THE **ASTRO** SOLUTION

ASTRO³ delivers all the key terms and core concepts for the **Introductory Astronomy** course.

Print + Online

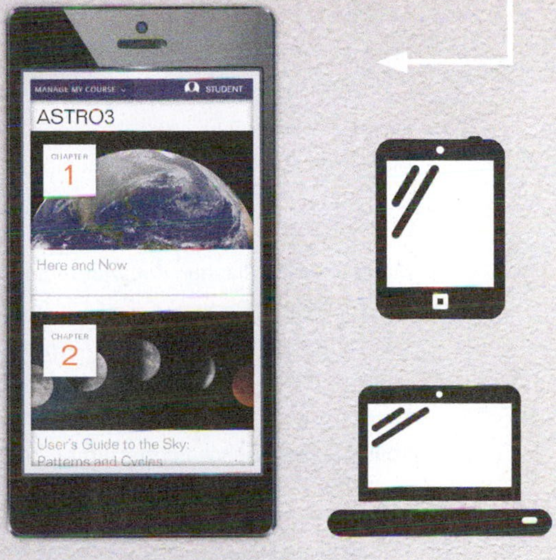

ASTRO Online provides the complete narrative from the printed text with additional interactive media and the unique functionality of **StudyBits**—all available on nearly any device!

What is a StudyBit™? Created through a deep investigation of students' challenges and workflows, the StudyBit™ functionality of **ASTRO Online** enables students of different generations and learning styles to study more effectively by allowing them to learn their way. Here's how they work:

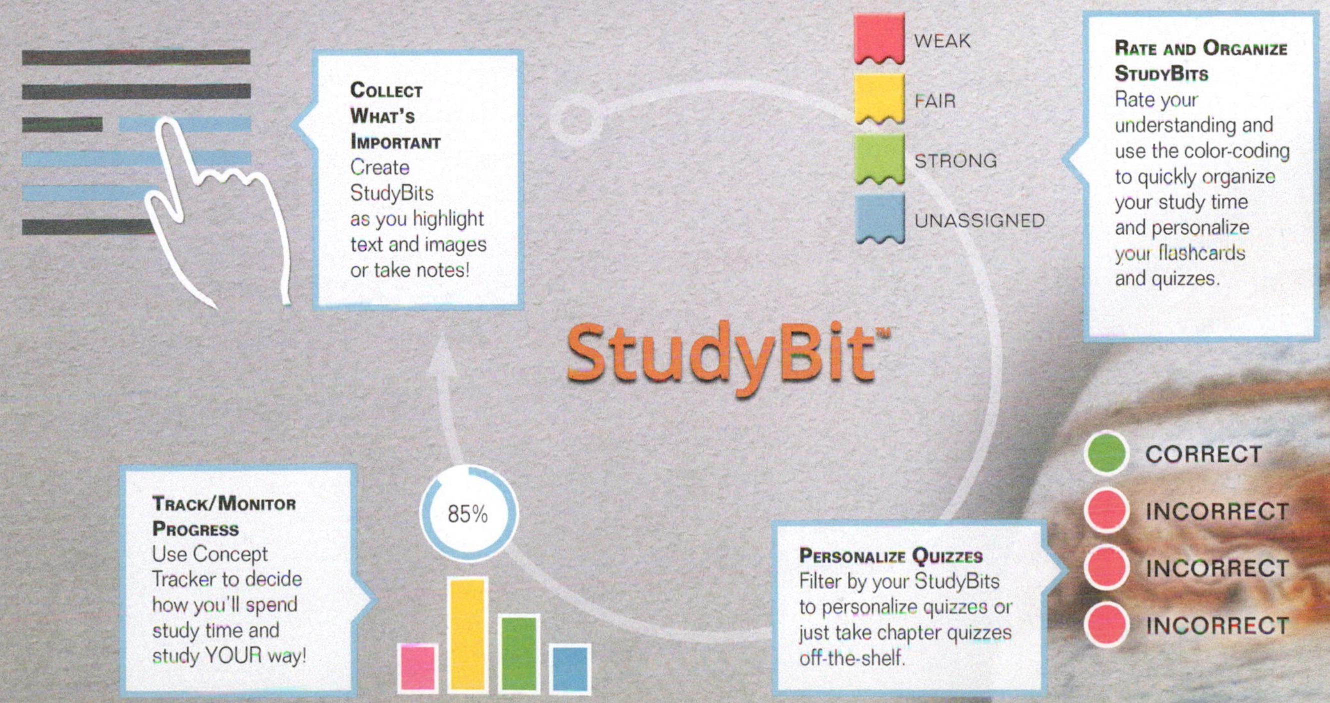

CENGAGE

ASTRO3

Michael A. Seeds, Dana Backman

Senior Vice President, General Manager: Balraj Kalsi

Product Manager: Clinton Kernen

Content/Media Developer: Colin Grover

Product Assistant: Eli Lewis

Marketing Manager: Heather Thompson

Sr. Content Project Manager: Martha Conway

Manufacturing Planner: Ron Montgomery

Sr. Art Director: Bethany Casey

Text Designer: Michael Stratton/ Stratton Design

Cover Designer: Lisa Kuhn/ Curio Press, LLC

Cover Image: Science Photo Library - VICTOR HABBICK VISIONS/Getty Images

Title Page and Back Cover Images:
Computer and tablet illustration: © iStockphoto.com/furtaev
Smart Phone illustration: © iStockphoto.com/dashadima
Feedback image: © Rawpixel.com/Shutterstock.com

Design Image for How Do We Know: © Suppakij1017/Shutterstock.com

Intellectual Property Analyst: Alexandra Ricciardi

Intellectual Property Project Manager: Reba Frederics

Production Service: SPi Global

For product information and technology assistance, contact us at
Cengage Customer & Sales Support, 1-800-354-9706

For permission to use material from this text or product, submit all requests online at **www.cengage.com/permissions**
Further permissions questions can be emailed to
permissionrequest@cengage.com

Library of Congress Control Number: 2016960912

Student Edition ISBN: 978-1-337-09751-2

Student Edition with Online ISBN: 978-1-337-09750-5

Cengage
200 Pier 4 Boulevard
Boston, MA 02210
USA

Cengage is a leading provider of customized learning solutions with employees residing in nearly 40 different countries and sales in more than 125 countries around the world. Find your local representative at **www.cengage.com.**

To learn more about Cengage platforms and services, register or access your online learning solution, or purchase materials for your course, visit **www.cengage.com**.

Printed in the United States of America
Print Number: 05 Print Year: 2020

SEEDS/BACKMAN

ASTRO³

BRIEF CONTENTS

Science Photo Library - VICTOR HABBICK VISIONS/Getty Images

CONTENTS

Part 1
EXPLORING THE SKY

Part 2
THE SOLAR SYSTEM

Part 3 THE STARS

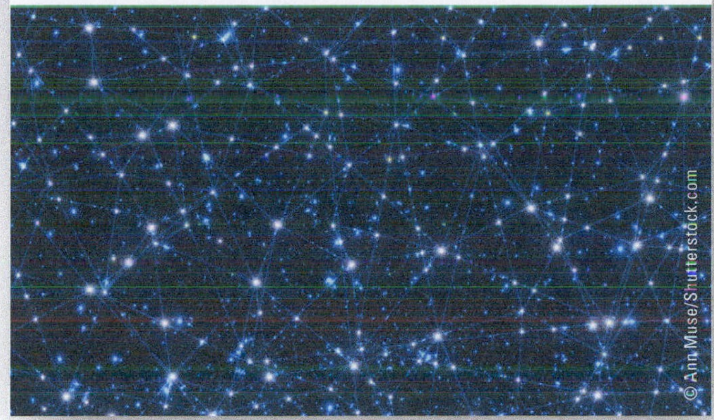

11 The Deaths of Stars 238

Part 4
THE UNIVERSE OF GALAXIES

12 The Milky Way Galaxy 272

13 Galaxies: Normal and Active 296

14 Modern Cosmology 322

Part 5
LIFE

15 Life on Other Worlds 348

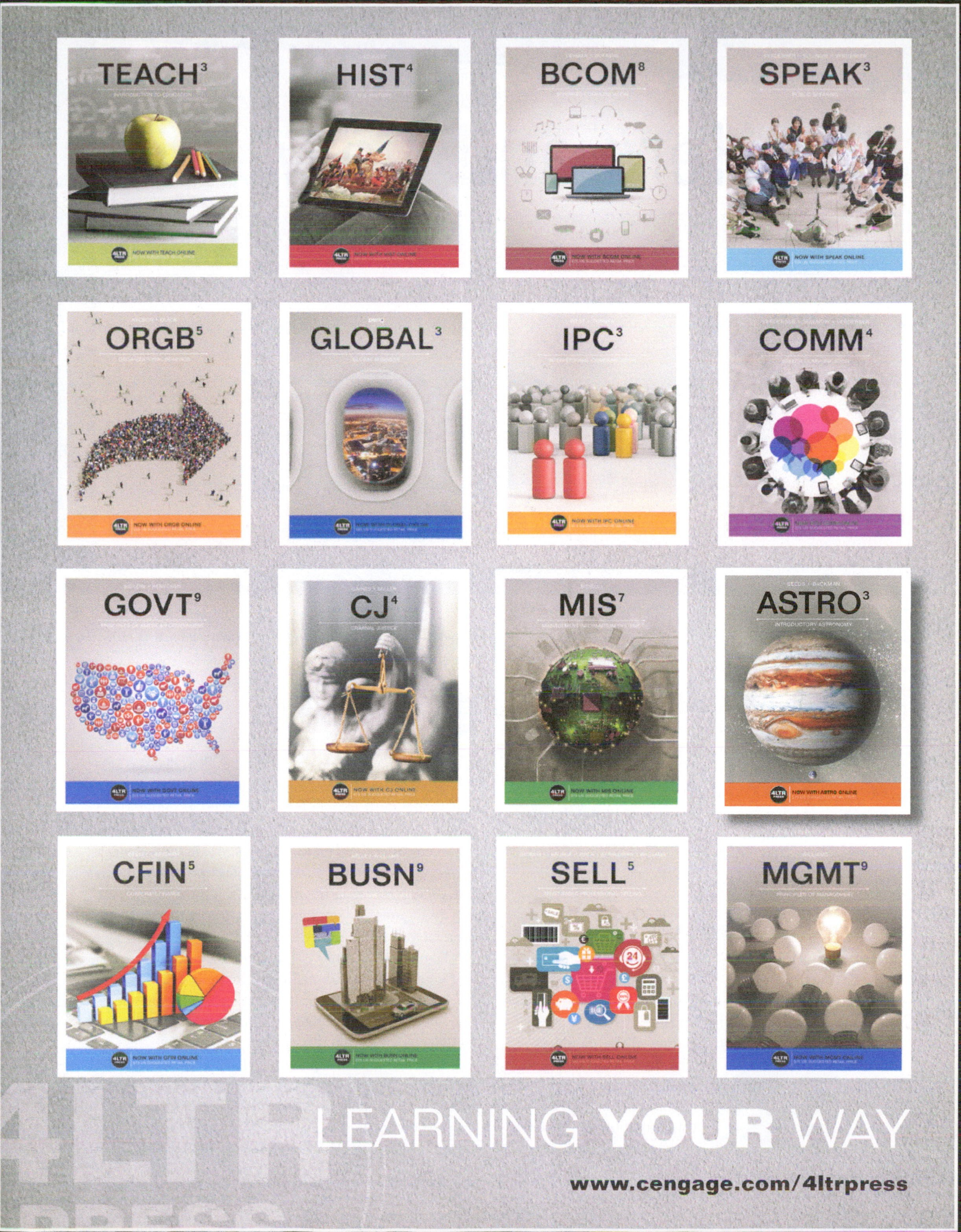

TEACH3
HIST4
BCOM8
SPEAK3
ORGB5
GLOBAL3
IPC3
COMM4
GOVT9
CJ4
MIS7
ASTRO3
INTRODUCTORY ASTRONOMY
CFIN5
BUSN9
SELL5
MGMT9
4LTR PRESS
LEARNING YOUR WAY
www.cengage.com/4ltrpress

1 Here and Now

The longest journey begins with a single step.
Laozi

NOAA/NASA GOES Project

After finishing this chapter go to **PAGE 9** for **STUDY TOOLS.**

GUIDEPOST

As you study astronomy, you learn about yourself. You are a planet-walker, and this chapter gives you a preview of what that means. The planet you live on whirls around a star that moves through a Universe filled with stars and galaxies. (See CO 1 for a view of Earth from space.) You owe it to yourself to know where you are located in the Universe and when you are living in its history because those are important steps to knowing what you are.

In this chapter, you will consider three important questions about astronomy:

- Where is Earth in the Universe?
- How does human history fit into the history of the Universe?
- Why study astronomy?

This chapter is a jumping-off point for your exploration of deep space and deep time. The next chapter continues your journey by looking at the night sky as seen from Earth. As you study astronomy, you will see how science gives you a way to know how nature works. Later chapters will provide more specific insights into how scientists study and understand nature.

field of view The area visible in an image. Usually given as the diameter of the region.

1-1 WHERE IS HERE?

You should begin with something familiar, like the size of yourself and your surroundings. Figure 1-1 shows a region about 16 meters (52 feet) across occupied by a human being, a sidewalk, and a few trees—all objects with sizes you can relate to. Each picture in the following sequence shows you a frame with a **field of view** that is 100 times wider than the preceding picture.

In Figure 1-2, your field of view has increased in size by a factor of 100, and you can now see an area about 1.6 kilometers (1 mile) in diameter. The 16-meter area of Figure 1-1 is quite small from this view. Now you can see an entire college campus and the surrounding streets and houses. This is still the world you know and can relate to the scale of your body.

Figure 1-3 has a span of 160 km (100 mi). Take a look at the picture and notice the red color. This is an infrared photograph, in which healthy green leaves and crops show up as red. Your eyes are sensitive to only a narrow range of colors. As you explore the Universe, you

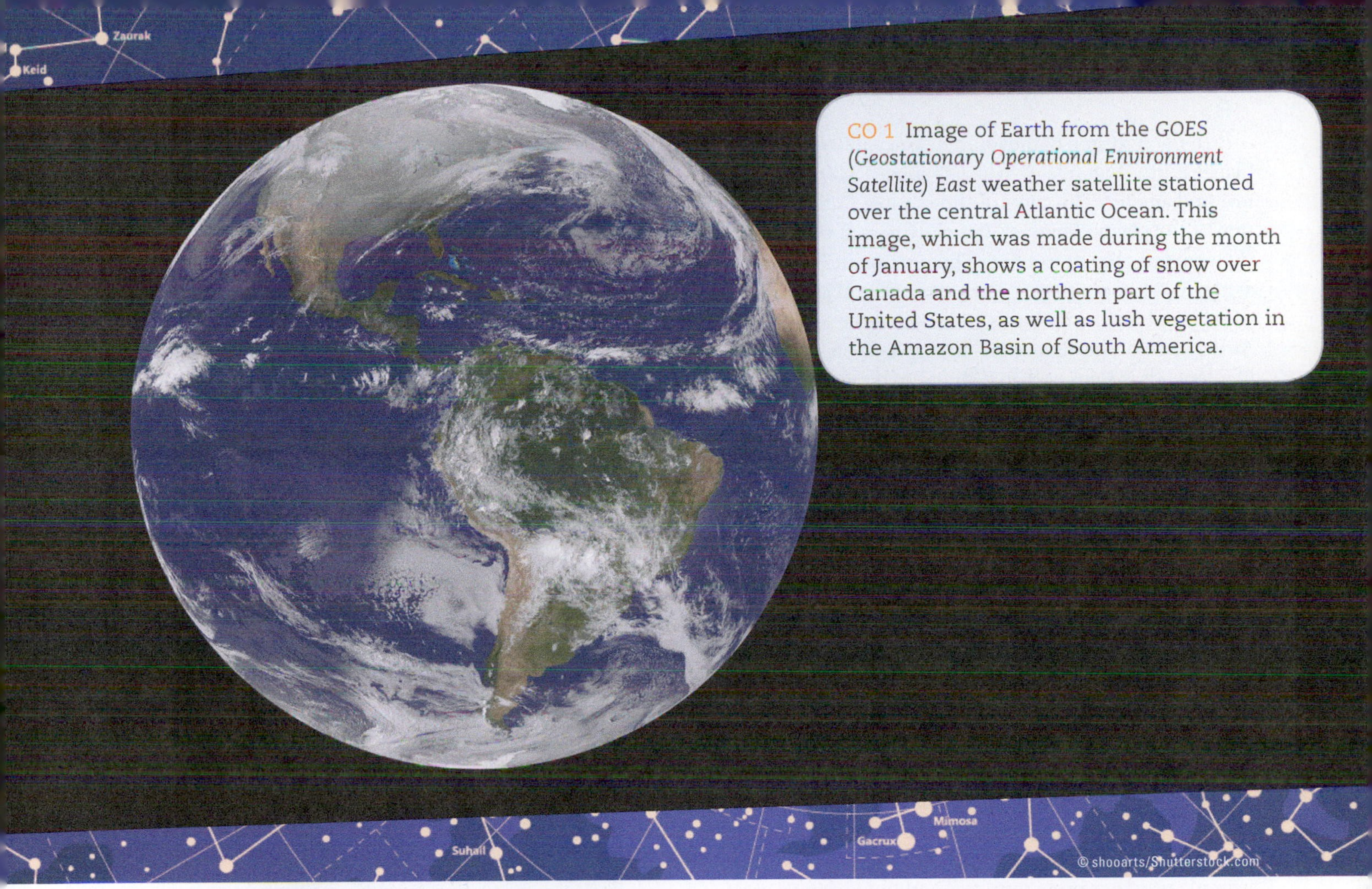

CO 1 Image of Earth from the GOES *(Geostationary Operational Environment Satellite) East* weather satellite stationed over the central Atlantic Ocean. This image, which was made during the month of January, shows a coating of snow over Canada and the northern part of the United States, as well as lush vegetation in the Amazon Basin of South America.

© shooarts/Shutterstock.com

FIGURE 1-1

Michael A. Seeds

FIGURE 1-2

USGS

This box ▪ represents the relative size of the previous frame.

will need to learn to use a wide range of "colors," from X-rays to radio waves, revealing sights invisible to your unaided eyes. Photographs in this book generally have labels noting which color or type of light was used to make the image.

The college campus is now invisible, and the patches of gray are towns and cities, including Wilmington, Delaware, visible in the lower right corner. At this scale, you can see natural features of Earth's surface: the Allegheny Mountains of southern Pennsylvania crossing the picture at the upper left, and the Susquehanna River flowing southeast into Chesapeake Bay. What look like white bumps are actually puffs of clouds. Mountains and valleys are only temporary features on Earth that are slowly, but constantly, changing. As you explore the Universe, you will come to see that it is also always evolving.

In the next step of the journey, you will see the entire planet Earth (Figure 1-4), which is about 13,000 km (8000 mi) in diameter. This picture shows most of the daylight side of the planet; however, the blurriness at the extreme right contains the sunset line. The rotation of Earth on its axis each 24 hours carries you eastward, and as you cross the sunset line into darkness you say that the Sun has set. At the scale of this figure, the atmosphere on which your life depends is thinner than a strand of thread.

Enlarge your field of view again by a factor of 100, and you see a region 1,600,000 km (1 million mi) wide (Figure 1-5). Earth is the small blue dot in the center, and the Moon, with a diameter of only about one-fourth that of Earth's, is an even smaller dot along its orbit. The relative sizes of Earth and Moon are shown in the inset at the bottom right of Figure 1-5. If you've had a high-mileage car, it may have traveled the equivalent of a trip to the Moon, which has an average distance from Earth of 380,000 km (240,000 mi). These numbers are so large that it is inconvenient to write them out. Astronomy is the science of big numbers, and you will use numbers much larger than these to describe the Universe.

Rather than writing out numbers, such as those in the previous paragraph, it is more convenient to write them in **scientific notation**. This is nothing more than a simple way to express numbers without writing a great many zeros. For example, in scientific notation, you can write 380,000 as 3.8×10^5.

When you once again enlarge your view by a factor of 100 (Figure 1-6), Earth, its Moon, and the Moon's orbit all lie in the small red box at lower left. This figure has a diameter of 1.6×10^8 km. Now you see the Sun and two other planets that are part of our Solar System. The **Solar System** consists of the Sun, its family of planets, and some smaller bodies, such as moons, asteroids, and comets.

Like Earth, Venus and Mercury are **planets**, small, nonluminous bodies that shine by reflecting sunlight. Venus is about the size of Earth, and Mercury is a bit larger than Earth's Moon. In this figure, they are both too small to be seen as anything but tiny dots. The Sun is a **star**, a

scientific notation A method for expressing very large or very small numbers by using powers of 10.

Solar System The Sun and all the objects gravitationally bound to it, including Earth, the other planets plus their moons, asteroids, comets, and so on.

planets A nonluminous body in orbit around a star, massive enough to be spherical and to have cleared its orbital zone of other objects.

star A globe of gas, held together by its own gravity and supported by the internal pressure of its hot gases, which generates energy by nuclear fusion.

FIGURE 1-3

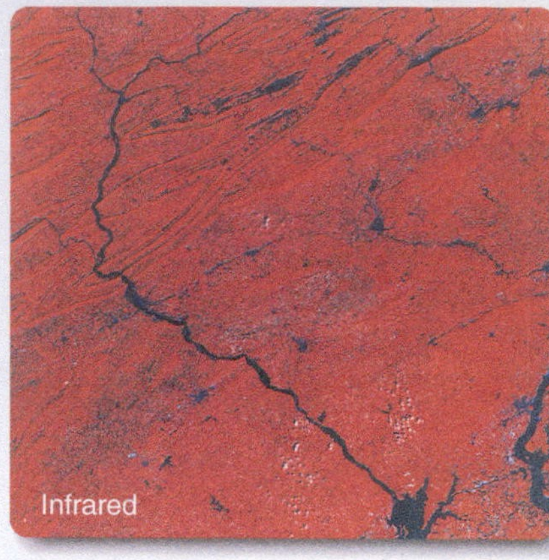

© NASA, Landsat infrared photograph

This box ▪ represents the relative size of the previous frame.

FIGURE 1-4

NASA

This box ▪ represents the relative size of the previous frame.

FIGURE 1-5

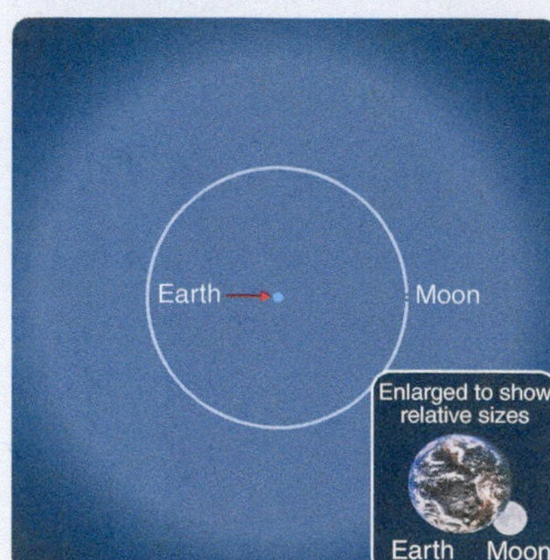

NASA

This box ▪ represents the relative size of the previous frame.

self-luminous ball of hot gas that generates its own energy. The Sun is about 100 times larger in diameter than Earth (inset), but it too is nothing more than a dot in this view. Earth orbits the Sun once a year.

Another way astronomers simplify descriptions and calculations using large numbers is to define new units. The average distance from Earth to the Sun is called the **astronomical unit (AU)**, a distance of 1.5×10^8 km (93 million mi). Using that unit you can then say, for example, that the average distance from Venus to the Sun is about 0.7 AU.

Your first field of view was only about 16 m (52 ft) in width. After just six steps, each enlarging by a factor of 100, you now see the entire Solar System, all the major planets, and their slightly elliptical orbits (Figure 1-7). Your view now is 1 trillion (10^{12}) times wider than in the first figure. The details of the previous figure are lost in the red square at the center of this figure. You see only the brighter, more widely separated objects as you back away. The Sun, Mercury, Venus, and Earth are so close together that you cannot separate them at this scale.

Mars, the next outward planet, lies only 1.5 AU from the Sun. In contrast, Jupiter, Saturn, Uranus, and Neptune are so far from the Sun that they are easy to find in this figure. Light from the Sun reaches Earth in only 8 minutes, but it takes more than 4 hours to reach Neptune. Pluto orbits mostly outside Neptune's orbit, but it is no longer considered a major planet.

When you again enlarge your view by a factor of 100, the Solar System becomes invisibly small (Figure 1-8). The Sun is only a point of light, and all the planets and their slightly elliptical orbits are now crowded into the small red square at the center. The planets are too small and reflect too little light to be visible so near the brilliance of the Sun.

Notice that no stars are visible in Figure 1-8 except for the Sun. The Sun is a fairly typical star, a bit larger than average, and it is located in a fairly normal neighborhood in the Universe. Although there are many billions of stars like the Sun, none is close enough to be visible in this figure. The stars are separated by average distances about 30 times larger than this view, which has a diameter of 11,000 AU. It is difficult to grasp the isolation of the stars. If the Sun were represented by a golf ball in New York City, the nearest star would be another golf ball in Chicago.

In Figure 1-9, your view has expanded to a diameter a bit over 1.1 million AU. The Sun is at the center, and you see a few of the nearest stars. These stars are so distant that it is not reasonable to give their distances in AU. Astronomers have defined a new larger unit of distance, the light-year. One **light-year (ly)** is the distance that light travels in 1 year, roughly 10^{13} km or 63,000 AU. The diameter of your view in Figure 1-9 is 17 ly. The nearest star to the Sun, Proxima Centauri, is 4.2 ly away. In other words, light from Proxima Centauri takes 4.2 years to reach Earth.

Although these stars are roughly the same size as the Sun, they are so far away that you cannot see them as anything but points of light. Even with the largest telescopes on Earth, you still see only points of light when you look at stars, and

FIGURE 1-6

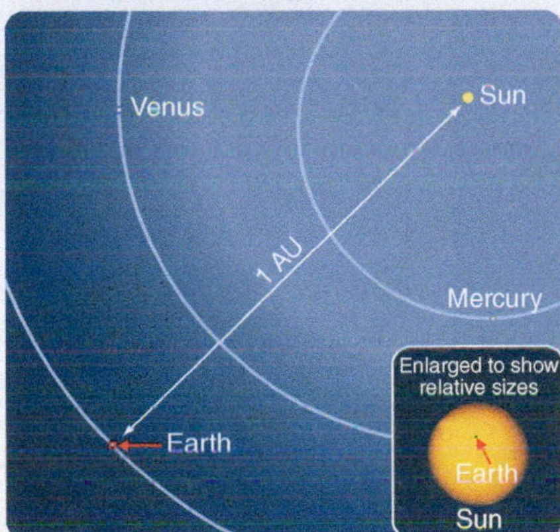

The small red box around Earth at lower left contains the entire field of view of Figure 1-5.

FIGURE 1-7

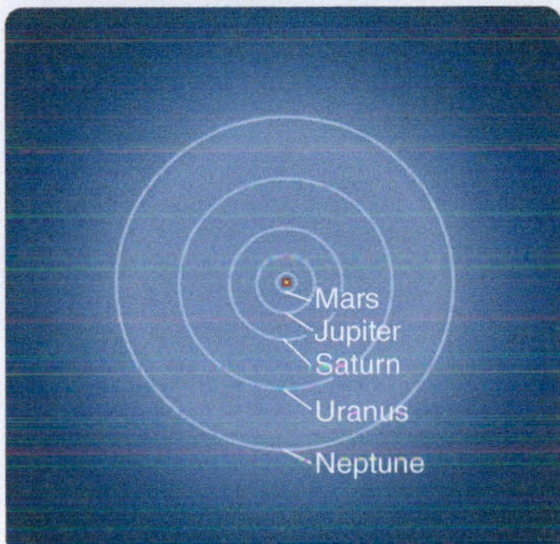

The small red box around the Sun at center contains the entire field of view of Figure 1-6.

astronomical unit (AU) Average distance from Earth to the Sun; 1.5×10^8 km (93 million mi).

light-year (ly) Unit of distance equal to the distance light travels in 1 year.

FIGURE 1-8

The small red box at the center contains the entire field of view of Figure 1-7.

FIGURE 1-9

This box ■ represents the relative size of the previous frame.

any planets that might circle those stars are much too small and faint to be visible. In Figure 1-9 the sizes of the dots represent not the sizes of the stars but their brightness. This is standard practice in astronomical diagrams, and it is also how starlight is recorded. Bright stars make larger spots in a photograph or electronic picture than faint stars. The size of a star image in a photograph tells you only how bright the star is, not how big it is.

galaxy A large system of stars, star clusters, gas, dust, and nebulae orbiting a common center of mass.

Milky Way The hazy band of light that circles our sky, produced by the glow of the stars and nebulae in our Galaxy.

Milky Way Galaxy The spiral galaxy containing our Sun, visible in the night sky as the Milky Way.

In Figure 1-10, you expand your field of view by another factor of 100, and the Sun and its neighboring stars vanish into the background of thousands of stars. This figure is 1700 ly in diameter. Of course, no one has ever journeyed thousands of light-years from Earth to look back and photograph the Sun's neighborhood, so this is a representative picture from Earth of a portion of the sky that can be used as a reasonable simulation. The Sun is faint enough that it would not be easily located in a picture at this scale.

Some things that are invisible in this figure are actually critically important. You do not see the thin gas that fills the spaces between the stars. Although those clouds of gas are thinner than the best vacuum produced in laboratories on Earth, it is those clouds that give birth to new stars. The Sun formed from such a cloud about 5 billion years ago. You will see more star formation in the next figure.

If you expand your view again by a factor of 100, you see our Galaxy (Figure 1-11). A **galaxy** is a great cloud of stars, gas, and dust bound together by the combined gravity of all the matter. In the night sky, you see our Galaxy from the inside as a great, cloudy band of stars ringing the sky as the **Milky Way**, and our Galaxy is called the **Milky Way Galaxy**. How does anyone know what the disk of the Milky Way Galaxy would look like from a vantage point tens of thousands of light years away? Astronomers use evidence to guide their explanations as they envision what our Galaxy looks like. Artists can then use those scientific descriptions to create a painting. Our Sun would be invisible in such a picture, but if you could see it, you

FIGURE 1-10

This box ■ represents the relative size of the previous frame.

How Do We Know? 1-1

The Scientific Method

How do scientists learn about nature? You have probably heard of the *scientific method* as the process by which scientists form hypotheses and test them against evidence gathered by experiment and observation. Scientists use the scientific method all the time and it is critically important, but they rarely think of it. It is such an ingrained way of thinking and understanding nature that it is almost invisible to the people who use it most.

A hypothesis is a suggestion about how nature works, and the evidence is reality. If a hypothesis is compared to the evidence and confirmed, it must be tested further; if it is contradicted, it must be discarded or revised. In that way, ideas get tested and refined to better describe how nature works.

For example, Gregor Mendel (1822–1884) was an Austrian abbot who liked gardening. He formed a hypothesis that offspring usually inherit traits from their parents not as a smooth blend, as most scientists of the time believed, but as discrete units according to strict mathematical rules. Mendel cultivated and tested more than 28,000 pea plants, noting which produced smooth peas and which produced wrinkled peas, and how that trait—and others—were inherited by successive generations. His studies confirmed his hypothesis and allowed the development of a series of laws of inheritance. Although the importance of his work was not recognized in his lifetime, Mendel is now called the "father of modern genetics".

NSphotostudio/Shutterstock.com

Whether peas are wrinkled or smooth is an inherited trait.

The scientific method is not a simple, mechanical way of grinding facts into understanding. It is actually a combination of many ways of analyzing information, finding relationships, and creating new ideas. A scientist needs insight and ingenuity to form and test a good hypothesis. Scientists use the scientific method almost automatically, forming, testing, revising, and discarding hypotheses sometimes minute-by-minute as they discuss new ideas. The scientific method is a way of thinking and a way of knowing about nature. The "How Do We Know?" essays in this book will introduce you to some of the methods scientists use.

FIGURE 1-11

Mark A. Garlick/space-art.co.uk

This box ▪ represents the relative size of the previous frame.

would find it about two-thirds of the way from the center to the edge, at about the location of the marker dot indicated in the figure. Our Galaxy contains more than 100 billion stars, and, like many others, has graceful **spiral arms** winding outward through the disk. You will discover in a later chapter that stars are born in great clouds of gas and dust as they pass through the spiral arms.

The visible portion of our Galaxy's disk is 80,000 ly in diameter. Only a century ago astronomers thought it was the entire Universe—an island Universe of stars in an otherwise empty vastness. Now, the Milky Way Galaxy is known to be not unique; it is a typical Galaxy in many respects, although

spiral arms Long spiral pattern of bright stars, star clusters, gas, and dust. Spiral arms extend across the disks of spiral galaxies.

larger than most. In fact, our Galaxy is only one of many billions of galaxies scattered throughout the Universe.

As you expand your field of view by another factor of 100, our Galaxy appears as a tiny luminous speck surrounded by other specks (Figure 1-12). This figure includes a region 17 million ly in diameter, and with each dot representing a galaxy. Notice that our Galaxy is part of a cluster of a few dozen galaxies. You will find that galaxies are commonly grouped together in clusters. Some of these galaxies have beautiful spiral patterns like our own Galaxy, but others do not. One of the questions you will investigate in a later chapter is what produces these differences among the galaxies.

Figure 1-13 represents a view with a diameter of 1.7 billion ly. It was made by combining observations with theoretical calculations. The figure shows clusters of galaxies connected in a vast network. Clusters are grouped into **superclusters**— clusters of clusters— and the superclusters are linked to form long filaments and walls outlining voids that seem nearly empty of galaxies. These filaments and walls appear to be the largest structures in the Universe.

Were you to expand your field of view one more time, you would probably see a uniform fog of filaments and voids. When you puzzle over the origin of these structures, you are at the frontier of human knowledge. The sequence of figures ends here because it has reached the limits of the best telescopes. Humanity's view does not extend as far as the region that would be covered by a figure 100 times larger than Figure 1-13.

superclusters A cluster of galaxy clusters.

1-2 WHEN IS NOW?

In later chapters, you will learn that the Universe began about 14 billion years ago. Since that time, galaxies have formed, and generation after generation of stars have formed and died. Our Sun and its planets formed only about 4.6 billion years ago. Humans have existed on Earth for only about 4 million years. That is only 0.004 billion years. Human civilization has existed for only 10,000 years (0.00001 billion years). Humanity is very new to the Universe.

How could anyone possibly know these secrets of nature? Science gives us a way to know how nature works. Throughout this book, you will find short essays titled "How Do We Know?"; these essays are designed to help you think not about *what* is known, but *how* it is known. To do that, they will explain a variety of scientific thought processes and procedures to help you understand how scientists know about the natural world.

FIGURE 1-12

Australian Astronomical Observatory/David Malin Images

This box ■ represents the relative size of the previous frame.

FIGURE 1-13

M. Seldner, B. L. Siebers, E. J. Groth, and P. J. E. Peebles, *Astronomical Journal* 82 (1977)

This box ■ represents the relative size of the previous frame.

What Are We? *Participants*

Astronomy will give you perspective on what it means to be here on Earth. This chapter has helped you locate yourself in space and time. Once you realize how vast our Universe is, Earth seems quite small, and people on the other side of the world seem like neighbors. Furthermore, in the entire history of the Universe, the human story is only a flicker of an eye blink. This may seem humbling at first, but you can be proud of how much we humans have understood in such a short time.

Not only does astronomy locate you in space and time, it places you in the physical processes that govern the Universe. Gravity and atoms work together to make stars, light the Universe, generate energy, and create the chemical elements in your body. The chapters that follow will show you how you fit into those cosmic processes.

Although you are very small and your kind have existed in the Universe for only a short time, you are an important part of something very large and very beautiful.

STUDY TOOLS 1

IN THE BOOK

- ☐ Tear Out the Review Card on *Here and Now*.

ASTRO ONLINE

Visit ASTRO Online at www.cengagebrain.com

- ☐ Interactive Reading
- ☐ Practice Quizzing
- ☐ Videos and Animations
- ☐ Exclusive Features

Questions and Problems

1

REVIEW QUESTIONS

1. What is the largest dimension of which you have personal knowledge? Have you run a mile? Hiked 10 mi? Run a marathon?
2. What is the difference between our Solar System and our Galaxy? Between our Galaxy and the Universe?
3. Why are light-years more convenient than miles, kilometers, or astronomical units for measuring certain distances?
4. Why is it difficult to detect planets orbiting other stars?
5. What does the size of the star image in a photograph tell you?
6. What is the difference between the Milky Way and the Milky Way Galaxy?
7. What are the largest known structures in the Universe?
8. How does astronomy help answer the question "What are we?"
9. **How Do We Know?** How does the scientific method give scientists a way to know about nature?

DISCUSSION QUESTIONS

1. You and three of your friends have won an all-expenses paid one-time-only round-trip first-class vacation to anywhere in the Universe, but only if the choice of destination is unanimous. Where do you want to go, and how do you convince your friends to agree?
2. Think back to the last time you got a new phone and had to figure out how it worked. In what way did you employ the scientific method, maybe without even realizing that you did?

PROBLEMS

1. The diameter of Earth across the equator is 7926 mi. Given that a mile equals 1.609 km, what is Earth's diameter in kilometers? In centimeters?
2. The diameter of the Moon across its equator is 3476 km. Given that a kilometer equals 0.6214 mi, what is the Moon's diameter in miles?
3. One astronomical unit is about 1.50×10^8 km. Explain why this is the same as 150×10^6 km.
4. Venus orbits 0.72 AU from the Sun. What is that distance in kilometers? (*Hint:* See Problem 3.)
5. Light from the Sun takes 8 minutes to reach Earth. How long does it take to reach Mars?
6. The Sun is almost 400 times farther from Earth than is the Moon. How long does light from the Moon take to reach Earth?
7. If the speed of light is 3.0×10^5 km/s, how many kilometers is 1 light-year? How many meters? (Note: One year contains 3.2×10^7 s.)
8. How long does it take light to cross the diameter of our Milky Way Galaxy?
9. The nearest large galaxy to our own is about 2.5 million ly away. How many meters is that? (*Hint:* See Problem 7.)
10. How many galaxies like our own would it take laid edge-to-edge to reach the nearest galaxy? (*Hint:* See Problem 9.)

LEARNING TO LOOK

1. In Figure 1-4, the division between daylight and darkness is at the right on the globe of Earth. How do you know this is the sunset line and not the sunrise line?
2. Look at Figure 1-6. How can you tell that Mercury does not follow a circular orbit?

3. Of the objects listed here, which would be contained inside the object shown in Figure 1-14?
   - Stars
   - Planets
   - Galaxy
   - Clusters
   - Spiral arms

FIGURE 1-14

Bill Schoening/NOAO/AURA/NSF

4. In Figure 1-15 which stars are brightest, and which are faintest? How can you tell? Why can't you tell which stars in this photograph are biggest or which have planets?

FIGURE 1-15

REU program/NOAO/AURA/NSF

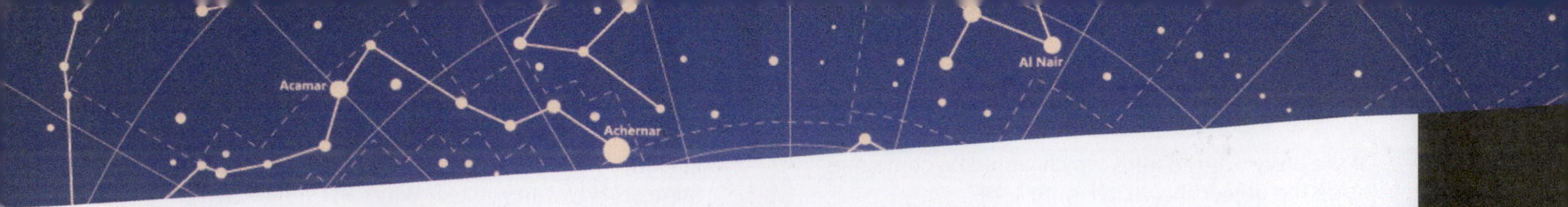

2 User's Guide to the Sky: Patterns and Cycles

The Southern Cross I saw every night abeam. The Sun every morning came up astern; every evening it went down ahead. I wished for no other compass to guide me, for these were true.

Captain Joshua Slocum *Sailing Alone Around the World*

Rami Daud/NASA

After finishing this chapter go to **PAGE 33** for **STUDY TOOLS.**

GUIDEPOST

The night sky is the rest of the Universe as seen from our planet. The previous chapter took you on a cosmic zoom out into the sky through deep space and deep time, setting the stage for the drama to come. In this chapter, you will continue your explorations by viewing the sky from Earth with your own unaided eyes. You will discover that the turning of Earth on its axis and its orbital motion around the Sun produce corresponding motions in the sky. You will see how the Sun, Moon, and planets move against the background of stars. Some of those motions have direct influences on your life and produce dramatic sights. (Look at CO 2 for a sequence of images of a lunar eclipse). As you explore, you will find answers to five important questions:

- How are stars and constellations named?
- How are the brightnesses of stars measured and compared?
- How does the sky appear to move as Earth rotates and revolves?
- What causes the seasons?
- How does the motion of the Moon produce phases and eclipses?

constellation One of the stellar patterns identified by name, usually of mythological gods, people, animals, or objects. Also, the region of the sky containing that star pattern.

As you study the sky and its motion, you will understand that Earth is a planet rotating on its axis. The next chapter will tell the story of how humans first understood that Earth is a planet.

2-1 THE STARS

On a dark night far from city lights, you can see a few thousand stars in the sky. Your observations can be summarized by naming individual stars and groups of stars and by specifying their relative brightness.

2-1a Constellations

All around the world, ancient cultures celebrated heroes, gods, and mythical beasts by naming groups of stars, called **constellations**, after them. You should not be surprised that the star patterns generally do not look like the creatures they are named after any more than Columbus, Ohio, looks like Christopher Columbus.

CO 2 A sequence of images spanning about 2 hours showing a lunar eclipse visible from the United States in September 2015. From left to right, the full moon is increasingly covered by Earth's shadow. In the image at far right, the Moon is shown fully shadowed, illuminated only by the orange glow of sunlight refracted through Earth's atmosphere. This was a rare "supermoon" eclipse that occurred when the Moon was closest in its orbit to Earth, and near the horizon for many observers, thereby appearing as large as possible.

The constellations named in Western culture originated in ancient civilizations of Mesopotamia, Babylon, Egypt, and Greece beginning as many as 5000 years ago. Of those ancient constellations, 48 are still in use. In those former times, a constellation was simply a loose grouping of bright stars, and many of the fainter stars were not included in any constellation. Regions of the southern sky not visible to observers living at northern latitudes also were not identified with any constellations. Constellation boundaries, when they were defined at all, were only approximate, so a star like Alpheratz could be thought of as both a part of Pegasus and also a part of Andromeda (Figure 2-1).

FIGURE 2-1

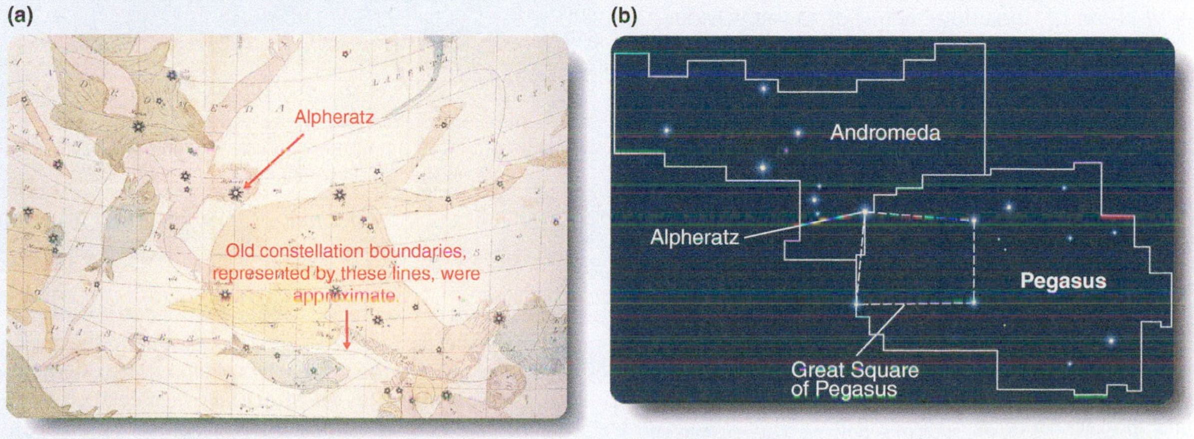

From Duncan Bradford, Wonders of the Heavens, Boston, John B. Russell 1837

(a) In antiquity, constellation boundaries were poorly defined, as shown on this map by the curving dotted lines that separate Pegasus from Andromeda. (b) Modern constellation boundaries are precisely defined by international scientific agreement.

In recent centuries, astronomers have added 40 modern constellations to fill gaps, and in 1928 the International Astronomical Union (IAU) established a total of 88 official constellations with clearly defined permanent boundaries that together cover the entire sky. Thus, a constellation now represents not a group of stars, but a section of the sky—a viewing direction—and any star within the region belongs only to that one constellation.

In addition to the 88 official constellations, the sky contains a number of less formally defined groupings called **asterisms**. The Big Dipper, for example, is an asterism you probably recognize; it is part of the constellation Ursa Major (the Great Bear). Another asterism is the Great Square of Pegasus that includes three stars from Pegasus and the previously mentioned star Alpheratz, now considered to be part of Andromeda only.

Although constellations and asterisms are named as if they were real groupings, most are made up of stars that are not physically associated with one another. Some stars may be many times farther away than others in the same constellation and moving through space in different directions. The only thing they have in common is that they lie in approximately the same direction as viewed from Earth, as in Figure 2-2.

2-1b Star Names

In addition to naming groups of stars, ancient astronomers named the brighter stars, and modern astronomers still use many of those names. The names of the constellations are in Latin or Greek, the languages of science in Medieval and Renaissance Europe. Most individual star names derive from ancient Arabic, much altered over centuries. For example, the name of Betelgeuse, the bright red star in Orion, comes from the Arabic phrase *yad al-jawza*, meaning "Hand of Jawza [Orion]." Aldebaran, the bright red eye of Taurus the bull, comes from the Arabic *al-dabar an*, meaning "the follower [of the Pleiades]."

Another way to identify stars is to assign Greek letters to the bright stars in a constellation in approximate order of brightness. Thus the brightest star is usually designated alpha (α), the second brightest beta (β), and so on. For many constellations, the letters follow the order of brightness, but some constellations, like Orion, are exceptions (Figure 2-3). A Greek-letter star name also includes the possessive form of the constellation name; for example, the brightest star in the constellation Canis Major is Alpha Canis Majoris. This name identifies the star and the constellation and gives a clue to the relative brightness of the star. Compare this with the ancient individual name for that star, Sirius, which tells you nothing about its location or brightness.

asterism A named grouping of stars that is not one of the recognized constellations.

magnitude scale The astronomical brightness scale. The larger the number, the fainter the star.

2-1c Star Brightness

Astronomers usually describe the brightness of stars using the **magnitude scale**. The ancient astronomers originally divided the stars into six brightness groups. The brightest were called first-magnitude stars. The scale continued downward to sixth-magnitude stars, the faintest visible to the human eye. Thus, the *larger* the magnitude number, the *fainter* the star. This makes sense if

FIGURE 2-2

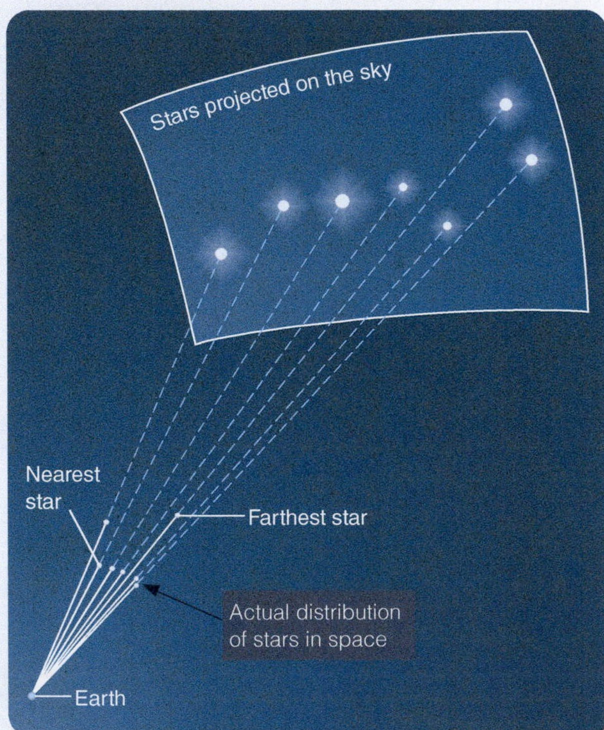

The stars you see in the Big Dipper are not at the same distance from Earth. You see the stars in a group in the sky because they lie in the same general direction as seen from Earth, not because they are all actually near each other. The sizes of the star dots in the star chart represent the apparent brightness of the stars.

FIGURE 2-3

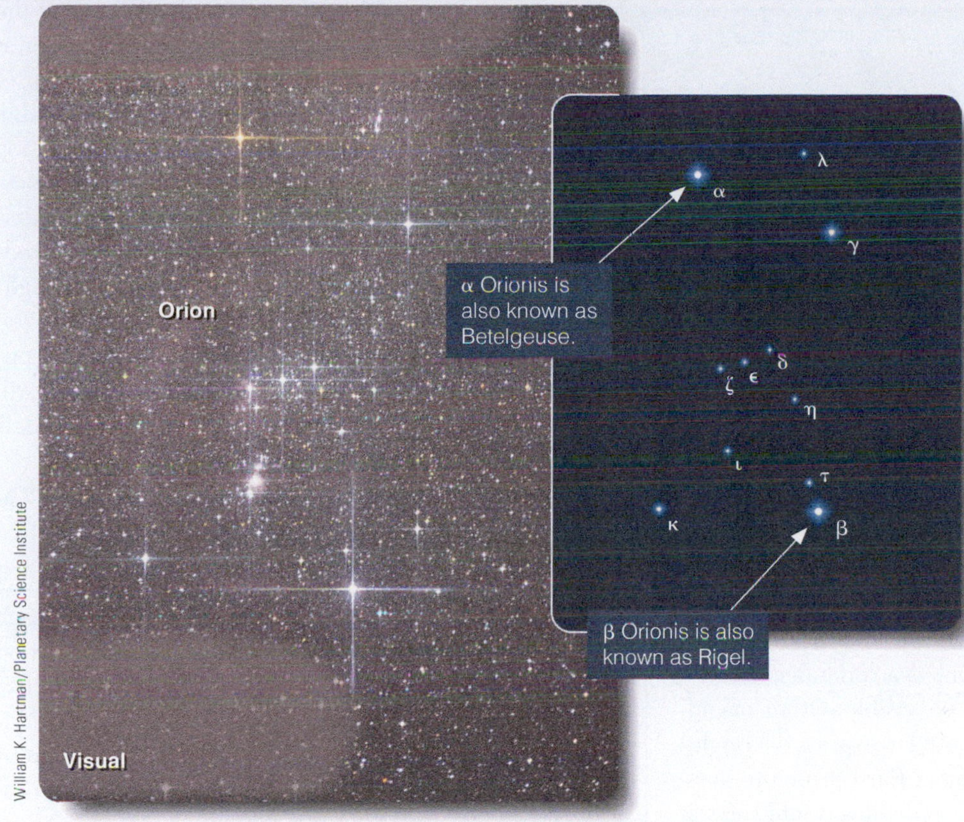

The stars in Orion do not quite follow the rule for assigning Greek letters in order of decreasing brightness. For example, β (Beta) is brighter than α (Alpha), and κ (Kappa) is brighter than η (Eta). Fainter stars do not have Greek letters or names, but if they are located inside the constellation boundaries, they are part of the constellation. The brighter stars in a constellation often also have individual names derived from Arabic. (The spikes on the star images in the photograph were produced by an optical effect in the telescope.)

you think of bright stars as first-class stars and the faintest stars visible as sixth-class stars.

The Greek astronomer Hipparchus (190–120 BCE) is believed to have compiled the first star catalog, and there is evidence he used the magnitude system in the catalog. About 300 years later (around 140 CE) the Egyptian-Greek astronomer Claudius Ptolemaeus (pronounced TAHL-eh-MAY-us; also known as Ptolemy, pronounced TAHL-eh-mee) definitely used the magnitude system in his own catalog, and successive generations of astronomers have continued to use the system. Star brightnesses expressed in this system are known as **apparent visual magnitudes (m_V)**, describing how the stars look to human eyes observing from Earth.

Brightness is quite subjective, depending on both the physiology of eyes and the psychology of perception. To be scientifically accurate you should refer to **flux**—a measure of the light energy from a star that hits one square meter in one second. (See **Magnitudes** in the Math Reference Cards MR.) With modern scientific instruments, astronomers can measure the flux of starlight with high precision and then use a simple mathematical relationship that relates light flux to apparent visual magnitude. Instead of saying that the star known by the charming name Chort (Theta Leonis) is about third magnitude, you can say its magnitude is 3.34. Thus, precise modern measurements of the brightness of stars are still connected to observations of apparent visual magnitude that go back to the time of Hipparchus.

Limitations of the apparent visual magnitude system have motivated astronomers to supplement it in various ways:

1. Some stars are so bright the magnitude scale must extend into negative numbers, as demonstrated in Figure 2-4. On this scale, Sirius, the brightest star in the sky, has a magnitude of −1.46.
2. With a telescope you can find stars much fainter than the limit for your unaided eyes. Thus the magnitude system has also been extended to numbers larger than sixth magnitude to include faint stars (Figure 2-4).
3. Although some stars emit large amounts of infrared

apparent visual magnitude (m_V) A measure of the brightness of a star as seen by human eyes from Earth.

flux A measure of the flow of energy through a surface. Usually applied to light.

FIGURE 2-4

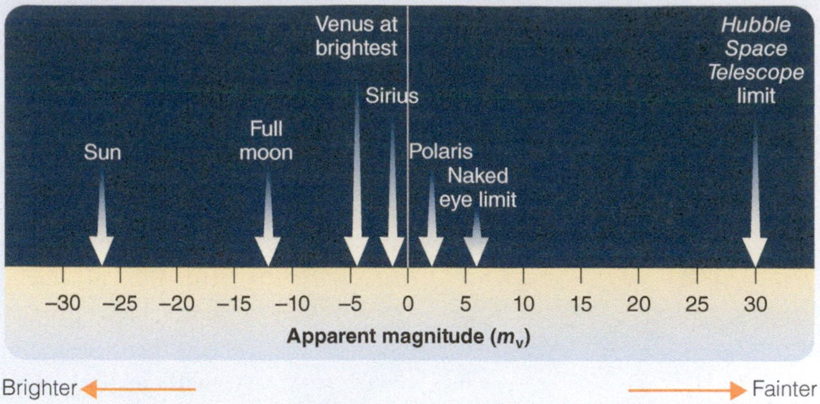

The scale of apparent visual magnitudes extends into negative numbers to represent the brightest objects and to positive numbers larger than 6 to represent objects fainter than the human eye can see.

or ultraviolet light, those types of radiation (discussed further in Chapters 4 and 5) are invisible to human eyes. The subscript "V" in m_V is a reminder that you are including only light that is visible. Other magnitude systems have been invented to express the brightness of invisible light arriving at Earth from the stars.

4. An apparent magnitude tells only how bright the star is as seen from Earth but doesn't tell anything about a star's true power output because the star's distance is not included. You can describe the true power output of stars with another magnitude system that will be described in Chapter 9.

2-2 THE SKY AND ITS MOTIONS

celestial sphere An imaginary sphere of very large radius surrounding Earth to which the planets, stars, Sun, and Moon seem to be attached.

scientific model A scientific metaphor; a concept that organizes thought about an aspect of nature without necessarily being literally true.

precession The slow change in orientation of the Earth's axis of rotation. One cycle takes nearly 26,000 years.

The sky above you seems to be a blue dome in the daytime and a sparkling ceiling at night. Learning to understand the sky requires that you first recall the perspectives of people who observed the sky thousands of years ago.

2-2a The Celestial Sphere

Ancient astronomers believed the sky was a great sphere surrounding Earth with the stars stuck on the inside like thumbtacks in a ceiling. Modern astronomers know that the stars are scattered through space at different distances, but it is still useful in some contexts for you to think of the sky as a great sphere enclosing Earth with stars all at one distance. The **celestial sphere** is an example of a **scientific model**, a common feature of scientific thought (See How Do We Know? 2-1.). You can use the celestial sphere as a convenient model of the sky. You will learn about more scientific models in the chapters that follow.

As you study **Concept Art 2A, "The Sky Around Us,"** notice three important points:

1. The sky appears to rotate westward around Earth each day, but that is a consequence of the eastward rotation of Earth. This produces day and night.
2. What you can see in the sky depends on your latitude. For example, Australians see many stars, constellations, and asterisms not visible from North America, but they never see the Big Dipper.
3. Astronomers measure distances across the sky as angles expressed in units of degrees and subdivisions of degrees called arc minutes and arc seconds.

2-2b Precession

In addition to causing the obvious daily motion of the sky, Earth's rotation is connected with a very slow celestial motion that can be detected only over centuries. More than 2000 years ago, Hipparchus compared positions of some stars with their positions recorded nearly two centuries previously and realized that the celestial poles and equator were slowly moving across the sky. Later astronomers understood that this motion is caused by a toplike motion of Earth known as **precession**.

If you have ever played with a gyroscope or top, you have seen how the spinning mass resists any sudden change in the direction of its axis of rotation. The more massive the top and the more rapidly it spins, the more it resists your efforts to twist it out of position. You may recall that even the most rapidly spinning top slowly swings its axis around in a circle. The weight of the top tends to make it tip over, and this combines with its rapid rotation to make its axis sweep

How Do We Know? 2-1

Scientific Models

How can a scientific model be useful if it isn't entirely true? A scientific model is a carefully devised conception of how something works, a framework that helps scientists think about some aspect of nature. The celestial sphere is a model that helps astronomers think about the motions of the sky.

Chemists, for example, use colored balls to represent atoms and sticks to represent the bonds between them, almost like Tinkertoys. Using these molecular models, chemists can see the three-dimensional shape of molecules and understand how the atoms interconnect.

The molecular model of DNA proposed by Watson and Crick in 1953 led to our modern understanding of the mechanisms of genetics. You have probably seen elaborate ball-and-stick models of DNA, but does the molecule really look like Tinkertoys? No, but the model is both simple enough and accurate enough to help scientists think about DNA.

A scientific model is not a statement of truth; it does not have to be precisely true to be useful. In an idealized model, some complex aspects of nature can be simplified or omitted. The ball-and-stick model of a molecule doesn't show the relative strength of the chemical bonds, for instance. A model gives scientists a way to think about some aspect of nature, but need not be true in every detail.

When you use a scientific model, it is important to remember the limitations of that model. If you begin to think of a model as true, it can be misleading instead of helpful. The celestial sphere, for instance, can help you think about the sky, but you must remember that it is only a model. The Universe is much larger and much more interesting than this ancient model of the heavens.

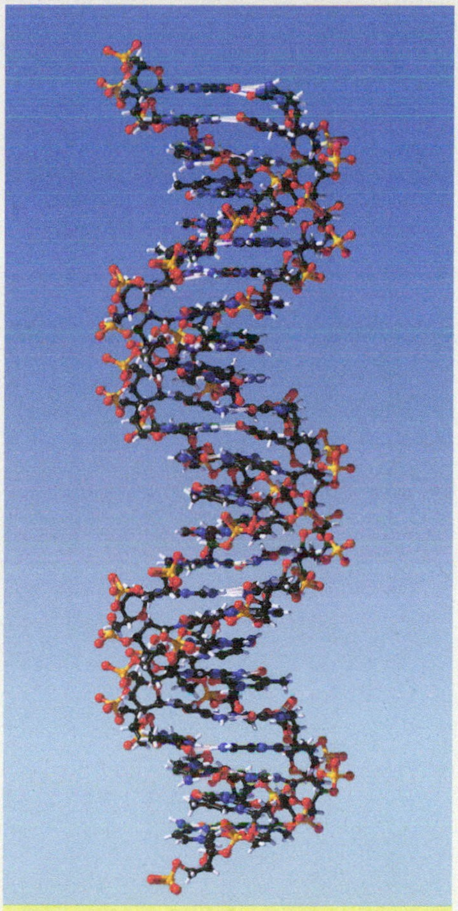

John Harwood/Photodisc/Getty Images

A ball-and-stick model of a short section of a DNA molecule.

out the shape of a cone. That motion is precession (Figure 2-7a). In later chapters, you will learn that many celestial bodies precess.

Earth spins like a giant top, but it does not spin upright in its orbit; its axis is tipped 23.4 degrees from vertical. Earth's large mass and rapid rotation keep its axis of rotation pointed toward a spot near the star Polaris, and the axis would remain pointed constantly in that direction except for the effect of precession.

Concept Art 2A
THE SKY AROUND US

1

The eastward rotation of Earth causes the Sun, Moon, and stars to move westward in the sky as if the **celestial sphere** were rotating westward around Earth. From any location on Earth you see only half of the celestial sphere, the half above the **horizon**. The **zenith** marks the top of the sky above your head, and the **nadir** marks the bottom of the sky directly under your feet. The drawing at right shows the view for an observer in North America. An observer in South America would have a dramatically different horizon, zenith, and nadir.

The apparent pivot points are the **north celestial pole** and the **south celestial pole** located directly above Earth's north and south poles. Halfway between the celestial poles lies the **celestial equator**. Earth's rotation defines the directions you use every day. The **north point** and **south point** are the points on the horizon closest to the celestial poles. The **east point** and the **west point** lie halfway between the north and south points. The celestial equator always touches the horizon at the east and west points.

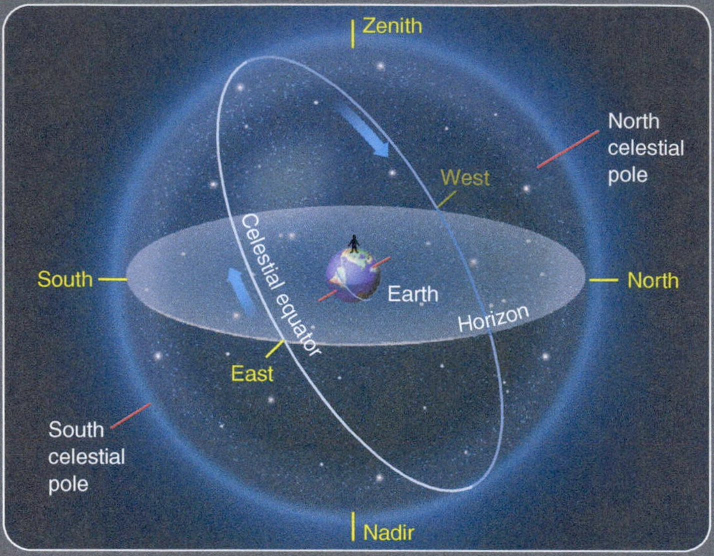

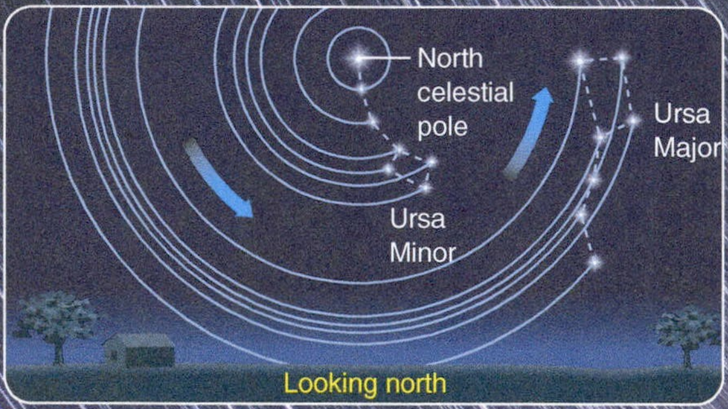

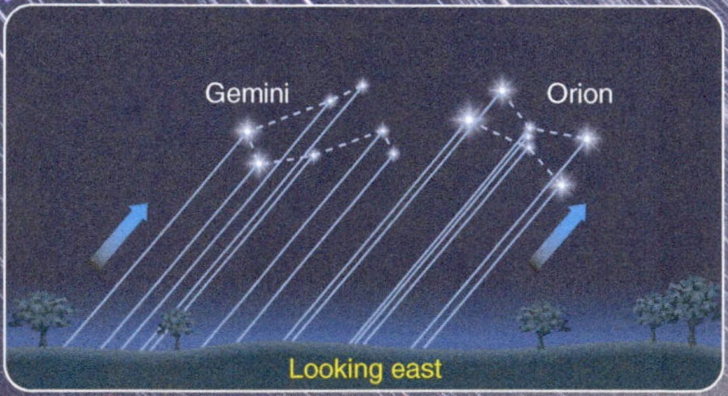

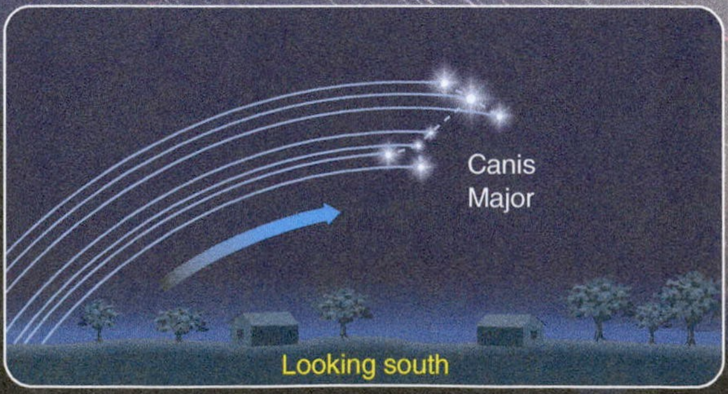

1a

This time exposure of about 30 minutes shows stars as streaks, called star trails, rising behind an observatory dome. The camera was facing northeast to take this photo. The motion you see in the sky depends on which direction you look, as shown at right. Looking north, you see the star Polaris, the North Star, located near the north celestial pole. As the sky appears to rotate westward, Polaris hardly moves, but other stars circle the celestial pole. Looking south from a location in North America, you can see stars circling the south celestial pole, which is invisible below the southern horizon.

P. Marenfeld & NOAO/AURA/NSF

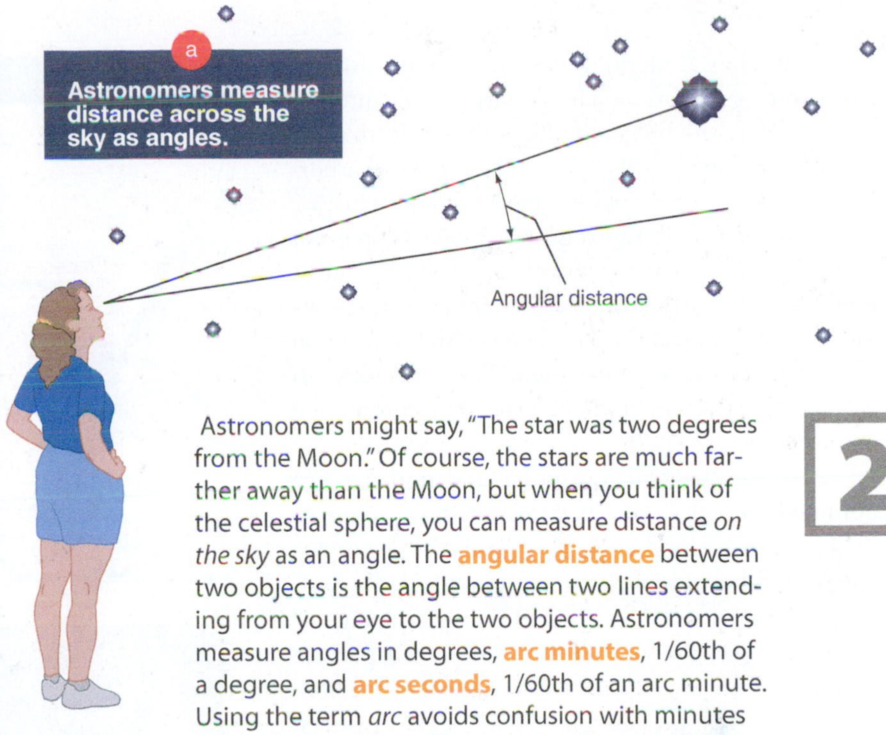

2

Astronomers might say, "The star was two degrees from the Moon." Of course, the stars are much farther away than the Moon, but when you think of the celestial sphere, you can measure distance *on the sky* as an angle. The **angular distance** between two objects is the angle between two lines extending from your eye to the two objects. Astronomers measure angles in degrees, **arc minutes**, 1/60th of a degree, and **arc seconds**, 1/60th of an arc minute. Using the term *arc* avoids confusion with minutes and seconds of time. The **angular diameter** of an object is the angular distance from one edge to the other. The Sun and Moon are each about half a degree in diameter, and the bowl of the Big Dipper is about 10 degrees wide.

3

What you see in the sky depends on your latitude as shown at left. Imagine that you begin a journey in the ice and snow at Earth's North Pole with the north celestial pole directly overhead. As you walk southward, the celestial pole moves toward the horizon, and you can see farther into the southern sky. The angular distance from the horizon to the north celestial pole always equals your latitude (L)—the basis for celestial navigation. As you cross Earth's equator, the celestial equator would pass through your zenith, and the north celestial pole would sink below your northern horizon.

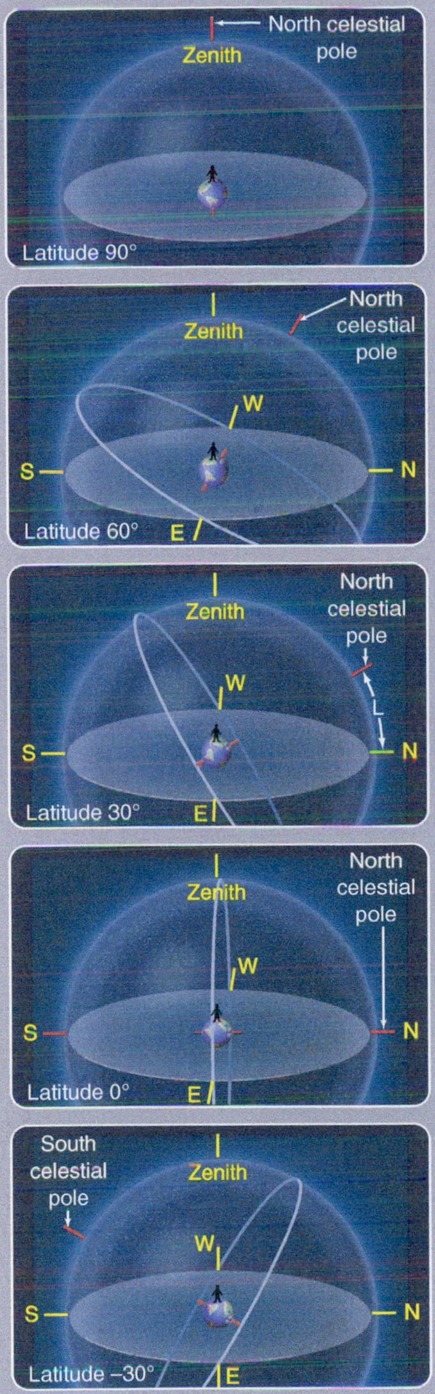

3a

Circumpolar constellations are those that never rise or set. From mid-northern latitudes, as shown at right, you see a number of familiar constellations circling Polaris and never dipping below the horizon. As the sky rotates, the pointer stars at the front of the Big Dipper always point toward Polaris. Circumpolar constellations near the south celestial pole never rise when seen from mid-northern latitudes. From a high latitude such as Norway, you would have more circumpolar constellations, and from Quito, Ecuador, located on Earth's equator, you would have no circumpolar constellations at all.

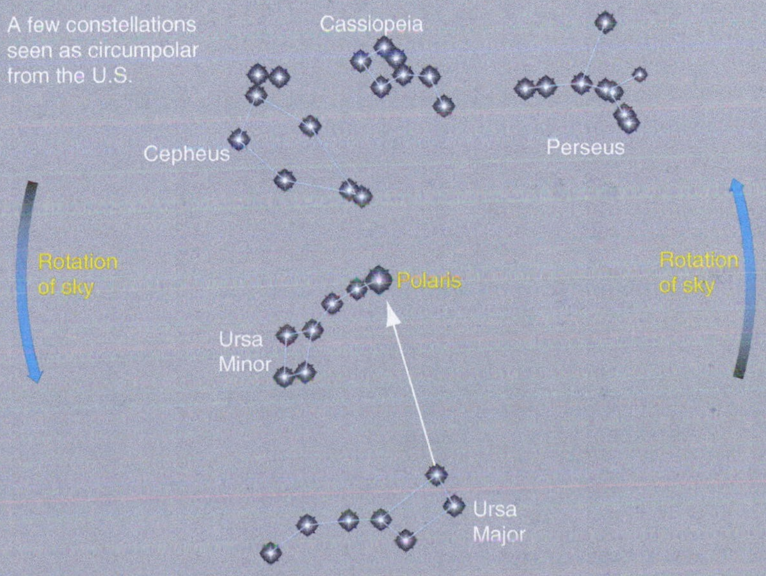

Earth has a slight bulge around its middle because of its rotation. The gravity of the Sun and Moon pull on the bulge, tending to twist Earth's axis "upright" relative to its orbit. If Earth were a perfect sphere, it would not be subjected to this twisting force. Notice that the analogy to a spinning top is not perfect; gravity tends to make a top fall over, but it tends to twist Earth upright. In both cases, the twisting of the axis of rotation combined with the rotation of the object causes precession. The precession of Earth's axis takes about 26,000 years for completing one cycle (Figure 2-7b).

Because the celestial poles and equator are defined by Earth's rotational axis, precession moves these reference marks. Figure 2-5c shows the apparent path followed by the north celestial pole over thousands of years. You would notice no change at all from night to night or year to year, but precise measurements reveal their slow apparent motion. Over centuries, precession has significant effects. Egyptian records show that 4800 years ago the north celestial pole was near the star Thuban (Alpha Draconis). The pole is now approaching Polaris and will be closest to it in about the year 2100. In another 12,000 years the pole will have moved to the apparent vicinity of the very bright star Vega (Alpha Lyrae).

FIGURE 2-5

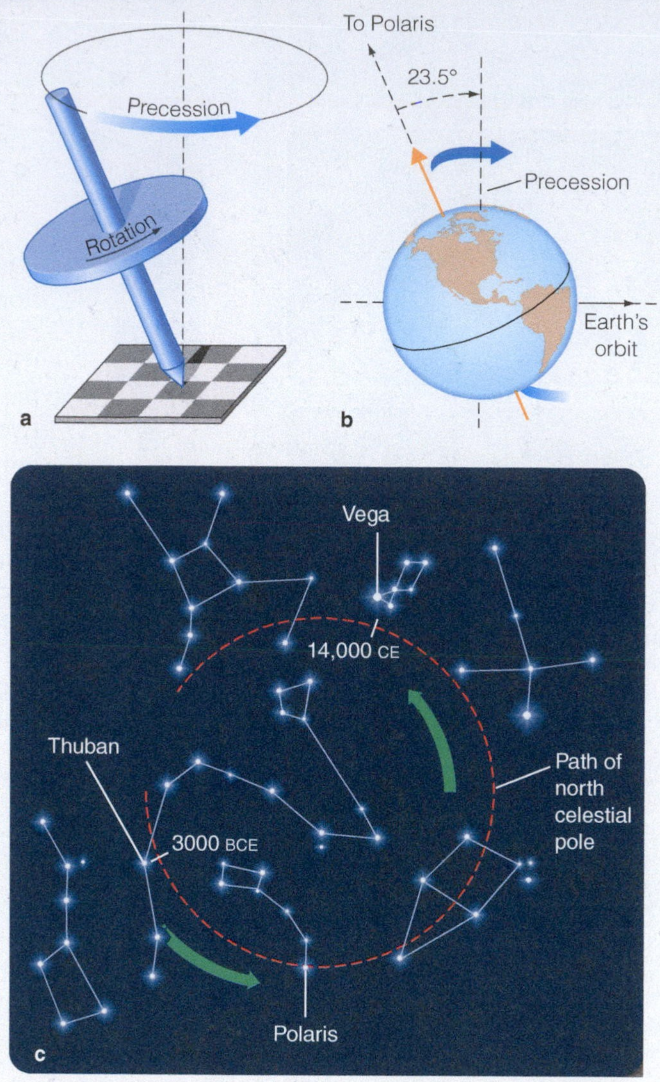

Precession. (a) The rotation axis of a spinning top precesses in a conical motion around the perpendicular to the floor because its weight tends to make it fall over. (b) Earth's axis precesses around the perpendicular to its orbit because the gravity of the Sun and Moon acting on Earth's equatorial bulge tend to twist it "upright." (c) Precession causes the north celestial pole to move slowly among the stars, completing a circle in about 26,000 years.

2-3 THE CYCLE OF THE SUN

The English language defines **rotation** as the turning of a body on its axis. **Revolution** means the motion of a body around a point outside the body. Earth rotates on its axis, and that produces day and night. Earth also revolves around the Sun, and that produces the yearly cycle.

2-3a Apparent Annual Motion of the Sun

Even in the daytime the sky is actually filled with stars, but the glare of sunlight fills our atmosphere with scattered light and you can see only the brilliant blue sky. If the Sun were fainter and you could see the stars in the daytime, you would notice that the Sun appears to be moving slowly eastward, relative to the background of the distant stars, as shown in Figure 2-6. This apparent motion is caused by the real orbital motion of Earth around the Sun. In early to mid-January, you would see the Sun in front of the constellation Sagittarius. By the beginning of March, you would

rotation Motion around an axis passing through the rotating body.

revolution Orbital motion about a point located outside the orbiting body.

FIGURE 2-6

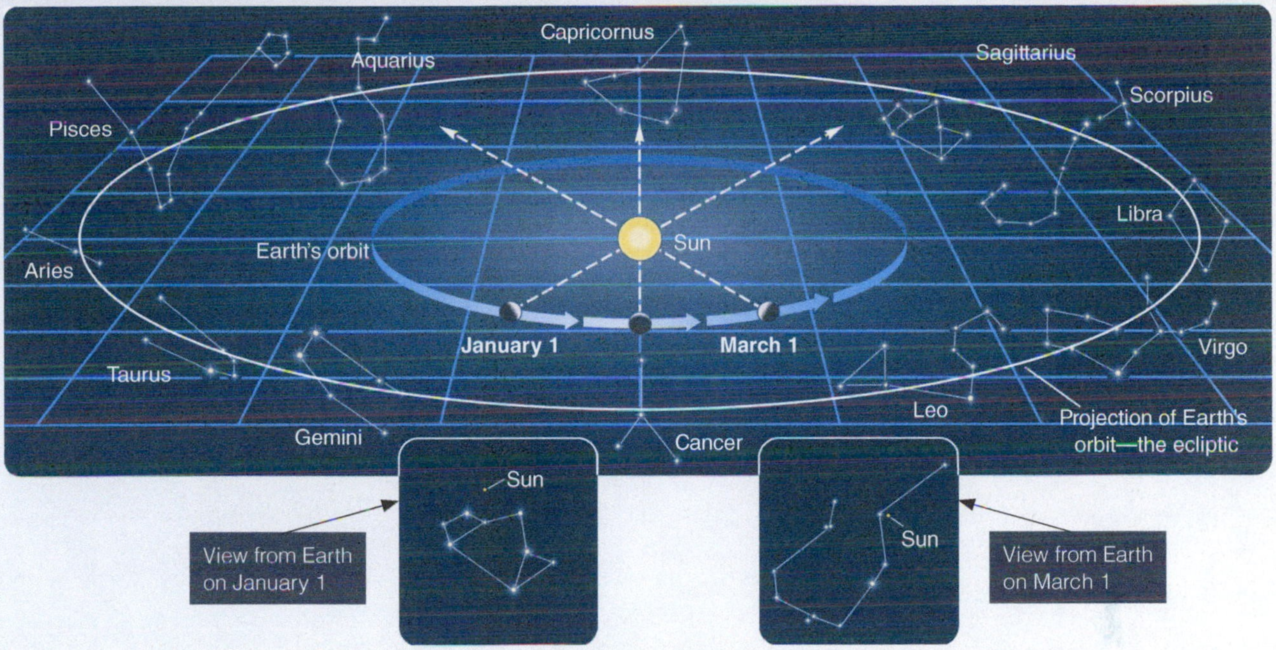

The motion of Earth around the Sun makes the Sun appear to move against the background of the stars. The circular orbit of Earth is thus projected on the sky as the ecliptic, the circular path of the Sun during the year as seen from Earth. If you could see the stars in the daytime, you would notice the Sun crossing in front of the distant constellations as Earth moves along its orbit.

see the Sun in front of Aquarius. Note that your angle of view in Figure 2-6 makes the Earth's orbit seem very elliptical when it is really almost a perfect circle.

Through the year, the Sun moves eastward among the stars following a line called the **ecliptic**, the apparent path of the Sun among the stars. Recall the concept of the celestial sphere; if the sky were a great screen, the ecliptic would be the shadow cast by the Earth's orbit. In other words, you can describe the ecliptic as the projection of the Earth's orbit on the celestial sphere. Earth circles the Sun in 365.26 days, and, consequently, the Sun appears to go around the sky once in the same period. You don't notice this motion because you cannot see the stars in the daytime, but the apparent motion of the Sun caused by a real motion of Earth has an important consequence that you do notice—the seasons.

2-3b Seasons

The seasons are caused by the revolution of Earth around the Sun combined with a simple fact you have already encountered: Earth's equator is tipped 23.4 degrees relative to its orbit. As you study **Concept Art 2B, "The Cycle of the Seasons,"** notice two important principles:

1. The seasons are *not* caused by variation in the distance between Earth and the Sun. Earth's orbit is nearly circular, so it is always about the same distance from the Sun.
2. The seasons *are* caused by changes in the amount of solar energy that Earth's Northern and Southern Hemispheres receive at different times of the year, resulting from the tip of the Earth's equator and axis relative to its orbit.

The seasons are so important as a cycle of growth and harvest that cultures around the world have attached great significance to the ecliptic. It marks the center line of the **zodiac** ("circle of animals"), and the motion of the Sun, Moon, and the five visible planets (Mercury, Venus, Mars, Jupiter, and Saturn) are the basis of the ancient superstition of astrology.

ecliptic The apparent path of the Sun around the sky.

zodiac A band centered on the ecliptic and encircling the sky.

Concept Art 2B

THE CYCLE OF THE SEASONS

1

You can use the celestial sphere to help you think about the seasons. The celestial equator is the projection of Earth's equator on the sky, and the ecliptic is the projection of Earth's orbit on the sky. Because Earth is tipped in its orbit, the ecliptic and equator are inclined to each other by 23.4° as shown at right. As the Sun moves eastward around the sky, it spends half the year in the southern half of the sky and half the year in the northern half. That causes the seasons.

The Sun crosses the celestial equator going northward at the point called the **vernal equinox**. The Sun is at its farthest north at the point called the **summer solstice**. It crosses the celestial equator going southward at the **autumnal equinox** and reaches its most southern point at the **winter solstice**.

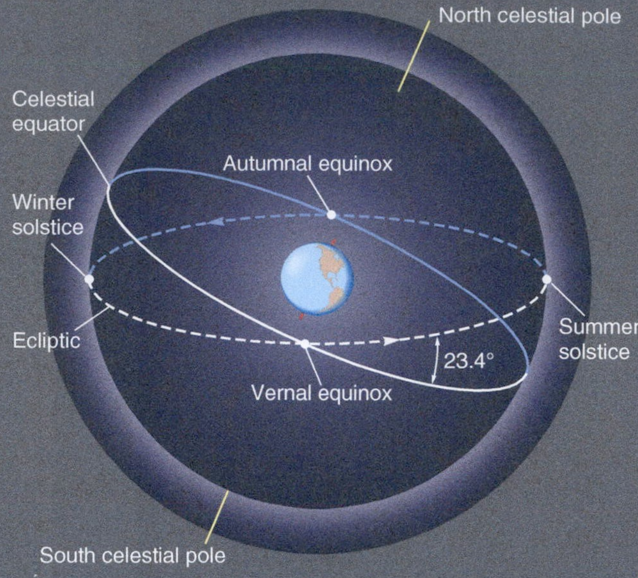

1a

The seasons are defined by the dates when the Sun crosses these four points, as shown in the table at the right. *Equinox* comes from the word for "equal"; the day of an equinox has equal amounts of daylight and darkness. *Solstice* comes from the words meaning "Sun" and "stationary." *Vernal* comes from the word for "green." The "green" equinox marks the beginning of spring in the Northern Hemisphere.

Event	Date*	N. Hemisphere
Vernal equinox	March 20	Spring begins
Summer solstice	June 22	Summer begins
Autumnal equinox	September 22	Autumn begins
Winter solstice	December 21	Winter begins

* Give or take a day due to leap years and other factors.

1b

On the day of the summer solstice in late June, Earth's Northern Hemisphere is inclined toward the Sun, and sunlight shines almost straight down at northern latitudes. At southern latitudes, sunlight strikes the ground at an angle and spreads out. North America has warm weather, and South America has cool weather.

Earth's axis of rotation points toward Polaris, and, like a top, the spinning Earth holds its axis fixed as it orbits the Sun. On one side of the Sun, Earth's Northern Hemisphere leans toward the Sun; on the other side of its orbit, it leans away. However, the direction of the axis of rotation does not change in the course of a year.

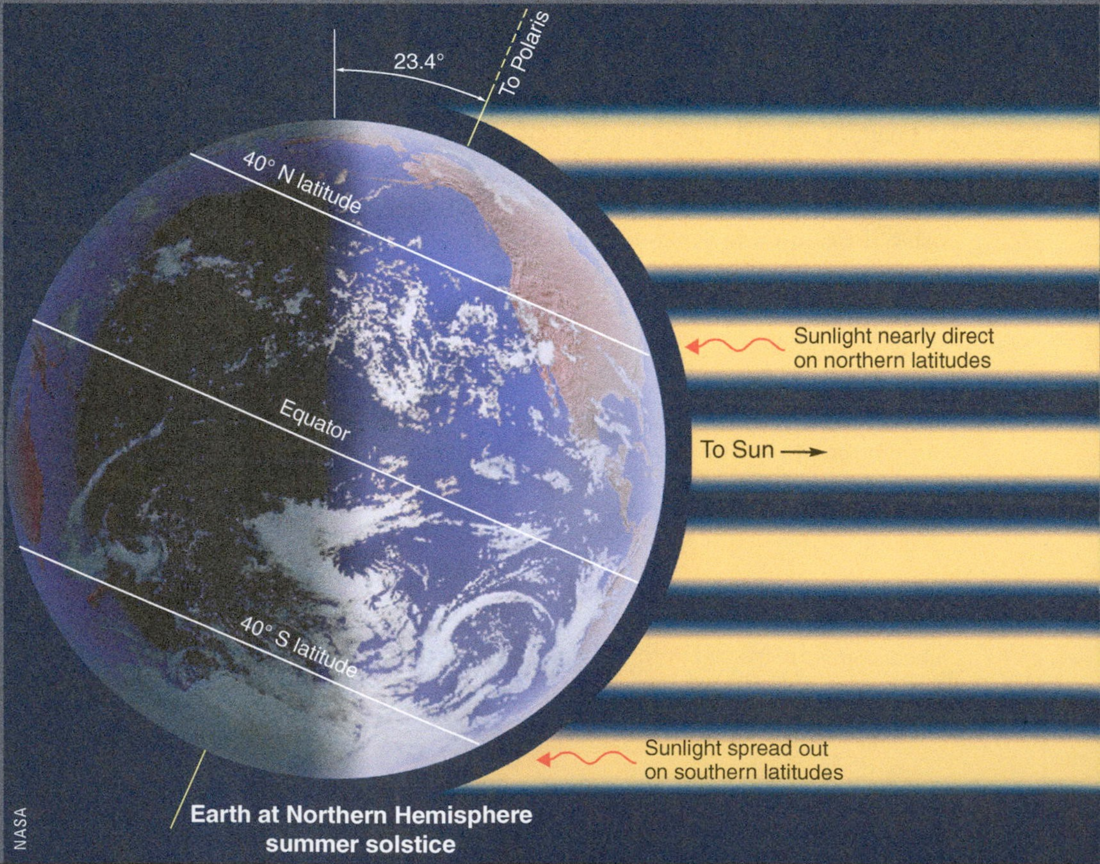

Earth at Northern Hemisphere summer solstice

1c Light striking the ground at a steep angle spreads out less than light striking the ground at a shallow angle. Light from the summer solstice Sun strikes northern latitudes from nearly overhead and is concentrated.

Light from the winter solstice Sun strikes northern latitudes at a much steeper angle and spreads out. The same amount of energy is spread over a larger area, so the ground receives less energy from the winter Sun.

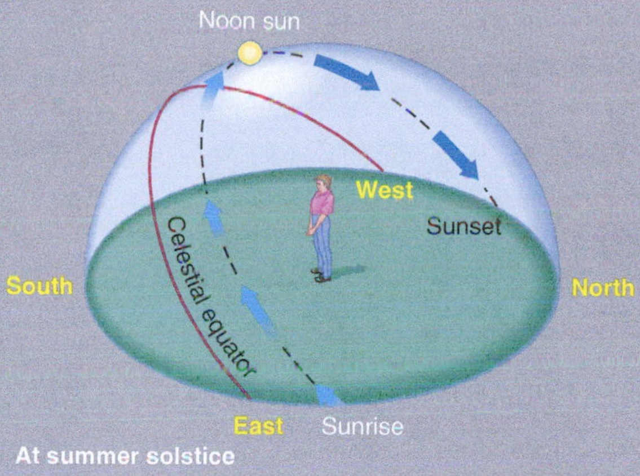

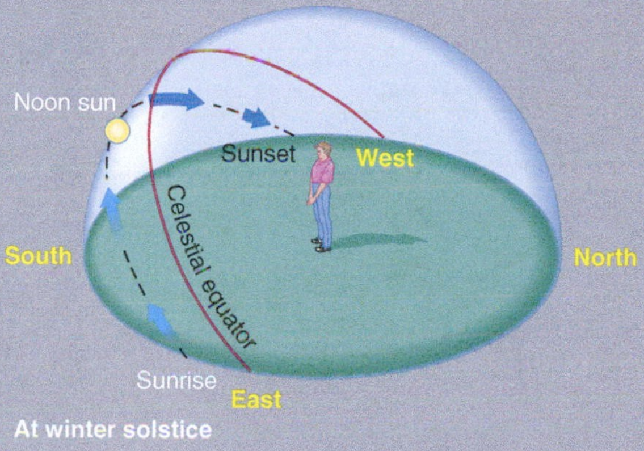

2

The two causes of the seasons are shown at right for someone in the Northern Hemisphere. First, the noon summer Sun is higher in the sky and the winter Sun is lower, as shown by the longer winter shadows. Thus winter sunlight is more spread out. Second, the summer Sun rises in the northeast and sets in the northwest, spending more than 12 hours in the sky. The winter Sun rises in the southeast and sets in the southwest, spending less than 12 hours in the sky. Both of these effects mean that northern latitudes receive more energy from the summer Sun, and summer days are warmer than winter days.

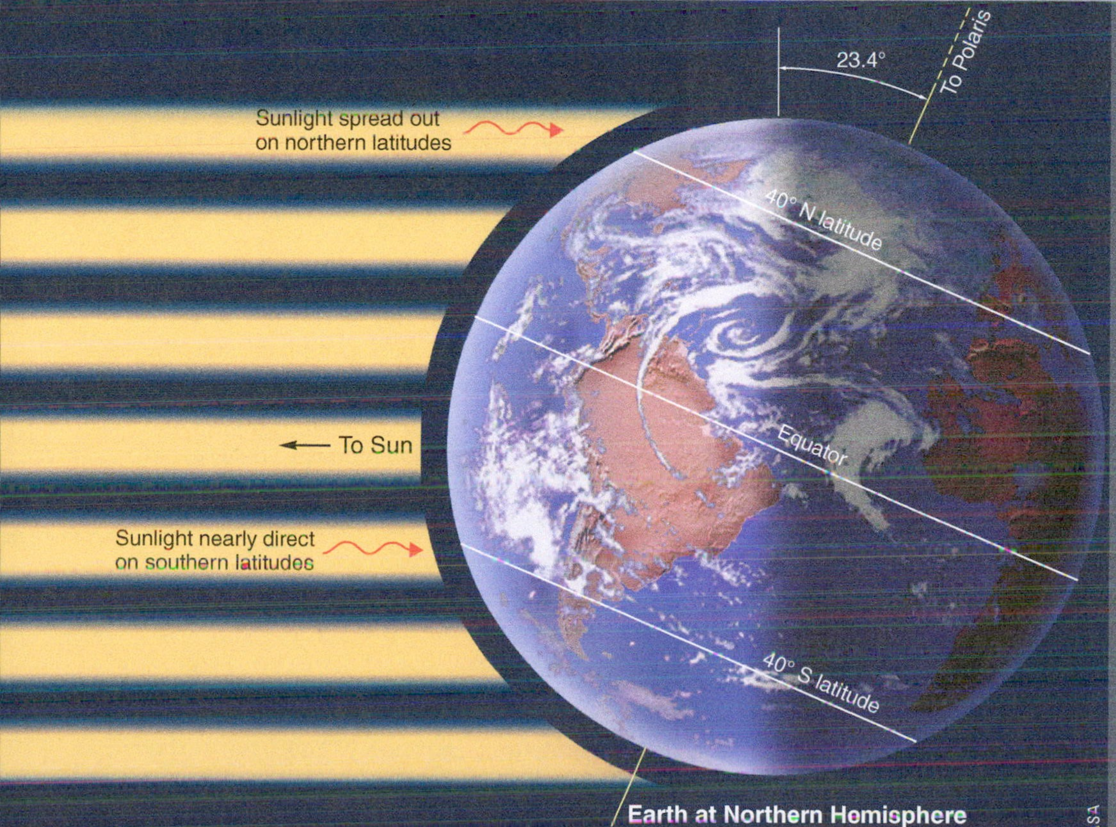

Earth at Northern Hemisphere winter solstice

1d On the day of the winter solstice in late December, Earth's Northern Hemisphere is inclined away from the Sun, and sunlight strikes the ground at an angle and spreads out. At southern latitudes, sunlight shines almost straight down and does not spread out. North America has cool weather and South America has warm weather.

Earth's orbit is only very slightly elliptical. About January 3, Earth is at **perihelion**, its closest point to the Sun, when it is only 1.7 percent closer than average. About July 5, Earth is at **aphelion**, its most distant point from the Sun, when it is only 1.7 percent farther than average. This small variation does not significantly affect the seasons.

FIGURE 2-7

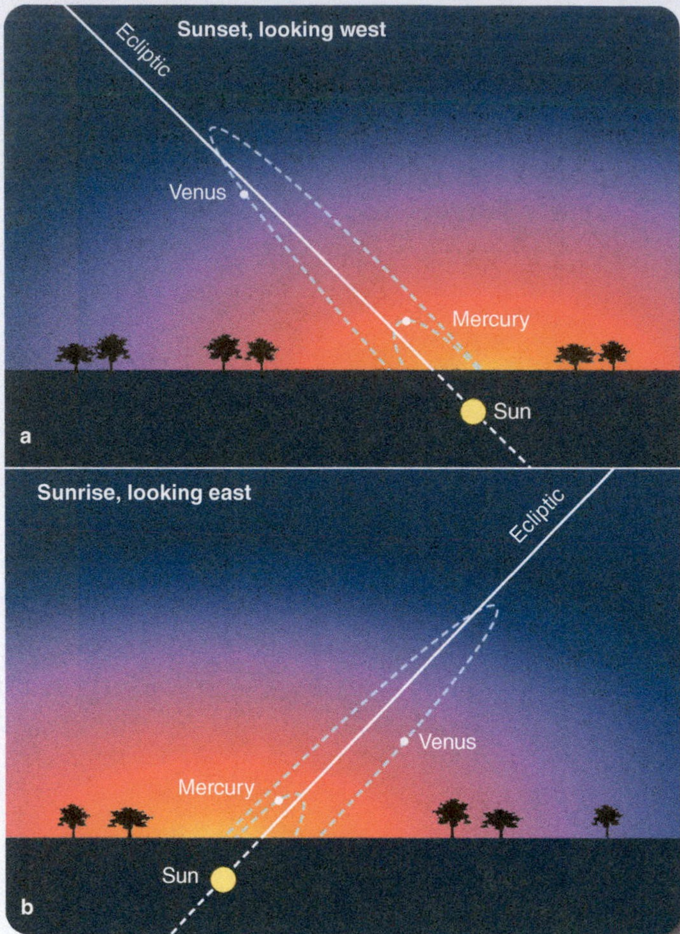

Mercury and Venus follow orbits that keep them near the Sun, so they are visible only soon after sunset or just before sunrise. Venus takes 584 days to move from morning sky to evening sky and back again, but Mercury zips around in only 116 days.

(See **How Do We Know 2-2**.) The signs of the zodiac are no longer important in astronomy. You can look for the planets along the ecliptic appearing like bright stars. Mars looks quite orange in color. Because Venus and Mercury orbit inside Earth's orbit, they never get far from the Sun and are visible in the west after sunset or in the east before sunrise, as you can see in Figure 2-7. Venus can be very bright, but Mercury is difficult to see near the horizon. By tradition, any planet in the sunset sky is called an **evening star**, and any planet in the dawn sky is called a **morning star**. Perhaps the most beautiful is Venus, which can become as bright as magnitude −4.7. As Venus moves around its orbit, it can dominate the western sky each evening for many weeks, but eventually its orbit appears to carry it back toward the Sun as seen from Earth, and it is lost in the haze near the horizon. A few weeks later you can see Venus reappear in the dawn sky as a brilliant morning star. Months later it will switch back to being an evening star once again.

evening star Any planet visible in the sky just after sunset.

morning star Any planet visible in the sky just before sunrise.

2-4 CYCLES OF THE MOON

The Moon orbits eastward around Earth once a month. Starting this evening, look for the Moon in the sky. If it is a cloudy night or if the Moon is in the wrong part of its orbit, you may not see it; but keep trying on successive evenings, and within a week or two you will see the Moon. Then, watch for the Moon on following evenings, and you will see it move along its orbit around Earth, cycling through its phases as it has done for billions of years.

2-4a Motion of the Moon

If you watch the Moon night after night, you will notice two things about its motion. First, you will see it moving relative to the background of stars; second, you will notice that the markings on its face don't change. These two observations will help you understand the motion of the Moon and the origin of the Moon's phases.

The Moon moves rapidly among the constellations. If you watch the Moon for just an hour, you can see it moving eastward against the background of stars by slightly more than its own apparent diameter. Each night when you look at the Moon, you will see it is roughly half the width of a zodiac constellation (about 13 degrees) to the east of its location the night before. This movement is the result of the motion of the Moon along its orbit around Earth.

2-4b The Cycle of Moon Phases

The changing shape of the illuminated part of the Moon as it orbits Earth is one of the most easily observed phenomena in astronomy. You have surely

How Do We Know? 2-2

Pseudoscience

What is the difference between a science and a pseudoscience? Astronomers have a low opinion of beliefs such as astrology, not only because they are groundless but also because they *pretend* to be a science. They are *pseudosciences*, from the Greek *pseudo*, meaning "false."

A **pseudoscience** is a set of beliefs that appears to include scientific ideas but fail to obey the most basic rules of science. For example, some years ago a claim was made that pyramidal shapes focus cosmic forces on anything underneath and might have healing properties. Supposedly, a pyramid made of paper, plastic, or other materials would preserve fruit, sharpen razor blades, and do other miraculous things. Many books promoted the idea of the special power of pyramids, and this idea led to a popular fad.

A key characteristic of science is that its claims can be tested and verified. In this case, simple experiments showed that any shape, not just a pyramid, protects a piece of fruit from airborne spores and allows it to dry without rotting. Likewise, any shape allows oxidation to improve the cutting edge of a razor blade. Because experimental evidence contradicted the claim and because supporters of that hypothesis declined to abandon or revise their claims, you can recognize pyramid power as a pseudoscience. The disregard of contradictory evidence and alternate explanations is a sure sign of a pseudoscience.

Pseudoscientific claims can be self-fulfilling. For example, some believers slept under pyramidal tents to improve their rest. There is no logical mechanism by which such a tent could affect a sleeper, but because people wanted and expected the claim to be true, they reported that they slept more soundly. Vague claims based on personal testimony that cannot be tested are another sign of a pseudoscience.

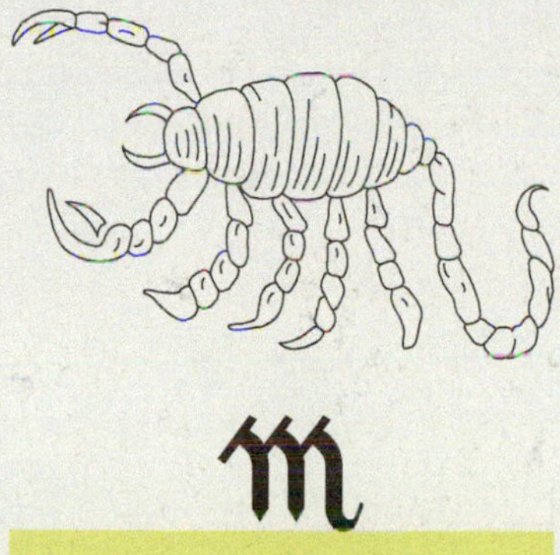

Astrology may be the oldest pseudoscience.

Astrology is probably the best-known pseudoscience. It has been tested over and over for centuries, and it doesn't work. It has been proven beyond a reasonable doubt that there is no connection between the positions of the Sun, Moon, and planets and people's personalities, or events in their lives. Nevertheless, many people believe in astrology despite contradictory evidence.

Pseudosciences appeal to our need to understand and control the world around us. Some such claims involve medical cures, ranging from using magnetic bracelets and crystals to focus mystical power to astonishingly expensive, illegal, and dangerous treatments for cancer. Logic is a stranger to pseudoscience, but human fears and needs are not.

seen the full moon rising dramatically or a thin crescent moon hanging in the evening sky. Study **Concept Art 2C, "The Phases of the Moon,"** and notice three important points:

1. The Moon always keeps the same side facing Earth, and you never see the far side of the Moon. "The man in the Moon" (some cultures see "the rabbit in the Moon" instead) is produced by familiar features on the Moon's near side.
2. The changing shape of the Moon as it passes through its cycle of phases is produced by sunlight illuminating different parts of the side of the Moon you can see. You always see the same side of the Moon looking down on you, but the shifting shadows make the "man in the Moon" change moods as the Moon cycles through its phases (see also Figure 2-8).
3. The orbital period of the Moon around the Earth is not the same as the length of a Moon phase cycle.

Concept Art 2C

THE PHASES OF THE MOON

1

As the Moon orbits Earth, it rotates to keep the same side facing Earth as shown at right. Consequently you always see the same features on the Moon, and you never see the far side of the Moon. A mountain on the Moon that points at Earth will always point at Earth as the Moon revolves and rotates.

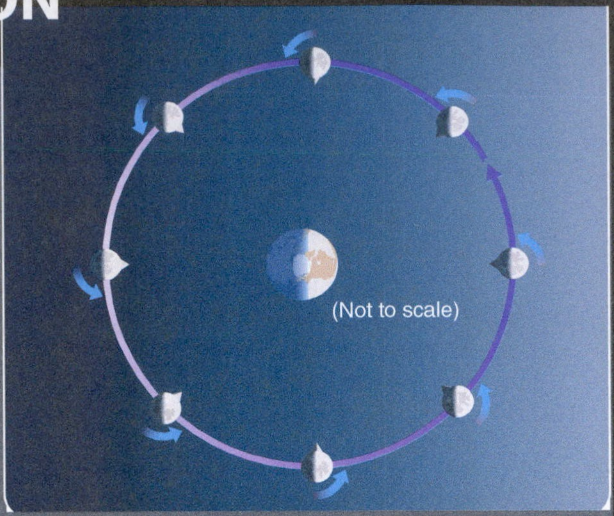

2

As seen at left, sunlight always illuminates half of the Moon. Because you see different amounts of this sunlit side, you see the Moon cycle through phases. At the phase called "new moon," sunlight illuminates the far side of the Moon, and the side you see is in darkness. At new moon, you can't see the Moon at all. At full moon, the side you see is fully lit, and the far side is in darkness. How much you see depends on where the Moon is in its orbit.

Notice that there is no such thing as the "dark side of the Moon." All parts of the Moon experience day and night in a monthlong cycle.

In the diagram at the left, you see that the new moon is close to the Sun in the sky, and the full moon is opposite the Sun. The time of day depends on the observer's location on Earth.

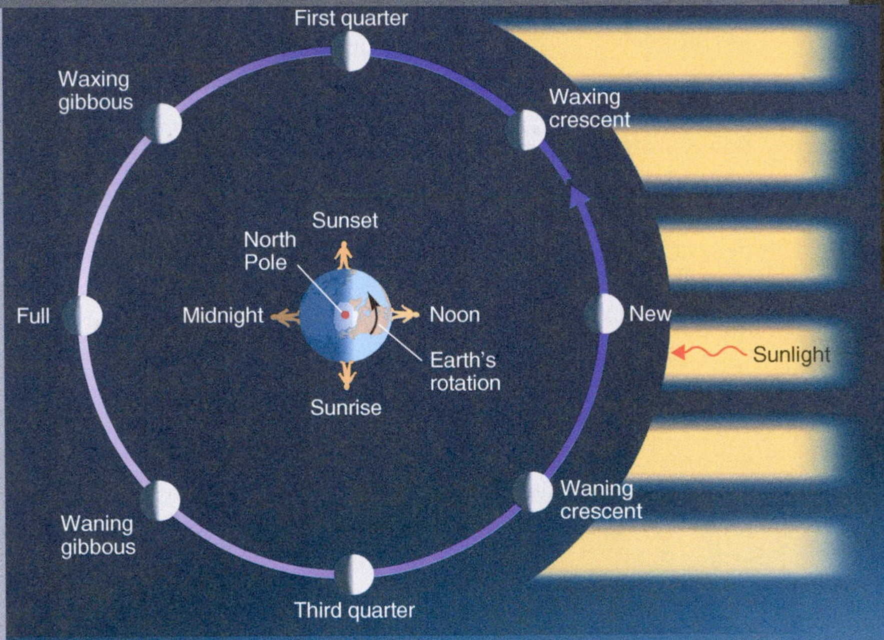

2a

The first two weeks of the cycle of the Moon are shown below according to its position at sunset on 14 successive evenings. As the Moon grows fatter from new to full, it is said to wax

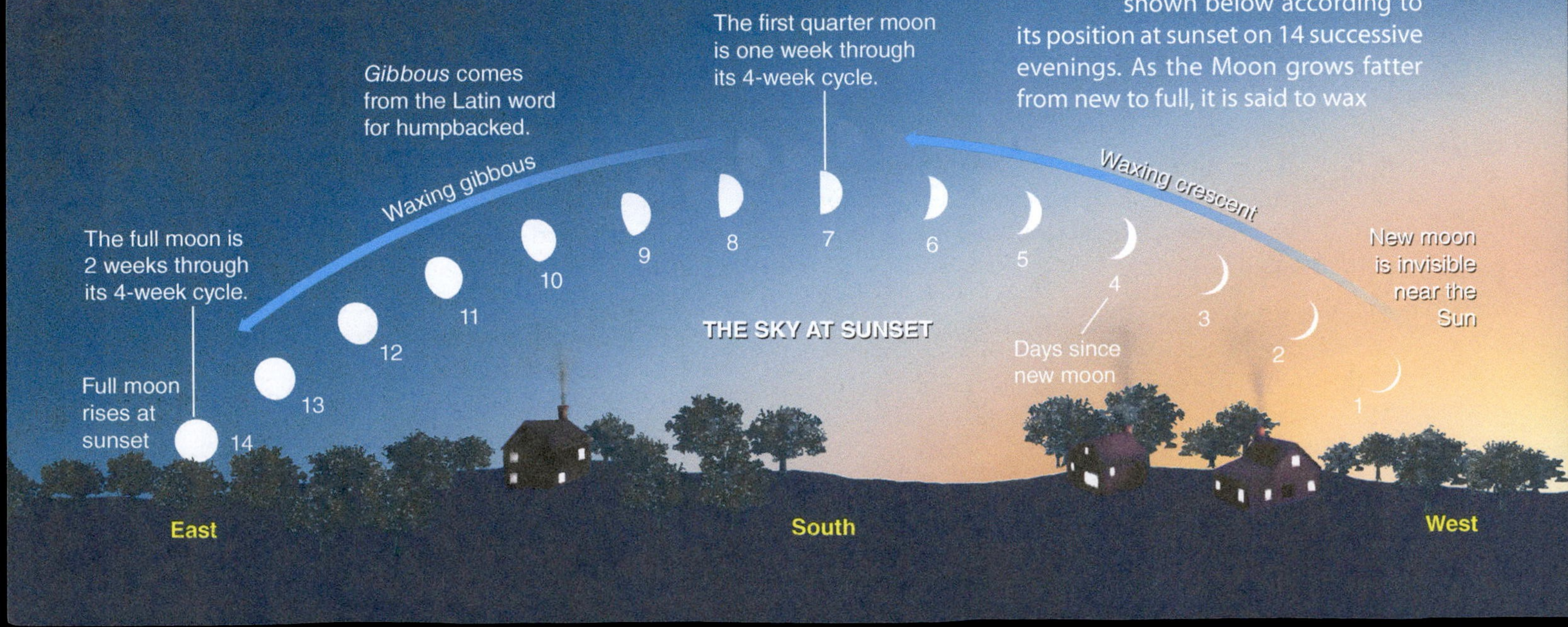

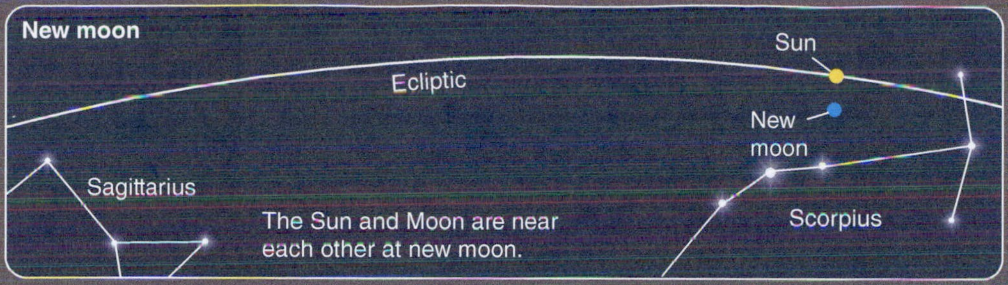

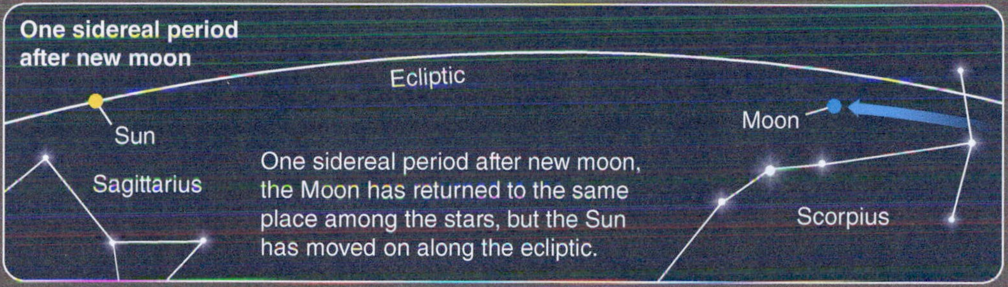

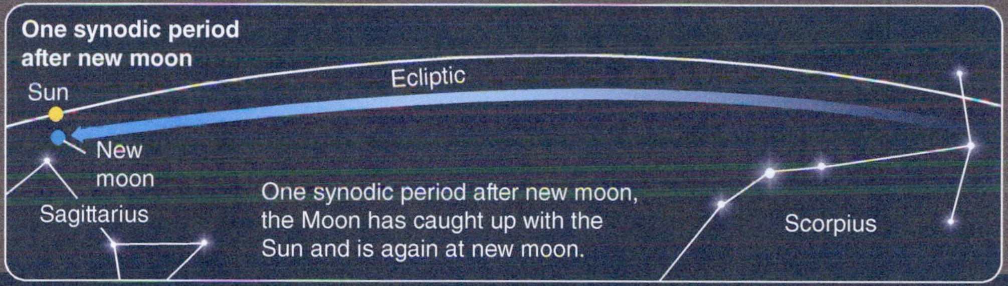

3

The Moon orbits eastward around Earth in 27.32 days, its **sidereal period** (pronounced si-DARE-ee-al). This is how long the Moon takes to circle the sky once and return to the same position among the stars.

A complete cycle of lunar phases takes 29.53 days, the Moon's **synodic period**. (*Synodic* comes from the Greek words for "together" and "path.")

To see why the synodic period is longer than the sidereal period, study the star charts at the right.

Although you think of the lunar cycle as being about 4 weeks long, it is actually 1.53 days longer than 4 weeks. The calendar divides the year into 30-day periods called months (literally "moonths") in recognition of the 29.53 day synodic cycle of the Moon.

a

To think about the changing phases of the Moon, imagine facing the southern sky, which is where people living in the Northern Hemisphere find the ecliptic. The Moon crosses from west to east night by night, following the ecliptic.

You can use the diagram on the opposite page to determine when the Moon rises and sets at different phases.

TIMES OF MOONRISE AND MOONSET

Phase	Moonrise	Moonset
New	Dawn	Sunset
First quarter	Noon	Midnight
Full	Sunset	Dawn
Third quarter	Midnight	Noon

2b The last two weeks of the cycle of the Moon are shown below by its position at sunrise on 14 successive mornings. As the Moon shrinks from full to new, it is said to wane.

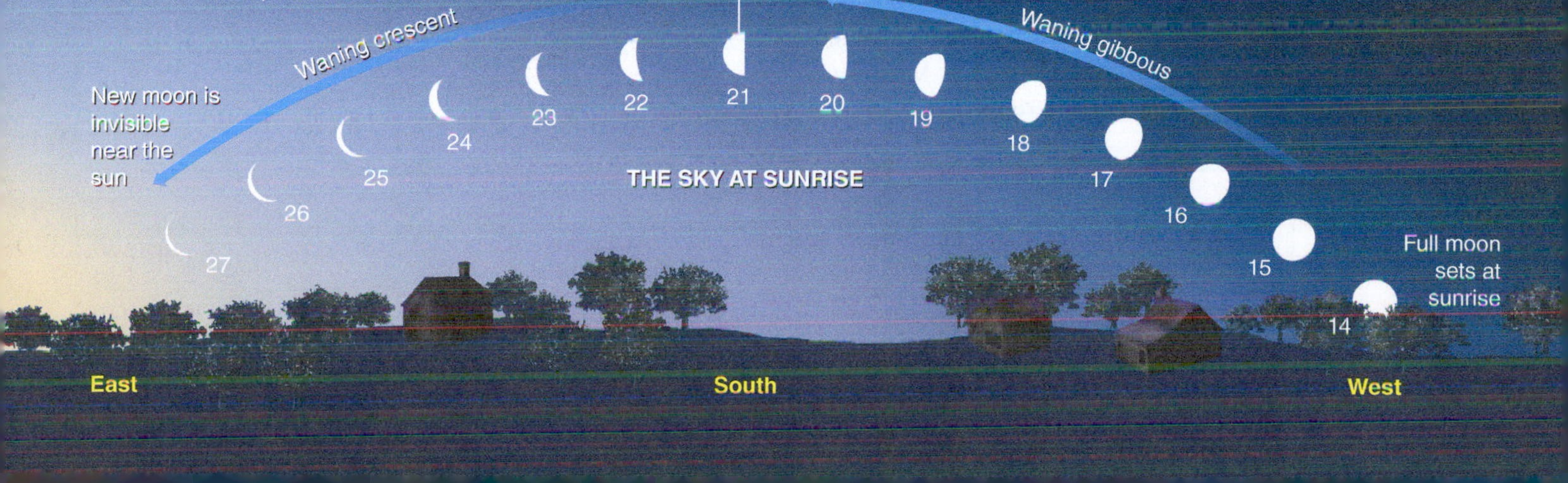

FIGURE 2-8

In this sequence of lunar phases, the Moon shrinks from a thin crescent at upper left to the last crescent at lower right. You see the same face of the Moon, the same mountains, craters, and plains, but the changing direction of sunlight produces the lunar phases.

2-4c Solar Eclipses

Eclipses are due to a seemingly complicated combination of apparent motions of the Sun and Moon, yet they are actually easy to predict once all the cycles are understood. Eclipses are also among the most spectacular of nature's sights you might witness.

From planet Earth you can see a phenomenon that is not visible from most planets. It happens that the Sun is 400 times larger than the Moon and, on the average, 390 times farther away, so the Sun and Moon have nearly equal apparent diameters. (See **The Small Angle Formula** in the Math Reference Cards MR.) Thus, the Moon is just about the right size to cover the bright disk of the Sun and cause a **solar eclipse**. In a solar eclipse, it is the Sun that is being hidden (eclipsed) and the Moon that is "in the way."

A shadow consists of two parts, as you can see in Figure 2-9. The **umbra** is the region of total shadow. For example, if you were in the umbra of the Moon's shadow, you would see no portion of the Sun. The umbra of the Moon's shadow usually just barely reaches Earth's surface and covers a relatively small circular zone (Figure 2-9a). Standing in that umbral zone, you would be in total shadow, unable to see any part of the Sun's surface. That is called a total eclipse, as seen in Figure 2-9b. If you moved into the **penumbra**, however, you would be in partial shadow but could also see part of the Sun peeking around the edge of the Moon. This is called a partial eclipse. Of course, if you are outside the penumbra, you see no eclipse at all.

Because of the orbital motion of the Moon and the rotation of Earth, the Moon's shadow sweeps rapidly across Earth in a long, narrow path of totality. If you want to see a total solar eclipse, you must be in the path of totality. When the umbra of the Moon's shadow sweeps over you, you see one of the most dramatic sights in the sky—the Sun, totally eclipsed.

The eclipse begins as the Moon slowly crosses in front of the Sun. It takes about an hour for the Moon to cover the solar disk, but as the last sliver of the Sun disappears, darkness arrives in a few seconds. Automatic streetlights come on, drivers turn on their cars' headlights, and birds go to roost. The sky usually becomes so dark you can even see the brighter stars. The darkness lasts

solar eclipse The event that occurs when the Moon passes directly between Earth and the Sun, blocking your view of the Sun.

umbra The region of a shadow that is totally shaded.

penumbra The portion of a shadow that is only partially shaded.

FIGURE 2-9

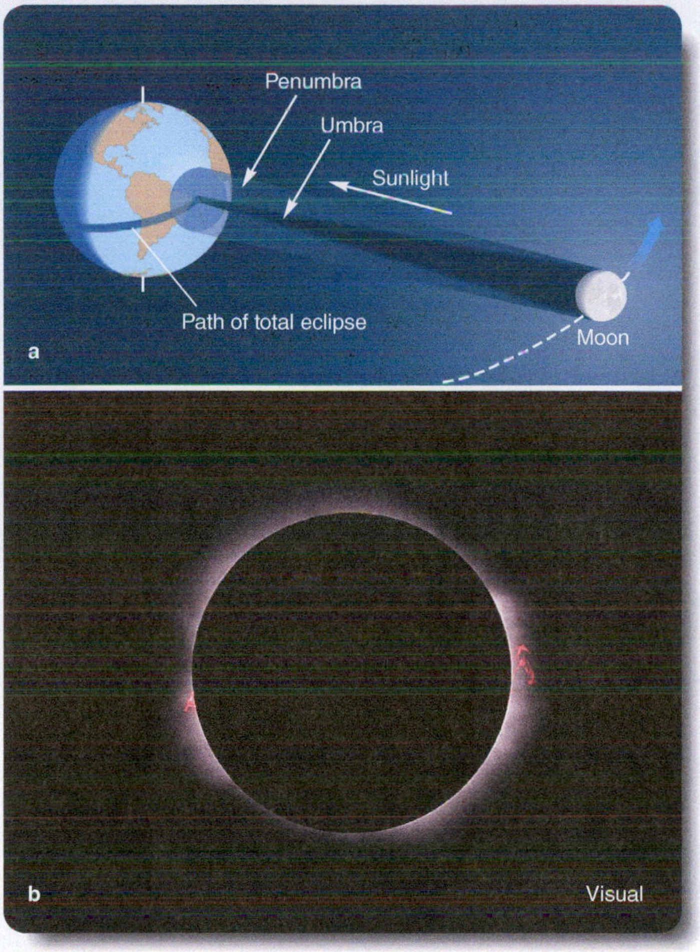

(a) The umbral shadow of the Moon sweeps over a narrow strip of Earth during a solar eclipse. (b) From a location inside the umbral shadow, you would see the Moon cover the bright surface of the Sun in a total solar eclipse. Note faint features of the Sun's atmosphere around the edge of the Moon's disk.

only a few minutes because the umbra is never more than 270 kilometer (170 miles) in diameter on the surface of Earth and sweeps across the landscape at over 1600 km/hr (1000 mph). On average, the period of totality lasts only 2 or 3 minutes and never more than 7.5 minutes. During totality you can see subtle features of the Sun's atmosphere, such as red flame-like projections that are visible only during those moments when the brilliant disk of the Sun is completely covered by the Moon. (The Sun's atmosphere will be discussed in more detail in Chapter 5.) As soon as part of the Sun's disk reappears, the fainter features vanish in the glare, and the period of totality is over. The Moon moves on in its orbit and in an hour the Sun is completely visible again.

Sometimes when the Moon crosses in front of the Sun it is too small to fully cover the Sun, and then you would witness an **annular eclipse**. That is a solar eclipse in which an annulus (meaning "ring") of the Sun's disk is visible around the disk of the Moon. The eclipse never becomes total; it never quite gets dark, and you can't see the faint features of the solar atmosphere. Annular eclipses occur because the Moon follows a slightly elliptical orbit around Earth. If the Moon is in the farther part of its orbit during totality, its apparent diameter will be less than the apparent diameter of the Sun, and thus you see an annular eclipse. Furthermore, Earth's orbit is slightly elliptical, so the Earth-to-Sun distance varies slightly, and consequently so does the apparent diameter of the solar disk, contributing to the effect of the Moon's varying apparent size.

If you plan to observe a solar eclipse, remember that the Sun is bright enough to burn your eyes and cause permanent damage if you look at it directly. This is true whether there is an eclipse or not. Solar eclipses can be misleading, tempting you to look at the Sun in spite of its brilliance and thus risking your eyesight.

During the few minutes of totality, the brilliant disk of the Sun is hidden, and it is safe to look at the eclipse, but the partial eclipse phases and annular eclipses can be dangerous (Figure 2-10). See Figure 2-11 for a safe way to observe the partially eclipsed Sun. Table 2-1 will allow you to determine when some upcoming solar eclipses will be visible from your location.

2-4d Lunar Eclipses

Occasionally you can see the Moon darken and turn copper-red in a **lunar eclipse**. A lunar eclipse occurs at full moon when the Moon moves through the shadow of Earth. Because the Moon shines only by reflected sunlight, you see the Moon

annular eclipse A solar eclipse in which the solar photosphere appears around the edge of the Moon in a bright ring, or annulus. Features of the solar atmosphere cannot be seen during an annular eclipse.

lunar eclipse The darkening of the Moon when it moves through Earth's shadow.

FIGURE 2-10

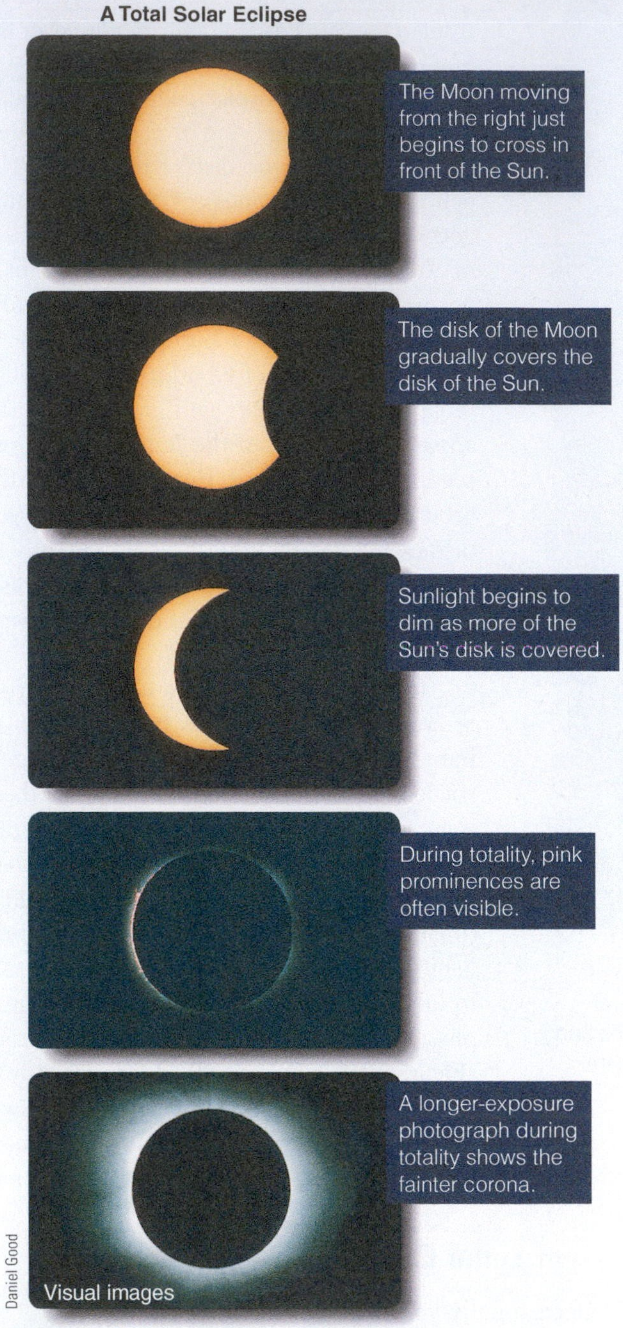

This sequence of photos shows the first half of a total solar eclipse.

FIGURE 2-11

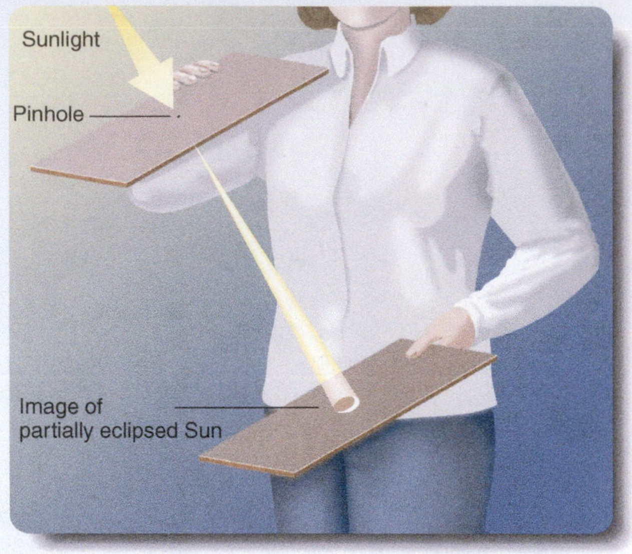

A safe way to view the partial phases of a solar eclipse. Use a pinhole in a card to project an image of the Sun on a second card. The greater the distance between the cards, the larger (and fainter) the image will be.

gradually darken as it enters the shadow. If you were on the Moon and in the umbra of Earth's shadow, you would see no portion of the Sun. If you moved into the penumbra, however, you would be in partial shadow and would see part of the Sun peeking around the edge of Earth so the sunlight would be dimmed but not extinguished. In a lunar eclipse, it is the Moon that is being hidden in the Earth's shadow and Earth that is "in the way" of the sunlight.

If the orbit of the Moon carries it through the umbra of Earth's shadow, you see a total lunar eclipse, demonstrated in Figure 2-12a. As you watch the Moon in the sky, it first moves into the penumbra and dims slightly; the deeper it moves into the penumbra, the more it dims. In about an hour, the Moon reaches the umbra, and you see the umbral shadow darken part of the Moon. It takes about an hour for the Moon to enter the umbra completely and become totally eclipsed. The period of total eclipse may last as long as 1 hour 45 minutes, though the timing of the eclipse depends on where the Moon crosses the shadow. Table 2-2 gives information about some upcoming lunar eclipses.

When the Moon is totally eclipsed, it does not disappear completely. Although it receives no direct sunlight, the Moon in the umbra does receive some sunlight refracted (bent) through Earth's atmosphere. If you were on the Moon during totality, you would not see any part of the Sun because it would be entirely hidden behind Earth. However, you would be able to see Earth's atmosphere illuminated from behind by the Sun. The red glow from this ring consisting of all the Earth's simultaneous sunsets and sunrises illuminates

TABLE 2-1 TOTAL AND ANNULAR ECLIPSES OF THE SUN, 2017 THROUGH 2024*

Date	Total/Annular (T/A)	Time of Mid-Eclipse** (UTC)	Maximum Length of Total or Annular Phase (Min:Sec)	Area of Visibility
2017 February 26	A	14:55	0:44	South America, Atlantic, Africa, Antarctica
2017 August 21***	T	18:27	2:40	Pacific, United States, Atlantic
2019 July 2	T	19:24	4:33	Pacific, South America
2019 December 26	A	05:19	3:39	Southeast Asia, Pacific
2021 June 10	A	10:43	3:51	North America, Arctic
2021 December 4	T	07:35	1:54	Antarctica, South Atlantic
2023 April 20	A/T****	04:18	1:16	Southeast Asia, Philippines, Indonesia, Australia
2023 October 14	A	18:00	5:17	United States, Central America, South America
2024 April 8	T	18:18	4:28	North America, Central America
2024 October 2	A	18:46	7:25	Pacific, South America

* There will be no total or partial solar eclipses in 2018.

** Times are Universal Time. Subtract 5 hours for Eastern Standard Time, 6 hours for Central Standard Time, 7 hours for Mountain Standard Time, and 8 hours for Pacific Standard Time. For Daylight Savings Time (mid-March through early November), add 1 hour to Standard Time.

*** The next major total solar eclipse visible from the United States will occur on August 21, 2017, when the path of totality will cross the United States from Oregon to South Carolina.

**** Hybrid eclipse: begins as annular, becomes total, ends as annular.

Source: NASA Goddard Space Flight Center

TABLE 2-2 TOTAL AND PARTIAL ECLIPSES OF THE MOON, 2017 THROUGH 2024

Date	Time of Mid-Eclipse (UTC)*	Length of Totality (Hr:Min)	Length of Eclipse** (Hr:Min)
2017 August 7	18:22	Partial	1:55
2018 January 31	13:31	1:16	3:23
2018 July 27	20:23	1:43	3:55
2019 January 21	05:13	1:02	3:17
2019 July 16	21:32	Partial	2:58
2021 May 26	11:20	0:15	3:07
2021 November 19	09:04	Partial	3:28
2022 May 16	04:13	1:25	3:27
2022 November 8	11:00	1:25	3:40
2023 October 28	20:15	Partial	1:17
2024 September 18	02:45	Partial	1:03

* Times are Universal Time. Subtract 5 hours for Eastern Standard Time, 6 hours for Central Standard Time, 7 hours for Mountain Standard Time, and 8 hours for Pacific Standard Time. For Daylight Savings Time (mid-March through early November), add 1 hour to Standard Time. Lunar eclipses that occur between sunset and sunrise in your time zone will be visible, and those at midnight will be best placed.

** Does not include penumbral phase.

Source: NASA Goddard Space Flight Center

FIGURE 2-12

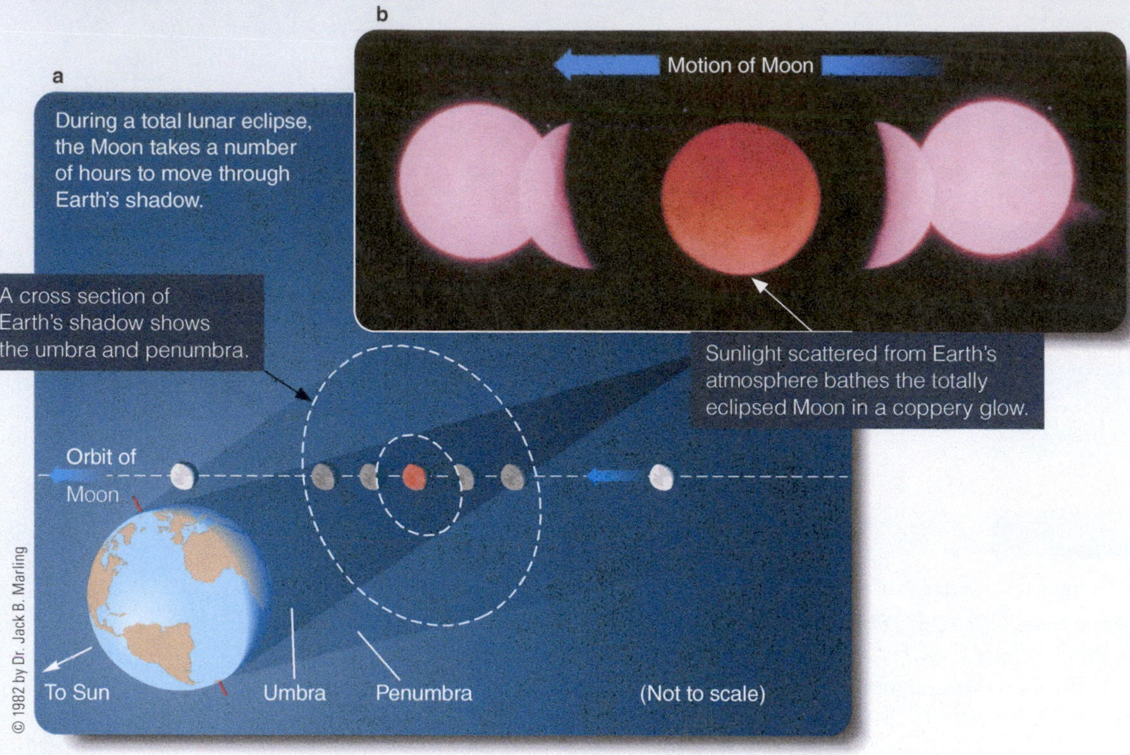

A longer exposure was used to record the Moon while it was totally eclipsed. The Moon's path appears curved in the photo because of photographic effects.

the Moon during totality and makes it glow coppery red, as shown in Figure 2-12b.

If the Moon passes a bit too far north or south of the center of Earth's shadow, it may only partially enter the umbra, and you see a partial lunar eclipse. The part of the Moon that remains outside the umbra in the penumbra receives some direct sunlight, and the glare is usually great enough to prevent your seeing the faint coppery glow of the part of the Moon in the umbra.

Lunar eclipses always occur at full moon but not at every full moon. The Moon's orbit is tipped about 5 degrees to the ecliptic, so most full moons cross the sky north or south of Earth's shadow and there is no lunar eclipse that month (see Figure 2-13). For the same reason, solar eclipses always occur during new moon but not at every new moon. The orientation of the Moon's orbit in space varies slowly and as a result solar and lunar eclipses repeat in a pattern called the

FIGURE 2-13

Umbral shadows of Earth and the Moon. Because of the tilt of the Moon's orbit relative to the ecliptic, it is easy for the shadows to miss their mark at full moon and at new moon and fail to produce eclipses. (The diameters of Earth and the Moon are exaggerated by a factor of 2 for clarity.)

Saros cycle lasting 18 years and 11⅓ days. Prehistoric peoples who understood the Saros cycle could predict eclipses without understanding what the Sun and Moon really were.

Although there are usually no more than one or two lunar eclipses each year, it is not difficult to see one. You need only be on the dark side of Earth when the Moon passes through Earth's shadow. That is, the eclipse must occur between sunset and sunrise at your location to be visible. Table 2-2 will allow you to determine when some upcoming lunar eclipses will be visible.

Saros cycle The 18-year, 11⅓-day period after which the pattern of lunar and solar eclipses repeats.

What Are We? Along for the Ride

We humans are planet-walkers. We live on the surface of a whirling planet, and on clear, dark nights we can look outward into space for billions of light-years.

Most of the stars we see are actually not very far way. They are only hundreds, not billions, of light-years from Earth. The vast majority of the stars in the Universe are too distant and, consequently, too faint to see with the unaided eye. For centuries, human cultures have labeled the visible stars with names and divided them into constellations. The constellation Hercules looked down on Plato and Aristotle just as it looks down on you.

Nearly all of the energy Earth receives comes from the Sun, and, as Earth spins on its axis and revolves around its orbit, the changing balance of sunlight produces the changing seasons. Winter and summer have cycled longer than there have been humans on Earth, and when you watch autumn leaves turn golden you are part of one of the great astronomical cycles of your planet.

Through it all, the Moon orbits Earth and cultures all through history have made it their timepiece. Your calendar is divided into months recognizing the cycle of the Moon, and your months are divided into four weeks recognizing the fourfold cycle of the phases.

Busy as you are, take a moment to look up and enjoy the cycles of the sky. They will remind you that you are a planet-walker.

STUDY TOOLS 2

IN THE BOOK

- ☐ Tear Out the Review Card on *User's Guide to the Sky: Patterns and Cycles*.

ASTRO ONLINE

Visit ASTRO Online at www.cengagebrain.com

- ☐ Interactive Reading
- ☐ Practice Quizzing
- ☐ Videos and Animations
- ☐ Exclusive Features

2 Questions and Problems

REVIEW QUESTIONS

1. What is the difference between an asterism and a constellation? Give some examples.
2. How does the Greek-letter designation of a star give you a clue to its brightness?
3. How did the magnitude system originate in the classification of stars by brightness?
4. What does the word *apparent* mean in *apparent visual magnitude*?
5. In what ways is the celestial sphere a scientific model?
6. If Earth did not rotate, could you define the celestial poles and celestial equator?
7. Where would you go on Earth to place a celestial pole at your zenith?
8. Why does the number of circumpolar constellations depend on the latitude of the observer?
9. If Earth did not rotate, could you still define the ecliptic? Why or why not?
10. Why are the seasons reversed in the Southern Hemisphere relative to the Northern Hemisphere?
11. Do the phases of the Moon look the same from every place on Earth, or is the Moon full at different times as seen from different locations?
12. What phase would Earth be in if you were on the Moon when the Moon was full? At first quarter? At waning crescent?
13. Why isn't there an eclipse at every new moon and at every full moon?
14. Why is the Moon red during a total lunar eclipse?
15. **How Do We Know?** How can a scientific model be useful if it isn't a correct description of nature?
16. **How Do We Know?** What are the main characteristics of a pseudoscience? Can you suggest other examples of pseudoscience not described in the textbook? Can you suggest other examples?

DISCUSSION QUESTIONS

1. Using stars from one or more of the "official" constellations, create an asterism that is significant to the culture of your school.
2. You discover an ancient document that lists the location of buried treasure as 36° 24' 11.9" N by 25° 25' 35.4" E. You look on a map to see approximately where that is, but when you start on your trip you find that your GPS is broken. How will you know when you're at the right latitude? Much harder question: How will you know when you're at the right longitude?
3. You are packing for a vacation on a planet orbiting another star that is much like the Sun. Why might you want some information about the planet's orbit size and axis tilt to know what to pack?
4. You happen to visit the Moon when some people on Earth see a total solar eclipse. Who has a more spectacular experience of this event, you or the friends you left behind back on Earth? Why?

PROBLEMS

1. If light from one star is 40 times brighter (has 40 times more flux) than light from another star, what is their difference in magnitudes? (*Hint:* See **Magnitudes** in the Math Reference Cards MR.)
2. If two stars differ by 8.6 magnitudes, what is their flux ratio? (*Hint:* See **Magnitudes** in the Math Reference Cards MR.)
3. Star A has a magnitude of 2.5; star B, 5.5; and star C, 9.5. Which is brightest? Which are visible to the unaided eye? Which pair of stars has an flux ratio of 16? (*Hint:* See **Magnitudes** in the Math Reference Cards MR.)
4. By what factor is sunlight brighter than moonlight? (*Hint:* Refer to Figure 2-4, and see **Magnitudes** in the Math Reference Cards MR.)
5. If you are at a latitude of 35 degrees north of Earth's equator, what is the angular distance from the northern horizon up to the north celestial pole? From the southern horizon down to the south celestial pole?
6. Given that Earth is 4.6 billion (4.6×10^9) years old, how many precessional cycles have occurred?
7. Identify the phases of the Moon if on March 20 the Moon were located at the position the Sun is located on: (a) March 20, (b) September 22, (c) June 22, and (d) December 21.
8. Identify the phases of the Moon if at sunset in the Northern Hemisphere the Moon were (a) near the eastern horizon, (b) high in the south, (c) in the southeast, (d) in the southwest.
9. Draw a diagram showing Earth, the Moon, and shadows during (a) a total solar eclipse, (b) a total lunar eclipse, (c) a partial lunar eclipse, and (d) an annular eclipse.
10. Phobos, one of the Moons of Mars, has an average diameter of 22.5 km and orbits 5980 km above the surface of the planet. What is the angular diameter of Phobos as seen from Mars? (*Hint:* See **The Small-Angle Formula** in the Math Reference Cards MR.)

LEARNING TO LOOK

1. To take the photos that are combined on the opening page of this chapter, was the photographer located on the day or night side of the Earth? Was the photographer in the Moon's umbra, penumbra, or both? How do you know?
2. Look at Concept Art 2C. Find the person looking at the third-quarter phase of the Moon at sunrise. What percentage of the near side of the Moon is illuminated? Likewise, what percentage is in the dark? Repeat the exercise for the new phase of the Moon.
3. Figure 2-14 shows a crescent Moon. Explain why the Moon could never look this way at night.

FIGURE 2-14

4. Figure 2-15 shows the annular eclipse of May 30, 1984. How is it different from the annular eclipse shown in Figure 2-9? Why do you suppose it is different?

FIGURE 2-15

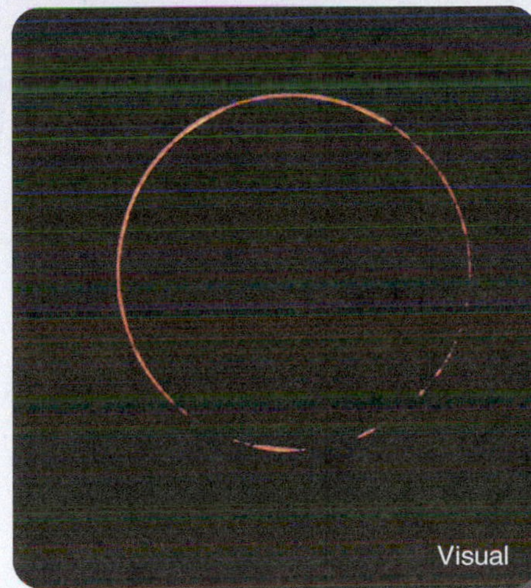

Laurence Marschall

3 The Origin of Modern Astronomy

How you would burst out laughing, my dear Kepler, if you would hear what the greatest philosopher of the Gymnasium told the Grand Duke about me . . .
From a letter by Galileo Galilei

After finishing this chapter go to **PAGE 57** for **STUDY TOOLS.**

GUIDEPOST

The preceding chapters gave you a modern view of Earth. You can now imagine how Earth, the Moon, and the Sun move through space and how that produces the sights you see in the sky. But how did humanity first realize that we live on a planet moving through space? This realization required the revolutionary overthrow of an ancient and honored idea of Earth's place.

By the 16th century, many astronomers were becoming uncomfortable with the idea that Earth sat at the center of a spherical Universe. In this chapter, you will discover how an astronomer named Copernicus directly challenged that idea, how Galileo Galilei changed the rules of the debate, and how Isaac Newton changed humanity's concept of nature. (See CO 3.) Here you will find answers to four important questions:

- How did classical philosophers describe Earth's place in the Universe?
- How did Copernicus revise that ancient concept?
- Why was Galileo condemned by the Inquisition?
- How did Isaac Newton change the way people thought about nature and themselves?

This chapter is not just about the history of astronomy. As the astronomers of the Renaissance struggled to understand Earth and the heavens, they invented a new way of understanding nature—a way of thinking that is now called science. Every chapter that follows will use the methods that were invented when Copernicus and others grappled with the ancient idea that Earth is the center of the Universe.

3-1 ASTRONOMY BEFORE COPERNICUS

To understand why Copernicus's work was so important, you first need to backtrack to ancient Greece and meet the two great authorities of ancient astronomy, the brilliant philosopher Aristotle and a later follower of Aristotle's principles, Claudius Ptolemy, whom you first met in Chapter 2 Section 2-1c in the context of measuring the brightness of stars.

Recall from Chapter 1 the differences among Solar System, galaxy, and Universe. The early astronomers discussed here didn't know there were other galaxies. They

CO 3 Galileo's telescope revealed such things as craters on the Moon, and he explained how that evidence could be used to test the prevailing Earth-centered model of the Universe. He was condemned by the Inquisition in 1633.

didn't know that we live inside a galaxy. They didn't even know that the stars were other objects like the Sun. They thought the entire Universe consisted of Earth, the Sun, the Moon, and five moving points of light they called planets. Furthermore, they did not know that the planets are composed of substances much like Earth's. They imagined that all of this was enclosed inside the celestial sphere that lay just beyond the most distant planet, Saturn. Their entire Universe was quite small, and they imagined that Earth lay at the center.

3-1a Aristotle's Universe

Philosophers of the ancient world attempted to deduce truth about the Universe by reasoning from **first principles**. A first principle was something that seemed obviously true to everyone and supposedly needed no further examination. That may strike you as peculiar; modern thinkers tend to observe how things work and then from that evidence make principles and conclusions that can always be reexamined. Before the Renaissance, however, reasoning from evidence (which you might call "scientific thinking") was not widespread.

Study **Concept Art 3A**, "**The Ancient Universe**," and notice three important ideas:

1. Ancient philosophers and astronomers accepted without question—as first principles— that heavenly objects must move on circular paths at constant speeds, and that the Earth was motionless at the center of the Universe. Although a few ancient writers mentioned the possibility that Earth might move, most of them did so in order to point out how that idea was "obviously" wrong.
2. As viewed by you from Earth, the planets seem to follow complicated paths in the sky, including episodes of "backward" motion that are difficult to explain in terms of motion on circular paths at constant speeds.
3. Finally, you can see how Ptolemy created an elaborate geometrical and mathematical model to explain details of the observed motions of the planets while assuming Earth was motionless at the center of the Universe.

first principle
Something that seems obviously true and needs no further examination.

Aristotle lived in Greece from the year 384 to 322 BCE. He believed as a first principle that the heavens were perfect. Because the sphere and circle were considered the only perfect geometrical figures, Aristotle also believed that all motion in the perfect heavens must be caused by the rotation of spheres carrying objects around in uniform circular motions. Aristotle's writings became so famous that he was known throughout the Middle Ages as "The Philosopher," and the geocentric universe of nested spheres that he devised dominated astronomy. His opinions on the nature of Earth and the sky were widely accepted for almost 2000 years.

Claudius Ptolemy, a mathematician who lived roughly 500 years after Aristotle, believed in the basic ideas of Aristotle's universe but was interested in practical rather than philosophical questions. For Ptolemy, first principles took second place to accuracy. He set about making an accurate mathematical description of the motions of the planets. Ptolemy weakened the first principles of Aristotle by moving Earth a little off-center in the model and inventing a way to slightly vary the planets' speeds. This made his model (published around the year 140) a better match to the observed motions.

Aristotle's universe, as embodied in the mathematics of Ptolemy's model, dominated ancient astronomy. At first the Ptolemaic model predicted positions of the planets with fair accuracy; but as centuries passed, errors accumulated, and Islamic and later European astronomers had to update the model, adjusting the sizes and locations of the circles and changing the rates of motion.

3-2 NICOLAUS COPERNICUS

Nicolaus Copernicus (originally, Mikolaj Kopernik) was born in 1473 in what is now Poland (Figure 3-1). At the time of his birth—and throughout his life—astronomy was based on Ptolemy's model of Aristotle's universe. In spite of many revisions, the Ptolemaic model was still a poor predictor of planet positions, but because of the authority of Aristotle, it was widely accepted. Moreover, because in Aristotle's philosophy the most perfect region was in the heavens and the most imperfect region was at Earth's center, the classical **geocentric universe** model matched the commonly held Christian view of the geometry of heaven and hell. Anyone who criticized Aristotle's model of the Universe was thereby also challenging belief in the locations of heaven and hell, risking at least criticism and perhaps a serious charge of heresy with a possible death penalty.

geocentric universe
A model universe with Earth at the center, such as the Ptolemaic universe.

heliocentric universe
A model universe with the Sun at the center, such as the Copernican universe.

3-2a Copernicus's Model

Copernicus was associated with the Roman Catholic Church throughout his life. His uncle, by whom he was raised and educated, was a bishop. After studying medicine and Church law in some of the major universities in Europe, Copernicus became a Church employee, serving as secretary and personal physician to his powerful uncle for 15 years and moving into quarters in the cathedral after his uncle died. Because of this connection to the Church and his fear of persecution, he hesitated to publish his revolutionary ideas that challenged the Ptolemaic model and, indirectly, belief in heaven and hell.

What were these revolutionary ideas? Copernicus believed that the Sun and not Earth was the center of the Universe and that Earth rotated on its axis and revolved around the Sun. Copernicus apparently began doubting Ptolemy's geocentric model during his college days. A **heliocentric universe** model (Sun centered, from the Greek word for Sun, *helios*) had been discussed occasionally before Copernicus's time, but Copernicus was the first person to produce a detailed model with substantial justifying arguments. Sometime before 1514, Copernicus wrote a short pamphlet summarizing his model and distributed it in handwritten form while he worked on his book.

B&Y Photography / Alamy Stock Photo

3-2b *De Revolutionibus*

Copernicus's book *De Revolutionibus Orbium Coelestium (On the Revolutions of Celestial Spheres)* was essentially finished by about 1530. He hesitated to publish, although other astronomers, and even church officials concerned about reform of the calendar, knew about his work, sought his advice, and looked forward to the book's publication. In 1542 Copernicus finally sent the manuscript for *De Revolutionibus* off to be printed. He died in 1543 before the printing was completed.

The most important idea in the book was that the Sun was the center of the Universe. That single innovation had an impressive consequence—the retrograde motion of the planets was immediately explained in a straightforward way without the epicycles that Ptolemy used.

In Copernicus's model, which you can see in Figure 3-3, Earth moves faster along its orbit than the planets that lie farther from the Sun. Consequently, Earth periodically overtakes and passes these planets. Imagine that you are a runner on a track moving along an inside lane. Runners well ahead of you appear to be moving forward relative to background scenery. As you overtake and pass slower runners in outside lanes, they fall behind, seeming to move backward for a few moments relative to the scenery. The same thing happens as Earth passes a planet such as Mars. Although Mars moves steadily along its orbit, as seen from Earth it seems to slow to a stop and move westward (retrograde) relative to the background stars as Earth passes it (Figure 3-2). Because the planets' orbits do not lie in precisely the same plane, a planet does not resume its eastward motion in precisely the same path it followed earlier. Instead, it describes a loop with a shape depending on the angle between the two orbital planes.

Copernicus's model was simple and straightforward compared with the multiple off-center circles of the Ptolemaic model. However, *De Revolutionibus* failed to immediately disprove the geocentric model for one critical reason—the Copernican model could not predict the positions of the planets any more accurately than the Ptolemaic model could.

Although Copernicus proposed a revolutionary idea in making the Solar System heliocentric, he was

FIGURE 3-1

Iryna1/Shutterstock.com

Nicolaus Copernicus (Latinized version of his birth name, Mikolaj Kopernik) pursued a lifetime career in the Church, but he was also a talented mathematician and astronomer. His work triggered a revolution in human thought.

FIGURE 3-2

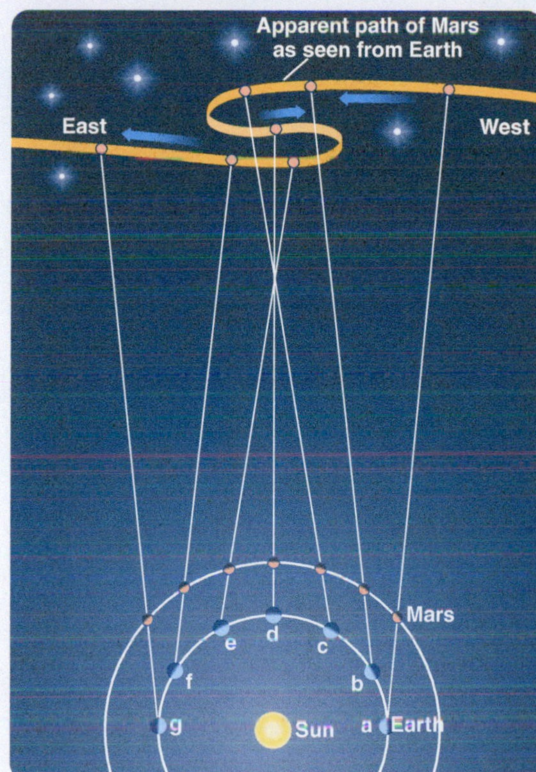

Earth and Mars are shown at equal intervals to show that as Earth overtakes Mars (a–c), Mars appears to slow its eastward motion. As Earth passes Mars (d), Mars appears to move westward. As Earth draws ahead of Mars (e–g), Mars resumes its eastward motion against the background stars.

Concept Art 3A

AN ANCIENT MODEL OF THE UNIVERSE

1

For 2000 years, the minds of astronomers were shackled by a pair of ideas. The Greek philosopher Plato argued that the heavens were perfect. Because the only perfect geometrical shape is a sphere, which carries a point around in a circle as it rotates, and because the only perfect motion is uniform motion, Plato concluded that all motion in the heavens must be made up of combinations of circles turning at uniform rates. This idea was called **uniform circular motion**.

Plato's student Aristotle argued that Earth was imperfect and lay at the center of the Universe. Such a model is known as a **geocentric universe**. His model contained 55 spheres turning at different rates and at different angles to carry the Moon, Mercury, Venus, the Sun, Mars, Jupiter, and Saturn across the sky.

Aristotle was known as the greatest philosopher in the ancient world, and for 2000 years his authority chained the minds of astronomers with uniform circular motion and geocentrism. See the model at right.

From *Cosmographica* by Peter Apian (1539)

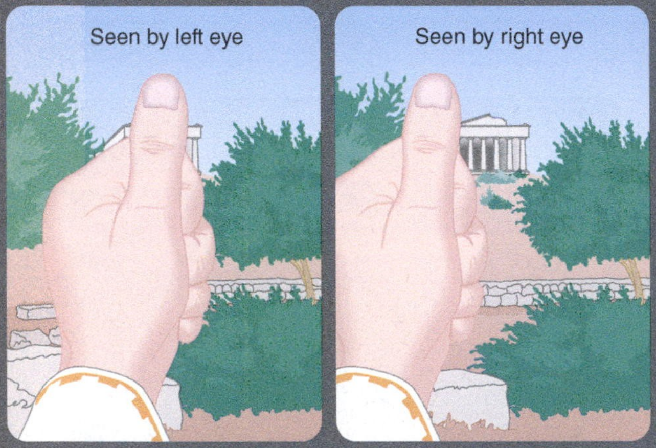

1a Ancient astronomers believed that Earth did not move because they saw no **parallax**, the apparent motion of an object because of the motion of the observer. To demonstrate parallax, close one eye and cover a distant object with your thumb held at arm's length. Switch eyes, and your thumb appears to shift position as shown at left. If Earth moves, ancient astronomers reasoned, you should see the sky from different locations at different times of the year, and you should see parallax distorting the shapes of the constellations. They saw no parallax, so they concluded Earth could not move. Actually, the parallax of the stars is too small to see with the unaided eye.

2

Planetary motion was a big problem for ancient astronomers. In fact, the word *planet* comes from the Greek word for "wanderer," referring to the eastward motion of the planets against the background of the fixed stars. The planets did not, however, move at a constant rate, and they could occasionally stop and move westward for a few months before resuming their eastward motion. This backward motion is called **retrograde motion**.

Every 2.14 years, Mars passes through a retrograde loop. Two successive loops are shown here. Each loop occurs further east along the ecliptic and has its own shape.

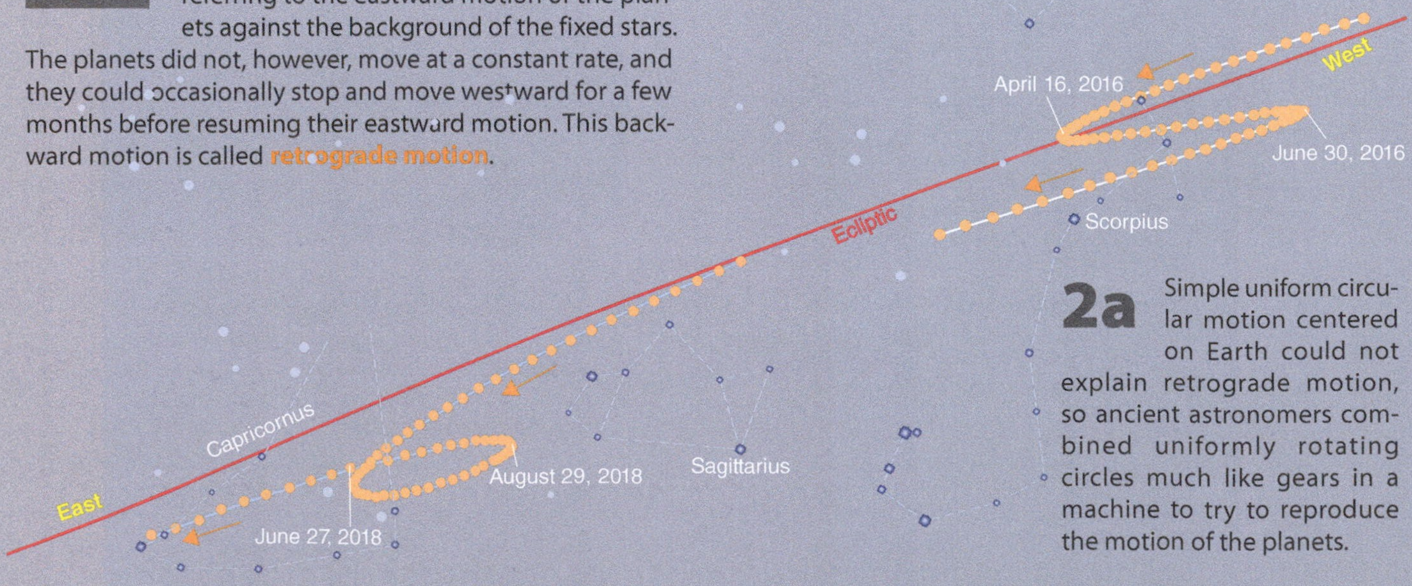

2a Simple uniform circular motion centered on Earth could not explain retrograde motion, so ancient astronomers combined uniformly rotating circles much like gears in a machine to try to reproduce the motion of the planets.

3 Uniformly rotating circles were key elements of ancient astronomy. Claudius Ptolemy created a mathematical model of the Aristotelian universe in which the planet followed a small circle called the **epicycle** that slid around a larger circle called the **deferent**. By adjusting the size and rate of rotation of the circles, he could approximate the retrograde motion of a planet. See illustration at right.

To adjust the speed of the planet, Ptolemy supposed that Earth was slightly off center and that the center of the epicycle moved such that it appeared to move at a constant rate as seen from the point called the **equant**.

To further adjust his model, Ptolemy added small epicycles (not shown here) riding on top of larger epicycles, producing a highly complex model.

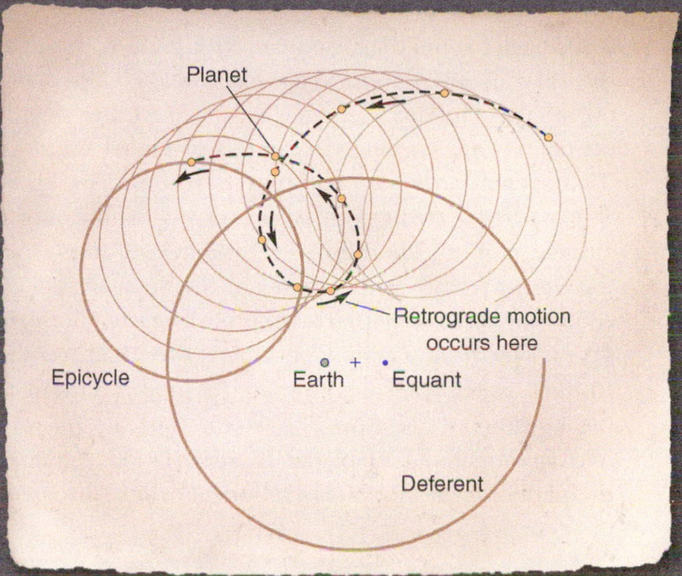

3a Ptolemy's great book *Mathematical Syntaxis* (about the year 140) contained the details of his model. Islamic astronomers preserved and studied the book through the Middle Ages, and they called it *Al Majisti (The Greatest)*. When the book was found and translated from Arabic to Latin in the 12th century, it became known as *Almagest*.

3b The Ptolemaic model of the Universe shown below was geocentric and based on uniform circular motion. Note that Mercury and Venus were treated differently from the rest of the planets. The centers of the epicycles of Mercury and Venus had to remain on the Earth–Sun line as the Sun circled Earth through the year.

Equants and smaller epicycles are not shown here. Some versions contained nearly 100 epicycles as generations of astronomers tried to fine-tune the model to better reproduce the motion of the planets.

Notice that this modern illustration shows rings around Saturn and sunlight illuminating the globes of the planets, features that could not be known before the invention of the telescope.

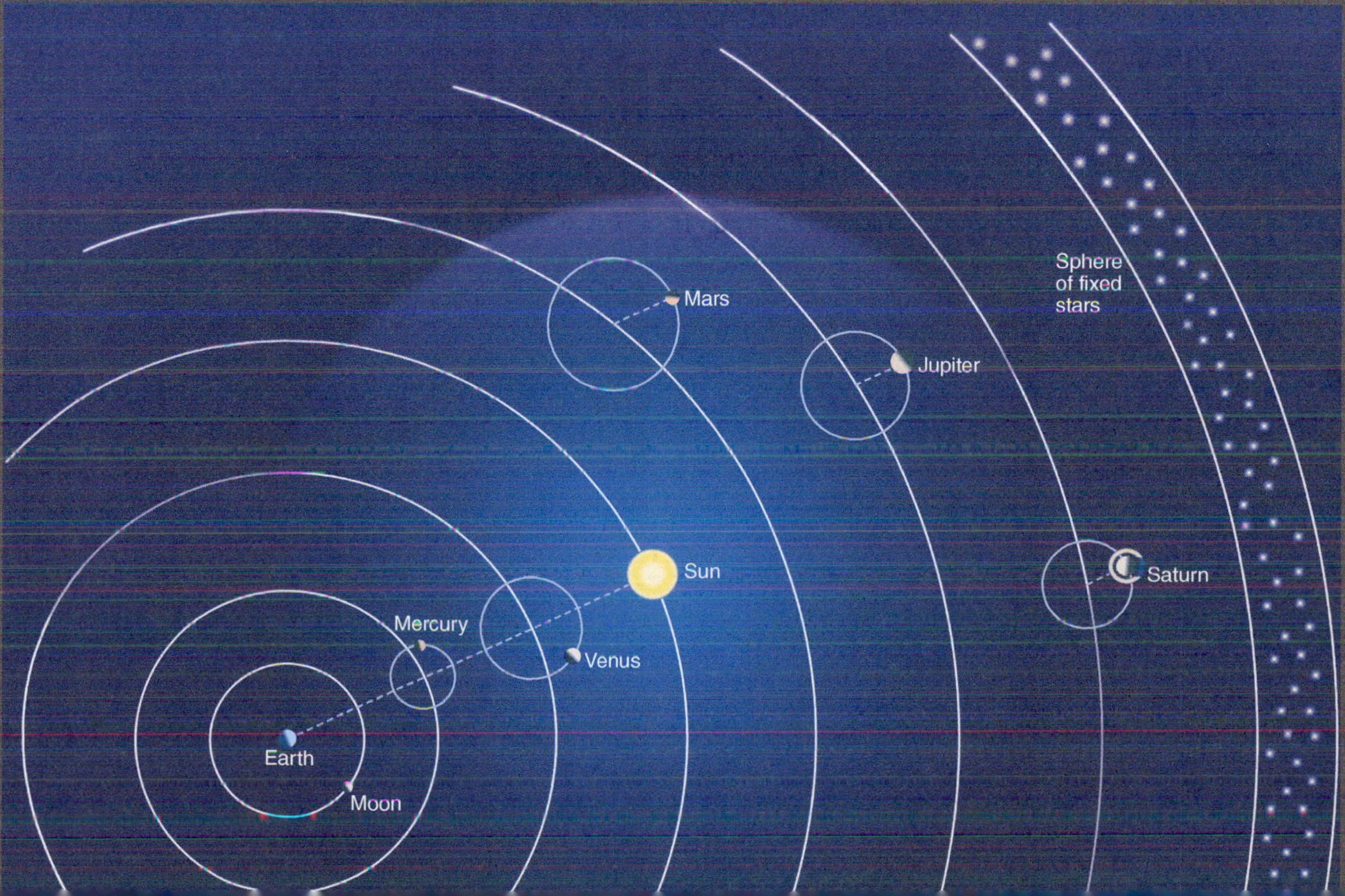

a classically trained astronomer with great respect for the old concept of uniform circular motion. Copernicus objected to Ptolemy's schemes for moving Earth slightly off-center and varying the speeds of planet motions. That seemed arbitrary and ugly to Copernicus, so he returned to a strong but incorrect belief in uniform circular motion. Therefore, even though his model put the Sun correctly at the center of the Solar System, it could not accurately predict the positions of the planets as seen from Earth. Copernicus even had to reintroduce small epicycles to match minor variations in the motions of the Sun, the Moon, and the planets. Astronomers today recognize those variations as due to the planets' real motions in elliptical orbits (discussed in Section 3-3).

You should notice the difference between the Copernican model and the Copernican hypothesis. The Copernican *model* is inaccurate. It includes uniform circular motion and thus does not precisely describe the motions of the planets. But the Copernican *hypothesis* that the Solar System is heliocentric is correct—the planets do circle the Sun, not Earth. Why that hypothesis gradually won acceptance is a question historians still debate. There are probably a number of reasons, including the revolutionary spirit of the times, but the most important factor may be the simplicity of the idea. For one thing, placing the Sun at the center of the Universe produced a symmetry among the motions of the planets that is elegant, pleasing to the eye and mind (Figure 3-3). In the Ptolemaic model, Mercury and Venus had to be treated

FIGURE 3-3

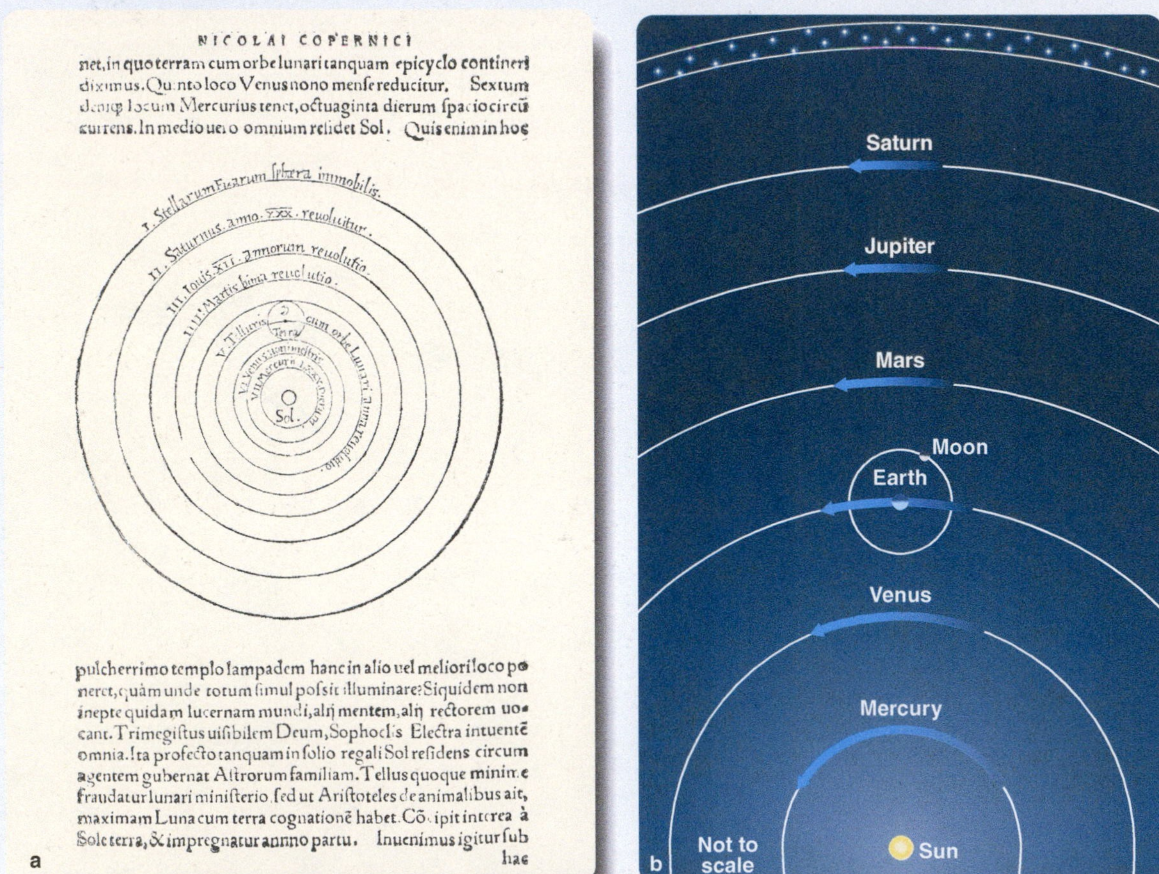

Yerkes Observatory

(a) The Copernican universe, as reproduced in his book *De Revolutionibus*. Earth and all the known planets revolve in separate circular orbits about the Sun (Sol) at the center. The outermost sphere carries the immobile stars of the celestial sphere. Notice the orbit of the Moon around Earth (Terra). (b) The model is simple not only in the arrangement of the planets but also in their motions. Orbital velocities (blue arrows) decrease from that of Mercury, the fastest, to that of Saturn, the slowest.

How Do We Know? 3-1

Scientific Revolutions

How do scientific revolutions occur? You might think from what you know of the scientific method that science grinds forward steadily as new hypotheses are tested against evidence and accepted or rejected. In fact, science sometimes leaps forward in scientific revolutions. The Copernican Revolution is often cited as the perfect example; in a few decades, astronomers rejected the 2000-year-old geocentric model and adopted the heliocentric model. Why does that happen? It's all because scientists are human.

The American philosopher of science Thomas Kuhn has referred to a commonly accepted set of scientific ideas and assumptions as a scientific **paradigm**. The pre-Copernican thinkers shared a geocentric paradigm that included uniform circular motion, geocentrism, and the perfection of the heavens. Although they were intelligent, they were prisoners of that paradigm. A scientific paradigm is powerful because it shapes your perceptions. It determines what you judge to be important questions and what you judge to be significant evidence. Consequently, the ancient philosophers could not recognize how their geocentric paradigms limited what they understood.

N A.Sharp, REU program/NOAO/AURA/NSF

Early philosophers believed the stars were attached to a solid sphere.

You will see here how the work of Copernicus, Galileo, and Kepler overthrew the geocentric paradigm. Scientific revolutions occur when the deficiencies of the old paradigm build up until finally a scientist has the insight to think "outside the box." Pointing out the failings of the old ideas and proposing a new paradigm with supporting evidence is like poking a hole in a dam; suddenly the pressure is released, and the old paradigm is swept away.

Scientific revolutions are exciting because they give you a dramatic new understanding of nature, but they are also times of conflict as new insights sweep away old ideas.

differently from the rest of the planets: Their epicycles had to remain centered on the Earth–Sun line. In Copernicus's model, all of the planets were treated the same. They all followed orbits that circled the Sun at the center.

Although astronomers throughout Europe read and admired *De Revolutionibus* and found Copernicus's astronomical observations and mathematics to have great value, few astronomers believed, at first, that the Sun actually was the center of the Solar System and that Earth moved. How the Copernican hypothesis was gradually recognized as correct has been called the Copernican Revolution, because it was not just the adoption of a new idea but a total change in the way astronomers and the rest of humanity thought about the place of Earth. (See **How Do We Know? 3-1**.)

The most important consequence of the Copernican hypothesis was not what it said about the Sun but what it said about Earth. By placing the Sun at the center, Copernicus made Earth move along an orbit like the other planets. By making Earth a planet, Copernicus revolutionized humanity's view of its place in the Universe and triggered a controversy that would eventually bring the astronomer Galileo Galilei

paradigm A commonly accepted set of scientific ideas and assumptions.

before the Inquisition, a controversy over the nature of scientific and religious truths that continues even today.

3-3 TYCHO BRAHE, JOHANNES KEPLER, AND PLANETARY MOTION

As astronomers struggled to understand the place of Earth, they also faced the problem of planetary motion. How exactly do the planets move? That problem was solved by a nobleman who built a fabulous observatory and a poor commoner with a talent for mathematics.

3-3a Tycho Brahe

The Danish nobleman Tycho Brahe is remembered in part for wearing false noses to hide a dueling scar from his college days (Figure 3-4). He was reportedly very proud of his noble station, so his disfigurement probably did little to improve his lordly disposition.

In 1572, astronomers were startled to see a new star (now called Tycho's supernova) appear in the sky. Aristotle had argued that the heavens were perfect and therefore unchanging, so astronomers concluded that the new star had to be nearer than the Moon. Tycho could not detect the new star's parallax (look back to Concept Art 3A), meaning it had to be far beyond the Moon; it was a change in the supposedly unchanging starry sphere.

When Tycho wrote a book about his discovery, the king of Denmark honored him with a generous income and the gift of an island, Hveen, where Tycho built a fabulous observatory. Tycho lived before the invention of the telescopes, so his observatory was equipped with wonderful instruments for measuring the positions of the Sun, the Moon, and the planets using the unaided eye and peering along sights. For 20 years, Tycho and his assistants carefully tracked the motions of the planets.

After the death of the Danish king, Tycho moved to Prague where he became the Imperial Mathematician to the Holy Roman Emperor Rudolph II. Tycho hired a few assistants including a German school teacher named Johannes Kepler. Just before Tycho died in 1601, he asked Rudolph II to make Kepler Imperial Mathematician. Thus the newcomer, Kepler, became Tycho's replacement (though at one-sixth Tycho's salary).

FIGURE 3-4

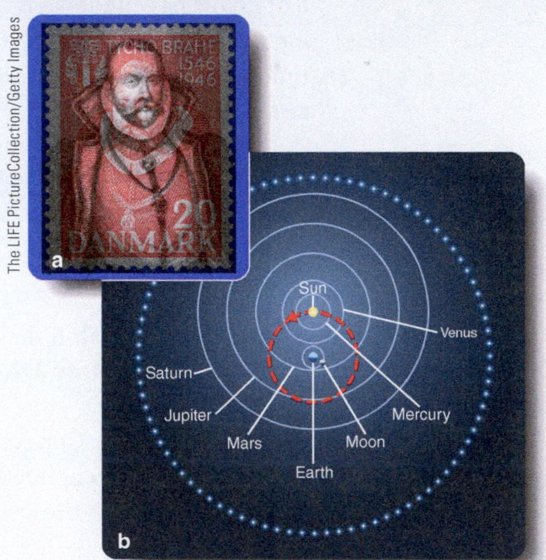

Tycho Brahe was, during his lifetime, the most famous astronomer in the world. Tycho's model of the Universe retained the first principles of classical astronomy; it was geocentric with the Sun and the Moon revolving around Earth, but the planets revolved around the Sun. All motion was along circular paths.

3-3b Johannes Kepler

No scientist could have been more different from Tycho Brahe than was Johannes Kepler. He was born in 1571, the oldest of six children in a poor family living in what is now southwest Germany. His father, who fought as a mercenary soldier for whomever could afford his fees, eventually disappeared. Kepler's mother was apparently an unpleasant and unpopular woman. She was accused of witchcraft in her later years, and Kepler defended her (successfully) in a trial that dragged on for 3 years. Kepler himself had poor health, even as a child, so it is surprising that he did well in school, winning promotion to a Latin school and eventually a scholarship to the university at Tübingen, where he studied to become a Lutheran pastor.

While still a college student, Kepler had become a believer in the Copernican hypothesis. During his last year of study, Kepler accepted a teaching job in the town of Graz, in what is now Austria, which allowed him to continue his studies in mathematics and astronomy. By 1596, the same year Tycho arrived in Prague, Kepler had learned enough to publish a book called *The Forerunner of Dissertations on the Universe, Containing the Mystery of the Universe*. The book is more commonly known by an abbreviation of its original Latin title: *Mysterium Cosmographicum* (Figure 3-5).

By modern standards, the book contains almost nothing of value. It begins with a long appreciation of Copernicus's model and then goes on to mystical

FIGURE 3-5

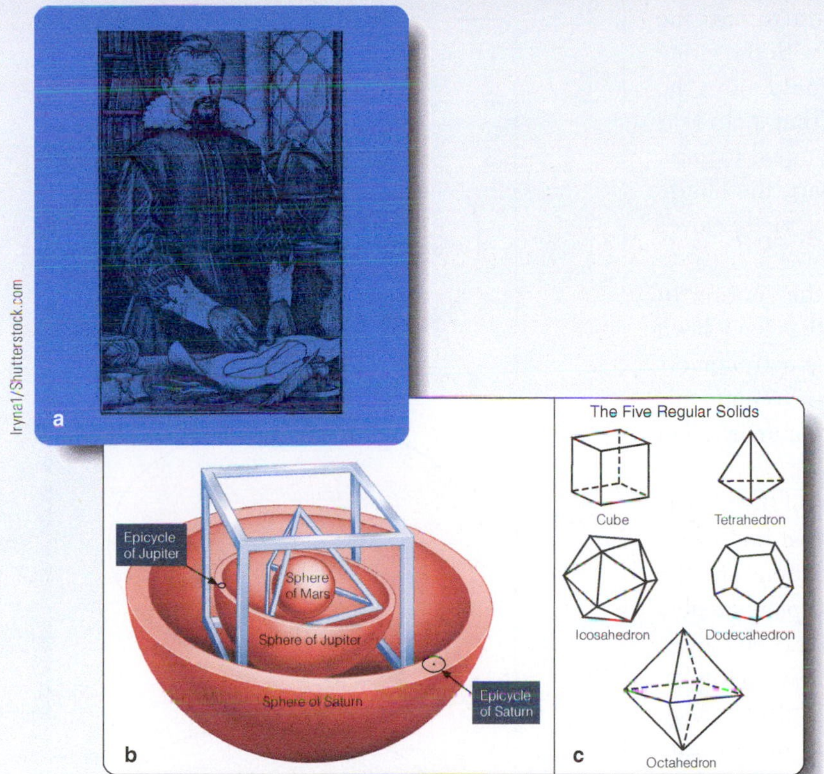

(a) Johannes Kepler was Tycho Brahe's successor. (b) This diagram, based on one drawn by Kepler, shows how he believed the sizes of the celestial spheres carrying the outer three planets—Saturn, Jupiter, and Mars—are determined by spacers (blue) consisting of two of the five regular solids. Inside the sphere of Mars, the remaining regular solids separated the spheres of Earth, Venus, and Mercury (not shown in this drawing). The Sun lay at the very center of this Copernican universe based on geometrical spacers. (c) The five regular solids are the tetrahedron, cube, octahedron, dodecahedron, and icosahedron, the only shapes with all faces of equal sizes and all equal angles between the faces.

speculation on the reason for the spacing of the planets' orbits. The second half has one virtue—as Kepler tried to understand planet orbits, he demonstrated that he was a talented mathematician and that he had become well versed in astronomy. He sent copies to Tycho and to Galileo, who both recognized Kepler's talent in spite of the mystical parts of the book.

Life was unsettled for Kepler in Graz because of the persecution of Protestants in that region, so when Tycho Brahe invited him to Prague in 1600, Kepler went eagerly, ready to work with the famous astronomer. Tycho's sudden death in 1601 left Kepler in a position to use Tycho's extensive records of observations to analyze the motions of the planets.

Kepler began by studying the motion of Mars, trying to deduce from the observations how the planet actually moved. By 1606, he had solved the mystery: The orbit of Mars is an ellipse, not a circle. Thus, he abandoned the ancient belief in the circular motion of the planets. But the mystery was even more complex. The planets do not move at uniform speeds along their elliptical orbits. Kepler recognized that they move faster when closer to the Sun and slower when farther away. Thus Kepler abandoned both uniform motion and circular motion and thereby finally solved the problem of planetary motion. Later, he discovered that the period of each planet's orbit is related to that orbit's radius. Kepler published his results in 1609 and 1619 in books called, respectively, *Astronomia Nova (New Astronomy)* and *Harmonices Mundi (The Harmony of the World)*.

3-3c Kepler's Three Laws of Planetary Motion

Although Kepler dabbled in the philosophical arguments of the day, he was a mathematician, and his triumph was the solution of the problem of the motion of the planets. The key to his solution was the **ellipse**. An ellipse is a figure drawn around two points, called the *foci*, in such a way that the distance from one focus to any point on the ellipse and back to the other focus equals a constant. This makes it easy to draw ellipses with two thumbtacks and a loop of string. Press the thumbtacks into a board, hook the string about the tacks, and place a pencil in the loop. As you can see in Figure 3-6a, if you keep the string taut as you move the pencil, it traces out an ellipse.

The geometry of an ellipse is described by two simple numbers. The **semi- major axis, *a***, is half of the longest diameter (Figure 3-6b). The **eccentricity, *e***, of an ellipse is half the distance between

ellipse A closed curve around two points called the foci, such that the total distance from one focus to the curve and back to the other focus remains constant.

semi-major axis, *a* Half of the longest diameter of an ellipse.

eccentricity, *e* A number between 1 and 0 that describes the shape of an ellipse; the distance from one focus to the center of the ellipse divided by the semi-major axis.

the foci divided by the semi-major axis. The eccentricity of an ellipse tells you its shape: If e is nearly equal to one, the ellipse is very elongated; if e is close to zero, the ellipse is more circular. To draw a circle with the string and tacks shown in Figure 3-6(a), you would move the two thumbtacks together, which shows that a circle is the same as an ellipse with eccentricity equal to zero. As you move the thumbtacks farther apart, the ellipse becomes flatter, and the value of its eccentricity moves closer to 1.

Kepler used ellipses to describe the motion of the planets in three fundamental rules that have been tested and confirmed so many times that astronomers now refer to them as "natural laws." They are commonly called Kepler's laws of planetary motion, summarized in Table 3-1 (See **How Do We Know? 3-2**.).

Kepler's first law states that the orbits of the planets around the Sun are ellipses with the Sun at one focus. Thanks to the precision of Tycho's observations and the sophistication of Kepler's mathematics, Kepler was able to recognize the elliptical shape of the orbits even though they are nearly circular. Of the planets known to Kepler, Mercury has the most elliptical orbit, which you can measure in Figure 3-7, but even it deviates only slightly from a circle.

Kepler's second law states that a line from the planet to the Sun sweeps over equal areas in equal intervals of time. This means that when the planet is closer to the Sun and the line connecting it to the Sun is shorter, the planet moves more rapidly to sweep over the same area that is swept over when the planet is farther from the Sun. Thus the planet in Figure 3-6c would move from point A' to point B' in one month, sweeping over the area shown. But when the planet is farther from the Sun, one month's motion would be shorter, from A to B although the total area swept out is the same.

The time that a planet takes to travel around the Sun once is its orbital period, P, and its average distance from the Sun equals the semi-major axis of its orbit, a. Kepler's third law tells us that these two quantities, orbital period and semi-major axis, are related: Orbital period squared is proportional to the semi-major axis cubed. For example, Jupiter's average distance from the Sun (which equals the semi-major axis of its orbit) is 5.2 AU. The semi-major axis cubed would be about 140.6, so the period must be the square root of 140.6, roughly 11.9 years.

It is important to notice that Kepler's three laws are **empirical**. That is, they

empirical Description of a phenomenon without explaining why it occurs.

FIGURE 3-6

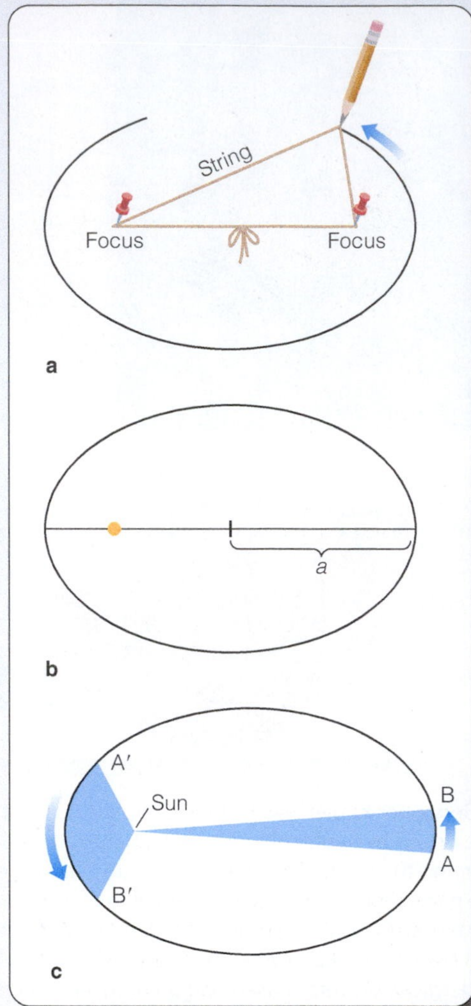

The geometry of elliptical orbits. (a) Drawing an ellipse with two tacks and a loop of string. (b) The semi-major axis, a, is half of the longest diameter. (c) Kepler's second law is demonstrated by a planet that moves from A to B in one month and from A′ to B′ in the same amount of time. The two blue sectors have the same area.

TABLE 3-1 KEPLER'S LAWS OF PLANETARY MOTION

I.	The orbits of the planets are ellipses with the Sun at one focus.
II.	A line from a planet to the Sun sweeps over equal areas in equal intervals of time.
III.	A planet's orbital period squared is proportional to its average distance from the Sun cubed: $P_y^2 = a_{AU}^3$

FIGURE 3-7

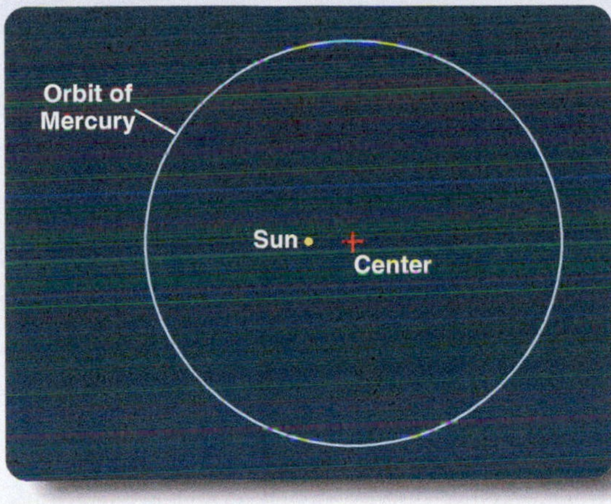

The orbits of the planets are nearly circular. You can measure the horizontal and vertical diameters of this orbit to detect its elliptical shape.

describe a phenomenon without explaining why it occurs. Kepler derived them from Tycho's extensive observations without referring to any first principles, fundamental assumptions, or theory. In fact, Kepler never knew what held the planets in their orbits or why they continued to move around the Sun in the ways he discovered.

3-4 GALILEO GALILEI

Galileo Galilei was born in the Italian city of Pisa in 1564 and studied medicine at the university there. His true love, however, was mathematics, and he eventually became professor of mathematics at the university at Padua, where he remained for 18 years. During this time, Galileo seems to have adopted the Copernican model, although he admitted in a 1597 letter to Kepler that he did not support that model publicly, fearing criticism.

Most people know two "facts" about Galileo, and both are wrong. Galileo did not invent the telescope, and he was not condemned by the Inquisition for believing that Earth moved around the Sun. As you learn about Galileo, you will discover that what was on trial were not just his opinions about the place of Earth but also the methods of science itself.

3-4a Telescopic Observations

It was the telescope that drove Galileo to publicly defend the heliocentric model. Galileo did not invent the telescope. It was apparently invented around 1608 by lens makers in Holland. Galileo, hearing descriptions in the fall of 1609, was able to build working telescopes in his laboratory. Galileo was also not the first person to look at the sky through a telescope, but he was the first person to observe the sky carefully and apply his observations to the main theoretical problem of the day—the place of Earth.

FIGURE 3-8

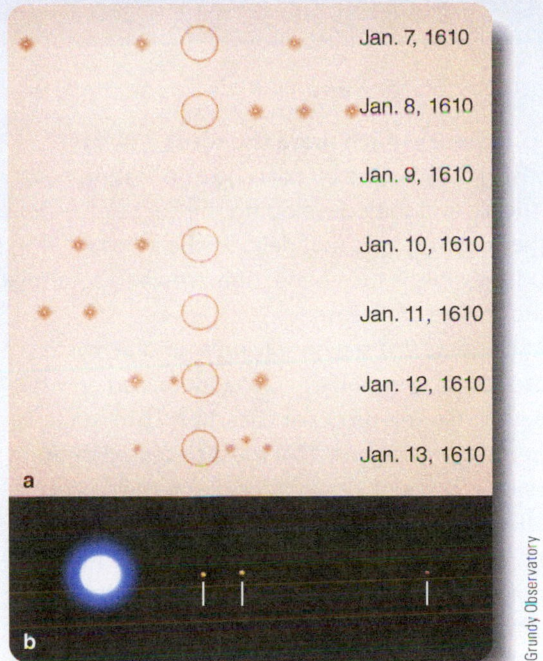

(a) On the night of January 7, 1610, Galileo saw three small "stars" near the bright disk of Jupiter and sketched them in his notebook. On subsequent nights (except January 9, which was cloudy), he saw that the stars were actually four Moons orbiting Jupiter. (b) This photograph, taken through a modern telescope, shows the overexposed disk of Jupiter and three of the four Galilean moons.

What Galileo saw through his telescopes was so amazing he rushed a small book into print, *Sidereus Nuncius (The Starry Messenger)*. In that book he reported two major discoveries about the Solar System. First, the Moon was not perfect. It had mountains and valleys on its surface, and Galileo used the shadows to calculate the height of the mountains. Aristotle's philosophy held that the Moon was perfect, but Galileo showed that it was not only imperfect but was even a world like Earth. Second, Galileo's telescope revealed four new "planets" circling Jupiter, planets that we know today as the Galilean moons of Jupiter, shown in Figure 3-8.

The moons of Jupiter supported the Copernican over the Ptolemaic model. Critics of Copernicus had

How Do We Know? 3-2

Hypothesis, Theory, and Law

Why is a theory much more than just a guess? Scientists study nature by devising and testing new hypotheses and then developing the successful ideas into theories and laws that describe how nature works. A good example is the connection between sour milk and the spread of disease.

A scientist's first step in solving a natural mystery is to propose a reasonable explanation based on what is known so far. This proposal, called a **hypothesis**, is an assertion or statement that must be tested through observation and experimentation. From the time of Aristotle, philosophers believed that food spoils as a result of the spontaneous generation of life—for example, mold growing out of drying bread. French chemist Louis Pasteur (1822–1895) hypothesized that microorganisms were not spontaneously generated but were carried through the air. To test his hypothesis he sealed an uncontaminated nutrient broth in glass, completely protecting it from the microorganisms on dust particles in the air. No mold grew, effectively disproving spontaneous generation. Although others had argued against spontaneous generation before Pasteur, it was Pasteur's meticulous testing of his hypothesis by experiment that finally convinced the scientific community.

A **theory** generalizes the specific results of well-confirmed hypotheses to give a broader description of nature, which can be applied to a wide variety of circumstances. For instance, Pasteur's specific hypothesis about mold growing in broth contributed to a broader theory that disease is caused by microorganisms transmitted from sick people to well people. This theory, called the germ theory of disease, is a cornerstone of modern medicine. It is a common misconception that the word theory means a tentative idea, a guess. As you can see, scientists actually use the word *theory* to mean an idea that is widely applicable and confirmed by abundant evidence.

hypothesis A conjecture, subject to further tests, that accounts for a set of facts.

theory A system of assumptions and principles applicable to a wide range of phenomena that has been repeatedly verified.

natural law A theory that has been so well confirmed that it is almost universally accepted as correct.

Sometimes, when a theory has been refined, tested, and confirmed so often that scientists have great confidence in it, it is called a **natural law**. Natural laws are the most fundamental principles of scientific knowledge. Kepler's laws of planetary motion are good examples.

Confidence is the key. In general, scientists have more confidence in a theory than in a hypothesis and the most confidence in a natural law. However, there is no precise distinction among hypotheses, theories, and laws, so use of these terms is sometimes a matter of tradition. For instance, some textbooks refer to the Copernican "theory" of heliocentrism, but it had not been well tested when Copernicus proposed it, and it is more rightly called the Copernican hypothesis. At the other extreme, Darwin's "theory" of evolution, containing many hypotheses that have been tested and confirmed over and over for nearly 150 years, might more correctly be called a natural law.

Franklin & Marshall College/Grundy Observatory/Mike Seeds

A fossil of a 500-million-year-old trilobite: Darwin's theory of evolution has been tested many times and is universally accepted in the life sciences, but by custom it is called Darwin's theory and not Darwin's law.

said Earth could not move because the Moon would be left behind; but Jupiter moved and kept its satellites. Galileo's discovery suggested that Earth, too, could move and keep its Moon. Also, Aristotle's philosophy included the belief that all heavenly motion was centered on Earth. Galileo showed that Jupiter's moons revolve around Jupiter, so there could be centers of motion other than Earth. Later, after the *Messenger* was published, Galileo noticed that Jupiter's inner-most moon had the shortest orbital period and the moons farther from Jupiter had proportionally longer periods. In this way, Jupiter's moons made up a harmonious system ruled by Jupiter, just as the planets in the Copernican universe were a harmonious system ruled by the Sun. This similarity didn't constitute proof, but Galileo saw it as an indication that the Solar System could be Sun centered and not Earth centered.

In the years of further exploration with his telescope, Galileo made additional fundamental discoveries (Figure 3-9). When he observed Venus, Galileo saw that it was going through phases like those of the Moon. In the Ptolemaic model, Venus moves around an epicycle centered on a line between Earth and the Sun. If that were true, it would always be seen as a crescent, like the model in Figure 3-10a. But Galileo saw Venus go through a complete set of phases, including full and gibbous, which proved that it did indeed revolve around the Sun (Figure 3-10b).

Sidereus Nuncius was popular and made Galileo famous. In 1611, Galileo visited Rome and was treated with great respect. He had friendly discussions with the powerful Cardinal Barberini, but because he was outspoken, forceful, and sometimes tactless he offended other important people who questioned his telescopic discoveries. Some critics said he was wrong, and others said he was lying. Some refused to look through a telescope lest it mislead them, and others looked and claimed to see nothing (hardly surprising given the awkwardness of

FIGURE 3-9

(a) Galileo is remembered as the great defender of Copernicanism. (b) Two of Galileo's telescopes, on display in a museum in Florence. Although he did not invent telescopes, Galileo will always be associated with them because they were the source of much of the observational evidence he used to try to understand the Universe.

Iryna1/Shutterstock.com

© Michael A. Seeds; Grundy Observatory, Franklin & Marshall College

FIGURE 3-10

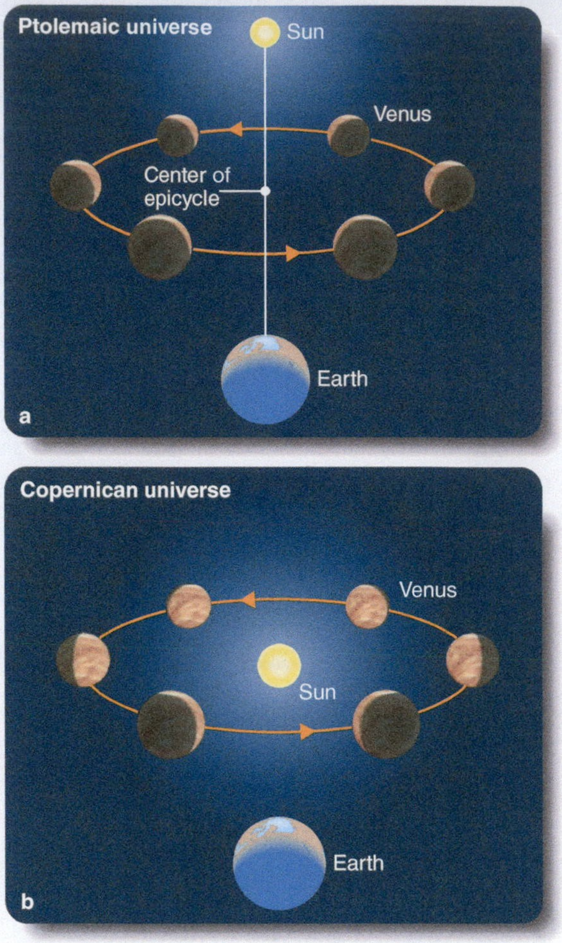

(a) If Venus moved in an epicycle centered on the Earth–Sun line, it would always appear as a crescent. (b) Galileo's telescope showed that Venus goes through a full set of phases, proving that it must orbit the Sun.

those first telescopes). When Galileo visited Rome again in 1616, Cardinal Bellarmine interviewed him privately and ordered him to cease public debate about models of the Universe, an order Galileo appears to have mostly followed.

The Inquisition (renamed in the 20th century the Congregation of the Holy Office) banned books relevant to the Copernican hypothesis, although *De Revolutionibus* itself was only suspended pending revision because it was recognized as useful for its predictions of planet positions. Everyone who owned a copy of the book was required to cross out certain statements and add handwritten corrections stating that Earth's motion and the central location of the Sun were only hypotheses and not facts, a situation that you will recognize as recurring today in connection with textbooks discussing biological evolution.

3-4b *Dialogo* and Trial

In 1623 Galileo's friend Cardinal Barberini became pope, taking the name Urban VIII. Galileo went to Rome in an attempt to have the 1616 order to cease debate lifted. Although that attempt was unsuccessful, Galileo began to write a massive defense of Copernicus's model, completing it in 1629. After some delay, Galileo's book was approved by both the local censor in Florence and the head censor of the Vatican in Rome. It was printed in 1632.

Called *Dialogo Sopra i Due Massimi Sistemi del Mondo (Dialogue Concerning the Two Chief World Systems)*, it confronts the ancient astronomy of Aristotle and Ptolemy with the Copernican model. Galileo wrote the book as a debate among three friends. Salviati is a swift-tongued defender of Copernicus; Sagredo is intelligent but largely uninformed; Simplicio is a dim-witted defender of Ptolemy. The book was a clear defense of Copernicus, and, either intentionally or unintentionally, Galileo exposed the pope's authority to ridicule. Urban VIII was fond of arguing that, as God was omnipotent, God could construct the Universe in any form while making it appear to humans to have a different form, and thus its true nature could not be deduced by mere observation. Galileo placed the pope's argument in the mouth of Simplicio. The pope took offense and ordered Galileo to face the Inquisition.

Galileo was interrogated by the Inquisition and threatened with torture. He must have thought of Giordano Bruno, a monk who was tried, condemned, and burned at the stake in Rome in 1600 for, among other offenses, supporting Copernicus. However, Galileo's trial did not center on his belief in Copernicus's model. *Dialogo* had been approved by two censors. Rather, the trial centered on a record of the meeting in 1616 between Galileo and Cardinal Bellarmine that included the statement that Galileo was "not to hold, teach, or defend in any way" the principles of Copernicus. Many historians believe that this document, which was signed neither by Galileo nor by Bellarmine nor by a legal secretary, was a forgery, or perhaps a draft that was never used. By this time Bellarmine was dead and could not testify about the meeting or the document.

The Inquisition condemned Galileo not for heresy but for disobeying the orders given him in 1616. In 1633, at the age of 70, kneeling before the Inquisition, Galileo read a recantation admitting his errors. Tradition has it that as he rose he whispered, *"E pur si muove"* ("Still it moves"), referring to Earth. Although he was sentenced to life imprisonment, he was actually confined at his villa for the next 10 years, perhaps through the intervention of the pope. He died there in 1642, 99 years after the death of Copernicus.

3-4c Two Ways to Understand the World

Galileo was tried and condemned on a charge you might call a technicality. Why then is his trial so important that historians have studied it for almost four centuries? Why have some of the world's greatest authors, including Bertolt Brecht, written about Galileo's trial? Why in 1992 did a commission established by Pope John Paul II report that the Inquisition had erred in condemning Galileo?

To understand the trial, you must recognize that it was the result of a conflict between two ways of understanding the Universe. Since the Middle Ages, European scholars had taught that the only path to true understanding was through religious faith. St. Augustine (354–430) wrote *"Credo ut intelligam,"* which can be translated as, "Believe in order to understand." Galileo and other scientists of the Renaissance, however, used their own observations as evidence to try to understand the Universe; when their observations contradicted religious authorities, they assumed their observations of reality were correct. (See **How Do We Know? 3-3**.) Galileo paraphrased Cardinal Baronius in saying, "The Bible tells us how to go to heaven, not how the heavens go." The significance of Galileo's trial is about the birth of modern science as a new way to understand the Universe.

How Do We Know? 3-3

Evidence as the Foundation of Science

Why is evidence critical in science? From colliding galaxies to the inner workings of atoms, scientists love to speculate and devise hypotheses, but all scientific knowledge is ultimately based on evidence from observations and experiments. Evidence is reality, and scientists constantly check their ideas against reality.

When you think of evidence, you probably think of criminal investigations in which detectives collect fingerprints and eyewitness accounts. In court, that evidence is used to try to understand the crime, but there is a key difference in how lawyers and scientists use evidence. A defense attorney can call a witness and intentionally fail to ask a question that would reveal evidence harmful to the defendant. In contrast, the scientist must be objective and not ignore any known evidence. The attorney is presenting only one side of the case, but the scientist is searching for the truth. In a sense, the scientist must deal with the evidence as both the prosecution and the defense.

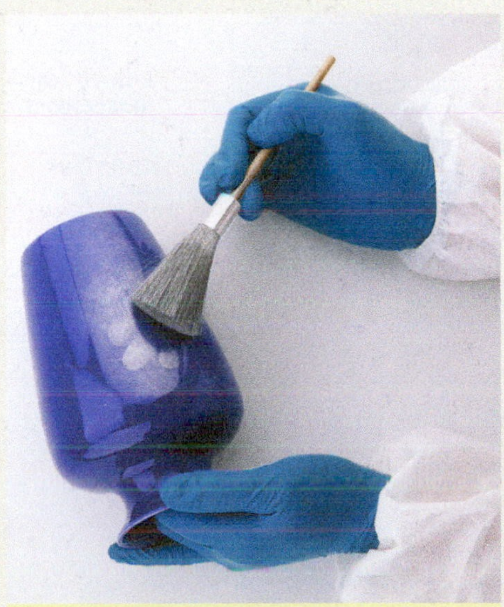
Dorling Kindersley/Getty Images

Fingerprints are evidence of past events.

It is a characteristic of scientific knowledge that it is supported by evidence. A scientific statement is more than an opinion or a speculation because it has been tested objectively against reality.

As you read about any science, look for the evidence in the form of observations and experiments. Every hypothesis or conclusion should have supporting evidence. If you can find and understand the evidence, the science will make sense. All scientists, from astronomers to zoologists, demand evidence. You should, too.

3-5 ISAAC NEWTON, GRAVITY, AND ORBITS

The birth of modern astronomy and of modern science dates from the 144 years between the publication of Copernicus's and Newton's chief works (in 1543 and 1687 respectively). The Renaissance is commonly taken to be the period approximately between 1350 and 1600, so the 144 years of this story lie at the climax of the European reawakening of learning in all fields (Figure 3-11).

The problem of the place of Earth was resolved by the Copernican Revolution, but the problem of planetary motion was only partly solved by Kepler's laws. For the last 10 years of his life, Galileo studied the nature of motion, especially the accelerated motion of falling bodies. Although he made some important progress, he was not able to relate his discoveries about motion to the heavens. That final step was taken by Isaac Newton.

3-5a Isaac Newton

Galileo died in January 1642. Some 11 months later, on Christmas day 1642, Isaac Newton was born in the English village of Woolsthorpe. Newton's life represented the flowering of the seeds planted by the previous four astronomers in this story, Copernicus, Tycho, Kepler, and Galileo.

Newton was a quiet child from a farming family, but his work at school was so impressive that his uncle financed his education at Trinity College, where he studied mathematics and physics. In 1665, plague swept through England, and the colleges were closed. During 1665 and 1666, Newton spent his time back home in Woolsthorpe, thinking and studying. It was during these years that he made most of his scientific discoveries. Among other things, he studied optics, developed three laws of motion, probed the nature of gravity, and invented calculus. The publication of his work in his book *Philosophia Naturalis Principia Mathematica*, know as

FIGURE 3-11

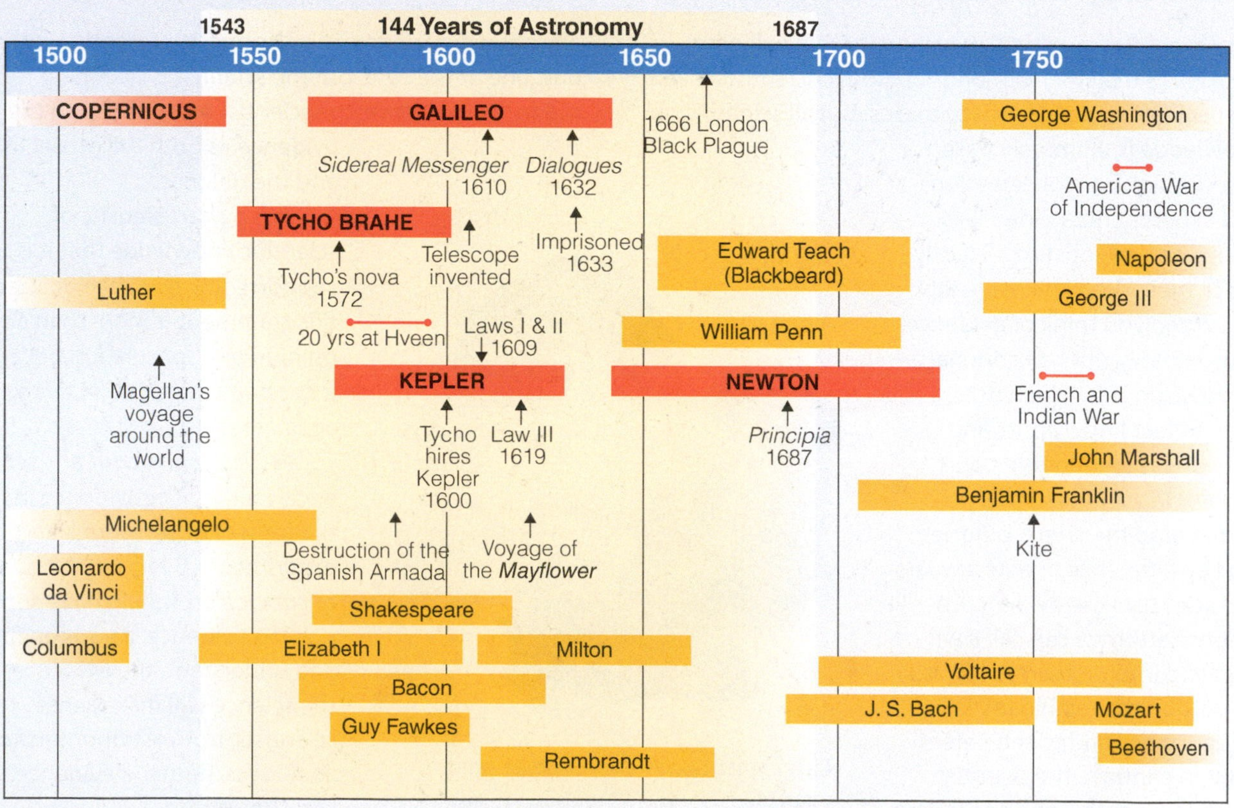

The five astronomers of this story lived at the culmination of the Renaissance.

the Principa (pronounced prin-KIP-ee-ah), in 1687 placed the fields of physics and astronomy on a new firm base.

It is beyond the scope of this book to analyze all of Newton's work, but his laws of motion and gravity had an important impact on the future of astronomy. From his study of the work of Galileo, Kepler, and others, Newton extracted three laws that relate the motion of a body to the forces acting on it (Table 3-2). These laws made it possible to predict exactly how a body would move if the forces were known.

TABLE 3-2 NEWTON'S THREE LAWS OF MOTION

I.	A body continues at rest or in uniform motion in a straight line unless acted upon by some force.
II.	A body's change of motion is proportional to the force acting on it, and is in the direction of the force.
III.	When one body exerts a force on a second body, the second body exerts an equal and opposite force back on the first body.

When Newton thought carefully about motion, he realized that some force must pull the Moon toward Earth's center. If there were no such force altering the Moon's motion, it would continue moving in a straight line and leave Earth forever. It can circle Earth only if Earth attracts it. Newton's insight was to recognize that the force that holds the Moon in its orbit is the same as the force that makes apples fall from trees.

Newtonian gravitation is sometimes called universal mutual gravitation. Newton's third law points out that forces occur in pairs. If one body attracts another, the second body must also attract the first. Thus, gravitation is mutual. Furthermore, gravity is universal. That is, all objects with mass attract all other masses in the Universe. The **mass** of an object is a measure of the amount of matter in the object, usually expressed in kilograms. Mass is not the same as **weight**. An object's weight is the force that Earth's gravity exerts on the object. Thus an object in space far from Earth might have no weight, but it would contain the same amount of matter and would thus have the same mass that it has on Earth.

Newton also realized that the distance between the objects is important. The gravitational force between two bodies depends not only on the masses of the bodies but also on the distance between them. He recognized that the force of gravity decreases as the square of the distance between the objects increases.

Specifically, if the distance from, say, Earth to the Moon were doubled, the gravitational force between them would decrease by a factor of 2^2, or 4. If the distance were tripled, the force would decrease by a factor of 3^2, or 9. This relationship is known as the **inverse square relation**. Newton guessed that gravity works by an inverse square relation because he had already discovered that light behaves this way (to be discussed in more detail in Chapter 9).

To summarize, the force of gravity attracting two objects to each other equals a constant times the product of their masses divided by the square of the distance between the objects. Gravity is universal: Your mass affects the planet Neptune and the galaxy M31, and every other object in the Universe, and their masses affect you—although not much, because they are so far away and your mass is relatively very small.

3-5b Orbital Motion

Newton's laws of motion and gravitation make it possible for you to understand why and how the Moon orbits Earth, the planets orbit the Sun, and to discover why Kepler's laws work.

To understand how an object can orbit another object, read **Concept Art 3B**, "**Orbiting Earth**." You need to see orbital motion as Newton did. Begin by noticing three important ideas:

1. An object orbiting Earth is actually falling (being accelerated) toward Earth's center. An object in a stable orbit continuously misses Earth because of its horizontal velocity.
2. Objects orbiting each other actually revolve around their mutual center of mass.
3. Notice the difference between closed orbits and open orbits. If you want to leave Earth never to return, you must give your spaceship a high enough velocity so it will follow an open orbit.

When the captain of a spaceship says to the pilot, "Put us into a circular orbit," the ship's computers must quickly calculate the velocity needed to achieve a circular orbit. That circular velocity depends only on the mass of the planet and the distance from the center of the planet. Once the engines fire and the ship reaches circular velocity, the engines can shut down. The ship is in orbit and will fall around the planet forever, as long as it is above the atmosphere's friction. No further effort is needed to maintain orbit, thanks to the laws Newton discovered (See **Circular Velocity** in the Math Reference Cards MR.).

mass A measure of the amount of matter making up an object.

weight The force that gravity exerts on an object.

inverse square relation A rule that the strength of an effect (such as gravity) decreases in proportion as the distance squared increases.

Concept Art 3B
ORBITS

1

You can understand orbital motion by thinking of a cannonball moving around Earth in a circular path. Imagine a cannon on a high mountain aimed horizontally as shown at right. A little gunpowder gives the cannonball a low velocity, and it doesn't travel very far before falling to Earth. More gunpowder gives the cannonball a higher velocity, and it travels farther. With enough gunpowder, the cannonball travels so fast it never strikes the ground. Earth's gravity pulls it toward Earth's center, but Earth's surface curves away from it at the same rate it falls. It is in orbit. The velocity needed to stay in a circular orbit is called **circular velocity**. Just above Earth's atmosphere, at an altitude of 200 km, circular velocity is 7790 m/s or about 17,400 miles per hour, and the orbital period is about 88 minutes.

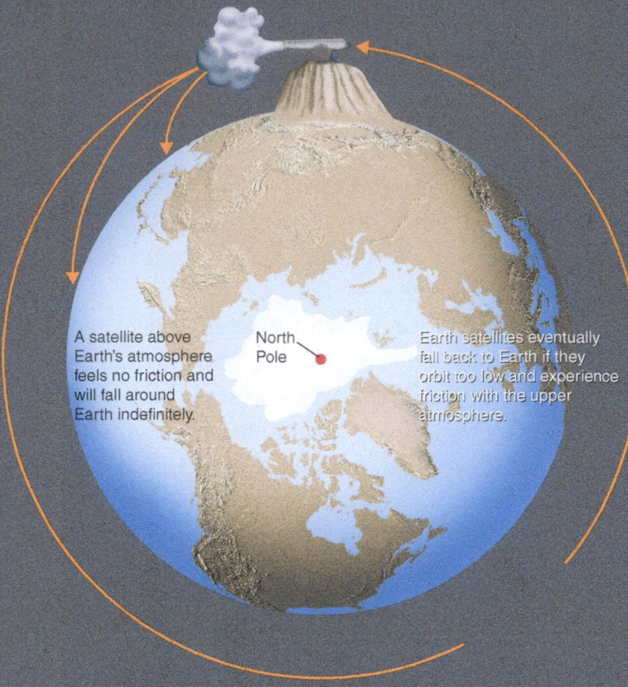

1a

A **geosynchronous satellite** orbits eastward with the rotation of Earth and remains above a fixed spot on the equator, which is ideal for communications and weather satellites.

A Geosynchronous Satellite

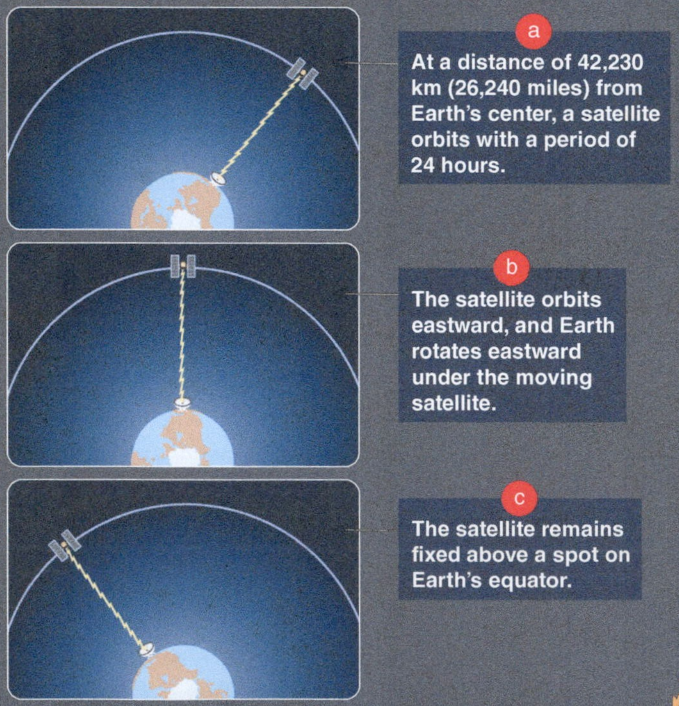

1b

According to Newton's first law of motion, the Moon should follow a straight line and leave Earth forever. Because it follows a curve, Newton knew that some force must continuously accelerate it toward Earth, gravity. Every second the Moon moves 1020 m (3350 ft) eastward and falls about 1.4 mm (1/18 in.) toward Earth. The combination of these motions produces the Moon's curved orbit. The Moon is falling all the time.

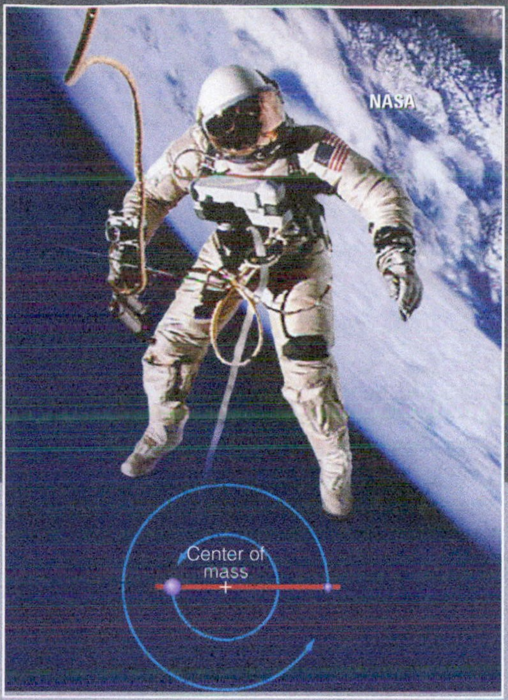

NASA

1c

Astronauts in orbit around Earth feel weightless, but they are not—to use a term from old science fiction movies—"beyond Earth's gravity." Like the Moon, the astronauts are accelerated toward Earth by Earth's gravity, but they travel fast enough along their orbits that they continually "miss the Earth." They are literally falling around Earth. Inside or outside a spacecraft, astronauts feel weightless because they and their spacecraft are falling at the same rate. Rather than saying they are weightless, you should more accurately say they are in free fall.

2

To be accurate you should not say that an object orbits Earth. Rather, the two objects orbit each other. Gravitation is mutual, and if Earth pulls on the Moon, the Moon pulls on Earth. The two bodies revolve around their common **center of mass**, the balance point of the system.

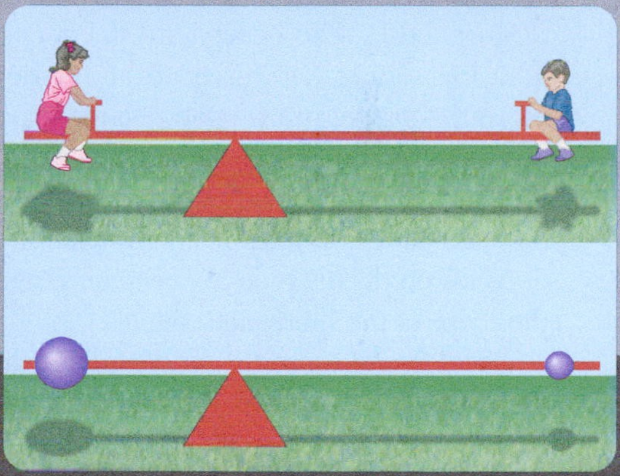

2a

Two bodies of different mass balance at their center of mass, which is located closer to the more massive object. As the two objects orbit each other, they revolve around their common center of mass as shown at right. The center of mass of the Earth–Moon system lies only 4670 km (2900 mi) from the center of Earth—inside Earth. As the Moon orbits the center of mass on one side, Earth swings around the center of mass on the opposite side.

3

Closed orbits are repeating cycles. The Moon and artificial satellites orbit Earth in closed orbits. Below, the cannonball could follow an elliptical or a circular **Closed orbits**. If the cannonball travels at the velocity needed to leave Earth permanently, called **escape velocity**, it will enter an **open orbit**. An open orbit does not return the cannonball to Earth; it will escape.

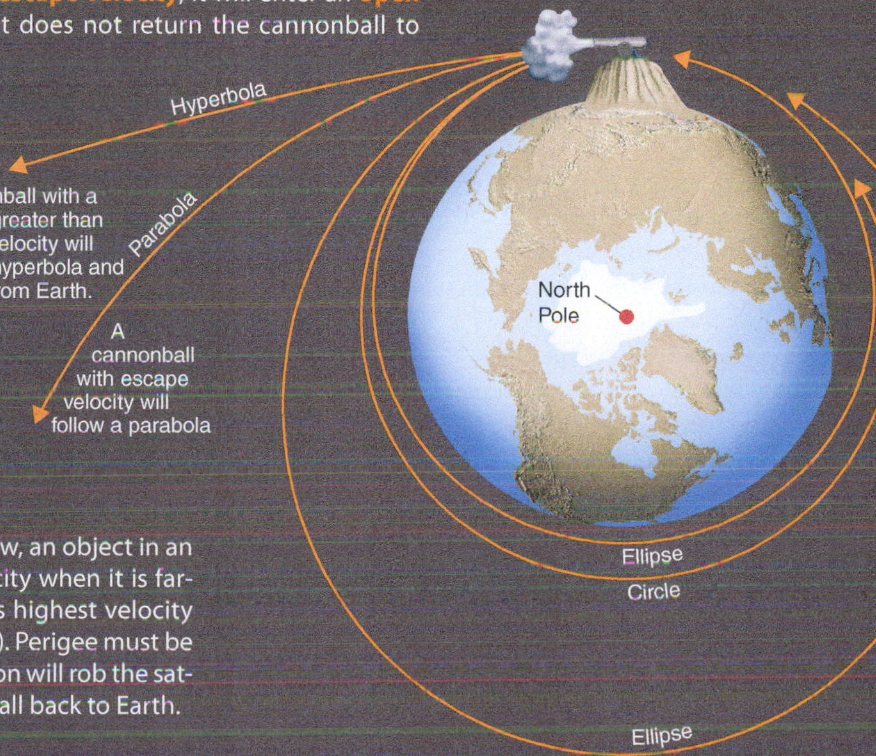

3a

As described by Kepler's second law, an object in an elliptical orbit has its lowest velocity when it is farthest from Earth (apogee), and its highest velocity when it is closest to Earth (perigee). Perigee must be above Earth's atmosphere, or friction will rob the satellite of energy and it will quickly fall back to Earth.

3-5c Tides: Gravity in Action

Newton understood that gravity is mutual—Earth attracts the Moon, and the Moon attracts Earth—and that means the moon's gravity can explain the ocean tides. But Newton also realized that gravitation is universal, and that means there is much more to tides than just Earth's oceans.

Tides are caused by small differences in gravitational force. As Earth and the Moon orbit around each other, they attract each other gravitationally. Because the side of Earth toward the Moon is a bit closer, the Moon pulls on it more strongly and that pulls up a bulge. Also, the Moon pulls on Earth a bit more than it pulls on Earth's far side and that produces a bulge on the far side (Figure 3-12). The oceans are deeper in these bulges, and as Earth rotates and carries you into a bulge, you see the tide creeping up the beach. Because there are two bulges, there are two high tides each day, although the exact pattern of tides at any given locality depends on details such as ocean currents, the shape of the shore, and so forth.

The Sun also produces tides on Earth, although they are smaller than lunar tides. At new and full moons, the lunar and solar tides add together to produce extra high and extra low tides that are called **spring tides**. At first- and third-quarter moons, the solar tides cancel out part of the lunar tides so that high and low tides are not extreme (see Figure 3-12). These are called **neap tides**.

Although the oceans flow easily into tidal bulges, the nearly rigid bulk of Earth flexes into tidal bulges and the plains and mountains rise and fall a few centimeters twice a day. Friction is gradually slowing Earth's rotation, and fossil evidence shows that Earth used to rotate faster. In the same way, Earth's gravity produces tidal bulges in the Moon, and, although the Moon used to rotate faster, friction has slowed it down and it now keeps the same side facing Earth.

Tides can also affect orbits. The rotation of Earth drags the tidal bulges slightly ahead of the Moon, and the gravitation of the bulges of water pull the Moon forward in its orbit. This makes the Moon's orbit

FIGURE 3-12

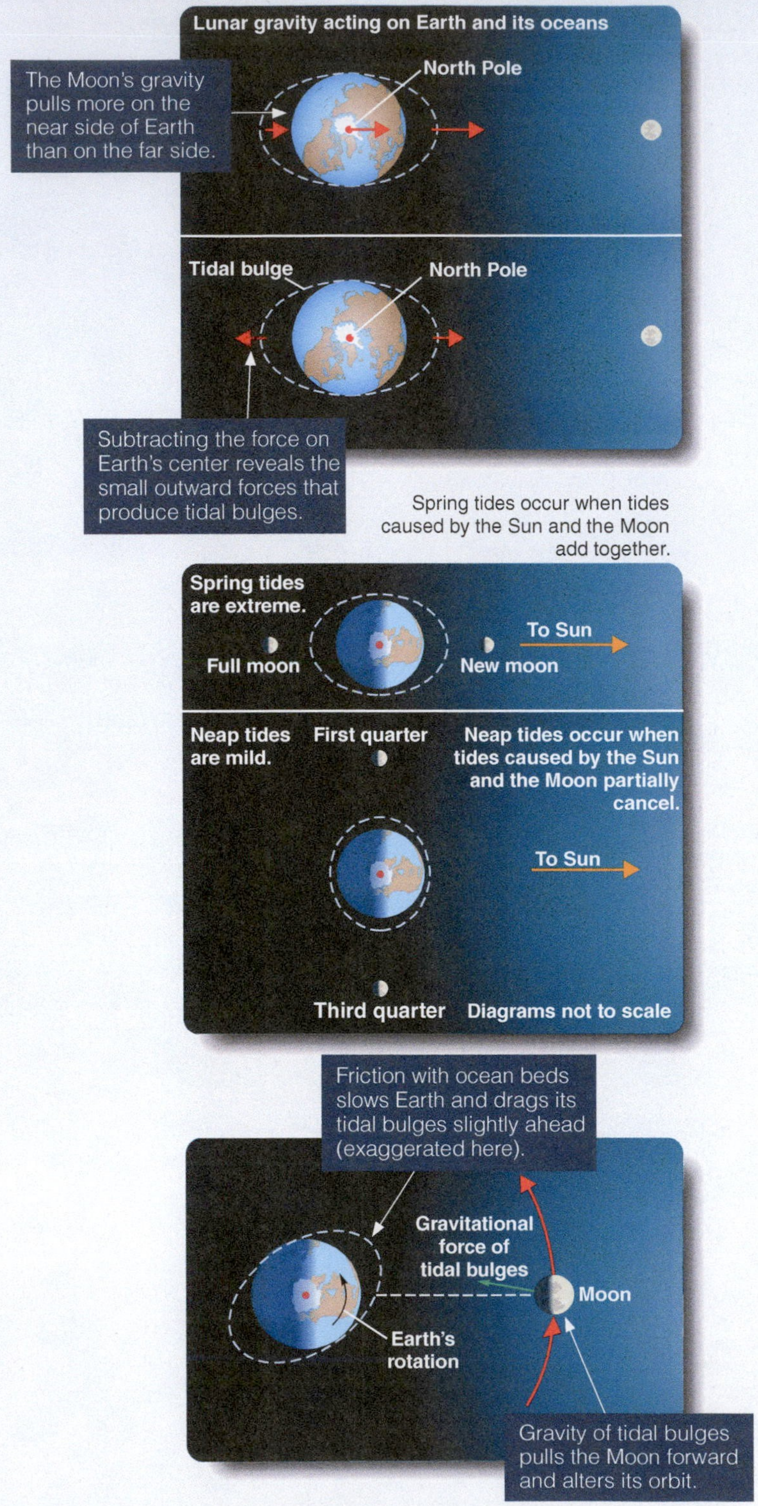

Tides are produced by small differences in the gravitational force exerted on different parts of an object. The side of Earth nearest the Moon feels a larger force than the side farthest away. Relative to Earth's center, small forces are left over, and they cause the tides. Both the Moon and the Sun produce tides on Earth. Tides can alter both an object's rotation and orbital motion.

grow larger by about 4 cm a year, an effect that astronomers can measure by bouncing lasers off reflectors left on the Moon by the Apollo astronauts.

3-5d Newton's Universe

Newton's insight gave the world a new conception of nature. His laws of motion and gravity were *general* laws that described the motions of all bodies under the action of external forces. In addition, the laws were *predictive* because they made possible specific calculations of predictions that could be tested by observation. For example, Newton's laws of motion can be used to derive Kepler's third law from the law of gravity. Newton's discoveries remade astronomy into an analytical science in which astronomers could measure the positions and motions of celestial bodies, calculate the gravitational forces acting on them, and predict their future motion.

One of the most often used and least often stated principles of science is *cause and effect*. You could argue that Newton's second law of motion was the first clear statement of that principle. Ancient philosophers such as Aristotle believed that objects moved because of innate tendencies; for example, objects made of earth or water had a natural tendency to move toward Earth at the center of the Universe. In contrast, Newton's second law says if an object changes its motion by an acceleration, then it must have been acted on by a force. The principle of cause and effect gives scientists confidence that every effect has a cause. If the Universe were not rational, then you could never expect to discover causes. Newton's second law of motion was arguably the first explicit statement that the behavior of the Universe is rational and depends on causes. (See "What Are We?")

spring tides Ocean tide of large range that occurs at full and new moon. Note that the term "spring" refers to leaping, not to the season of spring.

neap tides Ocean tide of small range occurring at first- and third-quarter moon.

What Are We? *Thinkers*

The scientific revolution began when Copernicus made humanity part of the Universe. Before Copernicus, people thought of Earth as a special place different from any of the objects in the sky, but in trying to explain the motions in the sky, Copernicus made Earth one of the planets. Galileo and those who brought him to trial understood the significance of making Earth a planet. It made Earth and humanity part of nature, part of the Universe.

Kepler showed that the planets move according to simple rules. We are not in a special place ruled by mysterious planetary forces. Earth, the Sun, and all of humanity are part of a Universe in which motions can be described by a few fundamental laws. If simple laws describe the motions of the planets, then the Universe is not ruled by mysterious influences as in astrology or the whims of the gods atop Mount Olympus. And if the Universe can be described by simple rules, then it is open to scientific study.

Before Copernicus, people felt they were special because they thought they were at the center of the Universe. Copernicus, Kepler, and Galileo showed that we are not at the center but nevertheless are part of an elegant and complex Universe. Astronomy tells us that we are special because we can study the Universe and eventually understand what we are. It also tells us that we are not only part of nature; we are the part of nature that can think about nature..

STUDY TOOLS 3

IN THE BOOK

- ☐ Tear Out the Review Card on *The Origin of Modern Astronomy*.

ASTRO ONLINE

Visit ASTRO Online at www.cengagebrain.com

- ☐ Interactive Reading
- ☐ Practice Quizzing
- ☐ Videos and Animations
- ☐ Exclusive Features

Questions and Problems

REVIEW QUESTIONS

1. Why did Greek philosophers conclude that the heavens were made up of perfect crystalline spheres moving at constant speeds?
2. Why did early astronomers conclude that Earth had to be motionless?
3. How did the Ptolemaic model explain retrograde motion?
4. In what ways were the models of Ptolemy and Copernicus similar?
5. Why did the Copernican hypothesis win gradual acceptance?
6. Why is it difficult for scientists to replace an old paradigm with a new paradigm?
7. Why did Tycho Brahe expect the new star of 1572 to show parallax? Why was the lack of parallax evidence against the Ptolemaic model?
8. Explain how Kepler's laws contradict uniform circular motion.
9. How do a hypothesis, a theory, and a law differ?
10. Review Galileo's telescopic discoveries and explain why they supported the Copernican model and contradicted the Ptolemaic model.
11. Galileo was condemned by the Inquisition, but Kepler, also a supporter of Copernicus, was not. Why not?
12. How do Newton's laws lead you to conclude that gravitation has to be universal?
13. Explain why you might describe the orbital motion of the Moon with the statement, "The Moon is falling."
14. **How Do We Know?** How does a paradigm affect the questions you ask and the answers you find acceptable?
15. **How Do We Know?** How would you respond to someone who said, "Oh, that's only a theory."
16. **How Do We Know?** Why would it be appropriate to refer to evidence as the "reality checks" in science?

DISCUSSION QUESTIONS

1. Science historian Thomas Kuhn has said that *De Revolutionibus* was a revolution-making book, but not a revolutionary book. How was that book in some ways classical and conservative?
2. Many historians suspect that Galileo offended Pope Urban VIII by putting the pope's favorite argument into the mouth of the character Simplicio in the book *Dialog Concerning the Two Chief World Systems*. How is the pope's argument a challenge to the principle of cause and effect?
3. How does the modern controversy over creationism and evolution reflect two ways of knowing about the physical world?

PROBLEMS

1. If you lived on Mars, which planets would describe retrograde loops? Which would sometimes be visible as crescent phases?
2. Galileo's telescope showed him that Venus has a large angular diameter (61 arc seconds) when it is a crescent and a small angular diameter (10 arc seconds) when it is nearly full. Use the small-angle formula to find the ratio of its maximum distance to its minimum distance. Is this ratio compatible with the Ptolemaic universe shown on the right-hand page of Concept Art 3A? (See The **Small-Angle Formula** in the Math Reference Cards MR.)
3. Galileo's telescopes were not of high quality by modern standards. He was able to see the Moons of Jupiter, but he never reported seeing features on Mars. Use the small-angle formula to find the maximum angular diameter of Mars when it is closest to Earth. How does that compare with the maximum diameter of Jupiter? (See **The Small-Angle Formula** in the Math Reference Cards MR.)
4. If a planet had an average distance from the Sun of 10 AU, what would its orbital period be?
5. If a space probe were sent into an orbit around the Sun that brought it as close as 0.5 AU to the Sun and as far away as 5.5 AU, what would its orbital period be?
6. Neptune orbits the Sun with a period of 164.8 years. What is its average distance from the Sun?
7. Venus's average distance from the Sun is 0.72 AU and Saturn's is 9.55 AU. Calculate the circular orbital velocity of Venus and Saturn around the Sun. (*Note*: The mass of the Sun is 1.99×10^{30} kg. An AU is 1.50×10^{11} m. Also, see the **Circular Velocity** formula in the Math Reference Cards MR.)
8. The circular velocity of Earth around the Sun is about 30 km/s. Do the velocity arrows in Figure 3-3 for Venus, Earth, and Saturn have correct relative sizes? (*Hint*: See Problem 7.)

9. What is the orbital velocity of an Earth satellite 42,230 km from Earth's center? How long does it take to circle its orbit once?

LEARNING TO LOOK

1. What three astronomical objects are represented in Figure 3-13? What are the two rings?
2. Why can the object shown in Figure 3-14 be bolted in place and used 24 hours a day without adjustment?
3. Why is it a little bit misleading to say that the astronaut in Figure 3-15 is weightless?

FIGURE 3-13

astronomical_objects

FIGURE 3-14

Larry Mulvehill/The Image Works

FIGURE 3-15

NASA/JSC

4 Light and Telescopes

The strongest thing that's given us to see with's
A telescope. Someone in every town
Seems to me owes it to the town to keep one.
Robert Frost, "The Star-Splitter"

After finishing this chapter go to **PAGE 77** for **STUDY TOOLS.**

NASA

GUIDEPOST

In earlier chapters of this book you looked at the sky the way the first astronomers did, with the unaided eye. In the previous chapter, you got a glimpse through Galileo's telescope that revealed amazing things about the Moon, Jupiter, and Venus. Now you can consider the telescopes, instruments, and techniques of modern astronomers.

Telescopes gather and focus light, so you need to study what light is and how it behaves on your way to understanding how telescopes work. You will learn about telescopes that capture invisible types of light such as radio waves and X-rays, enabling astronomers to put together a more complete view of celestial objects. (See CO 4.) This chapter will help you answer these five important questions:

- What is light?
- How do telescopes work?
- What are the powers and limitations of telescopes?
- What kind of instruments do astronomers use to record and analyze light gathered by telescopes?
- Why must some telescopes be located in space?

The science of astronomy is based on observations. Astronomers cannot visit distant galaxies and far-off worlds, so they have to study them using telescopes. Eleven chapters remain in your exploration, and every one will present information gained by astronomers using telescopes.

Natural light is a treasure that links you to the sky. The quest of astronomers is to study light from the Sun, planets, moons, asteroids, comets, stars, nebulae, and galaxies and extract information about their natures. Most celestial objects are very faint sources of light, so large telescopes, such as the ones shown in this chapter, are built to collect the greatest amount of light possible. Some types of telescopes gather light that is invisible to the human eye, but all telescopes work by the same basic principles. Some telescopes are used on Earth's surface, but others must go high into Earth's atmosphere, or even above the atmosphere into space, to work properly. This chapter focuses on telescopes, instruments, and techniques used to analyze light for astronomical research.

Astronomy is more than technology and scientific analysis, it helps us understand what we are. In the quotation that opens this chapter, the poet Robert Frost suggests that someone in every town should own a telescope to help us look upward and outward.

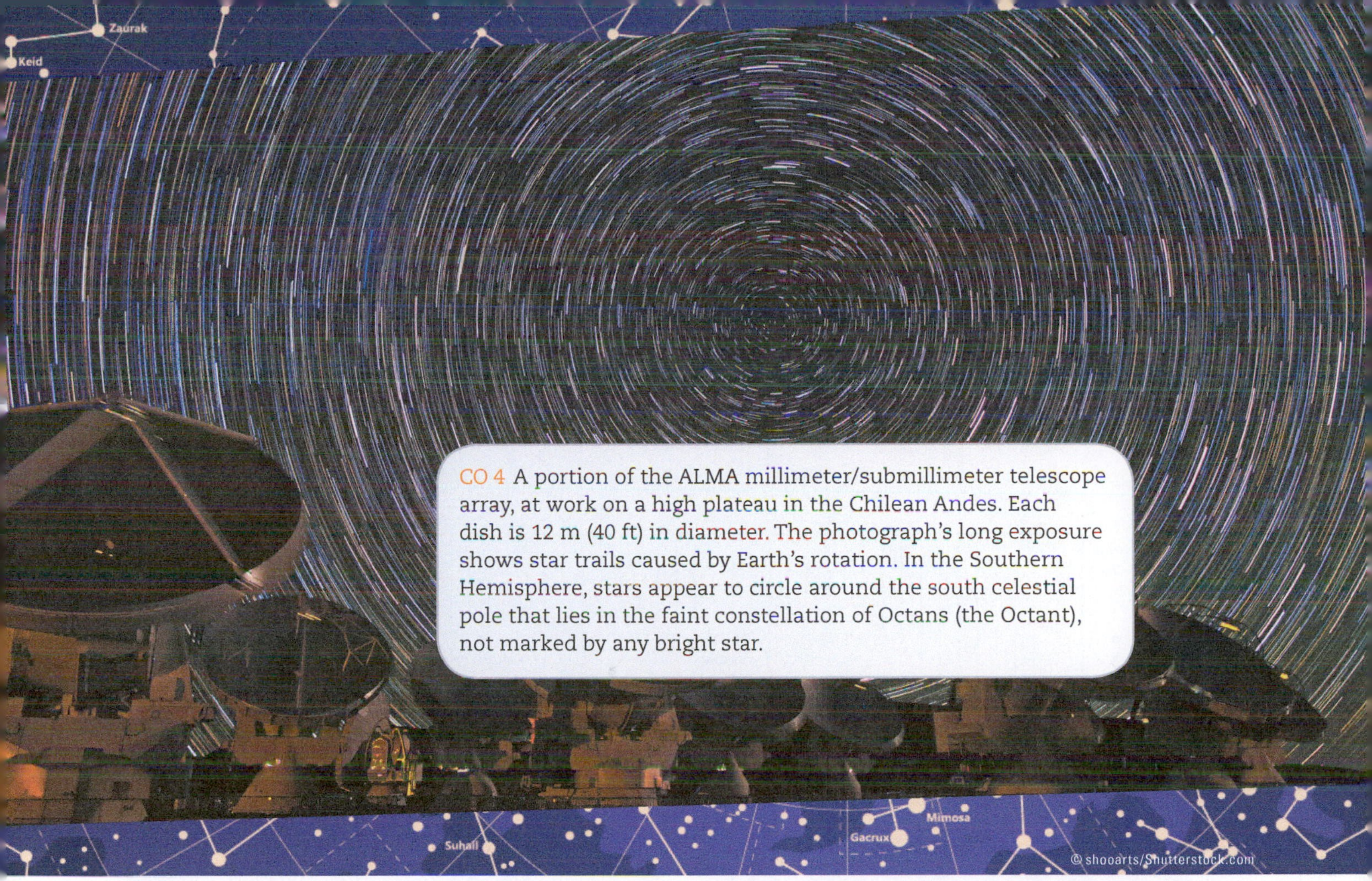

CO 4 A portion of the ALMA millimeter/submillimeter telescope array, at work on a high plateau in the Chilean Andes. Each dish is 12 m (40 ft) in diameter. The photograph's long exposure shows star trails caused by Earth's rotation. In the Southern Hemisphere, stars appear to circle around the south celestial pole that lies in the faint constellation of Octans (the Octant), not marked by any bright star.

4-1 RADIATION: INFORMATION FROM SPACE

Astronomers no longer study the sky by mapping constellations or charting the phases of the Moon. Modern astronomers analyze light using sophisticated instruments (Figure 4-1) and techniques to investigate the compositions, motions, internal processes, and evolution of celestial objects. To understand this, you need to learn about the nature of light.

4-1a Light as Waves and Particles

Have you ever noticed the colors in a soap bubble? If so, then you have seen one effect of light behaving as a wave. When that same light enters a digital camera, it behaves as a particle. How light behaves depends on how you treat it—light has both wavelike and particlelike properties. Sound is another type of wave that you have already experienced. Sound waves are air pressure disturbances that travel through the air from source to ear. Sound requires a solid, liquid, or gas medium to carry it; so, for example, in space outside a spacecraft, there can be no sound. In contrast, light

FIGURE 4-1

The northern Gemini telescope on Mauna Kea in Hawai'i stands over 19 m (60 ft) high when pointed straight up. Its main mirror is 8.1 m (26.5 ft) in diameter—larger than some classrooms.

FIGURE 4-2

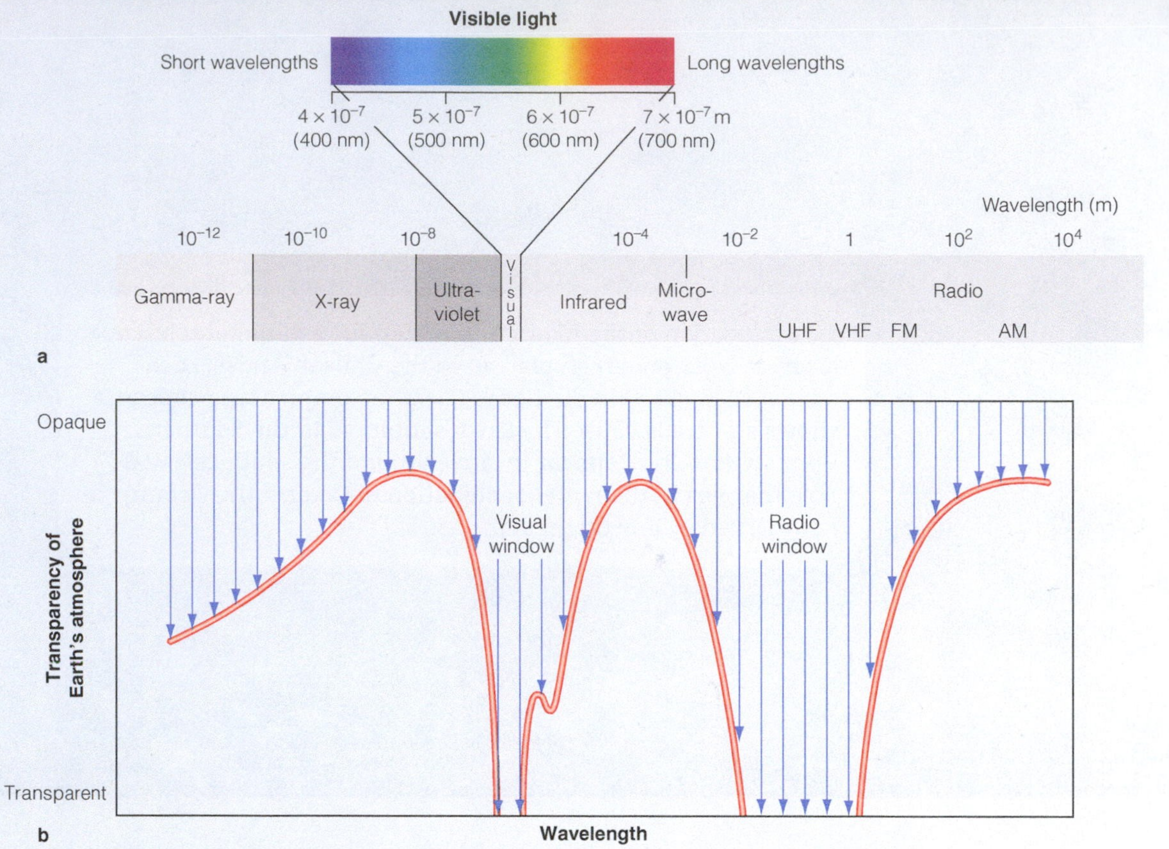

(a) The spectrum of visible light, extending from red to violet, is only part of the electromagnetic spectrum. (b) Most forms of light (electromagnetic radiation) are absorbed in Earth's atmosphere. Light can reach Earth's surface only through the visual and radio "windows."

is composed of a combination of electric and magnetic waves that can travel through empty space. Unlike sound, light waves do not require a medium and thus can travel through a vacuum.

Because light is made up of both electric and magnetic fields, it is referred to as **electromagnetic radiation**. Visible light is only one form of electromagnetic radiation. Electromagnetic radiation is a wave phenomenon, that is, it is associated with a periodically repeating disturbance (a wave) that carries energy.

Imagine waves in water: If you disturb a pool of water, waves spread across the surface. Now imagine placing a ruler parallel to the travel direction of the wave. The distance between peaks is the **wavelength**. The changing electric and magnetic fields of electromagnetic waves travel through space at about 300,000 km per second (186,000 mi per second). That is commonly referred to as the speed of light, but it is the speed of all electromagnetic radiation.

It may seem odd to use the word *radiation* when talking about light, but radiation really refers to anything that spreads outward from a source. Light radiates from a source, so you can correctly refer to light as a form of radiation.

electromagnetic radiation Changing electric and magnetic fields that travel through space and transfer energy from one place to another; examples are light and radio waves.

wavelength The distance between successive peaks or troughs of a wave, usually represented by a lowercase Greek lambda, λ.

4-1b The Electromagnetic Spectrum

The electromagnetic spectrum is, simply, the types of electromagnetic radiation arranged in order of increasing wavelength. Rainbows are spectra (plural of spectrum) of visible light. The colors of visible light have different wavelengths: Red has the longest wavelength and violet the shortest, as shown in the visible spectrum at the top of Figure 4-2.

The average wavelength of visible light is about 0.0005 mm. Two hundred light waves would fit end-to-end across the thickness of a sheet of paper. It is too awkward to measure such short distances in millimeters, so physicists and astronomers describe the wavelength of light using either the unit of **nanometer (nm)**, one-billionth of a meter (10^{-9} m), or the **Ångstrom (Å)**, equal to 10^{-10} m or 0.1 nm. The wavelength of visible light ranges between about 400 nm and 700 nm, or, equivalently, 4000 Å and 7000 Å. Some astronomers refer to wavelengths using units of microns (10^{-6} m), millimeters, centimeters, or meters. Figure 4-2 shows how the visible spectrum makes up only a small part of the electromagnetic spectrum.

Beyond the red end of the visible range lies **infrared (IR)** radiation, with wavelengths ranging from 700 nm to about 1 mm. Your eyes are not sensitive to this radiation, but your skin can sense some of it as heat. A heat lamp is nothing more than a bulb that gives off large amounts of infrared radiation. English astronomer William Herschel discovered infrared radiation, and thus, discovered that there is such a thing as invisible light (Figure 4-3).

FIGURE 4-3

World History Archive/Alamy Stock Photo

Depiction of Sir William Herschel in 1800 discovering that sunlight contains radiation detectable by thermometers but not by human eyes. He named that invisible light "infrared," meaning "below red."

Radio waves have even longer wavelengths than IR radiation. The radio radiation used for AM radio transmissions has wavelengths of a few hundred meters, while FM, television, and also military, governmental, and amateur radio transmissions have wavelengths from a few tens of centimeters to a few tens of meters. Microwave transmissions, used for radar and long-distance telephone communications, have wavelengths from about 1 mm to a few centimeters.

Look at the electromagnetic spectrum in Figure 4-2 and notice that electromagnetic waves with wavelengths shorter than violet light are called **ultraviolet (UV)**. Electromagnetic waves with wavelengths shorter than UV light are called **X-rays**, and the shortest are **gamma-rays**.

The distinction between these wavelength ranges is mostly arbitrary—they are simply convenient human-invented labels. For example, the longest-wavelength IR radiation and the shortest-wavelength microwaves are the same. Similarly, very short-wavelength UV light can be considered to be X-rays. Nonetheless, it is all electromagnetic radiation, and you could say we are "making light" of it all, because all these types of radiation are the same phenomenon as light: Some types your eyes can see, some types your eyes can't see.

Although light behaves as a wave, under certain conditions it also behaves as a particle. A particle of light is called a **photon**, and you can think of a photon as a minimum-sized bundle of electromagnetic waves. The amount of energy a photon carries depends on its wavelength. Shorter-wavelength photons carry more energy, and longer-wavelength photons carry less energy. A photon of visible light carries a small amount of energy, but an X-ray photon carries much more energy and a radio photon carries much less.

nanometer (nm) A unit of distance equaling one-billionth of a meter (10^{-9} m), commonly used to measure the wavelength of light.

Ångstrom (Å) A unit of distance commonly used to measure the wavelength of light. 1Å = 10^{-10} m.

infrared (IR) The portion of the electromagnetic spectrum with wavelengths longer than red light, ranging from 700 nm to about 1 mm, between visible light and radio waves.

ultraviolet (UV) The portion of the electromagnetic spectrum with wavelengths shorter than violet light, between visible light and X-rays.

X-rays Electromagnetic waves with wavelengths shorter than ultraviolet light.

gamma-rays The shortest-wavelength electromagnetic waves.

photon A quantum of electromagnetic energy that carries an amount of energy that depends inversely on its wavelength.

Astronomers are interested in electromagnetic radiation because it carries almost all available clues to the nature of planets, stars, and other celestial objects. Earth's atmosphere is opaque to most electromagnetic radiation, as shown by the graph at the bottom of Figure 4-2. Gamma-rays, X-rays, and some radio waves are absorbed high in Earth's atmosphere, and a layer of ozone (O_3) at an altitude of about 30 km (20 mi) absorbs almost all UV radiation. Water vapor in the lower atmosphere absorbs long-wavelength IR radiation. Only visible light, some short-wavelength infrared radiation, and some radio waves reach Earth's surface through what are called **atmospheric windows**. To study the sky from Earth's surface, you must look out through one of these "windows" in the electromagnetic spectrum.

4-2 TELESCOPES

Astronomers build optical telescopes to gather light and focus it into sharp images. This requires careful optical and mechanical designs, and it leads astronomers to build very large telescopes. To understand this, you need to learn the terminology of telescopes, starting with the different types of telescopes and why some are better than others.

atmospheric window Wavelength region in which our atmosphere is transparent—at visual, radio, and some infrared wavelengths.

refracting telescope A telescope that forms images by bending (refracting) light with a lens.

reflecting telescope A telescope that forms images by reflecting light with a mirror.

primary lens In a refracting telescope, the largest lens.

primary mirror In a reflecting telescope, the largest mirror.

eyepiece A short-focal-length lens used to enlarge the image in a telescope. The lens nearest the eye.

focal length The focal length of a lens or mirror is the distance from that lens or mirror to the point where it focuses parallel rays of light.

chromatic aberration A distortion found in refracting telescopes because lenses focus different colors at slightly different distances. Images are consequently surrounded by color fringes.

4-2a Two Kinds of Telescopes

Astronomical telescopes focus light into an image in one of two ways. Either (1) a lens bends (refracts) the light as it passes through the glass and brings it to a focus to form an image, or (2) a mirror—a curved piece of glass with a reflective surface—forms an image by bouncing light. Figure 4-4 demonstrates the difference between the two configurations. Because there are two ways to focus light, there are two types of astronomical telescopes. **Refracting telescopes** use a lens to gather and focus the light, whereas **reflecting telescopes** use a mirror.

FIGURE 4-4

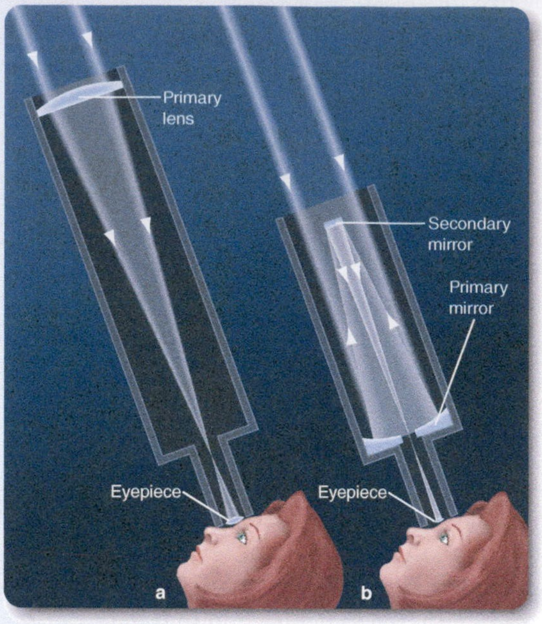

(a) A refracting telescope uses a primary lens to focus starlight into an image that is magnified by another lens called an eyepiece. The primary lens has a long focal length, and the eyepiece has a short focal length. (b) A reflecting telescope uses a primary mirror to focus the light by reflection. In this particular reflector design, called a Cassegrain telescope, a small secondary mirror reflects the starlight back down through a hole in the middle of the primary mirror to the eyepiece lens.

The main lens in a refracting telescope is called the **primary lens**, and the main mirror in a reflecting telescope is called the **primary mirror**. Both kinds of telescopes form a small, inverted image that is difficult to observe directly, so a lens called the **eyepiece** is used to magnify the image and make it convenient to view. The **focal length** is the distance from a lens or mirror to the image it forms of a distant light source such as a star. Creating the proper optical shape to produce a good focus is an expensive process. The surfaces of lenses and mirrors must be shaped and polished to have no irregularities larger than the wavelength of light. Creating the optics for a large telescope can take months or years; involve huge, precision machinery; and employ several expert optical engineers and scientists.

Refracting telescopes have serious disadvantages, most importantly, they suffer from an optical distortion that limits their usefulness. When light is refracted through glass, shorter wavelengths bend more than longer wavelengths. As a result, you see a color blur around every image. This color separation is called **chromatic aberration** and it

FIGURE 4-5

In most radio telescopes, a dish reflector concentrates the radio signal on the receiver. The signal is then amplified and recorded. Wire mesh is an adequate reflector for all but the shortest wavelength radio waves.

can be only partially corrected. Another disadvantage is that the glass in primary lenses must be pure and flawless because the light passes all the way through it. For that same reason, the weight of the lens can be supported only around its outer edge. In contrast, light reflects from the front surface of a reflecting telescope's primary mirror but does not pass through it, so reflecting telescopes have no chromatic aberration. Also, mirrors are less expensive to make than similarly sized lenses and the weight of telescope mirrors can be supported easily. These factors explain why every large astronomical telescope built since 1900 has been a reflecting telescope.

Optical telescopes gather visible light, but astronomers also build **radio telescopes** to gather radio radiation. Radio waves from celestial objects, like visible light waves, penetrate Earth's atmosphere and reach the ground. You can see in Figure 4-5 how the dish reflector of a typical radio telescope focuses the radio waves so their intensity can be measured. Because radio wavelengths are so long, the disk reflector does not have to be as perfectly smooth as the mirror of a reflecting optical telescope.

4-2b The Powers of a Telescope

Astronomers struggle to build large telescopes because a telescope can help human eyes in three important ways—these are called the three powers of a telescope. The two most important of these three powers depend on the diameter of the telescope. (See **The Powers of a Telescope** in the Math Reference Cards MR.)

Most celestial objects of interest to astronomers are faint, so you need a telescope that can gather large amounts of light to produce a bright image. **Light-gathering power** refers to the ability of a telescope to collect light. Catching light in a telescope is like catching rain in a bucket—the bigger the bucket, the more rain it catches. The light-gathering power is proportional to the *area* of the primary mirror, which is proportional to the square of the primary mirror's diameter. A telescope with a diameter of 2 m has four times (4×) the light-gathering power of a 1-m telescope. This is why astronomers use large telescopes and why telescopes are ranked by their diameters. One reason radio astronomers build big radio dishes is to collect enough radio photons, which have low energies, and concentrate them for measurement.

The second power of telescopes, **resolving power**, refers to the ability of the telescope to reveal fine detail. One consequence of the wavelike nature of light is that there is an inevitable small blurring called a **diffraction fringe** around every point of light in the image, and you cannot see any detail smaller than the fringe (Figure 4-6). Astronomers can't eliminate diffraction fringes, but the fringes are smaller in larger telescopes, and that means they have better resolving power and can reveal finer detail. (See **How Do We Know? 4-1**.) For example, a 2-m telescope has diffraction fringes ½ as large, and thus 2× better resolving power, than a 1-m telescope. The size of the diffraction fringes also depends on wavelength, and at the long wavelengths of radio waves, the fringes are large and the resolving power is poor. That's another reason radio telescopes need to be larger than optical telescopes.

One way to improve resolving power is to connect two or more telescopes

optical telescope Telescope that gathers visible light.

radio telescope Telescope that gathers radio radiation.

light-gathering power The ability of a telescope to collect light; proportional to the area of the telescope's primary lens or mirror.

resolving power The ability of a telescope to reveal fine detail. Depends on the diameter of the telescope's primary lens or mirror.

diffraction fringe Blurred fringe surrounding any image, caused by the wave properties of light. Because of this, no image detail smaller than the fringe can be seen.

FIGURE 4-6

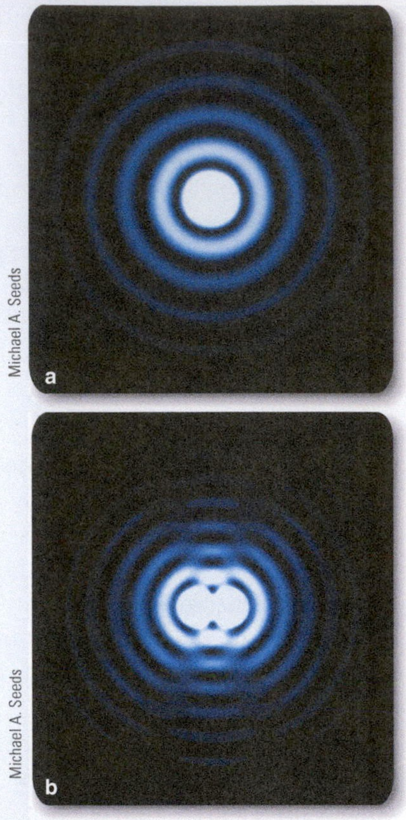

(a) Stars are so far away that their images are points, but the wavelike nature of light causes each star image to be surrounded with diffraction fringes, much magnified in this computer model. (b) Two stars close to each other have overlapping diffraction fringes and become difficult to detect separately.

FIGURE 4-7

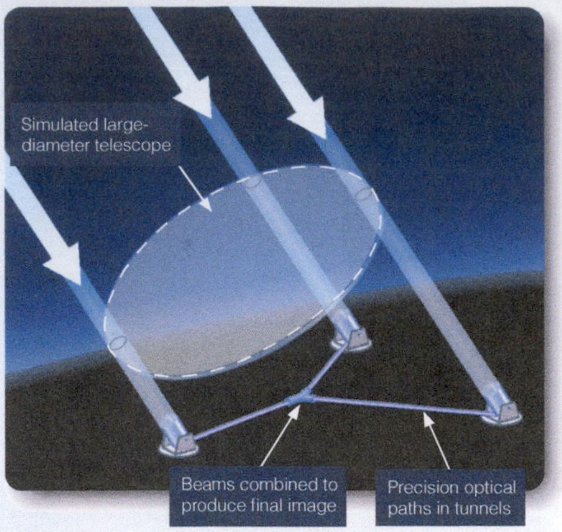

In an astronomical interferometer, smaller telescopes can combine their light to simulate a larger telescope with a resolution set by the separation of the smaller telescopes.

FIGURE 4-8

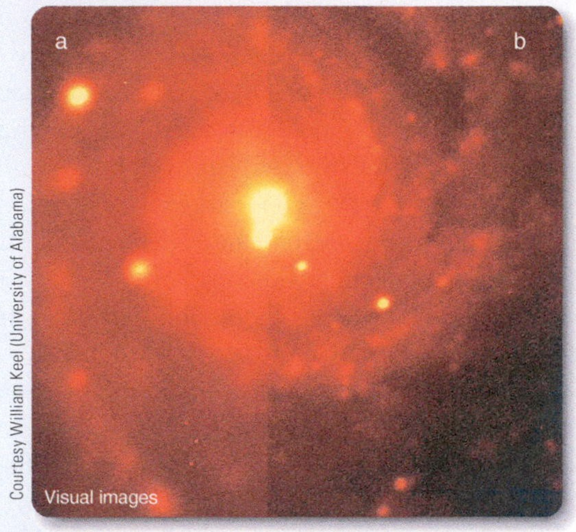

(a) The left half of this galaxy image is from a photograph recorded on a night of poor seeing. Small details are blurred. (b) The right half of this image is from a photo recorded on a night when the seeing was better—Earth's atmosphere above the telescope was steady. Much more detail is visible under good seeing conditions.

in an **interferometer**, which has a resolving power equal to that of a telescope as large as the maximum separation between the individual telescopes (Figure 4-7). The first interferometers were built by radio astronomers connecting radio dishes kilometers apart. Modern technology has allowed astronomers to connect optical telescopes to form interferometers with very high resolution.

interferometer Separated telescopes combined to produce a virtual telescope with the resolution of a much larger-diameter telescope.

seeing Atmospheric conditions on a given night. When the atmosphere is unsteady, producing blurred images, the seeing is said to be poor.

Aside from diffraction fringes, two other factors—optical quality and atmospheric conditions—limit resolving power. A telescope must contain high-quality optics to achieve its full potential resolving power. Even a large telescope shows little detail if its optical surfaces have imperfections. In addition, when you look through a telescope, you look through miles of turbulence in Earth's atmosphere, which makes images dance and blur, a condition that astronomers call ***seeing*** (Figure 4-8). A related

How Do We Know? 4-1

Resolution and Precision

What limits the detail you can see in an image? All images have limited resolution. You see this on your computer screen because images there are made up of picture elements, pixels. If your screen has large pixels, the resolution is low, and you can't see much detail. In an astronomical image, the size of a picture element is set by atmospheric seeing and by diffraction in the telescope. You can't see detail smaller than that resolution limit.

This limitation on detail in an image is related to the limited precision of any measurement. Imagine a zoologist trying to measure the length of a live snake by holding it along a meter stick. The wriggling snake is hard to hold, so it is hard to measure accurately. Also, meter sticks are usually not marked finer than millimeters. Both factors limit the precision of the measurement. If the zoologist said the snake was 43.28932 cm long, you might wonder if that were true. The resolution of this measurement technique does not justify the precision implied by all those digits.

Whenever you make a measurement you should ask yourself how precise that measurement can be. The precision of the measurement is limited by the resolution of the measurement technique, just as the amount of detail in a photograph is limited by its resolution. If you photograph a star, you would not be able to see details on its surface for a reason similar to why the zoologist can't measure the snake's length to high precision.

A high-resolution, visible-wavelength image of Mars made by the *Hubble Space Telescope* reveals details such as mountains, craters, and the south polar cap.

phenomenon is the twinkling of a star. The twinkles are caused by turbulence in Earth's atmosphere, and a star near the horizon, where you look through more air, will twinkle more than a star overhead. On a night when the atmosphere is unsteady, the stars twinkle, the images are blurred, and the seeing is bad. Even with good seeing, the detail visible through a large telescope is limited, not just by its diffraction fringes but by the steadiness of the air through which the observer must look.

A telescope performs best on a high mountaintop where the air is thin and steady, but even at good sites atmospheric turbulence spreads star images into blobs 0.5 to 1 arc seconds in diameter. The situation can be improved by a difficult and expensive technique called **adaptive optics**, in which rapid computer calculations adjust the telescope optics and partly compensate for seeing distortions.

This limitation on the amount of information in an image is related to the limitation of the accuracy of a measurement. All measurements have some built-in uncertainty, and scientists must learn to work within those limitations.

The third and, you may be surprised to learn, least important power of a telescope is **magnifying power**, the ability to make an image large. The magnifying power of a telescope equals the focal length of the primary mirror or lens divided by the focal length of the eyepiece. For example, a telescope with a primary mirror that has a focal length of 700 mm and an eyepiece with a focal length of 14 mm has a magnifying power of 50. Higher magnifying power does not necessarily show you more detail, because the amount of detail you can see in practice is limited by a combination of the seeing conditions and the telescope's resolving power and optical quality.

adaptive optics A computer-controlled optical system in an astronomical telescope used to partially correct for seeing.

magnifying power The ability of a telescope to make an image larger.

A telescope's primary function is to gather light and thus make faint things appear brighter, so the light-gathering power is the most important power and the diameter of the telescope is its most important characteristic. Light-gathering power and resolving power are fundamental properties of a telescope that cannot be altered, whereas magnifying power can be changed simply by changing the eyepiece. (Look again at **The Powers of a Telescope** in the Math Reference Cards MR.)

4-3 OBSERVATORIES ON EARTH—OPTICAL AND RADIO

Most major observatories are located far from big cities and usually on high mountains. Optical astronomers avoid cities because **light pollution**, the brightening of the night sky by light scattered from artificial outdoor lighting, can make it impossible to see faint objects. In fact, many residents of cities are unfamiliar with the beauty of the night sky because they can see only the brightest stars. Radio astronomers face a problem of radio interference analogous to light pollution. Weak radio signals from the cosmos are easily drowned out by human radio interference—everything from automobiles with faulty ignition systems to poorly designed transmitters in communication. To avoid this interference, radio astronomers locate their telescopes as far from civilization as possible. Hidden deep in mountain valleys, they are able to listen to the sky protected from human-made radio noise.

As you learned previously, astronomers prefer to place optical telescopes on mountains because the air there is thin and more transparent, but, most importantly they carefully select mountains where the airflow is usually not turbulent so the seeing is good. Building an observatory on top of a high mountain far from civilization is difficult and expensive, but the dark sky and good seeing make it worth the effort.

Examine **Concept Art 4A, "Modern Optical Telescopes,"** and you will notice three important points:

1. Research telescopes must focus their light to positions at which cameras and other instruments can be placed.
2. Small telescopes can use other focal arrangements that would be inconvenient in larger telescopes.
3. *Active optics*, computer control of the shape of a telescope's main mirrors, allows the use of thin, lightweight mirrors—either "floppy" mirrors or segmented mirrors. Reducing the weight of the mirror reduces the weight of the rest of the telescope, making it stronger and less expensive. Also, thin mirrors cool and reach a stable shape faster at nightfall, producing better images during most of the night.

sidereal tracking (pronounced "sih-dare-ee-al") The continuous movement of a telescope to keep it pointed at a star as Earth rotates.

light pollution The illumination of the night sky by waste light from cities and outdoor lighting, which prevents the observation of faint objects.

Telescopes located on the surface of Earth, whether optical or radio, must move continuously to stay pointed at a celestial object as Earth turns on its axis. This is called **sidereal tracking** ("sidereal" refers to the stars).

High-speed computers have allowed astronomers to build new, giant telescopes with unique designs (Figure 4-9). The European Southern Observatory has built the Very Large Telescope (VLT) in the remote foothills of the Andes Mountains of northern Chile. The VLT actually consists of four telescopes, each with a computer-controlled mirror 8.2 m (323 in., about 27 ft) in diameter and only 17.5 cm (6.9 in.) thick. The four telescopes can work singly or can combine their light to work as one large telescope. Italian and American astronomers have built the Large Binocular Telescope, which carries a pair of 8.4-m mirrors on a single mounting. The Gran Telescopio Canarias, located atop a volcanic peak, carries a segmented mirror 10.4 m in diameter. Other giant telescopes with effective diameters as large as 30 m are being planned or envisioned for the coming decades.

Ground-based telescopes usually are used by astronomers and technicians working in a control room in the same building or nearby, but some telescopes are now operated by astronomers thousands of miles from the observatory, and others are fully automated, operating without direct human supervision. Automation plus continuous improvement in computer speed and storage capacity has made possible huge surveys of the sky in which millions of objects have been observed or are planned for observation. For example, the Sloan Digital Sky Survey (SDSS) mapped the entire Northern Hemisphere sky, measuring the position and brightness of 100 million stars and galaxies at five ultraviolet, optical, and infrared wavelengths. Data from the SDSS are available for you to examine and manipulate at this website: skyserver.sdss.org. Some SDSS data are also at Microsoft's World Wide Telescope (www.worldwidetelescope.org) and Google Sky (www.google.com/sky/). The "citizen science" Galaxy Zoo site (www.galaxyzoo.org) allows volunteers (this could mean you) to classify galaxies based on their appearances in SDSS images.

FIGURE 4-9

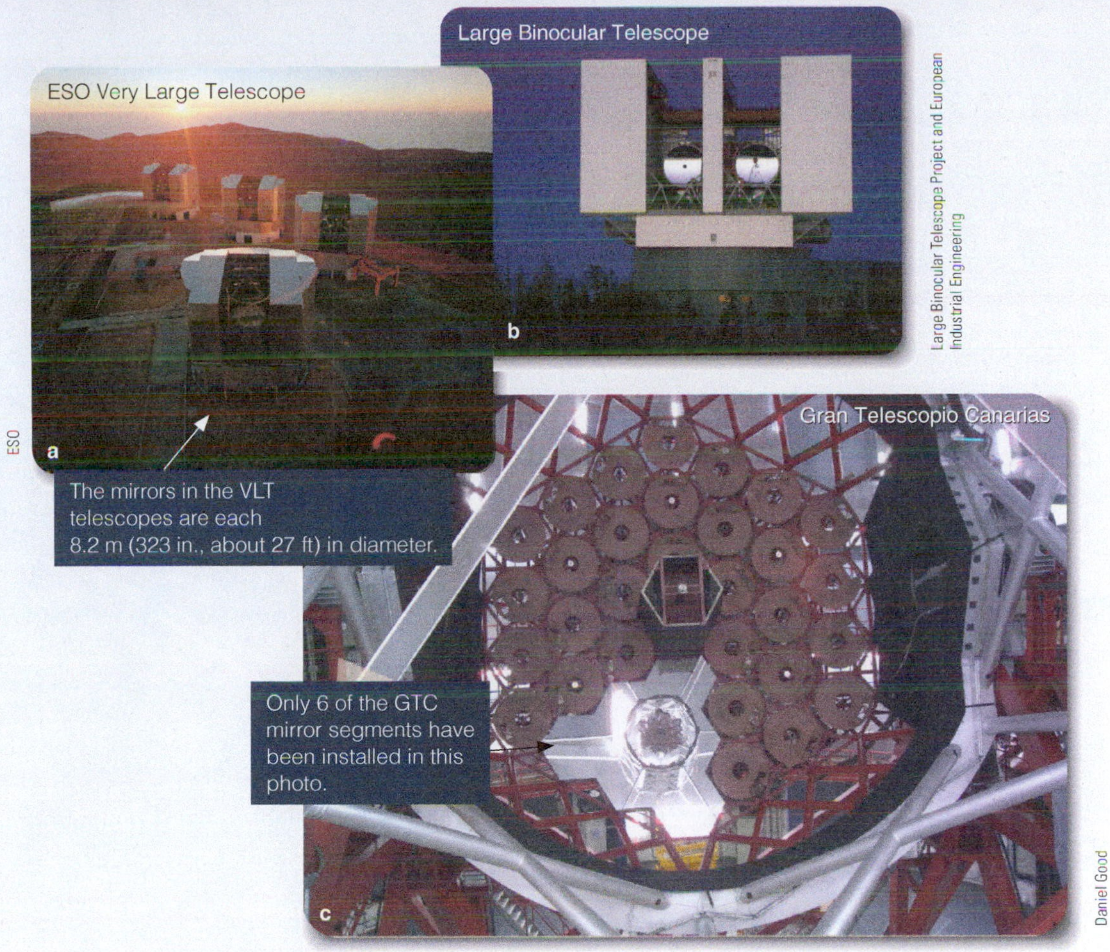

(a) The four telescopes of the European Very Large Telescope (VLT) are housed in separate domes at Paranal Observatory in Chile. (b) The Large Binocular Telescope (LBT) in Arizona carries two 8.4-m mirrors. The light gathered by the two mirrors can be analyzed separately or combined. The entire building rotates as the telescope moves. (c) The Gran Telescopio Canarias (GTC) on La Palma in the Canary Islands contains 36 hexagonal mirror segments in its 10.4-m primary mirror.

The largest fully steerable radio telescope in the world is at the National Radio Astronomy Observatory in West Virginia. The telescope has a reflecting surface 100 m in diameter made of 2,004 computer-controlled panels that adjust to maintain the shape of the reflecting surface. The largest radio dish in the world is 305 m (1,000 ft) in diameter, and is built into a mountain valley in Arecibo, Puerto Rico. The antenna hangs on cables above the dish, and, by moving the antenna, astronomers can point the telescope at any object that passes within 20 degrees of the zenith as Earth rotates (Figure 4-10a). China has constructed a 500-m-diameter radio telescope dubbed FAST, which, when operational, will have more than 2.5 times the collecting area of Arecibo's dish.

The Very Large Array (VLA) consists of 27 dishes spread in a Y-pattern across the New Mexico desert (Figure 4-10b). Operated as an interferometer, the VLA has the resolving power of a radio telescope up to 36 km (22 mi) in diameter. Such arrays are very powerful, and radio astronomers are now planning the Square Kilometer Array, which will consist of radio dishes spanning 6000 km (almost 4000 mi) and having a total collecting area of one square kilometer.

Concept Art 4A

MODERN OPTICAL TELESCOPES

1

The traditional telescopes described on this page are limited by complexity, weight, and Earth's atmosphere. Modern solutions are shown on the opposite page.

In larger telescopes, the light can be focused to a **prime focus** position high in the telescope tube as shown at the right. Although it is a good place to image faint objects, the prime focus is inconvenient for large instruments. A **secondary mirror** can reflect the light through a hole in the primary mirror to a **Cassegrain focus.** This focal arrangement may be the most common form of astronomical telescope.

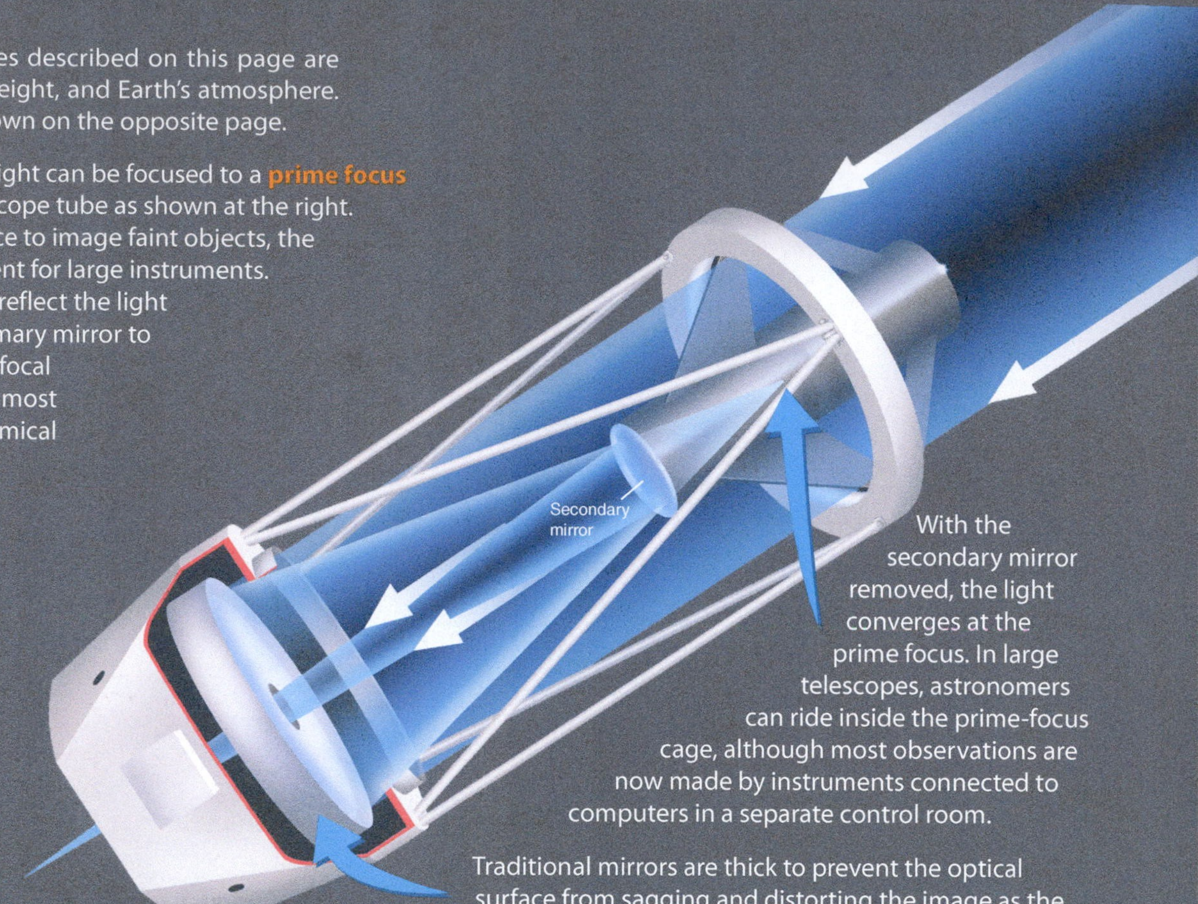

With the secondary mirror removed, the light converges at the prime focus. In large telescopes, astronomers can ride inside the prime-focus cage, although most observations are now made by instruments connected to computers in a separate control room.

The Cassegrain focus is convenient and has room for large instruments.

Traditional mirrors are thick to prevent the optical surface from sagging and distorting the image as the telescope is moved around the sky. Large mirrors can weigh many tons and are expensive to make and difficult to support. Also, they cool slowly at nightfall. Expansion and contraction in the cooling mirror causes distortion in the images.

1a

Smaller telescopes are often found with a **Newtonian focus,** the arrangement that Isaac Newton used in his first reflecting telescope. The Newtonian focus is inconvenient for large telescopes as shown at right.

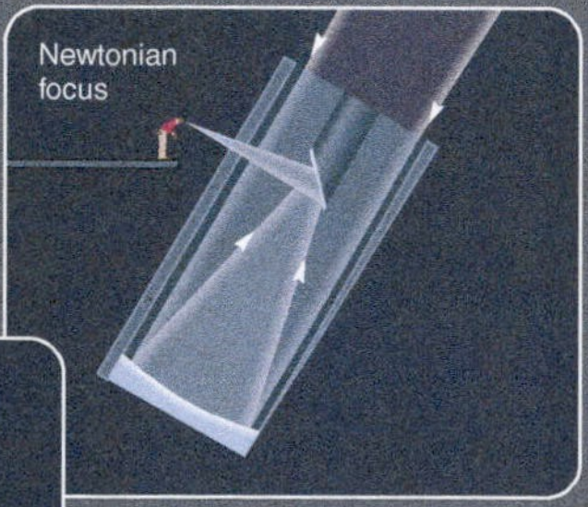

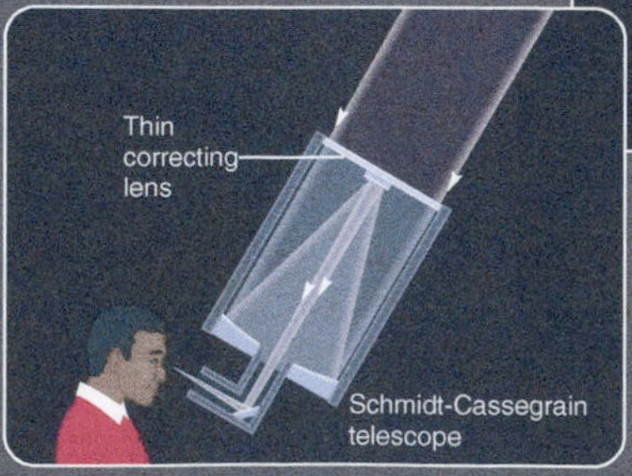

Courtesy of W. M. Keck Observatory

1b

Many small telescopes such as the one on your left use a **Schmidt-Cassegrain focus.** A thin correcting plate improves the image but is too slightly curved to introduce serious chromatic aberration.

1c

Shown below, the 4-meter Mayall Telescope at Kitt Peak National Observatory in Arizona can be used at either the prime focus or the Cassegrain focus. Note the human figure at lower right.

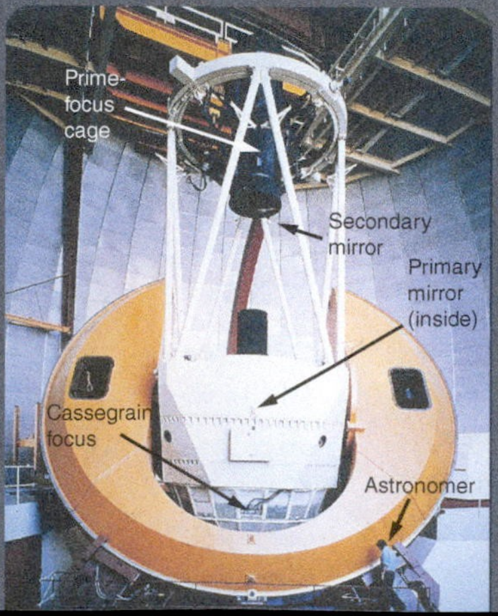

NOAO/AURA/NSF

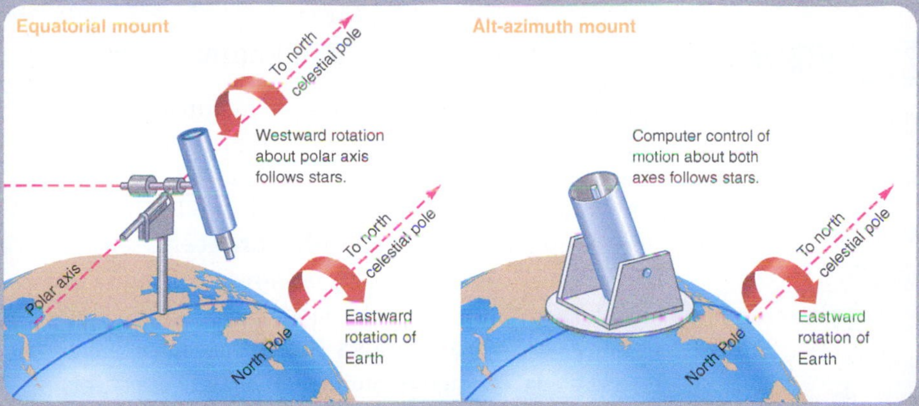

2

Telescope mountings must contain a **sidereal drive** to move the telescope smoothly westward, countering the eastward rotation of Earth. The earlier **equatorial mount** *(far left)* has a **polar axis** parallel to Earth's axis, but the mount used for the largest modern telescopes is **alt-azimuth mount** (altitude-azimuth; near left) moves like a cannon—up and down, left and right. Alt-azimuth mountings are simpler to build than equatorial mountings but require computer control to follow the stars.

3

Unlike traditional thick mirrors, thin mirrors, sometimes called "floppy" mirrors as shown at right, weigh less and require less massive support structures. Also, they cool rapidly at nightfall and there is less distortion from uneven expansion and contraction.

3a

Grinding a large mirror may remove tons of glass and take months, but new techniques speed the process. Some large mirrors are cast in a rotating oven that causes the molten glass to flow to form a concave upper surface. Grinding and polishing such a preformed mirror is much less time consuming.

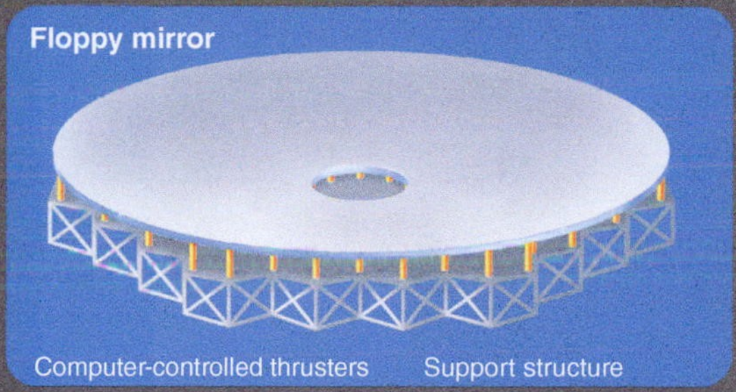

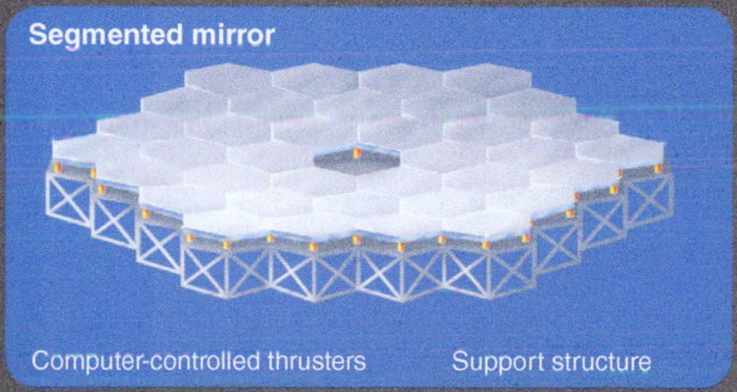

3b

Mirrors made of segments are economical because the segments can be made separately. The resulting mirror weighs less and cools rapidly.

3c

Both floppy mirrors and segmented mirrors sag under their own weight. Their optical shapes must be controlled by computer-driven thrusters behind the mirrors, a technique called **active optics**.

3d

The two Keck telescopes, each 10 meters in diameter, are located atop the extinct volcano Mauna Kea in Hawai'i. Their two primary mirrors are composed of hexagonal mirror segments, as shown at right.

NOAO/AURA/NSF

Courtesy of W. M. Keck Observatory

FIGURE 4-10

(a) The 305-m (1000-ft) radio telescope in Arecibo, Puerto Rico, is nestled in a naturally bowl-shaped valley. The receiver platform is suspended over the dish. (b) The Very Large Array (VLA) radio telescope uses 27 radio dishes that can be moved to different positions along a Y-shaped set of tracks across the New Mexico desert. They are shown here in the most compact arrangement. The individual dishes are usually combined to make an interferometer capable of very-high-resolution radio maps of celestial objects.

4-4 ASTRONOMICAL INSTRUMENTS AND TECHNIQUES

Just looking through a telescope doesn't tell you much. To learn about planets, stars, and galaxies, you must be able to analyze the light the telescope gathers. Special instruments attached to the telescope make that possible.

4-4a Imaging Systems and Photometers

The **photographic plate** was the first image-recording device used with telescopes. Brightness of objects imaged on a photographic plate can be measured with a lot of hard work, yielding only moderate precision. Astronomers also build **photometers**, sensitive light meters to measure the brightness of individual objects very precisely. Most modern astronomers use **charge-coupled devices (CCDs)** as both image-recording devices and photometers. A CCD is a specialized computer chip that can contain more than a million microscopic light detectors arranged in an array about the size of a postage stamp.

CCD and other types of **array detectors** can be used like a small photographic plate, but they have dramatic advantages over both photometers and photographic plates. CCDs can detect both bright and faint objects in a single exposure and are much more sensitive than a photographic plate. CCD images are **digitized**, or converted to numerical data, and can be read directly into a computer memory for later analysis. Although CCDs for astronomy are extremely sensitive and therefore expensive, less sophisticated CCDs are now used in commercial video and digital cameras. Infrared astronomers also use array detectors similar in operation to optical CCDs. At some other wavelengths, photometers are still used for measuring brightness of celestial objects.

The digital data representing an image from a CCD, or other array detector, are easy to manipulate to bring out details that would not otherwise be visible. For example, astronomical images are sometimes reproduced as negatives, with the sky white and the stars dark. This makes the faint parts of the image easier to see (Figure 4-11).

Astronomers also manipulate images to produce **false-color images** (also known as representational-color images) in which the colors represent different levels of intensity and are not related to the true colors of the object. For example, because humans can't see radio waves, astronomers must convert them into something perceptible. One way is to measure the strength of the radio signal at various places in the sky and draw a map in which contours mark areas of uniform radio intensity. Compare such a map to Figure 4-12a, a seating diagram for a baseball stadium in which the contours mark areas in which the seats have the same price. Contour maps are very common in radio astronomy and are often reproduced using false colors (Figure 4-12b).

FIGURE 4-11

Astronomical images can be manipulated to bring out difficult-to-see details. (a) The color photo of this galaxy is dark, and the dust clouds in the galaxy's central plane do not show very well. (b) The observing team produced this negative image to show the dust clouds more clearly.

4-4b Spectrographs

To analyze light in detail, you need to spread the light out according to wavelengths-into a spectrum, a task performed by a **spectrograph**. You can understand how this works by reproducing an experiment performed by Isaac Newton in 1666. Boring a hole in his window shutter, Newton admitted a thin beam of sunlight into his darkened bedroom. When he placed a prism in the path of the beam, the sunlight spread into a beautiful spectrum on the far wall. From this, and similar experiments, Newton concluded that white light was made of a mixture of all the colors.

Newton didn't think in terms of wavelength, but you can use that modern concept to see that the light passing through the prism is bent at an angle that depends on the wavelength (see Figure 4-13). Violet (short-wavelength) light bends the most, and red (long-wavelength) light the least. Thus, the white light entering the prism is spread into what is called a **spectrum**. A typical prism spectrograph contains more than one prism to spread the spectrum wider, plus lenses to guide the light into the prisms and to focus the light onto a photographic plate. Most modern spectrographs use a grating in place of a prism. A **grating** is a piece of glass with thousands of microscopic parallel lines scribed onto its surface. Different wavelengths of light reflect from the grating at slightly different angles, so white light is spread into a spectrum and can be recorded, often by a CCD camera. Recording the spectrum of a faint star or galaxy can require a long time exposure, so astronomers have developed multiobject spectrographs that can record the spectra of as many as 100 objects simultaneously. Multiobject spectrographs automated by computers have made large surveys of many thousands of stars and galaxies possible.

photographic plate The first image-recording device used with telescopes; it records the brightness of objects, but with only moderate precision.

photometer Sensitive astronomical instrument that measures the brightness of individual objects very precisely.

charge-coupled device (CCD) An electronic device consisting of a large array of light-sensitive elements used to record very faint images.

array detector Device for collecting and recording electromagnetic radiation using multiple individual detectors arrayed on the surface of a chip; for example, a CCD electronic camera.

digitized Converted to numerical data that can be read directly into a computer memory for later analysis.

false-color image A representation of graphical data with added or enhanced color to reveal detail.

FIGURE 4-12

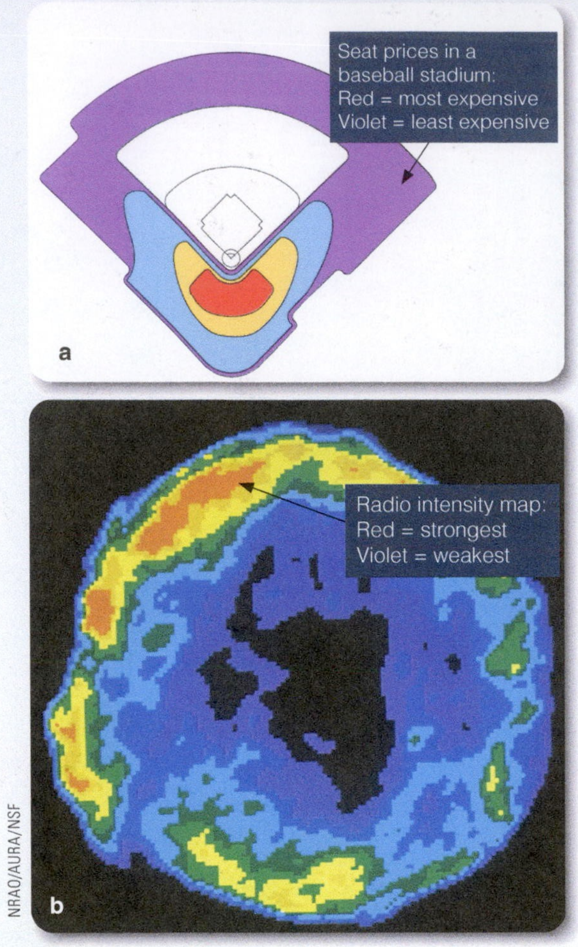

(a) A contour map of a baseball stadium shows regions of similar admission prices. The most expensive seats are those behind home plate. (b) A false-color radio map of Tycho's supernova remnant, the expanding shell of gas produced by the explosion of a star in 1572. The radio contour map has been color-coded to show intensity.

FIGURE 4-13

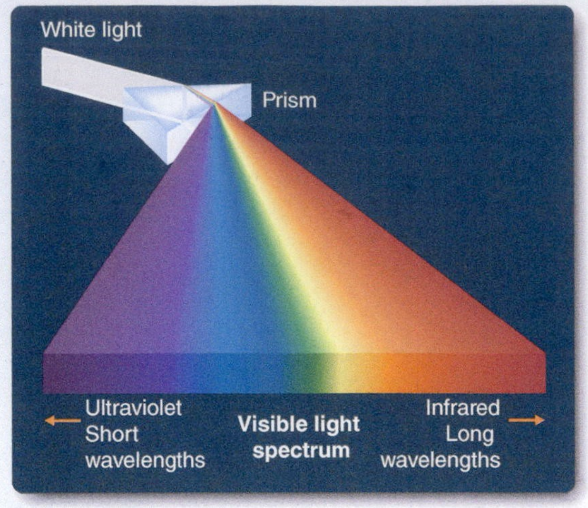

A prism bends light by an angle that depends on the wavelength of the light. Short wavelengths bend most and long wavelengths least. Thus, white light passing through a prism is spread into a spectrum.

spectrograph A device that separates light by wavelengths to produce a spectrum.

spectrum A range of electromagnetic radiation spread into its component wavelengths (colors); for example, a rainbow; also, representation of a spectrum as a graph showing intensity of radiation as a function of wavelength or frequency.

grating A piece of material in which numerous microscopic parallel lines are scribed. Light encountering a grating is dispersed to form a spectrum.

Because astronomers understand how light interacts with matter, a spectrum carries a tremendous amount of information, as you will see in more detail in Sections 5-4 and 5-5 of Chapter 5. A spectrograph is the astronomer's most powerful instrument. Astronomers are likely to remark, "We don't know anything about an object until we get a spectrum," and that is only a slight exaggeration. A spectrograph is the astronomer's most powerful instrument. Astronomers are likely to remark, "We don't know anything about an object until we get a spectrum," and that is only a slight exaggeration.

4-5 AIRBORNE AND SPACE OBSERVATORIES

You have learned about the observations that ground-based telescopes can make through the two atmospheric windows in the visible and radio parts of the electromagnetic spectrum. Most of the rest of the electromagnetic spectrum—infrared, ultraviolet, X-ray, and gamma-ray radiation—never reaches Earth's surface. To observe at these wavelengths, telescopes must fly above the atmosphere in high-flying aircrafts, rockets, balloons, and satellites. The only exceptions are observations that can be made in what are called the near-infrared and the near-ultraviolet, almost the same wavelengths as visible light.

4-5a The Ends of the Visual Spectrum

Astronomers can observe from the ground in the near-infrared just beyond the red end of the visible spectrum because some of this radiation leaks through the atmosphere in narrow, partially open atmospheric windows ranging in wavelength from 1200 nm to about 30,000 nm. Infrared astronomers usually describe wavelengths in micrometers or microns (10^{-6} m), so they refer to this wavelength range as 1.2 to 30 microns. In this range, most of the radiation is absorbed by water vapor and carbon dioxide molecules in Earth's atmosphere, so it is an advantage to place telescopes on the highest mountains where the air is especially thin and dry. For example, a number of important IR telescopes are located on the summit of Mauna Kea in Hawai'i at an altitude of 4200 m (13,800 ft) (Figure 4-14a).

The far-infrared range, which includes wavelengths longer than 30 microns, can tell us about planets, comets, star formation, and other low-temperature objects, but these wavelengths are absorbed high in the atmosphere. To observe in the far-infrared, telescopes must venture to high altitudes. Remotely operated IR telescopes suspended under balloons have reached altitudes as high as 41 km (25 mi). The Stratospheric Observatory for Infrared Astronomy (SOFIA), a Boeing 747SP aircraft that carries a 2.5-m (100-in.) telescope (Figure 4-14b), can get above more than 99 percent of the water vapor in Earth's atmosphere, allowing observations at IR wavelengths otherwise accessible only from space telescopes.

If a telescope observes at far-infrared wavelengths, it must be cooled. Infrared radiation is emitted by heated objects, and if the telescope is warm it will emit many times more IR radiation than that coming from a distant object. Imagine trying to look at a dim, moonlit scene through binoculars that are glowing brightly. In a telescope observing near-infrared radiation, only the detector, the element on which the infrared radiation is focused, must be cooled. To observe in the far-infrared, however, the entire telescope must be cooled.

At the short-wavelength end of the spectrum, astronomers can observe in the near-ultraviolet. Human eyes do not detect this radiation, but it can be recorded by photographic plates and CCDs. Wavelengths shorter than about 290 nm, including the far-ultraviolet, X-ray, and gamma-ray ranges, are completely absorbed by the ozone layer extending from 20 km to about 40 km above Earth's surface. No mountain is that high, and no balloon or airplane can fly that high, so astronomers cannot observe far-UV, X-ray, and gamma-ray radiation without going into space.

FIGURE 4-14

Richard Wainscoat, Aina Kai LLC

NASA / Jim Ross

(a) Optical, infrared, and submillimeter radio telescopes at 4200 m (13,800 feet) above sea level on Mauna Kea in Hawai'i. The high altitude, low atmospheric moisture, lack of nearby large cities, and location near the equator make this mountain one of the best places on Earth to build an observatory. (b) *SOFIA, the Stratospheric Observatory for Infrared Astronomy*, a joint project of NASA and the German Aerospace Center (DLR), flies at altitudes up to 14 km (45,000 ft), where it can collect light at IR wavelengths unobservable even from high mountaintops.

4-5b Telescopes in Space

The *Hubble Space Telescope* (or *HST*), named after Edwin Hubble, the astronomer who discovered the expansion of the Universe, is the most successful telescope ever to orbit Earth (Figure 4-15a). It was launched in 1990 and contains a 2.4-m (95-in.) mirror plus instruments that can observe at near-infrared, visual, and near-ultraviolet wavelengths. It is controlled from a research center on Earth and observes continuously. Nevertheless, there is time to complete only a fraction of the projects proposed by astronomers from around the world.

Most of the observations made by *HST* are at visual wavelengths, so its greatest advantage in being above Earth's atmosphere is the lack of seeing distortion. It can see faint objects and detect fine detail by concentrating light into sharp images.

The telescope is as big as a large bus and has been visited a number of times by the space shuttle so that

FIGURE 4-15

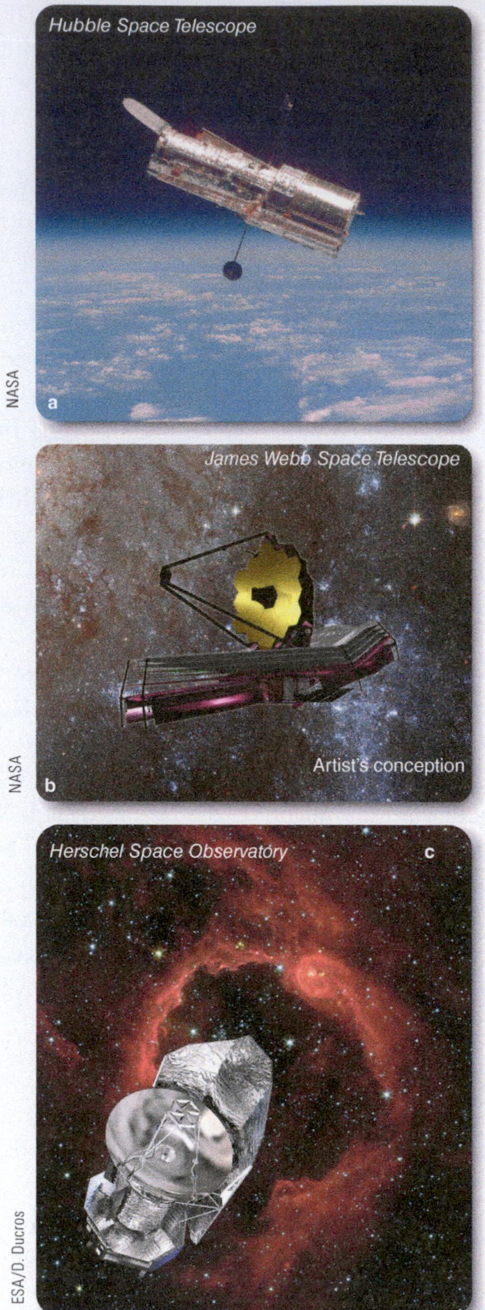

(a) The *Hubble Space Telescope* orbits Earth at an average of 570 km (355 mi) above the surface. In this image the telescope is viewing toward the upper left. (b) An artist's conception of *HST* eventual successor, the *James Webb Space Telescope (JWST)*. *JWST* will be located in solar orbit 1 million miles from Earth and will not have an enclosing tube, thus resembling a radio dish more than a conventional optical telescope. It will observe the Universe from behind a multilayered sun screen larger than a tennis court. (c) Artist's conception of the Herschel infrared space telescope that carried a 3-m mirror and instruments cooled almost to absolute zero.

FIGURE 4-16

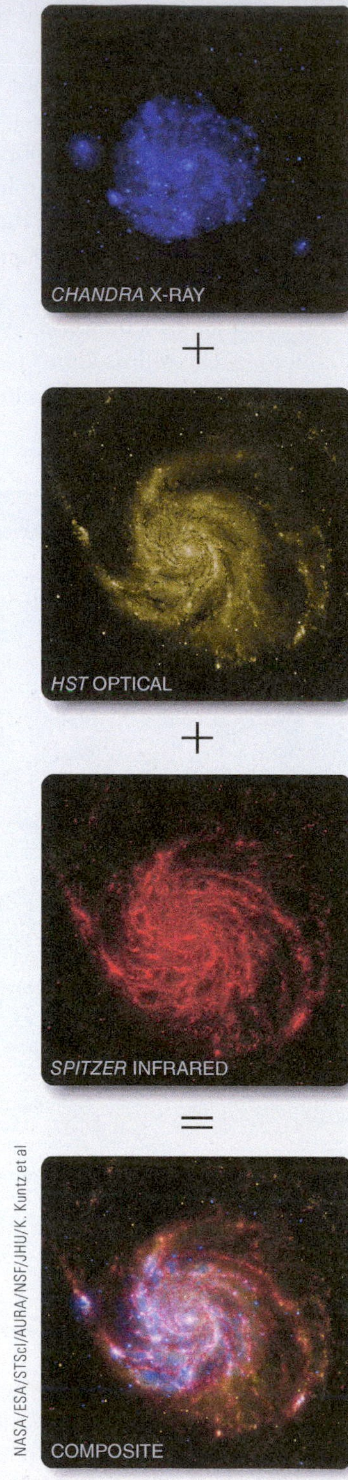

Images of the galaxy Messier 101 made by three space telescopes: the *Chandra X-ray Observatory*, the *Hubble Space Telescope*, and the *Spitzer Space Telescope*. Comparison and combination of these images yields abundant information about the life cycles of stars in a galaxy similar to our Milky Way.

astronauts could maintain its equipment and install new cameras and spectrographs. Astronomers hope that it will last until it is replaced by the *James Webb Space Telescope (JWST)*, expected to launch in the later half of 2018 (Figure 4-15b). The *JWST* telescope's primary mirror will have an effective diameter of 6.6 m (256 in.)

Telescopes that observe in the far-infrared must be protected from heat and must get above Earth's absorbing atmosphere. They have limited lifetimes because they must carry coolant to chill their optics. The most sophisticated of the IR telescopes put in orbit, the *Herschel Space Observatory* was cooled to −269°C (−453°F). Launched in 2009, it observed from behind a sunscreen. In fact, it could not observe from orbit around Earth because Earth is such a strong source of IR radiation, so the telescope was sent to a position 1.5 million km (1 million mi) from Earth in the direction away from the Sun. Named after the discoverer of infrared light (look back to Figure 4-3), *Herschel* made important discoveries concerning star and planet formation, interstellar chemistry, and distant galaxies during its 3-year mission.

High-energy astrophysics refers to the use of X- ray and gamma-ray observations of the sky. Making such observations is difficult but can reveal the secrets of processes such as the collapse of massive stars and eruptions of supermassive black holes. The largest X- ray telescope to date, the *Chandra X-ray Observatory*, was launched in 1999 and orbits a third of the way to the Moon. *Chandra* is named for the late Indian-American Nobel Laureate Subrahmanyan Chandrasekhar, who was a pioneer in many branches of theoretical astronomy. Focusing X-rays is difficult because they penetrate into most mirrors, so astronomers devised cylindrical mirrors in which the X-rays reflect from the polished inside of the cylinders and form images on special detectors. The telescope has made important discoveries about everything from star formation to monster black holes in distant galaxies (Figure 4-16).

One of the first gamma-ray observatories was the *Compton Gamma Ray Observatory*, launched in 1991. It mapped the entire sky at gamma-ray wavelengths. The European-built *INTEGRAL (International Gamma-Ray Astrophysics Laboratory)* satellite was launched in 2002 and has been very productive in the study of violent eruptions of stars and black holes. The *Fermi Gamma-ray Space Telescope*, launched in 2008 and operated by an international consortium led by the United States, is capable of mapping large areas of the sky to high sensitivity.

Modern astronomy has come to depend on observations that cover the entire electromagnetic spectrum. More orbiting space telescopes are planned, and they will be more versatile and more sensitive.

What Are We? Curious

The old saying "Curiosity killed the cat" is an insult to the cat and to curiosity. We humans are curious, and curiosity is a noble trait—the mark of an active, inquiring mind. At the limits of human curiosity lies the fundamental question, "What are we?" Telescopes extend and amplify our senses, but they also allow us to extend and amplify our curiosity about the Universe around us.

STUDY TOOLS 4

IN THE BOOK

- ☐ Tear Out the Review Card on *Light and Telescopes*.

ASTRO ONLINE

Visit ASTRO Online at www.cengagebrain.com

- ☐ Interactive Reading
- ☐ Practice Quizzing
- ☐ Videos and Animations
- ☐ Exclusive Features

Questions and Problems

4

REVIEW QUESTIONS

1. Why would you not plot sound waves in the electromagnetic spectrum?
2. If you had limited funds to build a large telescope, which type would you choose, a refractor or a reflector? Why?
3. Why do nocturnal animals usually have large pupils in their eyes? How is that related to the way astronomical telescopes work?
4. Why do optical astronomers often put their telescopes at the tops of mountains, while radio astronomers sometimes put their telescopes in deep valleys?
5. What are the advantages of making a telescope mirror thin? What problems might that cause? (*Hint:* What would happen to images of objects if a thin telescope mirror were to bend slightly?)
6. Small telescopes are often advertised as "200 power" or "magnifies 200 times." How would you improve such advertisements?
7. Why do radio telescopes have relatively poor resolving power?
8. The Moon has no atmosphere at all. What advantages would you have if you built an observatory on the lunar surface?
9. Why must telescopes observing at far-infrared wavelengths be cooled to low temperatures?
10. What purpose do the colors in a false-color image or map serve?
11. What might you detect with an X-ray telescope that you could not detect with an IR telescope?
12. How is the phenomenon of chromatic aberration related to how a prism spectrograph works?
13. Why would radio astronomers build identical radio telescopes in many different places around the world?
14. **How Do We Know?** How is the resolution of an astronomical image related to the precision of a measurement?

DISCUSSION QUESTIONS

1. When you think of a "nice night," what does the sky look like? You have a friend who is an astronomer. What would your friend think of your "nice night"? Does that depend on what type of astronomer he or she is (i.e., what wavelength of electromagnetic radiation that friend prefers to collect and study)?
2. Consider the "atmospheric windows" of Earth's atmosphere (Figure 4-2b). How do these match against the biological responses of animals and plants here on Earth? What do you think is the reason for that?

PROBLEMS

Note: for Problems 1 through 8, refer to **The Powers of a Telescope** in the Math Reference Cards.

1. The thickness of the plastic in dry cleaner bags is about 0.001 mm. How many wavelengths of red light is that? (*Hint:* You can estimate a typical wavelength for red light by looking at Figure 4-2a.)
2. Compare the light-gathering power of one of the 10-m Keck telescopes with that of a 0.5-m telescope.
3. How does the light-gathering power of one of the Keck telescopes compare with that of the human eye? Assume that the pupil of your eye can open to a diameter of about 0.8 cm in dark conditions.
4. What is the resolving power of a 25-cm (10-in.) telescope at a wavelength of 550 nm (in the middle of the visual band)? What do two stars 1.5 arc seconds apart look like through this telescope at that wavelength?
5. Most of Galileo's telescopes were only about 2 cm in diameter. Should he have been able to resolve the two stars mentioned in Problem 4?
6. How does the resolving power of the 5-m telescope on Mount Palomar near San Diego compare with that of the 2.5-m *Hubble Space Telescope*? Why does *HST* generally still outperform the Palomar 5-m telescope?
7. If you build a telescope with a focal length of 1.3 m, what focal length does the eyepiece need to give a magnification of 100 times?
8. Astronauts observing from a space station need a telescope with a light-gathering power 15,000 times that of the dark-adapted human eye (*Note:* see Problem 3), capable of resolving detail as small as 0.1 arc seconds at a wavelength of 550 nm, and having a magnifying power of 250. Design a telescope to meet their needs. Could you test your design by using it to observe stars from the surface of Earth?
9. A spy satellite orbiting 400 km above Earth is supposedly capable of counting individual people in a crowd in visual-wavelength images. Assume that the middle of the visual wavelength band is a wavelength of 550 nm. Assume an average person seen from above has a size of 0.7 m. Use the **Small-Angle Formula** and **The Powers of a Telescope** in the Math Reference Cards MR to estimate the minimum diameter of the telescope the satellite must carry.

LEARNING TO LOOK

1. The two images in Figure 4-17 show a star before and after an adaptive optics system attached to the telescope was switched on. What causes the distortion in the first image, and how do adaptive optics correct the image?

FIGURE 4-17

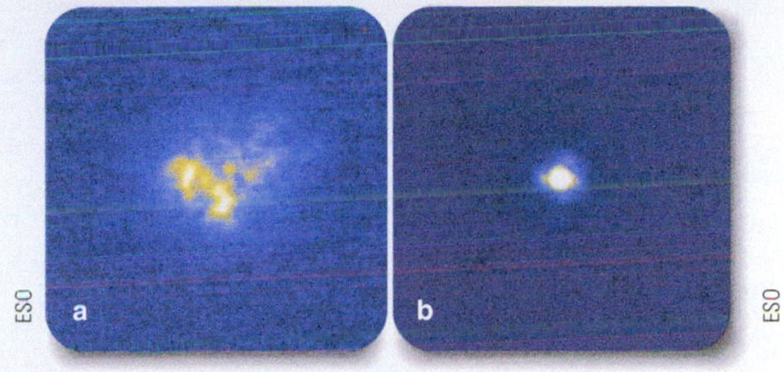

2. The star images in Figure 4-18 are tiny disks, but the diameter of these disks is not related to the diameter of the stars. Explain why the telescope can't resolve the diameter of the stars. What does cause the diameters of the star images?

FIGURE 4-18

3. The X-ray image in Figure 4-19 shows the remains of an exploded star. Explain why images recorded by telescopes in space are often displayed in representational ("false") color rather than in the "colors" (wavelengths) received by the telescope.

FIGURE 4-19

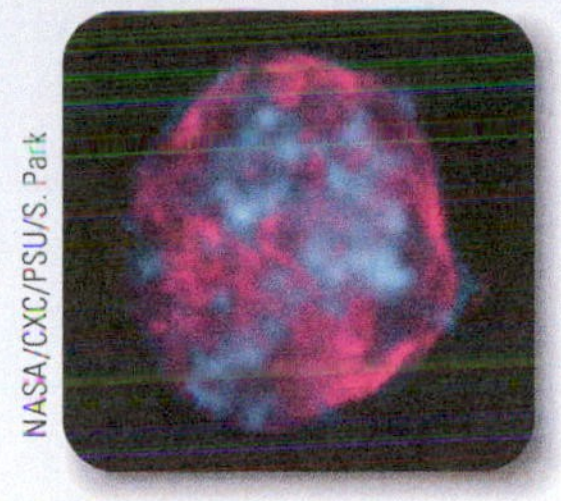

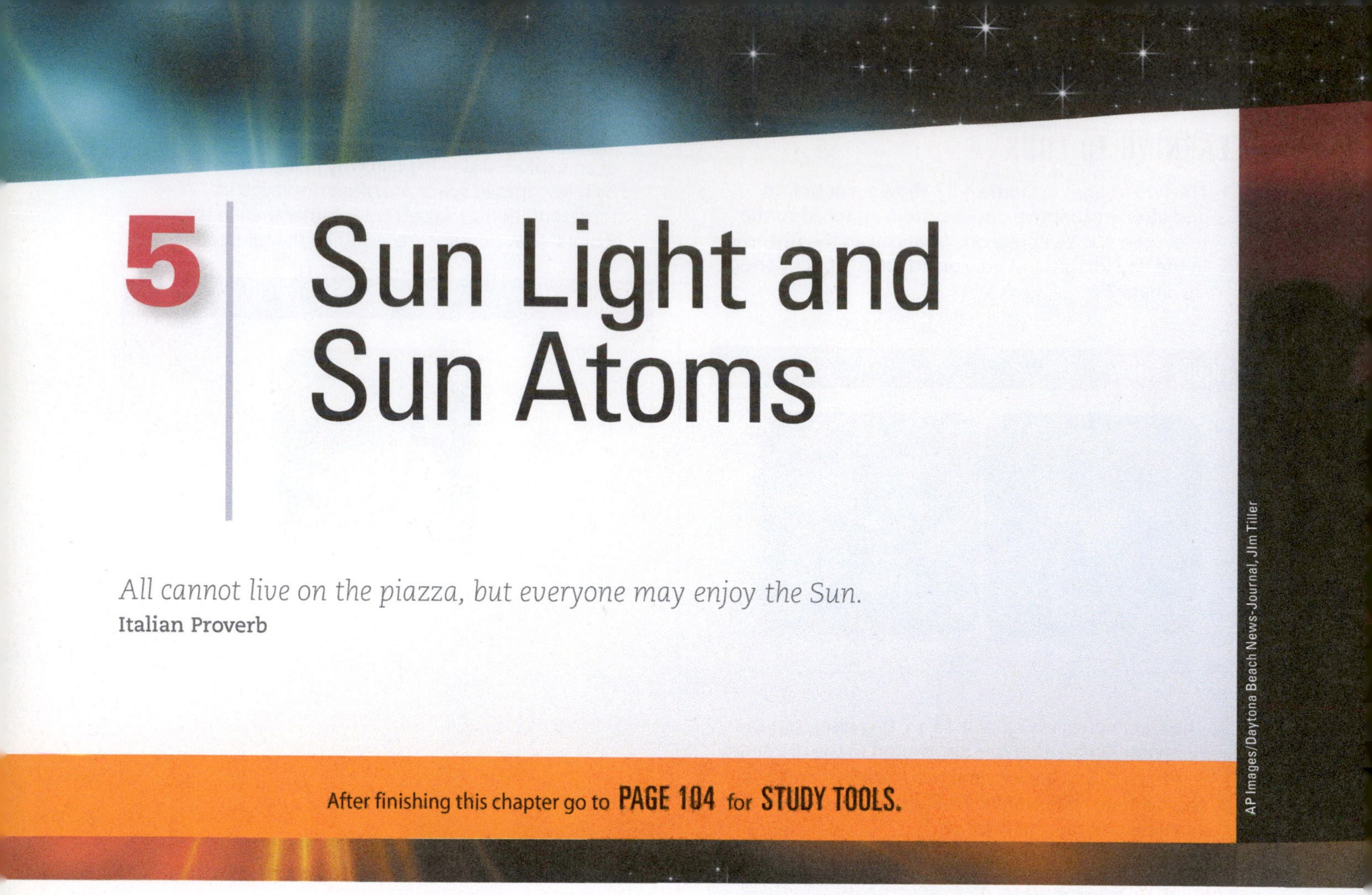

5 Sun Light and Sun Atoms

All cannot live on the piazza, but everyone may enjoy the Sun.
Italian Proverb

After finishing this chapter go to **PAGE 104** for **STUDY TOOLS.**

AP Images/Daytona Beach News-Journal, Jim Tiller

GUIDEPOST

This chapter marks a change in the way you will look at nature. Up to this point, you have been considering what you can see with your eyes alone or aided by telescopes. In this chapter, you will begin to understand how astronomical observations can be linked with physics experiments and theory to unlock celestial secrets that lie beyond what can be directly seen.

The Sun, which is the main source of light and warmth in our Solar System, has always been a natural subject of great human curiosity. (See CO 5.) In this chapter you will discover how careful analysis of sunlight, combined with understanding of how matter and light interact, paints a detailed picture of the Sun's properties. Here you will find answers to four important questions:

- How do atoms and light interact?
- What can you learn from spectra of celestial objects?
- What can you learn about the Sun by observing its surface and atmosphere?
- Why does the Sun go through 11- and 22-year cycles of activity?

Although this chapter considers only the star at the center of our Solar System, understanding how astronomers study the Sun leads you onward and outward among the planets, stars, and galaxies that fill the Universe, often using the rich information derived from their spectra.

Earthbound humans knew almost nothing about the Sun until the early 19th century, when the German optician Joseph von Fraunhofer studied the solar spectrum and found that it is interrupted by some 600 dark lines, representing colors that are missing from the sunlight Earth receives. As scientists realized that these **spectral lines** were related to the presence of various types of atoms in the Sun's atmosphere, a window finally opened to real understanding of the Sun's nature.

In this chapter, you will look through that window by considering how the Sun produces light and how atoms interact with light to make spectral lines. Once you understand that, you will know how astronomers have determined the chemical composition of the Sun, measured motions of gas on the Sun's surface and in its atmosphere, and detected magnetic fields that drive the Sun's cycle of activity.

spectral line A feature in a spectrum at a specific wavelength produced by the absorption or emission of light by certain atoms.

CO 5 The Sun is the main source of light and warmth on Earth's surface, so it is a natural object of human curiosity and awe. It is also the star that is most easily visible from Earth. Knowledge of the interaction of light and matter, which you studied in Chapter 4, helps unlock the secrets of the Sun and introduces you to the stars.

5-1 THE SUN: BASIC CHARACTERISTICS

In its general properties, the Sun is very simple. It is a great ball of hot gas held together by its own gravity. The tremendously hot gas inside the Sun has such a high pressure that it would explode were it not for its own confining gravity. The same gravity would make it collapse into a small, dense body were it not so hot inside. Like a soap bubble, the Sun is a simple structure balanced between opposing forces that, if unbalanced, would destroy it. These dramatic statements are also true for other stars, so you can study the Sun for insight into the rest of the stars in the Universe.

Another reason to study the Sun is that life on Earth critically depends on it. Very small changes in the Sun's luminosity can alter Earth's climate, and a larger change might make Earth uninhabitable. Nearly all of Earth's energy comes from the Sun—the energy in oil, gasoline, coal, and even wood is merely stored sunlight. Furthermore, the Sun's atmosphere of very thin gas extends past Earth's position, and changes in that atmosphere, such as eruptions or magnetic storms, can have a direct effect on Earth.

5-1a Distance and Size

When you watch the Sun set in the west, you see a glowing disk that is 150 million km (93 million mi) from Earth and has a diameter 109 times that of Earth's. How do you know? Recall from Chapter 3 Section 3-3c that, thanks to the work of Johannes Kepler in the 17th century, the relative sizes of the orbits of the planets were already known in Astronomical Units—one AU being equal to the average distance of the Sun from Earth. Recall also from Chapter 3, Concept Art 3A, that the true distance of a nearby object can be calculated from the size of the apparent shift in its position relative to the background as seen from two viewing positions. That shift is called *parallax*.

If you and another astronomer on the opposite side of Earth agree to observe the position of Venus at the same moment relative to more distant objects, you can measure the planet's parallax shift caused by you and your colleague's different observing locations. The distance of the planet in familiar "earthly" units then can be calculated with some trigonometry. With further analysis, using a map of the Solar System and Kepler's third law, you can find the real distance to the Sun. From the 17th century onward, astronomers made more and more accurate measurements of this type, especially on rare occasions

FIGURE 5-1

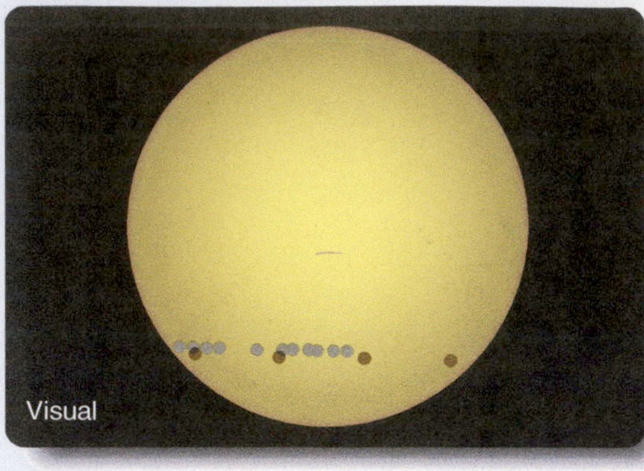

Comparison of Venus' path across the solar disk as observed from Australia (upper track) and Denmark (lower track). Measurement of the small parallax shift allows the distance to Venus to be calculated based on the known distance between the two observatories. Kepler's third law can then be used to find the distance to the Sun.

called **transits of Venus** most recently in 2008 and 2012, when that planet can be seen as a tiny dot directly between Earth and the Sun, as in Figure 5-1, so the edge of the Sun's disk acts as a convenient position marker. The distance from Earth to the Sun, combined with the Sun's easily measured angular size, gives you the Sun's diameter in familiar units. (Look at the Small Angle Formula in the Math Reference Cards MR.)

transits of Venus Rare occasions when Venus can be seen as a tiny dot directly between Earth and the Sun.

density Mass per volume.

atom The smallest unit of a chemical element, consisting of a nucleus containing protons and neutrons plus a surrounding cloud of electrons.

nucleus The central core of an atom containing protons and neutrons that carries a net positive charge.

proton A positively charged atomic particle contained in the nucleus of an atom.

neutron An atomic particle with no charge and about the same mass as a proton.

isotope A form of an atom with the same number of protons but a different number of neutrons.

electron Low-mass atomic particle carrying a negative charge.

5-1b Mass and Density

Once you know the distance to the Sun, then Newton's laws (Chapter 3, Section 3-5) tell you what the Sun's mass needs to be to produce the necessary amount of gravitational force to keep Earth and the other planets in their orbits at their observed speeds. If you know the mass and diameter of the Sun, you can make an easy calculation of the Sun's **density** (mass per volume). Look at **Celestial Profile 9** in Appendix C and notice the Sun's mass, equivalent to 333,000 times the mass of Earth, and its average density— only a little bit denser than water. Although the Sun is very large and very massive, the low density and high temperature together tell you that it must be gas from its surface to its center. When you look at the Sun you see only the outer layers, the surface and atmosphere, of this vast sphere of gas.

In the following sections of this chapter you will learn how, using the results of laboratory physics experiments on Earth, you can know facts as seemingly unknowable as the temperature and composition of the Sun—an object that has never been visited or touched by humans.

5-2 THE ORIGIN OF SUNLIGHT

The Sun's surface glows for the same reason and by the same process that makes the coils in your toaster glow when they are hot. You probably have also noticed that your toaster's coils glow different colors as they heat up. If they are not very hot, the coils are deep red, but as they heat up they grow brighter and yellower. Yellow-hot is hotter than red-hot but not as hot as white-hot. Different parts of the Sun's surface also glow slightly different colors depending on their temperatures.

5-2a Atoms and Subatomic Particles

Light from hot toaster coils and light from the Sun and other stars are all produced by moving electrons. As you know, an **atom** has a massive compact **nucleus** containing positively charged **protons**, usually accompanied by electrically neutral **neutrons**. **Isotopes** are versions of a given element with different numbers of neutrons. The nucleus of an ordinary atom is embedded in a large cloud of relatively low-mass, negatively charged **electrons**. These particles can also exist and move about unattached to an atom.

Charged particles, both protons and electrons, are surrounded by electric fields that they produce. Whenever you change the motion of a charged particle, the change in its electric field spreads outward at the speed of light as electromagnetic radiation. If you run a comb through your hair, you disturb electrons in both hair and comb, producing electric sparks and electromagnetic radiation, which you can sometimes hear as snaps and crackles if you are standing near a radio.

The Sun is hot, and there are plenty of electrons zipping around, colliding, changing directions and speeds, and thereby making electromagnetic radiation. Protons can also make electromagnetic radiation, but because electrons are less massive, usually it is electrons that do most of the moving around.

5-2b Temperature, Heat, and Blackbody Radiation

The particles inside any object—atoms linked together to form **molecules**, individual atoms, electrons inside atoms or wandering loose—are in constant motion, and in a hot object they are more agitated than in a cool object. You can refer to this agitation as *thermal energy*. When you touch a hot object, you feel **heat** as that thermal energy flows into your fingers. **Temperature** is simply a number related to the average speed of the particles, the intensity of the particle motion.

Astronomers and physicists express temperatures of the Sun and other objects on the **Kelvin temperature scale**. Zero degrees Kelvin (written 0 K) is **absolute zero** (−459.7°F), the temperature at which an object contains no thermal energy that can be extracted. Water freezes at 273 K and boils at 373 K. The Kelvin temperature scale is useful in astronomy because it is based on absolute zero and consequently is related directly to the motion of the particles in an object.

Now you can understand why a hot object glows. The hotter an object is, the more motion there is among its particles. The agitated particles, including electrons, collide with each other, and when electrons are accelerated, part of the energy is carried away as electromagnetic radiation. The radiation emitted by an opaque object is called **blackbody radiation**, a name translated from a German term that refers to the way a perfectly opaque emitter and absorber of radiation would behave. At room temperature, such a perfect absorber would look black, but at higher temperatures it would glow visibly. In astronomy, you will find the term *blackbody* actually refers to glowing objects.

Blackbody radiation is quite common. In fact, it is the type of light emitted by an ordinary incandescent light bulb. Electricity flowing through the filament of the bulb heats it to a high temperature, and it glows. You can also recognize the light emitted by a toaster coil as blackbody radiation. Many objects in the sky, including the Sun, primarily emit blackbody radiation because they are mostly opaque.

Two simple laws describe how blackbody radiation works. A hot object radiates at all wavelengths, but there is a **wavelength of maximum intensity** at which it radiates the most energy. **Wien's law** says that the hotter an object is, the shorter is the wavelength of its maximum intensity. This makes sense because in a hotter object, the particles travel faster, collide more violently, and emit more energetic photons, which have shorter wavelengths. This means that hot objects tend to emit radiation at shorter wavelengths and look bluer than cooler objects. Hot stars look bluer than cool stars.

The **Stefan-Boltzmann law** says that hotter objects emit more energy than cooler objects of the same size. That makes sense, too, because the hotter an object is, the more rapidly its particles move and the more violent and more frequent are the collisions that produce photons. In later chapters, you will use these two laws to understand stars and other objects.

Figure 5-2 shows plots of the intensity of radiation versus wavelength for three objects with different temperatures, illustrating both Wien's law and the Stefan- Boltzmann law. You can see how temperature determines the color of a glowing blackbody. The hotter object has its strongest intensity at shorter wavelengths, so it emits more blue light than red and thus looks blue. The cooler object emits more red than blue light and consequently looks red. Also, the total area under each curve is proportional to the total energy emitted, so you can see that the hotter object emits more total energy than the cooler objects.

Now you can understand why two famous stars, Betelgeuse and Rigel, have such different colors, which you saw in Chapter 2 in Section 2-1c. According to Wien's law, Betelgeuse is cooler than the Sun and so it looks red, but Rigel is hotter than the Sun and so it looks blue. A star with

molecule Two or more atoms bonded together.

heat Energy contained in matter in the form of agitation of its particles.

temperature A measure of the agitation among the atoms and molecules of a material.

Kelvin temperature scale A temperature scale using Celsius degrees and based on zero being equal to absolute zero.

absolute zero The theoretical lowest possible temperature at which a material contains no extractable heat energy; zero on the Kelvin temperature scale.

blackbody radiation Spectrum of radiation emitted by an ideal perfect radiator. The spectrum is continuous and its characteristics depend only on the object's temperature.

wavelength of maximum intensity The wavelength at which the radiation emitted by a glowing object is most intense.

Wien's law The hotter a glowing object is, the shorter is the wavelength of its maximum intensity, inversely proportional to its temperature.

Stefan-Boltzmann law Hotter objects emit more energy than cooler objects of the same size, in proportion to the fourth power of temperature.

FIGURE 5-2

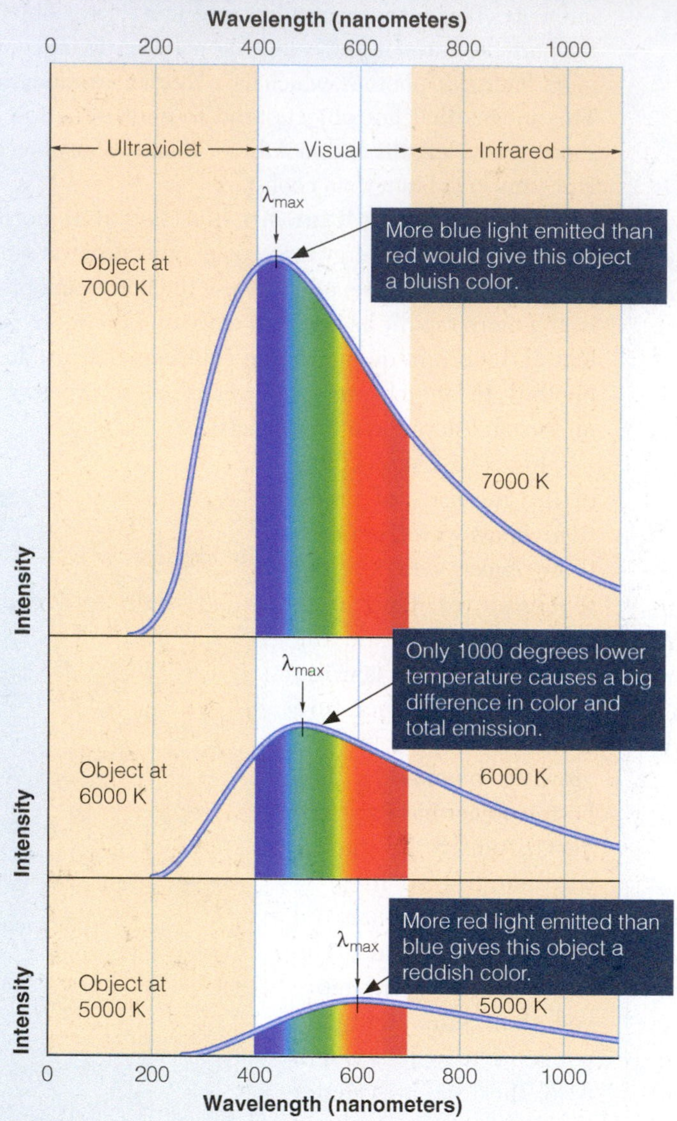

Graphs of blackbody radiation intensity versus wavelength for three objects at temperatures of 7000 K, 6000 K, and 5000 K, respectively (*top to bottom*). Comparison of the graphs demonstrates that a hot object radiates more total energy per unit area than a cooler object (Stefan-Boltzmann law) and that the wavelength of maximum intensity is shorter for hotter objects than for cooler objects (Wien's law). The hotter object here would look blue to your eyes, whereas the cooler object would look red.

the same temperature as the Sun would appear yellowish. According to the Stefan-Boltzmann law, Rigel also produces more energy from each square meter of its surface than does the Sun, which in turn produces more energy from each square meter than does Betelgeuse.

photosphere The bright visible surface of the Sun.

Notice that cool objects may emit little visible radiation but are still producing blackbody radiation. For example, the human body has a temperature of 310 K and emits blackbody radiation mostly in the infrared part of the spectrum. Infrared security cameras can detect burglars by the radiation they emit, and mosquitoes can track you down in total darkness by homing in on your infrared radiation. You have a wavelength of maximum intensity in the infrared part of the spectrum. At your temperature you almost never emit photons with other than infrared wavelengths. Look at **Blackbody Radiation** in the Math Reference Cards MR for more information and practice using Wien's law and the Stefan-Boltzmann law.

5-3 THE SUN'S SURFACE

The Sun's disk looks like a mostly smooth layer of gas. Although the Sun seems to have a real surface, it is not solid. In fact, the Sun is gaseous from its outer atmosphere right down to its center.

5-3a The Photosphere

The apparent surface of the Sun is called the **photosphere**. The photosphere is the layer in the Sun's atmosphere that is dense enough to emit plenty of light but not so dense that the light can't escape; so the photosphere is the source of most of the sunlight received by Earth. The photosphere is less than 500 km (300 mi) deep. If the Sun were represented by a model the size of a bowling ball, the photosphere would be no thicker than a layer of tissue paper wrapped around the ball. When you measure the amount of light with different wavelengths coming from the photosphere and use Wien's law, you find that the average temperature of the photosphere is about 5800 K.

Although the photosphere appears to be substantial, it is really a very low-density gas, even in its lowest and densest layers about 3000 times less dense than the air you breathe. To find gases as dense as the air at Earth's surface, you would have to descend about 15,000 km (10,000 mi) below the photosphere. With fantastically efficient insulation, you could fly a spaceship right through the photosphere. The bottom of the photosphere represents the depth where somebody outside the Sun would no longer be able to see the descending spaceship; conversely, onboard the ship, you would no longer be able to see the rest of the Universe.

There are regions of the photosphere that appear darker than the rest. These are called **sunspots**. They produce less light than equal-sized pieces of the normal photosphere, and their color is redder than the average, so from both Wien's law and the Stefan-Boltzmann law you can conclude that sunspots are cooler than the photosphere; in fact, they are usually about 1500 K cooler. Later in this chapter you will learn about the causes of sunspots and other solar dermatology problems.

5-3b Heat Flow in the Sun

At the temperature of 5800 K, every square millimeter of the Sun's photosphere must be radiating more energy than a 60-watt light bulb. Simple logic tells you that energy in the form of heat is flowing outward from the Sun's interior. With all that energy radiating into space, the Sun's surface would cool rapidly if energy did not flow up from inside to keep the surface hot.

As you learn more about the surface and atmosphere of the Sun, you will find many phenomena that are driven by this energy flow. Like a pot of boiling soup on a hot stove, the Sun is in constant activity as the heat comes up from below. In good photographs, the photosphere has a mottled appearance because it is made up of dark-edged regions called granules, and the visual pattern is called **granulation**, which you can see in Figure 5-3. Each granule is about the size of Texas and lasts for only 10 to 20 minutes before fading away. Faded granules are continuously replaced by new ones. The color and the amount of light from different portions of the granules show, by both Wien's law and the Stefan-Boltzmann law, that granule centers are a few hundred degrees hotter than the edges.

Astronomers recognize granulation as the surface effects of rising and falling currents of gas in and just below the photosphere. The Doppler effect that allows astronomers to measure speeds of these gas currents is discussed later in this chapter. The centers of granules are rising columns of hot gas, and the edges of the granules are cooler, sinking gas. The presence of granulation is clear evidence that energy is flowing upward through the photosphere by a process known as **convection**, which you can see displayed in Figure 5-3b.

Convection occurs when hot fluid rises and cool fluid sinks, as when, for example, a convection current of hot gas rises above a candle flame. You can watch liquid convection by adding a bit of cool nondairy creamer to an unstirred cup of hot coffee. The cool creamer sinks, warms, expands, rises, cools, contracts, sinks again, and so on, creating small regions on the surface of the coffee that mark the tops of convection currents. Viewed from above, these regions look much like solar granules.

FIGURE 5-3

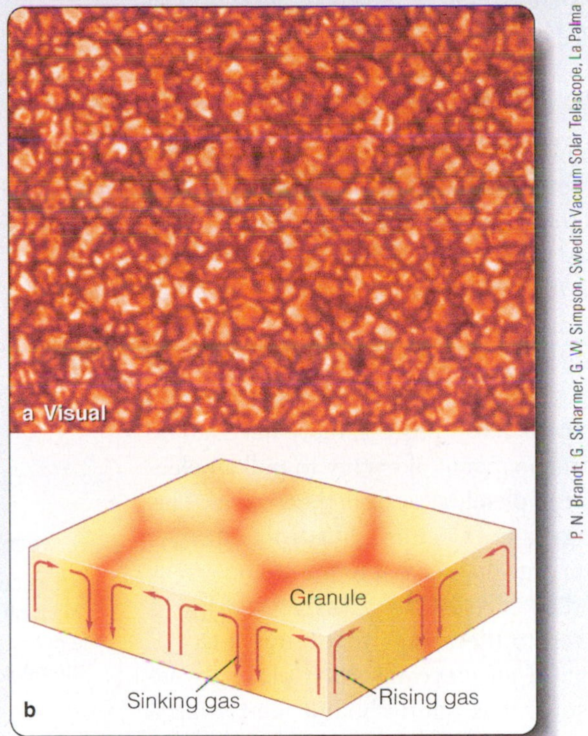

(a) This ultra-high-resolution image of the photosphere shows granulation. The largest granules here are about 1500 km across. (b) This model explains granulation as the tops of rising convection currents just below the photosphere. Heat flows upward as rising currents of hot gas and downward as sinking currents of cool gas. The rising currents heat the solar surface in small regions seen from Earth as granules.

5-4 LIGHT, MATTER, AND MOTION

If light did not interact with matter, you would not be able to see these words. In fact, you would not exist, because, among other problems, photosynthesis would be impossible, and there would be no grass, wheat, bread, beef, cheeseburgers, or any other kind of food. The interaction of light and matter makes your life possible, and it also makes it possible for you to understand the Universe.

You should begin your study of light and matter by considering electrons that are within atoms. As you

sunspot Relatively dark spot on the Sun that contains intense magnetic fields.

granulation The fine structure of bright grains with dark edges covering the Sun's surface.

convection Circulation in a fluid driven by heat. Hot material rises and cool material sinks.

learned in the previous chapter, electrons and other charged particles produce light when they change speed or direction of their motion.

5-4a Electron Shells

The electrons are bound to the atom by the attraction between their negative charge and the positive charge on the nucleus. This attraction is known as the **Coulomb force**, after the French physicist Charles-Augustin de Coulomb. A positive **ion** is an atom with missing electrons, meaning fewer electrons than protons. For an atom to go through the **ionization** process, there needs to be a certain amount of energy to pull an electron completely away from the nucleus. This energy is the electron's **binding energy**, the energy that holds it to the atom.

The size of an electron's orbit is related to the energy that binds it to the atom. If an electron orbits close to the nucleus, it is tightly bound, and a large amount of energy is needed to pull it away. Consequently, its binding energy is large. An electron orbiting farther from the nucleus is held more loosely, and less energy is needed to pull it away. That means it has less binding energy.

FIGURE 5-4

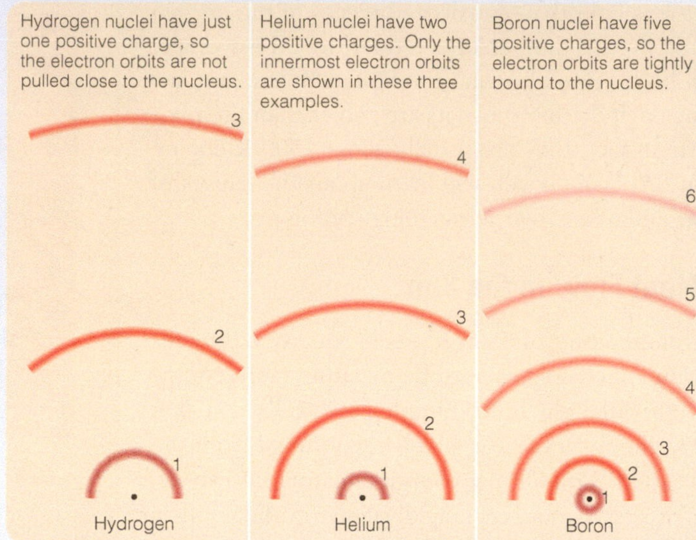

An electron in an atom may occupy only certain permitted orbits. Because each element has a different number of protons and therefore a different electrical charge in the nucleus attracting the electrons, each element has a different, unique pattern of permitted orbits.

Nature permits atoms only certain amounts (quanta) of binding energy, and the laws that describe how atoms behave are called the laws of **quantum mechanics**. Most of this chapter's discussions of the properties of atoms, blackbody radiation, and the interactions of light and matter are based on the laws of quantum mechanics (**How Do We Know? 5-1**).

Because atoms can only have certain amounts of binding energy, your model atom can have orbits of only certain sizes, called **permitted orbits**. These are like steps in a staircase: You can stand on the number-one step or the number-two step but not on the number-one-and-one-quarter step. The electron can occupy any permitted orbit but not orbits in between.

The arrangement of permitted orbits depends primarily on the charge of the nucleus, which in turn depends on the number of protons. The number of protons in the nucleus is unique to each element. Consequently, as shown in Figure 5-4 for hydrogen, helium, and boron, each kind of element has its own pattern of permitted orbits. Ionized forms of an element have orbital patterns quite different from their un-ionized forms. The arrangement of permitted orbits differs for every kind of atom and ion. Isotopes of an element have almost—but not quite—the same pattern of permitted electron orbits as each other because they have the same number of electrons while their nuclei have slightly different masses.

Coulomb force The electrostatic force of repulsion or attraction between charged bodies.

ion An atom that has lost or gained one or more electrons.

ionization The process in which atoms lose or gain electrons.

binding energy The energy needed to pull an electron away from its atom.

quantum mechanics The study of the behavior of atoms and atomic particles.

permitted orbit One of the unique orbits that an electron may occupy in an atom.

energy level One of a number of states an electron may occupy in an atom, depending on its binding energy.

5-4b The Excitation of Atoms

Each orbit in an atom represents a specific amount of binding energy, so physicists commonly refer to the orbits as **energy levels**. Using this terminology, you can say that an electron in its smallest and most tightly bound orbit is in its lowest permitted energy level. You could move the electron from one energy level to another by supplying enough energy to make up the difference between the two energy levels. It would be like moving a

How Do We Know? 5-1

Quantum Mechanics

How can you understand nature if it depends on the atomic world you cannot see? You can see objects such as stars, planets, aircraft carriers, and hummingbirds, but you can't see individual atoms. As scientists apply the principle of cause and effect, they study the natural effects they can see and work backward to find the causes. Invariably that quest for causes leads back to the invisible world of atoms.

Quantum mechanics is the set of rules that describe how atoms and subatomic particles behave. On the atomic scale, particles behave in ways that seem unfamiliar. One of the principles of quantum mechanics specifies that you cannot know simultaneously the exact location and motion of a particle. This is why physicists prefer to go one step beyond the simple atomic model that imagines electrons as particles following orbits, and instead describe the electrons in an atom as if they each were a cloud of negative charge. That's a better model.

© Antonio V. Oquias/Shutterstock.com

The world you see, including these neon signs, is animated by the properties of atoms and subatomic particles.

This raises some serious questions about reality. Is an electron really a particle at all? If you can't know simultaneously the position and motion of a specific particle, how can you know how it will react to a collision with a photon or another particle? The answer is that you can't know, and that seems to violate the principle of cause and effect.

Many of the phenomena you can see depend on the behavior of huge numbers of atoms, and quantum mechanical uncertainties average out. Nevertheless, the ultimate causes that scientists seek lie at the level of atoms, and modern physicists are trying to understand the nature of the particles that make up atoms. This is one of the most exciting frontiers of science.

flowerpot from a low shelf to a high shelf: The greater the distance between the shelves, the more energy you would need to raise the pot. The amount of energy needed to move the electron is the energy difference between the two energy levels.

If you move the electron from a lower energy level to a higher energy level, you can call the atom an **excited atom**. That is, you have added energy to the atom by moving its electron. If the electron falls back to the lower energy level, that energy is released. An atom can become excited by collision. If two atoms collide, one or both may have electrons knocked into higher energy levels. This happens very commonly in hot gas, where the atoms move rapidly and collide often.

Another way an atom can get the energy that moves an electron to a higher energy level is to absorb a photon (packet) of electromagnetic radiation. Only a photon with exactly the right amount of energy corresponding to the energy difference between the two levels can move the electron from one level to another. If the photon has too much or too little energy, the atom cannot absorb it, and the photon passes right by. Because the energy of a photon depends on its wavelength, only photons of certain wavelengths (certain colors) can be absorbed by a given kind of atom.

Figure 5-5 shows the lowest four energy levels of the hydrogen atom, along with three photons the atom could absorb. The longest-wavelength (reddest) photon has only enough energy to excite the electron from the first to the second energy level, but the shorter-wavelength (higher-energy, bluer) photons can excite the electron to higher levels. A photon with too much or too little energy cannot be absorbed. Because the hydrogen atom has many more energy levels than shown in Figure 5-5, it can absorb photons of many different wavelengths.

Atoms, like humans, cannot exist in an excited state forever. The excited atom is unstable and must eventually (usually within

excited atom An atom in which an electron has moved from a lower to a higher energy level.

FIGURE 5-5

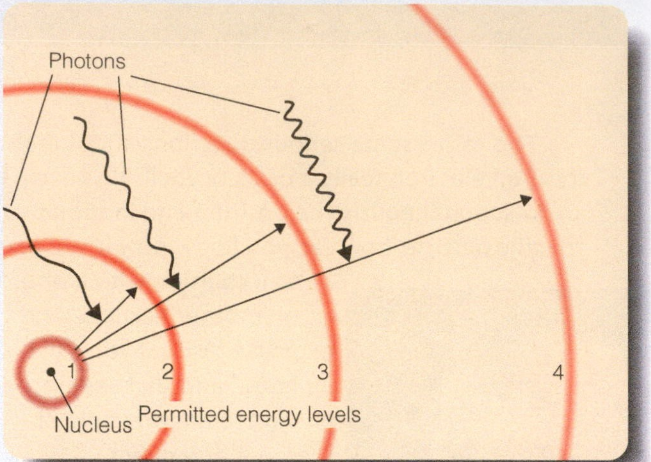

A hydrogen atom can absorb only those photons that have exactly the right energy to move the atom's electron to one of the higher-energy orbits. In this drawing, three photons with different wavelengths are shown along with the changes they respectively would produce in the orbit of the electron starting from the lowest orbital if they were absorbed.

10^{-9} to 10^{-6} seconds) give up the energy it has absorbed and return its electron to a lower energy level. The lowest energy level an electron can occupy is called the **ground state**.

When the electron drops from a higher to a lower energy level, it moves from a loosely bound level to one that is more tightly bound. The atom then has a surplus of energy—the energy difference between the levels—that it can emit as a photon. Study the sequence of events in Figure 5-6 to see how an atom can absorb and emit photons. Jumps of electrons from one orbit to another are sometimes called **quantum jumps**. In casual language that term has come to mean a huge change, but now you see that it in fact represents a very small change. The quantum jump represents a change of electron motion, so electromagnetic radiation is either released or absorbed in the process.

ground state The lowest permitted electron energy level in an atom.

quantum jumps Jumps of electrons from one orbit or energy state to another.

Doppler effect The change in the wavelength of radiation due to relative radial motion of source and observer.

Because each type of atom or ion has its unique set of energy levels, each type absorbs and emits photons with a unique set of wavelengths. As a result, you can identify the elements in a gas by studying the characteristic wavelengths of light absorbed or emitted. It is important to note that the wavelengths (colors) emitted and absorbed by jumping electrons are determined not by the starting or ending energy level of the jump, but by the difference between the energy levels.

The process of excitation and emission is a common sight in urban areas at night. A neon sign glows when atoms of neon gas in the glass tube are excited by electricity flowing through the tube. As the electrons in the electric current flow through the gas, they collide with the neon atoms and excite them. As you have seen, immediately after an atom is excited, its electron drops back to a lower energy level, emitting the surplus energy as a photon of a certain wavelength. The visible photons emitted by the most common electron jumps within excited neon atoms produce a reddish-orange glow. Street signs of other colors, erroneously also called "neon," contain other gases or mixtures of gases instead of pure neon.

5-4c Measuring Velocities: The Doppler Effect

The **Doppler effect** is an apparent change in the wavelength of radiation caused by relative motion of a source and observer. Astronomers use it to measure the speed of blobs of gas in the Sun's atmosphere toward or away from Earth, as well as speeds of entire stars and galaxies.

When astronomers talk about the Doppler effect, they are talking about small shifts in the wavelength

FIGURE 5-6

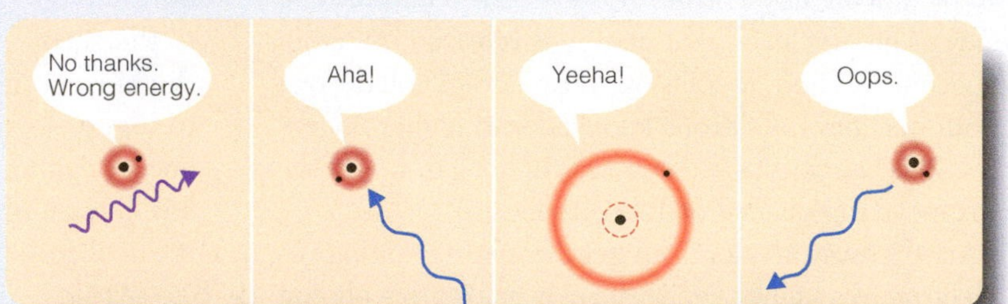

An atom can absorb a photon only if the photon has the correct amount of energy. The excited atom is unstable and within a fraction of a second returns to a lower energy level, reradiating the photon in a random direction.

of electromagnetic radiation. The Doppler effect, however, can occur for waves of all types—sound waves, for example. You probably hear the Doppler effect in sound every day without realizing what it is. The pitch of a sound is determined by its wavelength. Sounds with long wavelengths have low pitches, and sounds with short wavelengths have higher pitches. You hear the Doppler effect every time a car or truck passes you and the pitch of its engine noise or emergency siren seems to drop. Its sound is shifted to shorter wavelengths and higher pitches while it is approaching, and is shifted to longer wavelengths and lower pitches after it passes by, as shown in Figure 5-7a.

Understanding the Doppler effect for sound lets you understand the similar Doppler effect for light. Imagine a light source emitting waves continuously as it approaches you. The light will appear to have a shorter wavelength, making it slightly bluer. This is called a **blueshift**. A light source moving away from you has a longer wavelength and is slightly redder. This is a **redshift**. Redshifted and blueshifted spectra produced by a moving light source are illustrated in Figure 5-7b.

FIGURE 5-7

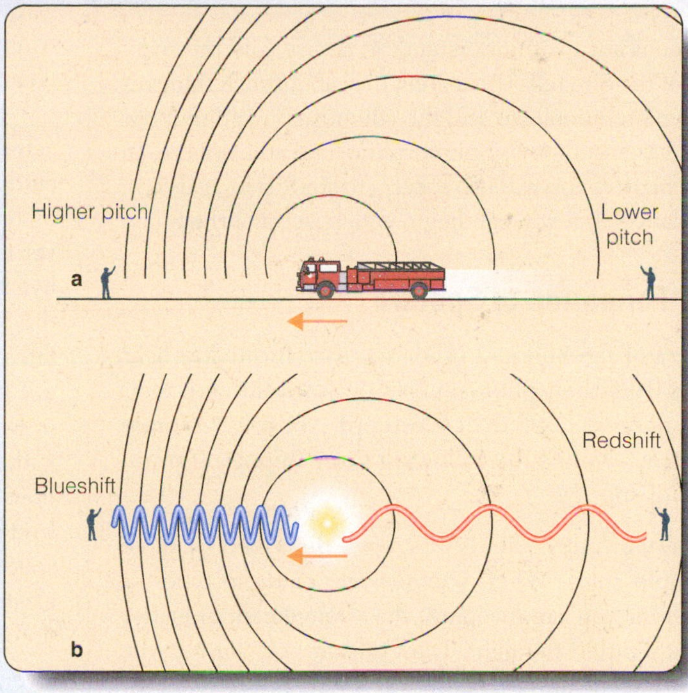

The Doppler effect. (a) The sound waves (*black circles*) emitted from a siren on an approaching truck will be received more often, and thus be heard with a higher pitch, than the sound waves from a stationary truck. The siren will have a lower pitch if it is going away from the observer. (b) A moving source of light emits waves that move outward (*black circles*). An observer toward whom the light source is moving observes a shorter wavelength (*a blueshift*); an observer for whom the light source is moving away observes a longer wavelength (*a redshift*).

The terms *redshift* and *blueshift* are used to refer to any range of wavelengths. The light does not actually have to be red or blue, and the terms apply equally to wavelengths in the radio, X-ray, or gamma-ray parts of the spectrum. "Red" and "blue" refer to the relative direction of the shift, not to actual color. Also, note that these shifts are much too small to noticeably change the color of a star, but they are easily detected by changes in the positions of narrow features in a star's spectrum such as spectral lines (discussed in detail in the next section). The Doppler shift, blue or red, reveals the relative motion of wave source and observer—you measure the same Doppler shift if the light source is moving and you are stationary, or if the light source is stationary and you are moving, as you can see in Figure 5-7b.

The amount of change in wavelength depends on the speed of the source. A moving car has a smaller sound Doppler shift than a jet plane, and a slow-moving star has a smaller light Doppler shift than one that is moving at high velocity. You can measure the speed of a star toward or away from you by measuring the size of the Doppler shift of its spectral lines.

Note that the Doppler shift is sensitive only to the part of the velocity directed away from you or toward you, called the **radial velocity** (V_r). A star moving perpendicular to your line of sight, to the left for instance, would have no blueshift or redshift because its distance from Earth would not be decreasing or increasing. Police radar guns use the Doppler effect to measure the speeds of cars. The police park next to the highway and aim their "hair dryers" directly along the road because they can measure only radial velocities, and they want to measure your full velocity along the highway. Look at the **Doppler Formula** in the Math Reference Cards MR for further information about the Doppler shift.

blueshift A Doppler shift toward shorter wavelengths caused by a velocity of approach.

redshift A Doppler shift toward longer wavelengths caused by a velocity of recession.

radial velocity (V_r) That component of an object's velocity directed away from or toward the observer.

5-5 THE SUN'S ATMOSPHERE

Science is a way of understanding nature, and the spectrum of the Sun tells you a great deal about such things as the Sun's temperature and the composition and motions of the gases in the solar photosphere and atmosphere. In later chapters you will use spectra to study stars, galaxies, and planets, but you can begin with spectra of the Sun.

5-5a Formation of Spectra

Spectra of the Sun and other stars are formed as light passes from their photospheres outward through their atmospheres. Look over **Concept Art 5A, "Atomic Spectra."** Notice the following three important properties of spectra:

1. There are three kinds of spectra described by three simple rules. When you see one of these types of spectra, you can recognize the arrangement of matter that emitted the light. Dark (absorption) lines in the Sun's spectrum are caused by atoms in the Sun's (or Earth's) atmosphere between you and the Sun's photosphere. The photosphere itself produces a blackbody (continuous) spectrum.
2. The wavelengths of the photons that are absorbed by a given type of atom are the same as the wavelengths of the photons emitted by that type of atom; both are determined by the electron energy levels in the atom. The emitted photons coming from a hot cloud of hydrogen gas have the same wavelengths as the photons absorbed by hydrogen atoms in the Sun's atmosphere. Although the hydrogen atom produces many spectral lines, called the Balmer lines, from the ultraviolet to the infrared, only three hydrogen lines, called the Balmer lines, are visible to human eyes.
3. Most modern astronomy books display spectra as graphs of intensity versus wavelength. Be sure you see the connection between dark absorption lines, bright emission lines, and dips and peaks in the graphed spectrum.

5-5b The Sun's Chemical Composition

Identifying the elements in the Sun's atmosphere by identifying the lines in its spectrum is a relatively straightforward procedure. For example, two dark absorption lines appear in the yellow region of the solar spectrum at the wavelengths 589.0 nm and 589.6 nm. The only atom that can produce this pair of lines is sodium, so the Sun must contain sodium. More than 90 elements in the Sun have been identified this way. The element helium was known in the Sun's spectrum first, before helium (from the Greek word *helios*, meaning "Sun") was found on Earth.

However, just because spectral lines that are characteristic of an element are missing, you cannot conclude that the element itself is absent. For example, the hydrogen Balmer lines are weak in the Sun's spectrum, yet more than 90 percent of the atoms in the Sun are hydrogen. The reason for this apparent paradox is that the Sun is too cool to produce strong hydrogen Balmer lines. At the Sun's photosphere temperature, atoms do not usually collide violently enough to knock electrons in hydrogen atoms into the second energy level, which is the necessary starting place for Balmer line absorptions. Spectral lines of other varieties of atoms (e.g., ionized calcium) are especially easy to observe in the Sun's spectrum because the Sun is the right temperature to excite those atoms to the energy levels that can produce visible spectral lines, even though those atoms are not very common in the Sun.

The effect of temperature on the visibility of spectral lines was first understood by Cecila Payne (later, Payne-Gaposchkin), who was an astronomer doing Ph.D. research work at Harvard Observatory in the 1920s. She used the new techniques of quantum mechanics to derive accurate chemical abundances for the Sun and other stars and so was the first person to know that the Sun is mostly composed of hydrogen, even though its visible-wavelength hydrogen spectral lines are only moderately strong.

Astronomers must use the physics that describe the interaction of light and matter to analyze a star's spectrum, taking into account the star's temperature, to calculate correctly the amounts of each element present in the star. Such results show that nearly all stars have compositions similar to the Sun—about 90 percent of the atoms are hydrogen, and about 9 percent are helium, with small traces of heavier elements (Table 5-1). It is fair to say that Cecilia Payne, whose thesis has been called the most important doctoral work in the history of astronomy, figured out the true chemical composition of the Universe. You will use her results in later chapters when you study the life stories of the stars, the history of our galaxy, and the origin of the Universe.

5-5c The Chromosphere

Above the photosphere lies the **chromosphere**. Solar astronomers define the lower edge of the chromosphere as lying just above the visible surface of the Sun with its

chromosphere Bright gases just above the photosphere of the Sun.

TABLE 5-1 THE MOST ABUNDANT ELEMENTS IN THE SOLAR ATMOSPHERE*

Element	Atomic Number	Fraction of Number of Atoms	Fraction of Mass
Hydrogen	1	0.926	0.749
Helium	2	0.073	0.237
Carbon	6	0.00023	0.0022
Nitrogen	7	0.000063	0.00071
Oxygen	8	0.00045	0.0059
Neon	10	0.000069	0.0011
Magnesium	12	0.000033	0.00064
Silicon	14	0.000032	0.00073
Sulfur	16	0.000014	0.00037
Iron	26	0.000027	0.0012

* Based on Lodders 2003, *Astrophysical Journal* 591, 1220

upper regions blending gradually with the atmosphere's outermost layer, the **corona**, which is described later in this chapter.

You can think of the chromosphere as being an irregular layer with an average depth less than Earth's diameter, as seen in Figure 5-8a. Because the chromosphere is roughly 1000 times fainter than the photosphere, you can see it with your unaided eyes only during a total solar eclipse when the Moon covers the brilliant photosphere (look back to Chapter 2 Section 2-4c). Then, the chromosphere flashes into view as a thin line of pink just above the photosphere. The word *chromosphere* comes from the Greek word *chroma*, meaning "color." The pink color is produced by combined light of three bright emission lines—the red, blue, and violet Balmer lines of hydrogen.

Astronomers know a great deal about the chromosphere from its spectrum. The chromosphere produces an emission spectrum, and Kirchhoff's second law (see **Concept Art 5A, "Atomic Spectra"**) tells you the chromosphere must be an excited, transparent, low-density gas viewed with a dark, cold background. The chromosphere's density relative to the density of the air you breathe on Earth's surface ranges from 1/10,000 just above the photosphere to 1/100 billion just below the corona.

Atoms in the lower chromosphere are ionized, and atoms in the higher layers of the chromosphere are even more highly ionized—that is, they have lost more electrons. From this, astronomers can find the temperature in different parts of the chromosphere. Just above the photosphere, the temperature falls to a minimum of about 4500 K and then rises to the extremely high temperatures of the corona.

Solar astronomers can take advantage of some of the physics of spectral line formation to study the chromosphere. A photon with a wavelength corresponding to one of the solar atmosphere's strong absorption lines is very unlikely to escape from deeper layers and reach Earth. A **filtergram** is a photograph made using light in one of those dark absorption lines. Those photons can

corona The faint outer atmosphere of the Sun, composed of low-density, high-temperature gas.

filtergram A photograph (usually of the Sun) taken in the light of a specific region of the spectrum—for example, an Hα (hydrogen Balmer-alpha) filtergram.

FIGURE 5-8

a) A cross-section at the edge of the Sun shows the relative thickness of the photosphere and chromosphere. Earth is shown for scale. On this scale, the disk of the Sun would be more than 1.5 m (5 ft) in diameter. The corona extends from the top of the chromosphere to great height above the photosphere. (b) This photograph, made during a total solar eclipse, shows only the inner part of the corona.

Concept Art 5A

ATOMIC SPECTRA

1

To understand how to analyze a spectrum, begin with a simple incandescent lightbulb. The hot filament emits blackbody radiation, which forms a **continuous spectrum**.

An **absorption spectrum** results when radiation passes through a cool gas. In this case you can imagine that the lightbulb is surrounded by a cool cloud of gas. Atoms in the gas absorb photons of certain wavelengths that are then missing from the observed spectrum, and you see dark **absorption lines** at those wavelengths. Such absorption spectra are also called **dark-line spectra**.

An **emission spectrum** is produced by photons emitted by an excited gas. You could see **emission lines** by turning your telescope aside so that photons from the bright bulb do not enter the telescope and the excited gas has a dark background. The photons you would see would be those emitted by the excited atoms near the bulb, and the observed spectrum is mostly dark with a few bright emission lines. Such spectra are also called **bright-line spectra**.

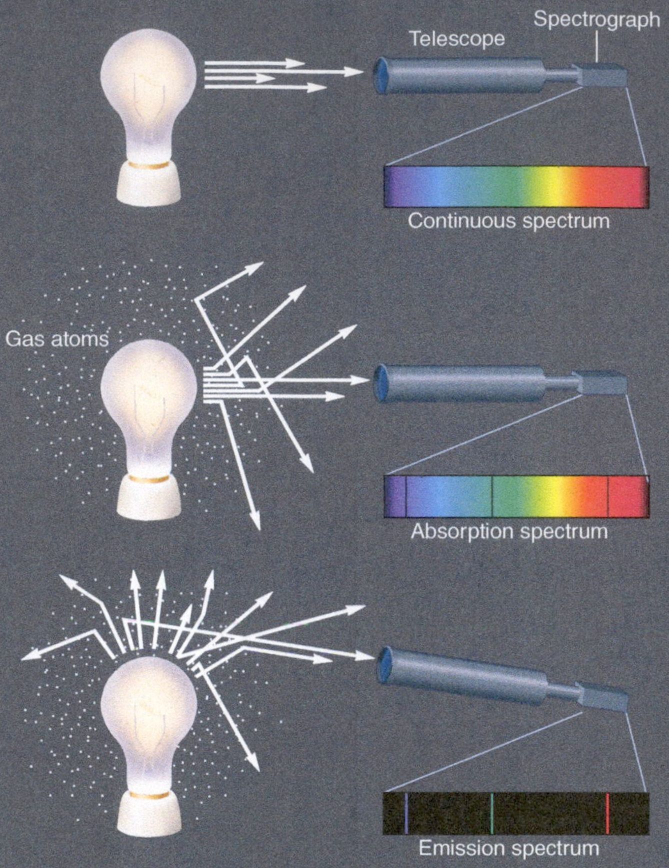

1a

The spectrum of a star is an absorption spectrum. The denser layers of the photosphere emit blackbody radiation. Gases in the atmosphere of the star absorb their specific wavelengths and form dark absorption lines in the spectrum.

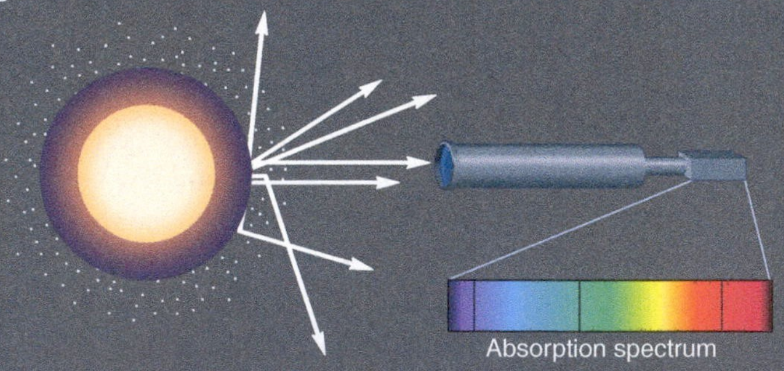

NOAO/AURA/NSF

Science & Society Picture Library/Getty Images

1b

In 1859, long before scientists understood atoms and electron energy levels, the German scientist Gustav Kirchhoff formulated three rules—now known as **Kirchhoff's laws**—describing the three types of spectra.

KIRCHHOFF'S LAWS

Law I: The Continuous Spectrum

A solid, liquid, or dense gas excited to emit light will radiate at all wavelengths and thus produce a continuous spectrum.

Law II: The Emission Spectrum

A low-density gas excited to emit light will do so at specific wavelengths and thus produce an emission spectrum.

Law III: The Absorption Spectrum

If light comprising a continuous spectrum passes through a cool, low-density gas, the result will be an absorption spectrum.

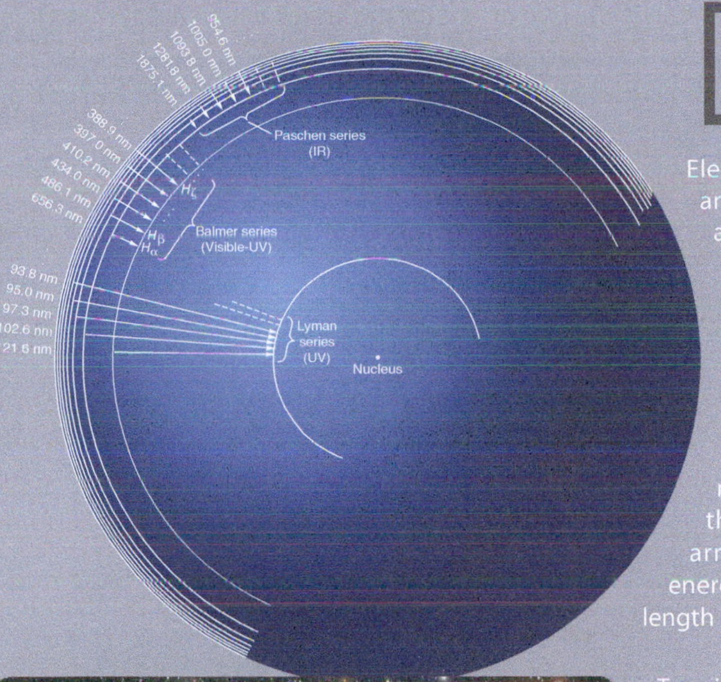

2

Electron orbits in the hydrogen atom are shown here as energy levels. When an electron makes a **transition** from one orbit to another, this means that the energy stored in the atom has changed. In this diagram, arrows pointed inward toward the nucleus represent transitions that result in the emission of a photon. If the arrows pointed outward, they would represent transitions that result from the absorption of a photon. Long arrows represent large amounts of energy and correspondingly short-wavelength photons.

2a

Transitions in the hydrogen atom can be grouped into series—the **Lyman series, Balmer series, Paschen series**, and so on, named after scientists who carefully investigated the spectra of hydrogen atoms. Transitions and the resulting spectral lines are identified by Greek letters. Only the first few transitions in the first three series are shown at left.

2b

In this drawing (*right*) of the hydrogen spectrum, emission lines in the infrared and ultraviolet are shown as gray. Only the first three lines of the Balmer series are visible to human eyes.

T.A. Rector/University of Alaska Anchorage; T. Abbott and NOAO/AURA/NSF

2c

Excited clouds of gas in space emit light at all of the Balmer wavelengths, but you see only the red, blue, and violet photons blending to create the purple-pink color typical of ionized hydrogen.

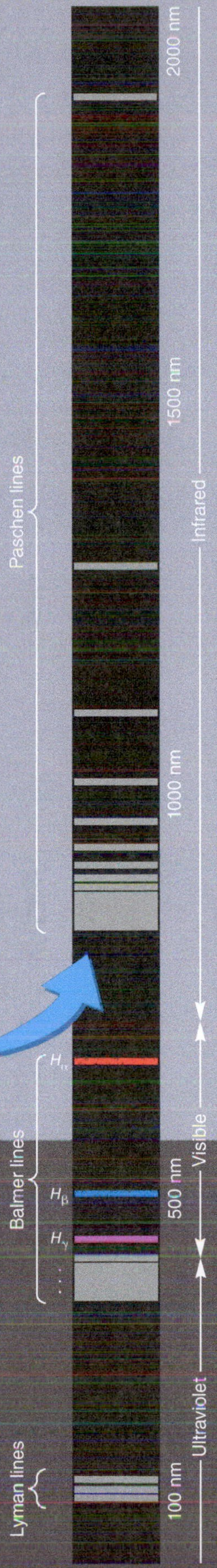

The shorter-wavelength lines in each series blend together.

3

Modern astronomers rarely work with spectra in the form of images of bands of light. Spectra are usually recorded digitally, so it is easy to represent them as graphs of intensity versus wavelength. Here, the artwork above the graph suggests the appearance of a stellar spectrum. The graph at right reveals details not otherwise visible and allows comparison of relative intensities. Notice that dark absorption lines in the spectrum appear as dips in the intensity graph.

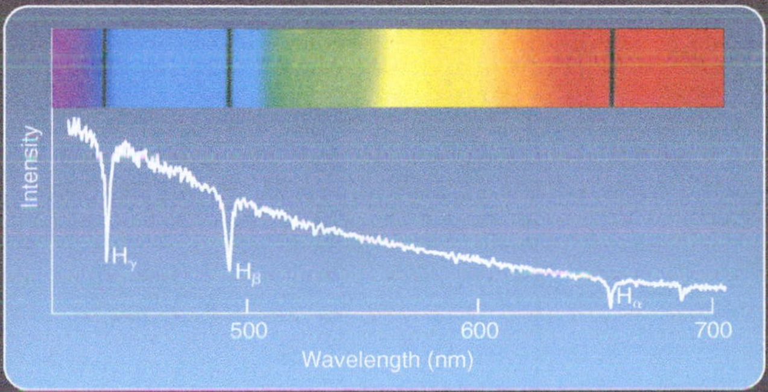

only come from high in the solar atmosphere. In this way, filtergrams reveal detail in the upper layers of the chromosphere. In a similar way, an image recorded in the far-ultraviolet or in the X-ray part of the spectrum reveals other structures in the hottest parts of the solar atmosphere (Figure 5-9).

Figure 5-10a shows a filtergram made at the wavelength of H_α, the hydrogen Balmer-alpha line. This image shows complex structure in the chromosphere, including long, dark **filaments** silhouetted against the brighter surface. **Spicules** are flamelike jets of gas extending upward into the chromosphere and lasting 5 to 15 minutes. Figure 5-10b shows how, at the limb of the Sun's disk, the spicules blend together and look like flames covering a burning prairie, but they are not flames at all. Spectra show that spicules are cooler gas from the lower chromosphere extending upward into hotter regions.

Spectroscopic analysis of the chromosphere shows that it is a low-density gas in constant motion where the temperature increases rapidly with height. Just above the chromosphere lies even hotter gas.

FIGURE 5-9

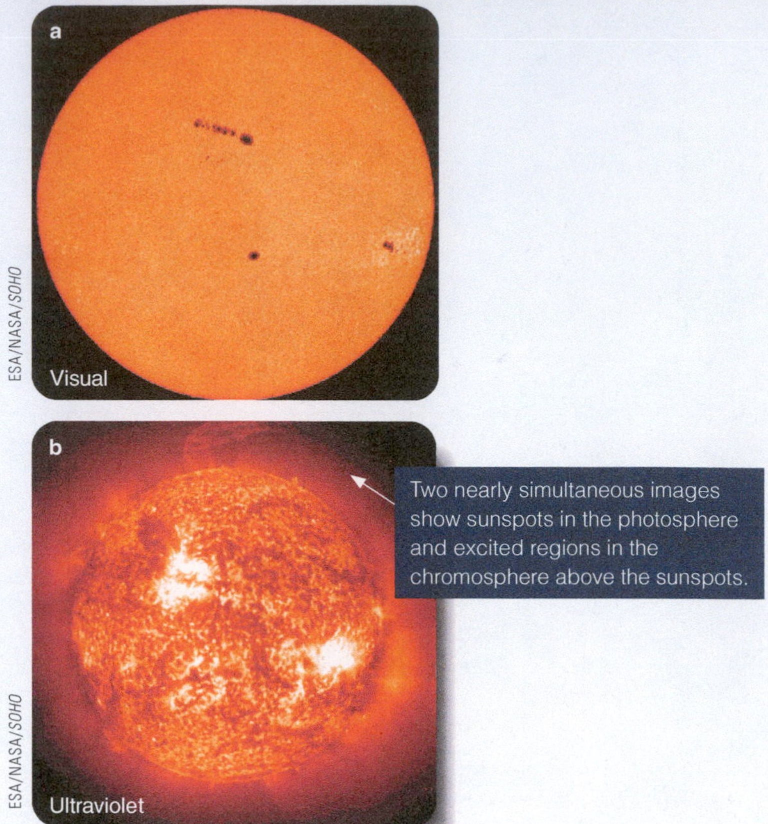

Simultaneous images of the solar photosphere and chromosphere. The ultraviolet image primarily shows material in the upper chromosphere that is much hotter than the photosphere. This pair of images shows that low-temperature sunspots have extra-hot active regions above them in the chromosphere.

5-5d The Corona

The outermost part of the Sun's atmosphere is called the *corona*, after the Greek word for "crown." The corona is so dim that it is not visible in Earth's daytime sky because of the glare of scattered light from the Sun's brilliant photosphere. During a total solar eclipse, however, when the Moon covers the photosphere, you can see the innermost parts of the corona, as shown in Figure 5-8b. Observations made with specialized telescopes called **coronagraphs** on Earth or in space can block the light of the photosphere and record the corona out beyond 20 solar radii, almost 10 percent of the way to Earth. Such images reveal that sunspots are linked with features in the chromosphere and corona (Figure 5-11).

filament A solar eruption, seen from above, silhouetted against the bright photosphere.

Spicule A small, flamelike projection in the chromosphere of the Sun.

coronagraph A telescope designed to capture images of faint objects such as the corona of the Sun that are near relatively bright objects.

The spectrum of the corona can tell you a great deal about the coronal gases and simultaneously illustrate how astronomers analyze a spectrum. Some of the light from the corona produces a continuous spectrum that lacks absorption lines. Superimposed on the corona's continuous spectrum are emission lines of highly ionized gases. In the lower corona, the atoms are not as highly ionized as they are at higher altitudes, and this tells you that the temperature of the corona rises with altitude. Just above the chromosphere, the coronal temperature is about 500,000 K, but in the outer corona the temperature can be 2,000,000 K or more. Despite that very high temperature, the corona does not produce much light because its density is very low, only 10^4 atoms / cm^3 or less. That is about 50 trillion times less dense than Earth's sea-level atmosphere.

Astronomers have wondered for years how the corona and chromosphere can be so hot. Heat flows from hot regions to cool regions, never from cool to

FIGURE 5-10

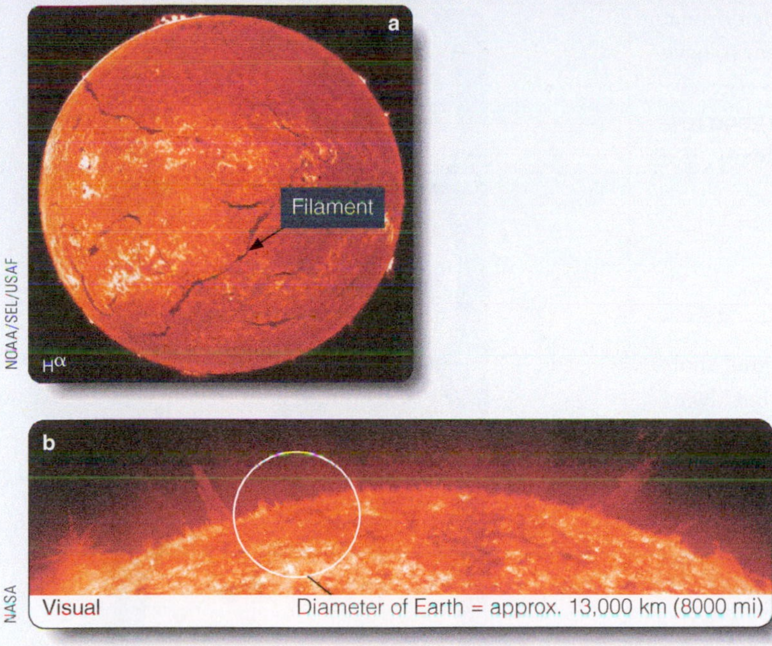

This hydrogen Balmer-alpha filtergram reveals complex structures in the chromosphere including long, dark filaments. (b) Seen at the edge of the solar disk, spicules look like a burning prairie, but they are not at all related to burning.

FIGURE 5-11

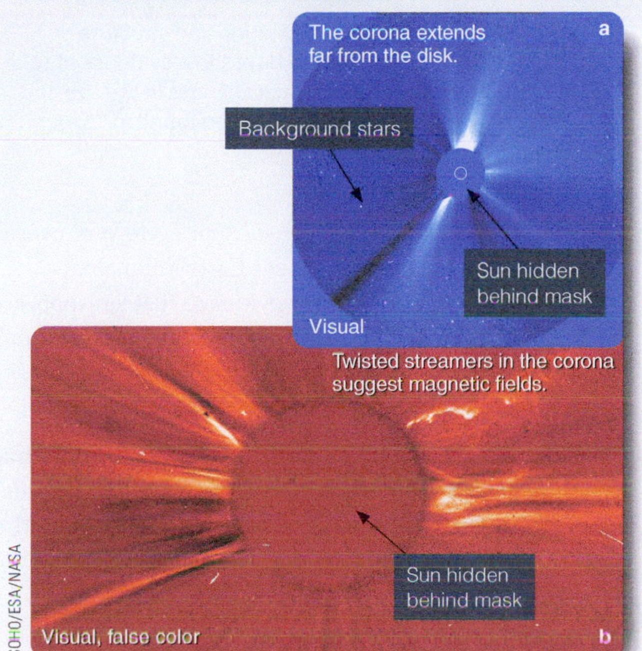

Coronagraph images reveal the corona by blocking light from the much brighter solar disk. The activity cycles and magnetic fields that play an important role in the photosphere and the chromosphere also affect the corona.

hot. So how can the heat from the photosphere, with a temperature of only 5800 K, flow out into the much hotter chromosphere and corona? Observations made by the *SOHO* satellite show a **magnetic carpet** of looped magnetic fields extending up through the photosphere. Remember that the gas of the chromosphere and corona has a very low density, so it can't resist movement in the magnetic fields. Turbulence below the photosphere seems to flick the magnetic loops back and forth and whip the gas about. That heats the gas. In this instance, energy appears to flow outward in the form of agitation of the magnetic fields. Solar magnetic phenomena will be discussed more thoroughly later in this chapter.

Not all of the Sun's magnetic field loops back toward the Sun; some of the field lines lead outward into space. Gas from the solar atmosphere follows along the magnetic fields that point outward and flows away from the Sun in a breeze called the **solar wind** that can be considered an extension of the corona. The low-density gases of the solar wind blow past Earth at 300 to 800 km/s with gusts as high as 1000 km/s (more than 2 million mph).

Because of the solar wind, the Sun is slowly losing mass, but this is a minor loss for an object as massive as the Sun. The Sun loses about 1 million tons per second, but that is only 10^{-14} of the solar mass per year. Later in life, the Sun, like many other stars, will lose mass rapidly. You will see in future chapters how this affects the evolution of stars.

Do other stars have chromospheres, coronae, and stellar winds like the Sun? Ultraviolet and X-ray observations suggest that the answer is yes.

magnetic carpet The network of small magnetic loops that covers the solar surface.

solar wind Rapidly moving atoms and ions that escape from the solar corona and blow outward through the Solar System.

The spectra of many stars contain emission lines in the far-ultraviolet that could only have formed in the low-density, high-temperature gases of a chromosphere or corona. Also, many stars are sources of X-rays that appear to have been produced by the high temperature gas in coronae. This observational evidence gives astronomers good reason to believe that the Sun, for all its complexity, is a typical star.

5-6 SOLAR ACTIVITY

The Sun is not quiet. It is home to slowly changing spots larger than Earth and rapid, vast eruptions that dwarf human imagination. All of these seemingly different forms of solar activity have one thing in common—magnetic fields.

5-6a Observing the Sun

Solar activity is often visible with even a small telescope, but you should be very careful about observing the Sun. It is not safe to look directly at the Sun, and it is even more dangerous to look at the Sun through any optical instrument such as a telescope, binoculars, or even the viewfinder of a camera. These concentrate sunlight and can cause severe injury. You can safely project an image of the Sun onto a screen, or you can use specially designed solar blocking filters.

In the early 17th century, Galileo observed the Sun and saw spots on its surface; day by day, he saw the spots moving across the Sun's disk. He correctly concluded that the Sun is a sphere and is rotating (Figure 5-12).

FIGURE 5-12

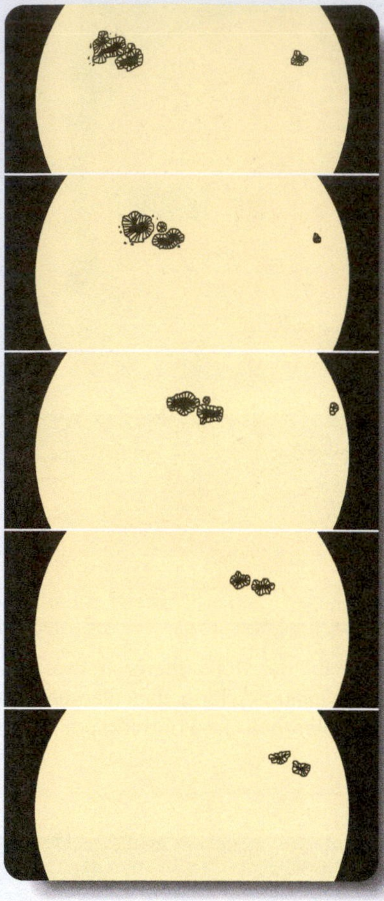

These sketches of the location and structure of sunspots on successive days show evidence of solar rotation as well as gradual changes in the size and structure of sunspots, as Galileo found in 1610.

5-6b Sunspots

The dark sunspots that you see at visible wavelengths only hint at the complex processes that go on in the Sun's atmosphere. To explore those processes, you must turn to the analysis of images and spectra at a wide range of wavelengths.

Study **Concept Art 5B, "Sunspots and the Solar Magnetic Cycle,"** and notice the following five important points:

1. Sunspots are cool spots on the Sun's surface caused by strong magnetic fields.
2. Sunspots follow an 11-year cycle not only in the number of spots visible but also in their locations on the Sun.
3. The Zeeman effect gives astronomers a way to measure the strength of magnetic fields on the Sun.
4. Characteristics of the sunspot cycle vary over centuries and appear to affect Earth's climate.
5. Finally, there is clear evidence that sunspots are part of a larger magnetic process that involves other layers of the Sun's atmosphere and parts of its interior in a 22-year cycle.

helioseismology The study of the interior of the Sun by the analysis of its modes of vibration.

The sunspot groups are merely the visible traces of magnetically active regions. But what causes this magnetic activity? The answer appears to be linked to the cyclical strengthening and weakening of the Sun's overall magnetic field.

5-6c Insight into the Sun's Interior

Almost no light emerges from below the photosphere, so you can't see into the solar interior. However, solar astronomers can use the vibrations in the Sun to explore its depths in a type of analysis called **helioseismology**. Random motions in the Sun constantly produce vibrations. Astronomers can detect these vibrations by observing Doppler shifts in the solar surface. These waves make the photosphere move up and down by small amounts— roughly plus or minus 15 km (10 mi). This covers the surface of the Sun with a pattern of rising and falling regions that can be mapped, as shown in Figure 5-13. In the Sun, a vibration with a period of 5 minutes is strongest, but other periods are observed ranging from 3 to 20 minutes. Just as geologists can study Earth's interior by analyzing vibrations from earthquakes, solar astronomers can use helioseismology to map the temperature, density, and rate of rotation in the Sun's invisible interior.

5-6d The Sun's Magnetic Cycle

Sunspots are magnetic phenomena, so the 11-year cycle of sunspots must be caused by cyclical changes in the Sun's magnetic field. To explore that idea, you can begin with the Sun's rotation.

The Sun does not rotate as a rigid body. It is a gas from its outermost layers down to its center, and some parts of the Sun rotate faster than other parts.

FIGURE 5-13

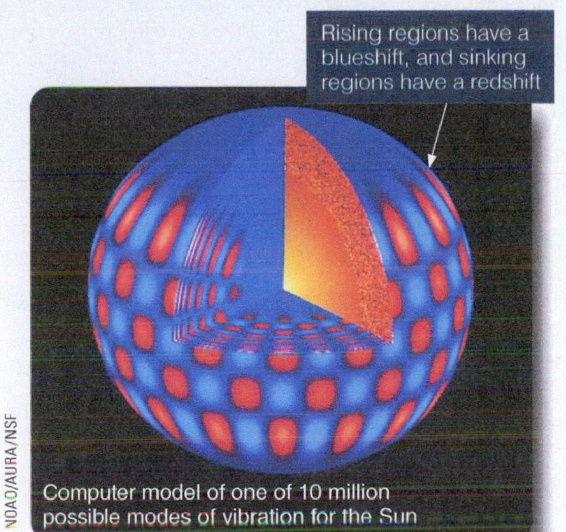

Model of one possible solar vibration mode, for comparison with helioseismology data. The Sun can vibrate in millions of different patterns or modes, and each mode corresponds to a different vibration wavelength that penetrates to a different level in the solar interior. By measuring Doppler shifts produced as the photosphere moves gently up and down, astronomers can map the inside of the Sun.

FIGURE 5-14

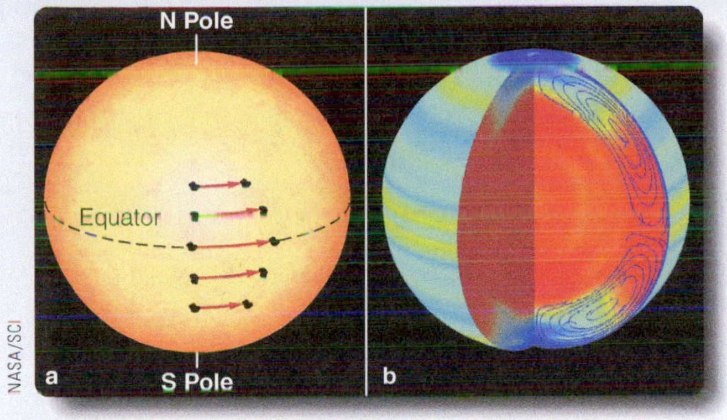

(a) In general, the photosphere of the Sun rotates faster at the equator than at higher latitudes. If you started five sunspots in a row, they would not stay lined up as the Sun rotates. (b) Detailed analyses of the Sun's interior rotation from helioseismology reveal regions of slow rotation (blue) and rapid rotation (red). Such studies show that differential rotation occurs inside the Sun as well as on its surface.

Figure 5-14a demonstrates how the equatorial region of the photosphere rotates faster than do regions at higher latitudes. At the equator, the photosphere rotates once every 24.5 days, but at latitude 45° one rotation takes 27.8 days. Helioseismology can map the rotation throughout the interior (Figure 5-14b). This phenomenon is called **differential rotation**, and it is clearly linked with the magnetic cycle.

The Sun's magnetic field appears to be powered by the energy flowing outward in the form of moving currents of gas. The gas is highly ionized, so it is a very good conductor of electricity. When an electrical conductor rotates rapidly and is stirred by convection, it can convert some of the energy flowing outward as convection into a magnetic field. This process is called the **dynamo effect**, and it is believed to operate also in the liquid metal of Earth's core to produce Earth's magnetic field. Helioseismology observations have found evidence that the Sun's magnetic field is generated by the dynamo effect at the bottom of the **convection zone** deep under the photosphere. The Sun's magnetic cycle is clearly related to how its magnetic field is created.

differential rotation The rotation of a body in which different parts of the body have different periods of rotation. This is true of the Sun, the giant planets like Jupiter, and the disk of the galaxy.

dynamo effect The process by which a rotating, convecting body of conducting matter, such as in Earth's core or in the Sun's convection zone, can generate a magnetic field.

convection zone A region inside a star where energy is carried outward as rising hot gas and sinking cool gas.

Concept Art 5B

SUNSPOTS AND THE SOLAR MAGNETIC CYCLE

1

The dark spots that appear on the Sun are only the visible traces of complex regions of activity. Evidence gathered over many years and at a wide range of wavelengths shows that sunspots are clearly linked to the Sun's magnetic field.

Spectra show that sunspots are cooler than the photosphere with a temperature of about 4200 K. The photosphere has an average temperature of about 5800 K. Because, according to the Stefan-Boltzmann law, the total amount of energy radiated by a surface depends on its temperature raised to the fourth power, sunspots look dark in comparison with the photosphere. Actually, a sunspot emits quite a bit of radiation. If the Sun were removed and only an average-size sunspot were left behind, it would be brighter than a full moon.

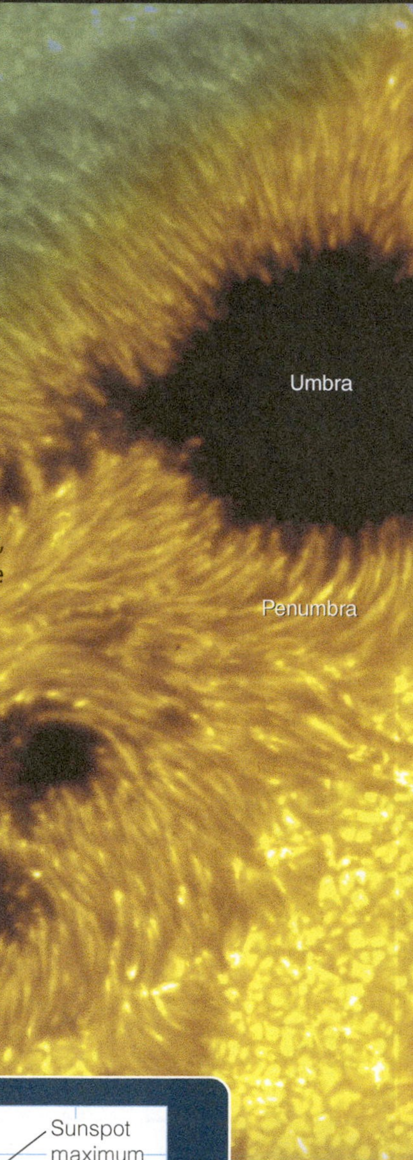

a A typical sunspot is about twice the size of Earth, but there is a wide range of sizes. They appear, last a few weeks to as long as 2 months, and then shrink away. Usually, sunspots occur in pairs or complex groups.

NASA

b Sunspots are not shadows, but astronomers refer to the dark core of a sunspot as its umbra and the outer, lighter region as the penumbra.

JAXA/NASA/Hinode

c Streamers above a sunspot suggest a magnetic field.

Hinode JAXA/NASA

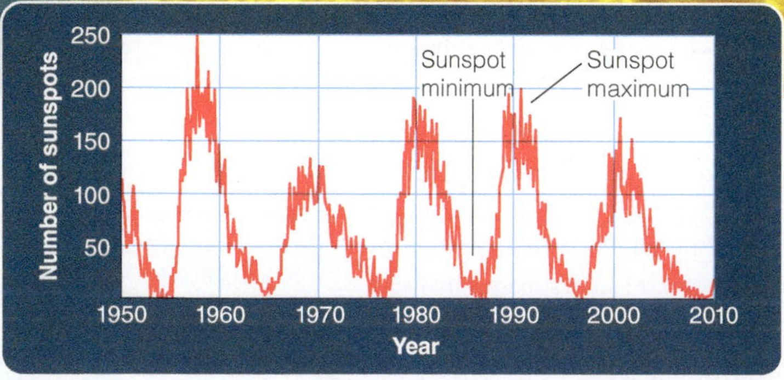

2

The number of spots visible on the Sun varies in a cycle with a period of 11 years. At maximum, there are often more than 100 spots visible. At minimum, there are very few or zero.

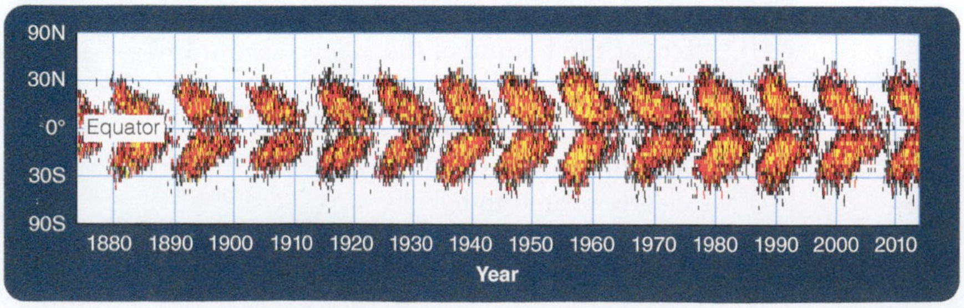

D. Hathaway NASA/MSFC

2a

Early in a cycle, spots appear at high latitudes north and south of the Sun's equator. Later in the cycle, new spots appear closer to the Sun's equator. If you plot the latitude of sunspots versus time, the graph looks like butterfly wings, as shown in this **Maunder butterfly diagram**, named after E. Walter Maunder of Greenwich Observatory.

3

Astronomers can measure magnetic fields on the Sun using the **Zeeman effect**, as shown below. When an atom is in a magnetic field, the electron energy levels are altered, and the atom is able to absorb photons with a greater variety of wavelengths than the same atom not in a magnetic field. In this spectrum you see single spectral lines split into multiple components, with the separation between the components proportional to the strength of the magnetic field.

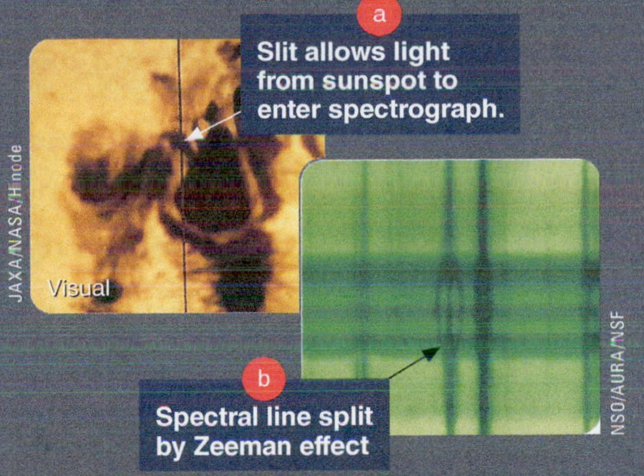

3a

Images of the Sun above show that sunspots contain magnetic fields a few thousand times stronger than Earth's. Such strong fields inhibit motions of ionized gas below the photosphere; consequently, convection is reduced below the sunspot, less energy is transported from the interior, and the Sun's surface at the position of the spot is cooler. Heat that is prevented from emerging at the sunspot's position is deflected and emerges around the sunspot, making the surrounding area hotter than the average photosphere. The deflected heat can be detected in ultraviolet and infrared images; the result is that the entire active region, including the sunspots, is actually emitting more energy than the same area of normal photosphere.

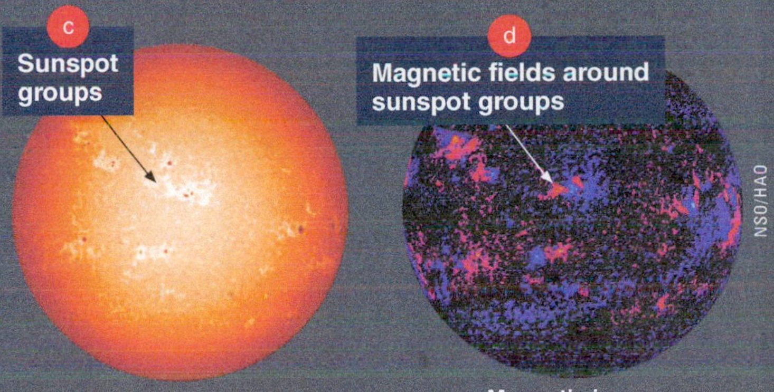

Ultraviolet filtergram

Magnetic image

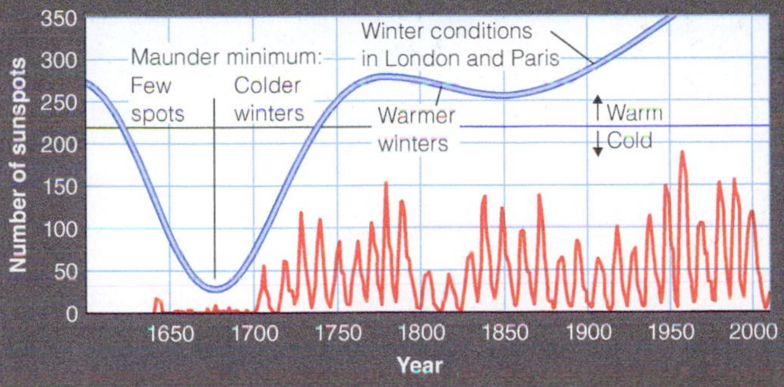

4

Historical records show that there were very few sunspots from about 1645 to 1715, a phenomenon known as the **Maunder minimum**. This coincides with the middle of a period called the "Little Ice Age," a time of unusually cool weather in Europe and North America from about 1500 to about 1850, as shown in the graph at left. Other such periods of cooler climate are known. Evidence suggests that there is a link between solar activity and the amount of solar energy Earth receives. This link has been confirmed by measurements made by spacecraft above Earth's atmosphere.

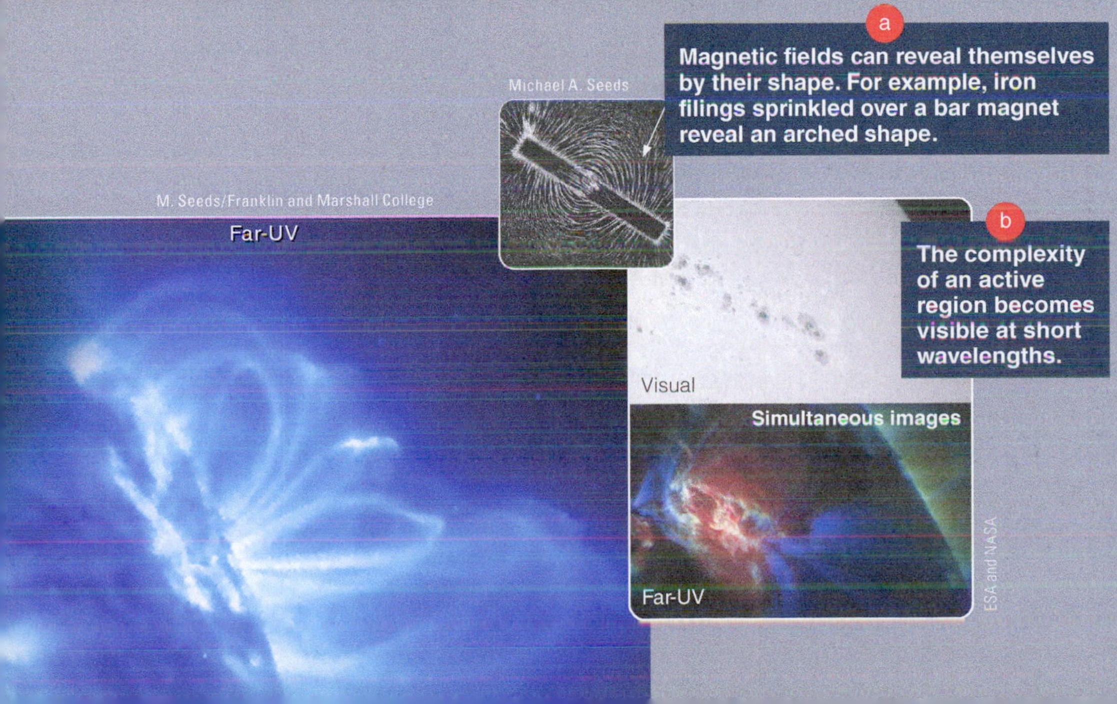

5

Observations at nonvisual wavelengths reveal that the chromosphere and corona above sunspots are violently disturbed in what astronomers call **active regions**. Spectrographic observations show that active regions contain powerful magnetic fields. If all wavelengths are included, Earth receives more radiation from the spotted, active Sun than at times of low activity.

Arched structures above an active region are evidence of gas trapped in magnetic fields.

FIGURE 5-15

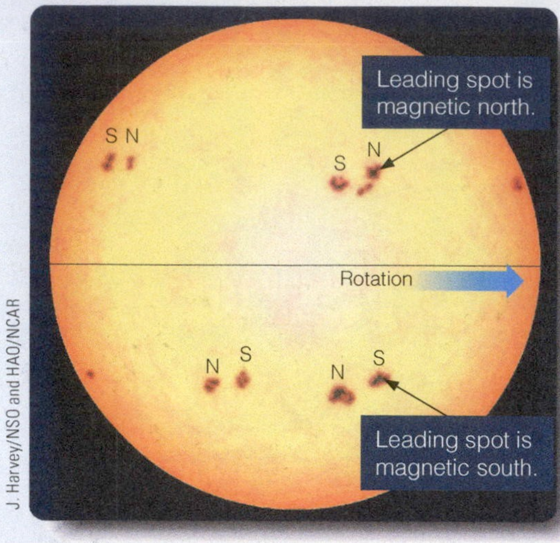

In sunspot groups, here simplified into pairs of major spots, the leading spot and the trailing spot have opposite magnetic polarity. Spot pairs in the Southern Hemisphere have reversed polarity from those in the Northern Hemisphere.

The magnetic behavior of sunspots provides an insight into how the magnetic cycle works. Sunspots tend to occur in groups or pairs, and the magnetic field around the pair resembles that around a bar magnet with one end magnetic north and the other end magnetic south. At any one time, sunspot pairs south of the Sun's equator have reversed polarity compared to those north of the Sun's equator. Figure 5-15 illustrates this by showing sunspot pairs south of the Sun's equator with magnetic south poles leading and sunspots north of the Sun's equator with magnetic north poles leading. At the end of an 11-year sunspot cycle, the new spots appear with reversed **magnetic polarity**.

5-6e The Babcock Model for Solar Activity

The Sun's magnetic cycle is not fully understood, but the **Babcock model** (named for its inventor) explains the magnetic cycle as a progressive tangling and then untangling of the solar magnetic field, demonstrated in Figure 5-16. Because the electrons in an ionized gas are free to move, the gas is a very good conductor of electricity, and any magnetic field in the gas is "frozen" (firmly embedded) in the gas. If the gas moves, the magnetic field must move with it. The Sun's magnetic field is frozen into its gases, and the differential rotation wraps this field around the Sun like a long string caught in a rotating wheel. Rising and sinking gas currents twist the field into ropelike tubes, which tend to float upward. Where these magnetic tubes burst through the Sun's surface, sunspot pairs occur.

The Babcock model also explains the reversal of the Sun's magnetic field from cycle to cycle. As the magnetic field becomes tangled, adjacent regions of the Sun's surface are dominated by magnetic fields that point in different directions. After about 11 years of tangling, the field

magnetic polarity Orientation and strength of a magnetic field's north and south poles.

Babcock model A model of the Sun's magnetic cycle in which the differential rotation of the Sun winds up and tangles the solar magnetic field. This is thought to be responsible for the sunspot cycle.

FIGURE 5-16

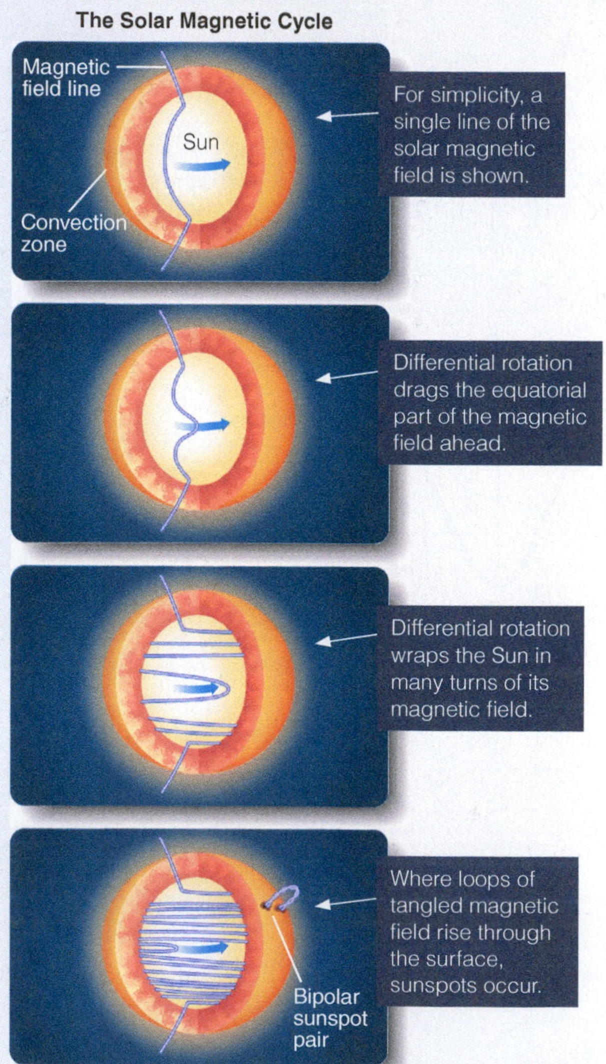

The Babcock model of the solar magnetic cycle explains the sunspot cycle as a primary consequence of the Sun's differential rotation gradually winding up and tangling the magnetic field near the base of the Sun's outer, convective layer.

becomes so complex that adjacent regions of the surface are forced to change their magnetic field directions to align with neighboring regions. The entire field quickly rearranges itself into a simpler pattern, and differential rotation begins winding it up to start a new cycle. The newly organized field is reversed, and the next sunspot cycle begins with magnetic north replaced by magnetic south. Consequently, the complete magnetic cycle is 22 years long, whereas the sunspot cycle is 11 years long.

This magnetic cycle seems to explain the Maunder butterfly diagram (look back to **Concept Art 5B, "Sunspots and the Sunspot Cycle"**). As a sunspot cycle begins, the twisted tubes of magnetic force first begin to float upward and produce sunspot pairs at higher latitude. In other words, the first sunspots in a new cycle appear far from the Sun's equator. Later in the cycle the field is more tightly wound, and tubes of magnetic force arch through the surface at lower latitudes, so sunspot pairs appear closer to the equator.

Notice the power of scientific models. Even though models of the sky (Chapter 2) and of atoms (Chapter 6) are only partly correct, they serve as organizing themes to guide further investigation. Similarly, the Babcock model is incomplete. For example, the solar cycle that should have begun around mid-2008 was delayed to late 2009. The subsequent 2013 cycle maximum was the weakest in 100 years in number of sunspots and amount of solar magnetic activity. The Babcock model does not easily account for such behavior. Nevertheless, the model provides a framework for organizing investigations of complex solar activity (**How Do We Know? 5-2**).

If the Sun is truly a representative star, you might expect to find similar magnetic cycles on other stars, but they are too distant for spots to be directly visible. Some stars, however, vary in brightness over a period of days, in a way revealing that they are marked with dark spots believed to resemble sunspots. Other stars have features in their spectra that vary over periods of years, suggesting

How Do We Know? 5-2

Confirmation and Consolidation

What do scientists do all day? The scientific method is sometimes portrayed as a kind of assembly line where scientists crank out new hypotheses and then test them through observation. In reality, scientists don't often generate entirely new hypotheses. It is rare that an astronomer makes an observation that disproves a long-held theory and triggers a revolution in science. Then what is the daily grind of science really about?

schankz/Shutterstock.com

A yellow jacket is a wasp from a nest containing a queen wasp.

Many observations and experiments confirm already-tested hypotheses. The biologist knows that all worker bees in a hive are sisters because they are all female, and they all had the same mother, the queen bee. A biologist can study the DNA from many workers and confirm that hypothesis. By repeatedly confirming a hypothesis, scientists build confidence in the hypothesis and may be able to extend it. Do all of the workers in a hive have the same father, or did the queen mate with more than one male drone?

Another aspect of routine science is consolidation, the linking of a hypothesis to other well-studied phenomena. A biologist can study yellow jacket wasps from a single nest and discover that the wasps, too, are sisters. There must be a queen wasp who lays all of the eggs in a nest. But in a few nests, the scientist may find two sets of unrelated sister workers. Those nests must contain two queens sharing the nest for convenience and protection. From a study of wasps, the biologist consolidates what is known about bees with what others have learned about wasps and reveals something new: that bees and wasps have evolved in similar ways for similar reasons.

Confirmation and consolidation allow scientists to build confidence in their understanding and extend it to explain more about nature.

Concept Art 5C

SOLAR ACTIVITY AND THE SUN-EARTH CONNECTION

1

Magnetic phenomena in the chromosphere and corona, like magnetic weather, result as constantly changing magnetic fields in the Sun's atmosphere trap ionized gas to produce beautiful arches and powerful outbursts. Some of this solar activity can affect Earth's magnetic field and atmosphere.

This ultraviolet image of the solar surface was made by the NASA *TRACE* spacecraft. It shows hot gas trapped in magnetic arches extending above active regions. At visual wavelengths, you would see sunspot groups in these active regions.

NASA/TRACE

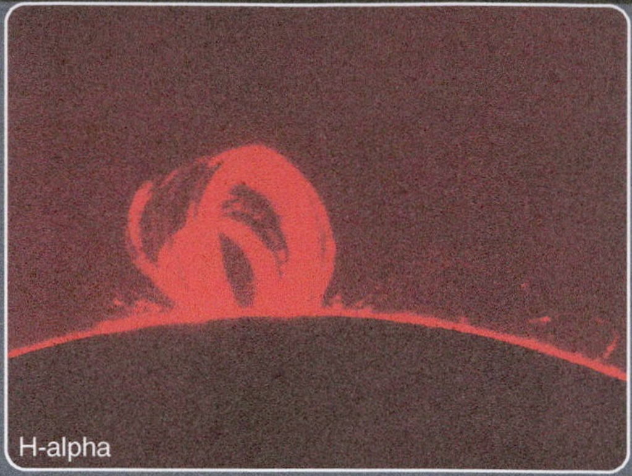

Sacramento Peak Observatory/NSO/AURA/NSF

1a

A **prominence** is composed of ionized gas trapped in a magnetic arch rising up through the photosphere and chromosphere into the lower corona. Seen during total solar eclipses at the edge of the solar disk, prominences look pink because of emission in the three hydrogen Balmer lines, H-alpha, H-beta, and H-gamma. The image above shows the arch shape suggestive of magnetic fields. Seen from above against the Sun's bright surface, prominences form dark **filaments**.

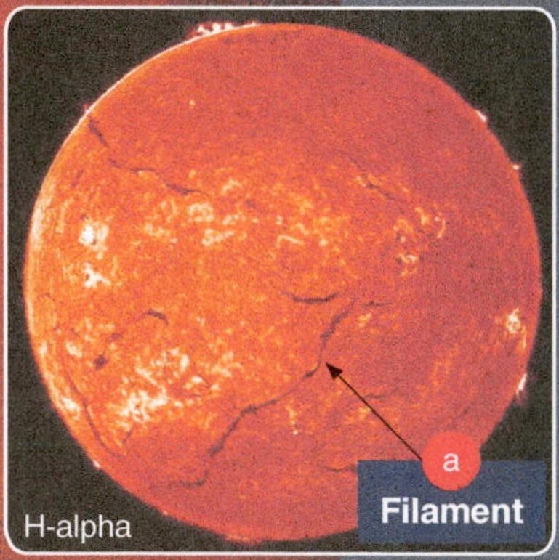

NOAA/SEL/USAF

1b

Quiescent prominences may hang in the lower corona for many days, whereas eruptive prominences burst upward in hours. The eruptive prominence below is many Earth diameters in length.

b

The gas in prominences may be 60,000 to 80,000 K, quite cold compared with the low-density gas in the corona, which may be as hot as a million Kelvin.

ESA/NASA/SOHO EIT

2

Solar **flares** rise to maximum in minutes and decay in an hour. They occur in active regions where oppositely directed magnetic fields meet and cancel each other in what are called **reconnection events**. Energystored in the magnetic fields is released as short-wavelength photons plus high-energy protons and electrons. X-ray and ultraviolet photons reach Earth in 8 minutes and increase ionization in our upper atmosphere, which can interfere with radio communications. Particles from flares reach Earth hours or days later as gusts in the solar wind, which can distort Earth's magnetic field and disrupt navigation systems. Solar flares can also cause surges in electrical power lines and damage to Earth satellites.

2a

At right, waves rush outward at 50 km/s from the site of a solar flare 40,000 times stronger than the 1906 San Francisco earthquake. The biggest solar flares can be a billion times more powerful than a hydrogen bomb.

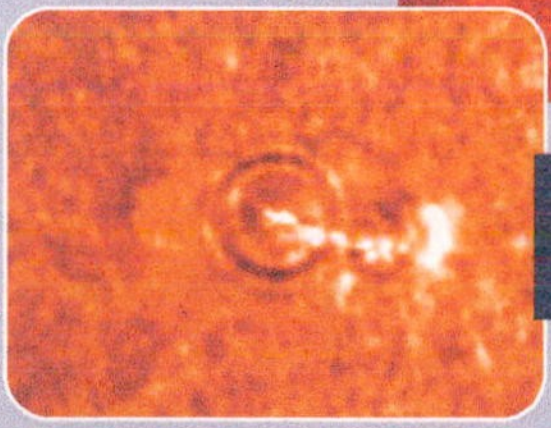

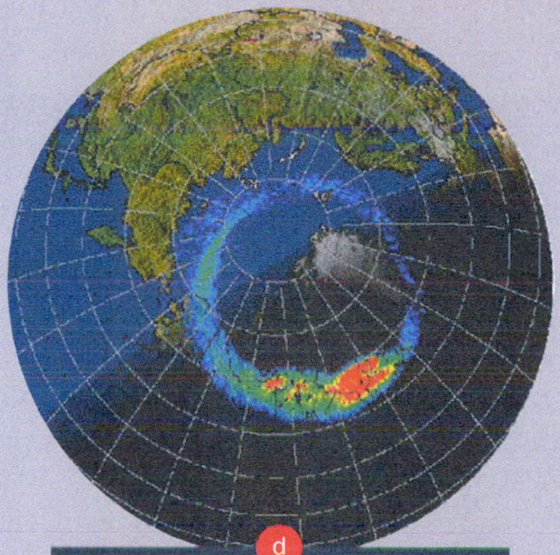

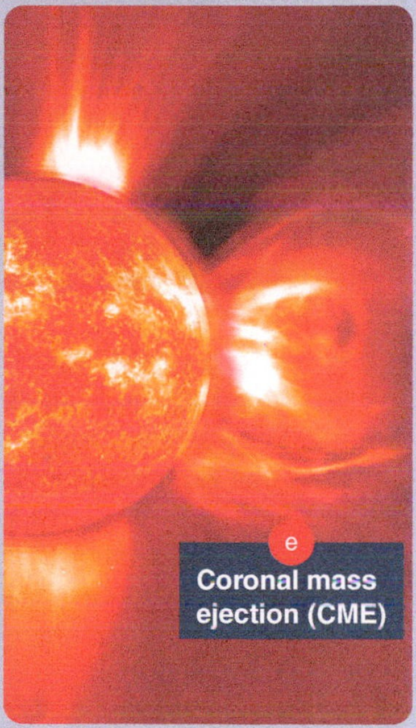

2b

The solar wind, enhanced by eruptions on the Sun, interacts with Earth's magnetic field and can create electrical currents with up to a million megawatts of power. Those currents flowing down into a ring around Earth's magnetic poles excite atoms in Earth's upper atmosphere to emit photons, as shown above. The emission results in glowing clouds and curtains of **auroras**.

2c

Reconnection events can release enough energy to blow large amounts of ionized gas outward from the corona in **coronal mass ejections (CMEs)**. If a CME strikes Earth, it can produce especially violent disturbances in Earth's magnetic field.

3

Much of the solar wind comes from **coronal holes**, where the magnetic field does not loop back into the Sun. These open magnetic fields allow ionized gas in the corona to flow away as the solar wind. The dark area in this X-ray image at right is a coronal hole.

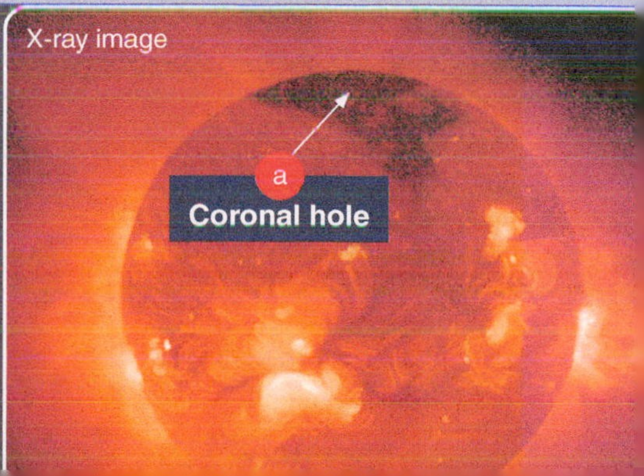

that they are subject to magnetic cycles much like the Sun's cycle. Once again, the evidence tells you that the Sun is a normal star.

5-6f Chromospheric and Coronal Activity

The solar magnetic fields extend high into the chromosphere and corona, where they produce beautiful and powerful phenomena. Study **Concept Art 5C, "Solar Activity and the Sun-Earth Connection,"** and notice the following three important points:

1. All solar activity is magnetic. You do not experience such events on Earth because Earth's magnetic field is weak and Earth's atmosphere is not ionized, so it is free to move independent of the magnetic field. On the Sun, however, the "weather" is a magnetic phenomenon.
2. Tremendous energy can be stored in arches of magnetic fields. These are visible near the edge of the solar disk as prominences, and, seen from above, as filaments. When that stored energy is released, it can trigger powerful eruptions; and, although these eruptions occur far from Earth, they can affect Earth in dramatic ways, including auroral displays. Aurorae, the eerie and pretty northern and southern lights, are produced when gases in Earth's upper atmosphere glow from energy delivered by the solar wind.
3. In some regions of the solar surface, the magnetic field does not loop back. High-energy gas from these regions flows outward and produces much of the solar wind.

You needed the entire range of physical principles presented in this chapter—parallax, Wien's law and the Stefan-Boltzmann law for blackbody radiation, atomic structure, Kirchhoff's laws for the formation of spectra, the Doppler effect, and the Zeeman effect—to realize how the Sun's surface temperature and composition are known, and to understand solar activity cycles and their effects on Earth. In a later chapter you will learn how the Sun fits into the context of other stars.

What Are We? Sunlight

We live very close to a star and depend on it for survival. All of our food comes from sunlight that was captured by plants on land or by plankton in the oceans. We either eat those plants directly or eat the animals that feed on those plants. Whether you had salad, seafood, or a cheeseburger for supper last night, you dined on sunlight, thanks to photosynthesis.

Almost all of the energy that powers human civilization came from the Sun through photosynthesis in ancient plants that were buried and converted to coal, oil, and natural gas. New technology is making energy from plant products such as corn, soy beans, and sugar. It is all stored sunlight. Windmills generate electrical power, and the wind blows because of heat from the Sun. Photocells make electricity directly from sunlight. Even our bodies have adapted to use sunlight to help manufacture vitamin D.

Our planet is warmed by the Sun, and without that warmth the oceans would be ice and much of the atmosphere would be a coating of frost. Books often refer to the Sun as "our Sun" or "our star." It is ours in the sense that we live beside it and by its light and warmth, but we can hardly say it belongs to us. It is closer to the truth to say that we belong to the Sun.

STUDY TOOLS 5

IN THE BOOK

☐ Tear Out the Review Card on *Sun Light and Sun Atoms*.

ASTRO ONLINE

Visit ASTRO Online at www.cengagebrain.com

☐ Interactive Reading

☐ Practice Quizzing

☐ Videos and Animations

☐ Exclusive Features

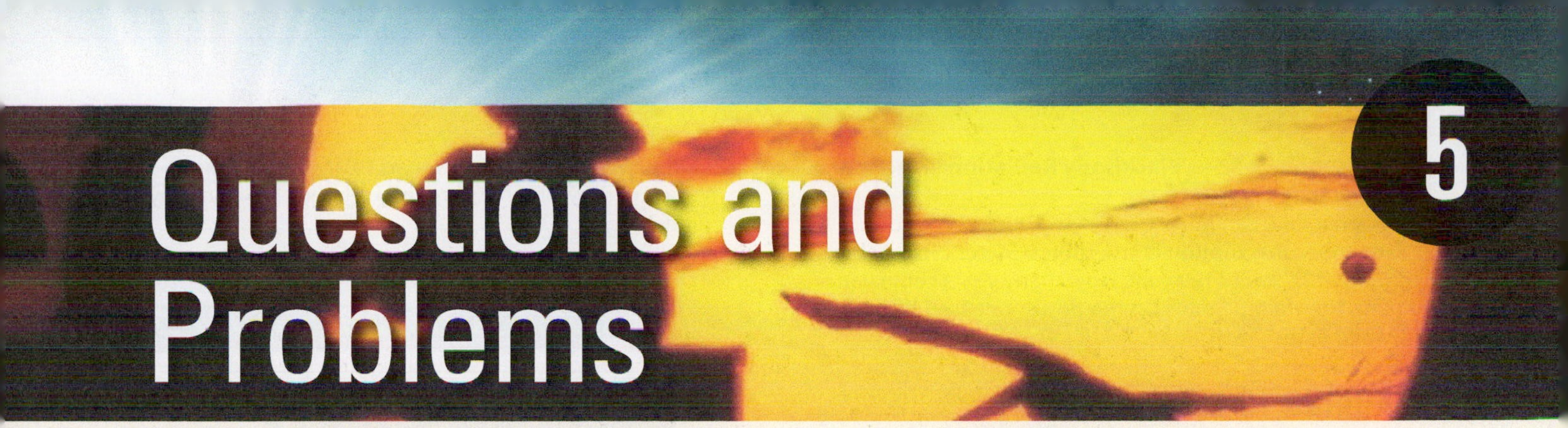

Questions and Problems

5

REVIEW QUESTIONS

1. Why is the binding energy of an electron related to the size of its orbit?
2. Explain why ionized calcium can form absorption lines, but ionized hydrogen cannot.
3. Describe two ways an atom can gain energy and become what is called "excited", with one or more electrons moved to higher orbitals.
4. Why do different atoms have different lines in their spectra?
5. Why does the amount of blackbody radiation emitted depend on the temperature of the object?
6. Why do hot stars look bluer than cool stars?
7. Why does the Doppler effect detect only radial velocity?
8. Why can't you see deeper into the Sun than the photosphere?
9. What evidence can you give that granulation is caused by convection?
10. How can astronomers detect structure in the chromosphere?
11. What evidence can you give that the corona has a very high temperature?
12. Why does hydrogen, which is abundant in the Sun's atmosphere, have relatively weak spectral lines, while calcium, which is not abundant, has very strong spectral lines?
13. How are astronomers able to explore the layers of the Sun below the photosphere?
14. What evidence can you give that sunspots and prominences are magnetic?
15. How does the Babcock model explain the sunspot cycle?
16. How can solar flares affect Earth?
17. **How Do We Know?** How is the world you see around you determined by a world you cannot see?
18. **How Do We Know?** What does "confirmation and consolidation of facts" mean? How does that extend scientific understanding?

DISCUSSION QUESTIONS

1. Imagine that you work in a police lab and you use a spectrograph to determine the chemical composition of samples from criminal investigations. Today, unfortunately, the power is out and none of your equipment works. Of the five basic human senses, which is the most useful in determining what a new incoming sample is made of?
2. Before Fraunhofer and others worked to observe and interpret spectra, most people were of the opinion that we would never know the composition of the Sun and stars. Can you think of any scientific question today that most people believe will probably never be answered?
3. What would the spectrum of an auroral display look like? Why?
4. What energy sources on Earth cannot be thought of as stored sunlight?
5. What observations would you make if you were ordered to set up a system that could warn astronauts in orbit of dangerous solar flares? Such a warning system exists.

PROBLEMS

1. If a star has a surface temperature of 20,000 K (2.00×10^4 K), at what wavelength will it radiate the most energy? (*Hint*: See **Blackbody Radiation** in the Math Reference Cards MR.)
2. Infrared observations of a star show that it is most intense at a wavelength of 2000 nm (2.00×10^3 nm). What is the temperature of the star's surface? (*Hint:* See **Blackbody Radiation** in the Math Reference Cards MR.)
3. If you triple the temperature of a blackbody, by what factor will the total energy radiated per second per square meter increase? (*Hint:* See **Blackbody Radiation** in the Math Reference Cards MR.)
4. Electron orbit transition *A* produces light with a wavelength of 500 nm. Transition *B* involves twice as much energy as *A*. What wavelength is the light it produces?
5. In a laboratory, the Balmer-beta line has a wavelength of 486.1 nm. If the line appears in a star's spectrum at 486.3 nm, what is the star's radial velocity? Is it approaching or receding? (*Hint:* See **The Doppler Formula** in the Math Reference Cards MR.)
6. The smallest detail visible with ground-based solar telescopes is about 1 arc second. How large a region does this represent on the Sun? (*Hint:* See **The Small-Angle Formula** in the Math Reference Cards MR. *Note:* The distance to the Sun is given in **Celestial Profile 9**, found in Appendix C.)
7. If a sunspot has a temperature of 4200 K and the average solar photosphere has a temperature of 5800 K, how many times brighter is a square meter of the photosphere compared to a square meter of the

sunspot? (*Hint:* See **Blackbody Radiation** in the Math Reference Cards MR.)

8. A solar flare can release 6.0×10^{25} Joules. The total energy consumption in the United States is now about 1.0×10^{20} Joules per year. How many years could you run the United States on the power released by a large solar flare?

LEARNING TO LOOK

1. The glowing gas cloud shown in Figure 5-17 contains mostly hydrogen excited to emit photons. What kind of spectrum would you expect this gas cloud to produce?

FIGURE 5-17

T. Rector, University of Alaska, and WIYN/NOAO/AURA/NSF

2. If the gas cloud in Figure 5-17 crosses in front of the star, and the gas cloud and star have different radial velocities, what would the combined spectrum of the gas cloud and the star look like?
3. Whenever there is a total solar eclipse, you can see something like Figure 5-18. Explain why the shape and extent of the glowing gases is observed to be different for each eclipse.

FIGURE 5-18

NSO/AURA/NSF

4. Figures 5-19 and 5-20 show two solar phenomena. What are they, and how are they related? How do they differ?

FIGURE 5-19

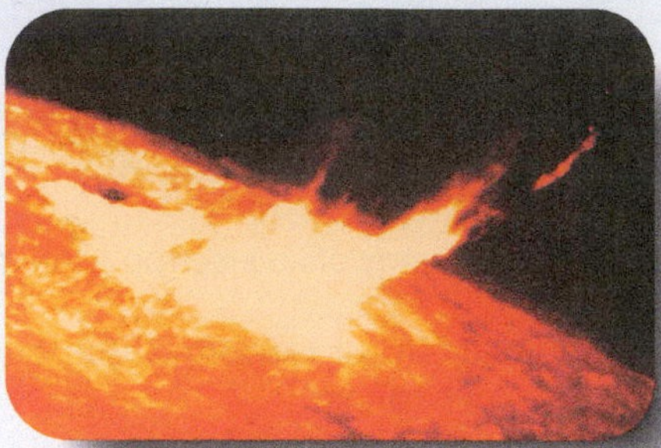

NSO/AURA/NSF

FIGURE 5-20

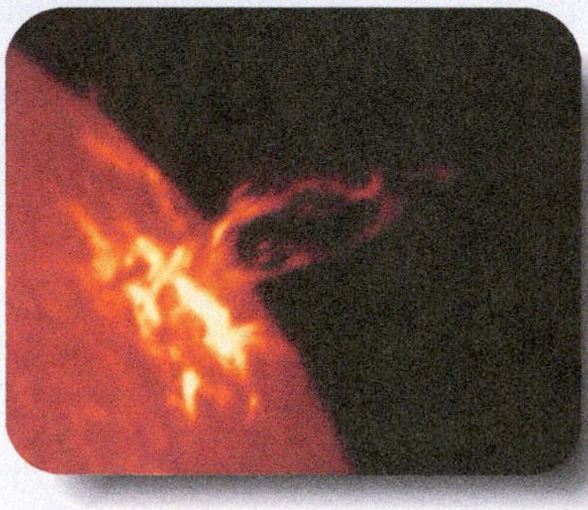

NSO/AURA/NSF

5. Figure 5-21, an image of the Sun was recorded in the extreme ultraviolet by the *SOHO* spacecraft. Explain the features you see.

FIGURE 5-21

SOHO/ESA & NASA

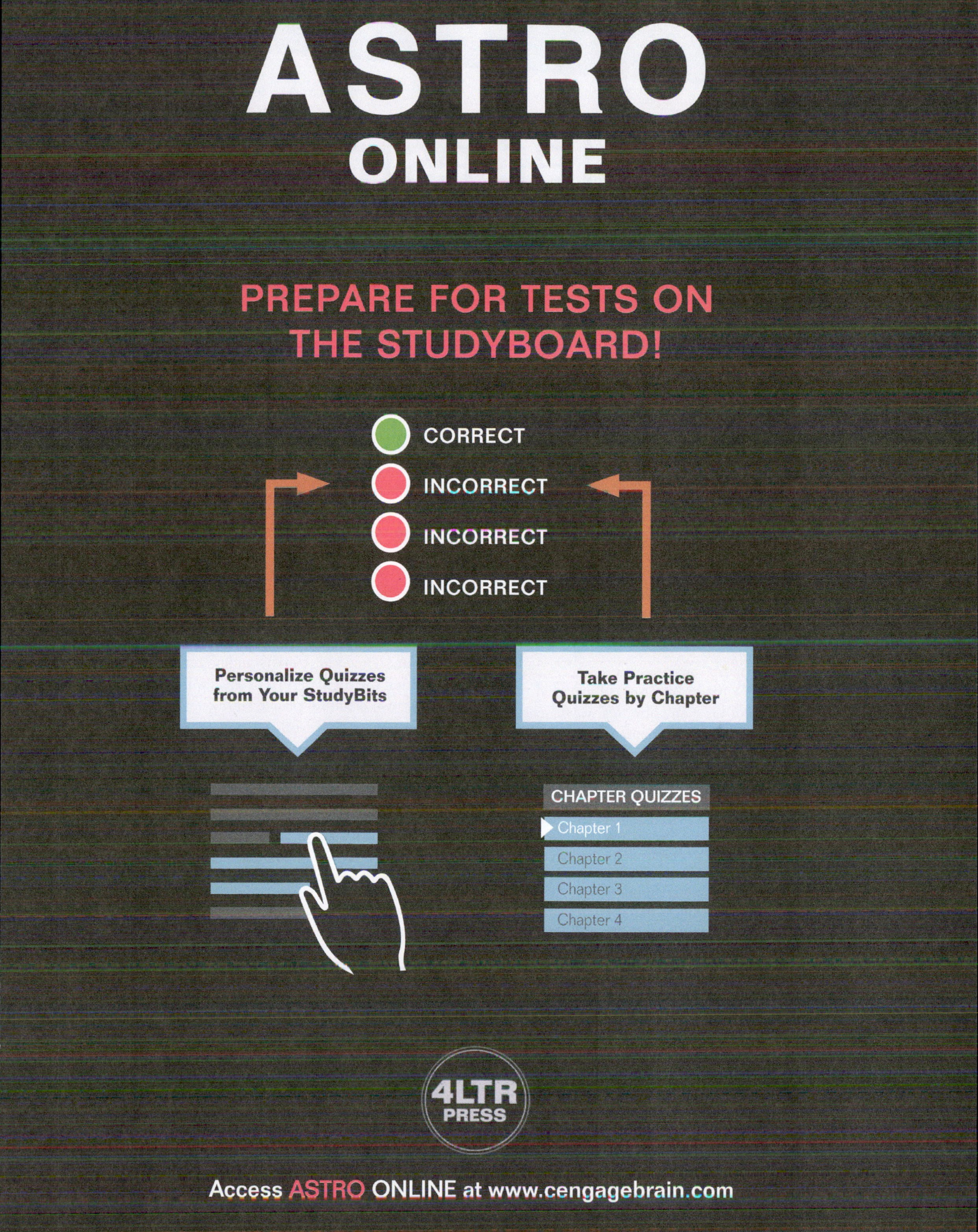
ASTRO
ONLINE
PREPARE FOR TESTS ON THE STUDYBOARD!
CORRECT
INCORRECT
INCORRECT
INCORRECT
Personalize Quizzes from Your StudyBits
Take Practice Quizzes by Chapter
CHAPTER QUIZZES
Chapter 1
Chapter 2
Chapter 3
Chapter 4
4LTR PRESS
Access ASTRO ONLINE at www.cengagebrain.com

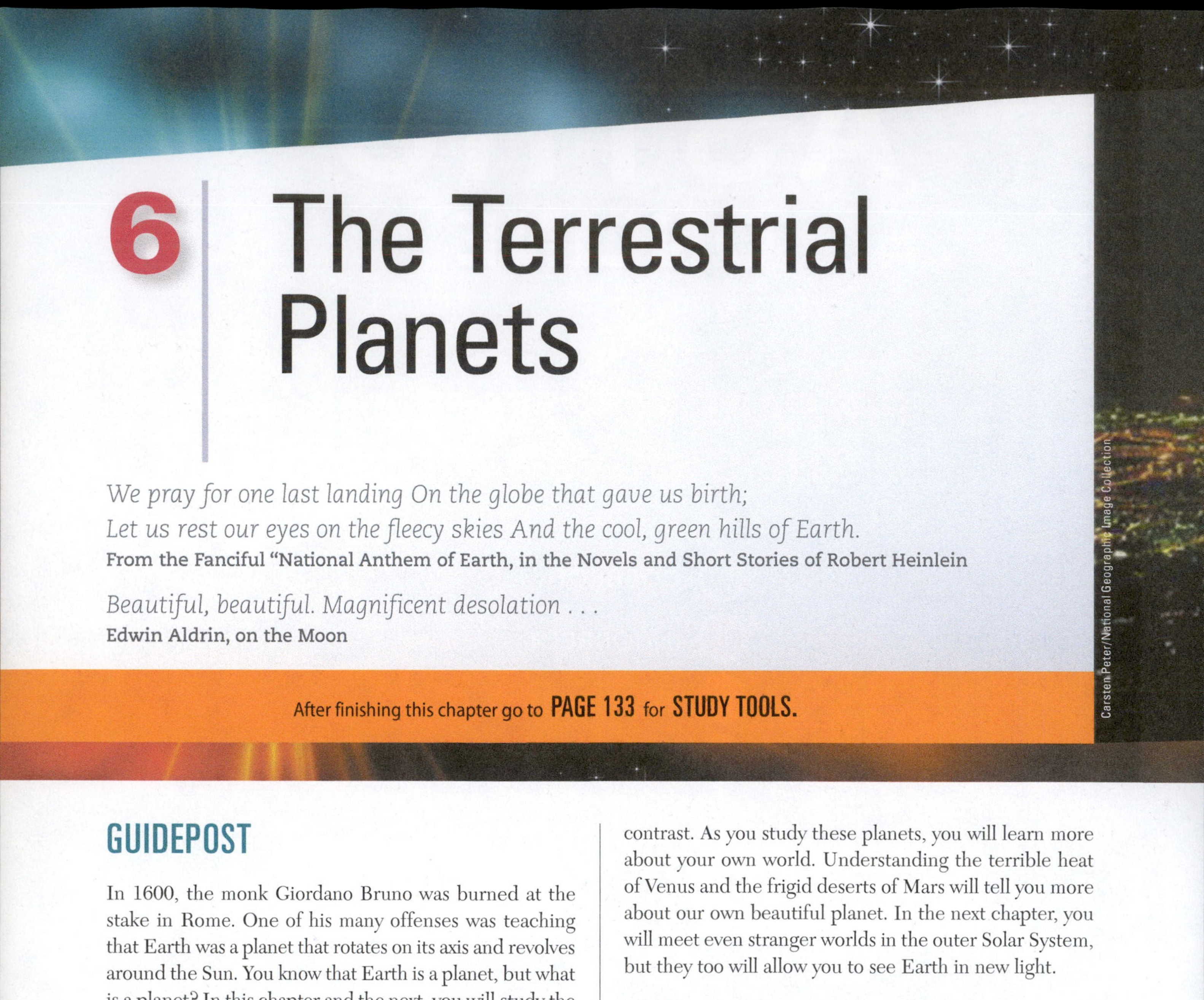

Carsten Peter/National Geographic Image Collection

6 The Terrestrial Planets

We pray for one last landing On the globe that gave us birth;
Let us rest our eyes on the fleecy skies And the cool, green hills of Earth.
From the Fanciful "National Anthem of Earth, in the Novels and Short Stories of Robert Heinlein

Beautiful, beautiful. Magnificent desolation . . .
Edwin Aldrin, on the Moon

After finishing this chapter go to **PAGE 133** for **STUDY TOOLS.**

GUIDEPOST

In 1600, the monk Giordano Bruno was burned at the stake in Rome. One of his many offenses was teaching that Earth was a planet that rotates on its axis and revolves around the Sun. You know that Earth is a planet, but what is a planet? In this chapter and the next, you will study the planets of our Solar System. You can start in this chapter with the planets that, like Earth, are made of rock and metal. (See CO 6.) Here you will find answers to four important questions:

- Why are some planets cratered, airless worlds while others have atmospheres and weather?
- How do size and distance from the Sun influence the evolution of planets?
- How does the history of the Moon illuminate the history of Earth and the other planets?
- What is the evidence that Venus and Mars were once more Earth-like?

comparative planetology Understanding planets by searching for and analyzing contrasts and similarities among them.

The worlds of our Solar System are truly un-Earthly, but you can learn about them by using **comparative planetology**, the study of planets by comparison and contrast. As you study these planets, you will learn more about your own world. Understanding the terrible heat of Venus and the frigid deserts of Mars will tell you more about our own beautiful planet. In the next chapter, you will meet even stranger worlds in the outer Solar System, but they too will allow you to see Earth in new light.

6-1 A TRAVEL GUIDE TO THE TERRESTRIAL PLANETS

In this chapter you will visit five worlds, and this preliminary section will be your guide to important features and comparisons.

6-1a The Scale of the Solar System

You can begin with the most general view of the Solar System. It is, in fact, almost entirely empty space (look back to Figure 1-7). Imagine reducing the scale of the Solar System until Earth is the size of a grain of table salt, about 0.3 millimeter (0.01 inch) in diameter. The Moon is a speck of pepper about 1 centimeter (0.4 in.) away, and the Sun is the size of a small plum located 4 meters (13 feet) from Earth. Jupiter is an apple seed 21 m (69 ft) from the Sun.

CO 6 Mount Etna erupting in 2001 above the Sicilian city of Catania. Humanity, a relative newcomer to Earth, has so far witnessed only a narrow range of the types of powerful activity that a rocky planet with a hot interior can display.

Neptune, at the edge of the planetary zone, is a large grain of sand more than 120 m (400 ft) from the central plum. Your model Solar System would be larger than two football fields and would need a powerful microscope to detect the asteroids orbiting between Mars and Jupiter. The planets are tiny specks of matter scattered around the Sun.

6-1b Revolution and Rotation

The planets revolve* around the Sun in orbits that lie close to a common plane. The orbit of Mercury, the closest planet to the Sun, is tipped 7.0° to Earth's orbit. The rest of the planets' orbital planes are inclined by no more than 3.4°. As you can see, the Solar System is basically flat and disk shaped.

The rotation of the Sun and the planets on their axes also seems related to the rotation of the disk. The Sun rotates with its equator inclined only 7.2° to Earth's orbit, and most of the other planets' equators are tipped less than 30°. The rotations of Venus and Uranus are peculiar, however. Venus rotates backward compared with the other planets, whereas Uranus rotates on its side with its equator almost perpendicular to its orbit.

The preferred direction of motion in the Solar System is counterclockwise, as seen from the north. All the planets revolve around the Sun in that direction. With the exception of Venus and Uranus, all the planets also rotate on their axes in that direction. Furthermore, nearly all of the moons in the Solar System, including Earth's Moon, orbit around their respective planets in that same direction.

6-1c Two Kinds of Planets

Perhaps the most striking clue to the origin of the Solar System comes from the obvious division of the planets into two groups, the small Earth-like worlds and the giant Jupiter-like worlds. The difference is so dramatic that you are led to say, "Aha, this must mean something!" In Chapter 8, you will use this as a critical clue as to how the Sun and the Solar System formed 4.6 billion years ago.

The inner four planets, Mercury, Venus, Earth, and Mars are small, dense worlds composed of rock

* Recall from Chapter 2 that the words *revolve* and *rotate* refer to different types of motion. A planet revolves around the Sun but rotates on its axis. Cowboys in the Old West didn't carry revolvers. They actually carried rotators. And you don't rotate your tires every 6 months, you actually revolve them.

FIGURE 6-1

Planets in comparison: Earth and Venus are similar in size, but their atmospheres and surfaces are very different. The Moon and Mercury are much smaller, and Mars is intermediate in size.

and iron. They are called **Terrestrial planets** because they resemble the rock and metal composition of Earth (*Terra* in Latin). The outer four planets, Jupiter, Saturn, Uranus, and Neptune are large, low-density worlds rich in hydrogen, water, and other low-density materials. They are called the **Jovian planets** because they are Jupiter-like (*Jove* in Latin). You will meet those Jovian worlds in the next chapter.

Terrestrial planet An Earth-like planet—small, dense, rocky and metallic.

Jovian planet A Jupiter-like planet with a large diameter and low density.

differentiation Separation of planetary material within a planet into layers according to density.

mantle The layer of dense rock and metal oxides that lies between the molten core and Earth's crust or a similar layer in another planet.

Figure 6-1 above compares the five worlds you are about to study. The first feature you should notice is diameter. The Moon is small, and Mercury is not much bigger. Earth and Venus are large and similar in size to each other, and Mars is a medium-sized world. You will discover that size is a critical factor in determining a world's personality: Small worlds tend to be internally cold and geologically dead, but larger worlds can be geologically active. (Note that the term *geology*, which originally referred only to Earth, can be used to refer to other worlds as well.)

6-1d Core, Mantle, and Crust

Differentiation is the separation of materials according to density, and the Terrestrial worlds are all differentiated, meaning that the denser materials have sunk into the interior and the lowest-density material has risen to the surface. They have rocky, lower-density crusts; higher-density metal cores; and **mantles**—the layers between their cores and crusts that are composed of dense rock.

The Moon and Mercury have craters all over their surfaces, and the evidence shows that most of these craters are very old. They must have formed when the Solar System was very young and contained lots of rocky debris orbiting the Sun among the planets. That debris hitting the planets produced an early period of crater formation known as the **heavy bombardment** and produced most of the craters in the Solar System. That debris was swept up long ago, so craters don't form often now. Smooth areas are younger and were formed by lava flows that covered up the ancient craters.

One important way you can study a planet is by following the energy. (See **How Do We Know? 6-1**.) The heat in the interior of a planet may be left over from the formation of the planet, or it may be heat generated by radioactive decay. In either case, the heat must flow outward toward the cooler surface where it is radiated into space. In flowing outward, the heat can cause convection currents in the mantle, magnetic fields, plate motions, quakes, faults, volcanism, mountain building, and more. Heat flowing upward through the cooler crust makes a large world like Earth geologically active. In contrast, the Moon and Mercury are so small that they cooled quickly and have little heat flowing outward. They are relatively inactive.

6-1e Atmospheres

When you look at airless Mercury and the Moon in Figure 6-1, you can see their craters and plains and mountains, but the surface of Venus is completely hidden by a cloudy atmosphere much thicker than Earth's. Mars, the medium-sized Terrestrial planet, has a relatively thin atmosphere.

You might ponder two questions. First, why do some worlds have atmospheres while others do not? You will discover that both size and temperature are important. The second question is more complex. Where did these atmospheres come from? To answer that question, you will have to study the geological histories of these worlds.

6-2 EARTH: THE ACTIVE PLANET

Earth is the basis for your comparative study of the Terrestrial planets, so you should pretend to visit it as if you didn't live here. It is an active planet with a molten interior and heat flowing outward that powers volcanism, earthquakes, and an active crust. Almost 75 percent of Earth's surface is covered by liquid water, and the atmosphere contains a significant amount of oxygen. (See **Celestial Profile 3** in Appendix C for Earth's data.)

6-2a Earth's Interior

How do Earth scientists know anything about the interior of Earth? As always, scientific knowledge is based on evidence.

Earth's mass divided by its volume tells you its average density, about 5.5 g/cm^3, but the density of Earth's rocky crust is only about half that. Clearly, a large part of Earth's interior must be made of materials denser than rock.

Each time an earthquake occurs, seismic waves travel through the interior and register on seismographs all over the world. Analysis of these waves shows that Earth's interior is divided into a metallic core, a dense rocky mantle, and a thin, low-density crust.

The core has a density of about 13 g/cm^3, denser than lead; mathematical models indicate it is composed of iron and nickel at a temperature of roughly 6000 K. The core of Earth is as hot as the gaseous surface of the Sun, but high pressure keeps the metal a solid near the middle of the core and a liquid in its outer parts. Two kinds of seismic waves show that the outer core is liquid. The ***P* waves** travel like sound waves, and they can penetrate a liquid, but ***S* waves** travel as a side-to-side vibration that can travel along the surface of a liquid but not through it. Scientists can deduce the size of the liquid core by observing where *S* waves get through and where they don't, as you can see in Figure 6-2.

Earth's magnetism gives you further clues about the core. The presence of a magnetic field is evidence that part of Earth's core must be a liquid metal. Convection currents stir the molten liquid, and because it is a very good conductor of electricity and is rotating as Earth rotates, it generates a magnetic field through the dynamo effect—a different version of the process that creates the Sun's magnetic field (look back to Chapter 5 Section 5-6d).

Earth's mantle is a deep layer of dense rock that lies between the molten core and the solid crust. Models indicate the mantle material has the properties of a solid but is capable of flowing slowly, like asphalt used in paving roads, which shatters if struck with a sledgehammer but bends slightly

heavy bombardment The intense cratering that occurred during the first 0.7 billion years of the Solar System's history.

***P* wave** A type of seismic wave involving compression and decompression of the material through which it passes.

***S* wave** A type of seismic wave involving lateral motion of the material through which it passes.

How Do We Know? 6-1

Understanding Planets: Follow the Energy

What causes change? One of the best ways to think about a scientific problem is to follow the energy. According to the principle of cause and effect, every effect must have a cause, and every cause must involve energy. Energy moves from regions of high concentration to regions of low concentration and, in doing so, produces changes. For example, coal burns to make steam in a power plant, and the steam passes through a turbine and then escapes into the air. In flowing from the burning coal to the atmosphere, the heat spins the turbine that spins a generator to make electricity.

Scientists commonly use energy as a key to understanding nature. A biologist might ask where certain birds get the energy to fly thousands of miles, and a geologist might ask where the energy comes from to power a volcano. Energy is everywhere, and when it moves, whether it is in birds or molten magma, it causes change. Energy is the "cause" in "cause and effect."

You have learned that the flow of energy from the inside of the Sun to its surface helps you know how the Sun and other stars work. The outward flow of energy supports the Sun against its own weight, drives convection currents that produce magnetic fields, and causes surface activity such as spots, prominences, and flares.

Heat flowing out of Earth's interior generates geological activity such as that at Yellowstone National Park.

You can also think of a planet by following the energy. The heat in the interior of a planet may be left over from the formation of the planet, or it may be heat generated by radioactive decay, but it must flow outward toward the cooler surface, where it is radiated into space. In flowing outward, the heat can cause convection currents in the mantle, magnetic fields, plate motions, quakes, faults, volcanism, mountain building, and more.

When you think about any world, be it a small asteroid or a giant planet, think of it as a source of heat that flows outward through that object's surface into space. If you can follow that energy flow, you can understand a great deal about the world. A planetary astronomer once said, "The most interesting thing about any planet is how its heat gets out."

under the weight of a truck. Just below Earth's crust, where the pressure is less than at greater depths, the mantle flows most easily.

The Earth's rocky crust is made up of low-density rocks. The crust is thickest under the continents, up to 60 kilometers (35 miles) thick, and thinnest under the oceans, where it is only about 10 km (6 mi) thick.

6-2b Earth's Active Crust

Earth's crust is composed of lower-density rock that floats on the mantle. The image of rock floating may seem odd, but recall that the rock of the mantle is denser than crust rocks. Also, just below the crust the mantle rock tends to be more fluid, so sections of low-density crust do indeed float on the mantle like great lily pads floating on a pond.

The motion of the crust and the erosive action of water make Earth's crust highly active and changeable. Read **Concept Art 6A, The Active Earth**, and notice the following three important points:

1. The motion of crustal plates produces much of the geological activity on Earth. Earthquakes, volcanism, and mountain building are linked to motions in the crust and the location of plate boundaries. While you are thinking about volcanoes, you can correct a common misconception. The molten rock that emerges from volcanoes comes from pockets of melted rock in the upper mantle and lower crust and not from the molten core.

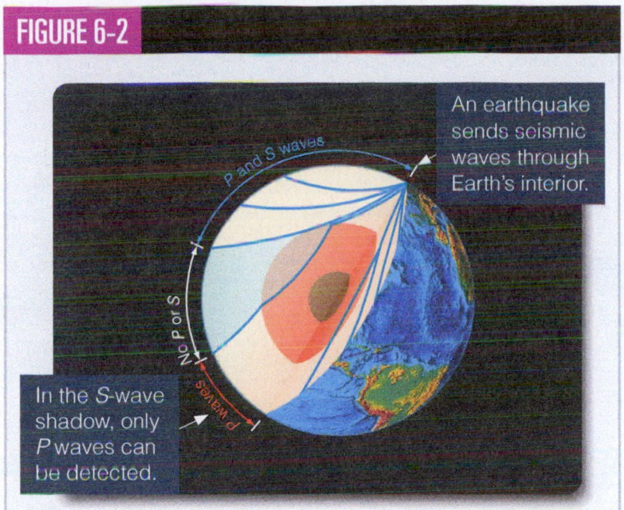

P and *S* waves give you clues to the structure of Earth's interior. No direct *S* waves from an earthquake reach the side of Earth opposite their source, indicating that Earth's core is liquid. The size of the *S* wave "shadow" tells you the size of the liquid outer part of the core.

2. The continents on Earth's surface have moved and changed over periods of hundreds of millions of years. A hundred million years is only 0.1 billion years, $\frac{1}{46}$ of the age of Earth, so sections of Earth's crust are in rapid motion from the perspective of geological time scales.
3. Notice that most of the geological features you know—mountain ranges, the Grand Canyon, and even the outline of the continents—are recent products of Earth's active surface.

The Earth's surface is constantly renewed, and small crystals called zircons are the oldest known Earth material. They are from Western Australia and are 4.4 billion years old. Most of the crust is much younger than that: The mountains and valleys you see around you are probably no more than a few tens, or hundreds, of millions of years old.

6-2c Earth's Atmosphere

When you think about Earth's atmosphere, you should consider three questions: How did it form? How has it evolved? How are humans changing it? Answering these questions will help you understand other planets as well as our own.

In Chapter 8 you will learn that the Sun and planets formed at the same time from a rotating nebula. Scientists once thought that Earth's **primary atmosphere**, its first atmosphere, was composed of gases such as hydrogen and methane captured from the nebula. Modern studies, however, indicate that the planets formed hot, so gases such as carbon dioxide, nitrogen, and water vapor would have been outgassed, meaning, released from the rock. The atmosphere you breathe today is a **secondary atmosphere** produced later in Earth's history.

Soon after Earth formed, it began to cool. Once it cooled enough, oceans began to form, and carbon dioxide began to dissolve in the water. Carbon dioxide is highly soluble in water— which explains the easy manufacture of carbonated beverages. As the oceans removed carbon dioxide from the atmosphere, it reacted with dissolved compounds in the ocean water to form silicon dioxide, limestone, and other mineral sediments. Thus, the oceans transferred the carbon dioxide from the atmosphere to the seafloor and left the air richer in other gases, especially nitrogen.

That removal of carbon dioxide is critical to Earth because an atmosphere rich in carbon dioxide can trap heat by a process called the **greenhouse effect**. When visible-wavelength sunlight shines through the glass roof of a greenhouse, it heats the interior. Infrared radiation from the warm interior can't get out through the glass, heat is trapped in the greenhouse, and the temperature climbs until the glass itself grows warm enough to radiate heat away as fast as sunlight enters, as seen in Figure 6-3a. (Of course, a real greenhouse also retains its heat because the walls prevent the warm air from mixing with the cooler air outside.) This is also called the "parked car effect," for obvious reasons.

Like the glass roof of a greenhouse, a planet's atmosphere can allow sunlight to enter and warm the surface. Note the similarity between the flow of light into the greenhouse in Figure 6-3a and into the atmosphere in Figure 6-3b. Carbon dioxide and other greenhouse gases such as water vapor and methane are opaque to infrared radiation, so an atmosphere containing enough of those greenhouse gases can trap heat and raise the temperature of a planet's surface.

It is a common misconception that the greenhouse effect is always bad. Without the greenhouse effect, Earth would be colder by at least 30°C (54°F), with a planetwide average temperature far below freezing. The problem is that human civilization is adding greenhouse gases to those that were already in

primary atmosphere A planet's first atmosphere.

secondary atmosphere A planet's atmosphere that replaces the primary atmosphere, for example by outgassing, impact of volatile-bearing planetesimals, or biological activity.

greenhouse effect The process by which an atmosphere traps heat and raises the temperature of a planetary surface.

Concept Art 6A

THE ACTIVE EARTH

1

Our world is an astonishingly active planet. Not only is it rich in water and therefore subject to rapid erosion, but its crust is divided into moving sections called *plates*. Where plates spread apart, lava wells up to form new crust; where plates push against each other, they crumple the crust to form mountains. Where one plate slides over another, you see volcanism. This process is called **plate tectonics**, referring to the Greek word for "builder." (An architect is literally an arch builder.)

c A rift valley forms where continental plates begin to pull apart. The Red Sea is forming as Africa starts to separate from the Arabian peninsula.

National Geophysical Data Center

a A typical view of planet Earth

William K. Hartmann/ Planetary Science Institute

Courtesy Janet Seeds

b Mountains are common on Earth, but they erode away rapidly because of the abundant water.

1a

Evidence of plate tectonics was first found in ocean floors, where plates spread apart and magma rises to form **midocean rises** made of rock called **basalt** that is typical of solidified lava. Radioactive dating shows that the basalt is younger near the midocean rise. Also, the ocean floor carries less sediment near the midocean rise. As Earth's magnetic field reverses back and forth, it is recorded in the magnetic fields frozen into the basalt. This produces a magnetic pattern in the basalt that shows that the seafloor is spreading away from the midocean rise.

1b

A **subduction zone** is a deep trench where one plate slides under another. Melting releases low-density magma that rises to form volcanoes such as those along the northwest coast of North America, including Mount Ranier.

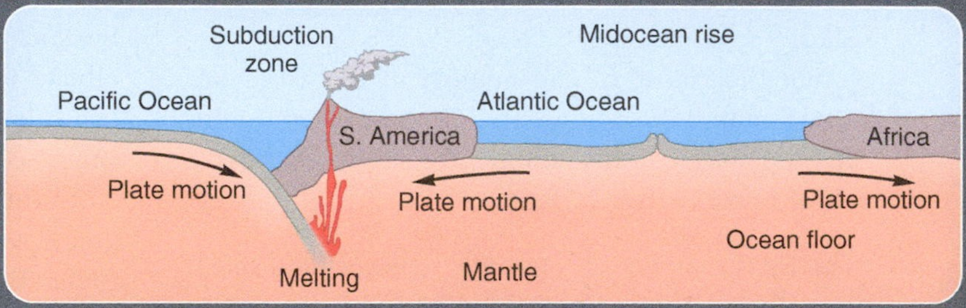

1c

Hot spots caused by rising magma in the mantle can poke through a plate and cause volcanism such as that in Hawai'i. As the Pacific plate has moved northwestward, the hot spot has punched through to form a chain of volcanic islands, which are now mostly worn below sea level. **Folded mountain ranges** can form where plates push against each other. For example, the Ural Mountains lie between Europe and Asia, and the Himalaya Mountains are formed by India pushing north into Asia. The Appalachian Mountains are the remains of a mountain range thrust up when North America was pushed against Europe and Africa.

1d The floor of the Pacific Ocean is sliding into subduction zones in many places around its perimeter. That process pushes up mountains such as the Andes and triggers earthquakes and active volcanism all around the Pacific in what is called the Ring of Fire. In places such as California, the plates slide past each other, causing frequent earthquakes.

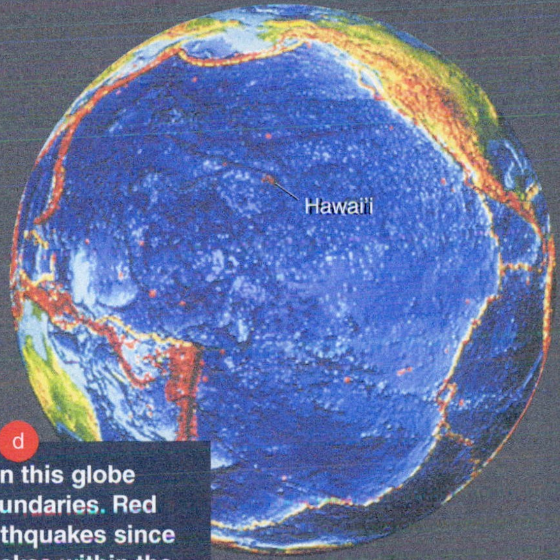

d Yellow lines on this globe mark plate boundaries. Red dots mark earthquakes since 1980. Earthquakes within the plate, such as those at Hawai'i, are related to volcanism over hot spots in the mantle.

Continental Drift

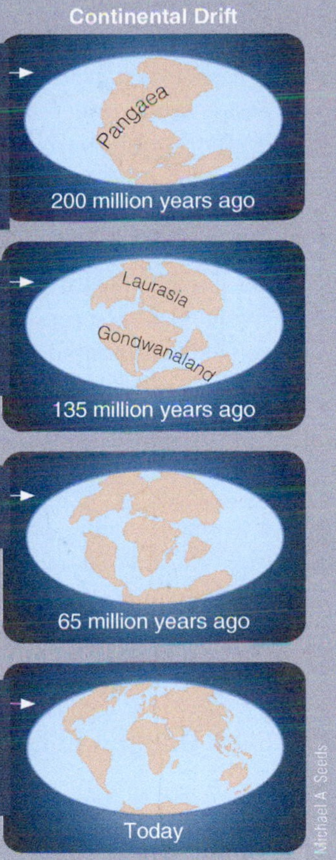

a Not long ago, Earth's continents came together to form one continent dubbed Pangaea by geoscientists.

b Pangaea broke into a northern and a southern continent.

c Notice India moving north toward Asia.

d The continents are still drifting on the highly plastic upper mantle.

2 The floor of the Atlantic Ocean is not being subducted. It is locked to the continents and is pushing North and South America away from Europe and Africa at about 3 cm per year in a motion called *continental drift*. Scientists can measure the motion by, for example, comparing the time for laser flashes sent from European and American observatories to bounce off mirrors left on the Moon by the *Apollo* astronauts. Roughly 200 million years ago, North and South America were joined to Europe and Africa. Evidence of that lies in similar fossils and similar rocks and minerals found in the matching parts of the continents. Notice how North and South America fit against Europe and Africa like a puzzle.

3 Plate tectonics push up mountain ranges and cause bulges in the crust, and water erosion wears the rock away. The Colorado River began cutting the Grand Canyon only about 10 million years ago when the Colorado plateau warped upward under the pressure of moving plates. That sounds like a long time ago, but it is only 0.01 billion years. A mile down, at the bottom of the canyon, lie rocks 0.57 billion years old, the roots of an earlier mountain range that stood as high as the Himalayas. It was pushed up, worn away to nothing, and covered with sediment long ago. Many of the geological features we know on Earth have been produced by relatively recent events.

the atmosphere. For 4 billion years, Earth's oceans and plant life have been absorbing carbon dioxide and burying it in the form of carbonates (such as limestone) and in carbon-rich deposits of coal, oil, and natural gas. In the past century or so, human civilization has begun digging up those fuels, burning them for energy, and releasing the carbon back into the atmosphere as carbon dioxide (Figure 6-3c). This process is steadily increasing the carbon dioxide concentration in the atmosphere and warming Earth's climate in what is called **global warming**. Evidence from proportions of carbon isotopes and oxygen in the atmosphere shows that most of the added CO_2 is the result of burning fossil fuels.

Global warming is a critical issue not just because it affects agriculture; it changes climate patterns that will warm some areas and cool other areas. It addition, the warming is melting permanently frozen ices in the polar caps, causing sea levels to rise. A rise of just a few feet will flood major land areas. When you visit Venus later in this chapter, you will see a planet dominated by a runaway greenhouse effect.

6-2d Oxygen in Earth's Atmosphere

When Earth was young, its atmosphere had no free oxygen. Oxygen is very reactive and quickly forms oxides in the soil, so metabolism of plants is needed to keep a steady supply of oxygen in the atmosphere. Photosynthesis makes energy for the plant by absorbing carbon dioxide and releasing free oxygen. About 2 to 2.5 billion years ago plants began to produce oxygen faster than chemical reactions could remove it from the atmosphere. Atmospheric oxygen then increased rapidly.

Because there is oxygen (O_2) in the atmosphere now, there is also a layer of ozone (O_3) at altitudes of 15 to 30 km (10 to 20 mi). Many people hold the common misconception that ozone is bad because they hear it mentioned as part of smog. Breathing ozone is bad for you, but you need the ozone layer in the upper atmosphere to protect you from harmful UV photons. Certain compounds called chlorofluorocarbons (CFCs) used in refrigeration and industry can destroy ozone when they leak into the atmosphere. Starting in the late 1970s, ozone concentrations began falling and the intensity of harmful UV radiation at Earth's surface increased year by year. Fortunately, as a result of these changes, international agreements banned most CFCs, and the ozone depletion eventually halted and now

global warming The gradual increase in the surface temperature of Earth caused by human modifications to Earth's atmosphere.

FIGURE 6-3

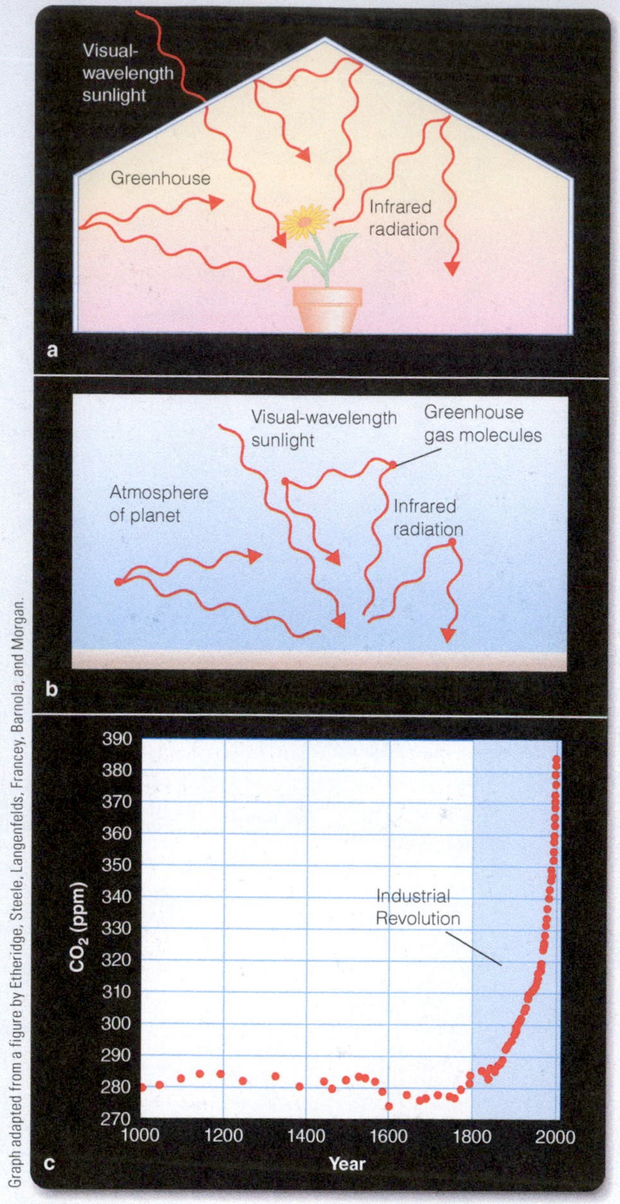

The greenhouse effect: (a) Visual-wavelength sunlight can enter a greenhouse and heat its contents, but the longer-wavelength infrared radiation cannot get out. (b) The same process can heat a planet's surface if its atmosphere contains greenhouse gases such as CO_2. (c) The concentration of CO_2 in Earth's atmosphere as measured in Antarctic ice cores remained roughly constant for thousands of years until the beginning of the Industrial Revolution around the year 1800. Since then it has increased by more than 40 percent. Evidence from proportions of carbon isotopes and oxygen in the atmosphere proves that most of the added CO_2 is the result of burning fossil fuels.

appears to have reversed. While this poses an immediate problem for public health on Earth, it is also of interest astronomically. When you visit Mars, you will see the effects of an atmosphere without ozone.

6-2e A Short Geological History of Earth

As Earth formed in the inner Solar System, it passed through three stages that also describe the histories of the other Terrestrial planets to varying extents. When you try to tell the story of each planet in our Solar System, you pull together all the known facts plus hypotheses and try to make them into a logical history of how the planet got to be the way it is. Of course, your stories will be incomplete because scientists don't yet understand all the factors affecting the histories of the planets (see Chapter 8).

The first stage of planetary evolution is *differentiation*, the separation of each planet's material into layers according to density. Some of that differentiation may have occurred very early as the heat released by infalling matter melted the growing Earth. Some of the differentiation, however, may have occurred later as radioactive decay released more heat and further melted Earth, allowing the denser metals to sink to the core.

The second stage, *cratering and giant basin formation*, could not begin until a solid surface formed. The heavy bombardment in the early Solar System cratered Earth just as it did the other Terrestrial planets. Some of the largest craters, called basins, were likely big enough to break through to the upper mantle where rocks are partly molten. That molten rock would have welled up through cracks to fill some of the basins and craters. On planets with liquid water, the basins may also have been the first oceans. As the debris of planet formation cleared away, the rate of impacts and crater formation fell to its present low rate.

The third stage, *slow surface evolution*, has continued for at least the past 3.5 billion years. Earth's surface is constantly changing as sections of crust slide over and past each other, push up mountains, and shift continents. In addition, moving air and water erode the surface and wear away geological features. Almost all traces of the earlier stages of differentiation and cratering have been destroyed by the active crust and erosion. Life apparently started on Earth around the beginning of the slow surface evolution stage, and the secondary atmosphere began to replace the primary atmosphere. That transition could be unique to Earth and might not have happened on the other Terrestrial planets.

Terrestrial planets pass through these stages, but differences in masses, temperature, and composition emphasize some stages over others and produce surprisingly different worlds.

6-3 THE MOON

You can't go for a stroll on the Moon without a spacesuit. There is no air, and the temperature difference from sunshine to shade is extreme. (See **Celestial Profile 10** in Appendix C for the Moon's data.)

6-3a Lunar Geology

You could visit two kinds of terrain on the Moon. The dark gray areas visible from Earth are the smooth lunar lowlands which astronomers named **maria** (plural of **mare**, pronounced mah-ray, the Latin word for *seas*). You could also visit the lighter-colored mountainous lunar highlands. The Moon looks quite bright in the night sky seen from Earth. In fact, the **albedo** of the near side of the Moon is only 0.12, meaning it reflects only 12 percent of the light that hits it. The Moon looks bright only in contrast to the night sky. In reality, it is a dark gray world.

Wherever you go on the Moon, you will find craters. The highlands are marked heavily by craters, but the smooth lowlands contain fewer craters. The craters on the Moon were formed by impacts, as evidenced by their distinguishing characteristics such as shape and the material (**ejecta**) they have spread across the Moon's surface. Craters range in size from giant basins hundreds of kilometers across to microscopic pits found in Moon rocks. Most of the craters on the Moon are old; they were formed long ago when the Solar System was young. This is clear from comparison of ages of Moon rocks collected from the heavily cratered highlands and the somewhat younger, less heavily cratered maria.

Twelve *Apollo* astronauts visited both the Moon's maria and highlands in six expeditions between 1969 and 1972 (Figure 6-4). Most of the rocks they found, both in the highlands and the maria, were typical of hardened lava. The maria are actually ancient basalt

mare One of the lunar lowlands filled by successive flows of dark lava, from the Latin for "sea." Plural: "maria."

albedo The ratio of the amount of light reflected from an object to the amount of light received by the object. Albedo equals 0 for perfect absorption and 1 for perfect reflection.

ejecta Pulverized rock scattered by meteorite impacts on a planetary surface.

FIGURE 6-4

Rocks returned from the Moon show that the Moon once had a deep magma ocean of molten rock, that after the surface solidified it was heavily fractured by cratering, and that it is now affected mainly by micrometeorites grinding away at surface rock.

lava flows. The highlands, in contrast, are composed of low-density rock, for example, **anorthosite**, a light-colored rock that contributes to the highlands' bright contrast with the dark maria and that would be among the first material to solidify and float to the top of molten rock. Many rocks found on the Moon are **breccias**, made up of fragments of broken rock cemented together under pressure. The breccias show how extensively the Moon's surface has been shattered by meteorites, as does the surface layer of powdery dust kicked up by the astronaut's boots.

anorthosite Aluminum and calcium-rich silicate rock found in the lunar highlands.

breccia Rock composed of fragments of older rocks bonded together.

large-impact hypothesis Hypothesis that the Moon formed from debris ejected during a collision between Earth and another planet-sized object.

6-3b The Origin of Earth's Moon

Over the past two centuries, astronomers developed three different hypotheses for the origin of Earth's Moon. The *fission hypothesis* proposed that the Moon broke from a rapidly spinning proto-Earth. The *condensation hypothesis* suggested that Earth and its Moon condensed from the same cloud of matter. The *capture hypothesis* suggested that the Moon formed elsewhere in the Solar System and was later captured by Earth. Each of these previous ideas had problems and failed to survive comparison with all the evidence

In the 1970s, a new hypothesis was proposed that combined some aspects of the three previous hypotheses. The **large-impact hypothesis** proposes that the Moon formed when an object estimated to have been at least as large as Mars smashed into the proto-Earth. Model calculations indicate that this collision would have ejected a disk of debris into orbit around Earth and that the debris could have quickly formed the Moon (Figure 6-5).

This would explain a number of phenomena. If the collision occurred off-center, it would have spun the

FIGURE 6-5

The Large-Impact Hypothesis

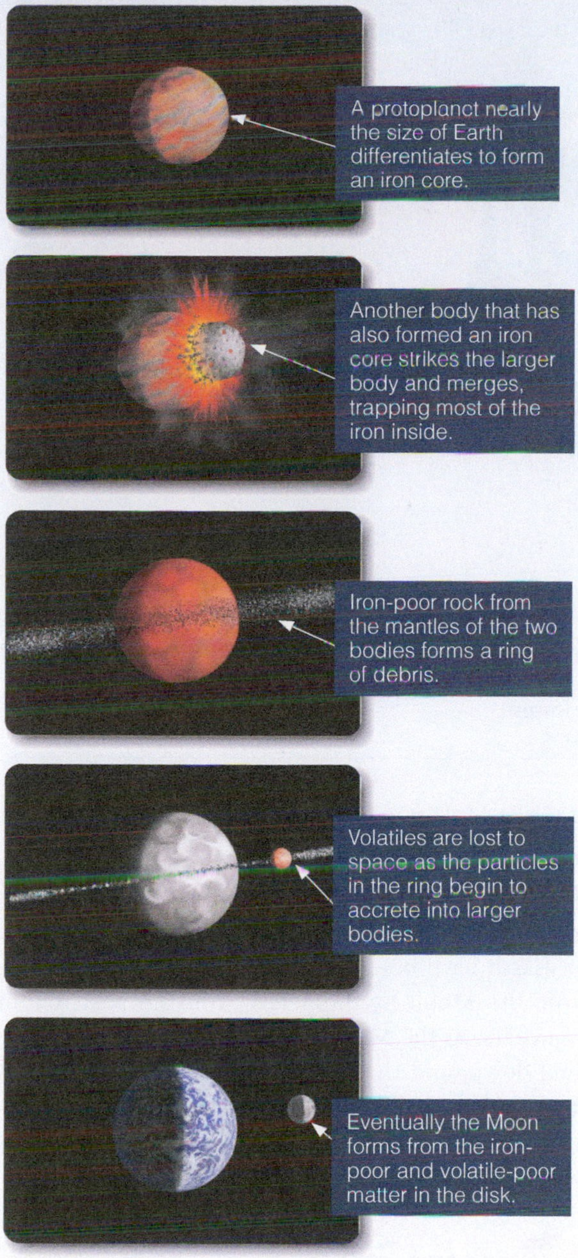

The large-impact hypothesis holds that when the Solar System was about 50 million years old, a massive collision produced the Moon in its orbit inclined to Earth's equator.

Earth–Moon system rapidly and would thus explain the present high angular momentum. If the proto-Earth and impactor had already differentiated, the ejected material would have been mostly iron-poor mantle and crust, which would explain the Moon's low density and iron-poor composition. Furthermore, the material would have lost much of its volatile components while it was in space, so the Moon also would have formed lacking those materials. Such an impact would have melted the proto-Earth, and the material falling together to form the Moon would have been heated enough to melt. This fits the evidence that the highland anorthosite in the Moon's oldest rocks formed by differentiation of large quantities of molten material. The large-impact hypothesis passes tests of comparison with the known evidence and is now considered likely to be correct.

6-3c A History of Earth's Moon

The history of Earth's Moon since its formation is dominated by a single fact—the Moon is small, only one-quarter the diameter of Earth. The escape velocity is low, so it has been unable to hold any atmosphere, and it cooled rapidly as its internal heat flowed outward into space.

The *Apollo* Moon rocks are the source of information about the timing of events during the Moon's history. For example, they show that the Moon must have formed in a mostly molten state. Planetary geologists now refer to the exterior of the newborn Moon as a **magma ocean**. Denser materials sank toward the center, and low-density minerals floated to the top to form a lowdensity crust. In this way the Moon partly differentiated. The radioactive ages of the Moon rocks show that the surface solidified about 4.4 billion years ago.

The second stage, cratering and basin formation, began as soon as the crust solidified, and the older highlands show that cratering was intense for approximately 0.7 billion years—during the heavy bombardment at the end of the Solar System's period of planet building. The Moon's crust was shattered, and the largest impacts formed giant **multiringed basins** hundreds of kilometers in diameter (Figure 6-6). There is some evidence indicating that roughly 4 billion years ago there was a sudden temporary increase in the cratering rate. Astronomers refer to this as the **late heavy bombardment** and hypothesize that it could have been caused by the final accretion and migration of Uranus and Neptune (see Chapter 8) scattering remaining debris across

magma ocean The exterior of the newborn Moon, a shell of molten rock hundreds of kilometers deep.

multiringed basin Large impact feature (crater) containing two or more concentric rims formed by fracturing of the planetary crust.

late heavy bombardment A sudden temporary increase in the Solar System's cratering rate that evidently occurred about 4 billion years ago.

FIGURE 6-6

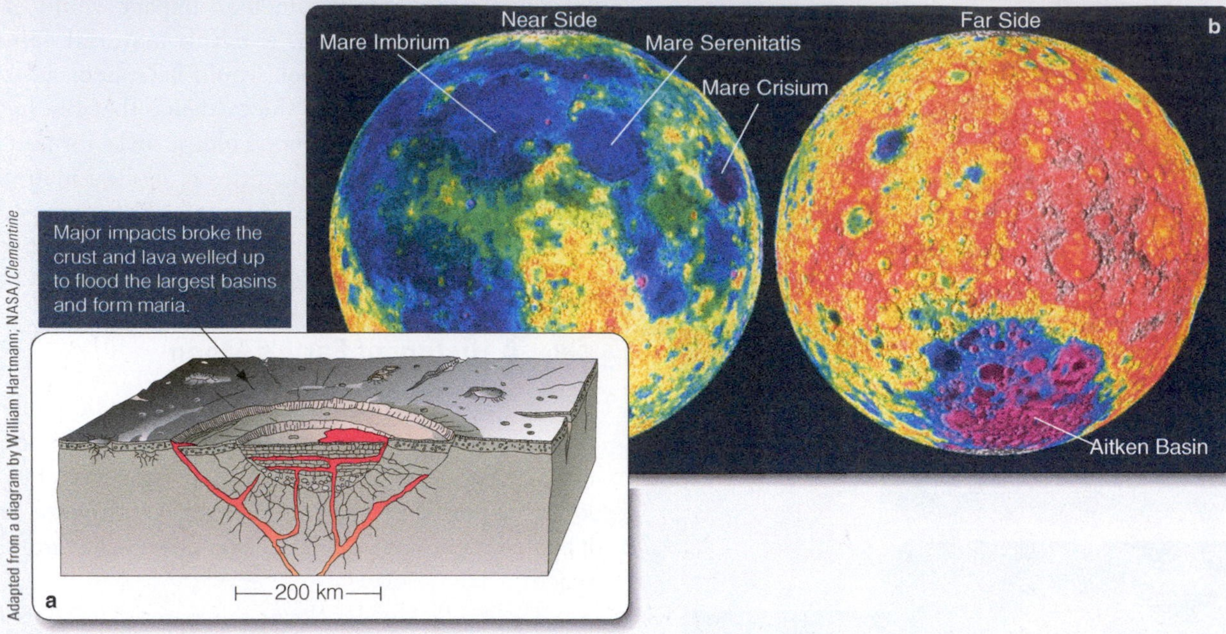

Much of the near side of the Moon is marked by large, generally circular lava plains called maria. The crust on the far side is thicker, and there is much less flooding. Even the huge South Pole-Aitken Basin contains little lava flooding. In these maps color marks elevation, with red the highest regions and purple the lowest.

the Solar System to collide with the Moon and planets. About 3.8 billion years ago the cratering rate fell rapidly to the current low rate.

The tremendous impacts that formed the lunar basins cracked the crust as deep as 10 km (6 mi) and led to flooding by lava. Though Earth's Moon cooled rapidly after its formation, radioactive decay heated the subsurface material, and part of it melted, producing lava that followed the cracks up into the giant basins (Figure 6-6a). Studies of rocks brought back from the Moon by *Apollo* astronauts show that the basins were flooded by successive lava flows of dark basalts from roughly 4 to 2 billion years ago, forming the maria. The Moon's crust is thinner on the side toward Earth, perhaps due to tidal effects. Consequently, while lava flooded the basins on the Earth-facing side, it was unable to rise through the thicker crust to flood the lowlands on the far side.

The third stage, slow surface evolution, was limited both because the Moon cooled rapidly and because it lacks water. Flooding on Earth included water, but the Moon has never had an atmosphere and thus has never had liquid water. With no air and no water, erosion is limited to the constant bombardment of **micrometeorites** and rare larger impacts. Indeed, a few meteorites found on Earth have been identified as Moon rocks ejected from the Moon by impacts within the past few million years. As the Moon lost its internal heat, volcanism died down, and the Moon became geologically dead. Its crust never divided into moving plates—there are no folded mountain ranges—and the Moon is now a "one-plate" planet.

micrometeorite Meteorite of microscopic size.

6-4 MERCURY

Like Earth's Moon, Mercury is small and nearly airless. (See **Celestial Profile 1** in Appendix C for Mercury's data.) Mercury cooled too quickly to develop plate tectonics, so you will find it also is a cratered, dead world.

6-4a Spacecraft at Mercury

Mercury orbits close to the Sun and is difficult to observe from Earth, so little was known about it until 1974–1975 when the *Mariner 10* spacecraft flew past Mercury three times and revealed a planet with a heavily cratered surface, much like that of Earth's Moon. A new spacecraft called *MESSENGER* made three flybys of Mercury in 2008 and 2009, taking impressively high-resolution images (Figure 6-7a) and measurements of the parts of the planet not covered by *Mariner 10's* cameras. *MESSENGER* went into orbit around Mercury in 2011, beginning a close-up investigation that lasted until it ran out of maneuvering fuel and crashed in 2015. Analysis shows that large areas have been flooded by lava and then cratered. The largest impact feature on Mercury is the Caloris Basin, a multiringed area 1550 km (960 mi) in diameter (Figure 6-7b).

Mercury is quite dense, and models indicate that it must have a large metallic core. In fact, the metallic core occupies about 70 percent of the radius of the planet. Mercury is essentially a metal planet with a thin rock mantle. *Mariner 10* photos revealed long curving ridges up to 3 km (2 mi) high and 500 km (300 mi) long, as shown in Figure 6-7c. The ridges even cut through craters, indicating that they formed after most of the heavy bombardment. Planetary scientists understand these ridges as evidence that long ago Mercury shrank as its massive core cooled and shrank, causing the crust to wrinkle and form the curving ridges.

6-4b A History of Mercury

Like Earth's Moon, Mercury is small, and that fact has determined much of its history. Not only is Mercury too small to retain an atmosphere, it has also lost much of its internal heat and thus is not geologically active.

Early in its formation, Mercury differentiated to form a metallic core and a rocky mantle. The presence of a magnetic field about 1 percent as strong as Earth's is evidence of a large metallic core. Astronomers thought for decades that Mercury has a lot of iron because it formed near the Sun where rock is less stable, but detailed calculations

FIGURE 6-7

(a) Mosaic of *MESSENGER* spacecraft images of the portion of Mercury that includes the Caloris multiringed basin. The color enhancement emphasizes composition differences. (b) Multiple rings of mountains around the rim of the Caloris Basin, photographed by the *Mariner 10* spacecraft in 1975. (c) The origin of the spider formation in the Caloris Basin photographed by *MESSENGER* is a puzzle.

show that Mercury contains even more iron than expected. Drawing on the large-impact hypothesis for the origin of Earth's Moon, scientists proposed that Mercury suffered a major impact that shattered and drove away much of the rocky mantle. However, data from *MESSENGER* indicate that the crust contains significant volatiles which would not have survived a giant impact. Therefore, a giant impact may not, after all, explain Mercury's large metal core.

In the second stage of planet formation, cratering battered the crust, and lava flows welled up to fill the lowlands, just as they did on the Moon. As the small world lost internal heat, its large metal core contracted, and its crust compressed and broke to form the long lobate scarp ridges, much as the peel of a drying apple wrinkles.

Lacking an atmosphere to erode it, Mercury has changed little since the last lava hardened, and it is now a "one-plate" planet like Earth's Moon.

6-5 VENUS

You might expect Venus to be much like Earth. It is 95 percent Earth's diameter, has a similar average density, and is only 30 percent closer to the Sun. (You can compare Earth's and Venus's data in detail by examining **Celestial Profiles 2 and 3** in Appendix C.) Unfortunately, the surface of Venus is perpetually hidden below thick clouds, and only in the past few decades have planetary scientists discovered that Venus is a hot desert world of volcanoes, lava flows, and impact craters lying at the bottom of a deep, hot atmosphere. No spacesuit is protective enough to allow you to visit the surface of Venus.

6-5a Venus's Atmosphere

In composition, temperature, and density, the atmosphere of Venus is more Hades than Heaven. The air is unbreathable, very hot, and very dense. How do planetary scientists know this? Because space probes have descended into the atmosphere and a few have landed on the surface.

In composition, the atmosphere of Venus is roughly 96 percent carbon dioxide. The rest is mostly nitrogen, with some argon, sulfur dioxide, and small amounts of sulfuric acid, hydrochloric acid, and hydrofluoric acid. There is only a tiny amount of water vapor. On the whole, the composition is deathly unpleasant, and most certainly smells bad too. Spectra show that the impenetrable clouds that hide the surface are made up of droplets of sulfuric acid and microscopic crystals of sulfur. This unbreathable atmosphere is 90 times denser than Earth's. The air you breathe is 1000 times less dense than water, but on Venus the air is only 10 times less dense than water. If you could survive the unpleasant composition and intense heat, you could strap wings on your arms and fly.

The surface temperature on Venus is hot enough to melt lead because the thick atmosphere creates a severe greenhouse effect. Sunlight filters down through the clouds and warms the surface, but infrared radiation cannot escape because the atmosphere is opaque to infrared. It is the overwhelming abundance of carbon dioxide that makes the greenhouse effect on Venus much more severe than on Earth.

6-5b Venus's Surface

Although the thick clouds and atmosphere on Venus are opaque to visible and infrared light, they are transparent to radio waves. Orbiting spacecraft mapped Venus by radar, revealing details as small as 100 m (about 300 ft) in diameter (Figure 6-8).

Radar maps of Venus are reproduced using arbitrary colors. Lowlands are colored blue in some maps, but this is misleading in a sense because there are no oceans on Venus. In other maps, the scientists have chosen instead to give Venus an overall orange tint because sunlight filtering down through the clouds bathes the landscape in a perpetual sunset glow. Some radar maps are colored gray, the natural color of the rocks.

Radar maps show that Venus is similar to Earth in one way but strangely different in other ways. Nearly 75 percent of Earth is covered by low-lying basaltic sea floors, and 85 percent of Venus is covered by basaltic lowlands. Of course on Venus the lowlands are not seafloors, and the remaining highlands are not the well-defined continents you see on Earth. Whereas Earth is dominated by plate tectonics, something different is happening on Venus.

The highland area Ishtar Terra, named for the Babylonian goddess of love, is about the size of Australia (Figure 6-8). At its eastern edge, the mountain called Maxwell Montes thrusts up 12 km (7.5 mi). For comparison, Everest, the tallest mountain on Earth, is 8.8 km (5.5 mi) high. The mountains bounding Ishtar Terra, including Maxwell, resemble folded mountain ranges, which suggests that some horizontal motion in the crust, as well as volcanism, has helped form the highlands.

FIGURE 6-8

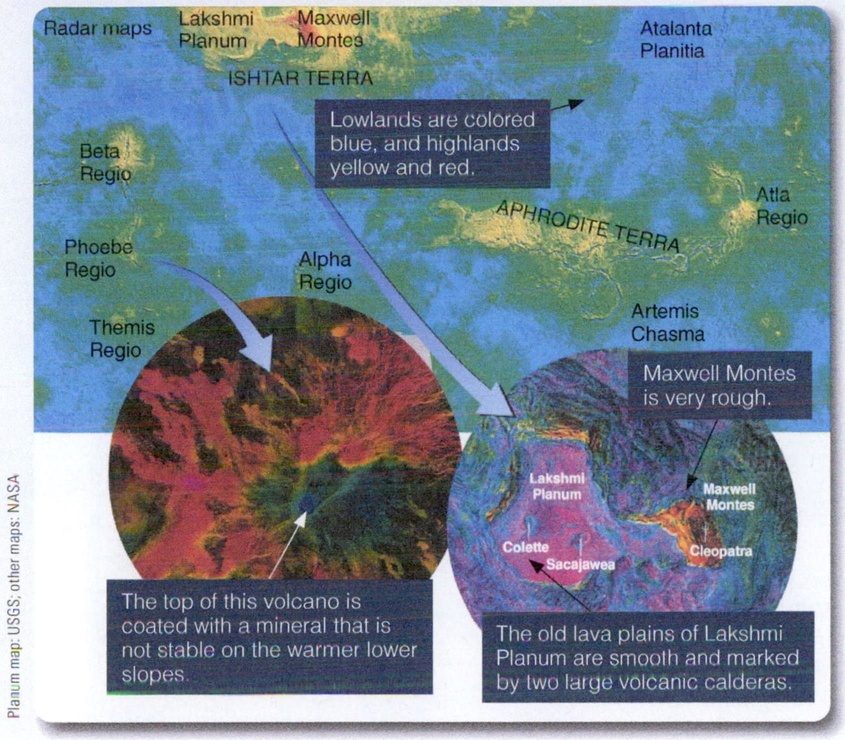

Three radar maps showing different aspects of Venus's surface. The main map at top shows elevation for most of the planet, omitting the polar areas. The inset map at left shows an electrical property of surface minerals related to chemical composition. The detailed map of Maxwell Montes and Lakshmi Planum inset at right is colored to show degree of terrain roughness, with purple representing smooth and orange rough.

Many features on Venus testify to its volcanic history. Long, narrow lava channels meander for thousands of kilometers. Radar maps reveal many volcanoes, faults, and sunken regions produced when magma below the surface drained away. Other volcanic features include the **coronae**, circular bulges up to 2600 km (1600 mi) in diameter bordered by fractures, volcanoes, and lava flows (Figure 6-9). These appear to be produced by rising convection currents of molten magma that push up under the crust. When the magma withdraws, the crust sinks back, and the circular fractures mark the edge of the original upwelling.

Radar images also show that Venus is marked by numerous craters. Meteorites big enough to make craters larger than 3 km in diameter have no trouble penetrating the thick atmosphere, but have formed only about 10 percent as many craters on Venus as on the maria of Earth's Moon. The number of craters shows that the crust of Venus is younger than the lunar maria but older than most of Earth's active surface. The average age of the surface of Venus is estimated to be roughly half a billion years. Clearly, geological processes cannot be renewing the surface as rapidly as they do on Earth.

Ten robot probes successfully landed on the surface of Venus in the 1970s and 1980s and managed to survive the heat and pressure for a few minutes or hours, while transmitting data to Earth. Some of those spacecraft analyzed the rock and snapped a few photographs (Figure 6-10). The surface rocks on Venus are dark gray basalts much like those in Earth's ocean floors. This evidence confirms that volcanism is important.

6-5c A History of Venus

To tell the story of Venus, you must draw together all the evidence and find hypotheses to explain two things; the thick carbon dioxide atmosphere, and the peculiar geology.

Calculations show that Venus and Earth have outgassed about the same amount of carbon dioxide, but Earth's oceans have dissolved most of that and converted it to sediments such as limestone. If all of the carbon in Earth's sediments and crust were dug up and converted back to carbon dioxide, our atmosphere would be about as dense as the air on Venus, and with similar composition. This suggests that the main difference between Earth and Venus is the lack of water on Venus to dissolve carbon dioxide.

Venus may have had oceans when it was young and much more Earth-like than at present, but, being closer to the Sun than Earth, it was warmer and the carbon dioxide in the atmosphere created a greenhouse effect that made the planet even

coronae On Venus, large round geological faults in the crust caused by the intrusion of magma below the crust.

FIGURE 6-9

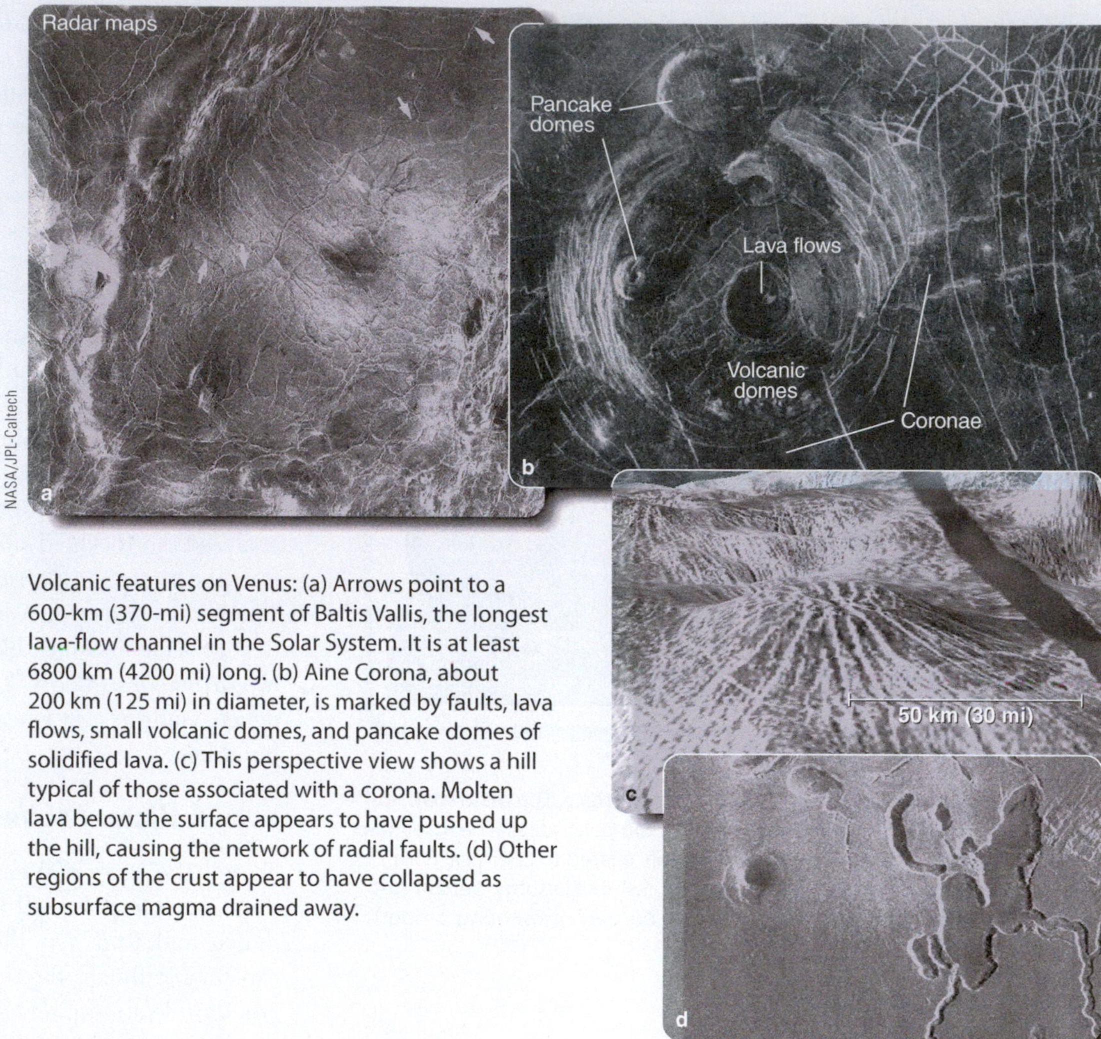

Volcanic features on Venus: (a) Arrows point to a 600-km (370-mi) segment of Baltis Vallis, the longest lava-flow channel in the Solar System. It is at least 6800 km (4200 mi) long. (b) Aine Corona, about 200 km (125 mi) in diameter, is marked by faults, lava flows, small volcanic domes, and pancake domes of solidified lava. (c) This perspective view shows a hill typical of those associated with a corona. Molten lava below the surface appears to have pushed up the hill, causing the network of radial faults. (d) Other regions of the crust appear to have collapsed as subsurface magma drained away.

warmer. That process could have dried up any oceans that did exist and prevented Venus from purging its atmosphere of carbon dioxide. In fact, evidence from the composition of Venus's atmosphere indicates that an ocean's worth of water might have been vaporized and lost. As carbon dioxide continued to be outgassed, the greenhouse effect grew even more severe. Thus, planetary scientists conclude that Venus was trapped in a **runaway greenhouse effect**.

runaway greenhouse effect A greenhouse effect so dramatic that it amplifies itself, becoming stronger with time.

The intense heat at the surface may have affected the geology of Venus by making the crust drier and more flexible so that it was unable to break into moving plates as on Earth. There is no sign of plate tectonics on Venus, but there is evidence that convection currents below the crust are deforming the crust to make coronae, push up mountains such as Maxwell, and create some folded mountains, like those around Ishtar Terra, by minor horizontal crust motions.

As you learned earlier, the small number of craters on the surface of Venus hints that the entire crust has been replaced within the past half-billion years or so. This may have occurred in a planetwide overturning as the old crust broke up and sank and lava flows created a new crust. Comparing Earth to Venus may eventually reveal more about how our own world's volcanism and tectonics work.

FIGURE 6-10

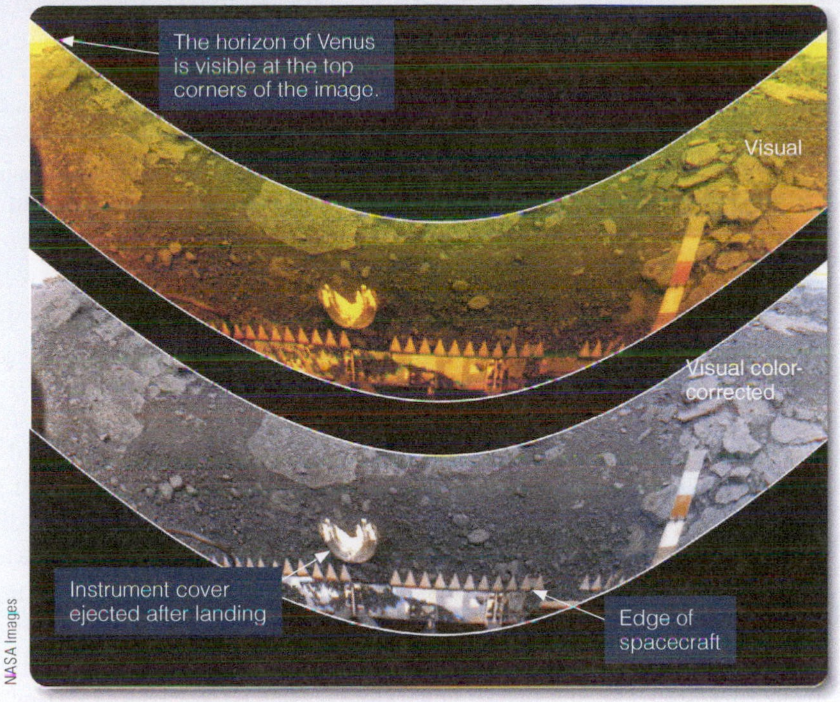

The *Venera 13* lander touched down on Venus in 1982 and carried a camera that swiveled from side to side to photograph the surface. The orange glow is produced by the thick atmosphere; when that is corrected to produce the view as it would be under white light, you can see that the rocks are dark gray. Isotopic analysis suggests they are basalts.

6-6 MARS

If you ever visit another world, Mars may be your best choice. You will need a heated, pressurized spacesuit, but Mars is not as inhospitable as the Moon (See **Celestial Profile 4** in Appendix C for Mars's data). It is also more interesting, with weather, complex geology, craters, volcanoes, and signs that water once flowed over its surface.

6-6a Mars's Atmosphere

The Martian air contains 95 percent carbon dioxide, 3 percent nitrogen, and 2 percent argon. That is similar to the composition of air on Venus, but the Martian atmosphere is very thin, less than 1 percent as dense as Earth's atmosphere and 1/10,000 as dense as Venus's.

There is very little water in the Martian atmosphere. Liquid water cannot survive on the surface of Mars because the air pressure is too low–any liquid water would immediately boil away. The polar caps appear to be composed of frozen water ice coated over by frozen carbon dioxide (dry ice). Whatever water is present on Mars is frozen either within the polar caps or as **permafrost** in the soil.

Although the present atmosphere of Mars is very thin, you will see evidence that the climate once permitted liquid water to flow over the surface, so Mars must have once had a thicker atmosphere. As a Terrestrial planet, it should have outgassed significant amounts of carbon dioxide, nitrogen, and water vapor; but because it was small it could not hold onto its atmosphere. The escape velocity on Mars is only 5 km/s (3 mi/s), less than half of Earth's, so it was easier for rapidly moving gas molecules to escape into space. Furthermore, if Mars had been colder, the gas molecules in its atmosphere would have been traveling more slowly and would not have escaped as easily.

Another process that has reduced Mars's atmosphere is the planet's lack of an ozone layer to screen out UV radiations–unbathing on Mars would be a fatal mistake. Solar UV photons can break atmospheric molecules up into smaller fragments. Water, for example, can be broken by UV into hydrogen and oxygen. The hydrogen then escapes into space, and the oxygen easily combines chemically with rocks and soil to form oxides. Iron oxide (rust) in the soil gives Mars its reddish landscape (Figure 6-11).

Mars may have had a substantial atmosphere when it was young, but it gradually lost much of that both by direct escape and by UV destruction. As a result, with a thin atmosphere that does not provide much greenhouse warming or permit liquid water to persist on the surface, Mars is now a cold, dry world.

permafrost Permanently frozen soil.

FIGURE 6-11

This image taken by the Mast Camera on the *Curiosity* rover highlights the interesting geology of Mount Sharp, a mountain inside Gale Crater where the rover landed. Prior to the rover's landing, observations from orbiting satellites indicated that the lower reaches of the mountain are rock layers containing water-bearing minerals.

6-6b Exploring Mars's Surface

Data recorded by orbiting satellites show that the Southern Hemisphere of Mars is a heavily cratered highland region up to 4 billion years old. The Northern Hemisphere is mostly a much younger lowland plain with few craters, visible in Figure 6-12.

Volcanism on Mars is dramatically evident in the Tharsis region, a highland region of volcanoes and lava flows bulging 10 km (6 mi) above the surrounding surface. A similar uplifted volcanic plain, the Elysium region, is more heavily cratered and eroded and appears to be older than the Tharsis bulge. The lack of impact craters on the slopes of some volcanoes in both Tharsis and Elysium suggests that there has been volcanic activity within the past few hundred million years, and maybe even more recently.

All of the volcanoes on Mars are **shield volcanoes**, very broad mountains with gentle slopes that are produced on Earth by hot spots penetrating upward through the crust. Shield volcanoes are not related to plate tectonics; in fact, the large shield volcanoes on Mars, the largest of which is Olympus Mons, provide evidence that plate tectonics has not been significant on that planet. Olympus Mons is 600 km (370 mi) in diameter at its base and rises 22 km (14 mi) high. In comparison, the largest volcano on Earth, also a shield volcano, is Mauna Loa in Hawai'i, rising only 10 km (6 mi) above its base on the seafloor (Figure 6-13a). On Earth, volcanoes like those that formed the Hawaiian Islands occur over rising currents of hot material in the mantle. Because the crust plate moves horizontally, a chain of volcanoes is formed instead of a single, large feature (Figure 6-13b). A lack of plate motion on Mars has allowed rising currents of magma to heat the crust repeatedly in the same places and build Olympus Mons, as well as other very large volcanic shields—especially in the Tharsis region (Figure 6-12).

When the crust of a planet is strained, it may break, producing faults and rift valleys. Near the Tharsis region is a great valley, Valles Marineris, which you can see in Figure 6-12, named after the *Mariner 9* spacecraft that first photographed it. The valley is a block of crust that has dropped downward along parallel faults. Erosion and landslides have further modified the valley into a great

shield volcanoe Wide, low-profile volcanic cone produced by highly liquid lava.

FIGURE 6-12

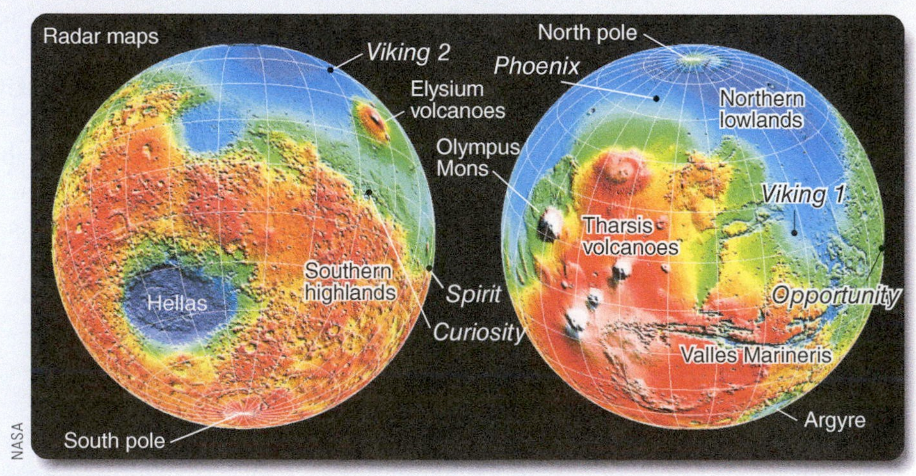

These hemisphere maps of Mars are color coded to show elevation. The northern lowlands lie about 4 km (2.5 mi) below the southern highlands. Volcanoes are very high (white), and the giant impact basins, Hellas and Argyre, are low. Note the depth of the canyon Valles Marineris. The two *Viking* spacecraft landed on Mars in 1976, *Pathfinder* in 1997, rovers *Spirit* plus *Opportunity* in 2004, the *Phoenix* lander in 2008, and *Curiosity* rover in 2012.

FIGURE 6-13

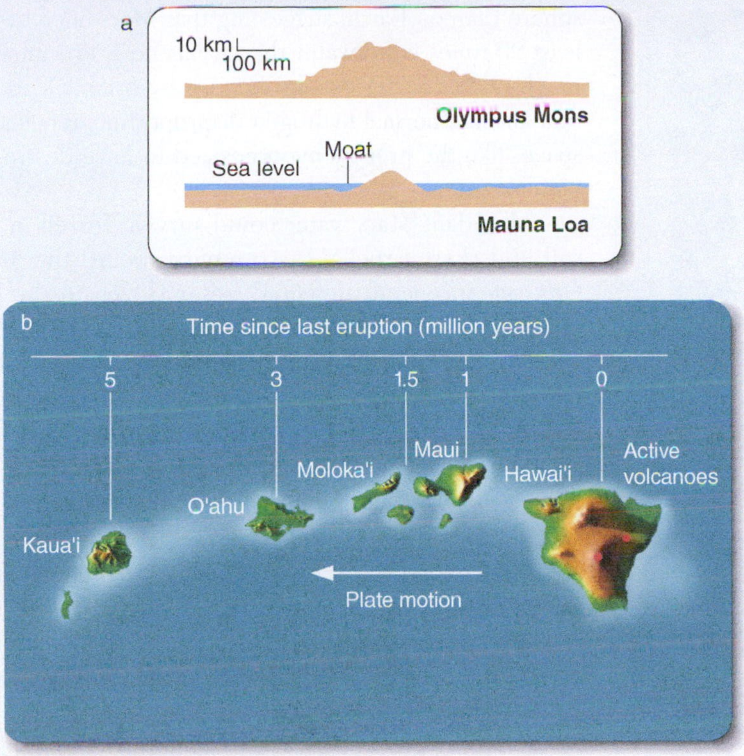

(a) Olympus Mons on Mars is much larger than Mauna Loa, the largest mountain on Earth. Mauna Loa is so heavy that it has sunk into Earth's crust, producing an undersea moat, but Olympus Mons has not, suggesting Mars's crust is much stronger than Earth's. (b) Volcanoes like Mauna Loa do not grow very large on Earth because the crust plates move horizontally and carry older volcanoes away from hot spots.

canyon. It is four times deeper, nearly ten times wider, and more than ten times longer than the Grand Canyon. The number of craters in the valley indicates that it is 1 to 2 billion years old, placing its origin sometime before the end of the most active volcanism in the Tharsis region.

6-6c Finding the Water on Mars

The quest for water on Mars is exciting because biologists conclude that life on Earth depends on water. You will learn in this section that Mars evidently had abundant liquid water on its surface for at least as long as it took life to arise on Earth. If conditions under which life started on Earth existed on Mars at the same time, then Mars provides a real opportunity for scientists to check their hypotheses about how life began on Earth.

Beginning in 1976 with the *Viking* program and continuing to the present day, spacecraft orbiting Mars such as *Mars Odyssey* (which reached Mars in 2001), *Mars Express* (2003), and *Mars Reconnaissance Orbiter* (2006 have returned photos showing that water once flowed over the surface. (All of those missions except *Viking* are still operating at the time of this writing, October 2016.) As you have learned, liquid water cannot exist now on the Martian surface because it would boil away in the extremely low atmospheric pressure, so evidence that water once flowed on Mars means that surface conditions there must have been quite different long ago.

Several types of formations hint at Martian surface water. **Outflow channels** appear to have been cut by massive floods carrying as much as 10,000 times the volume of water flowing down the Mississippi River. Such floods apparently swept away landscape features and eroded deep channels. The number of craters formed on top of the outflow channels show that they are billions of years old. **Valley networks** look like meandering river beds that probably formed over long periods. The valley networks are also located in the heavily cratered Southern Hemisphere, so they must be very old as well.

There are many other signs that there has been water on the Martian surface, some of which are geologically recent (Figure 6-14). Regions of collapsed terrain appear to be places where subsurface water has drained away. Gullies leading downhill appear to have been eroded recently, judging from their lack of craters, and a few gullies have been seen to appear between one Martian year and the next. In 2015, the *Mars Reconnaissance Orbiter* made spectroscopic observations confirming present-day seasonal groundwater flows.

Many flow features lead into the northern lowlands, and the smooth terrain there has been interpreted as an ancient ocean floor. Features along the edges of the lowlands are comparable to shorelines, and many planetary scientists conclude that the northern lowlands contained an ocean when Mars was younger. Did Mars once have enough water to fill

outflow channels Geological features on Mars and Earth caused by flows of vast amounts of water released suddenly.

Valley network A system of dry drainage channels on Mars that resembles the beds of rivers and tributary streams on Earth.

FIGURE 6-14

Panels a, b, d: NASA; Panel c: NASA/JPL-Caltech/MSSS

These visual-wavelength images made by *Viking* and *Mars Global Surveyor* orbiters show more features that suggest liquid water once flowed on the surface of Mars. Runoff channels are billions of years old, but some of the gullies have appeared since the first Earth-built spacecraft arrived.

a sea? Deuterium is three to five times more abundant relative to normal (light) hydrogen in the Martian atmosphere than on Earth, suggesting that Mars once had at least 20 times more water than it has now. Presumably, much of that water was broken up by solar UV radiation and the normal hydrogen disproportionately lost to space, like the process hypothesized to have destroyed Venus's water.

Abundant Mars water could survive frozen in the soil and crustal rocks. Instruments aboard the *Mars Odyssey* spacecraft detected water frozen in the soil over large areas of the planet. If you added a polar bear, changed the colors, and hid the craters, it would look like the broken pack ice on Earth's Arctic Ocean. *Mars Express* photographed these dust-covered formations near the Martian equator, and the shallow depth of the craters suggests that the ice is still there, just below the surface.

Much more of the ice on Mars may be hidden below the polar caps. Radar aboard the *Mars Express* orbiter was able to penetrate 3.7 km (2.3 mi) under the surface and map ice deposits in the crust of the planet's south polar region. There is enough water there, at least 90 percent pure, to cover the entire planet to a depth of 11 m (36 ft). The *Phoenix* probe landed in 2008 near Mars's north pole, finding water ice mixed with the soil not only as permafrost but also as small chunks of pure ice, indicating there once was standing water that froze in place.

Rovers *Spirit* and *Opportunity* landed in areas suspected of once having had water on their surfaces. (As of October 2016, *Opportunity* was still roving and communicating with Earth, more than 12 years after it landed.) Both rovers found evidence of past water, including small spherical concretions of the mineral hematite (dubbed "blueberries") that require abundant water to form (Figure 6-15a). In other rocks, *Opportunity* found layers of sediments with ripple marks and crossed layers showing they were deposited in moving water (Figure 6-15b). Chemical analysis of the rocks at both the *Opportunity* and *Spirit* sites detected sulfates much like Epsom salts plus bromides and chlorides. On Earth, these compounds are left behind when bodies of water such as desert lakes dry up. In its first year at work, the *Curiosity* rover discovered rounded rocks of the sort produced by running streams.

Mars has water, but it is hidden. When humans reach Mars, they will not need to dig far to find water in the form of ice. They can use solar power to break the water into hydrogen and oxygen. Hydrogen is fuel, and oxygen is the breath of life, so the water on Mars may prove to be buried treasure. Even more exciting is the knowledge

FIGURE 6-15

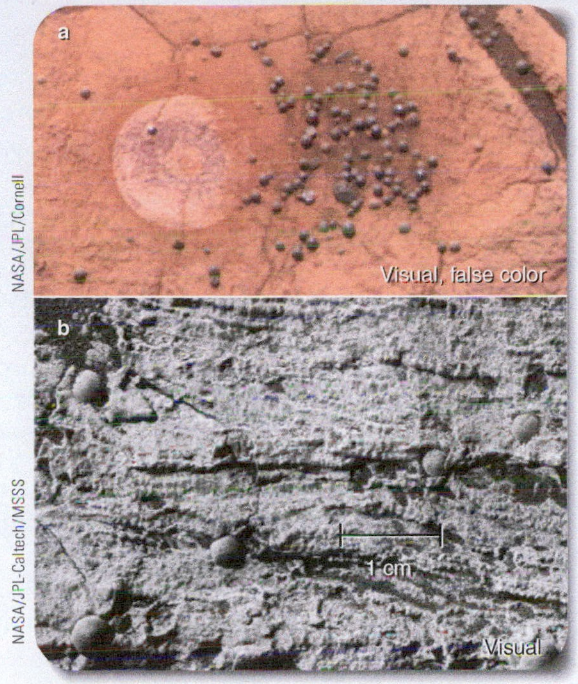

(a) Rover *Opportunity* photographed hematite concretions ("blueberries") weathered from rock. The round mark is a spot cleaned by the rover. The spheres appear to have grown as minerals collected around small crystals in the presence of water. Similar concretions are found on Earth. (b) The layers in this rock were deposited as sand and silt in rapidly flowing water. From the way the layers curve and cross each other, geologists can estimate that the water was at least 10 cm (4 in.) deep. A few "blueberries" are also visible in this image.

that Mars definitely once had bodies of liquid water on its surface. It is a desert world now, but someday an astronaut may scramble down an ancient Martian streambed, turn over a rock, and find a fossil. It is even possible that life started on Mars and has managed to persist as the planet gradually became less hospitable. For example, life may have hung on by retreating to limited warm and wet oases underground. You will learn in the final chapter about possible evidence of ancient life in one of the Martian meteorites.

6-6d A History of Mars

Did Mars ever have plate tectonics? Where did the water go? These fundamental questions challenge you to assemble the evidence and hypotheses for Mars and tell the story of its evolution. (See **How Do We Know? 6-2**.)

The history of Mars, like that of Venus, has been dominated by volcanism, but the differences are striking. Read **Concept Art 6B, When Good Planets Go Bad** and notice the following four important points:

1. The difference between Venus and Earth is not in the amount of CO_2 they have outgassed, but in the amount they have removed from their atmospheres. Being warmer and consequently having less liquid water on its surface sealed Venus's fate as a runaway greenhouse.
2. Venus is highly volcanic with a crust made up of lava flows that have covered over any older crust.
3. Mars is significantly smaller than Earth, with gravity too weak to prevent much of its atmosphere from leaking away. There is strong evidence of past water, but whatever is left is frozen in the soil or polar caps.
4. Mars is small and cooled quickly. Its crust has grown thick and volcanism has died to such a low level that it cannot replenish the gases lost to space.

The history of Mars is a case of arrested development. The planet began by differentiating into a crust, mantle, and core. Measurements from orbiting spacecraft reveal that it has a dense core but no planetwide magnetic field. The core must have cooled quickly and shut off the dynamo effect that would have produced a field.

The crust of Mars is now quite thick, as shown by the mass of the Olympus Mons volcano (Figure 6-13), but it was probably thinner in the past. Cratering may have broken or at least weakened the crust, triggering lava flows that flooded some basins. Mantle convection may have pushed up the Tharsis and Elysium volcanic regions and broken the crust to form Valles Marineris, but moving crustal plates never dominated Mars. There are no folded mountain ranges on Mars and no sign of plate boundaries. As the planet cooled, its crust grew thick and immobile.

At some point in the history of Mars, water was abundant enough to flow over the surface in great floods and might have filled oceans, but the age of plentiful liquid water evidently ended more than 3 billion years ago. The climate on Mars has changed as gases and water were lost to space and water was frozen into the soil as permafrost.

For Mars, the fourth stage of planetary development has been one of moderate activity and slow decline. Volcanoes still occasionally erupt, and gullies are eroded by melting permafrost, but this medium-sized planet has lost much of its internal heat. The atmosphere is thick and the surface is a forbiddingly dry, old desert.

Concept Art 6B

WHEN GOOD PLANETS GO BAD

1 VENUS: RUNAWAY GREENHOUSE

a **Venus and Mars haven't really gone wrong, but they have changed since they formed, and those changes can help you understand your own world.**

Venus and Earth have outgassed about the same amount of CO_2, but Earth's oceans have dissolved most of Earth's CO_2 and converted it to sediments such as limestone. If all of Earth's sedimentary carbon were dug up and converted back to CO_2, our atmosphere would be much like Venus's.

Because Venus was warmer when it formed, it had little if any liquid water to dissolve CO_2, and that produced a greenhouse effect that made the planet even warmer. The planet could not purge its atmosphere of CO_2, and as more was outgassed, Venus was trapped in a runaway greenhouse effect.

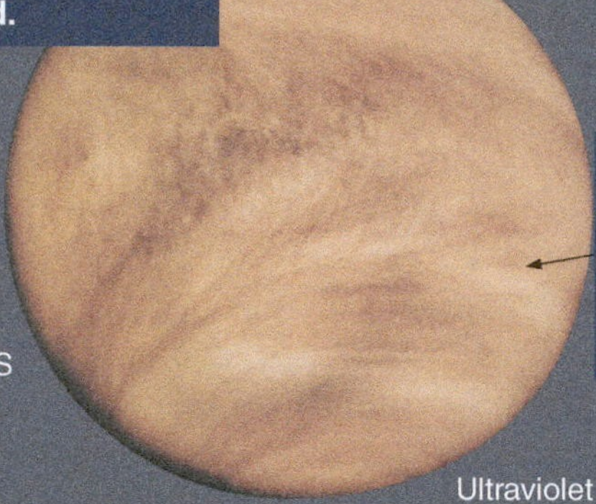

b **Only 0.7 AU from the Sun, Venus receives almost twice the solar energy per square meter that Earth does. Moved to Venus's orbit, Earth's surface would be 50°C (90°F) hotter.**

2

Lava covers Venus in layers of basalt, and volcanoes are common. Some of those volcanoes may be active right now whereas others are dormant. Traces of past lava flows show up in radar images including long narrow valleys cut by moving lava.

a **Baltis Vallis (arrows), at least 6800 km (4200 mi) long, is the longest lava flow channel in the Solar System.**

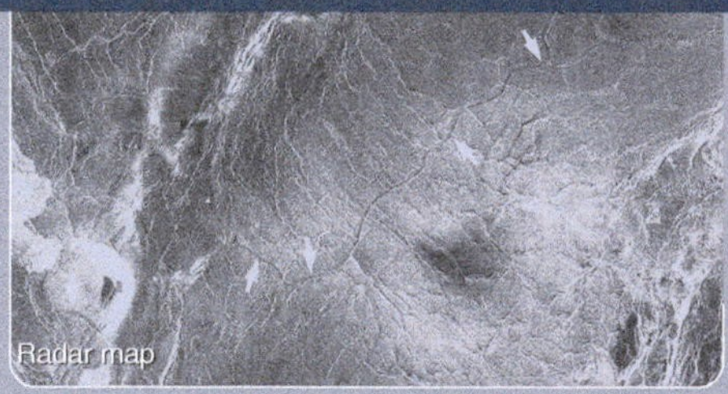

b **Even its thick atmosphere cannot protect Venus from larger meteorites, yet it has few craters. That must mean the surface is not old.**

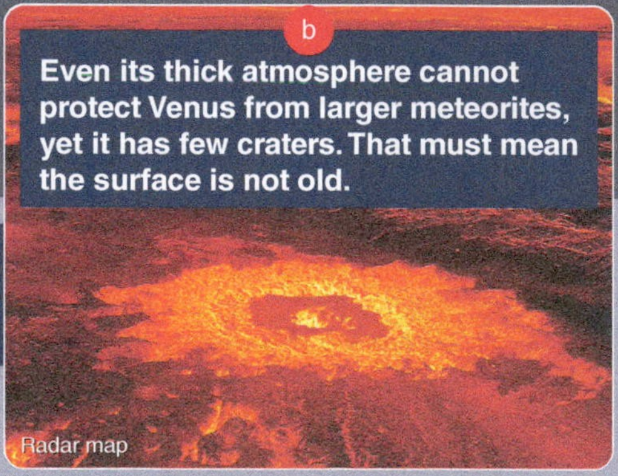

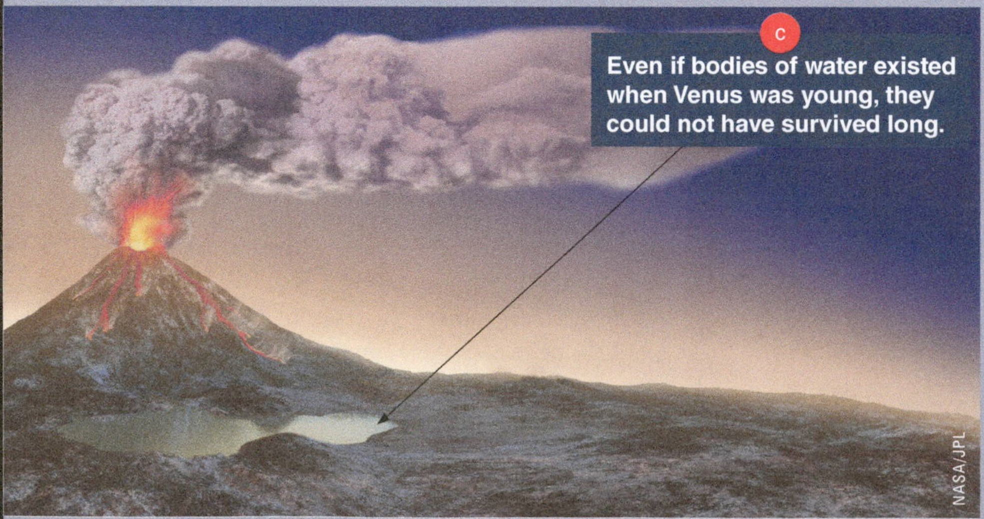

c **Even if bodies of water existed when Venus was young, they could not have survived long.**

2a

The crust of Venus must be no older than 0.5 billion years. One hypothesis is that an earlier crust broke up and sank in a sea of magma as fresh lava flows formed a new crust. Such resurfacing events might occur periodically on a hot, volcanic planet like Venus.

What could cause a resurfacing event? Models of the climate on Venus show that an outburst of volcanism could increase the greenhouse effect and drive the surface temperature up by as much as 100°C. This could soften the crust, increase the volcanism, and push the planet into a resurfacing episode. This type of catastrophe may happen periodically on Venus, or the planet may have had just one resurfacing event about half a billion years ago.

MARS: RUNAWAY REFRIGERATOR

3 When Mars was young, water was abundant enough to flow over the surface in streams and floods, and may have filled oceans. That age of liquid surface water ended more than 3 billion years ago. The climate on Mars has changed as atmospheric gases and water were lost to space and as water was frozen into the soil as permafrost.

a Evidence of surface water erosion on Mars

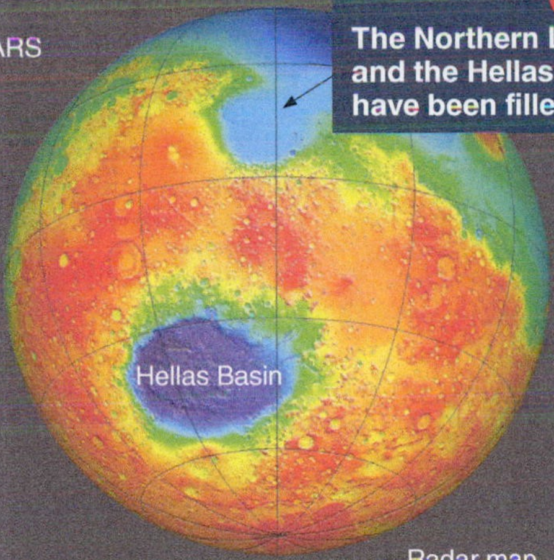

b The Northern Lowlands of Mars and the Hellas Basin may once have been filled with water.

Radar map

3a This distributary fan formed where an ancient stream flowed into a lake within a crater. Sediment deposited in the still lake water formed a lobed delta that later became sedimentary rock. Detailed analysis reveals that the stream changed course repeatedly, showing that the stream was not just a short-term flood.

4 Early in the history of Mars when its crust was thin, convection in the mantle could have pushed up volcanic regions such as Tharsis and Elysium, and limited plate motion could have produced Valles Marineris, but Mars cooled too fast, and its crust never broke into moving tectonic plates as did Earth's crust.

As its crust thickened, volcanism abated, and Mars lost the ability to replenish its waning atmosphere.

4a Mars has very large volcanoes, and their size shows that the crust has grown thick as the little planet has lost its heat to space. A thinner crust could not support such large volcanoes.

Olympus Mons volcano on Mars

a If the crust of Mars were made up of moving plates, its volcanoes could not have grown so big.

Computer-enhanced visual image

How Do We Know? 6-2

The Present Is the Key to the Past

How can we know what happened long ago if there weren't any witnesses? Geologists are fond of saying "The present is the key to the past." This implies that you can learn about the history of Earth by looking at the present condition of Earth's surface. The position and composition of various rock layers in the Grand Canyon, for example, tell you that the western United States was once at the floor of an ocean. This principle of geology is relevant today as you try to understand the history of other worlds such as Venus and Mars.

In the late 1700s, naturalists first recognized that the present gave them clues to the history of Earth. At the time that was astonishing because most people assumed Earth had no history. That is, they assumed either that Earth had been created in its present state as described in the Old Testament or that Earth was eternal. In either case, people commonly assumed that the hills and mountains they saw around them had always existed, more or less, as they were. By the 1700s, naturalists began to see evidence that the hills and mountains were not eternal but were the result of past processes and were slowly changing. That gave rise to the idea that Earth had a history.

NASA/JPL/Malin Space Science Systems

Layers of rock in the Martian crater Terby hint at a time when the crater was filled with a lake.

As those naturalists made the first attempts to thoughtfully and logically explain the nature of Earth by looking at the evidence, they were inventing modern geology as a way of understanding Earth. What Copernicus, Kepler, and Newton did for the heavens in the 1500s and 1600s, the first geologists did for Earth beginning in the 1700s. Of course, the invention of geology as the study of Earth led directly to the modern attempts to understand the geology of other worlds.

Geologists and astronomers share a common goal: They are attempting to reconstruct the past from the evidence. Whether you study Earth, Venus, or Mars, you are looking at the present evidence and trying to reconstruct the past history of the planet by drawing on observations and logic to test each step in the story. How did Venus get to be covered with lava, and how did Mars lose its atmosphere? The final goal of planetary astronomy is to draw together all of the available evidence (the present) to tell the story (the past) of how the planet got to be the way it is. Those first geologists in the 1700s would be fascinated by the stories planetary astronomers tell today.

6-6e Mars's Moons

Unlike Mercury or Venus, Mars has moons. Small and irregular in shape, Phobos is 28 × 23 × 20 km (17 × 14 × 12 mi) in diameter, and Deimos is 16 × 12 × 10 km (10 × 7.5 × 6 mi) in diameter. These moons are almost certainly captured asteroids (see Figure 6-16). These moons are so small they cannot pull themselves into spherical shape. Phobos and Deimos are not just small; they are tiny. An athletic astronaut who could jump 2 m (6 ft) high on Earth could jump almost 3 km (2 mi) on Phobos. However they formed, these Moons are so small that any interior heat would have leaked away very quickly, and there is no evidence of any internally driven geologic activity on either object.

Some futurists suggest that the first human missions to Mars may not land on the planet's surface but instead could build a colony on Phobos or Deimos. These plans involve speculation that these Moon and other asteroids may have water in deep interior rocks that colonists could use.

What Are We? Comfortable

Many planets in the Universe probably look like the Moon and Mercury—small, airless, and cratered. Some are made of stone; and some, because they formed farther from their parent star, are made mostly of ices. If you randomly visited a planet anywhere in the Universe, you would probably stand on a moonscape.

Earth is unusual but not rare. The Milky Way Galaxy contains more than 100 billion stars, and more than 100 billion galaxies are visible with existing telescopes. Most of those stars probably have planets, and although many planets may look like Earth's Moon and Mercury, there also must be plenty of Earth-like worlds.

As you look around your planet, you should feel comfortable living on such a beautiful planet, but it was not always such a nice place. The craters on the Moon and the Moon rocks returned by astronauts show that the Moon formed with a sea of magma. Mercury seems to have had a similar history, so Earth may have formed the same way. Its surface was once a seething ocean of liquid rock swathed in a hot, thick atmosphere, torn by eruptions of more rock, explosions of gas from the interior, and occasional impacts from space. The Moon and Mercury assure you that that is the way Terrestrial planets begin. Earth has evolved to become your home world, but Mother Earth has had a violent past.

FIGURE 6-16

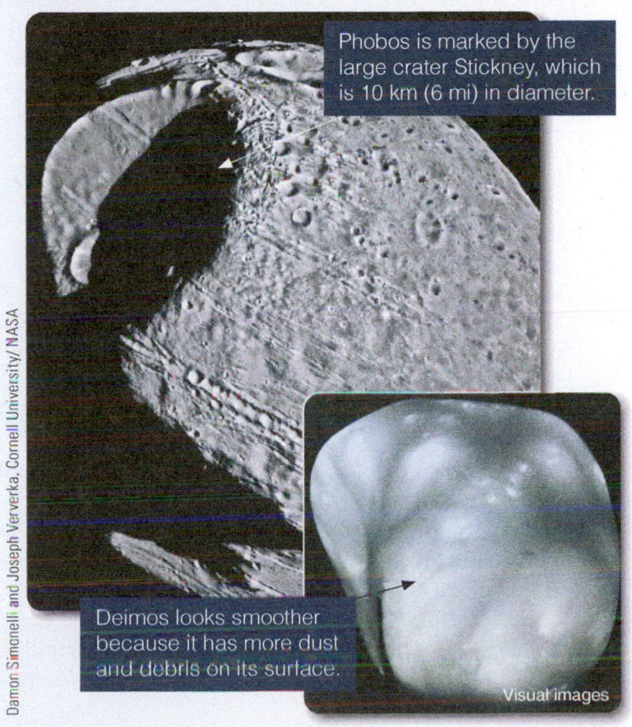

The moons of Mars are too small to pull themselves into spherical shapes. The two moons, shown here to scale, were named for the mythical attendants of the god of war, Mars. Phobos was the god of fear, and Deimos was the god of dread.

STUDY TOOLS 6

IN THE BOOK

- ☐ Tear Out the Review Card on *The Terrestrial Planets*.

ASTRO ONLINE

Visit ASTRO Online at www.cengagebrain.com

- ☐ Interactive Reading
- ☐ Practice Quizzing
- ☐ Videos and Animations
- ☐ Exclusive Features

6 Questions and Problems

REVIEW QUESTIONS

1. Why would you include the Moon in a comparison of the Terrestrial planets?
2. In what ways is Earth unique among the Terrestrial planets?
3. How do you know that Earth differentiated?
4. How are earthquakes in Hawai'i different from those in Southern California?
5. What characteristics must Earth's core have in order to generate a magnetic field?
6. How do island chains located in the centers of tectonic plates such as the Hawaiian Islands help you understand plate tectonics?
7. What has produced the oxygen in Earth's atmosphere?
8. How does the increasing abundance of CO_2 in Earth's atmosphere cause a rise in Earth's temperature?
9. Why would a decrease in the density of the ozone layer cause public health problems?
10. Why doesn't Earth have as many craters as the Moon?
11. Discuss the evidence and hypotheses concerning the origin of Earth's Moon.
12. What evidence indicates that plate tectonics does not occur on Venus? On Mars?
13. What evidence suggests that Venus has been resurfaced within the last half billion years?
14. Why are the atmospheres of Venus and Mars mostly carbon dioxide? Why is the atmosphere of Venus very dense but the atmosphere of Mars is very thin?
15. Why did Venus suffer from the runaway greenhouse effect but Earth has avoided it so far?
16. What evidence indicates that there has been liquid water on Mars?
17. Why do astronomers conclude that the crust on Mars must be thicker than Earth's crust?
18. **How Do We Know?** Why is heat flow the key to understanding a planet's surface activity?
19. **How Do We Know?** How is solving a crime that had no witnesses similar to figuring out what happened to Venus and Mars's oceans?

DISCUSSION QUESTIONS

1. Imagine that you visit a planet orbiting another star and discover oxygen in its atmosphere. What might you expect to find on its surface?
2. If liquid water is rare on the surfaces of planets, then would most Terrestrial planets have CO_2-rich atmospheres? Why or why not?
3. Old science-fiction paintings and drawings of colonies on the Moon often showed very steep, jagged mountains. Why did the artists assume that lunar mountains would be more jagged than mountains on Earth? Why are lunar mountains actually less jagged than mountains on Earth?
4. Imagine that you visit a planet orbiting another star and discovered that it is heavily cratered, but its small Moon is nearly crater-free. Why would that be a surprise? Speculate about what might have happened to those objects.

PROBLEMS

1. Assume *P* waves travel at 10 km/s and *S* waves travel at 5 km/s. If the *S* waves from an earthquake arrive at a seismographic station 10 minutes after the *P* waves, how far away was the earthquake from the station?
2. Earth's metallic core has about 55 percent of the planet's radius. What percentage of Earth's volume is taken up by the core?
3. If the Atlantic seafloor is spreading at 3.0 cm/year and is now 6400 km wide, how long ago were the continents in contact?
4. The Hawaiian-Emperor chain of undersea volcanoes is about 7500 km long, and the Pacific plate is moving 9.2 cm a year. How old is the oldest detectable volcano in the chain? What has happened to older volcanoes in the chain?
5. Calculate the age of the Grand Canyon as a percent of Earth's age.
6. Earth is four times larger in diameter than its Moon. How many times larger is it in surface area? In volume?
7. The smallest detail visible through Earth-based telescopes is about 1 arc second in diameter. What size object would this represent on Earth's Moon? (*Hint:* Use **The Small-Angle Formula** in the Math Reference Cards MR.)
8. Imagine that a spacecraft has landed on Mercury and is transmitting radio signals to Earth at a wavelength of 10.000 cm. When Mercury is seen from Earth in the evening sky, at its greatest angular distance east of the Sun, it is moving toward Earth at its maximum possible relative speed of 47.9 km/s. To what

wavelength must you tune your radio telescope to detect the signals? (*Hint:* Use **The Doppler Formula** in the Math Reference Cards MR. *Note:* The speed of light is 3.00×10^5 km/s.)

9. How long would it take radio signals to travel from Earth to Venus and back if Venus were at its nearest approach to Earth? Repeat the calculation for Venus at its farthest distance from Earth. (*Note:* The speed of light is 3.00×10^5 km/s. Useful data can be found in Appendix Table A-6.)
10. The smallest detail visible through Earth-based telescopes is about 1 second of arc in diameter. What size object would this represent on Mars at its closest approach to Earth? (*Hint:* Use the **The Small-Angle Formula** in the Math Reference Cards MR.) (*Note:* Useful data can be found in the Celestial Profiles in Appendix C.)
11. What is the maximum angular diameter of Phobos as seen from Earth? (*Hint:* Use **The Small-Angle Formula** in the Math Reference Cards MR.) (*Note*: Useful data can be found in the chapter text and in Appendix Table A-6.)
12. Phobos orbits Mars at a distance of 9376 km from the center of the planet and has a period of 0.3189 days. Assume Phobos's orbit is circular. Calculate the mass of Mars. (*Hint:* Use the formula for **Circular Velocity** in the Math Reference Cards MR. Make sure to convert relevant quantities to units of meters, kilograms, and seconds.)

LEARNING TO LOOK

1. Examine the globe of Earth shown in Concept Art 6A and look for volcanoes scattered over the Pacific Ocean. What is producing these volcanoes?
2. Olympus Mons on Mars is a very large volcano. In Figure 6-17 you can see multiple calderas superimposed at the top. What do those multiple calderas and the immense size of Olympus Mons indicate about the geology of Mars?

FIGURE 6-17

3. In what ways is Figure 6-18 a typical view of the surface of planet Earth? How is it unusual among planets in general?

FIGURE 6-18

William K. Hartmann/Planetary Science Institute

4. What do you see in Figure 6-19 that suggests heat is flowing out of Earth's interior?

FIGURE 6-19

USGS

5. In Figure 6-20, Astronaut Alan Bean works at the *Apollo 12* lander *Intrepid*. Describe the surface you see. What kind of terrain did they land on, for this, the second human landing on the Moon?

FIGURE 6-20

NASA

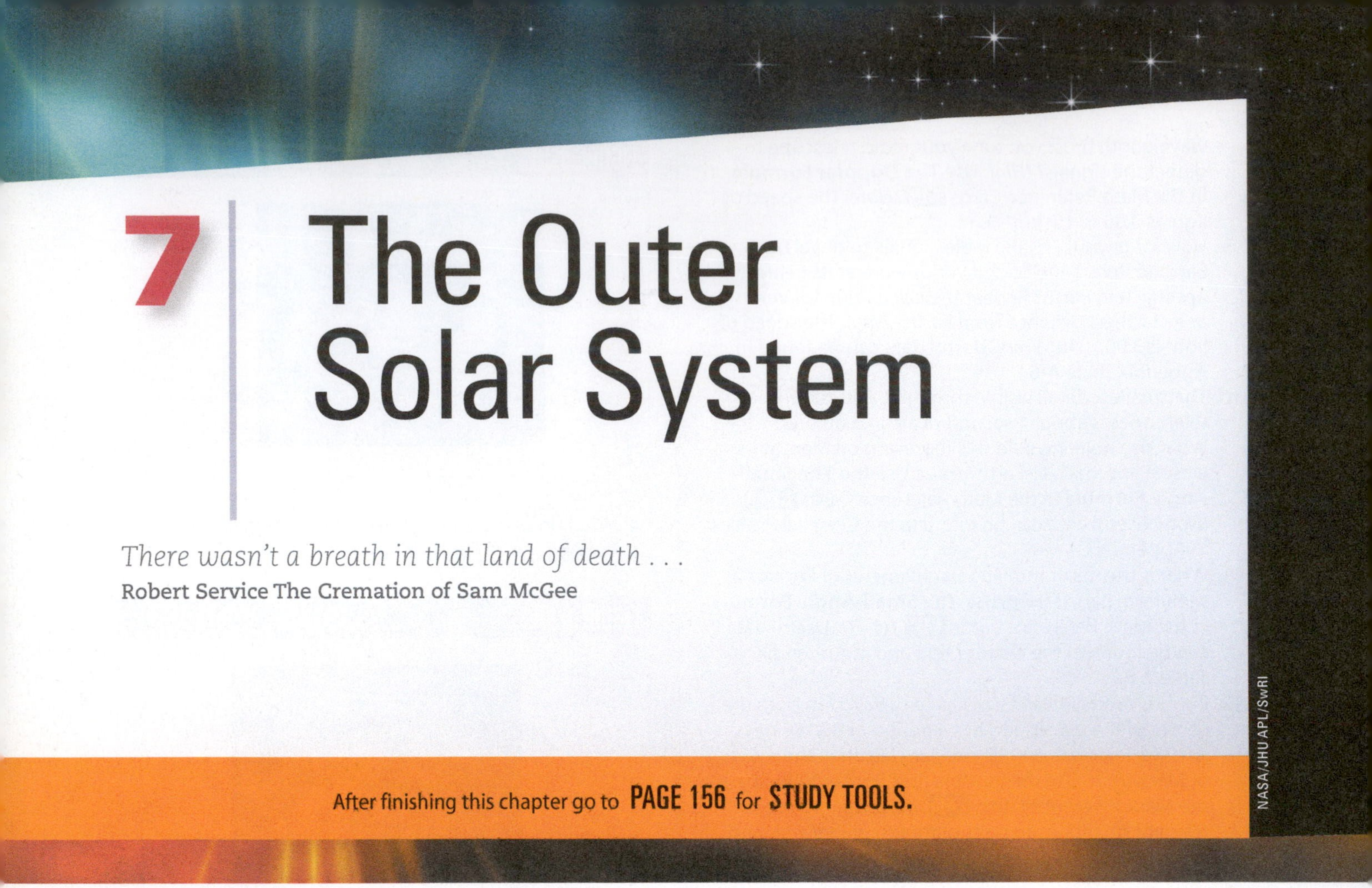

NASA/JHU APL/SwRI

7 The Outer Solar System

There wasn't a breath in that land of death . . .
Robert Service The Cremation of Sam McGee

After finishing this chapter go to **PAGE 156** for **STUDY TOOLS.**

GUIDEPOST

In the previous chapter, you studied how Earth and the other Terrestrial planets have evolved to their present forms. In this chapter, the Jovian (Jupiter-like) planets will be a challenge to your imagination. They are such alien worlds, they would be unbelievable if you didn't have direct observational evidence to tell you what they are like. In contrast, Pluto and the other bodies in the Kuiper Belt are not planets but small objects bearing intriguing clues about the planet construction process. (See CO 7.)

As you explore, you will find answers to four important questions:

- What are the properties of the Jovian planets?
- What is the evidence that some moons in the outer Solar System have been geologically active?
- How are planetary rings formed and maintained?
- What do Pluto and the other Kuiper Belt Objects tell us about the formation of the Solar System?

In this chapter and the previous chapter you can learn about the individual planets in detail, and some patterns will begin to emerge. In the next chapter you will discover that a comprehensive theory for the formation of the Solar System exists that manages to explain the overall characteristics of the planets as well as of the meteoroids, asteroids, and comets that also orbit the Sun. This theory predicts that planetary systems should be common around other stars, and you will learn that astronomers have found thousands of such systems.

The worlds of the outer Solar System can be studied from Earth, but much of what astronomers know has been radioed back to Earth from space probes. The *Pioneer* and *Voyager* probes flew past the outer planets in the 1970s and 1980s, the *Galileo* probe orbited Jupiter in the late 1990s, and the *Cassini/Huygens* orbiter and probe arrived at Saturn in 2004. The *New Horizons* probe flew past Pluto in 2015 and is now sailing deeper into the Kuiper Belt. The *Juno* orbiter arrived at Jupiter in 2016. Throughout this discussion, you will find images and data returned by these robotic explorers.

7-1 A TRAVEL GUIDE TO THE OUTER SOLAR SYSTEM

You are about to visit four worlds that are truly unearthly. This travel guide will warn you about what to expect.

CO 7 A true-color image of Pluto composed from data obtained when the *New Horizons* spacecraft was only a little farther from Pluto than the Moon is from Earth. Features as small as 2.2 km (1.4 mi) in size are resolved. A smooth plain with the shape of a Valentine heart is prominent at lower right.

7-1a Planets and Dwarf Planets

The outermost planets in our Solar System are Jupiter, Saturn, Uranus, and Neptune. They are often called the "Jovian planets," meaning they resemble Jupiter. In fact, they have their own separate personalities. Figure 7-1 compares these four worlds. Jupiter is the largest of the Jovian planets, more than 11 times the diameter of Earth. Saturn is slightly smaller, and Uranus and Neptune are quite a bit smaller than Jupiter. Pluto, not included in the illustration, is smaller than Earth's Moon but was considered a planet from the time of its discovery in 1930 until a decision by the International Astronomical Union (IAU) in 2006 that reclassified Pluto as a dwarf planet. You will learn about Pluto's characteristics, and reasons for the IAU decision, later in this chapter.

The other feature you will notice immediately is Saturn's rings. They are bright and beautiful and composed of billions of ice particles. Jupiter, Uranus, and Neptune also have rings, but they are not easily detected from Earth and are not visible in this figure. Nevertheless, as you visit these worlds you will be able to compare four different sets of planetary rings.

7-1b Atmospheres and Interiors

The four Jovian worlds have hydrogen-rich atmospheres filled with clouds. The atmospheres of the Jovian planets are not very deep; for example, Jupiter's atmosphere makes up only about 1 percent of its radius. On Jupiter and Saturn, you can see that the clouds form stripes that circle each planet. You will find traces of these same types of features on Uranus and Neptune, but they are not very distinct.

Models based on observations indicate that below their atmospheres Jupiter and Saturn are mostly liquid, so the old-fashioned term for these planets, the *gas giants*, should probably be changed to the *liquid giants*. Uranus and Neptune are sometimes called the *ice giants* because they are rich in water in both solid and liquid forms. Near their centers the Jovian planets probably have cores of dense material with the composition of rock and metal. None of the Jovian worlds has a definite solid surface on which you could walk.

You will read in Chapter 8 that the Jovian planets have low density because they are formed in the outer solar nebula where water vapor could freeze to form ice particles. The ice accumulated into protoplanets with density lower than the rocky Terrestrial planets and asteroids.

FIGURE 7-1

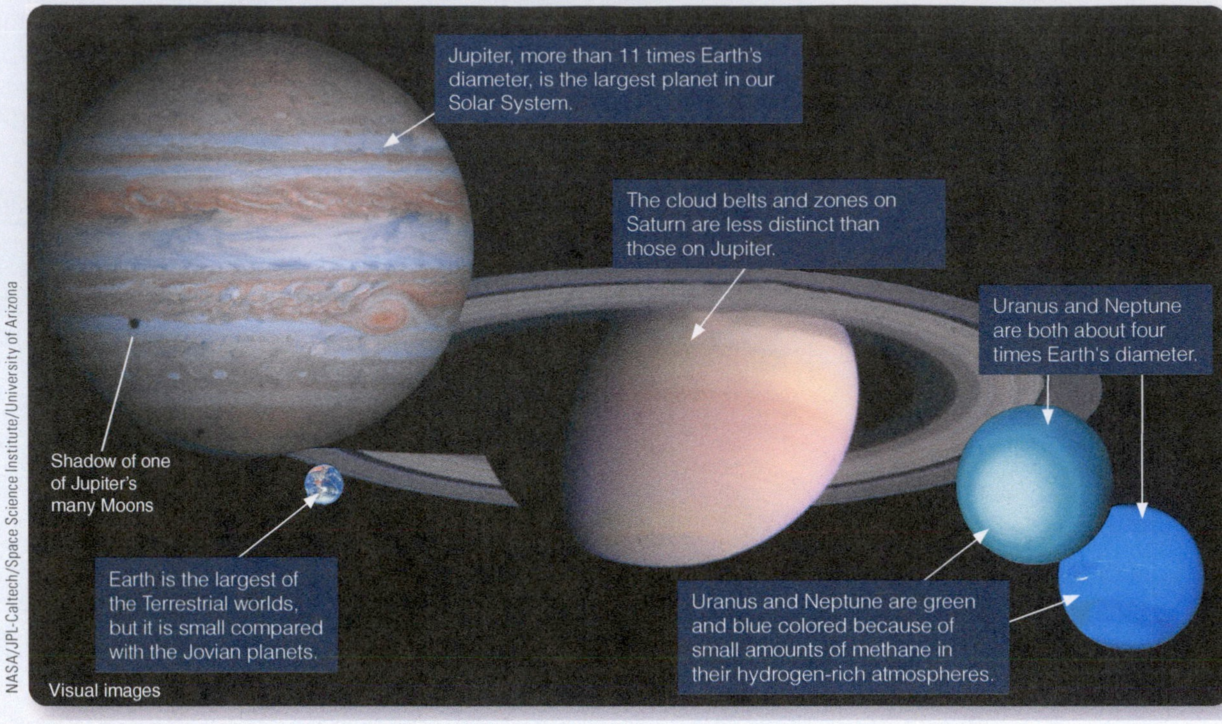

The principal worlds of the outer Solar System are the four massive but low-density Jovian planets, each much larger than Earth.

Once the Jovian planets grew massive enough, they could draw in even lower-density hydrogen and helium gas directly from the nebula by gravitational collapse.

7-1c Satellite Systems

You can't really land your spaceship on the Jovian planets, but you might be able to land on their moons. All of the outer Solar System planets have extensive moon systems. In many cases the moons interact gravitationally, mutually adjusting their orbits and also affecting the planetary ring systems. Some of the moons are geologically active now, while others show signs of past activity. Of course, geological activity requires heat flow from the interior, so you might wonder what could be heating the insides of these small objects.

7-2 JUPITER

Jupiter is named for the Roman king of the gods. It can be very bright in the night sky, and its cloud belts and four largest moons can be seen through even a small telescope. Jupiter is the largest and most massive of the Jovian planets, containing 71 percent of all the planetary matter in the entire Solar System. Just as you used Earth, the largest of the Terrestrial planets, as the basis for comparison with the others, you can examine Jupiter in detail as a standard in your comparative study of the other Jovian planets.

7-2a Jupiter's Interior

Jupiter is only 1.3 times denser than water. For comparison, Earth is more than 5.5 times denser than water. This gives astronomers a clue about the average composition of the planet's interior. Jupiter's shape also gives information about its interior. Jupiter and the other Jovian planets are all slightly flattened. A world with a large rocky core and mantle would not be flattened much by rotation, but an all-liquid planet would flatten significantly. Thus Jupiter's **oblateness**, the fraction by which its equatorial diameter exceeds its polar diameter, combined with its average density, helps astronomers

oblateness The flattening of a spherical body, usually caused by rotation.

model the interior. You can refer to **Celestial Profile 5** in Appendix C for more data about Jupiter.

Models indicate that the interior of Jupiter is mostly liquid hydrogen. However, if you jumped into Jupiter carrying a rubber raft expecting an ocean, you would be disappointed. The base of the atmosphere is so hot and the pressure is so high that there is no sudden boundary between liquid and gas. As you fall deeper and deeper through the atmosphere, you would find the gas density increased around you until you were sinking through a liquid, but you would never splash into a distinct liquid surface.

Under very high pressure, liquid hydrogen becomes **liquid metallic hydrogen**—a material that is a very good conductor of electricity. Most of Jupiter's interior is composed of this material. That large mass of conducting liquid, stirred by convection currents and spun by the planet's rapid rotation, drives the dynamo effect and generates a powerful magnetic field. Jupiter's field is more than 10 times stronger than Earth's. A planet's magnetic field deflects the solar wind and dominates a volume of space around the planet called the **magnetosphere**.

The strong magnetic field around Jupiter traps charged particles from the solar wind in radiation belts a billion times more intense than the Van Allen belts that surround Earth. The spacecraft that have flown through these regions received more than 4000 times the radiation than would have been lethal for a human.

At Jupiter's center, a so-called rocky core is hypothesized to contain heavier elements, such as iron, nickel, silicon, and so on. With a temperature four times hotter than the surface of the Sun and a pressure 50 million times Earth's air pressure at sea level, this material is unlike any rock on Earth. The term *rocky core* refers to the chemical composition, not to the properties of the material.

Careful infrared measurements of the heat flowing out of Jupiter reveal that the planet emits about twice as much energy as it absorbs from the Sun. This energy is understood to be heat left over from the formation of the planet. In Chapter 8, you will see that Jupiter should have grown very hot when it formed, and some of that heat remains in its interior. In 2016, the *Juno* spacecraft began orbiting Jupiter on a mission to make detailed measurements of Jupiter's gravitational field, magnetic field, and heat flow, aiming to better determine the planet's internal structure.

7-2b Jupiter's Complex Atmosphere

Study **Concept Art 7A, Jupiter's Atmosphere, Magnetosphere, and Rings** and, notice the following four important ideas:

1. The atmosphere is hydrogen rich, and the clouds are confined to a shallow layer.
2. The cloud layers are located at certain temperatures within the atmosphere where ammonia (NH_3), ammonium hydrosulfide (NH_4SH), and water (H_2O), respectively, can condense.
3. The belt–zone circulation pattern of colored cloud bands circling the planet like stripes on a child's ball is related to the high- and low-pressure areas found in Earth's atmosphere.
4. The large circular or oval spots seen in Jupiter's clouds are circulating storms that can remain stable for decades or even centuries.

7-2c Jupiter's Rings

Astronomers have known for centuries that Saturn has rings, but Jupiter's rings were not discovered until 1979, when the *Voyager 1* spacecraft sent back photos (see the second page of Concept Art 7A). Less than 1 percent as bright as Saturn's icy rings, Jupiter's rings are very dark and reddish, indicating that the rings are rocky material rather than icy.

Astronomers conclude that the ring particles are mostly microscopic. Photos of the rings show that they are very bright when illuminated from behind—in other words, they are scattering light forward. Large particles do not scatter light forward—a ring filled with basketball-size particles would look dark when illuminated from behind. **Forward scattering** of visible light tells you that the rings are mostly made of tiny grains with diameters approximately equal to the wavelengths of visible light, about the size of particles in cigarette smoke.

The rings orbit inside the **Roche limit**, the distance from a planet within which a moon cannot hold itself together by its own gravity. If a moon comes inside the Roche limit, tidal forces overcome the moon's gravity and pull the moon apart. Also, raw material for a moon cannot coalesce inside the Roche limit. The Roche limit is about 2.4 times

liquid metallic hydrogen A form of liquid hydrogen that is a good electrical conductor, inferred to exist in the interiors of Jupiter and Saturn.

magnetosphere The volume of space around a planet within which the motion of charged particles is dominated by the planetary magnetic field rather than the solar wind.

forward scattering The optical property of finely divided particles to preferentially direct light in the original direction of the light's travel.

Roche limit The minimum distance between a planet and a satellite that can hold itself together by its own gravity.

Concept Art 7A

JUPITER'S ATMOSPHERE, MAGNETOSPHERE, AND RINGS

NASA/JPL-Caltech

1 The visible clouds are not at the top of the atmosphere; an upper atmosphere of clear hydrogen and helium extends far above the cloud tops. Jupiter's atmosphere makes up only one percent of its radius. Major spots on Jupiter, like the Great Red Spot, are embedded circulating storms. They can remain stable for decades or even centuries.

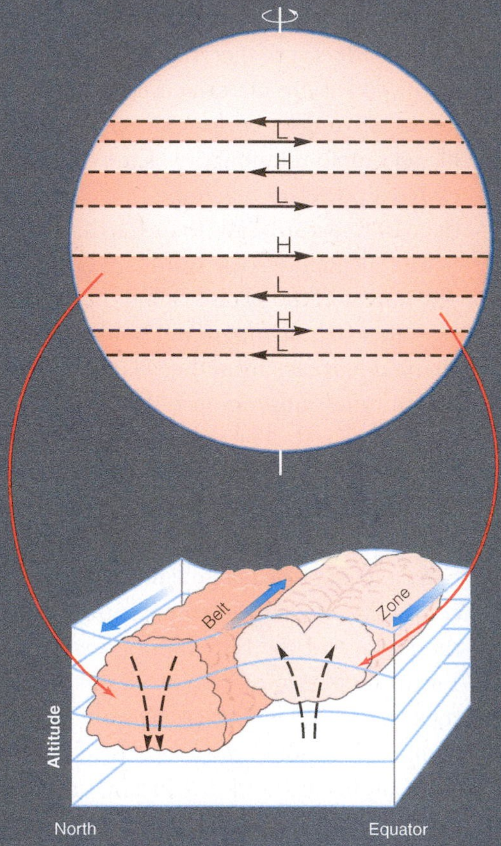

1a Cloud belts and zones circle Jupiter like stripes on a ball. This form of atmospheric circulation is belt–zone circulation. The poles and equator on Jupiter are about the same temperature, perhaps because of heat rising from the interior. Because of that, and also because of Jupiter's rapid rotation, instead of wave-shaped winds as on Earth, Jupiter's high- and low-pressure areas are stretched into belts and zones parallel to the equator. Zones are brighter than belts because rising gas forms clouds high in the atmosphere where sunlight is stronger and reflected back to outside viewers.

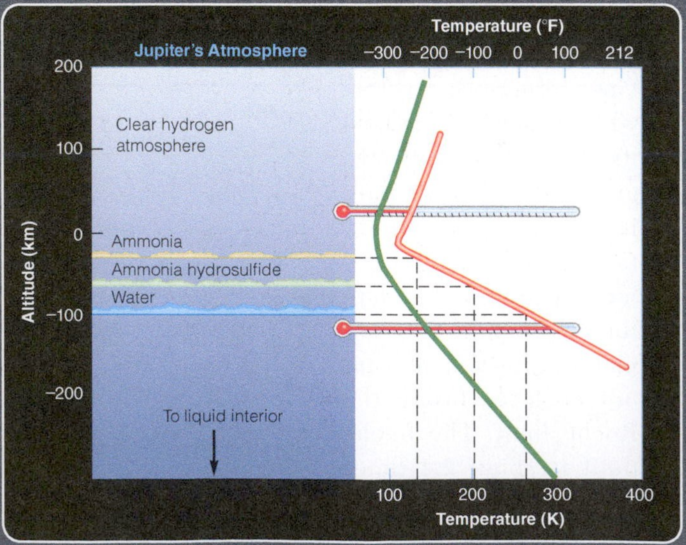

2 Jupiter's cloud layers lie at certain temperatures within the atmosphere where ammonia (NH_3), ammonium hydrosulfide (NH_4SH), and water (H_2O) can condense into droplets. If you could put thermometers into the different levels of the atmosphere, you would discover that the temperatures rise as you descend beneath the uppermost clouds, represented for Jupiter by the solid orange line. Because Saturn is farther from the Sun, its temperature is colder, shown by the green line. The cloud layers on Saturn form at the same temperature as the cloud layers on Jupiter, but they are deeper in Saturn's hazy atmosphere.

3

A planet's magnetic field deflects the solar wind and dominates a volume of space around the planet called the magnetosphere. Jupiter's magnetosphere is 100 times larger than Earth's. If you could see it in the sky, it would be 6 times larger than the full moon.

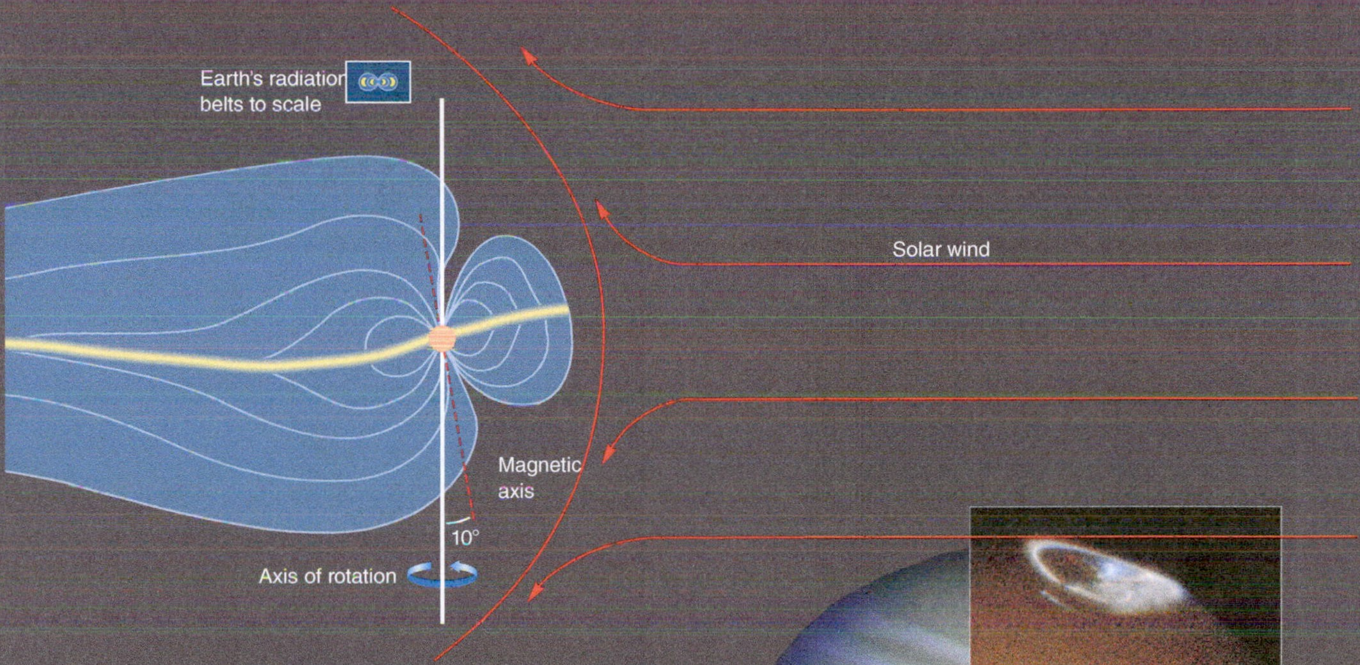

3a

As you saw with Earth in Chapter 5, interactions between Jupiter's magnetic field and the solar wind generate powerful electric currents that flow around the planet's magnetic poles. These aurorae on Jupiter are larger in diameter than Earth, and are confined to the north and south magnetic poles, as shown in these ultraviolet images.

John Clarke, University of Michigan, and NASA

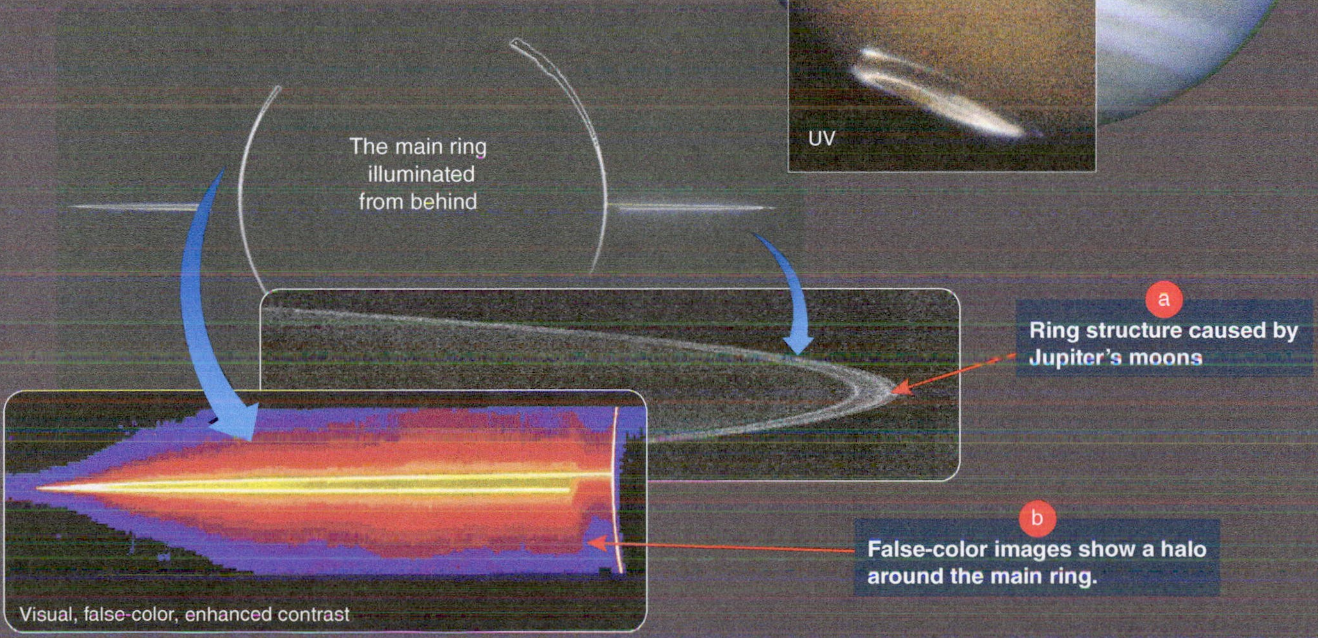

Panel a and c: NASA/JPL-Caltech/Univ. of Arizona;
Panel b: NASA/JPL-Caltech/Cornell Univ.

the planet's radius, depending somewhat on the relative densities of the planet and the moon material. Jupiter's rings, as well as those of Saturn, Uranus, and Neptune, lie inside the respective Roche limits for each planet.

Now you can understand Jupiter's dusty rings. If a dust speck gets knocked loose from a larger rock inside the Roche limit, the rock's gravity cannot hold the dust speck. For that same reason, the billions of dust specks in the rings can't pull themselves together to make a moon because of tidal forces inside the Roche limit.

You can be sure that Jupiter's ring particles are not old. The pressure of sunlight and the planet's powerful magnetic field alter the orbits of the particles. Images show faint ring material extending down toward the cloud tops; this, evidently, is dust grains spiraling into the planet. Dust is also destroyed by the intense radiation around Jupiter that grinds the dust specks down to nothing in a century or so. The rings you see today, therefore, can't be material left over from the formation of Jupiter—the rings of Jupiter must be continuously resupplied with new dust. Observations made by the *Galileo* spacecraft provide evidence that the source of ring material comes from micrometeorites eroding small moons orbiting near or within the rings.

The rings around Saturn, Uranus, and Neptune are also known to be short lived, and they also must be resupplied by new material, probably eroded from nearby moons. Besides supplying the Jovian rings with particles, moons confine the rings, keep them from spreading outward, and alter their shapes. You will explore these processes in detail when you study the rings of the other planets later in this chapter.

7-2d Jupiter's Family of Moons

Jupiter has four large moons and at least 60 smaller moons. Larger telescopes and modern techniques are rapidly finding more small moons orbiting each of the Jovian planets. Most of the small moons are probably captured asteroids. In contrast, the four largest moons (Figure 7-2), called the Galilean moons after their discoverer, Galileo, are clearly related to each other and probably formed with Jupiter.

The outermost Galilean moons, Ganymede and Callisto, are about the size of Mercury, 1.5 times the size of Earth's Moon. In fact, Ganymede is the largest Moon in the Solar System. Ganymede and Callisto have low densities of only 1.9 and 1.8 g/cm^3, respectively, meaning they must consist roughly of half rock and half ice. Observations of their gravitational fields by the *Galileo* spacecraft reveal that both moons have rocky, metallic cores and lower-density icy exteriors, meaning both moons are differentiated (see Chapter 6 Section 6-1d). Both moons interact with Jupiter's magnetic field in a way that shows they probably have mineral-rich layers of liquid water 100 kilometers or more below their icy crusts.

Callisto's surface, and most of Ganymede's surface, appear old because they are heavily cratered and very dark. The continuous blast of micrometeorites evaporates surface ice, leaving behind embedded minerals to form a dark skin like the grimy crust on an old snowbank, so surfaces get darker with age. More recent impacts dig up cleaner ice and leave bright craters, as you can see on Callisto in Figure 7-2. Ganymede has some younger, brighter grooved terrain believed to be systems of faults in the brittle crust. Some sets of grooves overlap other sets of grooves, suggesting extended episodes of geological activity.

FIGURE 7-2

Images of the Galilean moons of Jupiter, from left to right: Io, Europa, Ganymede, and Callisto. Note that Io and Europa are almost exactly the same size as Earth's Moon.

The density of the next moon inward, Europa, is 3 g/cm^3, high enough to mean that the moon is mostly rock with a thin icy crust. The visible surface is very clean ice, contains very few craters, has long cracks in the icy crust, and has complicated terrain that resembles blocks of ice in Earth's Arctic Ocean (Figure 7-3). The lack of craters tells you that Europa is an active world where craters are quickly erased. The pattern of folds on its surface suggests that the icy crust breaks as the moon is flexed by tides (see Section 3-5c of Chapter 3). Europa's gravitational influence on the *Galileo* spacecraft indicates that a liquid-water ocean, perhaps 200 km deep, lies below the 10- to 100-km-thick crust.

Images from spacecraft reveal that Io, the innermost of the four Galilean moons, has more than 150 volcanic vents on its surface (Figure 7-4). The active volcanoes throw sulfur-rich gas and ash high above the surface; the

FIGURE 7-3

Like icebergs on an arctic ocean, blocks of crust on Europa appear to have floated apart and rotated. The blue icy surface is stained brown by mineral-rich water venting from below the crust. White areas are ejecta from the impact that formed Pwyll crater.

FIGURE 7-4

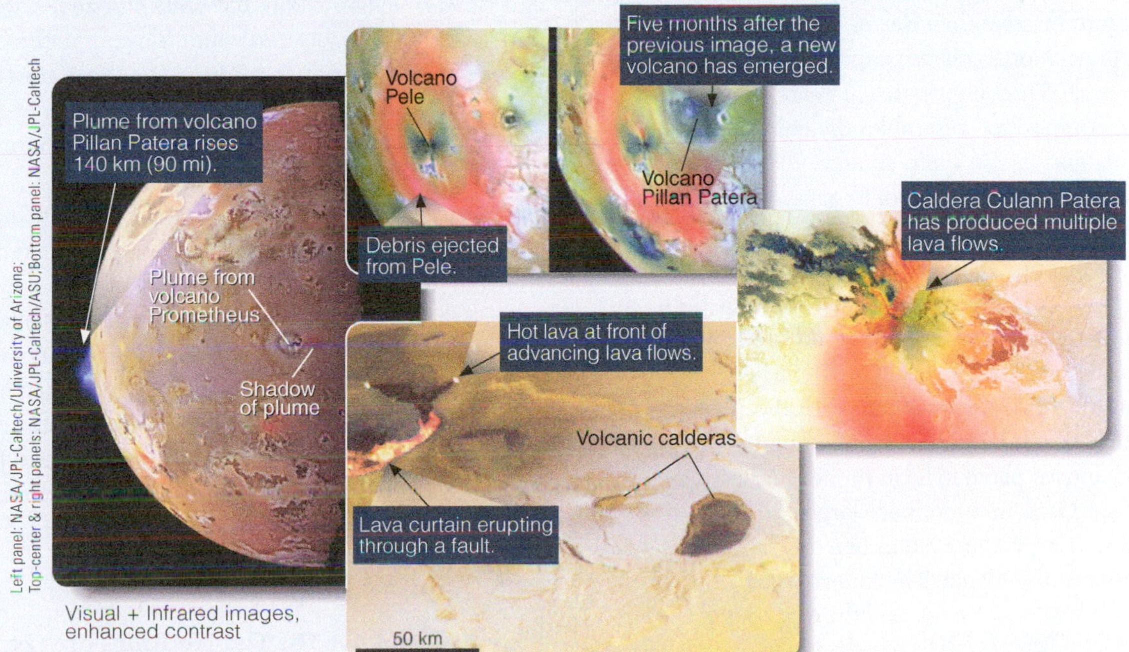

These representational-color images of volcanic features on Io were produced by combining visual and near-infrared images and digitally enhancing the colors. To human eyes, most of Io would look pale yellow and light orange.

ash falls back to bury the surface at a rate of a few millimeters a year. This explains why you see no impact craters on Io— they are covered up as fast as they form. Io's density is 3.6 g/cm^3, showing that it is not ice but rather rock and metal. Its gravitational influence on the passing *Galileo* spacecraft revealed that it is differentiated into a large metallic core, a rocky mantle, and a low-density crust.

The activity you see in the Galilean moons must be driven by energy flowing outward, yet these objects are too small to have remained hot from the time of their formation. Io's volcanism seems to be driven by **tidal heating**. Io follows a slightly eccentric orbit caused by its interactions with the other moons. Jupiter's gravitational field flexes Io with tides, and the resulting friction heats its interior. That heat flowing outward causes the volcanism. Europa is not as active as Io, but it too must have a heat source, presumably tidal heating. Ganymede is no longer active, but when it was younger it must have had internal heat to break the crust and produce the grooved terrain.

7-2e A History of Jupiter

Can you put all of the evidence together and tell the story of Jupiter? Creating such a logical argument from evidence and hypotheses is the ultimate goal of planetary astronomy.

Jupiter formed far enough from the Sun to incorporate large amounts of ices containing hydrogen, so it must have grown rapidly (see Chapter 8). Once it became about 10 to 15 times more massive than Earth, it could begin to grow further by gravitational collapse, capturing gas directly from the solar nebula. Thus, it grew rich in hydrogen and helium. Its present composition resembles the composition of the solar nebula and is also quite sunlike. Jupiter's gravity is strong enough to hold on to all of its gases, even hydrogen.

Most of the large family of moons may be captured asteroids, and Jupiter may still encounter a wandering asteroid or comet now and then. Some of these are deflected, some captured into orbit, and some, like comet Shoemaker-Levy 9 in 1994 and unidentified fragments in 2009 and 2012, actually collide with the planet. Dust blasted off of the inner moons by micrometeorites settles into the equatorial plane to form Jupiter's rings.

The four Galilean moons are large and seem to have formed like a mini Solar System in a disk of gas and dust around the forming planet. The innermost, Io, is densest, and the densities of the others decrease as you move away from Jupiter, similar to the way the densities of the planets decrease with distance from the Sun. Perhaps the inner moons incorporated less ice because they formed closer to the heat of the growing planet. You can recognize that tidal heating also has been important, and the intense warming of the inner moons could have driven off much of their ices. Thus, two processes together are probably responsible for the differences in compositions of the Galilean moons.

tidal heating The heating of a planet or satellite because of friction caused by flexing due to the gravitational influence of a nearby body.

7-3 SATURN

The Roman god Saturn, protector of the sowing of seed, was celebrated in a weeklong Saturnalia festival at the time of the winter solstice in late December. Early Christians took over the holiday to celebrate Christmas.

Saturn is most famous for its beautiful rings, which are easily visible through the telescopes of modern amateur astronomers. Large, Earth-based telescopes have explored the planet's atmosphere, rings, and moons. The two *Voyager* spacecrafts flew past Saturn in 1979, and the *Cassini* spacecraft went into orbit around Saturn in 2004 on an extended exploration of the planet, its rings, and its moons. (See **Celestial Profile 6** in Appendix C for specific data on Saturn.)

7-3a Planet Saturn

As you can see in Figure 7-1, Saturn has only faint belt–zone circulation, but *Voyager, Hubble Space Telescope*, and *Cassini* images show that belts and zones are present, and that the associated winds blow up to three times faster than on Jupiter (see **How Do We Know? 7-1**). Belts and zones on Saturn are less visible than on Jupiter because they occur deeper in the cold atmosphere, below a layer of methane haze.

Saturn is less dense than water (it would float!), suggesting that it is, like Jupiter, rich in hydrogen and helium. As photographs show, Saturn is the most oblate of the planets, and that evidence tells you that its interior is mostly liquid with possibly a small core of heavy elements. Because its internal pressure is lower, Saturn has less liquid metallic hydrogen than Jupiter. Perhaps that is why Saturn's magnetic field is 20 times weaker than Jupiter's. Like Jupiter, Saturn radiates more energy than it receives from the Sun, and models predict that it has a very hot interior.

7-3b Saturn's Rings

Study **Concept Art 7B, The Ice Rings of Saturn**, and notice the following three ideas:

1. The rings are made up of billions of ice particles, each in its own orbit around the planet. The ring particles you observe now can't be as old as Saturn. The rings

must be replenished now and then by impacts on Saturn's moons or other processes. The same is true of the rings around the other Jovian planets.

2. The gravitational effects of small moons can confine some rings in narrow strands or keep the edges of rings sharp. Moons can also produce waves in the rings that are visible as tightly wound ringlets.
3. The ring particles are confined in a thin layer in Saturn's equatorial plane, spread among small moons and confined by gravitational interactions with larger moons. The rings of Saturn, and the rings of the other Jovian worlds, are created and controlled by the planet's moons. Without the moons, there would be no rings.

7-3c Saturn's Family of Moons

Saturn has more than 60 known moons, many of which are small and all of which contain mixtures of ice and rock. It is probable that many are captured objects.

The largest of Saturn's moons, called Titan, is a bit larger than the planet Mercury. Its density suggests that it must contain a rocky core under a thick mantle of ices. Titan is so cold that its gas molecules do not travel fast enough to escape, so it has an atmosphere composed mostly of nitrogen with traces of argon and methane. Sunlight converts some of the methane into complex carbon-rich molecules that collect into small particles filling the atmosphere with orange smog, visible in

How Do We Know? 7-1

Funding for Basic Research

Who pays for science? Searching out scientific knowledge can be expensive, and that raises the question of funding. Some science has direct applications, and industry supports such research. For example, pharmaceutical companies have large budgets for scientific research leading to the creation of new drugs. But some basic science is of no immediate practical value, with no obvious commercial applications. Who pays the bill?

Should you consider a career as an industrial paleontologist? A paleontologist is a scientist who studies ancient life-forms by examining fossils of plant and animal remains. Except for the rare Hollywood producer about to release a dinosaur movie, corporations can't make a profit from the discovery of a new dinosaur. The practical-minded stockholders of a company will not approve major investments in such research. Consequently, digging up dinosaurs, like astronomy, is not well funded by industry.

Exploring other worlds is valuable; it helps us understand ourselves and our place in the Universe. Yet, sending the *Cassini* spacecraft to Saturn costs each U.S. citizen only 56¢ per year over the life of the project.

It falls to government institutions and private foundations to pay the bill for this kind of research. The Keck Foundation built two giant telescopes on Mauna Kea in Hawai'i with no expectation of financial return, and the National Science Foundation has funded thousands of astronomy research projects for the benefit of society.

The discovery of a new dinosaur or a new Galaxy is of no great financial value, but such scientific knowledge is not worthless. Its value lies in what it tells us about the world we live in. Such scientific research enriches our lives by helping us understand what we are. Ultimately, funding basic scientific research is a public responsibility that society must balance against other needs. There isn't anyone else to pick up the tab.

Concept Art 7B

THE ICE RINGS OF SATURN

1

The brilliant rings of Saturn are made up of billions of ice particles ranging from microscopic specks to chunks bigger than a house. Each particle orbits Saturn in its own circular orbit. Much of what planetary scientists know about the rings has been learned from the *Voyager 1* and *2* flybys and the *Cassini* orbiter. From Earth, astronomers see three rings labeled A, B, and C. *Voyager* and *Cassini* images reveal over a thousand ringlets within the rings.

Saturn's rings can't be leftover material from the formation of Saturn. The rings are made of ice particles, and the planet would have been so hot when it formed that it would have vaporized and driven away any icy material. Rather, the rings must be debris from collisions between Saturn's icy moons and passing comets or asteroids. Impacts large enough to scatter ice throughout the Saturn system are estimated to occur every 100 million years or so. The ice would quickly settle into the equatorial plane, and some would become trapped in rings.

Although the ice will tend to waste away because of meteorite impacts and damage from radiation in Saturn's magnetosphere, new impacts could replenish the rings with fresh ice. The bright, beautiful rings you see today may be only a temporary enhancement caused by an impact that occurred since the extinction of the dinosaurs.

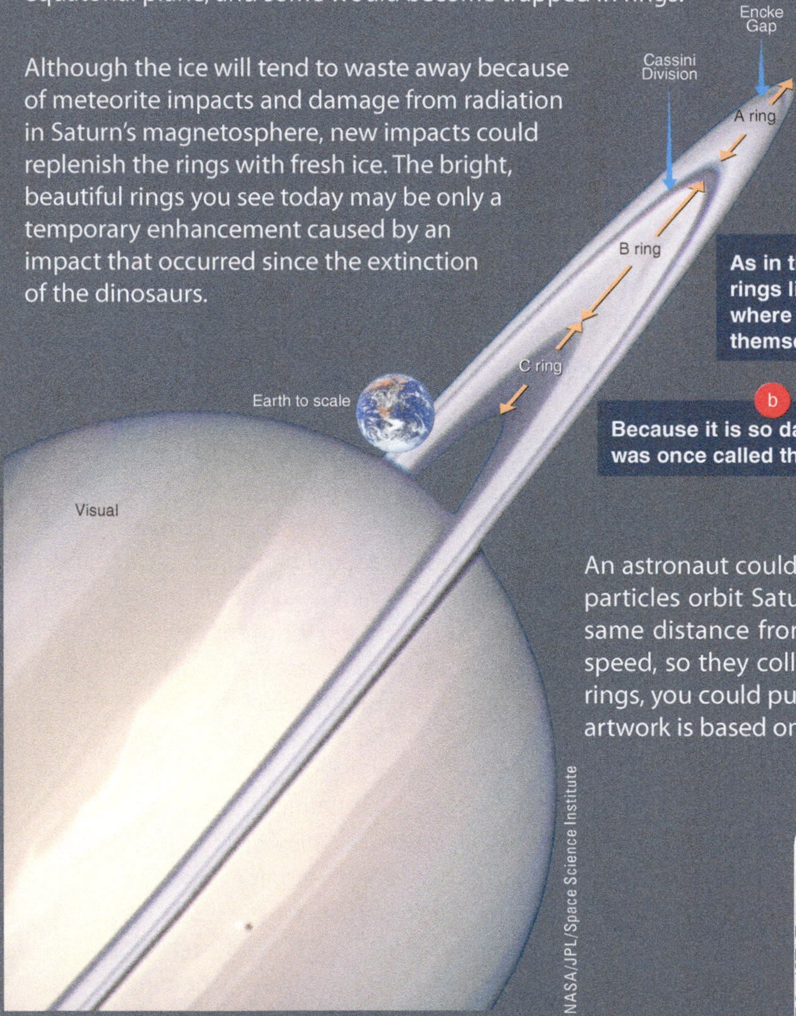

a **As in the case of Jupiter's ring, Saturn's rings lie inside the planet's Roche limit where the ring particles cannot pull themselves together to form a Moon.**

b **Because it is so dark, the C ring was once called the crepe ring.**

1a

An astronaut could swim through the rings. Although the particles orbit Saturn at high velocity, all particles at the same distance from the planet orbit at about the same speed, so they collide gently at low velocities. If you could visit the rings, you could push your way from one icy particle to the next. This artwork is based on a model of particle sizes in the A ring.

The C ring contains boulder-size chunks of ice, whereas most particles in the A and B rings are more like golf balls, down to dust-size ice crystals. Further, C ring particles are less than half as bright as particles in the A and B rings. *Cassini* observations show that the C ring particles contain less ice and more minerals.

2

Because of collisions among ring particles, planetary rings should spread outward. The sharp outer edge of the A ring and the narrow F ring are confined by **shepherd satellites** that gravitationally usher straying particles back into the rings.

Some gaps in the rings, such as the Cassini Division, are caused by resonances with moons. A particle in the Cassini Division orbits Saturn twice for each orbit of the moon Mimas. On every other orbit, the particle feels a gravitational tug from Mimas. These tugs always occur at the same places in the orbit and force the orbit to become slightly elliptical. Such an orbit crosses the orbits of other particles, which results in collisions, and that removes the particle from the gap.

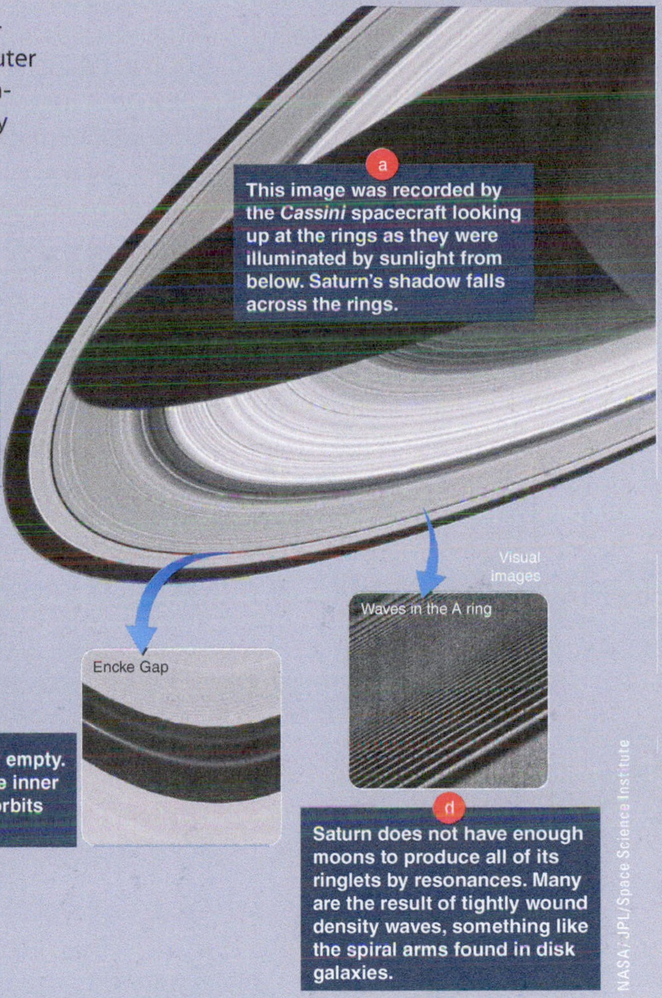

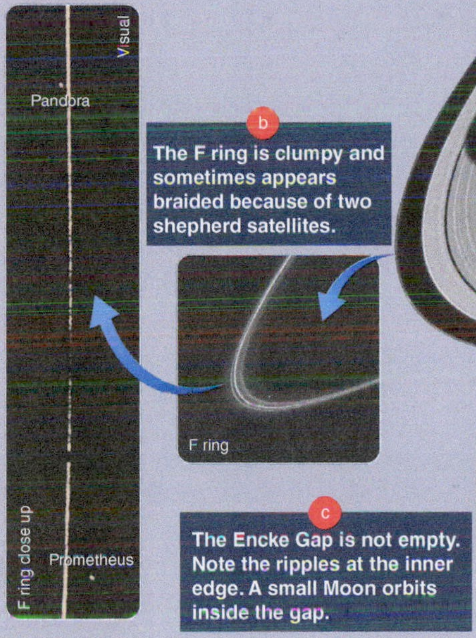

NASA/ JPL/ Space Science Institute

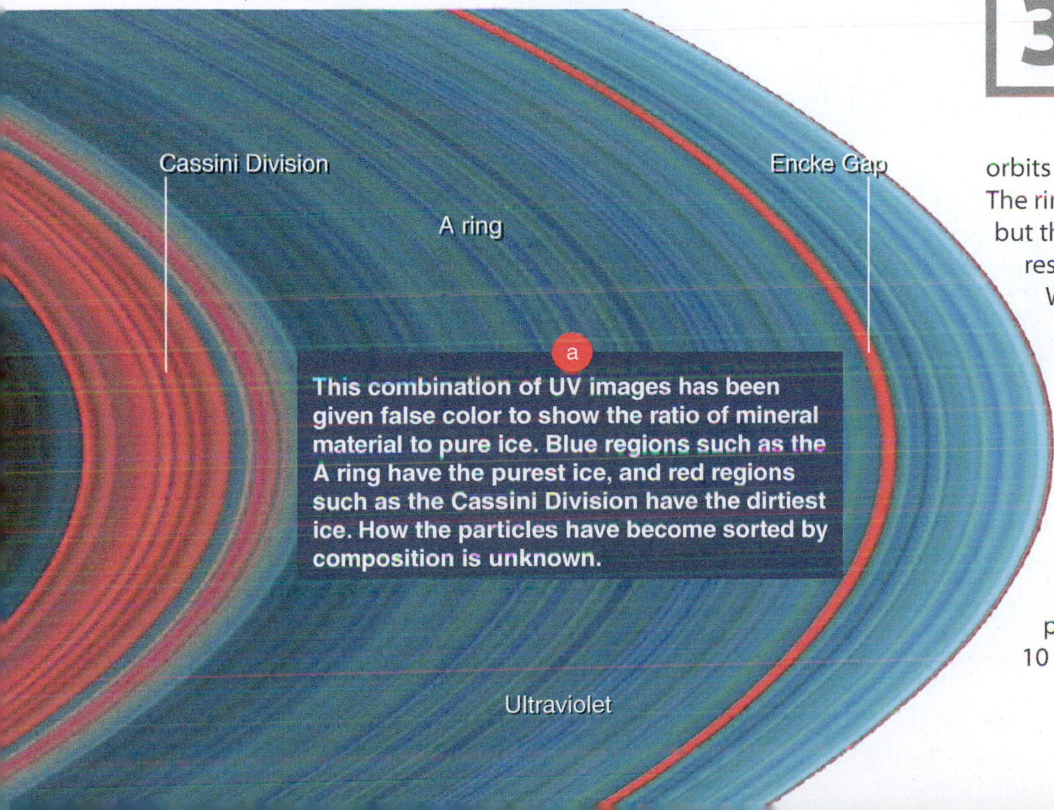

3

How do moons happen to be at just the right places to confine the rings? That puts the cosmic cart before the horse. The ring particles get caught in the most stable orbits among Saturn's innermost moons. The rings push against the inner moons, but those moons are locked in place by resonances with larger, outer moons. Without the moons, the rings would spread and dissipate.

Saturn's rings are a very thin layer of particles and nearly vanish when the rings turn edge-on to Earth. Although ripples in the rings may extend for hundreds of meters above and below the midplane, the sheet of particles may be only about 10 meters thick.

Figure 7-5. These particles are understood to slowly settle downward to the surface in the form of a dark organic goo, meaning it is composed of carbon-rich molecules.

Titan's surface is mainly composed of ices of water and methane at −180°C (−290°F). The *Cassini* spacecraft dropped the *Huygens* probe into the atmosphere of Titan, and it photographed dark drainage channels suggesting that liquid methane falls as rain, washes the dark goo off the higher terrain, and drains into the lowlands. Such methane downpours may be rare, however. No direct evidence of liquid methane was detected as the probe descended, but later radar and spectroscopic observations made by the *Cassini* orbiter detected lakes of methane and ethane. Infrared images suggest the presence of methane volcanoes that replenish the methane in the atmosphere, so Titan must have some internal heat source to power the activity.

Most of the remaining moons of Saturn are small and icy, have no atmospheres, are heavily cratered, and have dark, ancient surfaces. The moon Enceladus, however, shows signs of recent geological activity. Some parts of its surface contain 1000 times fewer craters than other regions, and infrared observations show that its south polar region is unusually warm and venting water and ice geysers (Figure 7-6). Evidently, a reservoir of liquid water lies only tens of meters below the surface. At some point in its history, this moon must have been caught in a resonance with another moon and was warmed by tidal heating. Enceladus appears to maintain the faint E ring that extends far beyond the visible rings. In 2009, astronomers detected infrared radiation from a dark ring 13 million km (8 million miles) in radius, beyond the orbits of most of Saturn's moons.

FIGURE 7-5

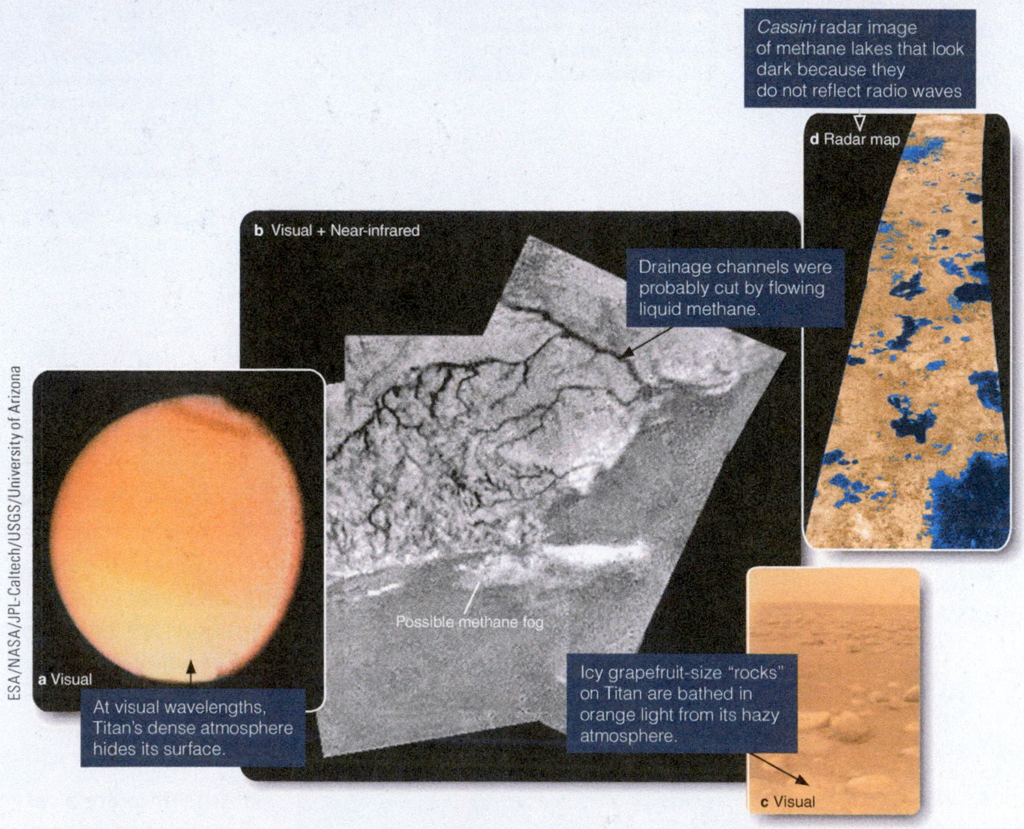

(a) As the *Huygens* probe descended through Titan's smoggy atmosphere, (b) it photographed the surface from an altitude of 8 km (5 mi). Although no liquid was present, dark drainage channels led into the lowlands. (c) Once the probe landed on the surface, it radioed back photos showing a level plain and chunks of ice smoothed by a moving liquid. (d) Radar images from the *Cassini* orbiter reveal lakes of liquid methane and ethane around the poles.

FIGURE 7-6

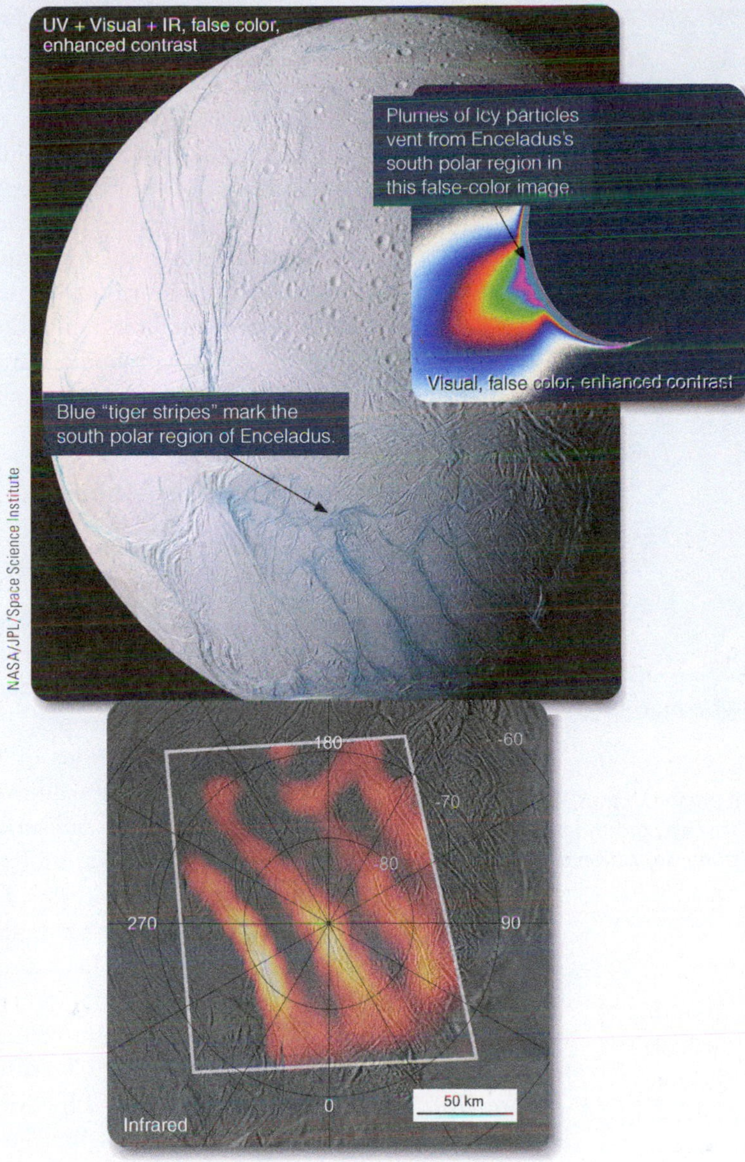

The bright, clean, icy surface of Enceladus does not look old. Some areas have few craters, and the numerous cracks and lanes of grooved terrain resemble the surface of Jupiter's moon Ganymede. Enceladus is venting water, ice, and organic molecules from geysers near its south pole. A thermal infrared image reveals internal heat leaking to space from the "tiger stripe" cracks where the geysers are located.

7-4 URANUS

Now that you are familiar with the gas giants in our Solar System, you will be able to appreciate how weird the ice giants, Uranus and Neptune, are. Uranus, especially, seems to have forgotten how to behave like a planet.

Uranus was discovered in 1781 by the scientist William Herschel, a German expatriate living in England. He named it *Georgium Sidus*, George's Star, after the English King George III. European astronomers, especially the French, refused to accept a planet named after an English king. They called it Herschel. Years later, German astronomer Johann Bode suggested Uranus, the oldest of the Greek gods.

7-4a Planet Uranus

Uranus is only one-third the diameter of Jupiter and only one-twentieth as massive, and, being four times farther from the Sun, its atmosphere is almost 100°C colder than Jupiter's.

Uranus never grew massive enough to capture large amounts of gas from the nebula as Jupiter and Saturn did, so it has much less hydrogen and helium. Its internal pressure is enough lower than Jupiter's that it should not contain any liquid metallic hydrogen. Models of Uranus, based in part on its density and oblateness, suggest that it has a small core of heavy elements and a deep mantle of partly solid water. Although referred to as ice, this material would not be anything like ice on Earth, given the temperatures and pressures inside Uranus. The mantle also contains rocky-composition material and dissolved ammonia and methane. Circulation in this electrically conducting mantle may generate the planet's peculiar magnetic field, which is highly inclined to its axis of rotation. Above this mantle lies a deep hydrogen and helium atmosphere.

Uranus rotates on its side, with its equator inclined about 98 degrees to its orbit. With an orbital period of 84 years, each of the four seasons lasts 21 years, and the winter–summer contrast is extreme. During a season when one of its poles is nearly pointed at the Sun

FIGURE 7-7

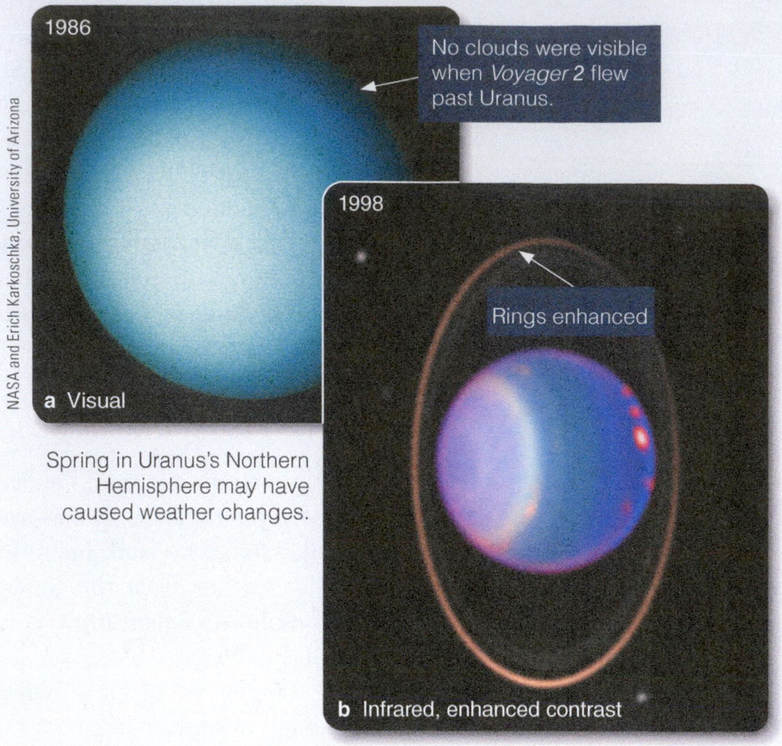

(a) If you had been riding onboard *Voyager 2* as it passed Uranus, the planet would have looked like a bland green-blue ball. (b) Later ground-based contrast-enhanced images reveal traces of belt–zone circulation deep in the atmosphere.

(a solstice), inhabitants of Uranus would never see the Sun rise or set. Compare this with seasons on Earth discussed in Section 2-3b of Chapter 2. Uranus's odd rotation may have been produced when it was struck by a very large object late in its formation, or by tidal interactions with the other giant planets as it migrated outward early in the history of the Solar System (see Chapter 8).

Voyager 2 photos, like the one in Figure 7-7a, show a nearly featureless ball. The atmosphere is mostly hydrogen and helium, but traces of methane absorb red light and thus make the atmosphere look blue or teal, depending on the image. There is no belt–zone circulation visible in the *Voyager 2* photographs, although extreme computer enhancement revealed a few clouds and bands around the south pole. In the decades since *Voyager 2* flew past Uranus, spring has come to the Northern Hemisphere of Uranus and autumn to the Southern Hemisphere. Images made by the *Hubble Space Telescope* as well as the most powerful Earth-based telescopes reveal changing clouds and cloud bands (Figure 7-7b).

Infrared measurements show that Uranus is radiating about the same amount of energy that it receives from the Sun, meaning it has much less heat flowing out of its interior than Jupiter or Saturn (or Neptune). This may account for its limited atmospheric activity. Astronomers are not sure why Uranus differs in this respect from the other Jovian worlds.

ovoids The large oval grooves found on Uranus's moon Miranda that indicate past geologic activity.

occultations The passage of a larger body in front of a smaller body.

7-4b Uranus's Moons

Until recently, astronomers could see only five moons orbiting Uranus. *Voyager 2* discovered 10 more small moons in 1986, and yet more have been found in images recorded by large, Earth-based telescopes.

The five major moons of Uranus are smaller than Earth's Moon and have old, dark, cratered surfaces. A few have deep cracks, produced, perhaps, when the interior froze and expanded. In some cases, liquid water "lava" appears to have erupted and smoothed over some regions. Miranda, the innermost moon, is only 14 percent the diameter of Earth's moon, but its surface is marked by grooves called **ovoids** (Figure 7-8). These may have been caused by internal heat driving convection in the icy mantle. By counting craters on the ovoids, astronomers conclude that the entire surface is old, and the Moon is no longer active.

7-4c Uranus's Rings

The rings of Uranus are dark and faint, contain little dust, are confined by shepherd satellites, and must be continuously resupplied with material from the moons. They are not easily visible from Earth—the first hint that Uranus had rings came from **occultations**, the passage of the planet in front of a star. Most of what astronomers know about these rings comes from observations by the *Voyager 2* spacecraft. Their composition appears to be water ice mixed with methane that has been darkened by exposure to radiation.

In 2006, astronomers found two new, very faint rings orbiting far outside the previously known rings of Uranus.

FIGURE 7-8

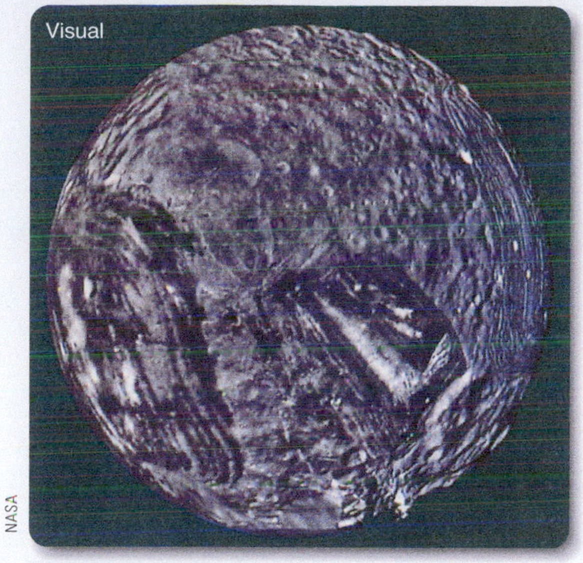

Evidence of geological activity on Uranus's Moon Miranda. The face of Miranda is marred by ovoids, which are believed to have formed when internal heating caused slow convection in the ice of the Moon's mantle. Note the 5 km (3 mi) high cliff at the lower edge of the Moon.

FIGURE 7-9

(a) Because Neptune's equator is inclined 28 degrees in its orbit, it experiences seasons. Because of the planet's orbital period, each season lasts about 40 years. (b) Since *Voyager 2* visited in 1989, spring has come to the Southern Hemisphere, and the weather has changed significantly. That is surprising because sunlight on Neptune is 900 times dimmer than on Earth.

The newly discovered satellite Mab appears to be the source of particles for the larger ring, and the smaller of the new rings is confined between the orbits of the moons Portia and Rosalind. Note that the International Astronomical Union (IAU) has declared that the newly discovered moons of Uranus are to be named after characters in Shakespeare's plays.

7-5 NEPTUNE

A British astronomer and a French astronomer independently calculated the existence and location of Neptune from irregularities in the motion of Uranus. British observers were too slow to respond; Neptune was discovered in 1846, and the French astronomer got the credit. Neptune looks like a tiny blue dot with no visible cloud features. Because of its blue color, astronomers named it after the god of the sea. In 1989, *Voyager 2* flew past and revealed some of Neptune's secrets.

7-5a Planet Neptune

Almost exactly the same size as Uranus, Neptune is calculated to have a similar interior: a small core of heavy elements lies within a slushy mantle of water, ices, and minerals (rock) below a hydrogen-rich atmosphere. Yet Neptune looks quite different; it is dramatically blue and has active cloud formations. Neptune has a dark blue tint because its atmosphere contains 1.5 times more methane than Uranus. Methane absorbs red photons better than blue and scatters blue photons better than red, giving Neptune a blue color and Uranus a green-blue color.

Atmospheric circulation on Neptune is much more dramatic than on Uranus. When *Voyager 2* flew by Neptune in 1989, the largest feature was the Great Dark Spot, which you can see in Figure 7-9. Roughly the size of Earth, the spot seemed to be an atmospheric circulation similar to Jupiter's Great Red Spot. Smaller spots were visible in Neptune's atmosphere, and photos showed they were circulating like hurricanes. More recently, the *Hubble Space Telescope* has photographed Neptune and found that the Great Dark Spot is gone, and new cloud formations have appeared. Evidently, the weather on Neptune is surprisingly changeable.

The atmospheric activity on Neptune is apparently driven by heat flowing from the interior plus some contribution by dim light from the Sun 30 AU away. Neptune may have more atmospheric activity than Uranus because it has more heat flowing out of its interior, for reasons that are unclear.

Like Uranus, Neptune has a highly inclined magnetic field that must be linked to circulation in the interior. In both cases, astronomers suspect that ammonia dissolved in the liquid water mantle makes the mantle a good electrical conductor and that convection in the water, coupled with the rotation of the planet, drives the dynamo effect and generates the magnetic field.

7-5b Neptune's Moons

Neptune has two moons that were discovered from Earth before *Voyager 2* flew past in 1989. The passing spacecraft discovered six more, very small moons. Since then, a few more small moons have been found by astronomers using large Earth-based telescopes.

The two largest moons have peculiar orbits. Nereid, about one-tenth the size of Earth's Moon, follows a large, elliptical orbit, taking about an Earth-year to circle Neptune once. Triton, about 80 percent the size of Earth's Moon, orbits Neptune backward—clockwise as seen from the north. These odd orbits suggest that the system was disturbed long ago in an interaction with some other body, such as a massive object.

Triton has an atmosphere of nitrogen and methane about 10^5 times less dense than Earth's, and a temperature of 36 K (−395°F). A significant part of Triton is ice, and deposits of nitrogen frost are visible at the southern pole (Figure 7-10). Many features on Triton suggest it has had an active past. It has few craters on its surface, but it does have long faults that appear to have formed when the icy crust broke, plus large basins that seem to have been flooded repeatedly by liquids from the interior. Even more interesting are the dark smudges visible in the southern polar cap that are interpreted as sunlight-darkened deposits of methane erupted out of liquid nitrogen geysers.

FIGURE 7-10

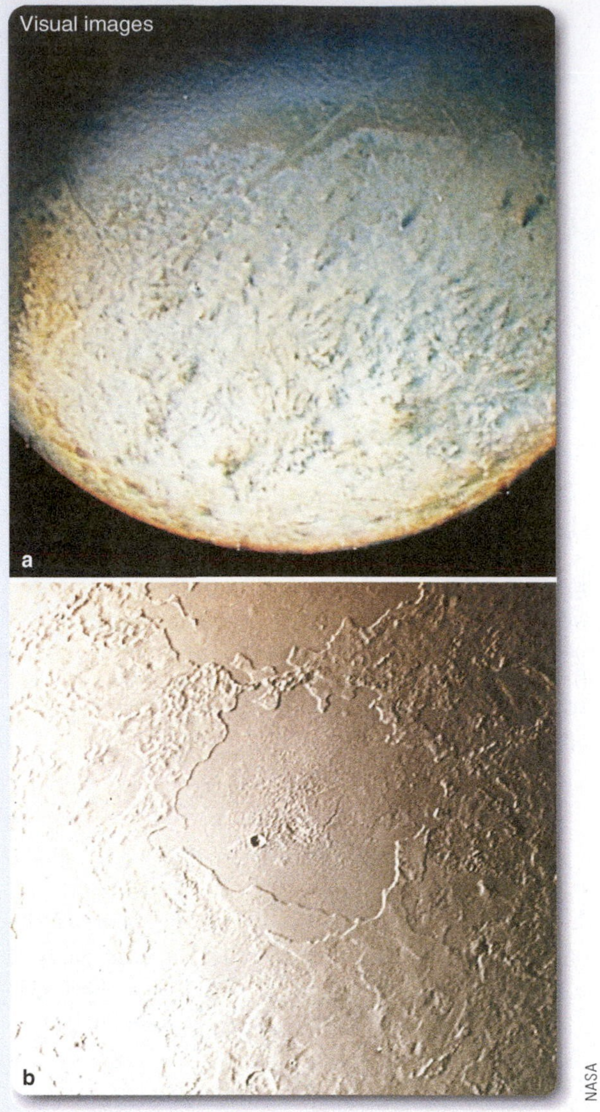

Visible-wavelength images of Neptune's moon Triton. (a) Triton's southern polar cap is composed of nitrogen frost. Note dark smudges caused by organic compounds sprayed from nitrogen geysers, and the absence of craters. (b) These round basins on Triton appear to have been repeatedly flooded by liquid from the interior.

7-5c Neptune's Rings

Neptune's rings are faint and very hard to detect from Earth, but they illustrate some interesting processes of comparative planetology (Figure 7-11). Neptune's rings are similar to those of Uranus but contain more small dust particles. The gravitational influence of one of Neptune's moons causes the outermost ring to be concentrated into short arcs. Neptune's ring system, like the others,

FIGURE 7-11

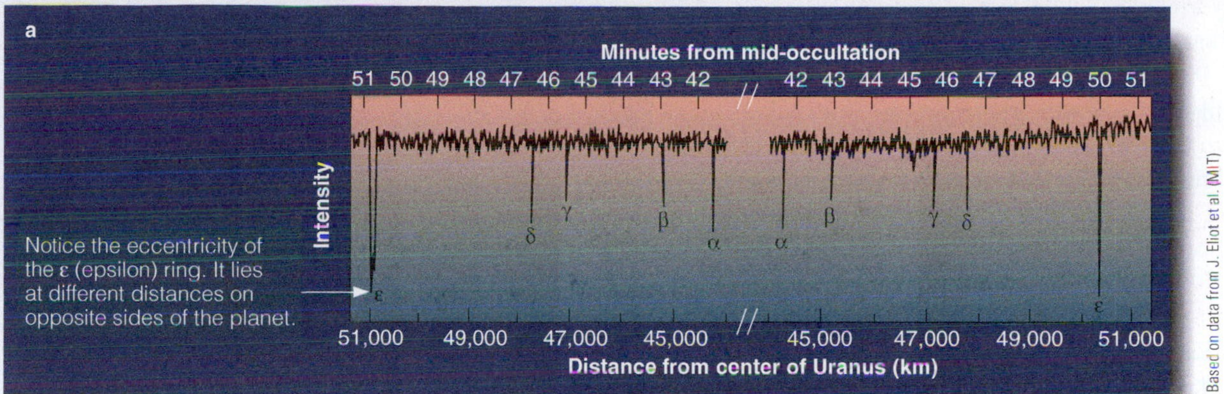

(a) The rings of Uranus were discovered in 1977, when Uranus crossed in front of a star. During this occultation, astronomers saw the star dim a number of times before and again after the planet crossed over the star. The dips in brightness were caused by rings circling Uranus. (b) The bright disk of Neptune is hidden behind the black bar in this *Voyager 2* image. Two narrow rings are visible, and a wider, fainter ring lies closer to the planet. More ring material is visible between the two narrow rings. The rings are bright in forward-scattered light, indicating that the rings contain significant amounts of very small dust particles with short lifetimes. Like the rings of Uranus, the rings of Neptune reflect very little light and probably contain methane-rich ice darkened by radiation.

is apparently resupplied by impacts on moons scattering debris that fall into the most stable places among the orbits of the moons.

7-6 PLUTO: PLANET NO MORE

Out on the edge of the Solar System orbits a family of small, icy worlds. Pluto was the first to be discovered, in 1930, but modern telescopes have found more.

You may have learned in school that there are nine planets in our Solar System, but, as mentioned previously, in 2006 the IAU voted to remove Pluto from the list of planets. Pluto is a very small, icy world—it isn't Jovian, and it isn't Terrestrial. Its orbit is highly inclined and so elliptical that Pluto actually comes closer to the Sun than Neptune at times. To understand Pluto's status, you must use comparative planetology to analyze Pluto and then compare it with its neighbors.

Pluto is very difficult to observe from Earth; its diameter is 2370 km (1470 mi), smaller than Australia and only two-thirds the diameter of Earth's Moon. In Earth-based telescopes, Pluto never looks like more than a faint point of light, and even in *Hubble Space Telescope* images, it shows little detail. Orbiting so far from the Sun, Pluto is cold enough to freeze most compounds you think of as gases, and spectroscopic observations have found evidence of nitrogen and methane ices on its surface. Pluto has a thin atmosphere of nitrogen and carbon monoxide with small amounts of methane.

Pluto has five known moons. Four are quite small, but Charon is relatively large, with half of Pluto's diameter. Charon orbits Pluto in an orbit inclined 118 degrees to the plane of Pluto's orbit around the Sun, which means Charon's motion is retrograde. Pluto and Charon are tidally locked facing each other, so Pluto's axis of rotation is inclined at the same large angle to its orbit and its rotation is also retrograde.

Knowing Charon's orbital period of 6.4 days and orbit semimajor axis of 19,600 km allows you to use Kepler's third law to calculate that the total mass of the system is only about 0.0024 Earth mass. Most of that mass is Pluto, which has about nine times the mass of Charon. Knowing the diameters and masses of Pluto and Charon allows determination that their densities are respectively about 1.9 and 1.7 g/cm^3. This indicates that both Pluto and Charon must contain about 30 to 40 percent ice and 60 to 70 percent rock.

The *New Horizons* spacecraft flew past Pluto in July 2015, radioing back impressive images (Figures 7-12 and CO 7) and other measurements. Close-up pictures show mountains up to 3.5 km (11,000 ft) high (Figure 7-12b). Methane and nitrogen ice would not be strong enough to stand that tall, so planetary scientists conclude that the mountains must be made of water ice that would be rock-hard at Pluto's temperature.

The lack of craters on and near those mountains indicates that this particular landscape is very young—no older than 100 million years, one of the youngest surfaces observed in the Solar System. Pluto has been active geologically recently, which means that heat must be flowing outward, although planetary scientists wonder what the internal heat source could be. Charon's orbit is perfectly circular, so tidal heating is not the explanation. Radioactive decay would seem to be the only plausible alternative, although that heat source should also have faded long ago. One planetary scientist on the *New Horizons* team remarked, "This may cause us to rethink what powers geological activity on many other icy worlds." Pluto turns out to be a world full of surprises.

FIGURE 7-12

(a) Enhanced-color images of Pluto (lower right) and Charon (upper left) taken by NASA's *New Horizons* spacecraft as it passed through the Pluto system on July 14, 2015. This composite image highlights the striking differences between Pluto and Charon. Pluto and Charon are shown with approximately correct relative sizes, but their separation is not to scale. (b) Just 15 minutes after its closest approach to Pluto, *New Horizons* captured this near-sunset view of the rugged, icy mountains and flat ice plains extending to Pluto's horizon. The backlighting reveals more than a dozen layers of haze in Pluto's tenuous but extended atmosphere. The scene is 380 km (230 mi) across.

7-6a What Defines a Planet?

To understand why Pluto is no longer considered a planet, you will need to consider the Kuiper Belt of small bodies orbiting on the outer edge of the Solar

System (see Chapter 8). Since 1992, astronomers have discovered roughly a thousand icy bodies orbiting beyond Neptune. There may be as many as 100 million objects in the Kuiper Belt larger than 1 km (0.6 mi) in diameter. They are understood to be icy bodies left over from the formation of the outer Solar System. Some of the Kuiper Belt Objects are quite large, and one, named Eris, has about the same diameter as Pluto, but is 27 percent more massive. Three other Kuiper Belt Objects found so far, Sedna, Orcus, and Quaoar (pronounced kwah-o-wahr), are half the size of Pluto or larger. Some of these objects have moons of their own. In that way, they resemble Pluto and its moons.

A bit of comparative planetology shows that Pluto is not related to the Jovian or Terrestrial planets; it is obviously a member of a newfound family of icy worlds that orbit beyond Neptune. These bodies must have formed at about the same time as the eight classical planets of the Solar System, but they did not grow massive enough to clear their orbital zones of remnant objects and remain embedded among a swarm of other objects in the Kuiper Belt.

One of the IAU's criteria for planet status is that an object must be large enough to dominate and gravitationally clear its orbital region of most, or all, other objects. Eris and Pluto, the largest objects found so far in the Kuiper Belt, and Ceres, the largest object in the asteroid belt, do not meet that standard. On the other hand, all three are large enough for their gravities to have pulled them into spherical shapes, so they are the prototypes of a new class of objects defined by the IAU as **dwarf planets**.

7-6b Pluto and the Plutinos

No, this section is not about a 1950s rock and roll band. It is about the history of the dwarf planets, and it will take you back 4.6 billion years to watch the outer planets form.

More than a hundred known Kuiper Belt Objects are caught with Pluto in a 3:2 resonance with Neptune. That is, they orbit the Sun twice while Neptune orbits three times. These Kuiper Belt Objects have been named **plutinos**. The plutinos formed in the outer solar nebula, but how did they get caught in resonances with Neptune? In Chapter 8, you will learn that models of the formation of the planets suggest that Uranus and Neptune may have formed closer to the Sun and that sometime later, gravitational interactions among the Jovian planets could have gradually shifted Uranus and Neptune outward. As Neptune migrated outward, its orbital resonances could have swept up small objects like a strange kind of snowplow. The Plutinos are caught in the 3:2 resonance, and other Kuiper Belt Objects are caught in other resonances. This appears to support the models that predict that Uranus and Neptune migrated outward.

The migration of the outer planets would have dramatically upset the motion of some of these Kuiper Belt Objects, and some could have been thrown inward where they could interact with the Jovian planets. Some of those objects may have been captured as moons, and astronomers wonder if moons, such as Neptune's Triton, could have started life as Kuiper Belt Objects. Other objects may have impacted bodies in the inner Solar System and caused the late heavy bombardment episode especially evident on the surface of Earth's Moon. The small frozen worlds on the fringes of the Solar System hold clues to the formation of Earth and the other planets 4.6 billion years ago.

dwarf planets A body that independently orbits the Sun or another star, is massive enough to pull itself into a spherical shape but not massive enough to clear out other bodies in and near its orbit; for example, Pluto, Eris, and Ceres.

plutinos One of the icy Kuiper Belt Objects that, like Pluto, is caught in a 3:2 orbital resonance with Neptune.

What Are We? Trapped

No one has ever been farther from Earth than the Moon. We humans have sent robotic spacecraft to visit most of the larger worlds in our Solar System, and we have found them strange and wonderful places, but no human has ever set foot on any of them. We are trapped on Earth.

Getting a human crew away from Earth's gravitational field is difficult and calls for very large rockets. The U.S. built rockets in the 1960s and early 1970s that could send astronauts to the Moon, but they stopped producing them. At the time of this writing (August 2016) the largest available rockets can only take astronauts just a few hundred kilometers above Earth's surface to the International Space Station. The United States and other nations are considering sending humans back to the Moon and eventually to Mars, but budget limitations have delayed specific plans. Does Earth's civilization have the resources to build spacecraft capable of carrying human explorers to other worlds? We'll have to wait and see.

Furthermore, we Earthlings have evolved to fit the environment on Earth. None of the planets or moons you explored in this chapter would welcome you. Lack of air, and extreme heat or cold are obvious problems, but, also, Earthlings have evolved to live with Earth's gravity. Astronauts in space for just a few weeks suffer biomedical problems because they are no longer in Earth's gravity. Living in a colony on Mars or the Moon might raise similar problems. Just getting to the outer planets would take decades of space travel; living for years in a colony on one of the Jovian moons under low gravity and exposed to the planet's radiation belts may be beyond the capability of the human body. We may be trapped on Earth not because we lack large enough rockets but because we need Earth's protection.

It seems likely that we need Earth more than it needs us. The human race is changing the world we live on at a startling pace, and some of those changes could make Earth less hospitable to human life. All of your exploring of un-Earthly worlds serves to remind you of the nurturing beauty of our home planet.

STUDY TOOLS 7

IN THE BOOK

- ☐ Tear Out the Review Card on *The Outer Solar System*.

ASTRO ONLINE

Visit ASTRO Online at www.cengagebrain.com

- ☐ Interactive Reading
- ☐ Practice Quizzing
- ☐ Videos and Animations
- ☐ Exclusive Features

7 Questions and Problems

REVIEW QUESTIONS

1. Why is Jupiter so much richer in hydrogen and helium than Earth?
2. How can Jupiter have a liquid interior and not have a definite liquid surface?
3. How does the dynamo effect account for the magnetic fields of Jupiter, Saturn, Uranus, and Neptune?
4. Why are the belts and zones on Saturn less distinct than those on Jupiter?
5. Why do astronomers conclude that none of the Jovian planets' rings can be left over from the formation of the planets?
6. How can a moon produce a gap in a planetary ring system?
7. Explain why the amount of geological activity on Jupiter's moons varies with distance from the planet.
8. What makes Saturn's F ring and the rings of Uranus and Neptune so narrow?
9. Why is the atmospheric activity of Uranus less than that of Saturn and Neptune?
10. Why do astronomers suspect that Saturn's moon Enceladus is geologically active?
11. What are the seasons on Uranus like?
12. Why are Uranus and Neptune, respectively, green-blue and blue?
13. What evidence is there that Neptune's moon Triton has been geologically active recently?
14. How do astronomers account for the origin of Pluto and the plutinos?
15. What evidence indicates that catastrophic impacts have occurred in the Solar System's past?
16. **How Do We Know?** Why would you expect research in archaeology to be less well funded than research in chemistry?

DISCUSSION QUESTIONS

1. Some astronomers argue that Jupiter and Saturn are unusual, while other astronomers argue that all planetary systems should contain one or two such giant planets. What do you think? Support your argument with evidence.
2. Why do all the Jovian planets have rings, but the Terrestrial planets apparently don't? If you wanted to search for a yet-undiscovered faint ring system, which planet would you examine first? Why?

PROBLEMS

1. What is the maximum angular diameter of Jupiter as seen from Earth? Repeat this calculation for Neptune. Relevant data can be found in the **Celestial Profiles** in Appendix C. (*Hint:* Use **The Small-Angle Formula** in the Math Reference Cards MR.)
2. What is the angular diameter of Jupiter as seen from Callisto? From Amalthea? Relevant data can be found in the **Celestial Profiles** in Appendix C and Appendix Tables A-6 and A-7. (*Hint:* Use **The Small-Angle Formula** in the Math Reference Cards MR.)
3. Measure the polar and equatorial diameters of Saturn in the photograph in **Celestial Profile 6**, Appendix C, and calculate the planet's oblateness using the definition given in this chapter.
4. If you observe light reflected from Saturn's rings, you should see a redshift at one edge of the rings and a blueshift at the other edge. If you observe a spectral line and see a difference in wavelength of 0.0560 nm between opposite edges of the rings, and the unshifted wavelength (observed in the laboratory) is 500 nm, what is the orbital velocity of particles at the outer edge of the rings? (*Hint:* Use the **Doppler Shift** formula in the Math Reference Cards MR. *Note:* The speed of light is 3.00×10^5 km/s.)
5. One way to recognize a distant planet is by its motion along its orbit. If Uranus circles the Sun in 84 years, how many arc seconds will it move in 24 hours? (For the purposes of this problem, ignore the motion of Earth.)
6. If Uranus's epsilon ring is 50 km wide and the orbital velocity of Uranus is 6.8 km/s, how long should the occultation last that you expect to observe from Earth when the ring crosses in front of the star? (For the purposes of this problem, ignore the motion of Earth.)
7. If Neptune's clouds have a temperature of 60 K, at what wavelength will they radiate the most energy? (*Hint:* Use **Wien's Law** in the Math Reference Cards MR.)
8. How long did it take radio commands to travel from Earth to *Voyager 2* as it passed Neptune? Relevant data can be found in Appendix Tables A-6 and A-7. Assume Earth and Neptune were as close as possible during the *Voyager 2* encounter. (*Note:* The speed of light is 3.00×10^5 km/s.)

9. What is the angular diameter of Pluto as seen from Charon, given that the average separation of Charon from Pluto is 19,600 km and the diameters of Pluto and Charon are, respectively, 2370 and 1210 km? (*Hint:* See **The Small-Angle Formula** in the Math Reference Cards MR.)
10. The orbital period of Charon is given in the chapter text. Given that the average distance of Charon from Pluto is 19,600 km, calculate the mass of Pluto. Assume the mass of Charon is negligible compared with the mass of Pluto. (*Hints:* Use the **Circular Orbit Velocity** formula in the Math Reference Cards MR; make sure to convert relevant quantities to units of meters, kilograms, and seconds.)

LEARNING TO LOOK

1. Figure 7-13 shows a segment of the surface of Jupiter's Moon Callisto. Why is the surface mostly dark? Why are some craters dark and some white? What does this image tell you about the history of Callisto?

FIGURE 7-13

NASA/JPL-Caltech/University of Arizona

2. The *Cassini* spacecraft recorded Figure 7-14, a photo of Saturn's A ring and Encke's division. What do you see in this photo that tells you about processes that confine and shape planetary rings?

FIGURE 7-14

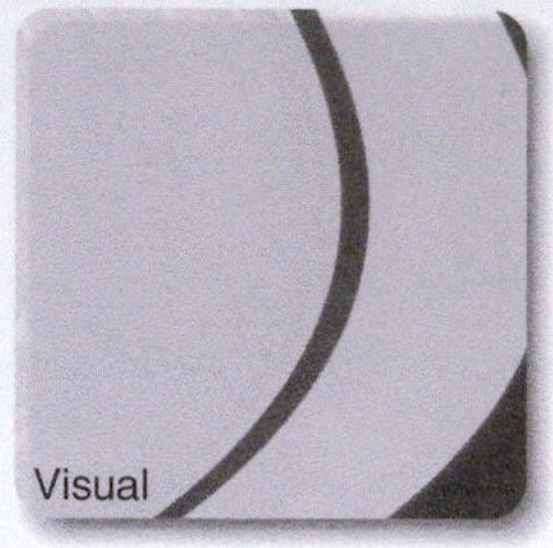

NASA/JPL-Caltech/SSI

3. Figures 7-15 and 7-16 show Uranus's Northern Hemisphere, one as it would look to the eye and the other through a red filter that enhances methane clouds. What do the atmospheric features tell you about circulation on Uranus?

FIGURE 7-15 AND 7-16

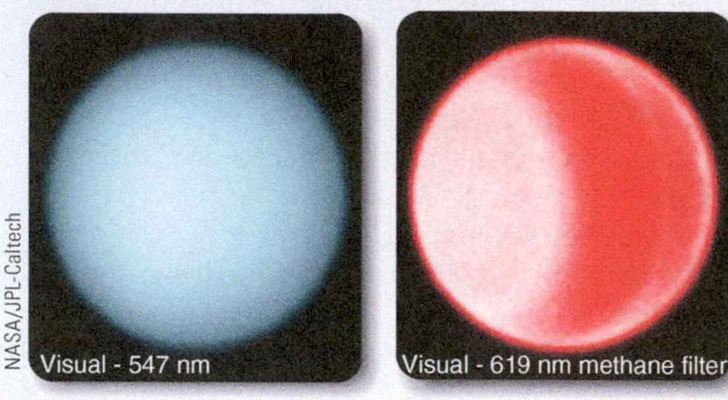

NASA/JPL-Caltech

NASA/ESA/STScI/AURA/NSF/H. Hammel

8 Origin of the Solar System and Extrasolar Planets

What place is this? Where are we now?
Carl Sandburg, *Grass*

NASA/JPL-Caltech

After finishing this chapter go to **PAGE 185** for **STUDY TOOLS.**

GUIDEPOST

You have learned how astronomers analyze light arriving at Earth from the rest of the Universe to try and unlock the secrets of celestial objects. The Sun, the center of our Solar System, was presented as a case study of what can be learned by studying the details of its spectrum. Then, you learned about the characteristics of the individual planets orbiting the Sun, Earth and its siblings, especially by using the technique of comparative planetology.

Now, having studied the Sun and the planets individually, it is time to draw all that information together and find answers to four important questions:

- What are the observed properties of the Solar System as a whole?
- How did the Solar System form?
- How do planets form?
- What do astronomers know about other planetary systems?

In the following six chapters you will travel far from our little Solar System, first exploring stars, then the Milky Way galaxy, then other galaxies, and finally the Universe as a whole. (See CO 8.) One important goal of this exploration is to place the Solar System and Earth in context.

The Solar System is your home in the Universe. Because humans are an intelligent species, we have the ability, and the responsibility, to wonder what we are. Our kind has inhabited this Solar System for several million years, but only within the past few hundred years have we begun to understand what a Solar System is.

8-1 SOLAR SYSTEM OVERVIEW

In order to make and then test hypotheses about how the Solar System formed, astronomers must search the present Solar System for evidence of its past (see **How Do We Know? 8-1**). In the next two sections you will survey the Solar System and compile a list of its most significant characteristics, potential clues to how it formed.

8-1a Size and Shape

Recall from Chapter 6 Section 6-1a that the Solar System's eight planets, Earth plus its siblings, are relatively tiny specks of matter scattered around the Sun. Imagine again a model of the Solar System in which 1 AU, the average distance between Earth and the Sun,

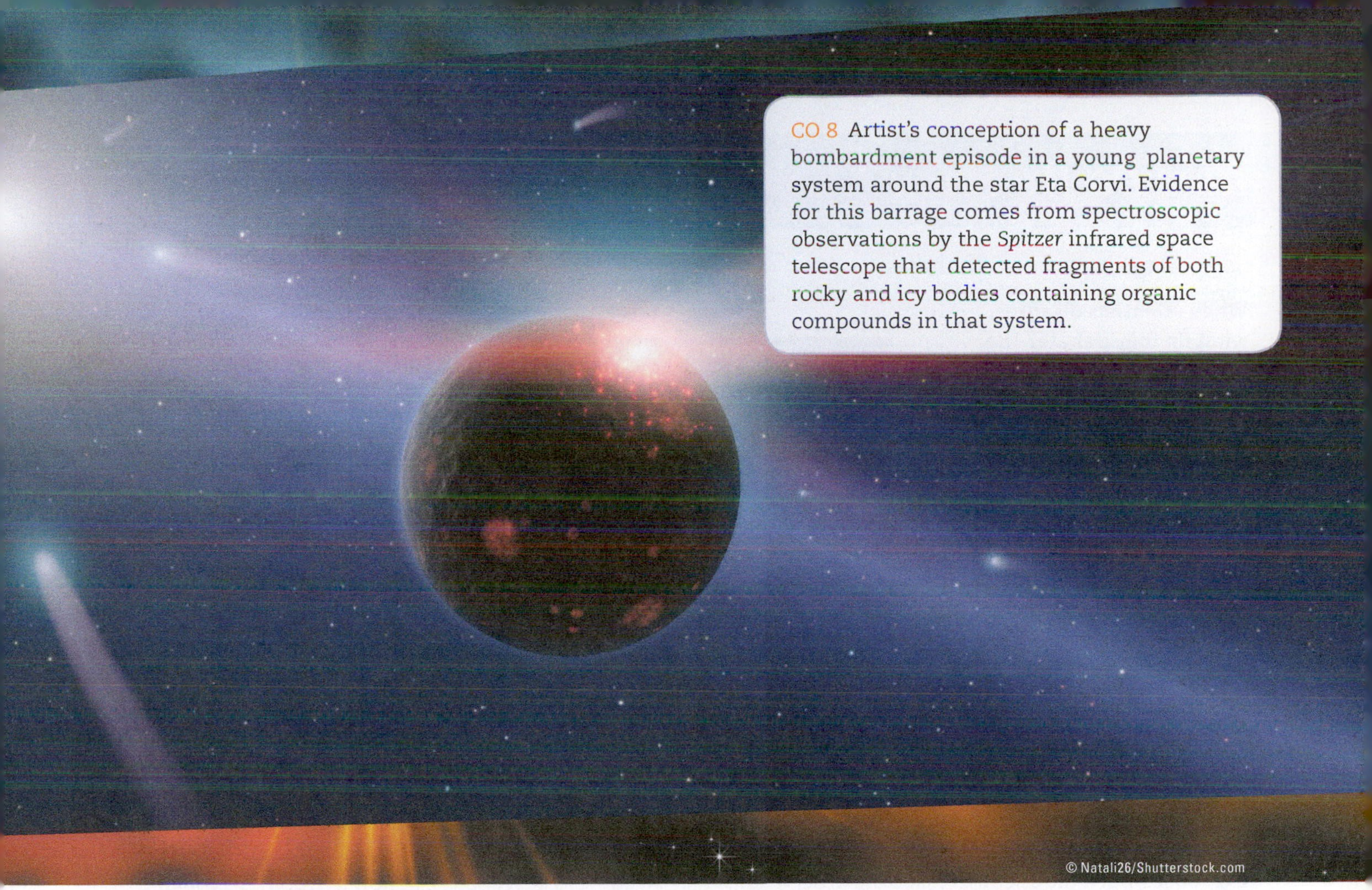

CO 8 Artist's conception of a heavy bombardment episode in a young planetary system around the star Eta Corvi. Evidence for this barrage comes from spectroscopic observations by the *Spitzer* infrared space telescope that detected fragments of both rocky and icy bodies containing organic compounds in that system.

is represented by 4 meters (13 feet). On that scale the Sun is a small plum, Earth is a grain of table salt, and Jupiter is an apple seed 21 m (69 ft) from the Sun. Neptune's orbit, marking the outer edge of the planetary zone, has a model radius of 120 m (400 ft). Also recall from Chapter 6 that the Solar System is relatively flat and disk-shaped, with the orbits of all the planets inclined less than 8 degrees to each other and to the Sun's equator. The planets all revolve around the Sun in the same direction, counterclockwise as seen from above Earth's north pole. With the exception of Venus and Uranus, all the planets rotate on their axes in that same direction, and nearly all of the Moons in the Solar System orbit around their planets in that same direction. With only a few exceptions, revolution and rotation in the Solar System both follow a common rule. Later in this chapter you will learn about hypotheses that may explain the exceptions to that rule.

8-1b Terrestrial and Jovian Planets

In the preceding two chapters you learned about the planets individually, but one important fact was obvious—the eight planets are divided into two categories: the small Earth-like worlds and the giant Jupiter-like worlds. The difference is so dramatic that it seemingly could not have occurred by chance; it must be significant. Read **Concept Art 8A, Two Kinds of Planets**, and notice the following three important points summarizing what you learned in Chapters 6 and 7.

1. The two kinds of planets are distinguished by their location. The four inner Terrestrial planets are quite different from the outer four Jovian planets.
2. Almost every solid surface in the Solar System is covered with craters.
3. The two groups of planets are also distinguished by individual properties such as rings, clouds, and Moons. Any hypothesis of the origin of the planets needs to explain these properties.

The division of the planets into two families is a clue to how our Solar System formed. The present properties of individual planets, however, don't tell everything you need to know about their origins.

The planets have all evolved since they formed. For further clues you can look at smaller objects that have remained largely unchanged since the birth of the Solar System.

How Do We Know? 8-1

Separating Facts from Hypotheses

When scientists disagree, what do they debate? The fundamental work of science is testing hypotheses by comparing them with facts. The facts are evidence of how nature works, and they represent reality. Hypotheses are attempts to explain how nature works. Scientists are very careful to distinguish between the two.

Scientific facts are those observations or experimental results of which scientists are confident. Ornithologists might note that fewer mountain thrushes are returning each spring to a certain mountain valley. Counting bird populations reliably is difficult and requires special techniques, but if the scientists made the observations correctly, they can be confident of their result and treat it as a fact.

To explain the declining population of thrushes, the scientists might consider a number of hypotheses, such as global warming or chemical pollution in the food chain. The ornithologists are free to combine or adjust their hypotheses to better explain the bird migration, but they are not free to adjust their facts.

Michael A. Seeds

In science, evidence is made up of facts that can range from precise numerical measurements to a qualitative description of a flower's shape.

Scientific facts are the hard pebbles of reality that can't be changed.

New facts can aggravate debates that are already politically charged. The declining number of mountain thrushes, for example, could be unwelcome news because addressing the root problem might cost taxpayer dollars or hurt local business interests. Nonscientists sometimes debate an issue by trying to adjust or even deny the facts, but scientists are not free to ignore a fact because it is unpopular or inconvenient. Scientists debate an issue by arguing about which hypothesis best applies, or how a hypothesis could be adjusted to fit the observed facts, but, once established, the facts themselves are not in question.

Whether scientists are measuring the density of an interstellar cloud or the size of a bird population, the final data become the reality against which hypotheses are tested. When Galileo said we should "read the book of nature," he meant we should consult reality as the final check on our understanding.

8-2 SPACE DEBRIS: ASTEROIDS, COMETS, AND METEORITES

As you will learn later in this chapter, astronomers have abundant evidence that the Sun plus Earth and Earth's sibling planets formed from a disk-shaped cloud called the solar nebula. (*Nebula* is the Greek word for cloud.) But, the Sun and the planets are not the only remains of the solar nebula. The Solar System is littered with three kinds of space debris: asteroids (Figure 8-1), comets, and meteoroids. Although these objects represent a tiny fraction of the mass of the system, they are a rich source of information about the origin of the planets.

asteroids Small, rocky world. Most asteroids orbit between Mars and Jupiter in the asteroid belt.

8-2a Asteroids

The **asteroids**, sometimes called minor planets, are small rocky worlds, most but not all of which orbit the Sun in a belt between the orbits of Mars and Jupiter. More than 400,000 asteroids have orbits that are charted, of which at least 2000 follow orbits that bring them into the inner Solar System where they can occasionally collide with a planet. You will learn in the next section that Earth has been struck

FIGURE 8-1

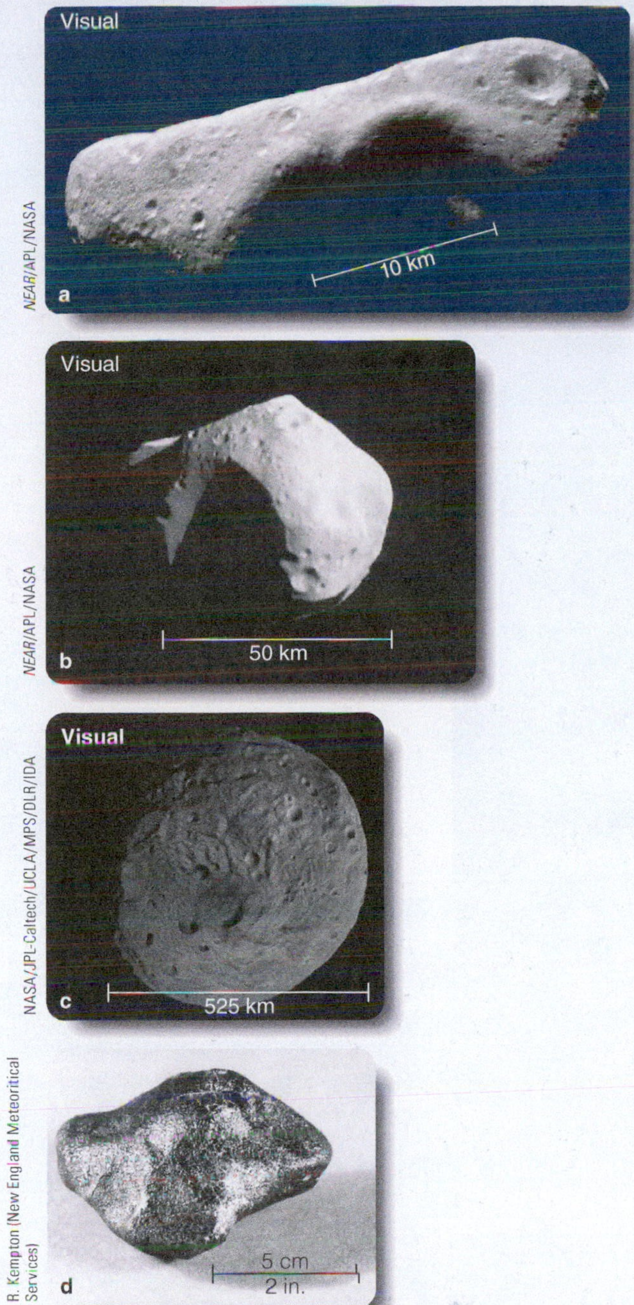

(a) The *Near Earth Asteroid Rendezvous (NEAR)* spacecraft visited the asteroid Eros in 2000 and found it to be heavily cratered by collisions and covered by a layer of crushed rock and dust. Eros appears to be solid rock. (b) The asteroid Mathilde, photographed by the *NEAR* spacecraft during a fly-by, has a very low density like some other known asteroids that may be rubble piles with large empty spaces between solid fragments. (c) The large asteroid Vesta has a spectrum resembling solidified lava. Vesta seems to have had internal heat and volcanic activity in its past, perhaps due to decay of radioactive minerals. This image of the asteroid Vesta, made from a distance of 2700 km (1700 mi) by NASA's *Dawn* spacecraft, shows Vesta's south polar region, which is dominated by the Rheasilvia impact basin. (d) One class of meteorites that is spectroscopically identical to Vesta is made of basalt, a volcanic rock.

many times in its history. Other asteroids share Jupiter's orbit, while some have been found beyond the orbit of Saturn.

About 200 asteroids are more than 100 kilometers (60 miles) in diameter, and tens of thousands are estimated to be more than 10 km (6 mi) in diameter. There are probably a million or more that are larger than 1 km (0.6 mi) and billions that are smaller. Because even the largest are only a few hundred kilometers in diameter, Earth-based telescopes can detect no details on their surfaces; even the *Hubble Space Telescope* can image only the largest features.

Photographs returned by robotic spacecraft and space telescopes show that asteroids are generally irregular in shape and battered by impact cratering. In fact, some asteroids appear to be rubble piles of broken fragments, and a few asteroids are known to be double objects, or to have small moons in orbit around them. These are understood by astronomers to be evidence of multiple collisions among the asteroids. A few larger asteroids show signs of volcanic activity on their surfaces that may have happened when those asteroids were young.

Astronomers recognize the asteroids as debris left over by a planet that failed to form at a distance of about 3 AU from the Sun. A good hypothesis for the formation of the Solar System should explain why a planet didn't form there, but instead left behind a belt of construction material.

8-2b Comets

In contrast to the rocky asteroids, the brightest **comets** are impressively beautiful objects, as you can see in Figure 8-2. Most comets are faint, however, and difficult to locate even at their brightest. A comet may take months to sweep through the inner Solar System,

comets One of the small, icy bodies that orbit the Sun and produce tails of gas and dust when they approach the Sun.

1 TWO KINDS OF PLANETS

The distinction between the Terrestrial planets and the Jovian planets is dramatic. The inner four planets, Mercury, Venus, Earth, and Mars, are **Terrestrial** (Earth-like) **planets**, meaning they are small, dense, rocky worlds with little or no atmosphere. The outer four planets, Jupiter, Saturn, Uranus, and Neptune, are **Jovian** (Jupiter-like) **planets**, meaning they are large, low-density worlds with thick atmospheres and liquid or ice interiors.

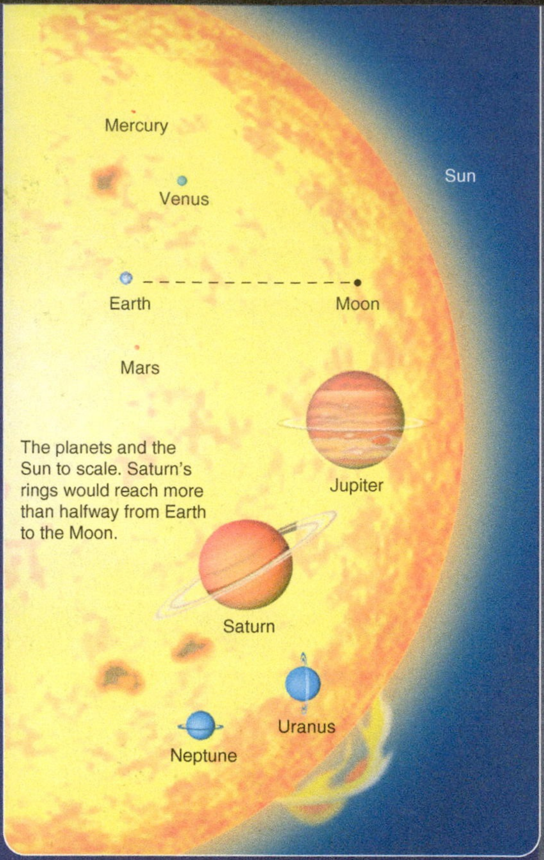

The planets and the Sun to scale. Saturn's rings would reach more than halfway from Earth to the Moon.

a Planetary orbits to scale. The Terrestrial planets lie close to the Sun, whereas the Jovian planets are spread far from the Sun.

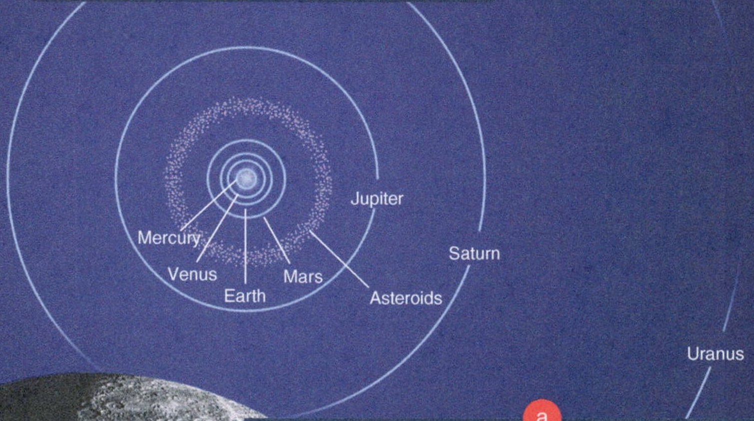

1a

Of the Terrestrial planets, Earth is the most massive, but the Jovian planets are much more massive. Jupiter contains over 300 Earth masses, Saturn nearly 100 Earth masses. Uranus and Neptune each contain about 15 Earth masses.

a Mercury is only 40 percent larger than Earth's Moon, and its weak gravity cannot retain a permanent atmosphere. Like the Moon, it is covered with craters from meteorite impacts.

Science/AAAS/NASA

© UC Regents/Lick Observatory

2

Craters are common on all of the surfaces in the Solar System that are strong enough to retain them. Earth has about 150 impact craters, but many more have been erased by erosion. Terrestrial planets, asteroids, comet nuclei, and nearly all of the moons in the Solar System are scarred by craters. Ranging from microscopic to hundreds of kilometers in diameter, these craters have been produced over the ages by meteorite impacts. When astronomers see a rocky or icy surface that contains few craters, they know that the surface is young.

3

The Terrestrial planets have densities like that of rock or metal. The Jovian planets all have low densities, and Saturn's average density is only about 70 percent that of water.

The atmospheres of the Jovian planets are turbulent, and some are marked by great storms such as the Great Red Spot on Jupiter, but the atmospheres are not deep. If Jupiter were shrunk to the size of a tennis ball, its atmosphere would be no deeper than the fuzz.

a **Mercury is so close to the Sun it is difficult to study from Earth. The *MESSENGER* spacecraft went into orbit around Mercury in 2011 and was able to take detailed photos of the planet's entire surface.**

b **The surface of Venus is not visible through its cloudy atmosphere, but radar maps reveal a dry world of craters and volcanoes.**

c **These five worlds are shown in proper relative size.**

d **Mars has a thin atmosphere and little water. Craters and volcanoes are common on its desert surface.**

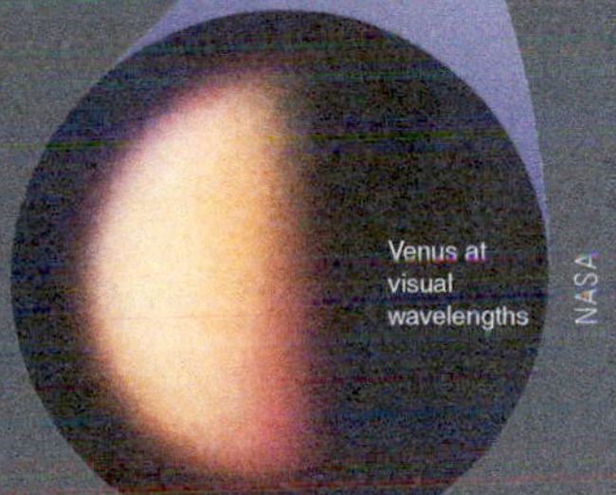

3a

The interiors of the Jovian planets contain small cores of heavy elements such as metals, surrounded by a liquid. Jupiter and Saturn contain hydrogen forced into a liquid state by the high pressure. Less-massive Uranus and Neptune contain heavy-element cores surrounded by partially solid water mixed with some rocks and minerals.

e **These Jovian worlds are shown in proper relative size.**

3b

The Jovian planets have extensive systems of satellites. For example, Jupiter is orbited by four large moons which were discovered by Galileo in 1610, and dozens of smaller moons discovered up to the present day.

f **The Terrestrial planets are drawn here to the same scale as the Jovian planets.**

3c

All four Jovian planets have ring systems. Saturn's rings are made of ice particles. The rings of Jupiter, Uranus, and Neptune are made of dark rocky particles. Terrestrial planets have no rings.

FIGURE 8-2

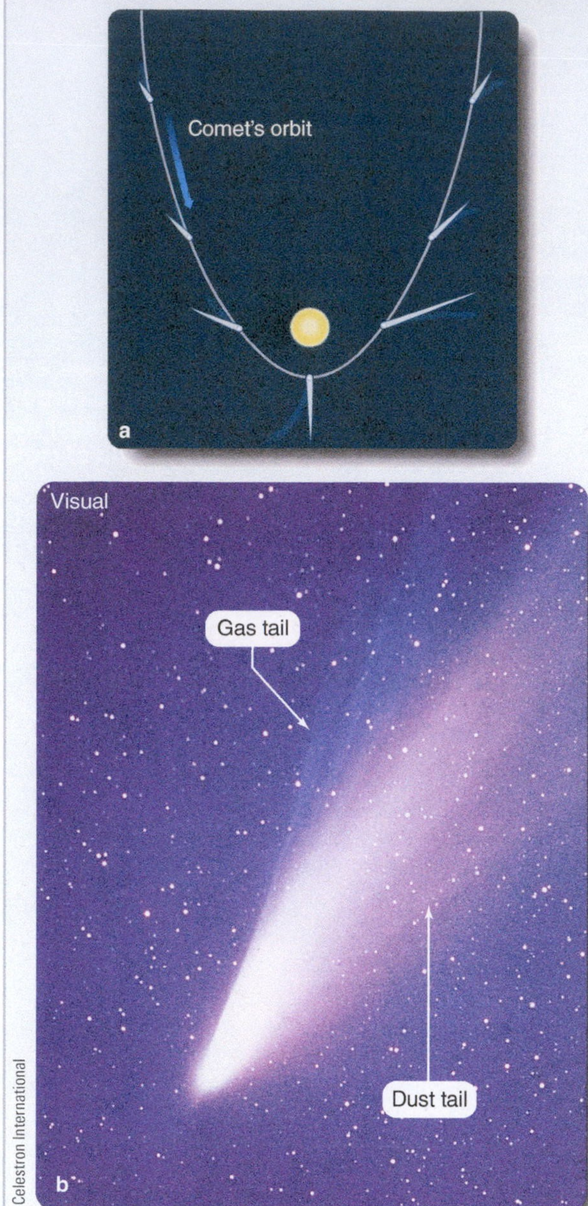

(a) A comet in a long, elliptical orbit becomes visible when the Sun's heat vaporizes its ices and pushes the gas and dust away in a tail pointing away from the Sun. (b) A comet may remain visible in the evening or morning sky for weeks as it moves through the inner Solar System. Comet West was in the sky during March 1976.

volatile Easily vaporized.

Kuiper Belt The collection of icy objects orbiting in a region from just beyond Neptune out to 50 AU or more from the Sun.

meteor A small bit of matter heated by friction to incandescent vapor as it falls into Earth's atmosphere.

during which time it appears as a glowing head with an extended tail of gas and dust.

The beautiful tail of a comet can be longer than an AU, but it is produced by an icy nucleus (sometimes described as a dirty snowball) only a few kilometers in diameter. The nucleus remains frozen and inactive while it is far from the Sun. As the nucleus moves along its elliptical orbit into the inner Solar System, the Sun's heat begins to vaporize the ices, releasing gas and dust. The pressure of sunlight and the solar wind pushes the gas and dust away, forming a long tail. The gas and dust respond differently to the forces acting on them so they sometimes separate into two separate subtails (Figure 8-2b). The motion of the nucleus along its orbit, the pressure of sunlight, and the outward flow of the solar wind cause the tails of comets to point always approximately away from the Sun (Figure 8-2a).

Comet nuclei contain ices of water and other **volatile** (easily vaporized) compounds such as carbon dioxide, carbon monoxide, methane, ammonia, and so on. These ices are the kinds of compounds that should have condensed from the outer solar nebula, and that makes astronomers think that comets are ancient samples of the gases and dust from which the outer planets formed.

Five spacecraft flew past the nucleus of Comet Halley when it passed through the inner Solar System in 1985 and 1986. Since then, spacecraft have visited the nuclei of several other comets. Images show that comet nuclei are irregular in shape and very dark, with jets of gas and dust spewing from active regions on the nuclei (Figure 8-3). In general, these nuclei are darker than a lump of coal, which suggests they have composition similar to certain dark, water- and carbon-rich meteorites that are described in the next section.

Since 1992, astronomers have discovered more than a thousand small, dark, icy bodies orbiting in the outer fringes of the Solar System beyond Neptune. This collection of objects is called the **Kuiper Belt** after the Dutch-American astronomer Gerard Kuiper (pronounced KIE-per), who predicted their existence in the 1950s. There are probably 100 million bodies larger than 1 km in the Kuiper Belt, and any successful hypothesis for how the Solar System formed should explain how they came to be where they are. Astronomers have evidence that some comets, those that move in the plane of the Solar System with relatively short orbital periods, come from the Kuiper Belt.

8-2c Meteors, Meteoroids, Meteorites

Unlike the stately comets, **meteors** flash across the sky in momentary streaks of light. They are commonly called "shooting stars." Of course, they are not stars but

FIGURE 8-3

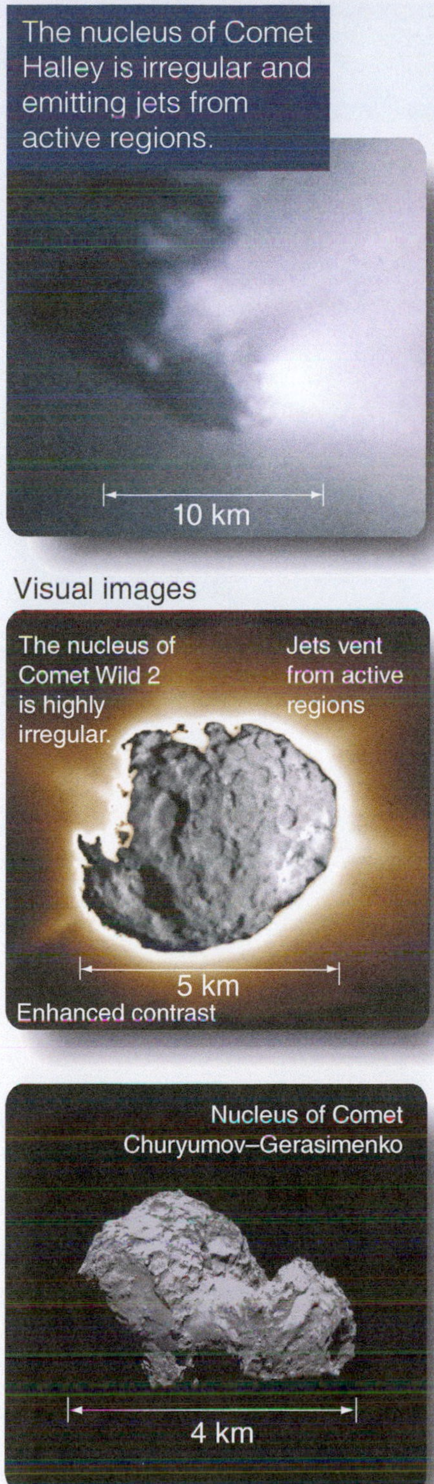

Visual-wavelength images made by passing spacecraft show comet nuclei as dark, irregular objects a few kilometers across. In some cases, jets of gas and dust are seen coming from active regions where sunlight vaporizes ices.

small bits of rock and metal falling into Earth's atmosphere and bursting into incandescent vapor about 80 km (50 mi) above the ground because of friction with the air. This vapor condenses to form dust, which settles slowly to the ground, adding about 40,000 tons per year to Earth's mass.

Technically, the word *meteor* refers to the streak of light in the sky. In space, before its fiery plunge, the object is called a **meteoroid**. Most meteoroids are the size of dust, grains of sand, or tiny pebbles. Almost all the meteors you see in the sky are produced by meteoroids that weigh less than 1 gram. Only rarely is a meteoroid massive and strong enough to survive its plunge, reach Earth's surface, and become what is called a **meteorite**.

Meteorites can be divided into three broad categories. *Iron* meteorites are solid chunks of iron and nickel. *Stony* meteorites are silicate masses that resemble Earth rocks. *Stony-iron* meteorites are mixtures of iron and stone. These types of meteorites are illustrated in Figure 8-4.

One type of stony meteorite called **carbonaceous chondrite** has a chemical composition that resembles a cooled lump of the Sun with the hydrogen and helium removed. These meteorites generally contain abundant volatile compounds including significant amounts of carbon and water, and may have similar composition to comet nuclei. Heating would have modified and driven off these fragile compounds, so carbonaceous chondrites must not have been heated since they formed. Astronomers conclude that carbonaceous chondrites, unlike the planets, have not changed since they formed, and thus give direct information about the early Solar System.

You can find evidence of the origin of meteors through one of the most pleasant observations in astronomy by watching a **meteor shower**, a display of meteors that are clearly related by a common origin. For example, the Perseid meteor shower occurs each year in August, and during the height of the shower you might see as many as 40 meteors per hour. The Perseid shower is so named because all its meteors appear to come

meteoroid A meteor in space before it enters Earth's atmosphere.

meteorite A meteor that survives its passage through the atmosphere and strikes the ground.

carbonaceous chondrite Stony meteorite that contains small glassy spheres called chondrules and volatiles. These chondrites may be the least altered remains of the solar nebula still present in the Solar System.

meteor shower A display of meteors that appear to come from one point in the sky, understood to be cometary debris.

FIGURE 8-4

The main types of meteorites have distinctive characteristics. (a) An iron meteorite, actually composed mostly of a mixture of iron and nickel. (b) An iron meteorite's Widmanstätten crystal pattern shows that this material cooled very slowly from a molten state and must have been inside a fairly large object. (c) A stony-iron meteorite. (d) A stony meteorite. (e) Carbonaceous chondrite meteorites are a rare type of stony meteorite that is rich in carbon, making the rock very dark.

from a point in the constellation Perseus. Meteor showers are seen when Earth passes near the orbit of a comet (Figure 8-5). The meteors in meteor showers must be produced by dust and debris released from the icy head of the comet. In contrast, the orbits of some meteorites have been calculated to lead back into the asteroid belt.

An important reason to mention meteorites here is for one specific clue they can give you concerning the solar nebula: Meteorites can reveal the age of the Solar System.

8-3 THE ORIGIN OF THE SOLAR SYSTEM

Astronomers have a hypothesis for the origin of our Solar System that is consistent with its characteristics, with what is now known about other planetary systems (described later in this chapter), and with observations of star formation (see Chapter 10). In fact, this hypothesis is so comprehensive and successful at explaining what we see that it can be considered to have graduated

FIGURE 8-5

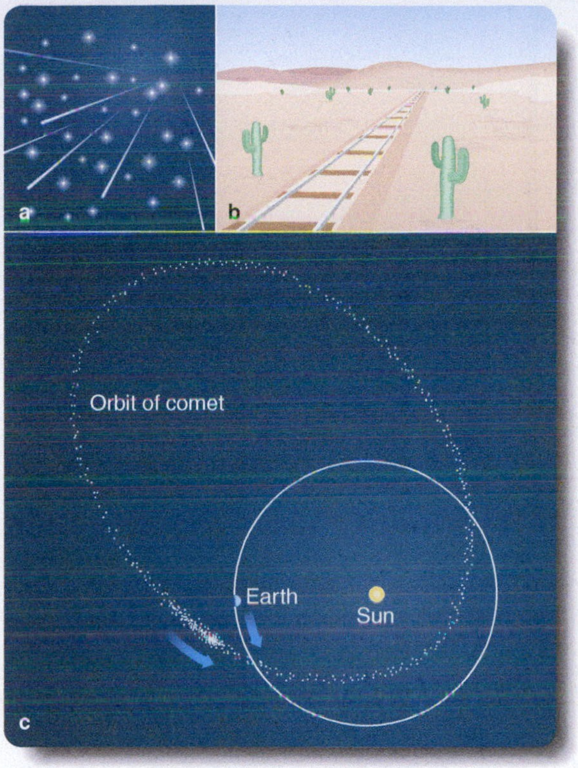

(a) Meteors in a meteor shower enter Earth's atmosphere along parallel paths, but perspective makes them appear to diverge from a single point in the sky. (b) Similarly, parallel railroad tracks appear to diverge from a point on the horizon. (c) Meteors in a shower are debris left behind as a comet's icy nucleus vaporizes. The rocky and metallic bits of matter spread along the comet's orbit. If Earth passes through such material, you can see a meteor shower.

from being just a hypothesis to being properly called a theory. Today, scientists are continuing to refine the details of that theory.

8-3a The Age of the Solar System

An important clue to how the Solar System formed is ages of the Earth, Moon, and other planets relative to each other and to the Sun. The most accurate way to find the age of a rocky body is to bring a sample into the laboratory and determine its age by analyzing the radioactive elements it contains.

When a rock solidifies, the process of cooling causes it to incorporate known proportions of the chemical elements. A few of these elements are radioactive and decay into other elements, called daughter elements or isotopes. The **half-life** of a radioactive element is the time it takes for half of the radioactive atoms to decay into the daughter elements. For example, potassium-40 decays into daughter isotopes calcium-40 and argon-40 with a half-life of 1.25 billion years. Another example, uranium-238, decays with a half-life of 4.5 billion years to lead-206 and other isotopes. As time passes, the abundance of a radioactive element in a rock gradually decreases, and the abundances of the daughter elements gradually increase (Figure 8-6). You can estimate the original abundances of the elements in the rock from rules of chemistry and observations of rock properties in general. Thus, measuring the present abundances of the parent and daughter elements allows you to find the age of the rock. This works best if you have several radioactive element "clocks" that can be used as independent checks on each other.

Of course, to find a radioactive age, you need a sample in the laboratory, and the only celestial bodies from which scientists have samples are Earth, the Moon, Mars, and meteorites. The oldest Earth rocks so far discovered and dated are tiny zircon crystals from Australia that are 4.4 billion years old. The surface of Earth is active, and the crust is continually destroyed and reformed from material welling up from beneath the crust (see Section 6-2b of Chapter 6). Consequently, the age of these oldest rocks tells you only that Earth is *at least* 4.4 billion years old.

Unlike Earth's surface, the Moon's surface is not being recycled by constant geologic activity, so you can guess that more of it might have survived unaltered since early in the history of the Solar System. The oldest Moon rocks brought back by the *Apollo* astronauts are 4.48 billion years old. That means the Moon must be at least 4.48 billion years old. Although no one has yet been to Mars, more than a dozen meteorites found on Earth have been identified by their chemical composition as having come from Mars. The oldest has an age of approximately 4.5 billion years. Mars must be at least that old.

The most important source for determining the age of the Solar System is meteorites. As you learned earlier, carbonaceous chondrite meteorites have compositions indicating that they have not been heated much, or otherwise altered, since they formed. They have a range of ages with a consistent and precise upper limit of 4.56 billion years. This figure is widely accepted as the age of the Solar System and is often rounded to 4.6 billion years. That is in agreement with the age of the Sun, which is estimated to be 5 billion years plus or minus 1.5 billion years using mathematical

half-life The time required for half of the radioactive atoms in a sample to decay.

FIGURE 8-6

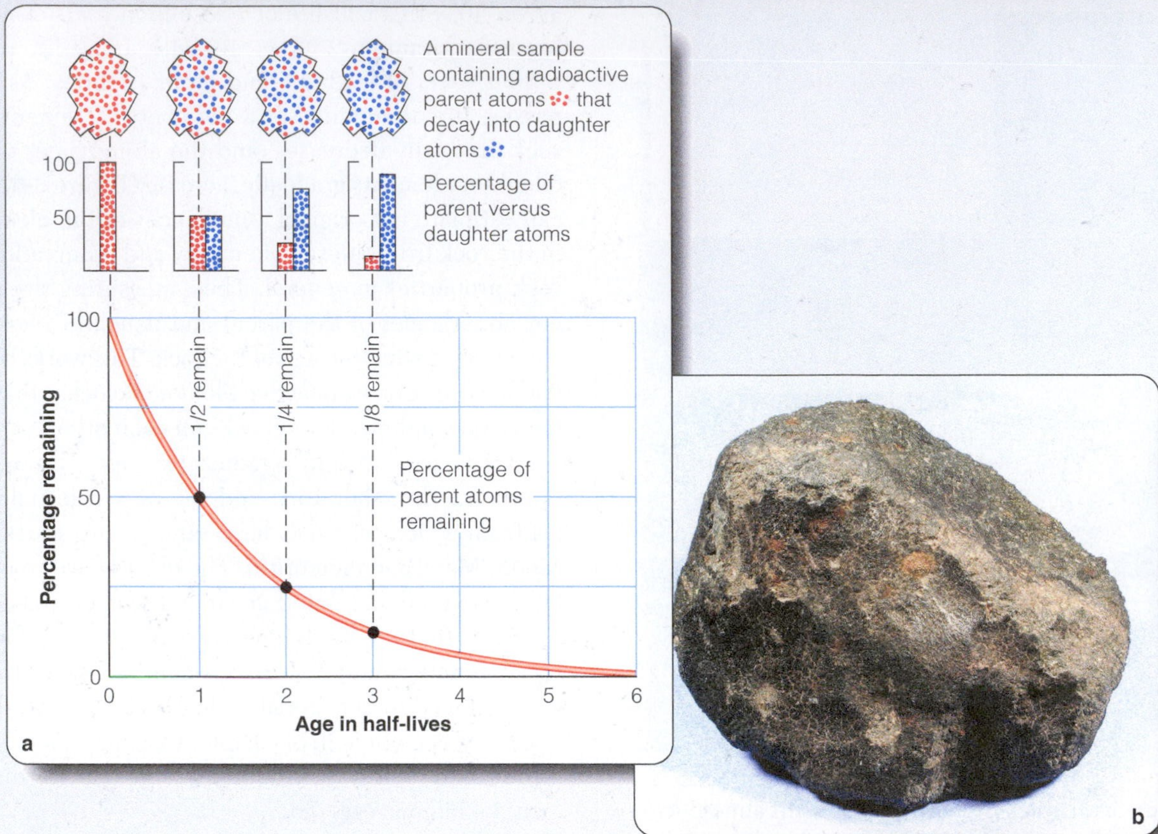

(a) The radioactive parent atoms (*red*) in a mineral sample decay into daughter atoms (*blue*). Half the radioactive atoms are left after one half-life, a fourth after two half-lives, an eighth after three half-lives, and so on. (b) Radioactive dating shows that this fragment of the Allende meteorite is 4.56 billion years old. It contains a few even older interstellar grains, which formed long before our Solar System did.

TABLE 8-1 CHARACTERISTIC PROPERTIES OF THE SOLAR SYSTEM

1. Disk shape of the Solar System
   - Orbits in nearly the same plane
   - Common direction of rotation and revolution
2. Two planetary types
   - Terrestrial—inner planets; high density
   - Jovian—outer planets; low density
3. Planetary rings and large satellite systems
   - Yes for Jupiter, Saturn, Uranus, and Neptune
   - No for Mercury, Venus, Earth, and Mars
4. Space debris—asteroids, comets, and meteoroids
   - Asteroids in inner Solar System, composition like Terrestrial planets
   - Comets in outer Solar System, composition like Jovian planets
5. Common age of about 4.6 billion years measured or inferred for Earth, the Moon, Mars, meteorites, and the Sun

models of the Sun's interior (see Chapter 10) that are completely independent of meteorite radioactive ages.

Apparently, all the bodies of the Solar System formed at about the same time some 4.6 billion years ago. You can include this fact in a list of Solar System properties that need to be explained by a complete theory of its formation (Table 8-1).

8-3b The Solar Nebula Theory

The **solar nebula theory** (Figure 8-8) supposes that planets form in rotating disks of gas and dust around young stars. Astronomers have found clear evidence that disks of gas and dust are common around young stars. This idea is so comprehensive and explains so many observations that it can be considered to have "graduated" from being just a hypothesis to being properly called a theory. Bipolar flows from protostars (see Chapter 10) were the first evidence of such disks, but now these disks can be imaged

solar nebula theory The theory that the planets formed from a spinning disk of material around the forming Sun.

FIGURE 8-7

A fanciful impression of the Solar System as if seen from outside. All the planets orbit in the same direction, in one plane, in approximately circular orbits. Comets, in contrast, normally have very eccentric orbits that are often inclined to the plane of the planets' orbits. These are all clues to how the Solar System formed. The planets are shown here roughly 2000 times larger than their true diameters relative to the sizes of their orbits.

directly by the *Hubble Space Telescope* (Figure 8-9) and other facilities. Our own planetary system formed in such a disk-shaped cloud around the Sun. When the Sun became luminous enough, the remaining gas and dust were blown away into space, leaving the planets orbiting the Sun.

According to the solar nebula theory, Earth and the other planets of our Solar System formed billions of years ago as the Sun condensed from the interstellar medium. This theory predicts that most stars should have planets because planet formation is a natural part of star formation, and therefore planets should be very common in the Universe, probably more common than stars.

8-3c Chemical Composition of the Solar Nebula

Everything astronomers know about the Solar System and star formation suggests that the solar nebula was a fragment of an interstellar gas cloud. Such a cloud would have been mostly hydrogen with some helium and minor traces of the heavier elements.

This is precisely what you see in the composition of the Sun (look back to Table 5-1). Analysis of the solar spectrum shows that the Sun is mostly hydrogen, with a quarter of its mass being helium and only about 2 percent being heavier elements. Of course, nuclear reactions have fused some hydrogen into helium, but that happens in the Sun's core and has not affected its surface composition.

FIGURE 8-8

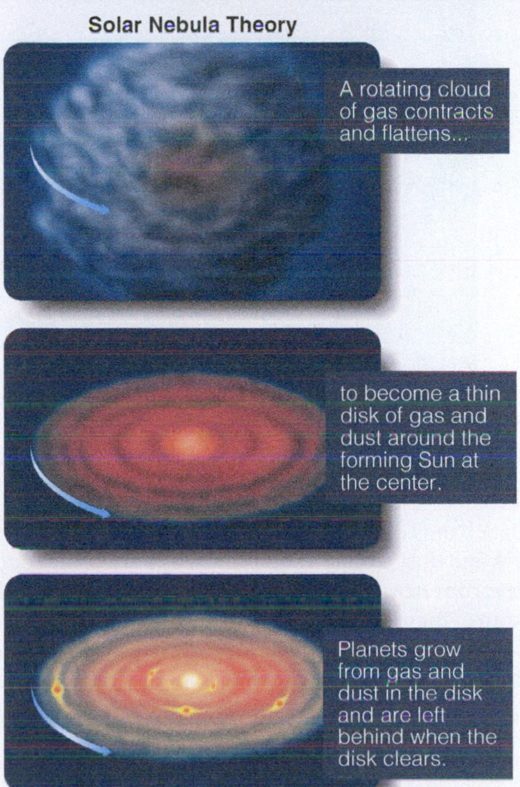

The solar nebula theory implies that the planets formed along with the Sun.

FIGURE 8-9

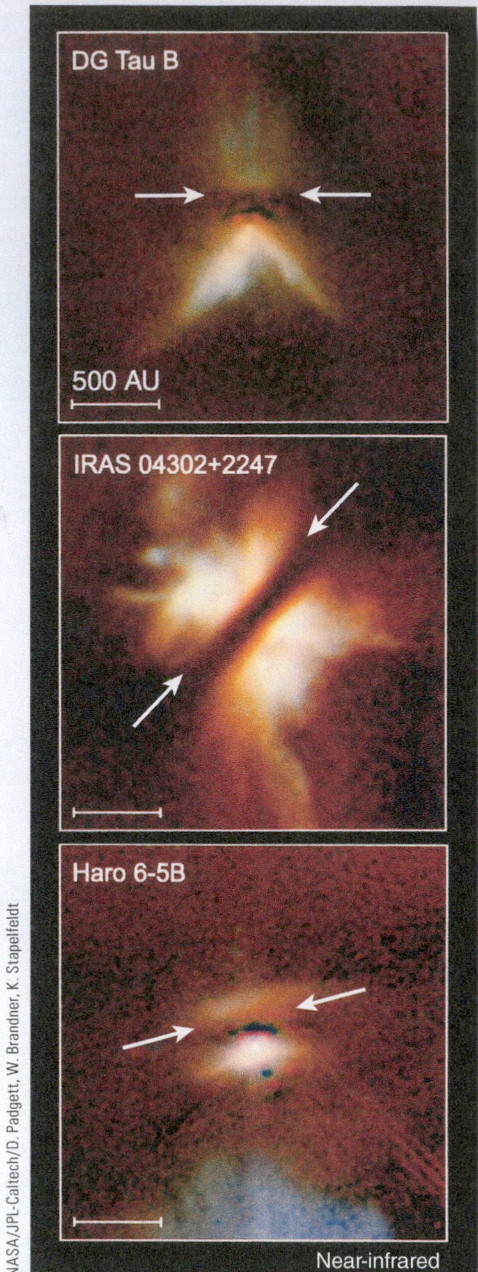

Dark bands (indicated by arrows) are edge-on disks of gas and dust around young stars seen in *Hubble Space Telescope* near-infrared images. Planets may eventually form in these disks. These systems are so young that material is still falling inward and being illuminated by light from the stars.

uncompressed densities The density a planet would have if its gravity did not compress it.

ice line Boundary in the forming Solar System beyond which water vapor could freeze to form ice.

That means the composition of the Sun's atmosphere revealed in its spectrum is essentially the composition of the solar nebula gases from which it formed.

You can see that same solar nebula composition reflected in the chemical compositions of the planets. The composition of Jupiter and the other Jovian planets resembles the composition of the Sun. Furthermore, if you allowed low-density gases to escape from a blob of Sun stuff, the remaining heavier elements would resemble the overall composition of Earth and the other Terrestrial planets, as well as carbonaceous chondrite meteorites.

8-3d Condensation of Solids

The key to understanding the process that converted the nebular gas into solid matter is the observed variation in density among Solar System objects. You have already noted that the four inner Earth-like planets are small, high-density objects, whereas the outer, Jupiter-like planets are low-density, giant planets. This division is due to the different ways gases condensed into solids in the inner and outer regions of the solar nebula.

Even among the four Terrestrial planets, you will find a pattern of slight differences in density. The **uncompressed densities**—the densities the planets would have if their gravity did not compress them, or to put it another way, the densities of their original construction material—can be calculated from the actual densities and masses of each planet. In general, the closer a planet is to the Sun, the higher is its uncompressed density.

This density variation is understood to have originated when the Solar System first formed solid grains. The kind of matter that condensed in a particular region would depend on the temperature of the gas there. In the inner regions, the temperature seems to have been 1500 K or so. The only materials that can form grains at this temperature are compounds with high melting points, such as metal oxides and pure metals, which are very dense, corresponding to the composition of Mercury. Farther out in the nebula it was cooler, and silicates (rocky material) could condense. These are less dense than metal oxides and metals, corresponding more to the compositions of Venus, Earth, and Mars. Somewhere farther from the Sun there was a boundary called the **ice line** beyond which the water vapor could freeze to form ice. Not much farther out, compounds such as methane and ammonia could condense to form other ices. Water vapor, methane, and ammonia were abundant in the solar nebula, so beyond the ice line, the nebula was filled with a blizzard of ice particles, and those ices have low densities like the Jovian planets.

The sequence in which different materials condensed from the gas at different distances from the Sun is called the **condensation sequence** (Table 8-2). This is really a hypothesis that says the planets, forming at different distances from the Sun, would have accumulated from different kinds of materials. The original chemical composition of the solar nebula should have been roughly the same throughout the nebula. The important factor was temperature: The inner nebula was hot, and only metals and rock could condense there, whereas the cold outer nebula could form lots of ices in addition to metals and rock. The ice line seems to have been between Mars and Jupiter, and it separates the formation of the dense Terrestrial planets from that of the lower-density Jovian planets.

8-3e Accretion of Planetesimals

In the development of a planet, three groups of processes operate to collect solid bits of matter—rock, metal, or ices—into larger bodies called **planetesimals** that eventually build the planets. The study of planet building is the study of these three groups of processes: condensation, accretion, and gravitational collapse. You encountered the condensation process in the previous section. Accretion and gravitational collapse will be described in detail in this section and the following section.

According to the solar nebula theory, planetary development in the solar nebula began with the growth of dust grains. These specks of matter, whatever their composition, grew from microscopic size first by condensation, then by accretion.

A particle grows by **condensation** when it adds matter, one atom or molecule at a time, from a surrounding gas. Snowflakes, for example, grow by condensation in Earth's atmosphere. In the solar nebula, dust grains were continuously bombarded by atoms of gas, some of which stuck to the grains.

The second process is **accretion**, the sticking together of solid particles. You may have seen accretion in action if you have walked through a snowstorm with big, fluffy flakes. If you caught one of those "flakes" on your mitten and looked closely, you would have seen that it was actually made up of many tiny, individual flakes that had collided as they fell and accreted to form larger particles. Model calculations indicate that in the solar nebula the dust grains were, on the average, no more than a few centimeters apart, so they collided frequently and accreted into larger particles.

There is no clear distinction between a very large grain and a very small planetesimal, but you can consider an object a planetesimal when its diameter approaches a kilometer or so, like the size of a typical small asteroid or comet.

Objects in the solar nebula larger than a centimeter were subject to new processes that tended to concentrate them. For example, collisions with the surrounding gas and with each other would have caused growing planetesimals to settle into a thin disk estimated to have been only about 0.01 AU thick in the central plane of the rotating nebula. This concentration of material would have made further planetary growth more rapid. Computer models show that the rotating disk of particles should have been gravitationally unstable and would have been disturbed by spiral density waves much like those found in spiral galaxies. This would have further concentrated the planetesimals and helped them coalesce into objects up to about 100 km (60 mi) in diameter.

Through these processes, the nebula became filled with trillions of solid particles ranging in size from

TABLE 8-2 THE CONDENSATION SEQUENCE

Temperature (K)	Condensate	Object; Estimated Temperature of Formation (K)
1500	Metal oxides	Mercury (1400)
1300	Metallic iron and nickel	
1200	Silicates	
1000	Feldspars	Venus (900)
680	Troilite (FeS)	Earth (600)
		Mars (450)
175	H_2O ice	Jovian (175)
150	Ammonia water ice	
120	Methane water ice	
65	Argon neon ice	Pluto (65)

condensation sequence The sequence in which different materials condense from the solar nebula depending on temperature and distance from the Sun.

planetesimal One of the small bodies that formed from the solar nebula and eventually grew into protoplanets. Asteroids and comets are remnants of the planetesimals.

condensation The growth of a particle by addition of material from surrounding gas, atom by atom.

accretion The sticking together of solid particles to produce a larger particle.

pebbles to small planets. As the largest began to exceed 100 km in diameter, new processes began to alter them, and a new stage in planet building began—the formation of protoplanets.

8-4 THE STORY OF PLANET BUILDING

You have learned that astronomers have several lines of evidence that the Solar System as a whole formed as a by-product of the Sun's formation, from the same cloud of raw material. And, astronomers observe disks around young stars that have the characteristics inferred for the original solar nebula from which the planets formed. As those disk nebulae cool down, computer models verified by observations predict that solid particles should condense and then accrete into small asteroid and comet-like bodies called planetesimals. Now the story of how those planetesimals became construction "bricks" for the planets can unfold, checked by observations of the properties of our Solar System and of planets around other stars.

8-4a Growth of Protoplanets

The coalescing of planetesimals eventually formed **protoplanets**, massive objects destined to become planets. As these larger bodies grew, new processes began making them grow faster and altered their physical structure.

If planetesimals collided at orbital velocities, it is unlikely that they would have stuck together. Typical orbital velocities in the Solar System are about 10 km/s (22,000 mph). Head-on collisions at this velocity would have vaporized the material. However, the planetesimals were all moving in the same direction in the nebular plane and didn't collide head-on. Instead, they merely "rubbed shoulders," so to speak, at low relative velocities. Such gentle collisions would have been more likely to combine them than to shatter them.

The largest planetesimals would grow the fastest because they had the strongest gravitational field and could more easily attract additional material. Computer models indicate that these planetesimals grew quickly to protoplanetary dimensions, sweeping up more and more material.

Protoplanets had to begin growing by accumulating solid material because they did not have enough gravity to capture and hold large amounts of gas. In the warm solar nebula, the atoms and molecules of gas were traveling at velocities much larger than the escape velocities of modest-size protoplanets. Because of that, in their early development, the protoplanets could only grow by attracting solid bits of rock, metal, and ice. Once a protoplanet approached a mass of 15 Earth masses or so, it could begin to grow by **gravitational collapse**, the rapid accumulation from the nebula of large amounts of infalling gas.

In its simplest form, the theory of Terrestrial protoplanet growth supposes that all the planetesimals in each orbital zone had about the same chemical composition. The planetesimals accumulated gradually to form a planet-size ball of material that was of homogeneous composition throughout. As the planet formed, heat accumulated in its interior both from the impacts of infalling planetesimals and from decay of short-lived radioactive elements. This heat eventually melted the planet and allowed it to differentiate. As you learned in Chapter 6 Section 6-1d, differentiation is the separation of material according to density. When the planet melted, the heavy metals such as iron and nickel settled to the core, while the lighter silicates floated to the surface to form a low-density crust. The story of planet formation from accretion of planetesimals followed by differentiation is shown in Figure 8-10.

This process depends on the presence of short-lived radioactive elements whose rapid decay would have released enough heat to melt the interior of planets. Astronomers know such radioactive elements were present because the oldest meteorites contain daughter isotope magnesium-26 that is produced by the decay of aluminum-26 with a half-life of only 0.74 million years.

There is evidence that not only protoplanets but also some planetesimals melted and differentiated. The Widmanstätten patterns of large crystals in iron meteorites (Figure 8-4b) indicate that the metal was once molten and deep enough inside some object that it took millions of years to cool and solidify. Also, the asteroid Vesta was once volcanic, producing the basalt lava flows seen on its surface.

If planets formed and were later melted by radioactive decay, gases released from the planet's interior would have formed an atmosphere. The creation of a planetary atmosphere from a planet's interior is called **outgassing**. Models of the formation of the Earth indicate that the

protoplanets Massive object, destined to become a planet, resulting from the coalescence of planetesimals in the solar nebula.

gravitational collapse The process by which a forming body such as a planet gravitationally captures gas rapidly from the surrounding nebula.

outgassing The release of gases from a planet's interior.

FIGURE 8-10

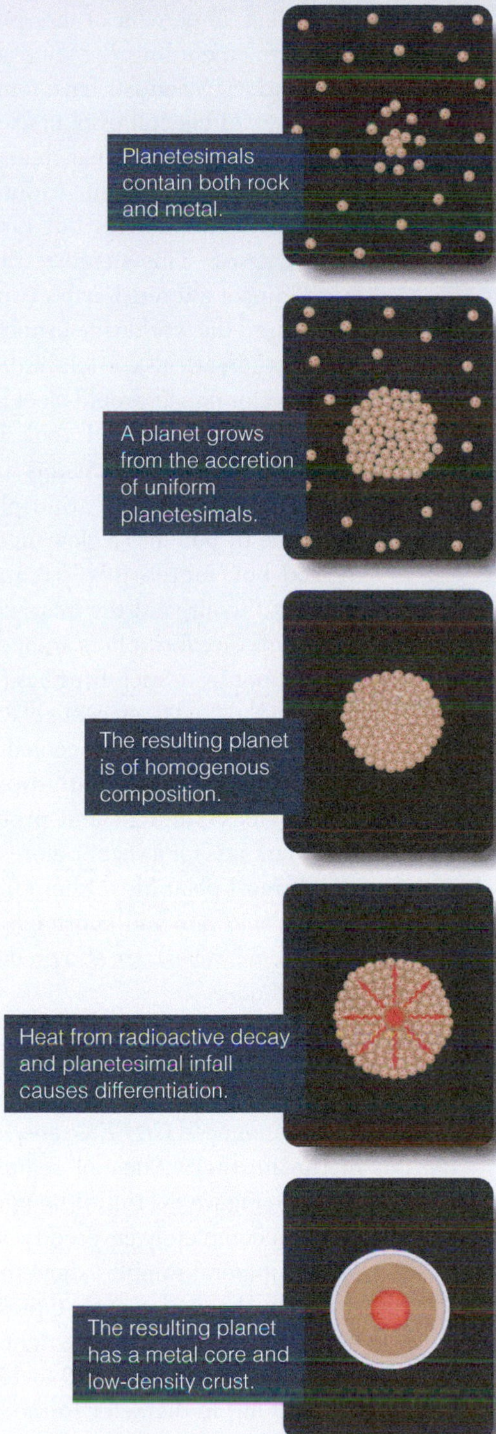

This simple model of planet building assumes planets formed from accretion and collision of planetesimals that were of uniform composition, containing both metals and rocky material, and that the planets later differentiated, meaning they melted and separated into layers by density and composition.

local planetesimals would not have included much water, so some astronomers now think that Earth's water and some of its present atmosphere accumulated late in the formation of the planets as Earth swept up volatile-rich planetesimals forming in the cooling solar nebula. Such icy planetesimals may have formed in the outer parts of the solar nebula and been scattered by encounters with the Jovian planets, creating a Solar System-wide bombardment of comets.

According to the solar nebula theory, the Jovian planets began growing by the same processes that built the Terrestrial planets, but the outer solar nebula not only contained solid bits of metals and silicates, it also included abundant ices. The Jovian planets grew rapidly and quickly became massive enough to grow by gravitational collapse, drawing in large amounts of gas from the solar nebula. Ices could not condense as solids at the locations of the Terrestrial planets, so those planets developed slowly and never became massive enough to grow by gravitational collapse.

The Jovian planets must have grown to their present size in less than about 10 million years, which is when astronomers calculate that the Sun become hot and luminous enough to blow away the gas remaining in the solar nebula. The Terrestrial planets grew from solids and not from the gas, so they could have continued to grow by accretion from the solid debris left behind when the gas was blown away. Model calculations indicate that the process of planet formation was almost completely finished by the time the Solar System was about 30 million years old.

8-4b Bombardment of the Planets

Astronomers have good reason to believe that comets and asteroids can hit planets. Meteorites hit Earth every day, and occasionally a large one can form a crater (Figure 8-11). Earth is marked by about 150 known meteorite craters. In a sense, this bombardment represents the slow continuation of the accretion of the planets. Earth's Moon, Mercury, Venus, Mars, and most of the Moons in the Solar System are covered with craters. A few of these craters have been formed recently by the steady rain of meteorites that falls on all the planets in the Solar System, but most of the craters appear to have been formed more than 4 billion years ago in what is called the heavy bombardment, when much of the remaining solid debris in the solar nebula was swept up by the planets. You learned in Chapter 6 Section 6-3c that the heavy bombardment may have had more than one episode.

FIGURE 8-11

(a) The Barringer Meteorite Crater (near Flagstaff, Arizona) is nearly 1.3 km (1 mi) in diameter and was formed about 50,000 years ago by the impact of an iron meteorite estimated to have been roughly 50 m (about 160 ft) in diameter. Notice the raised and deformed rock layers all around the crater. (b) Like all large-impact features, the Barringer Meteorite Crater has a raised rim and scattered ejecta. (c) This gravity map reveals a large crater and surrounding region of disturbed crust hidden below limestone near the village of Chicxulub in Mexico. The crater was formed about 65 million years ago by the impact of an asteroid or comet.

Sixty-five million years ago, at the end of the Cretaceous period, more than 75 percent of the species on Earth, including the dinosaurs, went extinct. Scientists have found a thin layer of clay all over the world that was laid down at that time, and it is rich in the element iridium—common in meteorites but rare in Earth's crust. This suggests that a large impact altered Earth's climate and caused the worldwide extinction.

Mathematical models indicate that a major impact would eject huge amounts of pulverized rock high above the atmosphere. As this material fell back, Earth's atmosphere would be turned into a glowing oven of red-hot meteorites streaming through the air, and the heat would trigger massive forest fires around the world. Soot from such fires has been found in the final Cretaceous clay layers. Once the firestorms cooled, the remaining dust in the atmosphere would block Sunlight and produce deep darkness for a year or more, killing off most plant life. Other effects, such as acid rain and enormous tsunamis (tidal waves), are also predicted by the models.

Geologists have located a crater at least 180 km (110 mi) in diameter centered near the village of Chicxulub (pronounced *CHEEK*-shoe-lube) in the northern Yucatán region of Mexico (Figure 8-11c). Although the crater is completely covered by sediments, mineral samples show that it contains shocked quartz typical of impact sites and that it is the right age. The impact of an object 10 to 15 km (6 to 10 mi) in diameter formed the crater about 65 million years ago, just when the dinosaurs and many other species died out. Most Earth scientists now consider this to be the scar of the impact that ended the Cretaceous period.

Earthlings watched in awe during 6 days in the summer of 1994 as

FIGURE 8-12

In 1992, Comet Shoemaker–Levy 9 passed within 1.3 planetary radii of Jupiter's center, well within its Roche limit, and tidal forces ripped the nucleus into more than 20 pieces. The fragmented pieces were as large as a few kilometers in diameter and spread into a long string of objects that looped away from Jupiter and then fell back to strike the planet, producing massive impacts over a period of 6 days in July 1994. Note that the impact flash in the upper right panel is larger than Earth.

20 or more fragments from the head of Comet Shoemaker–Levy 9 slammed into Jupiter and produced impacts equaling millions of megatons of TNT. Each impact created a fireball of hot gases and left behind dark smudges that remained visible for months afterward (Figure 8-12). Scars from smaller single impacts were observed on Jupiter in 2009 and in 2012. Major impacts on Earth occur less often because Earth is smaller, but they are inevitable.

We are sitting ducks. All of human civilization is spread out over Earth's surface and exposed to anything that falls out of the sky. Meteorites, asteroids, and comets bombard Earth, producing impacts that vary from dust settling on rooftops to blasts capable of destroying all life. In this case, the scientific evidence is conclusive and highly unwelcome. Statistically you are quite safe. The chance that a major impact will occur during your lifetime is so small that it is hard to estimate. But the consequences of such an impact are so severe that humanity should be preparing. One way to prepare is to find those **NEOs (Near-Earth Objects)** that could hit this planet, map their orbits in detail, and identify any that are dangerous.

NEO (Near-Earth Object) A small Solar System body (asteroid or comet) with an orbit near enough to Earth that it poses some threat of eventual collision.

8-4c The Jovian Problem

The solar nebula theory has been very successful in explaining the formation of the Solar System. But there are some problems, and the Jovian planets are the troublemakers.

The gas and dust disks around newborn stars don't last long. In Chapter 10 you will see images of dusty gas disks around the young stars in the Orion Nebula. Those disks are being evaporated by the intense ultraviolet radiation from hot stars within the nebula. Astronomers estimate that most stars form in clusters containing some massive stars, so this evaporation process must happen to most disks. Even if a disk did not evaporate quickly, model calculations predict that the gravitational influence of the crowded stars in a cluster should quickly strip away the outer parts of the disk. This is a troublesome observation because it seems to mean that planet-forming disks around young stars are unlikely to last longer than a few million years, and many must evaporate within 100,000 years or so. That's not long enough to grow a Jovian planet by the processes in the solar nebula theory. Yet, as you will learn in a later section of this chapter, Jovian planets are common in the Universe.

A modification to the solar nebula theory has come from mathematical models of the solar nebula. The results show that the rotating gas and dust of the solar nebula could have become unstable and formed outer planets by direct gravitational collapse. That is, massive planets may have been able to form directly from the nebula gas without first accreting a dense core from planetesimals. Jupiters and Saturns can form in these models within a few hundred years. If the Jovian planets formed in this way, they could have formed quickly, long before the solar nebula disappeared.

This new insight into the formation of the outer planets may help explain the formation of Uranus and Neptune. They are so far from the Sun that accretion could not have built them rapidly. It is hard to understand how they could have reached their present mass in a region where the material should have been sparse and orbital speeds are slow. Theoretical calculations show that Uranus and Neptune might instead have formed closer to the Sun, in the region of Jupiter and Saturn, and then moved outward by gravitational interactions with the other planets or with planetesimals in the Kuiper Belt. In any case, the formation of Uranus and Neptune is part of the Jovian problem.

The original solar nebula theory proposes that the planets formed by accreting a solid core and then, if they became massive enough, by gravitational collapse of nebula gas. The new theories suggest that some of the outer planets could have skipped the core accretion phase.

evolutionary theory An explanation of a phenomenon involving slow, steady processes of the sort seen happening in the present day.

catastrophic theory An explanation of a phenomenon involving special, sudden, perhaps violent, events.

8-4d Explaining the Characteristics of the Solar System

Now you have learned enough to put all the pieces of the puzzle together and explain the distinguishing characteristics of the Solar System in Table 8-1.

The disk shape of the Solar System is inherited from the solar nebula. The Sun and the planets revolve and rotate in the same direction because they are formed from the same rotating gas cloud. The orbits of the planets lie in the same plane because the rotating solar nebula collapsed into a disk, and the planets formed in that disk.

The solar nebula theory calls on continuing evolutionary processes to gradually build the planets. Scientists call this type of explanation an **evolutionary theory**. In contrast, a **catastrophic theory** invokes special, sudden, even violent, events. For example, as you learned earlier, Uranus rotates on its side and Venus rotates backward. Both these peculiarities could have been caused by off-center impacts of massive planetesimals on each planet as they were forming—an explanation of the catastrophic type. On the other hand, computer models suggest that the Sun can produce tides in the thick atmosphere of Venus that eventually could have resulted in the reversal of that planet's rotation—an explanation of the evolutionary type.

The second item in Table 8-1, the division of the planets into Terrestrial and Jovian worlds, can be understood through the condensation sequence. The Terrestrial planets formed in the inner part of the solar nebula, where the temperature was high and only compounds such as metals and silicates could condense to form solid particles. That produced the small, dense Terrestrial planets. In contrast, the Jovian planets formed in the outer solar nebula, where the lower temperature allowed the gas to form large amounts of ices, perhaps three times more ices than silicates. That allowed the Jovian planets to grow rapidly and become massive, low-density worlds. Also, Jupiter and Saturn are so massive they eventually were able to grow even larger by drawing in gas directly from the solar nebula. The Terrestrial planets could not do this because they never became massive enough.

The **heat of formation** (the energy released by infalling matter) was tremendous for these massive planets. Jupiter must have grown hot enough to glow with a luminosity of about 1 percent that of the present Sun. However, because it never got hot enough to start nuclear fusion as a star would, it never generated its own energy.

Jupiter is still hot inside. In fact, both Jupiter and Saturn radiate more heat than they absorb from the Sun, so they are evidently still cooling.

A glance at the Solar System suggests that you should expect to find a planet between Mars and Jupiter at the present location of the asteroid belt. Mathematical models show Jupiter grew into such a massive planet that it was able to gravitationally disturb the motion of nearby planetesimals. The bodies that should have formed a planet between Mars and Jupiter were broken up, thrown into the Sun, or ejected from the Solar System due to the gravitational influence of massive Jupiter. The asteroids seen today are the last remains of those rocky planetesimals.

The comets, in contrast, are evidently the last of the icy planetesimals. Some may have formed in the outer solar nebula beyond Neptune, but many probably formed among the Jovian planets, where ices could condense easily. Mathematical models show that the massive Jovian planets could have ejected some of these icy planetesimals into the far outer Solar System, to a region called the **Oort cloud**, from which comets come with very long periods and orbits highly inclined to the plane of the Solar System.

The icy Kuiper Belt Objects, including Pluto, appear to be ancient planetesimals that formed in the outer Solar System but were never incorporated into a planet. They orbit slowly far from the light and warmth of the Sun and, except for occasional collisions, have not changed much since the Solar System was young.

In Table 8-1, point 3 notes that all four Jovian worlds have ring systems, and you can understand this by considering the large mass of these worlds and their remote location in the Solar System. A large mass makes it easier for a planet to hold onto orbiting ring particles; and, being farther from the Sun, the ring particles are not as easily swept away by the pressure of sunlight and the solar wind. It is hardly surprising, then, that the Terrestrial planets, low-mass worlds located near the Sun, have no planetary rings.

The last entry in Table 8-1 is the common ages of Solar System bodies, and the solar nebula theory has no difficulty explaining that characteristic. The theory predicts that the planets formed at the same time as the Sun and should all have roughly the same age.

8-4e The History of the Atoms in Your Body

Here is a good place to pause, step back for some perspective, and summarize an amazing story that will be told across the next six chapters, providing context for the formation of Earth and the Sun. By the time the Universe was a few minutes old, the protons, neutrons, and electrons in your body had come into existence (see Chapter 14). You are made of very old matter.

Although those particles formed quickly, they were not linked together to form the atoms that are common today. Most of the matter was hydrogen, and about 25 percent of the mass was helium. Very few of the heavier atoms were made in the Big Bang. Although your body does not contain helium, it does contain many of those ancient hydrogen atoms unchanged since the Universe began.

Within a few hundred million years after the Big Bang, matter began to collect to form galaxies containing billions of stars. You have learned how nuclear reactions inside stars combine low-mass atoms, such as hydrogen, to make heavier atoms. Generations of stars cooked the original particles, fusing them into atoms such as carbon, nitrogen, and oxygen (see Chapters 11 and 12). Those are common atoms in your body. Even the calcium atoms in your bones were assembled inside stars.

Most of the iron in your body was produced by nuclear fusion reactions during explosions called *supernovae* that occur at the ends of the lives of stars more massive than the Sun, or by decay of radioactive atoms in the clouds of matter ejected by some supernovae. You will encounter these titanic explosions and their results in Chapter 11. Other atoms heavier than iron, for example, iodine that is critical in the function of your thyroid gland, plus atoms of other elements that are uncommon enough to be expensive such as gold, silver, and platinum, also are mostly or entirely made during the violent deaths of rare massive stars.

Our Galaxy contains at least 100 billion stars, including the Sun. It formed from a cloud of gas and dust about 5 billion years ago, and the atoms in your body were part of that cloud. How the Sun took shape, how the cloud gave birth to the planets, how the atoms in your body found their way onto Earth and into you is the story of this chapter. As you explore the origin of

heat of formation In planetology, the heat released by infalling matter during the formation of a planetary body.

Oort cloud The hypothetical source of comets, a swarm of icy bodies understood to lie in a spherical shell extending to 100,000 AU from the Sun.

the Solar System, keep in mind the great chain of origins that created the atoms. As the geologist Preston Cloud remarked, "Stars have died that we might live."

8-5 PLANETS AROUND OTHER STARS

Are there planets orbiting other stars? Yes, that is now certain. Are there planets like Earth? The first such objects have now been discovered.

In Chapter 10 you will learn about dense disks of gas and dust around stars that are forming. Astronomers have evidence that these disks resemble the solar nebula and may have planets forming within them. For example, at least 50 percent of the stars in the Orion Nebula are encircled by dense disks of gas and dust with more than enough mass to make planetary systems like ours. The Orion starforming region is only a few million years old, so it does not seem likely that planets have finished forming there yet. Nevertheless, the important point for astronomers is that disks of gas and dust that could become planetary systems are a common feature of star formation.

8-5a Debris Disks

In addition to these planet-forming disks around young stars, infrared astronomers have found cold, low-density dust disks around older stars such as Vega and Beta Pictoris. Although much younger than the Sun, these stars are on the main sequence and have completed their formation, so they are clearly in a later stage than the newborn stars in Orion. These low-density disks generally have even lower- density inner zones where planets may have formed, as you can see in Figure 8-13. Such tenuous dust disks are sometimes called **debris disks** because they are understood to be dusty debris released in collisions among small bodies such as comets, asteroids, and Kuiper Belt Objects. Our own Solar System contains such dust, and astronomers have evidence the Sun has an extensive debris disk of cold dust extending far beyond the orbits of the planets.

Few planets orbiting stars with debris disks have been detected so far, but the presence of dust particles with lifetimes shorter than the ages of the stars assures you that remnant planetesimals such as asteroids and comets are present as the sources of the dust. If planetesimals are there, then you can expect that there are also planets orbiting those stars. Many of the debris disks have details of structure and shape that are probably caused by the gravity of planets orbiting within, or at the edges of, the debris (Figure 8-13).

Notice the difference between the two kinds of disks that astronomers have found. The low-density dust disks, such as the one around Beta Pictoris, are produced by dust from collisions among remnant planetesimals such as comets, asteroids, and Kuiper Belt Objects. Such disks are evidence that planetary systems have already formed.

In contrast, the dense disks of gas and dust such as those seen around the stars in Orion are sites where planets could be forming right now.

debris disk A disk of dust found by infrared observations around some stars. The dust is debris from collisions among asteroids, comets, and Kuiper Belt Objects.

extrasolar planet A planet orbiting a star other than the Sun.

8-5b Extrasolar Planets

A planet orbiting another star is called an **extrasolar planet**. Such a planet would be quite faint and difficult to detect so close to the glare of its star. But there are ways to find these planets. To see how, all you have to do is imagine walking a dog.

You will remember that Earth and its Moon orbit around their common center of mass, and two stars in a binary system orbit around their center of mass. When a planet orbits a star, the star moves very slightly as it orbits the center of mass of the planet–star system. Think of someone walking a poorly trained dog on a leash; the dog runs around pulling on the leash, and even if it were an invisible dog, you could plot its path by watching how its owner was jerked back and forth. Astronomers can detect a planet orbiting another star by watching how the star moves as the planet tugs on it.

The first planet orbiting a Sun-like star detected this way was discovered in 1995. It orbits the star 51 Pegasi. As the planet circles the star, the star wobbles slightly, and this very small motion of the star is detectable as Doppler shifts in the star's spectrum, as shown in Figure 8-14a. This is the same technique used to study spectroscopic binary stars (see Chapter 9). From the motion of the star and estimates of the star's mass, astronomers can deduce that the planet has half the mass of Jupiter and orbits only 0.05 AU from the star. Half the mass of Jupiter amounts to 160 Earth masses, so this is a large planet, larger than Saturn. Note also that it orbits very close to its star.

Astronomers were not surprised by the announcement that a planet orbited 51 Pegasi; for years they had

FIGURE 8-13

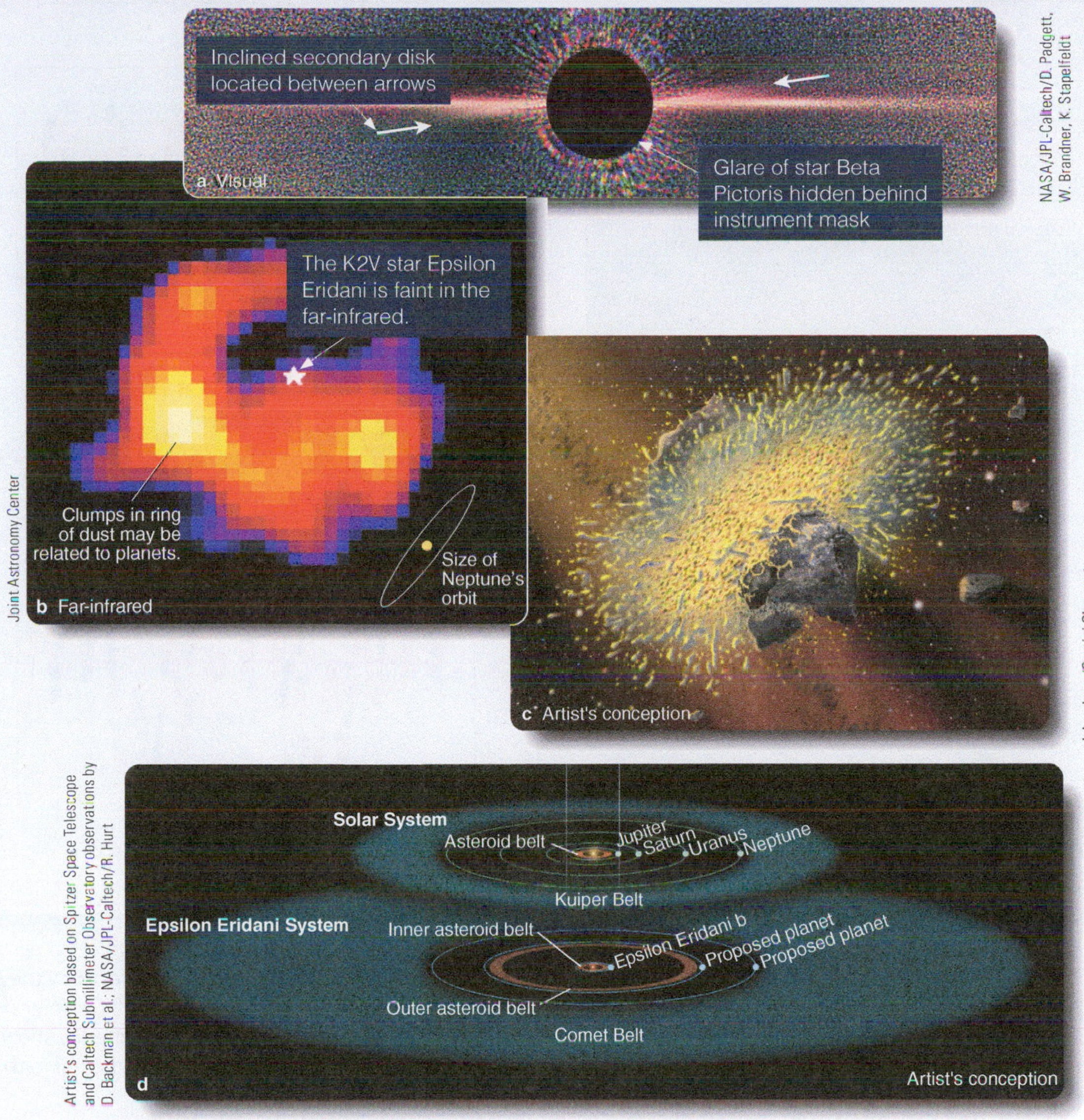

Dust disks around other stars. (a) Hiding the bright star behind a mask reveals a dust disk around Beta Pictoris. A second fainter disk may mark the orbital plane of a massive planet. (b) At far-infrared wavelengths, dust disks can be brighter than the star. (c) Asteroid collisions are rare but produce lots of dust, which is blown away quickly. The presence of such dust is evidence of planetesimals. (d) Star Epsilon Eridani has dust disks comparable to those in our Solar System. In both cases, planets control the extent of the dust. Dust in debris belts around older main-sequence stars indicates ongoing collisions of remnant planetesimals such as asteroids and comets. Such activity in our Solar System is ultimately driven by the gravitational influence of planets. The locations of debris belt edges may be defined by adjacent orbits of planets. The inferred architecture of the Epsilon Eridani planetary system is shown in comparison with our Solar System.

assumed that many stars had planets. Nevertheless, they greeted the discovery with typical skepticism (see **How Do We Know? 8-2**). That skepticism led to careful tests of the data and further observations that confirmed the discovery. In fact, more than 400 planets have been discovered in this way, including at least four planets orbiting the star Upsilon Andromedae (Figure 8-14b) and five orbiting 55 Cancri— true planetary systems. More than 500 such multiple-planet systems have been found.

Another way to search for planets is to look for changes in the brightness of a system when an orbiting planet crosses in front of or behind its parent star.

FIGURE 8-14

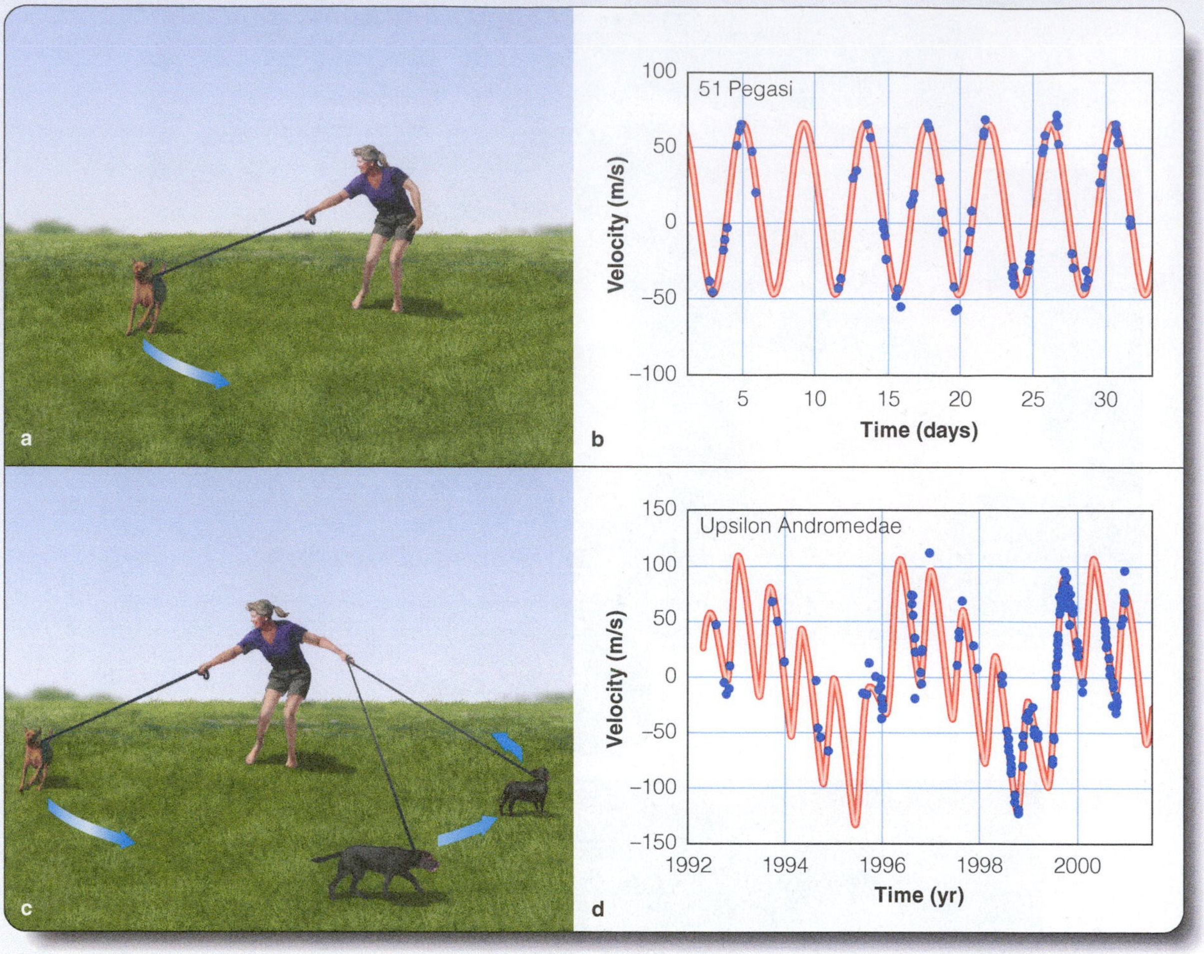

(a) A person walking a lively dog is tugged off course by the dog. (b) The star 51 Pegasi is pulled back and forth by the gravity of the planet that orbits it every 4.2 days. The wobble is detectable in precision observations of the star's Doppler shifts. (c) A person walking multiple dogs has a complicated motion. (d) Doppler shifts of the star Upsilon Andromedae show the combined effects of at least four planets orbiting it. The influence of its shortest-period planet has been removed in this graph to reveal more clearly the orbital influences of the other three planets.

These brightness changes are very small, but detectable, and can be used to measure sizes and in some cases temperatures of the planets directly. This is the same technique used to study eclipsing binary stars (see Chapter 9). The *Spitzer* infrared space telescope confirmed the existence of two extrasolar planets by observing them pass behind their parent stars, finding them to be very hot and about the size of Jupiter.

hot Jupiters A massive and presumably Jovian-composition planet that orbits close to its star and consequently has a high temperature.

The *Kepler Space Telescope*, launched in 2009, "stared" continuously at 150,000 stars during its 4-year primary mission, searching for small decreases in brightness as planets pass in front of them. The *Kepler* team has identified more than 2000 candidate planets, some smaller in diameter than Earth. One system discovered by *Kepler* and confirmed by ground-based Doppler shift measurements has six super-Earth planets, more massive than Earth but less massive than Uranus (Figure 8-15a).

The planets discovered so far tend to be massive and have short periods—they have been dubbed "**hot Jupiters**." Lower-mass planets (or longer-period planets) are harder to detect. Lower-mass planets don't tug on their stars very much, and present-day spectrographs can't detect the very small velocity changes that these gentle tugs produce. Planets with longer periods

FIGURE 8-15

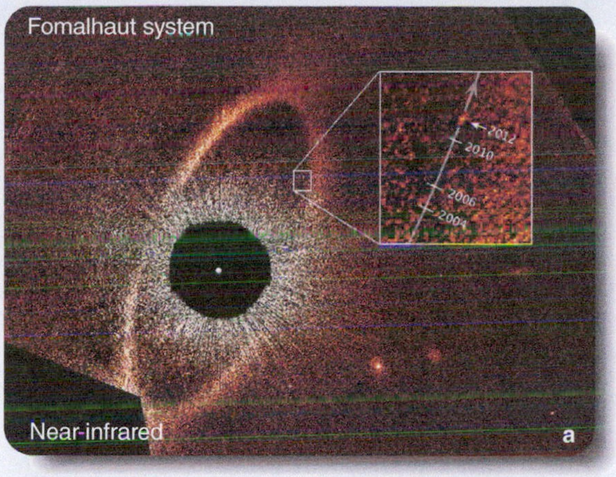

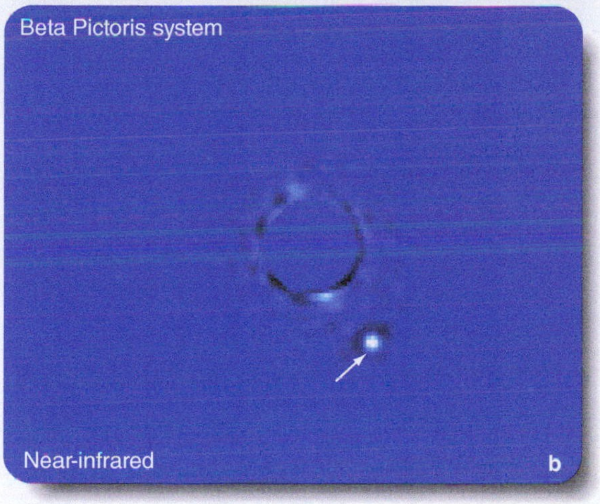

(a) Images of a Jovian-sized planet orbiting about 120 AU from the star Fomalhaut (Alpha Piscis Austrinus), at the inside edge of that star's previously known debris disk/ring. The inset shows motion of the object between 2004 and 2012, at the correct rate for a planet at that position orbiting a star of that mass. (b) Image of a Jovian-size planet orbiting 9 AU from the star Beta Pictoris. In both images, the central circular blank region represents a combination of hardware and software masks implemented to block light from the central stars that are much brighter than the planets.

are harder to detect because Earth's astronomers have not been making high-precision observations for a long enough time yet. Jupiter takes almost 12 years to circle the Sun once, so it will take years for astronomers to see the longer-period wobbles produced by planets lying farther from their stars. You should not be surprised that the first extrasolar planets discovered are massive and have short periods, so they are the easiest to detect. This is called a selection effect.

The new planets may seem odd for another reason. In our own Solar System, the large planets formed farther from the Sun where the solar nebula was colder and ices could condense. How could big planets form so near their stars? Model calculations predict that planets that form in an especially dense disk of matter could spiral inward as they sweep up gas or planetesimals. That means it is possible for a few planets to become the massive, short-period planets that are detected most easily.

Another puzzle is that many of the newly discovered extrasolar planets have eccentric orbits, and a few others appear to have orbits inclined at large angles to the equators of their parent stars. Simple interpretation of the solar nebular theory would predict that planets generally should have nearly circular orbits that lie approximately in their star's equatorial planes, as do the planets in our Solar System. Theorists point out, however, that planets in some young planetary systems can interact with each other and can be thrown into eccentric or highly inclined orbits. This effect is probably rare in planetary systems, but astronomers have found some of these extreme systems more easily because they have easily detected wobbles.

Actually getting an image of a planet orbiting another star is about as easy as photographing a bug crawling on the bulb of a searchlight miles away. Planets are small and dim and get lost in the glare of the stars they orbit. Nevertheless, as you can see in Figure 8-15, astronomers managed to image planets around the star HR 8799 using a combination of the telescopes on the ground and in space.

The discovery of extrasolar planets gives astronomers added confidence in the solar nebula theory. The theory predicts that planets are common, and astronomers are finding them orbiting many stars.

How Do We Know? 8-2

Scientists: Courteous Skeptics

What does it mean to be skeptical, yet also open to new ideas? "Scientists are just a bunch of skeptics who don't believe in anything." That is a common misconception among people who don't understand the methods and goals of science. Yes, scientists are skeptical about new ideas and discoveries, but they do hold strong beliefs about how nature works. Scientists are skeptical not because they want to disprove everything but because they are searching for the truth and want to be sure that a new description of nature is reliable before it is accepted.

Another common misconception is that scientists automatically accept the work of other scientists. On the contrary, scientists skeptically question every aspect of a new discovery. They may wonder if another scientist's instruments were properly calibrated or whether the scientist's mathematical models are correct. Other scientists will want to repeat the work themselves using their own instruments to see if they can obtain the same results. Every observation is tested, every discovery is confirmed, and only an idea that survives many of these tests begins to be accepted as a scientific truth.

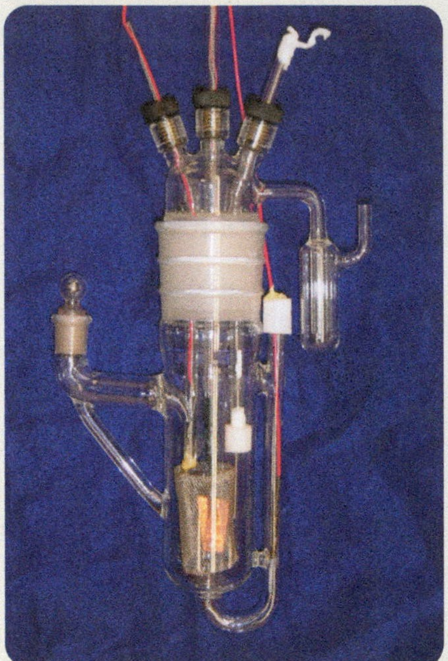
Photo by Steven Krivit of a device at the U.S Navy SPAWAR Systems Center in San Diego

A laboratory cell for the investigation of cold fusion claims that were thereby found to be false.

Scientists are prepared for this kind of treatment at the hands of other scientists. In fact, they expect it. Among scientists it is not bad manners to say, "Really, how do you know that?" or "Why do you think that?" or "Show me the evidence!" And it is not just new or surprising claims that are subject to such scrutiny. Even though astronomers had long expected to discover planets orbiting other stars, when a planet was finally discovered circling 51 Pegasi, astronomers were skeptical. That was not because they thought the observations were necessarily flawed but because this is how science works.

The goal of science is to tell stories about nature. Some people use the phrase "telling a story" to describe someone who is telling a fib. But the stories that scientists tell are exactly the opposite; perhaps you could call them "antifibs" because they are as true as scientists can make them. Skepticism eliminates stories with logical errors, flawed observations, or misunderstood evidence and tends eventually to leave standing only the stories that best describe nature. Skepticism is not a refusal to hold beliefs. Rather, it is a way for scientists to find and keep those natural principles that are worthy of trust.

What Are We? Planet-Walkers

The matter you are made of came from the Big Bang, and it has been cooked into a wide variety of atoms inside stars. Now you can see how those atoms came to be part of Earth. Your atoms were in the cloud of gas that formed the Solar System 4.6 billion years ago, and nearly all of that matter contracted to form the Sun, but a small amount left behind in a disk formed planets. In the process, your atoms became part of Earth.

You are a planet-walker, and you have evolved to live on the surface of Earth. Are there other beings like you in the Universe? Now you know that planets are common, and you can reasonably suppose that there are more planets in the Universe than there are stars. However complicated the formation of the Solar System was, it is a common process, so there may indeed be more planet-walkers living on other worlds.

STUDY TOOLS 8

IN THE BOOK

- ☐ Tear Out the Review Card on *Origin of the Solar System and Extrasolar Planets*

ASTRO ONLINE

Visit ASTRO Online at www.cengagebrain.com

- ☐ Interactive Reading
- ☐ Practice Quizzing
- ☐ Videos and Animations
- ☐ Exclusive Features

Questions and Problems

8

REVIEW QUESTIONS

1. What do Widmanstätten patterns (see Figure 8-6) indicate about the history of iron meteorites?
2. Why do astronomers refer to carbonaceous chondrites as unmodified or "primitive" material?
3. How do observations of meteor showers reveal one of the sources of meteoroids?
4. What evidence indicates that the asteroids are mostly fragments of larger bodies?
5. What evidence indicates that some asteroids have had geologically active surfaces?
6. How is the composition of meteorites related to the formation and evolution of asteroids?
7. What is the evidence that the Solar System formed about 4.6 billion years ago?
8. According to the solar nebula theory, why is the Earth's orbit nearly in the plane of the Sun's equator?
9. Why does the solar nebula theory predict that planetary systems are common?
10. What is the difference between condensation and accretion?
11. What does the term *differentiated* mean when applied to a planet? Would you expect to find that planets are usually differentiated? Why?
12. What processes cleared the nebula away and ended planet building?
13. Why is almost every solid surface in our Solar System scarred by craters?
14. Why would the formation of Jovian planets be difficult to understand if gas and dust disks around forming stars have short lifetimes in astronomical terms?
15. How does the solar nebula theory explain the dramatic density difference between the Terrestrial and Jovian planets?
16. What produced the helium now present in the Sun's atmosphere? In Jupiter's atmosphere? In the Sun's core?
17. What produced the iron and heavier elements such as gold and silver in Earth's core and crust?
18. What is the difference between the dense hot disks seen around some stars and the low-density cold disks seen around some other stars?
19. What are the methods by which extrasolar planets have been discovered?
20. Why is the existence of "hot Jupiters" puzzling? What is the current hypothesis for how they formed?
21. **How Do We Know?** Why are scientists free to adjust their hypotheses but not their facts?
22. **How Do We Know?** How can scientists be open to new ideas yet also properly skeptical?

DISCUSSION QUESTIONS

1. If you could visit another planetary system while the planets are forming, would you expect to see the condensation sequence at work, or do you think that process was likely unique to our Solar System? How do the properties of the extrasolar planets discovered so far affect your answer?
2. Do you expect that most planetary systems in the Universe have analogs to our Solar System's asteroid belt and Kuiper Belt? Would all planetary systems show signs of an age of heavy bombardment?
3. If the solar nebula hypothesis is correct, do you think there are more planets in the Universe than stars? Why or why not?

PROBLEMS

1. If a trillion (10^{12}) asteroids, each 1 km in diameter, were assembled into one body, how large would it be? Compare that to the size of Earth. (*Note*: The volume of a sphere is $\frac{4}{3}\pi r^3$.)
2. The asteroid Vesta has a mass of 2.6×10^{20} kg and an average radius of 260 km. What is the circular orbit velocity at its surface? The fastest major league pitchers can throw a baseball at 105 mph (47 m/s). Could they throw a baseball into orbit around Vesta? (*Hint*: Use the **Circular Orbit Velocity** formula in the Math Reference Cards MR.)
3. If you saw Comet Halley when it was 0.7 AU from Earth and it had a visible tail 5 degrees long, how long was the tail in kilometers? Suppose that the tail was not perpendicular to your line of sight. Is your first answer too large or too small? (*Hint*: Use **The Small-Angle Formula** in the Math Reference Cards MR.)
4. The mass of an average comet's nucleus is about 10^{14} kg. If the Oort cloud contains 2×10^{12} comet nuclei, what is the mass of the cloud in Earth masses? Compare that with Jupiter's mass. (*Note*: Relevant data can be found in Appendix C, Celestial Profiles 3 and 5.)

5. If you observed the Solar System from the nearest star (distance = 1.3 parsecs), what would the maximum angular separation be between Earth and the Sun? (*Note*: 1 pc is 2.1×10^5 AU.) (*Hint*: Use **The Small-Angle Formula** in the Math Reference Cards MR.)
6. What is the smallest-diameter crater you can identify in the photo of Mercury on the first page of Concept Art 8A? (*Hint*: See Appendix Table A-6, "Properties of the Planets," to find the radius of Mercury in kilometers.)
7. A sample of a meteorite has been analyzed, and the result shows that out of every 1000 nuclei of potassium-40 originally in the meteorite, only 200 have not decayed. How old is the meteorite? (*Hint*: See Figure 8-6a.)
8. What composition might you expect for a planet that formed in a region of the solar nebula where the temperature was about 100 K?
9. Suppose that Earth grew to its present size in 1 million years through the accretion of particles averaging 100 grams each. On the average, how many particles did Earth capture per second? (*Notes:* The mass of Earth is 6.0×10^{27} grams, and 1 year is 3.2×10^7 seconds.)

LEARNING TO LOOK

1. What do you see in Figure 8-16 that tells you the size of planetesimals when the Solar System was forming?

FIGURE 8-16

R. Kempton, New England Meteoritical

2. What do you see in Figure 8-17, the nucleus of Comet Borrelly, that tells you how comets produce their comae and tails?

FIGURE 8-17

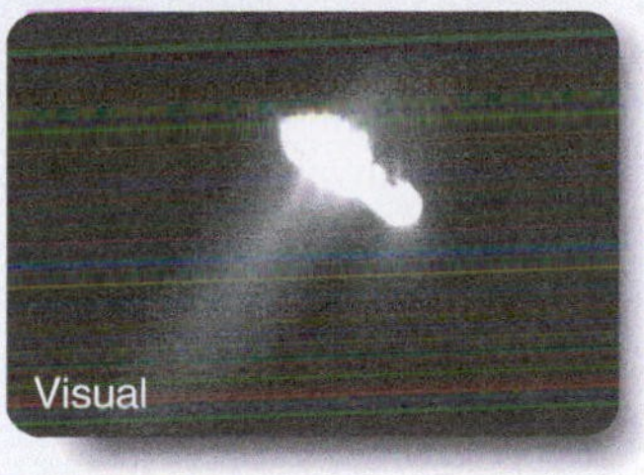

NASA

3. What do you see in Figure 8-18 that indicates this planet formed far from the Sun?

FIGURE 8-18

NASA/JPL-Caltech/SSI

4. Why do astronomers conclude that the surface of Mercury, shown in Figure 8-19, is old? When did the majority of those craters form?

FIGURE 8-19

NASA

9 The Family of Stars

[Love] is the star to every wandering bark,
Whose worth's unknown, although his height be taken.
William Shakespeare
Sonnet 116

ESO/Y. Beletsky

After finishing this chapter go to **PAGE 208** for **STUDY TOOLS.**

GUIDEPOST

To discover the properties of stars, astronomers use telescopes and instruments such as photometers, cameras, and spectrographs in clever ways to learn the secrets hidden in starlight. The result is a family portrait of the stars. Knowing the distances to stars is the key to knowing most of their other properties, but measuring those distances is very difficult.

Here you will find answers to the following five important questions about stars:

- How far away are the stars?
- How much energy do stars make?
- How are temperatures of stars determined?
- How big are stars?
- How much mass do stars contain?

With this chapter, you leave our Sun and Solar System behind and begin your study of the billions of stars that dot the sky. In a sense, stars are the basic building blocks of the Universe. If you hope to understand what the Universe is, what our Sun is, what our Earth is, and what we are, you need to understand stars. Once you know how to find the basic properties of stars, you will be ready to trace the history of the stars from birth to death, a story that begins in Chapter 10.

Shakespeare compared love to a star that can be seen easily and even used for guidance, but whose real nature is utterly unknown. He was born the same year as Galileo, and like everyone else at the time, had no idea what stars actually are. To understand the history of the Universe, the origin of Earth, and the nature of our human existence, you need to discover what people in Shakespeare's time did not know—the real nature of the stars.

In this chapter, you will learn how difficult it is to determine the real properties of a star. When you look at a star even through a telescope, you see only a point of light. Real understanding of stars requires careful analysis of starlight. (See CO 9.) This chapter concentrates on five goals: knowing how far away stars are, how much energy they emit, what their surface temperatures are, how big they are, and how much mass they contain. By the time you finish this chapter, you will be well-acquainted with the family of stars.

CO 9 Star cluster NGC 4755, also known as the Jewel Box, is located in the constellation Crux. This image was taken using the European Southern Observatory's (ESO's) Very Large Telescope (VLT) on Cerro Paranal in Chile. Several blue supergiant stars and one red supergiant star can be seen. The cluster is about 6500 ly (2000 pc) away.

9-1 STAR DISTANCES

Distance is the most important and the most difficult measurement in astronomy, although astronomers have several different ways to find the distances to stars. Each way ultimately depends on a simple and direct geometrical method that is much like the method surveyors would use to measure the distance across a river they cannot cross. You can begin by reviewing that method and then apply it to stars.

9-1a The Surveyor's Triangulation Method

To measure the distance across a river, a team of surveyors begins by driving two stakes into the ground. The distance between the stakes is called the *baseline*. The surveyors then choose a landmark on the opposite side of the river, a tree perhaps, thus establishing a large triangle marked by the two stakes and the tree, as shown in Figure 9-1. Using their instruments, the surveyors sight the tree from the two ends of the baseline and measure the two angles on their side of the river. Knowing two angles of this large triangle and the length of the baseline side between them, the surveyors can find the distance across the river by simple trigonometry. For example, if the baseline is 50 meters and the angles are 66 degrees and 71 degrees, they would calculate that the distance from the baseline to the tree is 64 m.

FIGURE 9-1

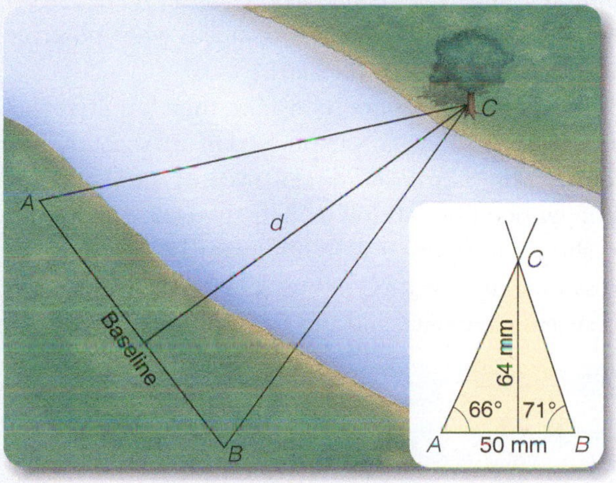

You can find the distance *d* across a river by measuring the length of the baseline and the angles *A* and *B* and then constructing a scale drawing of the triangle (1 mm = 1 m) or using trigonometry to calculate it.

9-1b The Astronomer's Triangulation Method

To find the distance to a star, astronomers use a very long baseline, the size of Earth's orbit, as you can see in Figure 9-2. If you take a photograph of a nearby star and then wait 6 months, Earth will have moved halfway around its orbit. You can then take another photograph of the star from that slightly different location in space. When you examine the photographs, you will discover that the star is not in exactly the same place in the two photographs.

In Chapter 3, Concept Art 3A, you learned that *parallax* is the term that refers to the common experience of an apparent shift in the position of a foreground object due to a change in the location of the observer's viewpoint. Your thumb, held at arm's length, appears to shift position against a distant background when you look first with one eye and then with the other. In that case, the baseline is the distance between your eyes, and the parallax is the angle through which your thumb appears to move when you switch eyes. The farther away you hold your thumb, the smaller the parallax. Recall from Section 5-1 of Chapter 5 that Venus shows a parallax when observed from different locations on Earth.

FIGURE 9-2

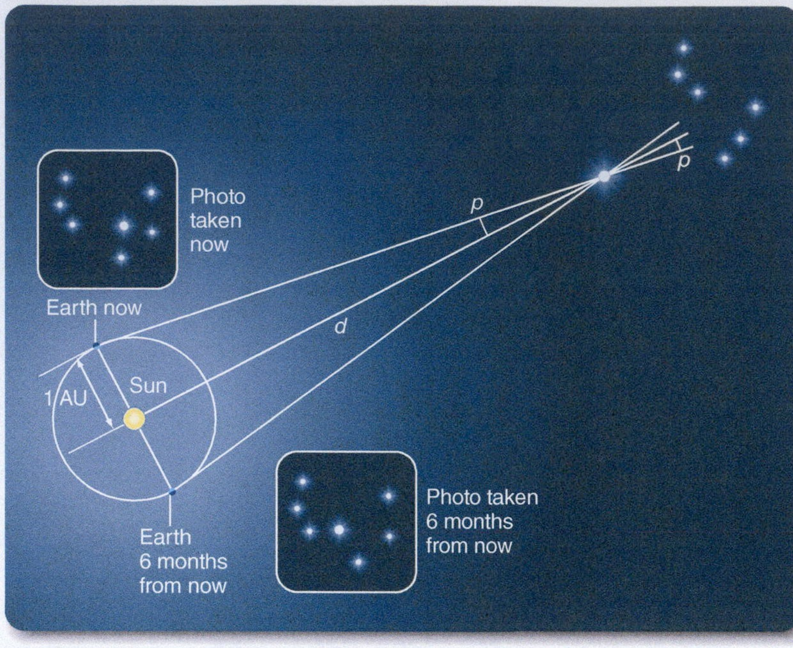

You can measure the parallax of a nearby star by photographing it from two points around Earth's orbit. For example, you might photograph it now and again after 6 months, when Earth's position has shifted by 2 AU.

Because the stars are so distant, their parallaxes are very small angles, usually expressed in arc seconds. The quantity that astronomers call **stellar parallax (*p*)** is conventionally defined as half the total shift of the star shown in Figure 9-2, in other words, the shift seen across a baseline of 1 AU rather than 2 AU. Astronomers measure the parallax, and surveyors measure the angles at the ends of the baseline, but both measurements tell the same thing—the shape and size of the triangle and thus the distance to the object in question. (Look at **Parallax and Distance** in the Math Reference Cards MR.)

stellar parallax (*p*) The small apparent shift in position of a nearby star relative to distant background objects due to Earth's orbital motion.

parsec (pc) The distance to a hypothetical star whose parallax is 1 arc second.

Measuring the parallax *p* is very difficult because it is such a small angle. The visible star nearest the Sun, Alpha Centauri, has a parallax of only 0.76 arc second, and more distant stars have even smaller parallaxes. To see how small these angles are, imagine a dime 2 miles away from you. That dime covers an angle of about 1 arc second. Stellar parallaxes are so small that the first successful measurement of one did not happen until 1838, more than 200 years after the invention of the telescope.

The distances to the stars are so large that it is not convenient to use kilometers or Astronomical Units. When you measure distance via parallax, it is better to use the unit of distance called a **parsec (pc)**. The word *parsec* was created by combining *par*allax and arc *sec*ond. One parsec equals the distance to an imaginary star that has a parallax of 1 arc second. A parsec is 2.06×10^5 AU, which equals roughly 3.26 ly (light-years). Parsec units are used more often than light-years by astronomers because parsecs are more directly related to the process of measurement of star distances. However, there are instances in which the light-year is also useful. You will notice that the following chapters use either parsecs or light-years as convenience and custom dictate.

The blurring caused by Earth's atmosphere makes star images appear to be about 1 arc second in diameter, and that makes it difficult to measure parallax. Even if you average together many observations made from Earth's surface, you cannot measure parallax with an uncertainty smaller than about 0.002 arc second. Thus, if you measure a parallax of 0.02 arc second from an observatory on the ground, your uncertainty will be about 10 percent. If you consider, as astronomers generally do, that 10 percent is the maximum acceptable level of uncertainty, then ground-based astronomers can't accurately measure parallaxes that are smaller than about 0.02 arc second. That parallax corresponds to a distance of about 50 pc (165 ly).

In 1989, the European Space Agency launched the satellite *Hipparcos* to measure stellar parallaxes from above the blurring effects of Earth's atmosphere. That small space telescope observed for 4 years, and the data were used to produce two parallax catalogs in 1997. One catalog contains 120,000 stars with parallaxes 20 times more accurate than ground-based measurements. The other catalog contains more than a million stars with parallaxes as accurate as ground-based parallaxes. Knowing accurate distances from the *Hipparcos* observations has given astronomers new insights into the nature of stars.

The European Space Agency launched the *Gaia* spacecraft in 2013. It is designed to measure the parallaxes of a billion stars as faint as apparent visual magnitude +20. This will allow the first real three-dimensional map of our Galaxy; an initial catalog of results was released in 2016.

FIGURE 9-3

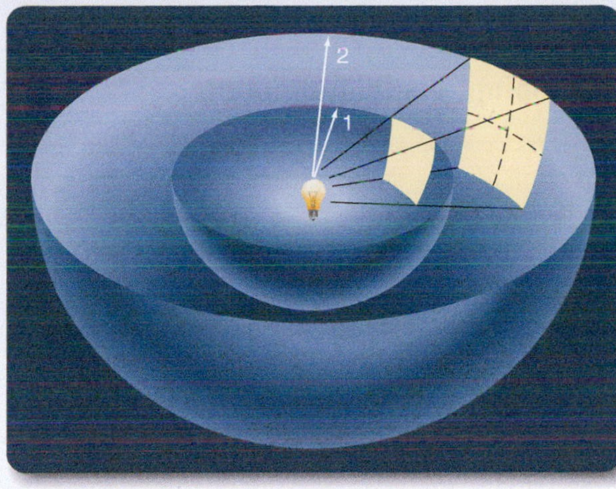

The inverse square relation: A light source is surrounded by imaginary spheres with radii of 1 unit and 2 units. The light falling on an area of 1 m^2 on the inner sphere spreads to illuminate an area of 4 m^2 on the outer sphere. Thus, the brightness of the light source is inversely proportional to the square of the distance.

9-2 APPARENT BRIGHTNESS, INTRINSIC BRIGHTNESS, AND LUMINOSITY

Your eyes tell you that some stars look brighter than others, and in Section 2-1c of Chapter 2 you learned about the apparent magnitude scale that refers to stellar apparent brightness. The scale of apparent magnitudes only tells you how bright stars appear to you on Earth, however. To know the true nature of a star you need to know its **intrinsic brightness**, a measure of the amount of light the star produces. An intrinsically very bright star might appear faint if it is far away. Thus, to know the intrinsic brightness of a star, you must take into account its distance from you.

9-2a Brightness and Distance

When you look at a bright light, your eyes respond to the visual-wavelength energy falling on your retinas, which tells you how bright the object appears. Thus, brightness is related to the flux of energy entering your eye. As you may recall from Chapter 2 Section 2-1c, astronomers and physicists define flux as the energy in joules (J) per second falling on 1 square meter. Recall that a joule is about the amount of energy released when an apple falls from a table onto the floor. A flux of 1 joule per second is also known as 1 watt. The wattage of a light bulb tells you its *intrinsic* brightness. Compare that with the *apparent* brightness of a light bulb, which depends on its distance from you.

If you placed a screen 1 square meter near a light bulb, a certain amount of flux would fall on the screen. If you moved the screen twice as far from the bulb, the light that previously fell on the screen would be spread to cover an area four times larger, and the screen would receive only one-fourth as much light. If you tripled the distance to the screen, it would receive only one-ninth as much light. Thus, the flux you receive from a light source is inversely proportional to the square of the distance to the source. This is known as the inverse square relation, which you can see in Figure 9-3. You first encountered the inverse square relation in Section 3-5a of Chapter 3, where it was applied to the strength of gravity.

Now you can understand how the apparent brightness of a star depends on its distance. If astronomers know the

intrinsic brightness
A measure of the amount of light a star produces.

apparent brightness of a star and its distance from Earth, they can use the inverse square law to correct for distance and find the intrinsic brightness of the star.

9-2b Absolute Visual Magnitude

If all the stars were the same distance away, you could compare one with another and decide which was intrinsically brighter or fainter. Of course, the stars are scattered at different distances, and you can't move them around to line them up for comparison. If, however, you know the distance to a star, you can use the inverse square relation to calculate the brightness the star would have at some standard distance. Astronomers use 10 pc as the standard distance and refer to the intrinsic brightness of the star as its **absolute visual magnitude (M_V)**, which is the apparent visual magnitude that star would have if it were 10 pc away. The subscript V tells you it is a visual magnitude, referring only to the wavelengths of light your eye can see. Other magnitude systems are based on other parts of the electromagnetic spectrum, such as infrared and ultraviolet radiation.

The Sun's absolute magnitude is easy to calculate because its distance and apparent magnitude are well known. The absolute visual magnitude of the Sun is about +4.8. In other words, if the Sun were only 10 pc (33 ly) from Earth, not a great distance in astronomy, it would have an apparent magnitude of +4.8 and look no brighter to your eye than the faintest star in the handle of the Little Dipper. (Look at **Absolute Magnitude and Distance** in the Math Reference Cards MR.)

This path to find the distance to stars has led you to absolute magnitude, a measure of the intrinsic brightness of the stars. You are now ready to reach the next of your five goals in this chapter.

9-2c Luminosity

The second goal for this chapter is to find out how much energy the stars emit. With the absolute magnitudes of the stars in hand, you can now compare stars using our Sun as a standard. The intrinsically brightest stars have absolute magnitudes of about −8.3. If such a star were 10 pc away from Earth, it would be more than 25 times as bright as Venus at its brightest. Such stars emit more than 100,000 times as much light at visual wavelengths than the Sun.

Absolute visual magnitude refers to visible light, but you want to know the total output including all types of radiation. Hot stars emit a lot of ultraviolet radiation that you can't see, and cool stars emit plenty of infrared radiation. To add in the energy you can't see, astronomers make a mathematical correction that depends on the temperature of the star. With that correction, astronomers can find the total electromagnetic energy output of a star, which they refer to as its **luminosity (L)**.

Astronomers know the luminosity of the Sun because they can send satellites above Earth's atmosphere and measure the amount of energy arriving from the Sun, adding up radiation of every wavelength, including the types blocked by the atmosphere. Of course, they also know the distance from Earth to the Sun very accurately, which is necessary to calculate luminosity. The luminosity of the Sun is about 3.8×10^{26} watts (joules per second).

You can express a star's luminosity in two ways. For example, you can say that the star Capella (Alpha Aurigae) is 150 times more luminous than the Sun. You can also express this in real energy units by multiplying by the luminosity of the Sun. The luminosity of Capella is about 5.7×10^{28} watts.

When you look at the night sky, the stars look much the same, yet your study of distances and luminosities reveals an astonishing fact. The most luminous stars emit 10 billion times more energy per second than the least luminous. Clearly, the family of stars is filled with interesting characters.

absolute visual magnitude (M_V) Intrinsic brightness of a star. The apparent visual magnitude the star would have if it were 10 pc away.

luminosity (L) The total amount of energy a star radiates per second at all wavelengths.

9-3 STAR TEMPERATURES

Your third goal in this chapter is to learn about the temperatures of stars. The surprising fact is that stellar spectral lines can be used as a sensitive star thermometer.

From the discussion of blackbody radiation in Section 5-2b of Chapter 5, you know that temperatures of stars can be estimated from their color—red stars are cool and blue stars are hot. The relative strengths of various spectral lines, however, give much greater accuracy in measuring star temperatures. Recall from Section 5-3 of Chapter 5 that, for stars, the term *surface* refers to the photosphere, which is the limit of our vision into the star from outside, but is not an actual solid surface. Stars typically have surface temperatures of a few thousand or tens of thousands of degrees Kelvin. As you will discover in Chapter 10, the centers of stars are much hotter than their surfaces—many millions of degrees hotter—but the spectra tell only about the outer layers from which the light you see departed.

As you learned in Section 5-5 of Chapter 5, hydrogen Balmer absorption lines are produced by hydrogen atoms with electrons initially in the second energy level. The strength of these spectral lines can be used to gauge the temperature of a star because scientists know from lab experiments with gasses and radiation, and also from theoretical calculations, that:

- If the surface of a star is as cool as the Sun or cooler, there are few violent collisions between atoms to excite the electrons, and most atoms will have their electrons in the ground (lowest) state. These atoms can't absorb photons in the Balmer series. As a result, you should expect to find weak hydrogen Balmer absorption lines in the spectra of very cool stars.
- In the surface layers of stars hotter than about 20,000 K, there are many violent collisions between atoms, exciting electrons to high energy levels or knocking the electrons completely out of most atoms so they become ionized. In this situation, few hydrogen atoms will have electrons in the second energy level to form Balmer absorption lines. As a result, you should also find weak hydrogen Balmer absorption lines in the spectra of very hot stars.
- At an intermediate temperature, roughly 10,000 K, the collisions have the correct amount of energy to excite large numbers of electrons into the second energy level. With many atoms excited to the second level, the gas absorbs Balmer-wavelength photons well and thus produces strong hydrogen Balmer lines.

FIGURE 9-4

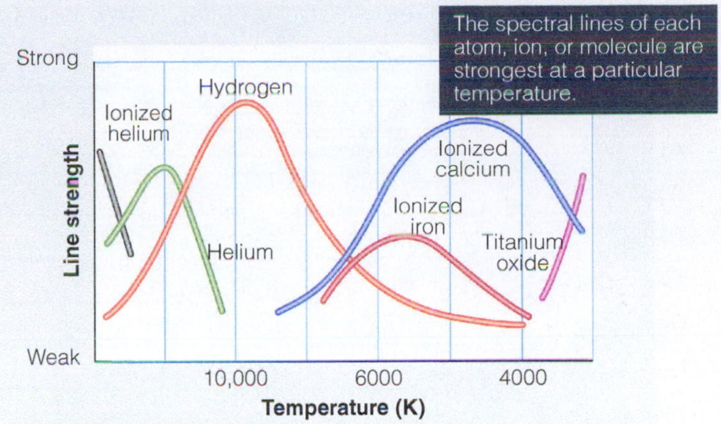

The strength of spectral lines can tell you the temperature of a star. Using the lines of several atoms or molecules gives you the most accurate results.

The strength of the hydrogen Balmer lines thus depends on the temperature of the star's surface layers. Both hot and cool stars have weak Balmer lines, but medium-temperature stars have strong Balmer lines.

Figure 9-4 shows the relationship between gas temperature and strength of spectral lines for hydrogen and other substances. Each type of atom or molecule produces spectral lines that are weak at high and low temperatures and strong at some intermediate temperature, although the temperature at which the lines reach maximum strength is different for each type of atom or molecule. Theoretical calculations of the type first made by Cecilia Payne (look back again to Chapter 5 Section 5-5) can predict just how strong various spectral lines should be for stars of different temperatures. Astronomers can determine a star's temperature by comparing the strengths of its spectral lines with the predicted strengths.

From stellar spectra, astronomers have found that these temperatures range from about 2300 K to about 50,000 K. Compare these with the surface temperature of the Sun, which is about 5800 K.

9-3a Temperature Spectral Types

You have seen that the strengths of spectral lines depend on the surface (photosphere) temperature of the star. From this you can predict that all stars of a given temperature should have similar spectra. Learning to recognize the pattern of spectral lines produced in the atmospheres of stars of different temperatures means there is no need to do a full analysis every time each type of spectrum is encountered. Time can be saved by classifying stellar spectra rather than analyzing each spectrum individually. Astronomers classify stars by the lines and bands in their spectra, as shown in Table 9-1. For example, if it has weak Balmer lines and lines of ionized helium, it must be an O star.

The star classification system now used by astronomers was devised at Harvard during the 1890s and 1900s. One of the astronomers there, Annie J. Cannon, personally inspected and classified the spectra of more than 250,000 stars. The spectra were first classified in groups labeled A through Q, but some groups were later dropped, merged with others, or reordered.

The final classification scheme includes seven main **spectral classes** or types that are still

spectral classes A star's label in the temperature classification system based on the appearance of the star's spectrum.

FIGURE 9-5

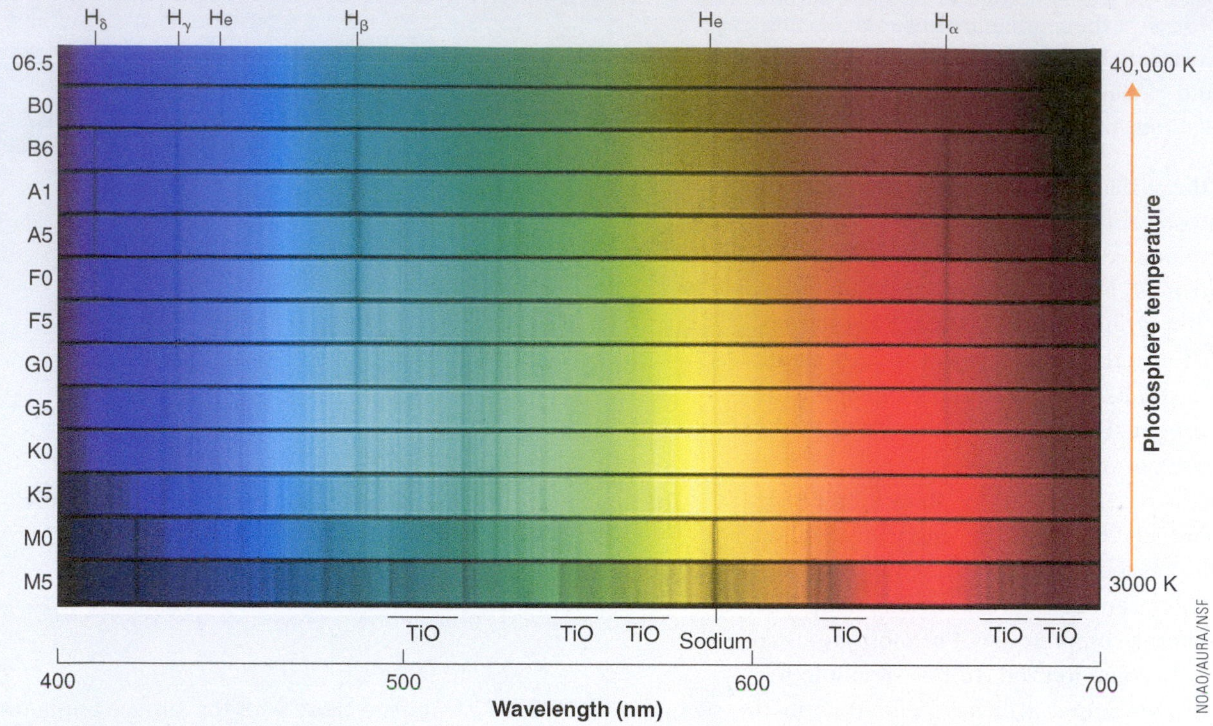

Color photographs of stellar spectra ranging from hot O stars at the top to cool M stars at the bottom: The hydrogen Balmer lines are strongest at spectral type A0, but the two closely spaced lines of sodium in the yellow are strongest for very cool stars. Helium lines appear only in the spectra of the hottest stars. Notice that the helium line visible in the top spectrum has nearly, but not exactly, the same wavelength as the sodium lines visible in cooler stars. Bands produced by the molecule titanium oxide (TiO) are strong in spectra of the coolest stars.

used today: O, B, A, F, G, K, and M. This set of star types, called the **spectral sequence**, is important because it is a temperature sequence. The O stars are the hottest, and the temperature continues to decrease down to the M stars, the coolest. For further precision, astronomers divide each spectral class into 10 subclasses. For example, spectral class A consists of the subclasses A0, A1, A2, . . . A8, A9. Next come F0, F1, F2, and so on. These finer divisions define a star's temperature to a precision of about 5 percent. The Sun, for example, is not just a G star, but a G2 star, with a temperature of 5800 K.

TABLE 9-1 TEMPERATURE SPECTRAL TYPES

Spectral Class	Approximate Temperature (K)	Hydrogen Balmer Lines	Other Spectral Features	Unaided Eye Example
O	40,000	Weak	Ionized helium	Meissa (O8)
B	15,000	Medium	Neutral helium	Achernar (B6)
A	8500	Strong	Ionized calcium weak	Sirius (A1)
F	6600	Medium	Ionized calcium weak	Canopus (F0)
G	5500	Weak	Ionized calcium medium	Sun (G2)
K	4100	Very weak	Ionized calcium strong	Aldebaran (K5)
M	3000	Very weak	TiO strong	Betelgeuse (M2)

spectral sequence The arrangement of spectral classes (O, B, A, F, G, K, and M) ranging from hot to cool.

Generations of astronomy students have remembered the spectral sequence by using mnemonics such as "Oh Boy, An F Grade Kills Me," or "Only Bad Astronomers Forget Generally Known Mnemonics."

Figure 9-5 shows color images of 13 stellar spectra ranging from the hottest at the top to the coolest at the bottom. Notice how spectral features change gradually from hot to cool stars. Although these spectra are attractive, astronomers rarely work with spectra as color images. Rather, they display spectra as graphs of intensity versus wavelength with dark absorption lines as dips in the graph in Figure 9-6. Such graphs show more detail than photographs. Notice also that the overall curves are similar to blackbody curves. The wavelength of maximum is in the infrared for the coolest stars and in the ultraviolet for the hottest stars.

Compare Figures 9-5 and 9-6 and notice how the strength of spectral lines depends on temperature, as indicated in the previous discussion regarding Figure 9-4. It is straightforward to determine a star's temperature from the details of its spectrum.

The study of spectral types is more than a century old, but astronomers continue to discover and define new types. The **L dwarfs**, found in 1998, are cooler and fainter than M stars. They are understood to be objects smaller than stars but larger than planets and are called **brown dwarfs**, which you will learn more about in a later chapter. The spectra of M stars contain bands produced by metal oxides, such as titanium oxide (TiO), but L dwarf spectra contain bands produced by molecules such as iron hydride (FeH). The **T dwarfs** are an even cooler and fainter type of brown dwarfs than L dwarfs. Their spectra show absorption by methane (CH_4) and water vapor. In 2011, astronomers using highly sensitive infrared detectors on space telescopes and large ground-based telescopes discovered a class of objects with temperatures below 500 K that are labeled **Y dwarfs**.

FIGURE 9-6

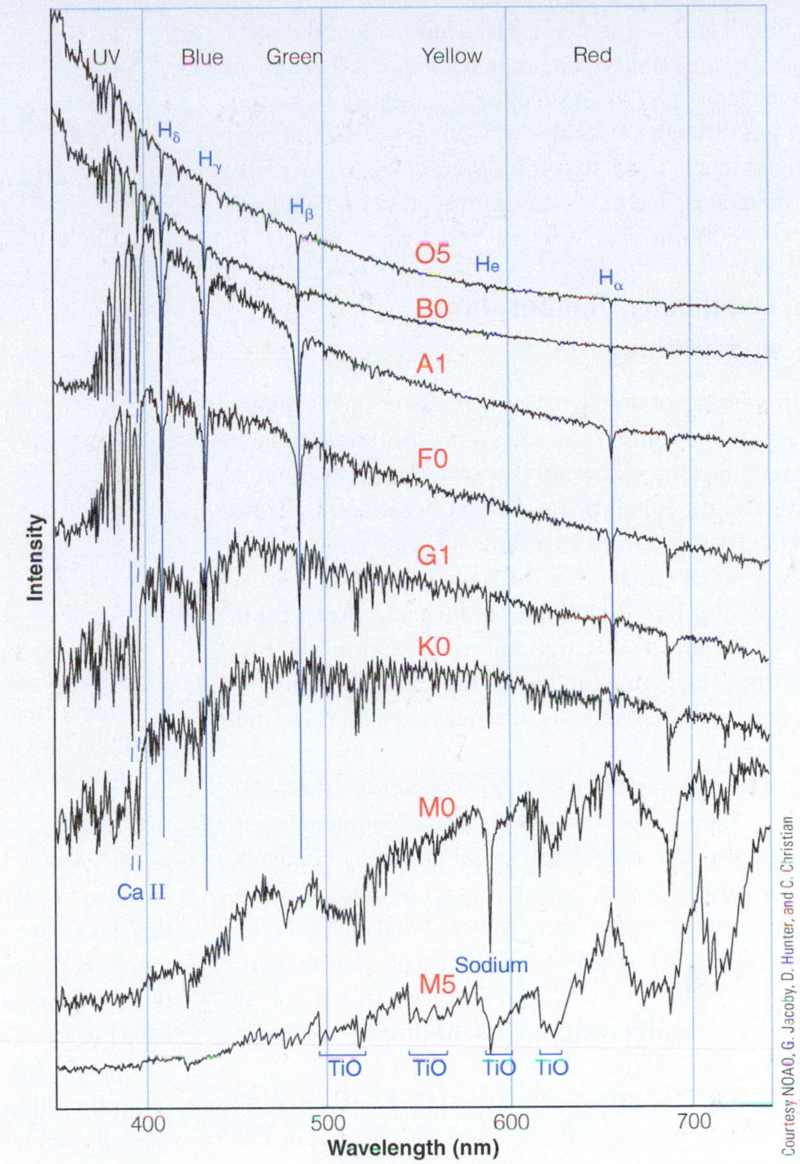

Spectra are often represented as graphs of intensity versus wavelength, with dark absorption lines appearing as sharp dips in the curves. This set of spectra displays hot stars at the top and cool stars at the bottom. Hydrogen Balmer lines are strongest at spectral type A0, whereas lines of ionized calcium (Ca II) are strongest in K stars. Titanium oxide (TiO) bands are strongest in M stars. Compare these spectra with the information in Figures 9-4 and 9-5, and Table 9-1.

L dwarfs, T dwarfs Spectral classes of brown dwarf stars with lower surface temperatures and luminosities than M dwarfs.

brown dwarf A very cool, low-luminosity star whose mass is not sufficient to ignite nuclear fusion.

Y dwarf A substellar object with temperature below 500 K, having inferred properties intermediate between brown dwarfs and Jovian planets.

9-4 STAR SIZES

Your fourth goal in this chapter is to learn about the sizes of stars. Do they all have the same diameter as the Sun, or are some larger and some smaller? You certainly can't see their sizes through a telescope; the images of the stars are much too small for you to resolve their disks and measure their diameters. There is a way, however, to find out how big stars really are.

9-4a Luminosity, Temperature, and Radius

The luminosity of a glowing object like a star is determined by its temperature (Stefan-Boltzmann law, Chapter 5 Section 5-2b) and its surface area. For example, you can eat dinner by candlelight because the candle flame has a small surface area, and, although it is very hot, it cannot radiate much heat because it is small; it has a low luminosity. However, if the candle flame were 12 feet tall, it would have a very large surface area from which to radiate, and, although it might be the same temperature as a normal candle flame, its luminosity would drive you from the table.

In a similar way, a star's luminosity is proportional to its surface area. A hot star may not be very luminous if it has a small surface area, but it could be highly luminous if it were larger. Even a cool star could be luminous if it had a large surface area. You can use stellar luminosities to determine the diameters of stars if you can separate the effects of temperature and surface area. (Look at **Luminosity, Temperature, and Radius** in the Math Reference Cards MR.)

The **Hertzsprung–Russell (H–R) diagram**, named after its originators, Ejnar Hertzsprung in the Netherlands and Henry Norris Russell in the United States, is a graph that separates the effects of temperature and surface area on stellar luminosities and enables astronomers to sort the stars according to their diameters.

Before discussing the details of the H–R diagram, let's look at a similar diagram you might use to compare automobiles. You can plot a diagram such as that shown in Figure 9-7 to show horsepower versus weight for various makes of cars. In so doing, you will find that in general the more a car weighs, the more horsepower it has. Most cars fall somewhere along the normal sequence of cars running from heavy, high-powered cars to light, low-powered models. You might call this the main sequence of cars. Some cars, however, have much more horsepower than normal for their weight—the sport or racing models—and the economy models have less power than normal for cars of the same weight.

Hertzsprung–Russell (H–R) diagram A plot of the intrinsic brightness versus the surface temperature of stars. It separates the effects of temperature and surface area on stellar luminosity.

FIGURE 9-7

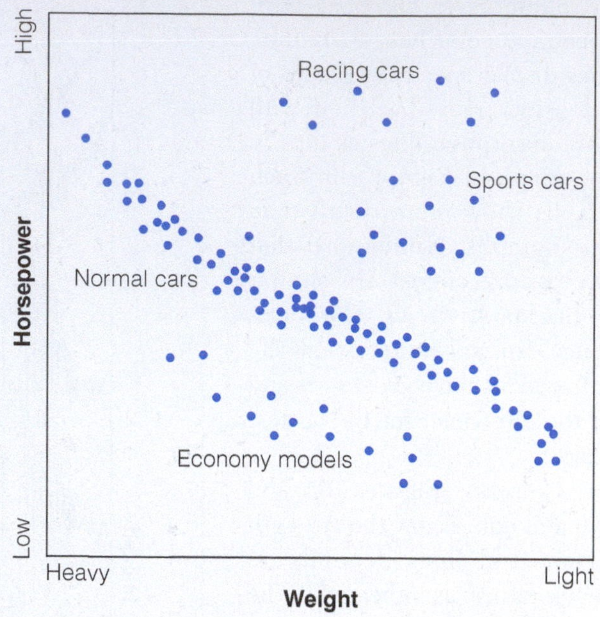

You could analyze automobiles by plotting their horsepower versus their weight and thus reveal relationships among various models. Most would lie somewhere along the main sequence of "normal" cars.

Just as the diagram in Figure 9-7 helps you understand the different kinds of autos, so the H–R diagram can help you understand different kinds of stars. An H–R diagram, seen in Figure 9-8, has luminosity on the vertical axis and temperature on the horizontal axis. A star is represented by a point on the graph that tells you its luminosity and temperature. Note that in astronomy the symbol $\odot$ refers to the Sun. Thus $L_\odot$ refers to the luminosity of the Sun, $T_\odot$ refers to the temperature of the Sun, and so on. The H–R diagram in Figure 9-8 also contains a scale of spectral types across the top. The spectral type of a star is determined by its temperature, so you can use either spectral type or temperature as the horizontal axis of an H–R diagram.

Astronomers use H–R diagrams so often that they usually skip the words "the point that represents the star." Rather, they will say that a star is located in a certain place in the diagram. Of course, they mean the point that represents the luminosity and temperature of the star and

FIGURE 9-8

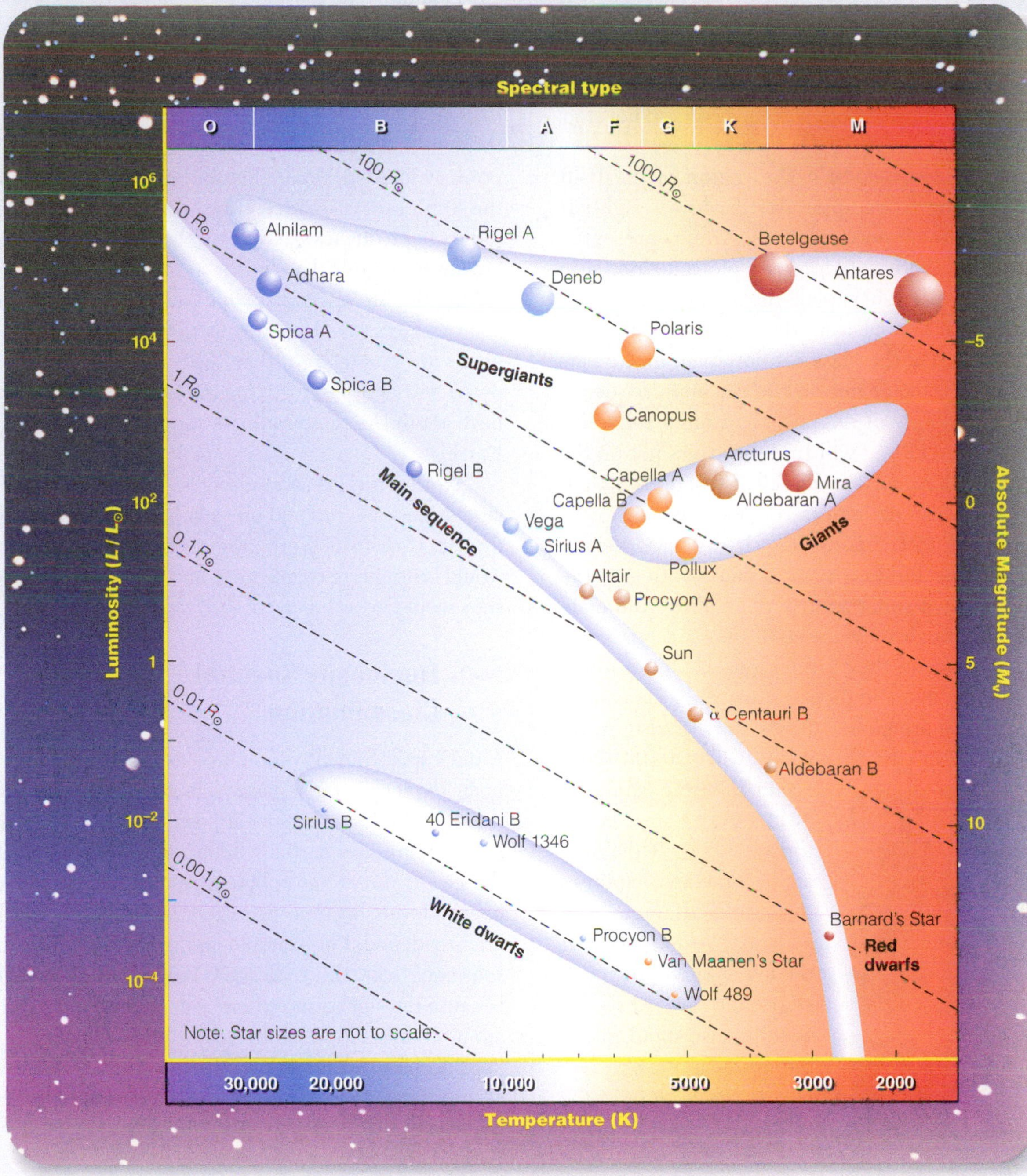

An H–R diagram showing the luminosity and temperature of many well-known stars. (Individual stars that orbit each other are designated A and B, for example Spica A and Spica B.) The dashed lines are lines of constant radius; star sizes on this diagram are not to scale. To visualize the size of the largest stars, imagine that the Sun is the size of a tennis ball. Then, the largest supergiants would be the size of a sports stadium and white dwarfs the size of grains of sand.

not the star itself. The location of a star in the H–R diagram has nothing to do with the location of the star in space. Furthermore, a star may move in the H–R diagram as it ages and its luminosity and temperature change, but such motion in the diagram has nothing to do with the star's motion in space.

In an H–R diagram, the location of a point representing a star tells you a great deal about the star.

Points near the top of the diagram represent very luminous stars, and points near the bottom represent very-low-luminous stars. Also, points near the right edge of the diagram represent very cool stars, and points near the left edge of the diagram represent very hot stars. Notice in Figure 9-8 how the artist has used color to represent temperature. Cool stars are red and hot stars are blue.

The **main sequence** is the region of the H–R diagram running from upper left to lower right. It includes roughly 80 percent of all stars, represented by a curved line with dots for stars plotted along it in Figure 9-8. As you might expect, the hot main-sequence stars are more luminous than the cool main-sequence stars. Look again at the classification diagram for cars in Figure 9-7. Vehicles not on the car main sequence have different kinds of engines than main-sequence cars. In Chapter 11 you will find that stars not on the main sequence have different nuclear reactions as their power sources than do main-sequence stars.

In addition to temperature, size is important in determining the luminosity of a star. Notice in the H–R diagram that some cool stars lie above the main sequence. Although they are cool, they are luminous, and that must mean they are larger—have more surface area—than main-sequence stars of the same temperature. These are called **giant** stars, and they are roughly 10 to 100 times larger than the Sun. The **supergiant** stars at the top of the H–R diagram are as large as 1000 times the Sun's diameter. In contrast, **red dwarfs** at the lower end of the main sequence are not only cool but also small, giving them low luminosities.

At the bottom of the H–R diagram are the "economy models," stars that are very low in luminosity because they are very small. The **white dwarfs** lie in the lower left of the H–R diagram, and although some white dwarfs are among the hottest stars known, they are so small they have very little surface area from which to radiate, and that limits them to low luminosities.

main sequence The region of the H–R diagram running from upper left to lower right, which includes roughly 90 percent of all stars generating energy by nuclear fusion.

giant Large, cool, highly luminous star in the upper right of the H–R diagram, typically 10 to 100 times the diameter of the Sun.

supergiant Exceptionally luminous star whose diameter is 100 to 1000 times that of the Sun.

red dwarf A faint, cool, low-mass, main-sequence star.

white dwarf Dying star at the lower left of the H–R diagram that has collapsed to the size of Earth and is slowly cooling off.

luminosity class A category of stars of similar luminosity, determined by the widths of lines in their spectra.

A simple calculation can be made to draw precise lines of constant radius across the H–R diagram based on the luminosities and temperatures at each point as shown in Figure 9-8. For example, locate the line labeled 1 $R_\odot$ (1 solar radius) and notice that it passes through the point representing the Sun. Any star whose point is located along this line has a radius equal to the Sun's radius. Notice also that the lines of constant radius slope downward to the right, because cooler stars are always fainter than hotter stars of the same size, following the Stefan-Boltzmann Law. These lines of constant radius show dramatically that the supergiants and giants are much larger than the Sun. In contrast, white dwarf stars fall near the line labeled 0.01 $R_\odot$. They all have about the same radius—approximately the size of Earth!

Notice the great range of sizes among stars. The largest stars are 100,000 times larger than the tiny white dwarfs. If the Sun were a tennis ball, the white dwarfs would be grains of coarse sand, and the largest supergiant stars would be as big as football fields.

9-4b Luminosity Spectral Classification

A star's spectrum also contains clues as to whether it is a main-sequence star, a giant, or a supergiant. The larger a star is, the less dense its atmosphere is. The widths of spectral lines are partially determined by the density of the gas: If the atoms collide often in a dense gas, their energy levels become distorted, and the spectral lines are broadened. For example, in the spectrum of a main-sequence star, the hydrogen Balmer lines are broad because the star's atmosphere is dense and the hydrogen atoms collide often. In the spectrum of a giant star, the lines are narrower (Figure 9-9), because the giant star's atmosphere is less dense and the hydrogen atoms collide less often. In the spectrum of a supergiant star, the Balmer lines are very narrow.

Thus, an astronomer can look closely at a star's spectrum and tell roughly how big it is. Size categories are called **luminosity classes**, because the size of the star is the dominating factor in determining luminosity. Supergiants, for example, are very luminous because they are very large.

The luminosity classes are represented by the Roman numerals I through V, with supergiants further subdivided into types Ia and Ib. For example, you can distinguish between a luminous supergiant (Ia) such as Rigel (Beta Orionis) and a regular supergiant (Ib) such as Polaris, the North Star (Alpha Ursa Minoris). The star Adhara (Epsilon Canis Majoris) is a luminous giant (II), Aldebaran (Alpha Tauri) is a giant (III), and Altair (Alpha Aquilae) is a subgiant (IV). The Sun is a main-sequence star (V). The luminosity class notation appears after the spectral type, as in G2 V for the Sun. White dwarfs don't enter into this classification because their spectra are very different from the other types of stars.

FIGURE 9-9

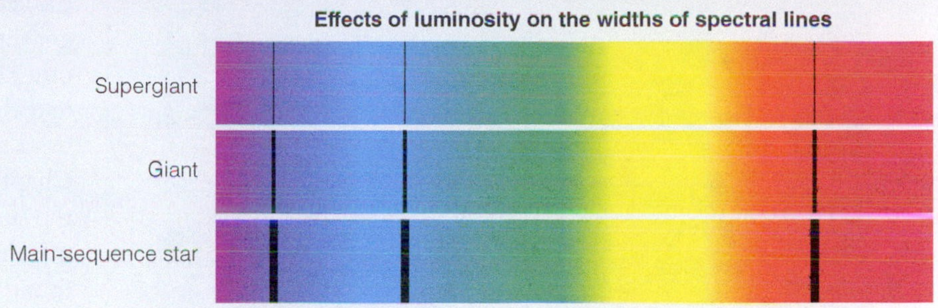

These model spectra show how the widths of spectral lines reveal a star's luminosity class. Supergiants have very narrow spectral lines, and main-sequence stars have broad lines. Certain spectral lines are more sensitive to this effect than others; careful inspection of a star's spectrum can determine its luminosity classification.

The approximate positions of the main sequence, giant, and supergiant luminosity classes are shown on the H–R diagram in Figure 9-8. Luminosity classification is subtle and not very accurate, but it is an important technique in modern astronomy because it provides clues to distance.

Most stars are too distant to have measurable parallaxes, but astronomers can find the distances to these stars if they can record the stars' spectra and determine their luminosity classes. From spectral type and luminosity class, astronomers can estimate the star's absolute magnitude, compare with its apparent magnitude, and compute its distance. Although this process finds distance and not true parallax, it is called **spectroscopic parallax**.

For example, Betelgeuse (Alpha Orionis) is classified M2 Ib, and its apparent magnitude averages about +0.4. (Betelgeuse is somewhat variable.) You can plot this star in an H–R diagram shown in Figure 9-8, where you would find that a temperature class of M2 and luminosity class of Ib (supergiant) corresponds to an absolute visual magnitude of about −6.0, which means a luminosity of about 30,000 $L_{\odot}$. That information, combined with the star's apparent brightness, allows astronomers to estimate that Betelgeuse is about 190 pc (see **Absolute Magnitude and Distance** in the Math Reference Cards **MR**). A combination of parallax measurements made by the *Hipparcos* satellite and radio telescopes yields a distance of 197 pc with an uncertainty of 24 percent. Spectroscopic parallax does give a good first estimate of the distances of stars so far away that their parallax can't easily be measured.

9-5 STAR MASSES—BINARY STARS

Your fifth goal is to find out how much matter stars contain, that is, to know their masses. Do they all have about the same mass, or is there a wide range, as there is for luminosity?

Gravity is the key to determining mass. Matter produces a gravitational field. Astronomers can figure out how much matter a star contains if they watch an object such as another star move through the star's gravitational field.

Finding the masses of stars involves studying **binary stars**, pairs of stars that orbit each other. Many of the familiar stars in the sky are actually pairs of stars orbiting each other, as shown in Figure 9-10. Binary systems are common; more than half of all stars are members of binary star systems. Few, however, can be analyzed completely. Many are so far apart that their periods are much too long for practical mapping of their orbits. Others are so close together they are not visible as separate stars.

9-5a Binary Star Orbits

The key to finding the mass of a binary star is an understanding of orbital motion (see Section 3-5b of Chapter 3). Each star in a binary system moves in its own

spectroscopic parallax The method of determining a star's distance by comparing its apparent magnitude with its absolute magnitude as estimated from its spectrum.

binary stars Pairs of stars that orbit around their common center of mass.

FIGURE 9-10

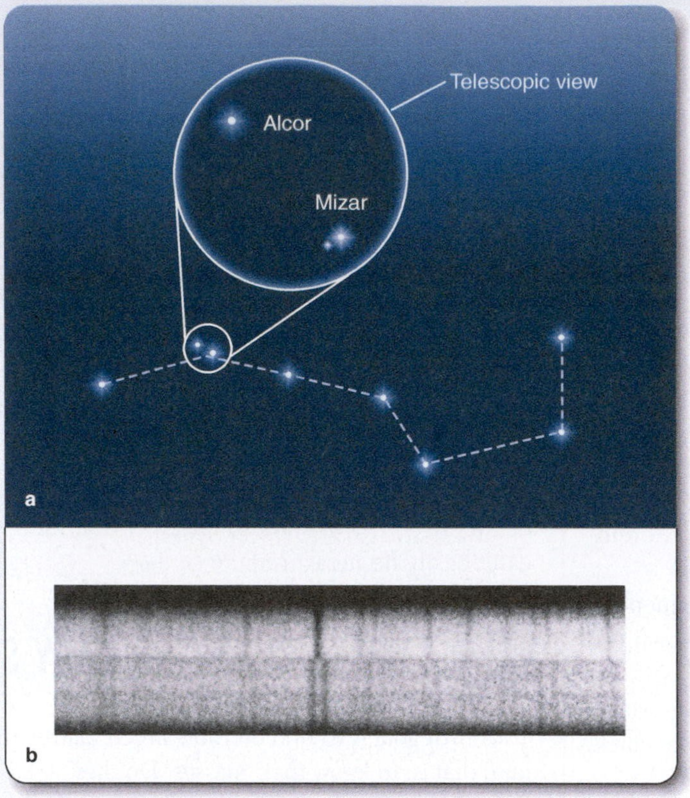

The Observatories of the Carnegie Institution of Washington

(a) At the bend of the handle of the Big Dipper lies a pair of stars, Mizar (Zeta Ursae Majoris) and Alcor (80 Ursae Majoris). Through a telescope you can discover that Mizar has a fainter companion and so is a member of a visual binary system. Adaptive optics observations reveal a faint close companion of Alcor, not pictured in this diagram. (b) Spectra of Mizar sometimes show single spectral lines (*top strip*) and other times show double lines (*bottom strip*), indicating that the main star of the visual binary system is itself a spectroscopic binary system rather than a single star.

orbit around the system's center of mass, the balance point of the system (look back to Chapter 3, **Concept Art 3B**). If one star is more massive than its companion, then the massive star is closer to the center of mass and travels in a smaller orbit, while the lower-mass star whips around in a larger orbit (Figure 9-11). The ratio of the masses of the stars in the binary system portrayed in Figure 9-11 is M_A/M_B, which equals r_B/r_A, the inverse of the ratio of the radii of the orbits. For example, if one star in a binary system has an orbit twice as large as the other star's orbit, then it must be half as massive. Getting the ratio of the masses is easy, but that doesn't tell you the individual masses of the stars, which is what you really want to know. That takes one more step.

To find the total mass of a binary star system, you must know the size of the orbits and the orbital period—the length of time the stars take to complete one orbit. The smaller the orbits are and the shorter the orbital period is, the stronger the stars' gravity must be to hold each other in orbit. From the sizes of the orbits and the orbital period, astronomers can use Kepler's third law (see Section 3-3c of Chapter 3) to figure out how much mass the stars contain in total. Combining that information with the ratio of the masses found from the relative sizes of the orbits reveals the individual masses of the stars. (Look at **The Masses of Binary Stars** on the Math Reference Cards MR.)

Finding the mass of a binary star system is easier said than done. One difficulty is that the true sizes of the star orbits must be measured in units, such as meters or Astronomical Units, in order to find the masses of the stars in units such as kilograms or solar masses. Measuring the true sizes of the orbits in turn requires knowing the distance to the binary system. Therefore, the only stars whose masses astronomers know for certain are in binary systems with orbits that have been determined and distances from Earth that have been measured. Other complications are that the orbits of the two stars may be elliptical; also, the plane of their orbits can be tipped at an angle to your line of sight, distorting the apparent shapes of the orbits. Astronomers must find ways to correct for these complications. Notice that finding the masses of binary stars requires a number of steps to get from what can be observed to what astronomers really want to know—the masses. Constructing such sequences of steps is an important part of science. (See **How Do We Know? 9-1**.)

9-5b Three Kinds of Binary Systems

Although there are many different kinds of binary stars, three types are especially important for determining stellar masses. Studying binary stars is also preparation for knowing how to find planets around stars other than our Sun, because a star with a planet orbiting around it is like a binary star system with one very small component. Each type of binary star system corresponds to a different technique for finding planets, as you learned in Section 8-5 of Chapter 8.

FIGURE 9-11

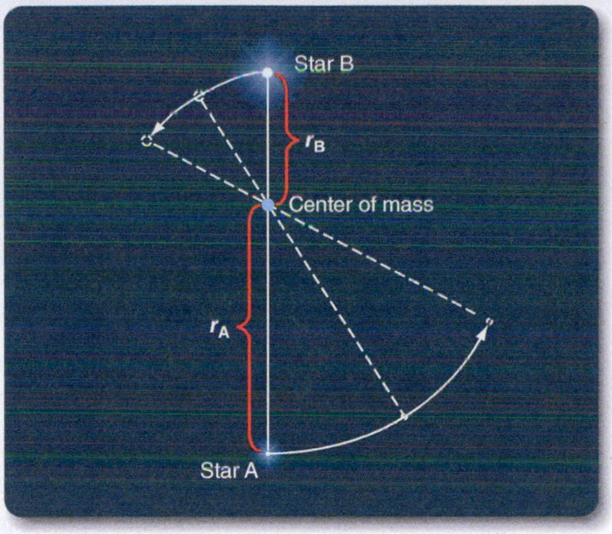

As stars in a binary star system revolve around each other, the line connecting them always passes through the center of mass, and the more massive star is always closer to the center of mass.

FIGURE 9-12

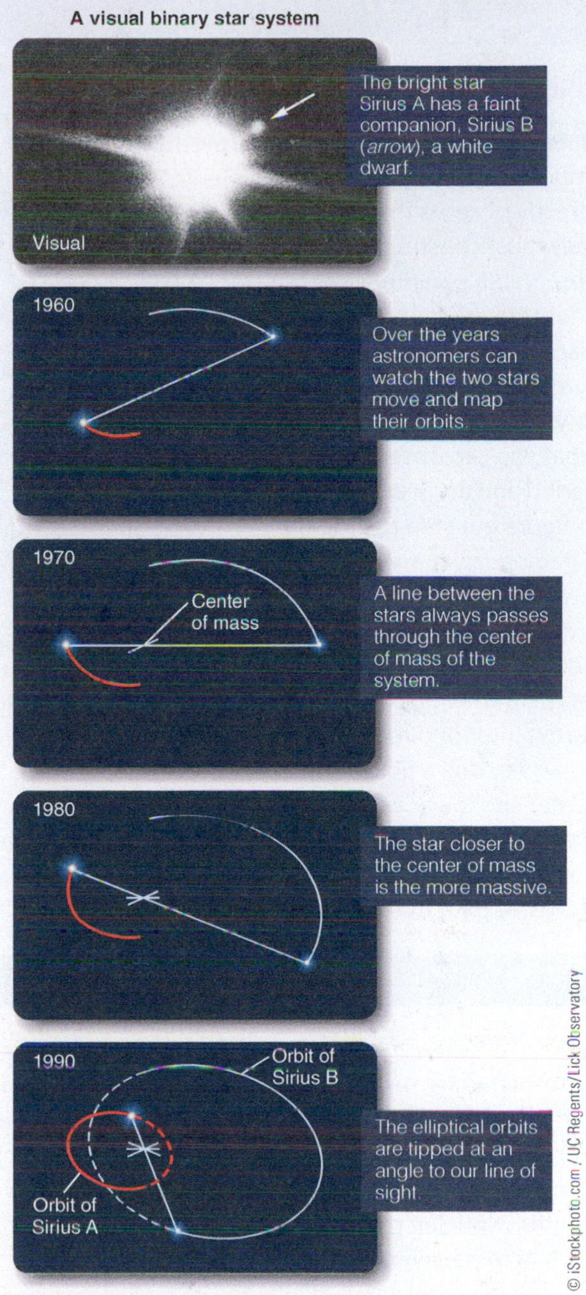

Measuring the orbital motion of Sirius A and Sirius B allows calculation of their individual masses.

In a **visual binary system**, the two stars are separately visible in the telescope and astronomers can watch the stars orbit each other over periods of years or decades, as the series of illustrations of Sirius A and B in Figure 9-12 demonstrates. From that, astronomers can find the orbital period and, if the distance of the system from Earth can be found, the size of the orbits. That is enough to find the masses of the stars.

Many visual binaries have such large orbits their orbital periods are hundreds or thousands of years, and astronomers have not yet seen them complete an entire orbit. Also, many binary stars orbit so close to each other they are not visible as separate stars. Such systems can't be analyzed as a visual binary.

If the stars in a binary system are close together, a telescopic view, limited by diffraction and atmospheric seeing, shows a single point of light. Only by looking at a spectrum, which is formed by light from both stars and contains spectral lines from both, can astronomers tell that there are two stars present and not one. Such a system is a **spectroscopic binary system**. Familiar examples of spectroscopic binary systems are the stars Mizar and Alcor in the handle of the Big Dipper (Figure 9-10b).

Figure 9-13 shows a pair of stars orbiting each other; the circular orbit appears elliptical because you see it nearly edge-on. If this were a true spectroscopic binary system, you would not see the separate stars. Nevertheless, the Doppler shift would tell you there

visual binary system A binary star system in which the two stars are separately visible in the telescope.

spectroscopic binary system A star system in which the stars are too close together to be visible separately. We see a single point of light, and only by taking a spectrum can we determine that there are two stars.

How Do We Know? 9-1

Chains of Inference

How do scientists determine something they can't measure directly? Sometimes scientists cannot directly observe the things they really want to study, so they must construct chains of inference that connect observable parameters to the unobservable quantities they want to know. You can't measure the mass of a star directly, so you must find a way to use what you can observe, orbital period and angular separation, to figure out step by step the parameters you need to calculate the mass.

Consider another example. Geologists can't measure the temperature and density of Earth's interior directly. There is no way to drill a hole to Earth's center and lower a thermometer or recover a sample. However, the speed of vibrations from a distant earthquake depends on the temperature and density of the rock they pass through. Geologists can't measure the speed of the vibrations deep inside Earth; but they can measure the delays in the arrival times at different locations on the surface, and that allows them to work their way back to the speed and, finally, the temperature and density.

USGS

San Andreas fault: A chain of inference connects earthquakes to conditions inside Earth.

Chains of inference can be nonmathematical. Biologists studying the migration of whales can't follow individual whales for years at a time, but they can observe them feeding and mating in different locations; take into consideration food sources, ocean currents, and water temperatures; and construct a chain of inference that leads back to the seasonal migration pattern for the whales.

This chapter contains a number of chains of inference. Almost all fields of science use chains of inference. When you can link the observable parameters step by step to the final conclusions, you gain a strong insight into the nature of science.

were two stars orbiting each other. As the two stars move in their orbits, they alternately approach toward and recede from Earth and their spectral lines are Doppler shifted alternately toward blue and then red wavelengths. Noticing pairs of spectral lines moving back and forth across each other would alert you that you were observing a spectroscopic binary.

Although spectroscopic binaries are very common, they are not as useful as visual binaries. Astronomers can find the orbital period easily, but they can't find the true size of the orbits because there is no way to find the angle at which the orbits are tipped. That means they can't find the true masses of a spectroscopic binary. All they can find is a lower limit to the masses.

eclipsing binary system A binary star system in which the stars cross in front of each other as seen from Earth.

light curve A graph of brightness versus time commonly used in analyzing variable stars and eclipsing binaries.

If the plane of the orbits is nearly edge-on to Earth, then the stars can cross in front of each other as seen from Earth. When one star moves in front of the other, it blocks some of the light, and the star is eclipsed. Such a system is called an **eclipsing binary system**.

Seen from Earth, the two stars are not visible separately. The system looks like a single point of light. But, when one star moves in front of the other star, part of the light is blocked, and the total brightness of the point of light decreases. Figure 9-14 shows a smaller star moving in an orbit around a larger star, first eclipsing the larger star and then being eclipsed as it moves behind. The resulting variation in the brightness of the system is shown as a graph of brightness versus time, a **light curve**.

The light curves of eclipsing binary systems contain plenty of information about the stars, but the curves can be difficult to analyze. Figure 9-14 shows an idealized example of such a system. Once the light curve of an eclipsing binary system has been accurately observed,

FIGURE 9-13

A Spectroscopic Binary Star System

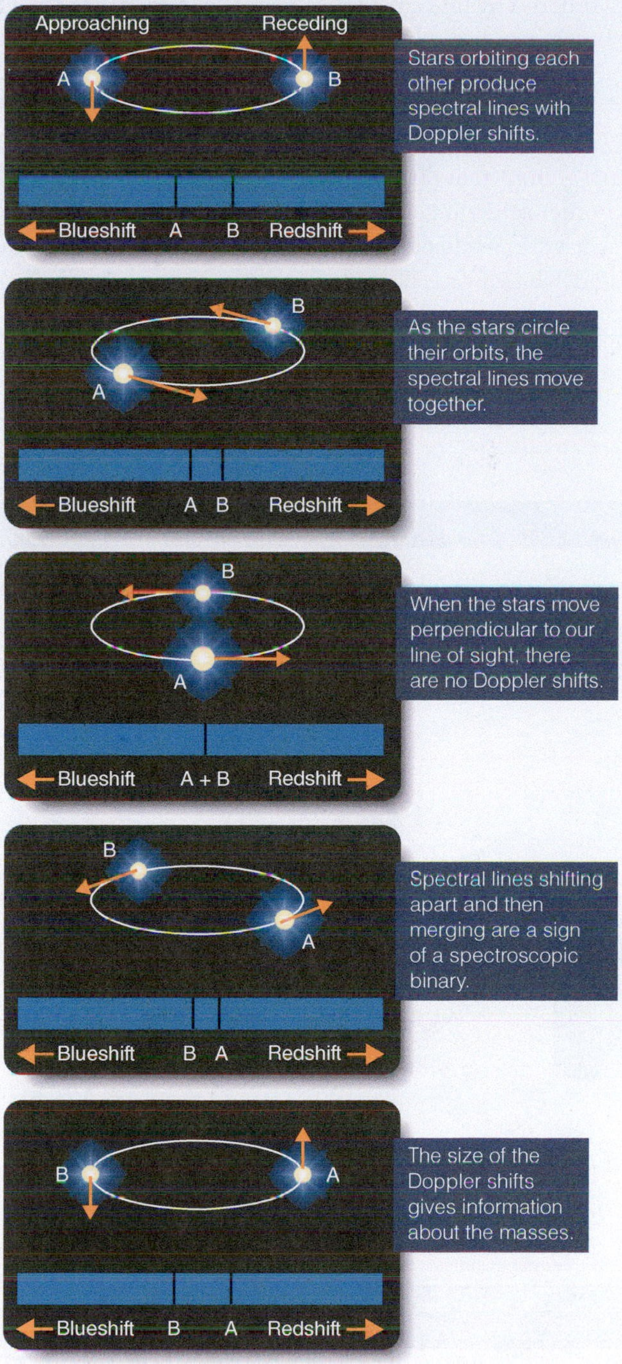

From Earth, a spectroscopic binary looks like a single point of light, but the Doppler shifts in its spectrum reveal the orbital motion of the two stars.

FIGURE 9-14

An Eclipsing Binary Star System

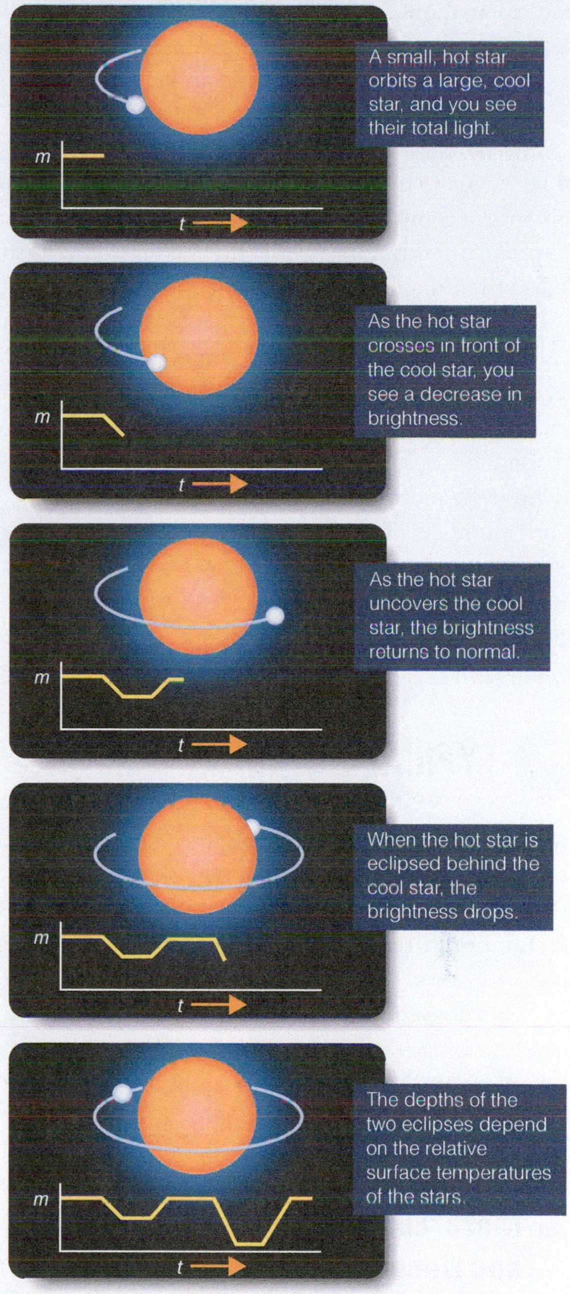

From Earth, an eclipsing binary looks like a single point of light, but changes in brightness reveal that two stars are eclipsing each other. The light curve, shown here as magnitude versus time, combined with Doppler shift information from spectra, can reveal the size and mass of the individual stars.

astronomers can construct a chain of inference that leads to the masses of the two stars. They can find the orbital period easily and can get spectra showing the Doppler shifts of the two stars. They can find the orbital speed because they don't have to correct for the inclination of the orbits; you know the orbits are nearly edge-on, or there would not be eclipses. From that, astronomers can find the size of the orbits and the masses of the stars.

Earlier in this chapter you learned that luminosity and temperature can be used to determine the radii of stars, but eclipsing binary systems give a way to check those calculations by measuring the sizes of a few stars directly. The light curve shows how long it takes for the stars to cross in front of each other, and multiplying these time intervals by the orbital speeds gives the diameters of the stars. There are complications due to the inclination and eccentricity of orbits, but often these effects can be taken into account, so observations of an eclipsing binary system can directly tell you not only the masses of its stars but also their diameters.

From the study of binary stars, astronomers have found that the masses of stars range from roughly 0.1 solar mass to nearly 100 solar masses. The most massive stars ever found in a binary system have masses of about 150 solar masses. A few other stars may be even more massive, but they are not members of binary systems, so astronomers can only estimate their masses from models based on their luminosities, temperatures, and other characteristics.

9-6 TYPICAL STARS

You have achieved the five goals set at the start of this chapter. You know how to find the distances, luminosities, temperatures, diameters, and masses of stars. Now you can put those data together to paint a family portrait of the stars. As in human family portraits, both similarities and differences are important clues to the history of the family.

9-6a Mass, Luminosity, and Density

The H–R diagram is filled with patterns that give you clues as to how stars are born, how they age, and how they die. When you add your data, you see traces of those patterns.

If you label an H–R diagram with the masses of the stars determined by observations of binary star systems, you will discover that the main-sequence stars are ordered by mass, which you can see in Figure 9-15. The most massive main-sequence stars are the hot stars. As you run your eye down the main sequence, you will find lower-mass stars, and the lowest- mass stars are the coolest, faintest main-sequence stars.

Stars that do not lie on the main sequence are not in order according to mass. Some giants and supergiants are massive, while others are no more massive than the Sun. All white dwarfs have about the same mass, usually in the narrow range of 0.5 to about 1.0 solar mass.

Because of the systematic ordering of mass along the main sequence, these main-sequence stars obey a **mass–luminosity relation**—the more massive a star

FIGURE 9-15

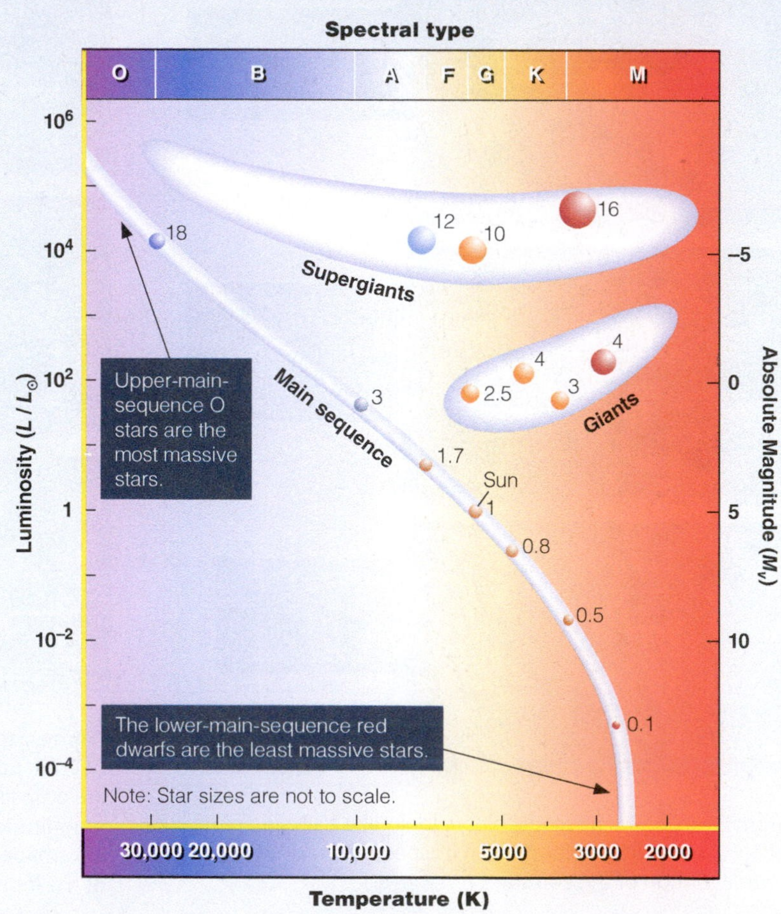

The masses of the plotted stars are labeled on this H–R diagram. Notice that the masses of main-sequence stars decrease systematically from upper left to lower right in the diagram but that masses of giants and supergiants are not arranged in any ordered pattern.

mass–luminosity relation The more massive a main-sequence star is, the more luminous it is.

is, the more luminous it is (Figure 9-16). In fact, the mass-luminosity relation can be expressed as: Luminosity is proportional to mass to the 3.5 power. For example, a star with a mass of 4.0 $M_\odot$ can be expected to have a luminosity of about $4^{3.5}$ or 128 $L_\odot$. Giants, supergiants, and white dwarfs do not follow the mass–luminosity relation. In the next chapters, the main sequence mass–luminosity relation will help you understand how main sequence stars generate their energy. (Look at **The Mass–Luminosity Relation** in the Math Reference Cards MR.)

Though mass alone does not reveal any pattern among giants, supergiants, and white dwarfs, density does. Once you know a star's mass and diameter, you can calculate its average density by dividing its mass by its volume. Stars are not uniform in density but are most dense at their centers and least dense near their surfaces. The center of the Sun, for instance, is about 100 times as dense as water; its density at the bottom of the photosphere is about 3400 times less dense than Earth's atmosphere at sea level. A star's average density is intermediate between its central and surface densities. The Sun's average density is approximately 1 gram per cubic centimeter—about the density of water.

Main-sequence stars have average densities similar to the Sun's density. As you learned earlier in the discussion about luminosity classification, giant stars are much larger in diameter than the main-sequence stars but not much larger in mass, so giants have low average densities, ranging from 0.1 to 0.01 g/cm^3. The enormous supergiants have still lower densities, ranging from 0.001 to 0.000001 g/cm^3. These densities are thinner than the air you breathe, and if you could insulate yourself from the heat, you could fly an airplane through these stars. Only near the center would you be in any danger; astronomers calculate that the material there is very dense.

The white dwarfs have masses about equal to the Sun's but are very small—only about the size of Earth. Thus, the matter is compressed to densities of 3 million g/cm^3 or more. On Earth, a teaspoonful of this material would weigh about 15 tons.

Density divides stars into three groups. Most stars are main-sequence stars with densities similar to the Sun's. Giants and supergiants are very-low-density stars, and white dwarfs are high-density stars. You will see in later chapters that these densities reflect different stages in the evolution of stars.

FIGURE 9-16

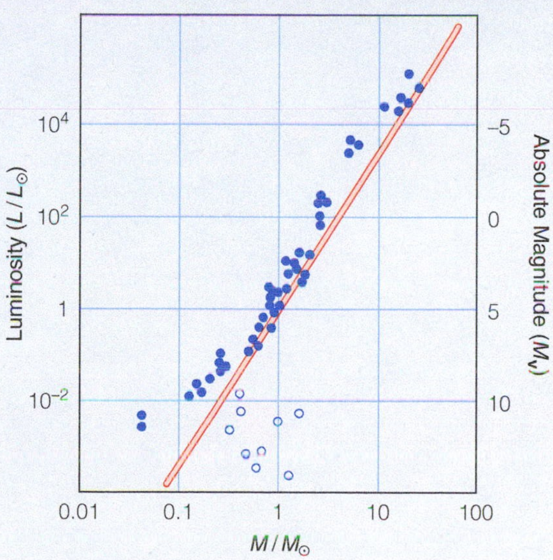

The mass–luminosity relation shows that the more massive a main-sequence star is, the more luminous it is. The open circles represent white dwarfs, which do not obey the relation. The red line represents the equation $L = M^{3.5}$.

9-6b Surveying the Stars

If you want to know what the average person thinks about a certain subject, you take a survey. If you want to know what the average star is like, you can survey the stars. Such surveys reveal important relationships among the family of stars. Over the years, many astronomers have added their results to the growing collection of data on star distances, luminosities, temperatures, sizes, and masses. They can now analyze those data to search for relationships among these and other parameters.

Modern astronomers are deeply involved in extensive surveys. Remember the earlier mention of the *Hipparcos* satellite that surveyed the entire sky measuring the parallax of more than a million stars? Powerful computers to control instruments and analyze data make such immense surveys possible. For example, the Sloan Digital Sky Survey mapped a quarter of the sky, measuring the position and brightness of 100 million stars and galaxies. Also, the Two Micron All Sky Survey (2MASS) mapped the entire sky at three near-infrared wavelengths. A number of other sky surveys are underway. Astronomers will "mine" these mountains of data for decades to come.

What could you learn about stars from a survey of the stars near the Sun? The evidence astronomers have is that the Sun is in a typical place in the Universe. Therefore, such a survey could reveal general characteristics

Concept Art 9A

THE FAMILY OF STARS

1 What is the most common type of star? What types of stars are rare? To answer these questions, you need to make a census of the stars. In doing so you collect information on their spectral classes, luminosity classes, and distances. Your survey of the family of stars produces some surprising demographic results.

1a You could survey the stars by observing every star within 62 pc of Earth. A sphere 62 pc in radius encloses a million cubic parsecs. Such a survey would tell you how many stars of each type are found within a volume of a million cubic parsecs.

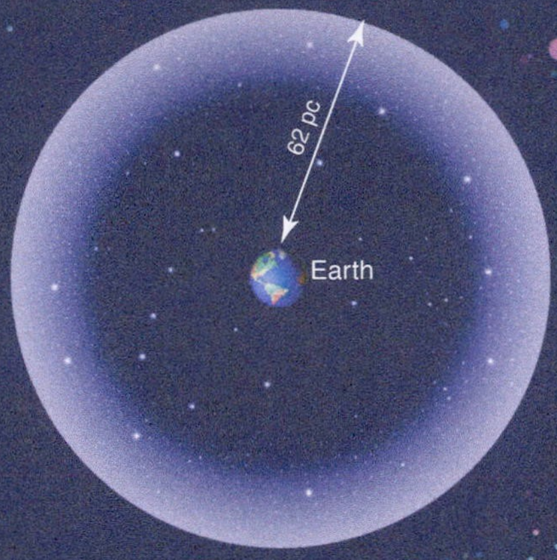

2 Your survey faces two difficulties:

1. The most luminous stars are so rare you find only a few in your survey region. In fact, there is not even one O star within 62 pc of Earth.
2. Lower-main-sequence M stars—sometimes called red dwarfs—and white dwarfs are so faint they are hard to locate even when they are only a few parsecs from Earth. Finding every one of these stars in your survey sphere is a difficult task, but if you don't, your survey is incomplete.

The star chart in the background of these two pages shows most of the constellation Canis Major; stars are represented as dots with colors assigned according to spectral class. The brightest stars in the sky tend to be the rare, highly luminous stars, which appear bright even though they are far away. Most stars are of very low luminosity, so nearby stars tend to be very faint red dwarfs.

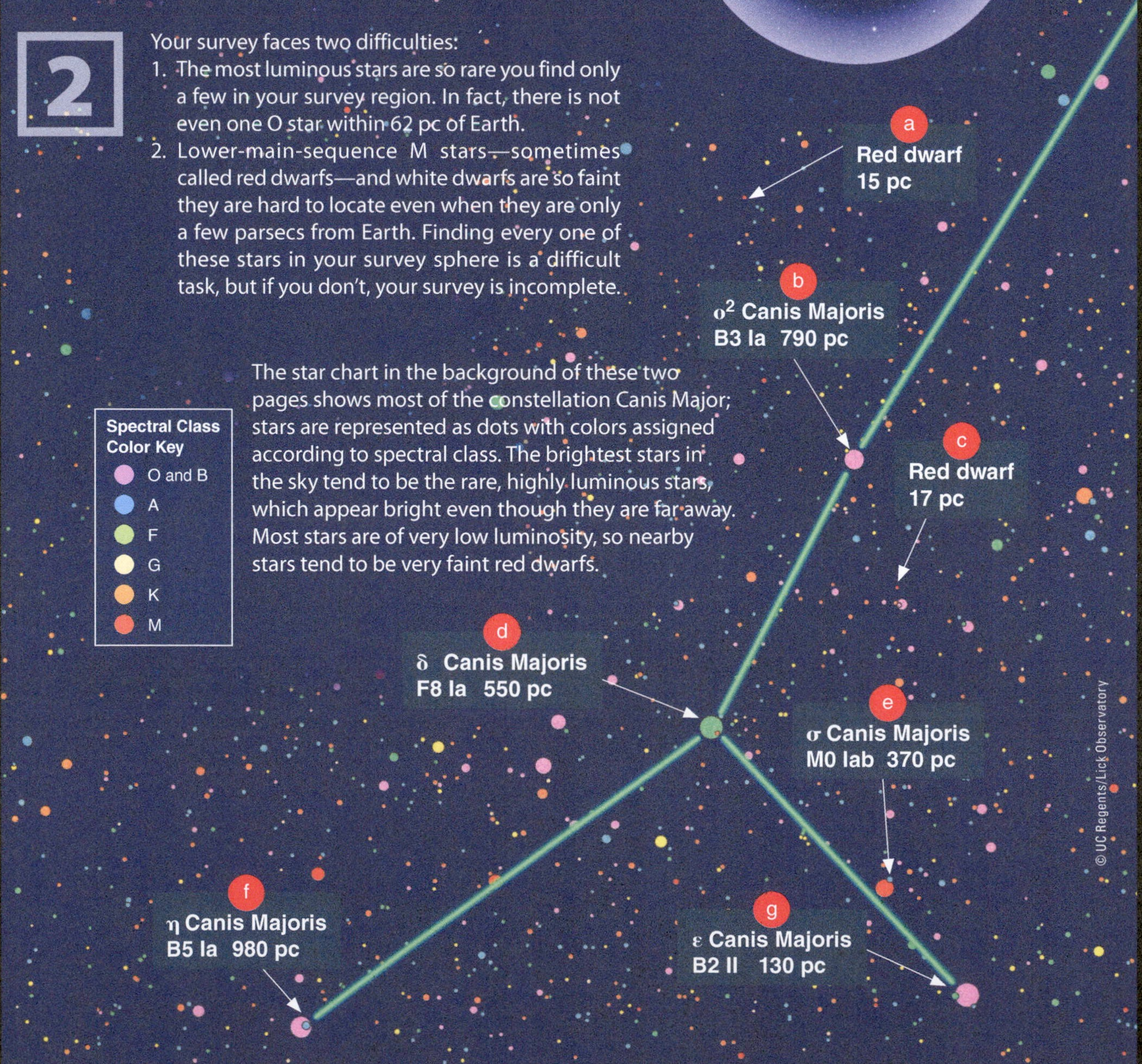

2a Supergiants, giants, and O and B main-sequence stars are so rare their bars are not visible in this graph.

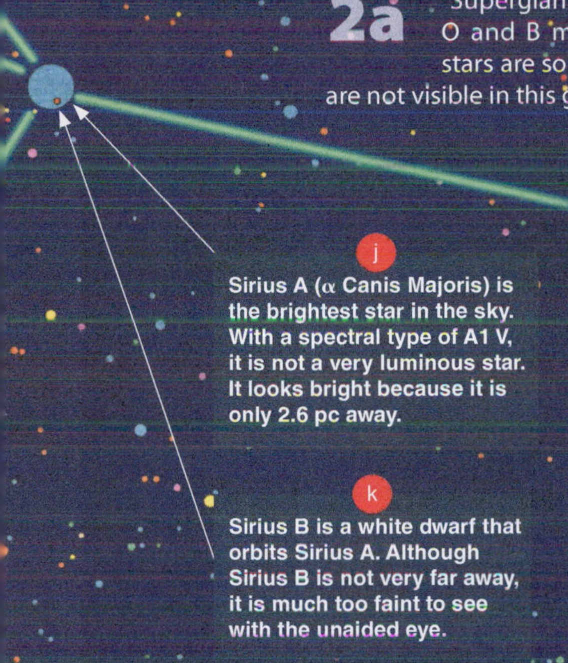

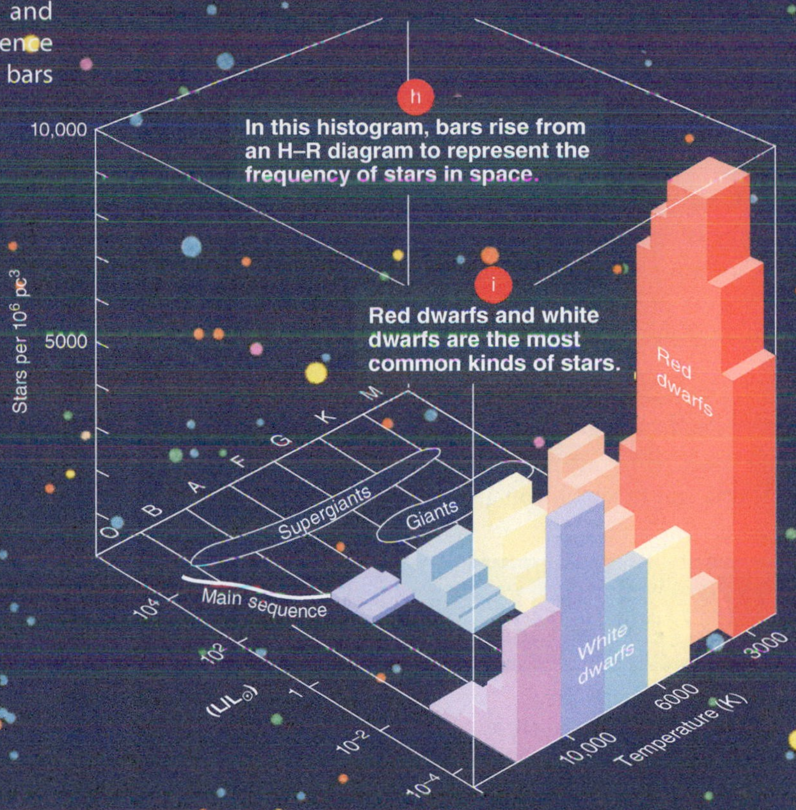

3 Luminous stars are rare but are also easy to see even at great distances. Most stars are very-low-luminous objects. Not a single white dwarf or red dwarf is bright enough to see with the unaided eye. See the H–R diagrams at right.

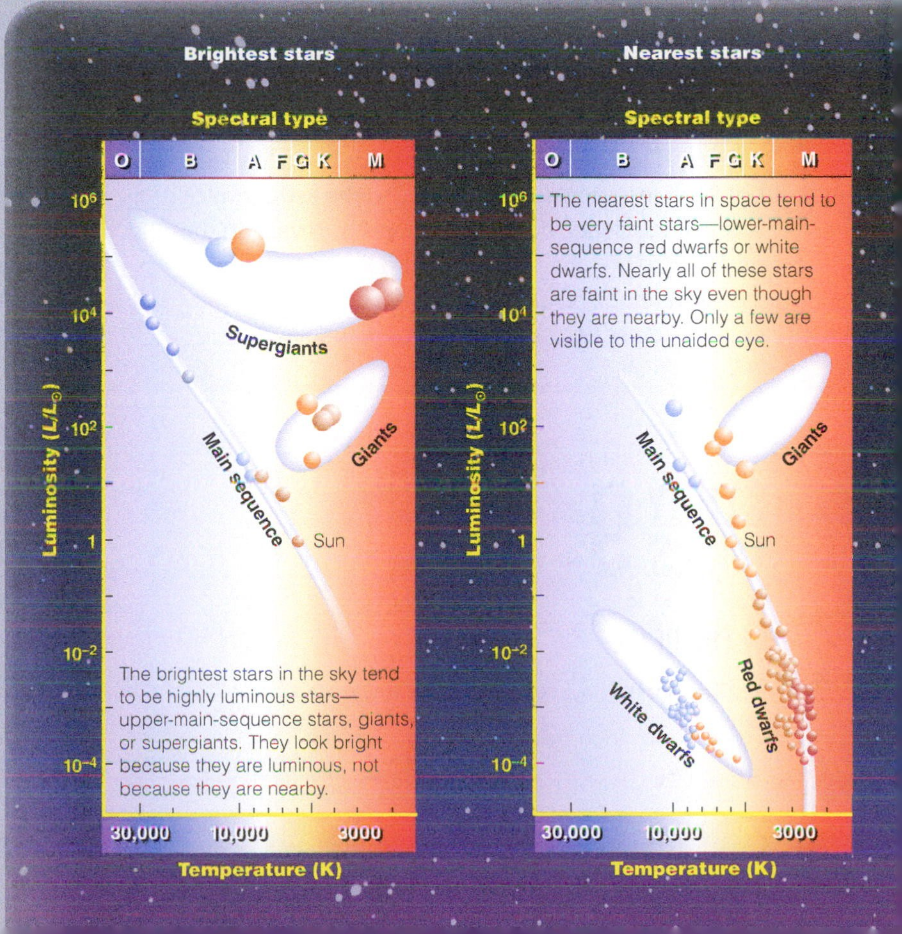

of the stars. Study **Concept Art 9A**, **"The Family of Stars,"** and notice the following three important points:

1. Making a good survey is difficult because you must be sure you get a good sample. If you don't survey enough stars, or if you miss some kinds of stars, your survey might produce biased results.
2. Luminous stars are rare. The most common types of stars are the low-luminous red dwarfs.
3. A careful survey reveals that what you see in the sky is deceptive. Most of the bright stars in the sky are not typical stars but instead are highly luminous giants and supergiant stars that can be seen from great distances.

The night sky is a beautiful carpet of stars. Some are giants and supergiants, and some are dwarfs. The family of stars is rich in its diversity. Pretend space is like a very clear ocean in which you are swimming. When you look at the night sky you are seeing mostly "whales," the rare very large and luminous stars, mostly far away. If instead you cast a net near your location, you would mostly catch "sardines," very-low-luminous M dwarfs that make up most of the stellar population of the Universe.

In this chapter, you explored the basic properties of stars. Once you found the distance to the stars, you were able to find their luminosities. Knowing their luminosities and temperatures gives you their diameters. Studying binary stars gives you their masses. These are all rather straightforward and simple data. But you have now discovered a puzzling situation. The largest and most luminous stars are so rare you might joke that they hardly exist, and the average stars are such small, low-mass things they are hard to see even if they are near Earth in space. Why does nature make stars in this peculiar way? To answer that question, you can explore the birth, life, and death of stars. That quest begins in Chapter 10.

What Are We? Medium

Humans are medium creatures, and experience medium things. You can see trees and flowers and small insects, but you cannot see the beauty of the microscopic world without ingenious instruments and special methods.

Similarly, you can sense the grandeur of a mountain range, but larger objects, such as stars, are too big for our medium-scale senses. You have to use your ingenuity and imagination to experience the truth of such large objects. That is what Science allows us to do. We live between the microscopic world and the astronomical world, and science enriches our lives by revealing the parts of the Universe beyond our daily experience.

Experience is fun, but it is very limited. You may enjoy a flower by admiring its color and shape and by smelling its fragrance. But the flower is more wonderful than your direct experience can reveal. To truly appreciate the flower you need to understand it—how complex it truly is, how it serves its plant, and how the plant came to create such a beautiful blossom.

Humans have a natural drive to understand as well as experience. You have experienced the stars in the night sky, and now you are beginning to understand them as objects ranging from hot blue O stars to cool red M dwarfs. It is natural for you to wonder why these stars are so different. As you investigate that question in the following chapters, you will discover that although you have medium-scale senses, you can understand the stars.

STUDY TOOLS 9

IN THE BOOK

- ☐ Tear Out the Review Card on *The Family of Stars*.

ASTRO ONLINE

Visit ASTRO Online at www.cengagebrain.com

- ☐ Interactive Reading
- ☐ Practice Quizzing
- ☐ Videos and Animations
- ☐ Exclusive Features

Questions and Problems

9

REVIEW QUESTIONS

1. Why are Earth-based parallax measurements limited to the nearest stars?
2. Why was the *Hipparcos* satellite able to make more accurate parallax measurements than ground-based telescopes?
3. What do the words *absolute* and *visual* mean in the definition of absolute visual magnitude?
4. What does luminosity measure that is different from what absolute visual magnitude measures?
5. Why are Balmer lines strong in the spectra of medium-temperature stars and weak in the spectra of hot and cool stars?
6. Why are titanium oxide features visible in the spectra of only the coolest stars?
7. Explain the interrelationships among Table 9-1, Figure 9-4, Figure 9-5, and Figure 9-6.
8. Why does the luminosity of a star depend on both its radius and its temperature?
9. How can you be sure that giant stars really are larger than main-sequence stars?
10. What evidence shows that white dwarfs must be very small?
11. What observations would you make to classify a star according to its luminosity? Why does that method work?
12. Why does the orbital period of a binary star depend on its mass?
13. What observations would you make to study an eclipsing binary star, and what would those measurements tell you about the component stars?
14. Why don't astronomers know the inclination of a spectroscopic binary? How do they know the inclination of an eclipsing binary?
15. How do the masses of stars along the main sequence illustrate the mass–luminosity relation?
16. Why is it difficult to find out how common the most luminous stars are? The least luminous stars?
17. What is the most common type of star?
18. If you look only at the brightest stars in the night sky, what type of star are you likely to be observing? Why?
19. **How Do We Know?** What is the missing link in the chain of inference leading from observations of spectroscopic binaries to the masses of stars?

DISCUSSION QUESTIONS

1. Your research requires you to discover the 10 stars nearest to the Sun. How would you select your sample to check? In other words, what measurements would you make, and how would you analyze those measurements, to choose stars likely to be nearby?
2. Can you think of classification systems that are used to simplify our understanding of what would otherwise be complex topics? Consider foods, movies, cars, grades, and clothes.
3. Your marketing agency has been tasked with rebranding the Sun as something other than an average star. What slogans can you come up with that are fairly accurate, given what you know about the Sun compared with other stars?

PROBLEMS

1. If a star has a parallax of 0.050 arc second, what is its distance in pc? In ly? In AU?
2. If a star has a parallax of 0.016 arc second and an apparent magnitude of 6, how far away is it, and what is its absolute magnitude?
3. Complete the following table of apparent visual magnitudes, absolute visual magnitudes, distances, and parallaxes:

m_v	M_v	d (pc)	p (arc seconds)
—	7	10	—
11	—	1000	—
—	−2	—	0.025
4	—	—	0.040

4. Determine the temperatures of the following stars based on their spectra. Use Figure 9-4.
   a. medium-strength Balmer lines, strong helium lines
   b. medium-strength Balmer lines, weak ionized calcium lines
   c. strong TiO bands
   d. very weak Balmer lines, strong ionized calcium lines
5. To which spectral classes do the stars in Problem 4 belong?

6. If a main-sequence star has a luminosity of 400 $L_\odot$, what is its spectral type? (*Hint:* See Figure 9-8.)
7. If a star has an apparent magnitude equal to its absolute magnitude, how far away is it in parsecs? In light-years?
8. An O8 V star has an apparent magnitude of 11. Use the method of spectroscopic parallax to estimate the distance to the star. (*Hints:* Refer to one of the H–R diagrams in the chapter; use the **Absolute Magnitude-Distance** formula in the Math Reference Cards MR.)
9. At the position of Earth, the total flux of Sunlight at all wavelengths, called the solar constant, is 1366 watts per m^2. Use that to find the total luminosity of the Sun. Make your calculations in two steps. First, use $4\pi R^2$ to calculate the surface area in m^2 of a sphere surrounding the Sun with a radius of 1 AU. Second, multiply by the solar constant to find the total solar energy passing through the sphere in 1 second. (This assumes that the Sun's luminosity is emitted equally in all directions.) The result will be the luminosity of the Sun. Compare your result with the data in **Celestial Profile 9,** Appendix C.
10. In the following table, which star is brightest in apparent magnitude? Most luminous in absolute magnitude? Largest? Farthest away?

Star	Spectral Type	m_v
a	G2 V	5
b	B1 V	8
c	G2 Ib	10
d	M5 III	19
e	White dwarf	15

11. What is the total mass of a visual binary system if the average separation of the stars is 8 AU and their orbital period is 20 years? (*Hint:* See the **The Masses of Binary Stars** feature in the Math Reference Cards MR.)
12. If an eclipsing binary has a period of 32 days, and the two components are in circular orbits around their center of mass with speeds, respectively, of 154 km/s and 77 km/s, what are the circumferences of the two orbits? The separation between the two stars? The total mass of the system? The mass of each star? (*Hint:* See the **Circular Velocity** formula in the Math Reference Cards MR.)
13. What is the luminosity of a 4-solar-mass main-sequence star? Of a 8-solar-mass main-sequence star? Can you easily determine the luminosity of a 4-solar-mass red giant star? Why or why not? (*Hint:* See the **The Mass–Luminosity Relation** feature in the Math Reference Cards MR.)

LEARNING TO LOOK

1. If all of the stars in Figure 9-17 are members of the same star cluster, they all have about the same distance from Earth. Then why are three of the brightest much redder than the rest? What kind of star are they?

FIGURE 9-17

10 Structure and Formation of Stars

Jim he allowed [the stars] was made, but I allowed they happened. Jim said the moon could'a laid them; well, that looked kind of reasonable, so I didn't say nothing against it, because I've seen a frog lay most as many, so of course it could be done.

Mark Twain, The Adventures of Huckleberry Finn

After finishing this chapter go to **PAGE 235** for **STUDY TOOLS.**

ESO

GUIDEPOST

In Chapter 9, you discovered the wide range of differences within the family of stars. In this chapter, you will combine observations and hypotheses to understand how nature makes stars, and learn the answers to the following five important questions:

- How do stars remain stable?
- How do stars make energy?
- How do the luminosities and lifetimes of stars depend on their masses?
- How do astronomers study the gas and dust between the stars, called the interstellar medium?
- How do stars form from the interstellar medium?

Perhaps more important than these factual questions is the key question you should ask in any scientific context: What's the evidence? Astronomers have evidence about how stars are born, how they remain stable, and how they make energy. Testing hypotheses against evidence is the basic skill required of all scientists, and you will find opportunities to use it repeatedly in the five chapters that follow.

The stars are not eternal. When you look at the sky, you see hundreds of points of light. (See CO 10.) Each is an object like the Sun held together by its gravity and generating tremendous energy in its core through nuclear reactions. The stars you see tonight are the same stars your parents, grandparents, and great-grandparents saw. Stars hardly change at all in a human lifetime, but they are not eternal. Stars are born, and stars die. This chapter begins that story.

How can you know what the internal processes and life cycles of stars are when you can't see inside them and humans don't live long enough to see them evolve? The answer lies in the methods of science. By constructing hypotheses about how nature works and then testing those hypotheses against evidence from observations, you can unravel some of nature's greatest secrets.

In this chapter, you will learn how the flow of energy from inside to the surfaces of the stars balances gravity, making the stars stable, and how nuclear reactions inside stars generate that energy. You will also see how gravity creates new stars from the thin dusty gas between the stars.

10-1 STELLAR STRUCTURE

If there is a single idea in stellar astronomy that can be called crucial, it is the concept of balance. In this section you will discover that stars are simple, elegant

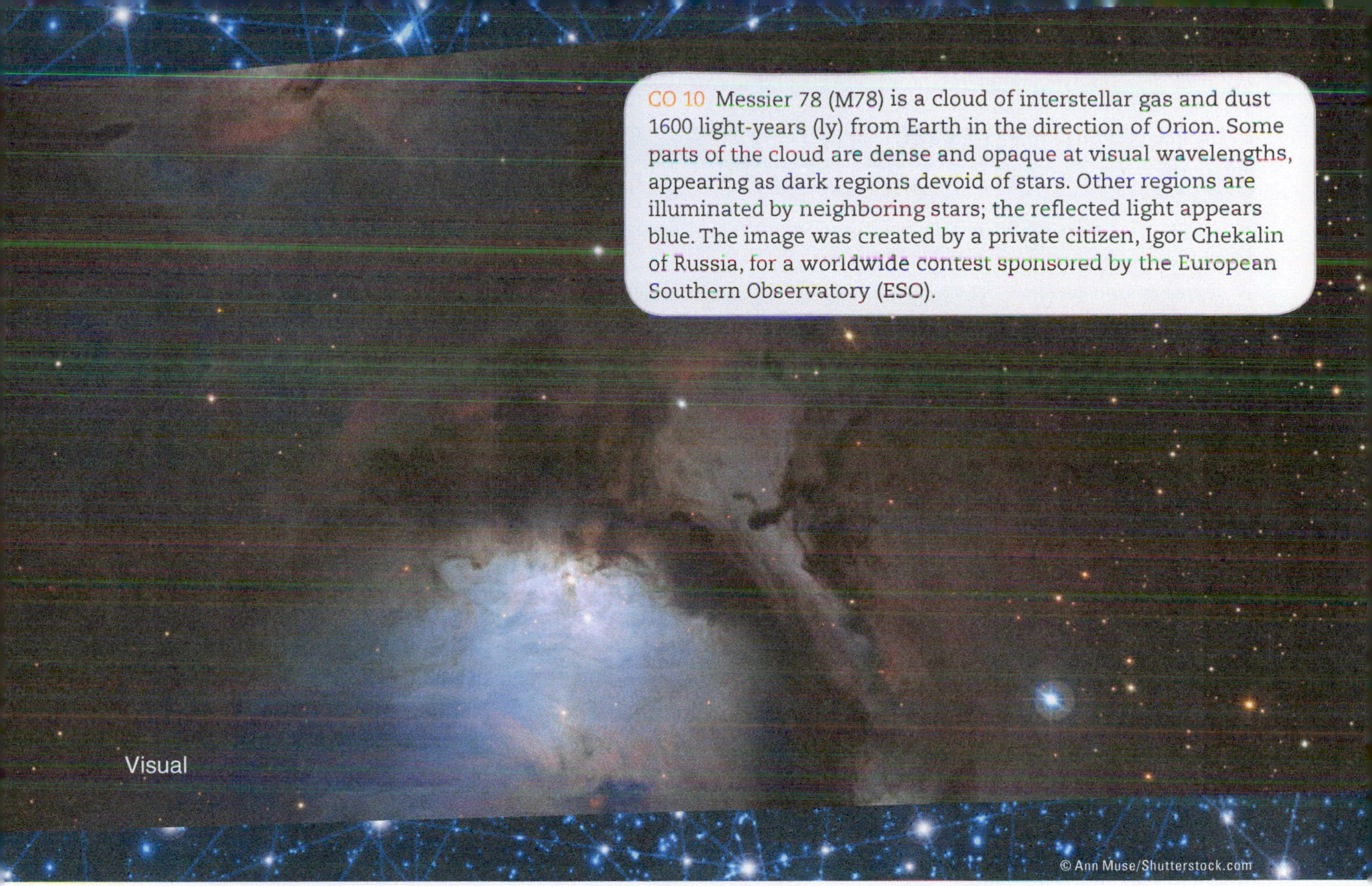

CO 10 Messier 78 (M78) is a cloud of interstellar gas and dust 1600 light-years (ly) from Earth in the direction of Orion. Some parts of the cloud are dense and opaque at visual wavelengths, appearing as dark regions devoid of stars. Other regions are illuminated by neighboring stars; the reflected light appears blue. The image was created by a private citizen, Igor Chekalin of Russia, for a worldwide contest sponsored by the European Southern Observatory (ESO).

power sources held together by their own gravity, balanced by the support of their internal heat and pressure.

Using the basic concept of balance, you can consider the structure of a star. What is meant here by structure is the variation in temperature, density, pressure, and so on between the surface of the star and its center. It will be easier to think about stellar structure if you imagine that the star is divided into concentric shells like those in an onion. Knowing the four basic laws of stellar structure that are described in the next section will also help you understand stars. With these laws you can consider the temperature, pressure, and density in each shell. These helpful shells exist only in the imagination; stars do not actually have such separable layers.

10-1a The Laws of Conservation of Mass and Energy

The first two laws of stellar structure have something in common—they are both what astronomers and physicists call *conservation* laws. Conservation laws say that certain things cannot be created out of nothing or vanish into nothing. Such conservation laws are powerful aids to help you understand how nature works.

The law of ***conservation of mass*** (Table 10-1) is a basic law of nature that can be applied to the structure of stars. It says that the total mass of a star must equal the sum of the masses of its shells. This is like saying the weight of a cake must equal the sum of the weight of its layers.

The law of ***conservation of energy*** (Table 10-1) is another basic law of nature. It says that the amount of energy flowing out of the top of a layer in the star must be equal to the amount of energy coming in at the bottom, plus whatever energy is generated within the layer. That means that the energy leaving the surface of the star, its luminosity, must equal the sum of the energies generated in all the layers inside the star. This is like saying that all the new cars driving out of a factory must equal the sum of all the cars made on each of the production lines.

These two laws may seem so familiar or so obvious that they hardly need to be stated, but they are important clues to the structure of stars. The third law of stellar structure is a law about balance.

TABLE 10-1 THE FOUR LAWS OF STELLAR STRUCTURE

I. Conservation of mass	Total mass equals the sum of the masses in all the layers.
II. Conservation of energy	Total luminosity equals the sum of energy generated in all of the layers.
III. Hydrostatic equilibrium	The weight on each layer is balanced by the pressure in that layer.
IV. Energy transport	Energy moves from hot to cool regions by conduction, radiation, or convection.

10-1b Hydrostatic Equilibrium

The weight of each layer of a star must be supported by the layer below. This is another way to state the third law in Table 10-1. (The words *down* or *below* are conventionally used to refer to regions closer to the center of a star.) Picture a pyramid of people in a circus stunt—the people in the top row do not have to hold up anybody else; the people in the next row down are holding up the people in the top row; and so on. In a star that is stable, the deeper layers must support the weight of all of the layers above. The inside of a star is made up of gas, so the weight pressing down on a layer must be balanced by gas pressure in that layer. If the pressure is too low, the weight from above will compress and push down the layer; and if the pressure is too high, the layer will expand and lift the layers above.

This balance between weight and pressure is called **hydrostatic equilibrium** (Table 10-1). *Hydro* (from the Greek word for water) tells you the material is a fluid, which by definition includes the gases of a star, and *static* tells you the fluid is stable, neither expanding nor contracting.

Figure 10-1 shows this hydrostatic balance in the imaginary layers of a star. The weight pressing down on each layer is shown by lighter red arrows, which grow larger with increasing depth because the weight grows larger. The pressure in each layer is shown by darker red arrows, which must grow larger with increasing depth to support the weight above.

The pressure in a gas depends on the temperature and density of the gas. Near the surface, there is little weight pressing down, so the pressure does not need to be high for stability. Deeper in the star, the pressure must be higher, which means that the temperature and density of the gas must also be higher. Hydrostatic equilibrium tells you that stars must have high temperature, pressure, and density inside to support their own weight and be stable.

Although the law of hydrostatic equilibrium can tell you some things about the inner structure of stars, you need one more law to completely describe a star. You need to know how energy flows from the inside to the outside.

FIGURE 10-1

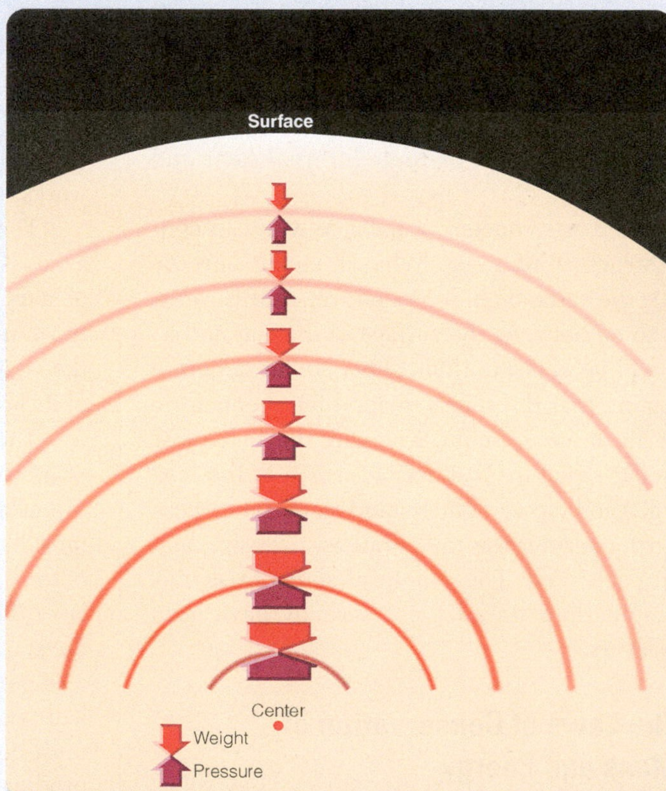

The principle of hydrostatic equilibrium says that the pressure in each layer of a star must balance the weight on that layer. Consequently, as the weight increases from the surface of the star to its center, the pressure must also increase.

hydrostatic equilibrium Stable balance of forces in a fluid (liquid or gas).

10-1c Energy Transport

Now look at the fourth law in Table 10-1. The surface of a star radiates light and heat into space and would quickly cool if that energy were not replaced. Because the inside of the star is hotter than the surface, energy must flow outward to the surface, where it radiates away. This flow of energy through each shell determines its temperature, which, as you saw previously, determines how much weight that shell can balance.

The law of ***energy transport*** (Table 10-1) says that energy must flow from hot regions to cooler regions either by conduction, convection, or radiation.

Conduction is the most familiar form of heat flow. If you hold the bowl of a spoon in a candle flame, the handle of the spoon grows warmer as heat, in the form of motion among the atoms of the spoon, is conducted from atom to atom up the handle. Conduction requires close contact between the atoms, so it is not an important cause of heat flow in normal stars because radiation or convection are much more efficient, even in their centers; it is only important in rare types of stars with extremely high densities.

The transport of energy by *radiation* is another familiar experience. Put your hand near a candle flame, and you can feel the heat. What you actually feel are infrared photons—packets of energy— radiated by the flame and absorbed by your hand. Radiation is the principal means of energy transport in the interiors of most stars. Photons are absorbed and reemitted in random directions repeatedly as energy works its way from the hot interior toward the cooler surface.

The flow of energy by radiation depends on how difficult it is for photons to move through the gas. The **opacity** of the gas, its resistance to the flow of radiation, depends strongly on its temperature: A hot gas is more transparent than a cool gas. If the opacity is high, radiation cannot flow through the gas easily, and it backs up like water behind a dam. When enough heat builds up, the gas begins to churn as hot gas rises upward and cool gas sinks downward. This heat-driven circulation of a fluid is *convection*, the third way energy can move in a star. You are familiar with convection; the rising wisp of smoke above a candle flame is carried by convection. Energy is carried upward in these convection currents as rising hot gas and also as sinking cool gas. The cross section of the Sun in Figure 10-3 shows the locations of the radiative and convective zones.

The four laws of stellar structure, when properly understood, not only tell you about the insides of stars, but also how stars are born, how they live, and how they die.

FIGURE 10-3

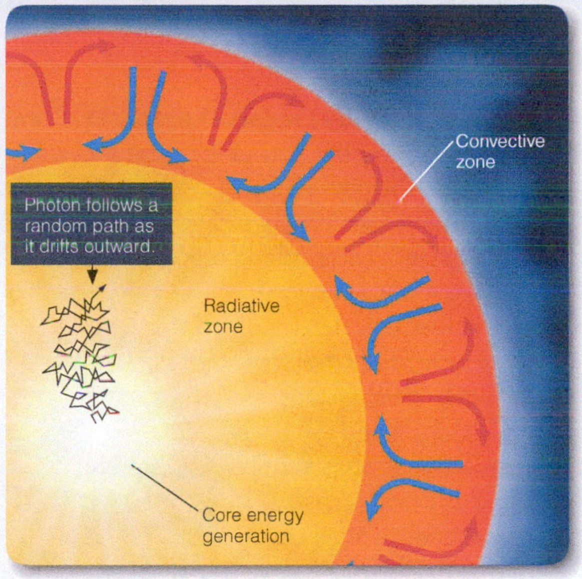

A cross section of the Sun. Near the center, nuclear fusion reactions sustain high temperatures. Energy flows outward through the radiative zone as photons are randomly defected over and over by electrons. In the cooler, more opaque outer layers, the energy is carried by rising convection currents of hot gas (red arrows) and sinking currents of cooler gas (blue arrows).

FIGURE 10-2

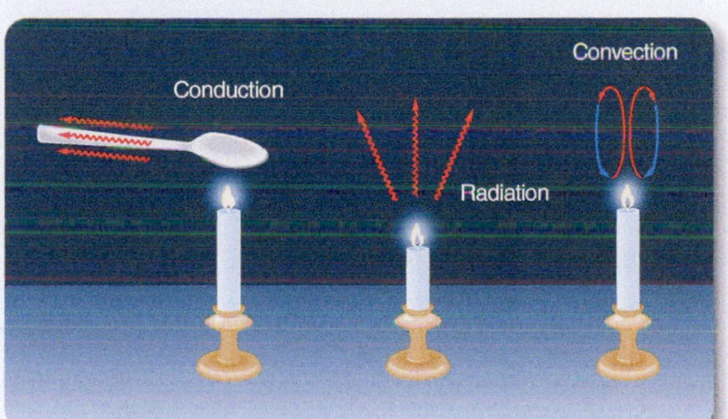

The three modes by which energy may be transported from the flame of a candle, as shown here, are the three modes of energy transport within a star.

10-1d Stellar Models

The laws of stellar structure, described in general terms in the previous

opacity The resistance of matter to the passage of radiation.

How Do We Know? 10-1

Mathematical Models

How can scientists study aspects of nature that cannot be observed directly? One of the most powerful tools in science is the mathematical model, a group of equations or a computer program carefully designed to mimic the behavior of objects and processes that scientists want to study. Astronomers build mathematical models of stars to study the structure hidden deep inside them. Models can speed up the slow evolution of stars and slow down the rapid processes that generate energy. Stellar models are based on only four equations, but other models are much more complicated and may require many more equations.

For example, scientists and engineers designing a new airplane don't just cross their fingers, build it, and ask a test pilot to try it out. Long before any metal parts are made, mathematical models are created to test whether the wing design will generate enough lift, whether the fuselage can support the strain, and whether the rudder and ailerons can safely control the plane during takeoff, flight, and landing. Those mathematical models are put through all kinds of tests: Can a pilot fly with one engine shut down? Can the pilot recover from sudden turbulence? Can the pilot land in a crosswind? By the time the test pilot rolls the plane down the runway for the first time, the mathematical models have flown many thousands of miles.

The Boeing Company

Before any new airplane flies, engineers build mathematical models to test its stability.

Scientific models are only as good as the assumptions that go into them and must be compared with the real world at every opportunity. If you are an engineer designing a new airplane, you can test your mathematical models by making measurements in a wind tunnel. Models of stars are much harder to test against reality, but they do predict some observable things. Stellar models predict the existence of a main sequence, the mass–luminosity relation, the observed numbers of giant and supergiant stars, and the shapes of H–R diagrams. Without mathematical models, astronomers would know little about the lives of the stars, and designing new airplanes would be a very dangerous business.

sections, can be written as mathematical equations. By solving those equations, astronomers can make a mathematical model of the inside of a star. (See **How Do We Know? 10-1**.) To start one of these models requires knowing the type of information you learned in Chapter 9 about the external properties of a star: its diameter, luminosity, mass, surface temperature, atmospheric density, and composition.

If you want to build a model of a star, you could divide the star into, say, 100 concentric shells and then write down the four equations of stellar structure for each shell. You would then have 400 equations that would have 400 unknowns, namely, the temperature, density, mass, and energy flow in each shell. Solving 400 equations simultaneously is not easy, and the first such solutions, done by hand before the invention of electronic computers, took months of work. Now a properly programmed computer can solve the equations in a few seconds and print a table of numbers that represent the conditions in each shell of the star. Such a table is called a **stellar model**.

stellar model A table of numbers representing the conditions in various layers within a star.

The table shown in Figure 10-4 is a model of the Sun. The bottom line, for radius equal to 0.00, represents the center of the Sun, and the top line, for radius equal to 1.00, represents the surface. The other lines tell

FIGURE 10-4

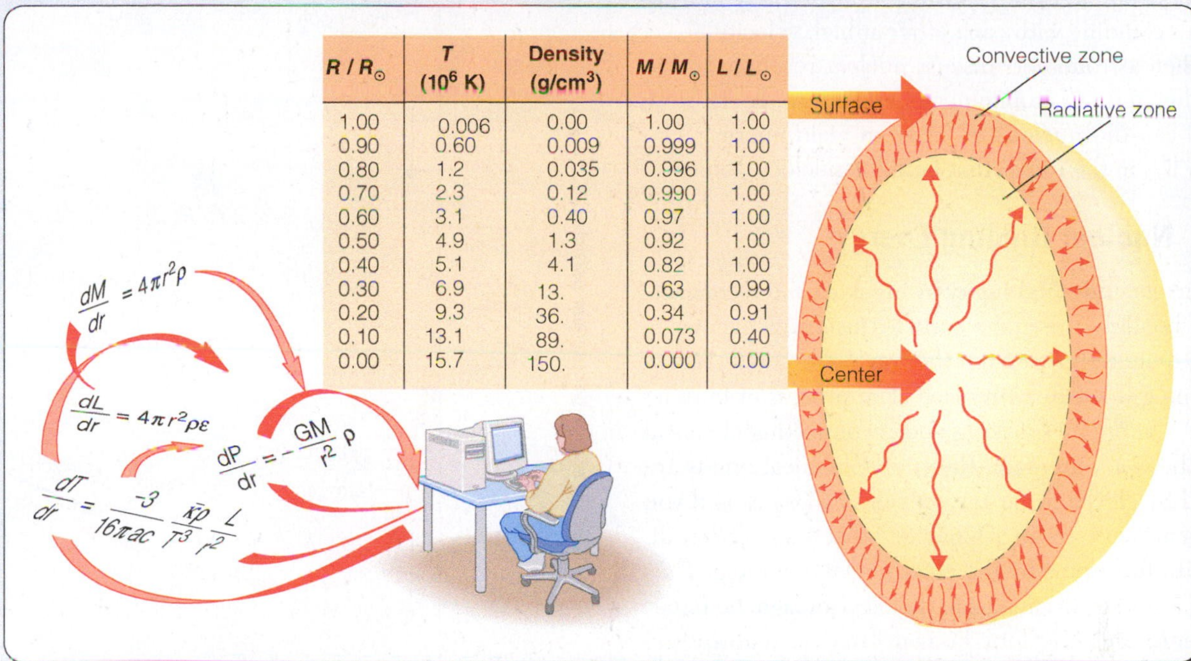

$R / R_\odot$	T (10^6 K)	Density (g/cm^3)	$M / M_\odot$	$L / L_\odot$
1.00	0.006	0.00	1.00	1.00
0.90	0.60	0.009	0.999	1.00
0.80	1.2	0.035	0.996	1.00
0.70	2.3	0.12	0.990	1.00
0.60	3.1	0.40	0.97	1.00
0.50	4.9	1.3	0.92	1.00
0.40	5.1	4.1	0.82	1.00
0.30	6.9	13.	0.63	0.99
0.20	9.3	36.	0.34	0.91
0.10	13.1	89.	0.073	0.40
0.00	15.7	150.	0.000	0.00

A stellar model is a table of numbers that represent conditions inside a star. Such tables can be computed using four laws of stellar structure, shown here in equation form, including the principle of hydrostatic equilibrium and the law of energy transport. The table in this figure describes the present-day Sun.

you the temperature and density in each shell, the mass inside each shell, and the fraction of the Sun's luminosity flowing outward through the shell. The bottom line tells you that the temperature at the center of the Sun must be more than 15 million Kelvin in order for the Sun to be stable. At such a high temperature, the gas is very transparent, and energy flows as radiation. Nearer the surface, the temperature is lower, the gas is more opaque, and energy moves by convection. This is confirmed by evidence of convection visible on the Sun's surface as granulation (look back to Chapter 5, Section 5-3b).

Stellar models also let you look into a star's past and future. In fact, you could use models as time machines to follow the evolution of stars over billions of years. To look into a star's future, for instance, you could use a stellar model to determine how fast the star uses its fuel in each shell. As the fuel is consumed, the chemical composition of the gas changes, the opacity changes, and the amount of energy generated declines. By calculating the rate of these changes, you could predict what the star will look like at some time in the future. You could repeat that process over and over and follow the evolution of the star step by step as it ages.

Stellar astronomy has made great advances since the 1960s, when high-speed computers that could calculate models of the structure and evolution of stars began to be available. The summary of star formation later in this chapter is based on thousands of models. You will continue to rely on theoretical models as you study main-sequence stars later in this chapter, and the deaths of stars in Chapter 11.

10-2 NUCLEAR FUSION IN THE SUN AND STARS

Astronomers often use the wrong words to describe energy generation in the Sun and stars. Astronomers will say, "The star ignites hydrogen burning." You normally use the word *ignite* to mean "catch on fire" and *burn* to mean "consume by fire." What goes on inside stars isn't really burning in the usual sense.

The Sun is a normal star, and it is powered by nuclear reactions that occur near its center. Later in this section you will learn some of the evidence for this. The energy produced keeps a star's interior hot, and the gas is totally ionized. That is, the electrons are not attached to the atomic nuclei, and the gas is a soup of rapidly moving particles colliding with each other at high velocity.

When astronomers discuss nuclear reactions inside stars, they refer to atomic nuclei and not to atoms. How exactly can the nucleus of an atom yield energy? The answer lies in the forces that hold the nuclei together.

10-2a Nuclear Binding Energy

The Sun generates its energy by breaking and reconnecting the bonds between the particles *inside* atomic nuclei. That is quite different from the way we generate energy by burning wood in a fireplace. The process of burning extracts energy by breaking and reconnecting chemical bonds between atoms in the fuel. Chemical bonds are formed by electrons on the outsides of atoms, and you learned in Section 5-4b of Chapter 5 that the electrons are bound to the atoms by the electromagnetic force. Thus chemical energy originates in the electromagnetic force.

There are only four known forces in nature: the force of gravity, the electromagnetic force, and the strong and weak **nuclear forces**. The weak nuclear force is involved in the radioactive decay of certain kinds of nuclear particles, and the strong force binds together atomic nuclei. Thus, nuclear energy originates in the strong nuclear force.

Nuclear power plants on Earth generate energy through **nuclear fission** reactions that split uranium nuclei into less massive fragments. The isotope of uranium used in power plants has a total of 235 protons and neutrons in its nuclei, which it splits into a range of fragments containing roughly half as many particles. Because the fragments produced are more tightly bound than the uranium nuclei, energy is released during uranium fission, as illustrated in Figure 10-5.

Stars make energy in **nuclear fusion** reactions that combine light nuclei into heavier nuclei. The most common reaction, including that in the Sun, fuses hydrogen nuclei (single protons) into helium nuclei (two protons and two neutrons). Because the nuclei produced are more tightly bound than the original nuclei, energy is released. Notice in Figure 10-5 that both fusion and fission reactions move downward in the diagram, meaning they release the binding energy of nuclei and create more tightly bound nuclei.

nuclear forces The two forces of nature that only affect the particles in the nuclei of atoms.

nuclear fission Reactions that break the nuclei of atoms into fragments.

nuclear fusion Reactions that join the nuclei of atoms to form more massive nuclei.

FIGURE 10-5

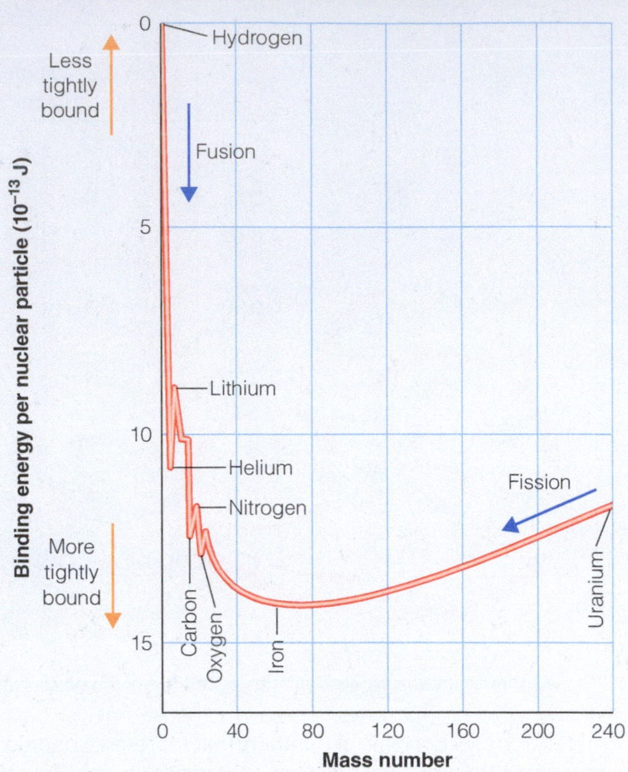

The orange curve in this graph shows the binding energy per particle, the energy that holds particles inside an atomic nucleus. The horizontal axis shows the atomic mass number of each element, the number of protons and neutrons in the nucleus. Both fission and fusion nuclear reactions move downward in the diagram (arrows), meaning the nucleus produced by a reaction is more tightly bound than the nuclei that went into the reaction. Iron has the most tightly bound nucleus, so no nuclear reactions can begin with iron and release energy.

10-2b Hydrogen Fusion

The Sun fuses four hydrogen nuclei to make one helium nucleus. Because one helium nucleus contains 0.7 percent less mass than four hydrogen nuclei, some mass vanishes in the process. In fact, that mass is converted to energy, and you could figure out how much by using Einstein's famous equation, $E = mc^2$. For example, converting 1 kilogram (2.2 pounds) of matter completely into energy would produce an enormous amount of energy, 9×10^{16} joules, comparable to the amount of energy a big city like Philadelphia uses in a year.

Making one helium nucleus releases only a small amount of energy, hardly enough to raise a housefly one-thousandth of an inch into the air. Only by concentrating

FIGURE 10-6

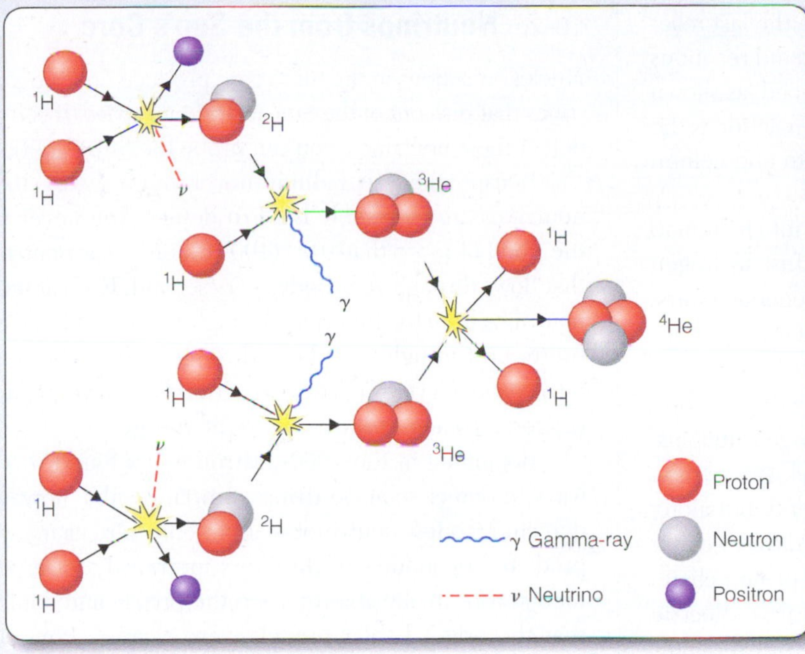

The proton–proton chain combines four protons (at far left) to produce one helium nucleus (at right). Energy is produced mostly as gamma-rays (γ) and as positrons, which combine with electrons and convert their mass into more gamma-rays. Neutrinos (ν) escape without heating the gas.

many reactions in a small region can nature produce significant amounts of energy. The Sun has a voracious appetite and needs 10^{38} reactions per second, transforming 4 million tons of mass into energy every second, just to replace the energy pouring into space from its surface. That might sound as if the Sun is losing mass at a furious rate, but in its entire 11-billion-year main-sequence lifetime, the Sun will convert only about 0.1 percent of its mass into energy.

Fusion reactions can occur only when the nuclei of two atoms get very close to each other. Because atomic nuclei carry positive charges, they repel each other with an electrostatic force called the Coulomb force. Physicists commonly refer to this repulsion between nuclei as the **Coulomb barrier**. To overcome this barrier, atomic nuclei must have violent collisions that are rare unless the gas is very hot, in which case the nuclei move at high speeds. (You recall that an object's temperature is just a measure of the average speed with which its particles move.)

Thus nuclear reactions in the Sun take place only near the center, where the gas is hot and dense. A high temperature ensures that collisions between nuclei are violent enough to overcome the Coulomb barrier, and a high density ensures that there are enough collisions, and thus enough reactions, to produce enough energy to keep the Sun stable.

You can symbolize this process with a simple nuclear reaction:

$$4\,{}^1\text{H} \rightarrow {}^4\text{He} + \text{energy}$$

In this equation, ^{1}H represents a proton, the nucleus of a hydrogen atom, and ^{4}He represents the nucleus of a helium atom. The superscripts indicate the total numbers of protons and neutrons in each nucleus.

The actual steps in the process are more complicated than this simple summary suggests. Instead of waiting for four hydrogen nuclei to collide simultaneously, which would be a highly unlikely event, the process can proceed by steps in a series of reactions called the proton–proton chain.

The **proton–proton chain** is a series of three nuclear reactions that build a helium nucleus by adding protons one at a time. This process is efficient at temperatures above 10,000,000 K. The Sun's energy is created this way. Recall from the previous section that models of the interior of the Sun based on its overall stability indicate the central temperature is about 16,000,000 K.

The three reactions in the proton–proton chain are as follows:

$$^1\text{H} + {}^1\text{H} \rightarrow {}^2\text{H} + e^+ + \nu$$
$$^2\text{H} + {}^1\text{H} \rightarrow {}^3\text{He} + \gamma$$
$$^3\text{He} + {}^3\text{He} \rightarrow {}^4\text{He} + {}^1\text{H} + {}^1\text{H}$$

In the first reaction, two hydrogen nuclei (two protons) combine, and one changes into a neutron, to result in a heavy hydrogen nucleus called **deuterium**, while emitting a particle called a *positron* (a positively charged electron, symbolized by e^+) and another called a **neutrino** (ν). In the second reaction, the heavy hydrogen nucleus absorbs

Coulomb barrier The electrostatic force of repulsion between bodies of like charge, commonly applied to atomic particles.

proton–proton chain A series of three nuclear reactions that builds a helium atom by adding together protons. The main energy source in the Sun.

deuterium An isotope of hydrogen in which the nucleus contains a proton and a neutron.

neutrino A massless atomic particle that travels at or nearly at the speed of light.

another proton and, with the emission of a gamma-ray (γ), becomes a lightweight helium nucleus (^3He). Finally, two light helium nuclei combine to form a normal helium nucleus and two hydrogen nuclei. Because the last reaction needs two ^3He nuclei, the first and second reactions must occur twice for each ^4He to be produced, as shown in Figure 10-6. The net result of this chain reaction is the transformation of four hydrogen nuclei into one helium nucleus plus energy.

All main-sequence stars fuse hydrogen into helium to generate energy. The Sun and smaller stars fuse hydrogen by the proton–proton chain. Upper-main-sequence stars, more massive than the Sun, fuse hydrogen by a more efficient process called the **CNO (carbon-nitrogen-oxygen) cycle**. The CNO cycle begins with a carbon nucleus and transforms it first into a nitrogen nucleus, then into an oxygen nucleus, and then back to a carbon nucleus. The carbon is unchanged in the end, but along the way four hydrogen nuclei are fused to make a helium nucleus plus energy, just as in the proton–proton chain. A carbon nucleus has six times more positive electric charge than hydrogen, so the Coulomb barrier is higher than for combining two protons. Temperatures higher than 16,000,000 K are required to make the CNO cycle work: The center of the Sun is not quite hot enough for this reaction. In stars more massive than about 1.1 solar masses, the cores are hot enough and the CNO cycle dominates over the slower proton–proton chain.

In both the proton–proton chain and the CNO cycle, energy appears in the form of gamma-rays, positrons, and neutrinos. The gamma-rays are photons that are absorbed by the surrounding gas before they can travel more than a few centimeters. This heats the gas. The positrons produced in the first reaction combine with free electrons, and both particles vanish, converting their mass into more gamma-rays. Thus, the positrons also help keep the center of the star hot. The neutrinos, however, are particles that travel at nearly the speed of light and almost never interact with other particles. The average neutrino could pass unhindered through a lead wall more than 1 ly thick. Consequently, the neutrinos do not help heat the gas but race out of the star, carrying away roughly 2 percent of the energy produced.

It is time to ask the critical question that lies at the heart of science: What is the evidence to support this theoretical explanation of how the Sun and other stars make energy? The search for that evidence will introduce you to one of the great triumphs of modern science.

CNO (carbon-nitrogen-oxygen) cycle A series of nuclear reactions that use carbon as a catalyst to combine four hydrogen nuclei to make one helium nucleus plus energy, effective in stars more massive than the Sun.

10-2c Neutrinos from the Sun's Core

Nuclear reactions in the Sun's core produce floods of neutrinos that rush out of the Sun and off into space. If you can detect these neutrinos, you can probe the Sun's interior.

Because neutrinos almost never interact with atoms, neutrinos are extremely hard to detect. You never feel the flood of more than 10^{14} (100 trillion) solar neutrinos that flows through your body every second. Even at night, neutrinos from the Sun rush through Earth as if it weren't there, up through your bed, through you, and onward into space. Certain particle reactions, however, can be triggered by a neutrino of the right energy.

Beginning in the 1960s, astronomers found several ways to detect solar neutrinos. Initial results, however, detected too few neutrinos—about one-third as many as predicted by models of the Sun's interior. Later experiments were finally able to solve the puzzle and confirm that the original solar models were correct. Physicists know of three kinds of neutrinos, which they call "flavors." The first neutrino experiments could detect (or taste) only one flavor, electron-neutrinos. Theory hinted that the electron-neutrinos produced in the core of the Sun might change back and forth (oscillate) among the three flavors as they rushed out through the Sun and across space to Earth. Observations from the facility shown in Figure 10-7 confirmed this explanation. Two-thirds of the electron-neutrinos produced in the Sun transform into muon- and tau-neutrinos, which most detectors cannot count, before they reach Earth. You can be confident that models of the Sun's interior are basically correct because they predict the observed rate at which neutrinos come from the Sun's core.

10-2d The Pressure–Temperature Thermostat

Nuclear reactions in stars manufacture energy and heavy atoms under the supervision of a natural thermostat that keeps the reactions from erupting out of control. That thermostat is the relation between gas pressure and temperature.

In a star, the nuclear reactions generate just enough energy to balance the inward pull of gravity. Consider what would happen if the reactions began to produce too much energy. The star balances gravity by generating energy, so the extra energy flowing out of the star would force it to expand. The expansion would lower the central temperature and density and slow the nuclear reactions

FIGURE 10-7

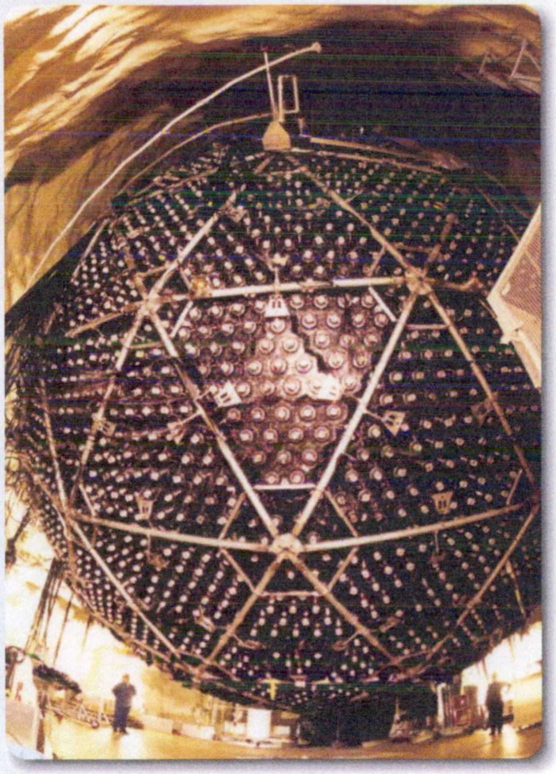

The *Sudbury Neutrino Observatory* is a 12 m diameter globe containing water rich in deuterium (heavy hydrogen) in place of ordinary hydrogen. Buried 2100 m (6800 ft) deep in an Ontario mine, it can detect all three flavors of neutrinos and confirms that neutrinos oscillate.

until the star regained stability. Thus the star has a built-in regulator that keeps the nuclear reactions from occurring too rapidly.

The same thermostat keeps the reactions from dying down. Suppose the nuclear reactions began making too little energy. Then the star would contract slightly, increasing the central temperature and density and increasing the nuclear energy generation.

The overall stability of a star depends on the relation between gas pressure and temperature. If the material of the star has the property of normal gases—for which an increase or decrease in temperature produces a corresponding change in pressure—then the nuclear reaction pressure–temperature thermostat can function properly and contribute to the stability of the star. In the next section of this chapter, you will see how this thermostat accounts for the relation between mass and luminosity for main-sequence stars.

10-3 MAIN-SEQUENCE STARS

You can understand the deepest secrets of the stars by looking at the most obvious feature of the H–R diagram—the main sequence. Roughly 90 percent of all the stars that make energy by nuclear fusion are main-sequence stars, and they obey a simple relationship that is the key to understanding the life and death of stars.

10-3a The Mass–Luminosity Relation

You learned in Section 9-6a of Chapter 9 that observations of the temperature and luminosity of stars show that main-sequence stars obey a simple rule—the more massive a star is, the more luminous it is. That rule is the key to understanding the stability of main-sequence stars. In fact, the mass–luminosity relation is predicted by the theories of stellar structure, giving astronomers direct observational confirmation of those theories.

To understand the mass–luminosity relation, you can consider both the law of hydrostatic equilibrium, which says that pressure balances weight, and the pressure–temperature thermostat, which regulates energy production. A star that is more massive than the Sun has more weight pressing down on its interior, so the interior must have a high pressure to balance that weight. That means the massive star's automatic pressure–temperature thermostat must keep the gas in its interior hot and the pressure high. A star less massive than the Sun has less weight on its interior and thus needs less internal pressure; therefore, its pressure–temperature thermostat is set lower. In other words, massive stars are more luminous because they must make more energy to support their larger weight. If they were not so luminous, they would not be stable.

10-3b Brown Dwarfs

The mass–luminosity relation also tells you why the main sequence has a lower end, a minimum mass. Objects with masses less than about 0.08 of the Sun's mass cannot raise their central temperature high enough to sustain hydrogen fusion. Called *brown dwarfs*, such objects are only about 10 times the diameter of Earth—about the same size as Jupiter—and may still be warm from the processes of formation, but they do not generate energy by hydrogen fusion. They have contracted as much as they can and are slowly cooling off.

Brown dwarfs fall in the gap between low-mass M stars and massive planets like Jupiter. They would look dull orange or red to your eyes—which is why they are labeled "brown"—because they emit most of their energy in the infrared. The warmer brown dwarfs fall in spectral class L and the cooler in spectral class T (look back to Chapter 9 Section 9-3a). Because they are so small and cool, brown dwarfs are very low-luminous objects and thus are difficult to find. Nevertheless, hundreds are known, and they may be as common as M stars.

10-3c Lives of Main-Sequence Stars

While a star is on the main sequence, it is stable, so you might think its life would be totally uneventful. But a main-sequence star balances its gravity by fusing hydrogen, and as the star gradually uses up its fuel, that balance must change. Thus, even the stable main-sequence stars are changing as they consume their hydrogen fuel.

Recall that hydrogen fusion combines four nuclei into one. As a main-sequence star fuses its hydrogen, the total number of particles in its interior decreases. Each newly made helium nucleus can exert only the same pressure as a hydrogen nucleus. Because the gas has fewer nuclei, its total pressure is less. This unbalances the gravity–pressure stability, and gravity squeezes the core of the star more tightly. As the core contracts, its temperature increases and the nuclear reactions run faster, releasing more energy. This additional energy flowing outward forces the outer layers to expand. Therefore, as a main-sequence star slowly turns hydrogen into helium in its core, the core contracts and heats up while the outer parts of the star become larger, the star becomes more luminous, and its surface cools down.

These gradual changes during the lifetimes of main-sequence stars mean that the main sequence is not a sharp line across the H–R diagram, but rather a band. Stars begin their stable lives fusing hydrogen on the lower edge of this band, which is known as the **zero-age main sequence (ZAMS)**, but gradual changes in luminosity and surface temperature move the stars upward and slightly to the right, as shown in Figure 10-8. By the time they reach the upper edge of the main sequence, they have exhausted nearly all the hydrogen in their centers. If you precisely measure and plot the luminosity and temperature of main-sequence stars on the H–R diagram, you will find them at various positions within the main-sequence band, indicating how much hydrogen has been converted to helium in their cores. Therefore, you can use the position of a star in the band, combined with stellar evolution models, as one way to estimate the star's age.

zero-age main sequence (ZAMS) The location in the H–R diagram where stars first reach stability as hydrogen-burning stars.

FIGURE 10-8

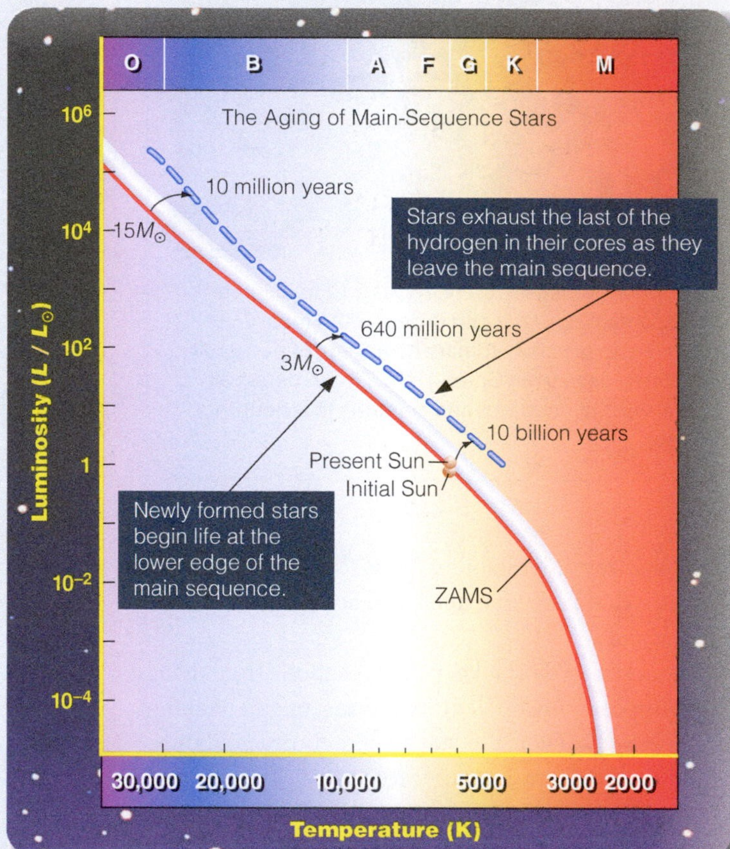

A newborn star reaches stability with properties that place it on the H–R diagram at the lower edge of the main-sequence band, along the line called the zero-age main sequence (ZAMS). As a star converts hydrogen in its core into helium, it changes slowly away from the ZAMS and across the main sequence, becoming slightly more luminous and cooler. Once a star consumes all of the hydrogen in its core, it can no longer remain a stable main-sequence star. More massive stars age rapidly, but less massive stars use up the hydrogen in their cores more slowly and live longer main-sequence lives.

The Sun is a typical main-sequence star; and, as it undergoes these gradual changes, Earth will suffer. When the Sun began its main-sequence life about 5 billion years ago, it was only about 70 percent as luminous as it is now, and by the time it leaves the main sequence in another 6 billion years, the Sun will have twice its present luminosity. Long before that, the rising luminosity of the Sun will raise Earth's average temperature, melt the polar caps, modify Earth's climate, and ultimately boil away and destroy Earth's oceans. Life on Earth will probably not survive these changes in the Sun, but humans have a billion years or more to prepare.

The average star spends 90 percent of its life on the main sequence. This explains why 90 percent of all true stars (stars powered by nuclear fusion) are main-sequence stars—you are most likely to see a star during that long, stable period while it is on the main sequence. The amount of time a star spends on the main sequence depends on its mass (Table 10-2). Massive stars consume fuel rapidly and live short lives, but low-mass stars conserve their fuel and shine for billions of years. For example, a 25-solar-mass star will exhaust its hydrogen and die in only about 4 million years. The Sun has enough fuel to last about 11 billion years. The red dwarfs, although they have little fuel, use it up very slowly and may be able to survive for 100 billion years or more.

10-4 THE BIRTH OF STARS

The key to understanding star formation is the correlation between young stars, short-lived stars, and interstellar clouds of gas and dust. Where you find the youngest groups of stars, as well as stars with very short lifetimes, you also find large clouds of gas and dust illuminated by the hottest and brightest of the new stars, as you can see in Figure 10-9. This should lead you to suspect that stars form from such clouds.

TABLE 10-2 LIFETIMES OF MAIN-SEQUENCE STARS

Spectral Type	Mass (Sun = 1)	Luminosity (Sun = 1)	Years on Main Sequence
O5	40	400,000	1×10^6
B0	15	13,000	11×10^6
A0	3.5	80	440×10^6
F0	1.7	6.4	3×10^9
G0	1.1	1.4	8×10^9
K0	0.8	0.46	17×10^9
M0	0.5	0.08	56×10^9

10-4a The Interstellar Medium

A common misperception is to imagine that space is empty, or a vacuum. In fact, if you glance at Orion's sword on a winter evening, you can see the Great Nebula in Orion, a glowing cloud of gas and dust. That cloud and others are prominent examples of the low-density gas and dust between the stars called the **interstellar medium (ISM)**.

About 75 percent of the mass of interstellar gas is hydrogen, and 25 percent is helium; there are also traces of carbon, nitrogen, oxygen, calcium, sodium, and heavier atoms. Notice that this is very similar to the composition of the Sun (refer to Table 5-1) and other stars. Roughly 1 percent of the ISM's mass is made up of microscopic particles called **interstellar dust**. The dust particles are about the size of the particles in cigarette smoke. The dust seems to be made mostly of carbon and silicates (rocklike minerals) mixed with or coated with frozen water and other ices, plus some organic compounds.

This interstellar material is not uniformly distributed through space; it consists of a complex tangle of cool, dense clouds pushed and twisted by currents of hot, low-density gas. Although the cool clouds contain only 10 to 1000 atoms/cm^3 (fewer than in any laboratory vacuum on Earth), astronomers refer to them as dense clouds in contrast with the hot, low-density gas that fills the spaces between clouds. That thin gas contains only about 0.1 atom/cm^3.

How do you know what the properties of the interstellar medium are? As you just learned, the ISM is in some cases easily visible as clouds of gas and dust, as in, for example, the Great Nebula in Orion. Astronomers call such a cloud a **nebula** from the Latin word for "cloud." Such nebulae (plural) give clear evidence of an interstellar medium. Read **Concept Art 10A, Three Kinds of Nebulae**, and notice the following three important points:

1. First, very hot stars can excite clouds of gas and dust to emit light, and this reveals that the clouds contain mostly hydrogen gas at very low densities.
2. Second, where dusty clouds reflect the light of stars you see evidence that the dust in the clouds is made up of very small particles.

interstellar medium (ISM) The gas and dust distributed between the stars.

interstellar dust Microscopic solid grains in the interstellar medium.

nebula A relatively dense cloud of interstellar gas and dust.

FIGURE 10-9

Young stars are found in clouds of gas and dust from which they have been born. The nebula N44 is 170,000 ly from Earth in a nearby galaxy, and the Horsehead Nebula is only about 1500 ly distant in our Galaxy. Gas in both nebulae is excited to glow by hot, young stars. Interstellar dust is visible as dark, twisting clouds seen against the bright background gas.

3. The third thing to notice is that some dense clouds of gas and dust are dark and are detectable only where they are silhouetted against background regions filled with stars or bright nebulae.

If a cloud is less dense, the starlight may be able to penetrate it, and stars can be seen through the cloud; but the stars look dimmer because the dust in the cloud scatters some of the light. Because shorter wavelengths are scattered more easily than longer wavelengths, the redder photons are more likely to make it through the cloud, and the stars appear redder than they should for their respective spectral types. This effect is called **interstellar reddening** and is illustrated in Figure 10-10. The same physical process makes the setting Sun look red. The fact that distant stars are dimmed and reddened by intervening gas and dust is clear evidence of an interstellar medium.

interstellar reddening The process in which dust scatters blue light out of starlight and makes the stars look redder.

Although you have been imagining individual nebulae, the thin gas and dust of the interstellar medium also fills the spaces between the nebulae. You can see evidence of that because the gas forms interstellar absorption lines in the spectra of distant stars, as shown in Figure 10-11. As starlight travels through the interstellar medium, gas atoms of elements such as calcium and sodium absorb photons of certain wavelengths, producing absorption lines. You can be sure those lines originate in the interstellar medium because they appear in the spectra of O and B stars—stars too hot to have visible calcium and sodium absorption lines in their own atmospheres. Also, the narrowness of the interstellar lines indicates they could not have been formed in the hot atmospheres of stars—the line widths indicate a very small range of Doppler shifts, meaning the atoms are moving at speeds corresponding to temperatures of only 10 to 50 K.

FIGURE 10-10

(a) Interstellar reddening makes stars seen through a cloud of dust look redder than they should because shorter wavelength light is more easily scattered. (b) No stars are detected at visual wavelengths through a dense cloud except near the edges. (c) At longer, near-infrared wavelengths, many stars can be detected behind the cloud.

Observations at nonvisible wavelengths provide valuable evidence about the interstellar medium. Infrared observations can directly detect dust in the interstellar medium. Although the dust grains are very small and very cold, there are huge numbers of grains in a cloud, and each grain emits infrared radiation at long wavelengths, as shown in Figure 10-12a. Furthermore, some molecules in the cold gas emit in the infrared, so infrared observations can detect very cold clouds of gas. Also, infrared light penetrates ISM dust better than shorter-wavelength radiation, so astronomers can use infrared cameras to look into and through interstellar clouds that are

FIGURE 10-11

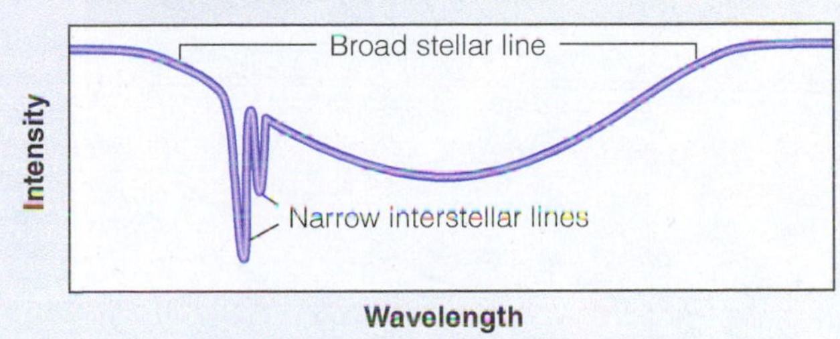

Interstellar absorption lines can be recognized easily because they are much narrower than the spectral lines produced in stellar atmospheres. The multiple interstellar lines in this spectrum are produced by separate interstellar clouds with slightly different radial velocities and are seen superimposed on a broad stellar line.

Concept Art 10A

THREE KINDS OF NEBULAE

1

Emission nebulae are produced when a hot star excites the gas near it to produce an emission spectrum. The star must be hotter than about spectral type B1 (25,000 K). Cooler stars do not emit enough ultraviolet radiation to ionize the gas. Emission nebulae have a distinctive pink color produced by the blending of the red, blue, and violet Balmer lines. Emission nebulae are also called **H II regions**, following the convention of using Roman numerals to indicate an atom's ionization state. Thus, H I means neutral hydrogen; H II is ionized.

In an H II region, the ionized nuclei and free electrons are mixed. When a nucleus captures an electron, the electron falls down through the atomic energy levels, emitting photons at specific wavelengths. Spectra indicate that the nebulae have compositions much like that of the Sun—mostly hydrogen. Emission nebulae have densities of 100 to 1000 atoms per cubic centimeter, thinner than the best vacuums produced in laboratories on Earth.

2

A **reflection nebula** is produced when starlight scatters from the dust in a nebula. Consequently, the spectrum of a reflection nebula is just the reflected spectrum of the starlight, atmospheric absorption lines included. Gas is surely present in a reflection nebula, but it is not excited enough to emit photons. See image below.

2a Reflection nebulae look blue for the same reason the sky looks blue. Short wavelengths scatter more easily than long wavelengths. See sketch below.

Reflection nebulae NGC 1973, 1975, and 1977 are just north of the Orion Nebula. The pink glow from ionized hydrogen is an emission nebula deep inside the complex of reflection nebulae.

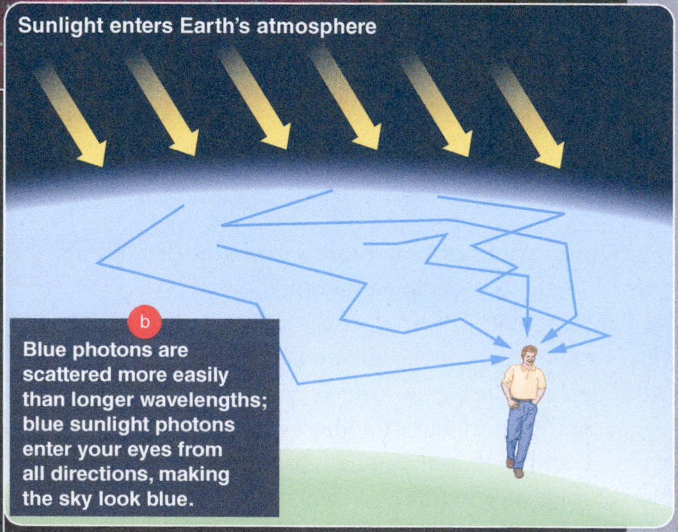

Blue photons are scattered more easily than longer wavelengths; blue sunlight photons enter your eyes from all directions, making the sky look blue.

2b The blue color of reflection nebulae at left shows that the dust particles must be very small to preferentially scatter the blue photons. Interstellar dust grains must have typical diameters ranging from about 10 to 1000 nm (0.01 to 1 micron). This particular reflection nebula includes an emission nebula.

2c The hottest stars in the Pleiades star cluster are spectral type B6, not hot enough to ionize hydrogen in the ISM. Instead, the brightest stars produce a reflection nebula as their light is scattered from interstellar dust in a cloud that the cluster happens to be passing through.

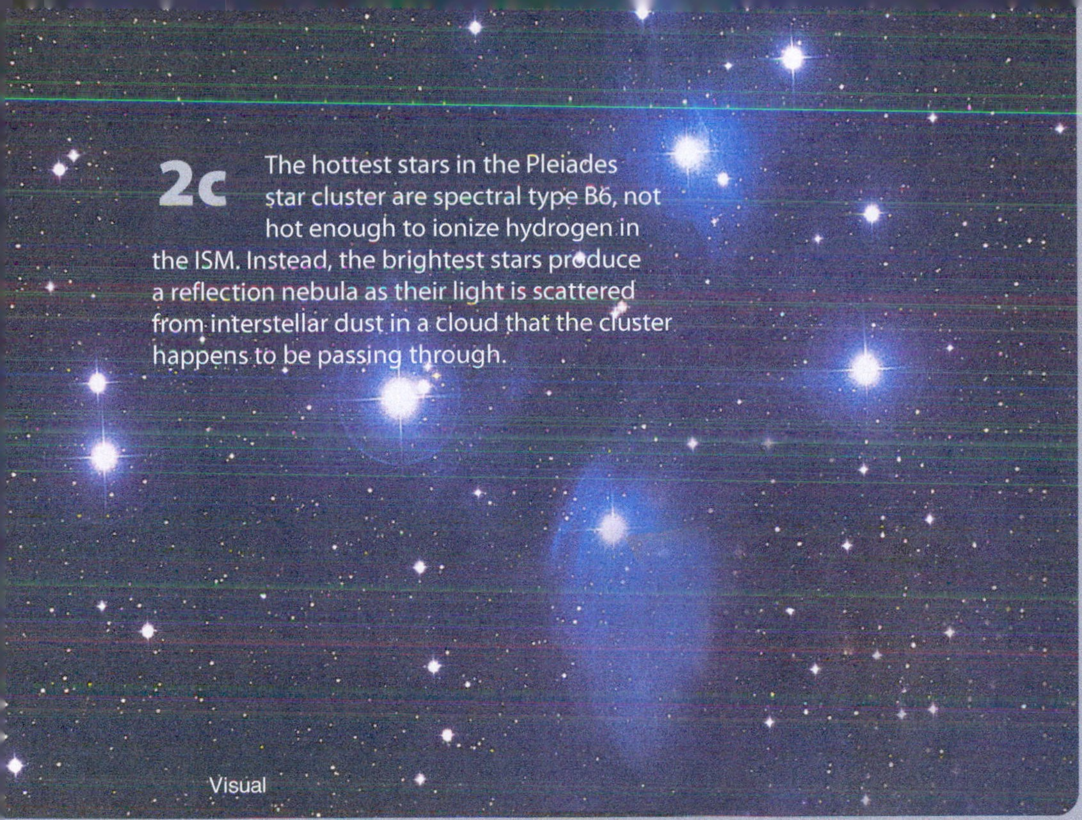

Australian Astronomical Observatory/David Malin Images

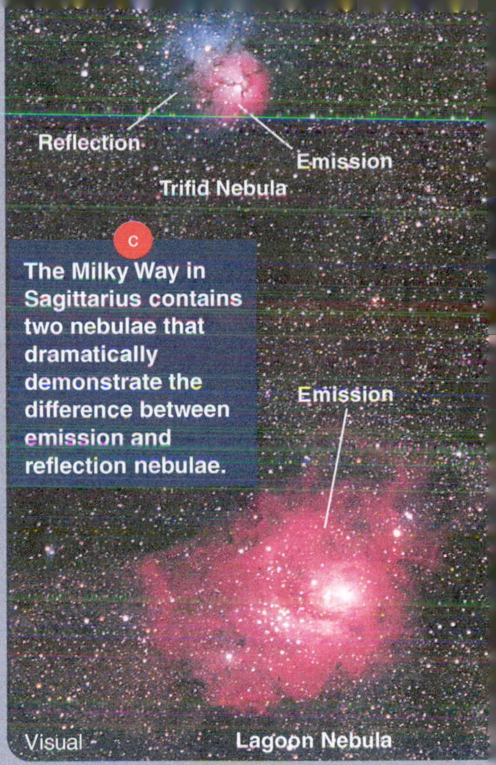

c The Milky Way in Sagittarius contains two nebulae that dramatically demonstrate the difference between emission and reflection nebulae.

Daniel Good

3 **Dark nebulae** are clouds of gas and dust dense enough to obstruct the view of more distant stars. Their shapes, as shown at left and lower left, are generally irregular, suggesting that, in addition to the effects of nearby stars, there are breezes and currents pushing through the ISM.

Australian Astronomical Observatory/David Malin Images

a Twisted by intense light from nearby stars, this dark nebula is visible because it obscures more distant stars.

Australian Astronomical Observatory/David Malin Images

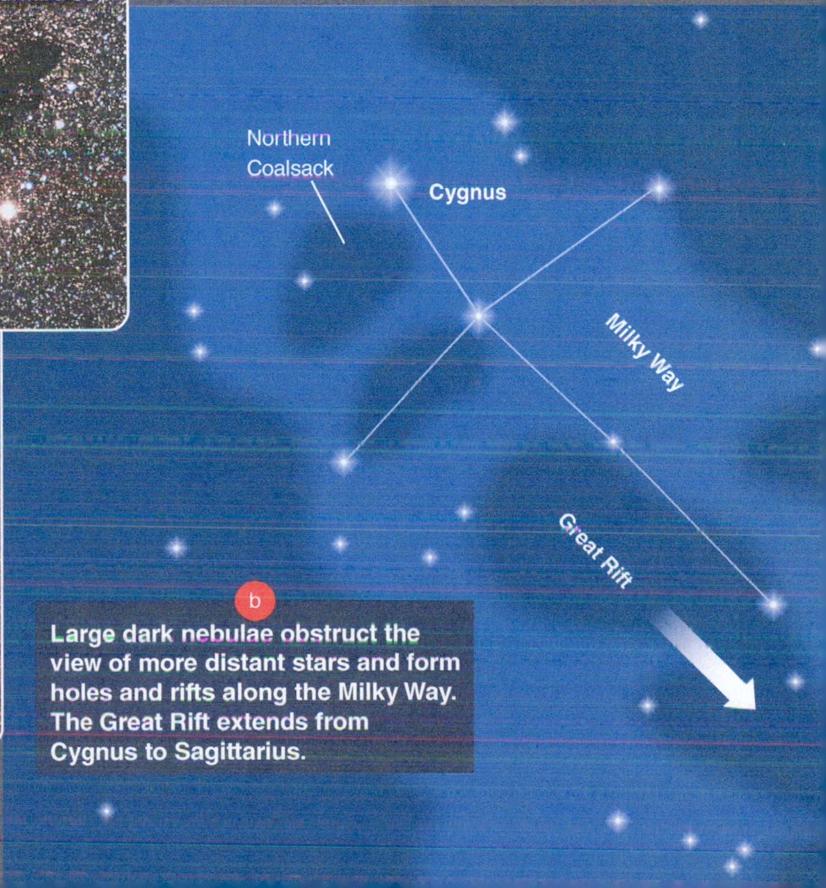

b Large dark nebulae obstruct the view of more distant stars and form holes and rifts along the Milky Way. The Great Rift extends from Cygnus to Sagittarius.

opaque to visible light. X-ray observations can detect regions of very hot gas apparently produced by exploding stars, like those in the constellation Cygnus (Figure 10-12b). Radio observations reveal the emissions of specific molecules in the interstellar medium—the equivalent of emission lines in visible light. Such studies show that some of the atoms in space have linked together to form molecules.

There is no shortage of evidence concerning the interstellar medium. Clearly, space is not empty, and from the correlation between young stars and clouds of gas and dust you can suspect that stars form from these clouds.

10-4b Formation of Stars from the Interstellar Medium

To study the formation of stars, you can continue comparing hypothesis with evidence. The theory of gravity predicts that the combined gravitational attraction of the atoms in a cloud of gas will shrink the cloud, pulling every atom toward the center. That might lead you to expect that every cloud would eventually collapse and become a star; however, the thermal energy in the cloud resists collapse. Interstellar clouds are very cold; but, even at a temperature of only 10 K, the average particle moves at about 0.33 km/s (740 mph). That much thermal motion would make the cloud drift apart if gravity were too weak to hold it together.

Other factors can help a cloud resist its own gravity. Observations show that clouds are turbulent with currents of gas pushing through and colliding with each other. Also, magnetic fields in clouds may resist being squeezed. The thermal motion of the atoms, turbulence in a cloud, and magnetic fields resist gravity, and only the densest clouds are likely to contract.

The densest interstellar clouds contain from 10^3 to 10^5 atoms/cm^3 and have temperatures as low as 10 K. They include a few hundred thousand to a few million solar masses. In such dense clouds, hydrogen can exist as molecules (H_2) rather than as atoms. These very densest parts of the ISM are called **molecular clouds**, and the largest are called giant molecular clouds. Although hydrogen molecules cannot be detected by radio telescopes, the clouds can be mapped by the emission lines of carbon monoxide molecules (CO) present in small amounts in the gas. By now you may have realized the main point: Stars can form inside molecular clouds when the densest parts of the clouds become unstable and contract under the influence of their own gravity.

molecular clouds A dense interstellar gas cloud in which atoms are able to link together to form molecules such as H_2 and CO.

shock wave A sudden change in pressure that travels as an intense sound wave.

Most clouds do not appear to be gravitationally unstable and will not contract to form stars on their own. However, a stable cloud colliding with a **shock wave** (the astronomical equivalent of a sonic boom) can be compressed and disrupted into fragments. Theoretical calculations show that some of these fragments can become dense enough to collapse under the influence of their own gravity and form stars, as shown in Figure 10-13.

Supernova explosions (exploding stars described in Chapter 11) produce shock waves that compress

FIGURE 10-12

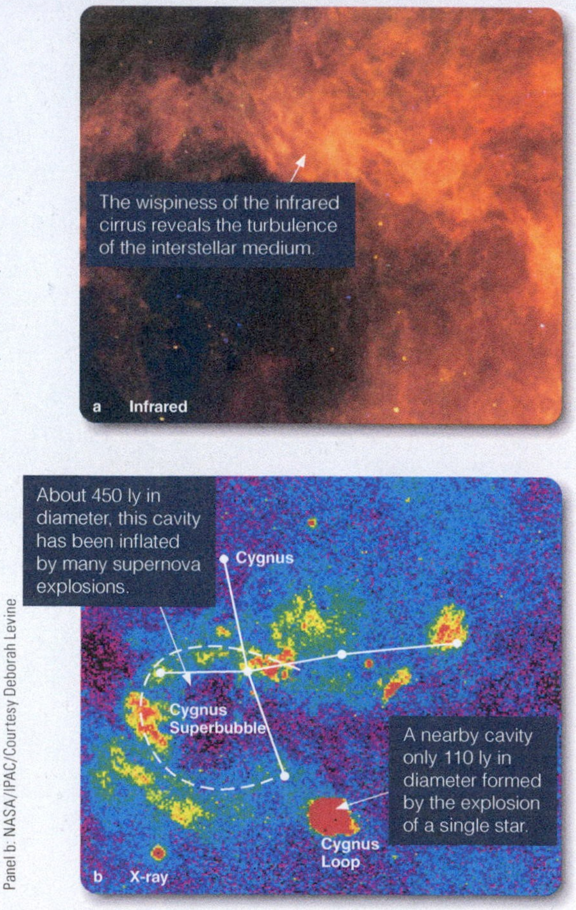

Panel a: Courtesy of Steve Snowden and Max Planck Institute for Extraterrestrial Physics, Germany; Panel b: NASA/IPAC/Courtesy Deborah Levine

Infrared and X-ray observations reveal different parts of the interstellar medium. (a) This far-infrared image shows the infrared cirrus, wispy clouds of very cold dust and gas spread all across the sky. (b) The Cygnus Superbubble, detectable in X-rays, triggers star formation where it pushes into the surrounding gas.

the interstellar medium, and recent observations show young stars forming at the edges of such shock waves. Another source of shock waves may be the birth of very hot stars. A massive star is so luminous and hot that it emits vast amounts of ultraviolet photons. When such a star is born, the sudden blast of light, especially ultraviolet radiation, can ionize and drive away nearby gas, forming a shock wave that could compress nearby clouds and trigger further star formation. Even the collision of two interstellar clouds can produce a shock wave and trigger star formation. Some of these processes are shown in Figure 10-14.

Although these are important sources of shock waves, the dominant trigger of star formation in our Galaxy may be the spiral pattern itself. In Chapter 1, you learned that our Galaxy contains spiral arms. As interstellar clouds encounter these spiral arms, the clouds are compressed, and star formation can be triggered (see Chapter 12).

Once begun, star formation can spread like a grass fire. Both high-mass and low-mass stars form together, but when the massive stars form, their intense radiation or eventual supernova explosions push back the surrounding gas and compress it. This compression in turn can trigger the formation of more stars, some of which will be massive. Thus, a few massive stars can drive a continuing cycle of star formation in a giant molecular cloud.

A collapsing cloud of gas does not form a single star. Because of instabilities, the cloud breaks into fragments, producing perhaps thousands of stars. A stable group of stars that formed and are held together by their combined gravity is called a **star cluster**. In contrast, a **stellar association** is a group of stars that formed together but are not gravitationally bound to one another. The stars in an association drift away from each other in a few million years. The youngest associations are rich in young stars, including O and B stars. O and B stars have lifetimes so short in astronomical terms (Table 10-2) that they must be located close to their birthplaces, so these stars serve as brilliant signposts pointing to regions of star formation.

FIGURE 10-13

Shock Wave Triggers Star Formation

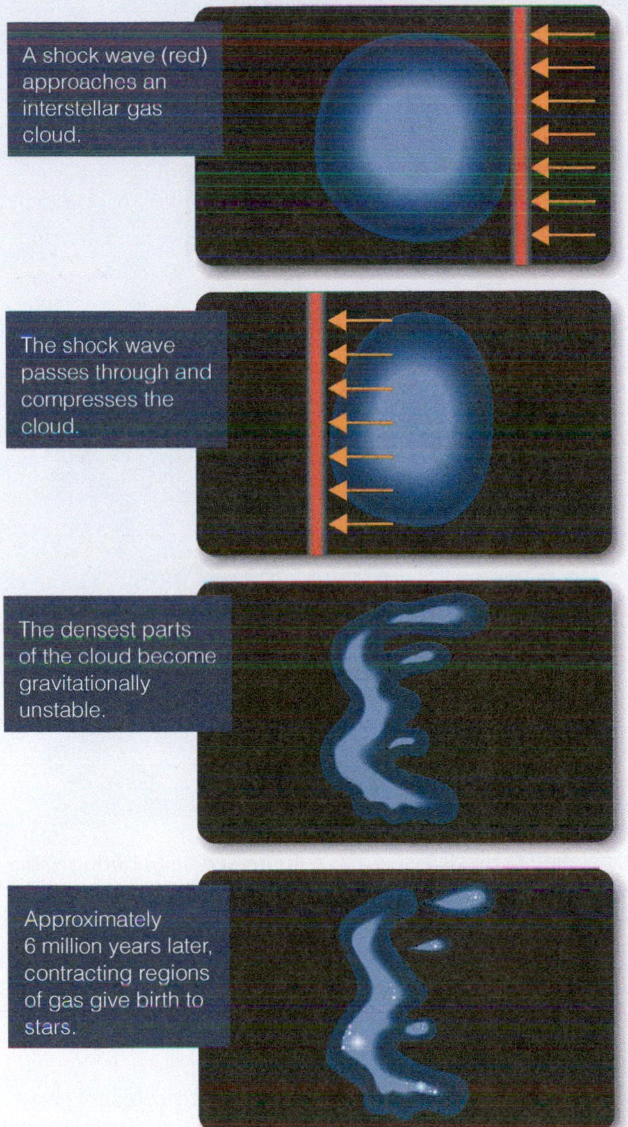

In this summary of a computer model, an interstellar gas cloud is triggered into collapsing and forming clusters of new stars by a passing shock wave. The events summarized here might span about 6 million years.

10-4c Formation of Protostars

You might be wondering how the unimaginably cold gas of an interstellar cloud can heat up to form a star. The answer is gravity.

Once part of a cloud is triggered to collapse, gravity draws each atom toward the center. At first the atoms fall unopposed; they hardly ever collide with each other. In this free-fall contraction, the atoms pick up speed as they fall until, by the time the gas becomes dense enough for the atoms to collide often,

star cluster A group of stars that formed together and orbit a common center of mass.

stellar association A group of stars that formed together but are not gravitationally bound to one another.

FIGURE 10-14

(a) The blast of light and ultraviolet radiation from an earlier generation of massive stars has compressed neighboring gas and triggered the formation of more stars. Those stars are now triggering the birth of a third generation. (b) The explosion of a supernova a few million years ago produced a shock wave that is triggering star formation where it slams into nearby clouds of gas and dust.

they are traveling very fast. Collisions convert the inward velocities of the atoms into random motions, and you remember that temperature is a measure of the random velocities of the atoms in a gas. Thus, the temperature of the gas goes up. As the internal temperature climbs, the gas becomes ionized, changing into a mixture of positively charged atomic nuclei and free electrons.

The initial collapse of the gas forms a dense core of gas, and as more gas falls in, a warm protostar develops, buried deep in the dusty gas cloud that continues to contract, although much more slowly than free-fall. A protostar is an object that will eventually become a star. Theory predicts that protostars are luminous red objects larger than main-sequence stars, with temperatures ranging from a few hundred to a few thousand degrees Kelvin.

Throughout its contraction, the protostar converts its gravitational energy into thermal energy. Half of this thermal energy radiates into space, but the remaining half raises the internal temperature. When the center gets hot enough, nuclear reactions begin generating enough energy to replace the radiation, leaving the surface of the star. The protostar then halts its contraction and becomes a stable, main-sequence star.

The time a protostar takes to contract from a cool interstellar gas cloud to a main-sequence star depends on its mass. The more massive the star, the stronger its gravity and the faster it contracts. The Sun took about 30 million years to reach the main sequence, but a 15-solar-mass star can contract in only 100,000 years. Conversely, a star of 0.2 solar mass takes 1 billion years to reach the main sequence.

FIGURE 10-15

NASA/ESA/STScI/AURA/NSF/J. Morse/ Science Source

Visible light image of the Herbig–Haro object HH 47, a jet of gas about half a light-year long flowing to the right away from a protostar hiding in the nebula to the left. The wavy nature of the jet suggests that it is perturbed by a second body, perhaps a binary companion to the new star.

10-4d Observations of Star Formation

By understanding what the interstellar medium is like, and by understanding how the laws of physics work, astronomers have been able to work out the story of how stars must form. But you can't accept a scientific theory without testing it, and that means you must compare the theory with the evidence.

The protostar stage is predicted to be less than 0.1 percent of a star's total lifetime, so, although that is a long time in human terms, you cannot expect to find many stars in the protostar stage. Furthermore, protostars form deep inside clouds of dusty gas that absorb any light the protostar might emit. Only when the protostar is hot enough to drive away its enveloping cloud of gas and dust does it become easy to observe at wavelengths your eye can see.

Although astronomers cannot easily observe protostars at visible wavelengths, protostars can be detected in the infrared. Protostars are cooler than stars and radiate as blackbodies, predominantly at infrared wavelengths. The dust of the surrounding interstellar cloud absorbs light from the protostar and grows warm, and that warm dust radiates additional large amounts of infrared radiation. Infrared radiation also can penetrate the dust to reach us. Thus, infrared observations reveal many bright sources of radiation that are protostars buried in interstellar clouds.

You can be sure that star formation is going on right now because astronomers find regions containing massive stars with lifetimes so short they must still be in their birthplaces. Small, very dense zones of gas and dust called **Bok globules** within larger nebulae probably represent the very first stages of star formation. Observations of jets coming from hidden protostars show that protostars are often surrounded by disks of gas and dust, as shown in Figure 10-15. The spinning disks focus the jets into **bipolar flows** that point in opposite directions along the axis of rotation. These jets appear to produce small flickering nebulae called **Herbig–Haro objects**. The same regions contain many stars with temperatures and luminosities that correspond to model predictions of the properties of very young stars. These include **T Tauri stars**, protostars with about the mass of the Sun that show signs of active chromospheres as you might expect from young, rapidly rotating stars with strong magnetic dynamos (look back to Chapter 5 Section 5-6d). T Tauri stars are also often surrounded by thick disks of gas and dust that are very bright at infrared wavelengths.

These observations all confirm the theoretical models of contracting protostars. Although star formation still holds many puzzles, the general process seems clear. In at least some cases, interstellar gas clouds are compressed by passing shock waves, and the clouds' gravity, acting unopposed, draws the matter inward to form protostars.

Now you know how to make a star: Just find some way to compress more than 0.08 solar mass of interstellar material into a small enough region of space, and gravity will do the rest of the job.

Bok globule Small, dark cloud only about 1 ly in diameter that contains 10 to 1000 solar masses of gas and dust, thought to be related to star formation.

bipolar flows Jets flowing in opposite directions away from a central object.

Herbig–Haro object A small nebula that varies irregularly in brightness, evidently associated with star formation.

T Tauri star A young medium-mass star surrounded by gas and dust, understood to be contracting toward the main sequence.

Concept Art 10B

STAR FORMATION IN THE ORION NEBULA

1

The visible Orion Nebula shown below is a pocket of ionized gas on the near side of a vast, dusty molecular cloud that fills much of the southern part of the constellation Orion. The molecular cloud can be mapped by radio telescopes. To scale, the cloud would be many times larger than this page. As the stars of the Trapezium were born in the cloud, their radiation has ionized the gas and pushed it away. Where the expanding nebula pushes into the larger molecular cloud, it is compressing the gas (see diagram at right) and may be triggering the formation of the protostars that can be detected at infrared wavelengths within the molecular cloud.

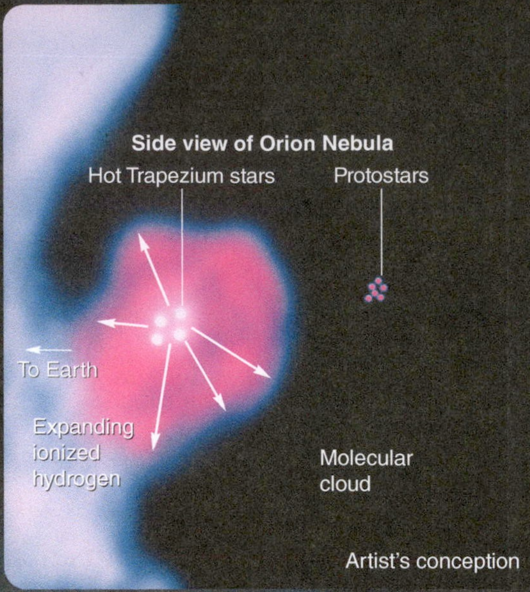

Hundreds of stars lie within the nebula, but only the four brightest, those in the Trapezium, are easy to see with a small telescope. A fifth star, at the narrow end of the Trapezium, can be visible on nights of good seeing.

The cluster of stars in the nebula is less than 2 million years old. This means the nebula is similarly young.

Panel a: A.Fujii; Panel b: NASA/IRAS/IPAC, Courtesy D.Levine

NASA, ESA, M. Robberto, STScI and the Hubble Space Telescope Orion Treasury Project Team

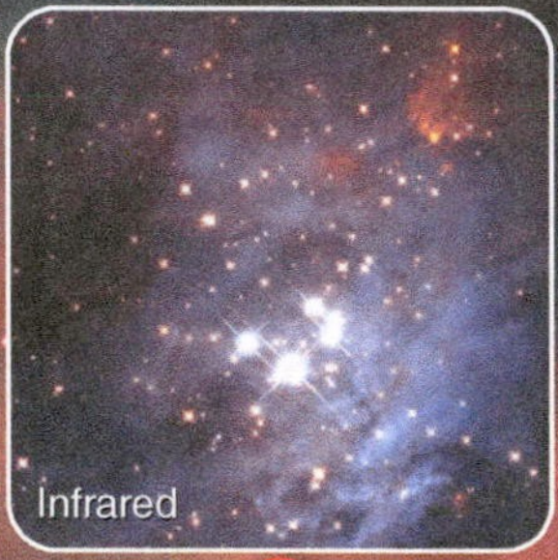

NASA/JPL-Caltech/n.Flagey & A. Noriega-Crespo

a

The near-infrared image above reveals more than 50 low-mass, cool protostars.

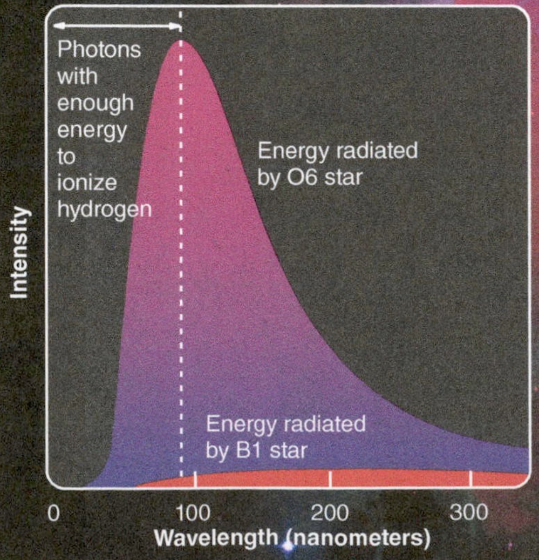

2

Of all the stars in the Orion Nebula, just one is hot enough to ionize the gas. Only photons with wavelengths shorter than 91.2 nm can ionize hydrogen. The second-hottest stars in the nebula are B1 stars, and they emit little of this ionizing radiation. The hottest star, however, is an O6 star that has 30 times the mass of the Sun. At a temperature of 40,000 K, it emits plenty of photons with wavelengths short enough to ionize hydrogen. Remove that one star, and the nebula's emission would turn off.

NASA and The Hubble Heritage Team (STScI)

b

Small dark clouds called Bok globules, named after astronomer Bart Bok, are found in and near star-forming regions. The one pictured here is part of nebula NGC 1999 near the Orion Nebula. Typically about 1 light year in diameter, they contain from 10 to 1000 solar masses.

3 The infrared image from the *Spitzer Space Telescope* reveals extensive nebulosity surrounding the visible Orion Nebula. Red and orange show the locations of warm dust that has been heated by starlight, green shows hot dust and ionized gas, and blue shows light coming directly from stars.

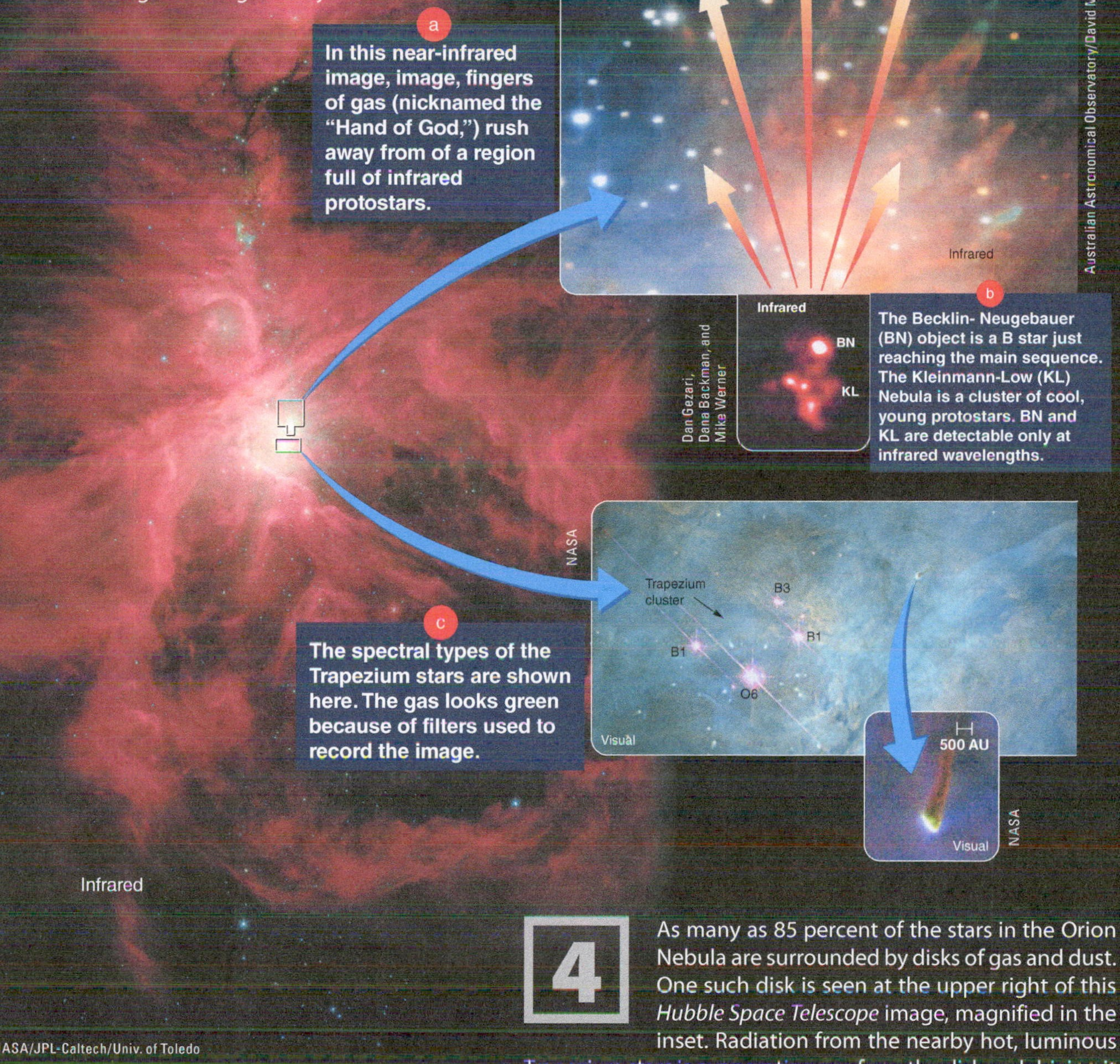

4 As many as 85 percent of the stars in the Orion Nebula are surrounded by disks of gas and dust. One such disk is seen at the upper right of this *Hubble Space Telescope* image, magnified in the inset. Radiation from the nearby hot, luminous Trapezium stars is evaporating gas from the disk and driving it away to form an elongated nebula.

10-4e The Orion Nebula

On a clear winter night, you can see with your unaided eye the Great Nebula of Orion as a fuzzy wisp in Orion's sword. With binoculars or a small telescope it is striking, and through a large telescope it is breathtaking. At the center lie four brilliant blue-white stars known as the Trapezium, the most luminous of a cluster of a few hundred stars. Surrounding the stars are the glowing filaments of a nebula more than 8 pc (26 ly) across. Like a great thundercloud illuminated from within, the churning currents of gas and dust suggest immense power. The significance of the Orion Nebula lies hidden, figuratively and literally, beyond the visible nebula. The region is ripe with star formation.

10-4f Evidence of Young Stars

You should not be surprised to find star formation in Orion. The constellation is a brilliant landmark in the winter sky because it is marked by hot blue stars. These stars are bright not because they are nearby but because they are tremendously luminous. These O and B stars cannot live more than a few million years, so you can conclude that they must have been born astronomically recently, near where they are seen now. Furthermore, the constellation contains large numbers of T Tauri stars, which are known to be young. Orion is rich with young stars.

The history of star formation in the constellation of Orion is written in its stars. The stars at Orion's west shoulder are about 12 million years old, whereas the stars of Orion's belt are about 8 million years old. The stars of the Trapezium at the center of the Great Nebula are no older than 2 million years. Apparently, star formation began near the west shoulder, and the massive stars that formed there triggered the formation of the stars you see in Orion's belt. That star formation may have triggered the formation of the stars you see in the Great Nebula. Like a grass fire, star formation has swept across Orion from northwest to southeast.

Examine **Concept Art 10B, Star Formation in the Orion Nebula**, and notice the following four important points:

1. The visible nebula is only a small part of a vast, dusty molecular cloud. You see the nebula because the larger stars born within it have ionized the nearby gas and driven it outward, breaking out of the much larger molecular cloud.
2. A single, very hot and short-lived O star is almost entirely responsible for ionizing the gas and making the nebula glow. This massive star has already become a main-sequence star while its smaller siblings are still protostars.
3. Infrared observations reveal ongoing star formation deep in the molecular cloud behind the visible nebula.
4. Finally, notice that many stars visible in the Orion Nebula are surrounded by disks of gas and dust. Such disks do not last long and are clear evidence that the stars are very young.

Observations of stars in the Orion nebula show that some of the gas and dust in the cloud from which a star forms settles into a disk. Make special note of these disks, because planets may form in them. In Chapter 8, you studied how our Solar System formed in the disk of gas and dust that encircled the young Sun.

What Are We? Myth-Makers and Explainers

On cold winter nights when the sky is clear and the stars are bright, Jack Frost paints icy lacework across your windowpane. That's a fairy tale, of course, but it is a graceful evocation of the origin of frost. We humans are explainers, and one way to explain the world around us is to create myths.

An ancient Aztec myth tells the story of the origin of the Moon and stars. The stars, known as the Four Hundred Southerners, and the Moon-goddess Coyolxauhqui, plotted to murder their unborn brother, the great war god Huitzilopochtli. Hearing their plotting, he leaped from the womb fully armed, hacked Coyolxauhqui into pieces, and chased the stars away. You can see the Four Hundred Southerners scattered across the sky, and each month you can see the Moon being chopped into pieces as it passes through its phases.

Stories like these explain the origins of things and can make our Universe seem more understandable. Science is a natural extension of our need to explain the world. The stories have become sophisticated scientific hypotheses and theories that are tested over and over against reality, but we humans build those theories for the same reason people used to tell myths.

STUDY TOOLS 10

IN THE BOOK

- ☐ Tear Out the Review Card on *Structure and Formation of Stars*.

ASTRO ONLINE

Visit ASTRO Online at www.cengagebrain.com

- ☐ Interactive Reading
- ☐ Practice Quizzing
- ☐ Videos and Animations
- ☐ Exclusive Features

Questions and Problems

10

REVIEW QUESTIONS

1. Describe the principle of hydrostatic equilibrium as it relates to the internal structure of a star.
2. How does the CNO cycle differ from the proton–proton chain? How is it similar?
3. How does the pressure–temperature thermostat control the nuclear reactions inside stars?
4. Step by step, explain how energy flows from the center of the Sun to Earth.
5. Why is there a mass–luminosity relation?
6. Why is there a lower limit to the mass of a main-sequence star?
7. Why does a star's life expectancy depend on its mass?
8. Why do distant stars look redder than their spectral types suggest?
9. What evidence can you cite that the interstellar medium contains both gas and dust?
10. What evidence visible to human eyes can you cite that the spaces between the stars are not totally empty?
11. How do the spectra of H II regions differ from the spectra of reflection nebulae? Why?
12. How is the blue color of a reflection nebula related to the blue color of the daytime sky?
13. Why are interstellar lines so narrow?
14. What factors resist the contraction of a cloud of interstellar matter?
15. How can a molecular cloud be triggered to form stars?
16. How does the contraction of a gas cloud increase its temperature?
17. What evidence is there that (a) star formation has occurred recently? (b) protostars really exist? (c) the Orion region is actively forming stars?
18. How does the geometry of bipolar flows and Herbig–Haro objects support the hypothesis that protostars are surrounded by rotating disks?
19. **How Do We Know?** How can mathematical models help you understand natural processes that occur in locations or with time scales that make them impossible to observe directly?

DISCUSSION QUESTIONS

1. Long ago, astronomers, philosophers, and poets thought that the stars were eternal and unchanging. Is there any observation they could have made, or any line of reasoning they could have followed, that might have led them to conclude that stars change and don't live forever?
2. How does hydrostatic equilibrium relate to hot-air ballooning?
3. When you see distant streetlights through smog (smoke plus fog), they look dimmer and redder than they do normally. In contrast, when you see the same streetlights through fog or falling snow, they look dimmer but not redder. Use your knowledge of the interstellar medium to compare the relative sizes of the particles in smog, fog, and snowstorms relative to the wavelength of visible light.
4. If you could see a few stars behind a dark nebula, how would you expect their spectra to differ from similar stars in front of the dark nebula?

PROBLEMS

1. Count all of the ^{1}H and ^{4}He nuclei in Figure 10-6 and explain how the proton–proton chain can be summarized by $4\,^{1}\text{He} \rightarrow {}^{4}\,\text{He} + \text{energy}$.
2. How much energy is produced when the proton–proton chain converts 1 kg of hydrogen into helium?
3. How many times longer can the Sun spend on the main sequence than a 15-solar-mass star? (*Hint:* See Table 10-2.)
4. Extinction dims starlight by about 1 magnitude per 1000 pc. What fraction of photons survives a trip of 1000 pc? (*Hint:* Consider the definition of the magnitude scale in Chapter 2. Also see **Magnitudes** in the Math Reference cards MR.)
5. The density of air in a child's balloon 20 cm in diameter is roughly the same as the density of air at sea level, 10^{19} particles/cm^3. To how large a diameter would you have to expand the balloon to make the gas inside the same density as the interstellar medium, about 1 particle/cm^3? (*Note:* The volume of a sphere is $\frac{4}{3}\pi r^3$.)
6. The dust in a molecular cloud has a temperature of about 50 K. At what wavelength does it emit the maximum energy? (*Hint:* See **Blackbody Radiation** in the Math Reference Cards MR.)
7. A giant molecular cloud is 30 pc in diameter and has a density of 1000 hydrogen atoms/cm^3. What is its total mass in kilograms? (*Note:* The volume of a sphere is $\frac{4}{3}\pi r^3$, and the mass of a hydrogen atom is about 1.7×10^{-27} kg, to two digits' precision.)
8. The expanding bubble of hot gas inflated by the cluster of new stars in its center, shown in Figure 10-9,

has a diameter of about 70 ly. If the bubble is 170,000 ly from Earth, what is the observed diameter of the bubble in arc seconds? (*Hint:* See **The Small-Angle Formula** in the Math Reference Cards MR.)

9. If a giant molecular cloud is 50 pc in diameter and a shock wave can sweep through it in 2 million years, how fast is the shock wave going in kilometers per second?
10. If a giant molecular cloud has a mass of 10^{35} kg and it converts 1 percent of its mass into stars during a single encounter with a shock wave, how many stars can it make? Assume the stars each contain 1 solar mass. (*Note:* The mass of the Sun can be found in **Celestial Profile 9**, Appendix C.)
11. The hottest star in the Orion Nebula has a surface temperature of 40,000 K. At what wavelength does it radiate the most energy? (*Hint:* See **Wien's Law** in the Math Reference Cards MR.)
12. An O star with a lifetime of 0.5 million years drifts away from its birthplace in the center of a molecular cloud at a speed of 10 km/s. If the cloud has a diameter of 10 pc, can the O star escape the cloud while the star is still on the main sequence?
13. If a protostellar disk is 200 AU in radius and the disk plus the forming star together contain 2 solar masses, what is the orbital speed at the outer edge of the disk in kilometers per second? (*Hint:* See the formula for **Circular Orbital Velocity** in the Math Reference Cards MR.)
14. If a contracting protostar is five times the radius of the Sun and has a temperature of only 2000 K, how luminous will it be? (*Hint:* See the relation between **Stellar Luminosity, Radius, and Temperature** in the Math Reference Cards MR.)
15. If a T Tauri star is the same temperature as the Sun but is 10 times more luminous, what is its radius? (*Hint:* See the relation between **Stellar Luminosity, Radius, and Temperature** in the Math Reference Cards MR.)

LEARNING TO LOOK

1. Figure 10-16 shows a dark globule of dusty gas. What do you think that globule would look like if you could see it from the other side?

FIGURE 10-16

NASA and the Hubble Heritage Team, STScI/AURA

2. The star in Figure 10-17 appears to be ejecting a jet of gas. What is happening to this star?

FIGURE 10-17

NASA/ESA/STScI/AURA/NSF

11 The Deaths of Stars

ESO

Natural laws have no pity.

Robert Heinlein

The Notebooks of Lazarus Long

After finishing this chapter go to **PAGE 268** for **STUDY TOOLS.**

GUIDEPOST

The preceding chapter described how stars are born, and how they resist their own gravity by fusing nuclear fuels in their centers to keep their internal pressure high. But they can last only as long as their fuel supply.

In this chapter you will learn that a star's mass determines what happens when the star runs out of fuel, and what type of "corpse" it will become after its death. You also will learn the surprising fact that star deaths are actually important to Earth life because massive stars create the atomic elements of which you and everything on Earth are made. If stars didn't die, you would not exist.

Some of the matter that was once in stars becomes trapped in different types of remnants known as white dwarfs, neutron stars, and black holes. Other matter from dying stars escapes back into the interstellar medium and is incorporated into new stars and planets. The deaths of stars are part of a great cycle of creation and destruction that includes our Sun, our planet, and us.

As you consider the fates of the stars, you will find answers to the following five important questions:

- What happens to a star when it uses up the last of the hydrogen in its core?
- What is the evidence that stars really evolve?
- What happens if an evolving star is in a binary star system?
- How does theory predict the existence of neutron stars and black holes?
- What is the evidence that neutron stars and black holes really exist?

This chapter ends the story of individual stars, but it does not end the story of stars. In the next chapter, you will begin exploring the giant communities in which stars live—the galaxies.

Gravity is patient—so patient it can kill stars. Occasionally, astronomers see a formerly faint star suddenly appear much brighter in the sky, then fade away after a few weeks or months. You will learn that what is seemingly a new star in the sky is either a **nova**, an eruption on the surface of a stellar remnant, or a **supernova**, the violent

nova From the Latin, meaning "new," a sudden and temporary brightening of a star making it appear as a new star in the sky, evidently caused by an explosion of nuclear fuel on the surface of a white dwarf stellar remnant.

supernova A "new star" in the sky that is up to 50,000 times more luminous than a normal nova and longer lasting, evidently the result of an explosion destroying a massive star.

© Ann Muse/Shutterstock.com

explosive death of an aging massive star. A few novae (plural of nova) are discovered each year. Supernovae (plural) are so rare that only one or two occur each century in our entire Galaxy, but astronomers surveying thousands of galaxies can see many supernovae flare per year.

In Chapter 10, you saw that stars resist their own gravity by generating energy through nuclear fusion. The energy keeps their interiors hot, and the resulting high pressure balances gravity and prevents the star from collapsing. Stars, however, have limited fuel. When they exhaust their fuel, gravity wins, and the stars die.

The mass of a star is critical in determining its fate. Massive stars use up their nuclear fuel at a furious rate and die after only a few million years. In contrast, the lowest-mass stars use their fuel sparingly and might be able to live hundreds of billions of years. Stars with different masses lead different lives and die in dramatically different ways.

11-1 GIANT STARS

A main-sequence star generates its energy by nuclear fusion reactions that combine hydrogen to make helium. A star remains on the main sequence for a time span equal to 90 percent of its total existence as an energy-generating star. When the hydrogen is exhausted, however, the star begins to evolve rapidly. It swells into a giant star and then begins to fuse helium into heavier elements. A star can remain in this giant stage for only about 10 percent of its total lifetime; then, it must die. The giant-star stage is the first step in the death of a star.

11-1a Expansion into a Giant

The nuclear reactions in a main-sequence star's core produce helium. Helium can fuse into heavier elements only at temperatures higher than 100,000,000 K, and no main-sequence star has a core that hot, so helium accumulates at the star's center like ashes in a fireplace. Initially, this helium ash has little effect on the star, but as hydrogen is exhausted and the stellar core becomes almost pure helium, the star's ability to generate nuclear energy is reduced. Because the energy generated at the center is what opposes gravity and supports the star, the core begins to contract as soon as the energy generation starts to decline.

Although the core of helium ash cannot generate nuclear energy, it can grow hotter because it contracts and converts gravitational energy into thermal energy. The rising temperature heats the unprocessed hydrogen just outside the core—hydrogen that previously was not hot enough to fuse. When the temperature of the surrounding hydrogen becomes high enough, hydrogen fusion begins in a spherical layer, called a shell, surrounding the exhausted core of the star. Like a grass fire burning outward from an exhausted campfire, the hydrogen-fusion shell creeps outward, leaving helium ash behind and increasing the mass of the helium core.

Figure 11-1 illustrates how the flood of energy produced by the hydrogen-fusion shell pushes toward the surface, heating the outer layers of the star and forcing them to expand dramatically. Stars like the Sun become **giant stars** 10 to 100 times the current diameter of the Sun, and the most massive stars become **supergiant stars** as much as 1000 times larger than the Sun.

The expansion of a star to giant or supergiant size cools the star's outer layers, and so the stars move toward the upper right in the H–R diagram. Look back to Figure 9-8 and notice that some of the most familiar stars, such as Aldebaran, Betelgeuse, and Polaris, are giants or supergiants.

Although the energy output of the hydrogen-fusion shell can force the envelope of the star to expand, it cannot stop the contraction of the helium core. Because the core is not hot enough to fuse helium, gravity squeezes it to a relatively tiny size. If you represented a typical giant star as being the size of a baseball stadium, its helium core would be only about the size of a baseball, yet would contain about 10 percent of the star's mass.

11-1b Helium Fusion

As a star becomes a giant, fusing hydrogen in a shell, the inert core of helium ash contracts and grows hotter. When the core finally reaches a temperature of 100,000,000 K, helium nuclei can begin fusing to make carbon nuclei.

The ignition of helium in the core changes the structure of the star. The star now makes energy in two locations by two different processes, helium fusion in the core and hydrogen fusion in the surrounding shell. The energy flowing outward from the core halts the contraction of the core, and at the same time the star's envelope contracts and grows hotter. Consequently, the point that represents the star in the H–R diagram moves downward, corresponding to lower luminosities, and to the left, corresponding to higher surface temperatures (Figure 11-2), to a region above the main sequence called the **horizontal branch**. Astronomers sometimes refer to stars with those characteristics as "yellow giants."

Helium fusion produces carbon and oxygen that accumulate in an inert core. Once again, the core contracts and heats up, and a helium-fusion shell ignites below the hydrogen-fusion shell. The star now makes energy in two fusion shells, so it quickly expands and its

FIGURE 11-1

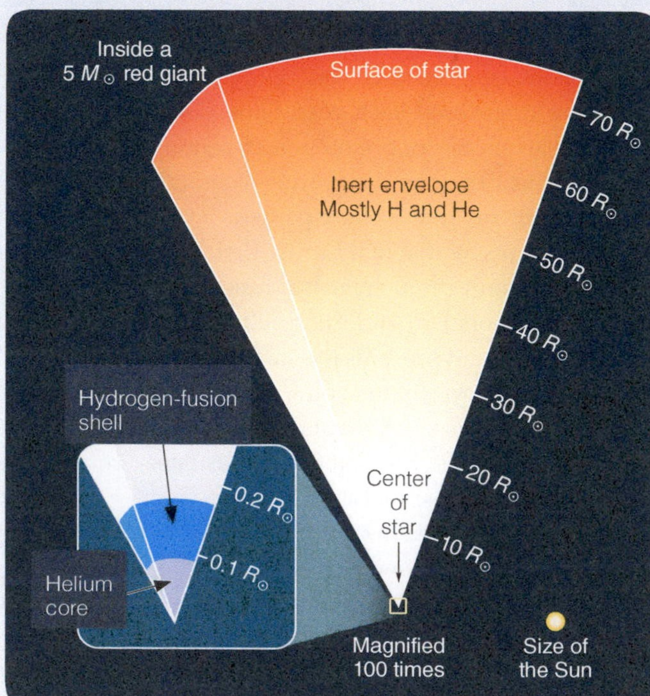

When a star runs out of hydrogen at its center, the core contracts to a small size, becomes very hot, and begins nuclear fusion in a shell (blue). The outer layers of the star expand and cool. The red giant star shown here has an average density much lower than the air at Earth's surface. Here $M_\odot$ stands for the mass of the Sun, and $R_\odot$ stands for the radius of the Sun.

giant star Large, cool, highly luminous star in the upper right of the H–R diagram, typically 10 to 100 times the diameter of the Sun.

supergiant star Exceptionally luminous star whose diameter is 100 to 1000 times that of the Sun.

horizontal branch The location in the H–R diagram of giant stars fusing helium in their cores.

FIGURE 11-2

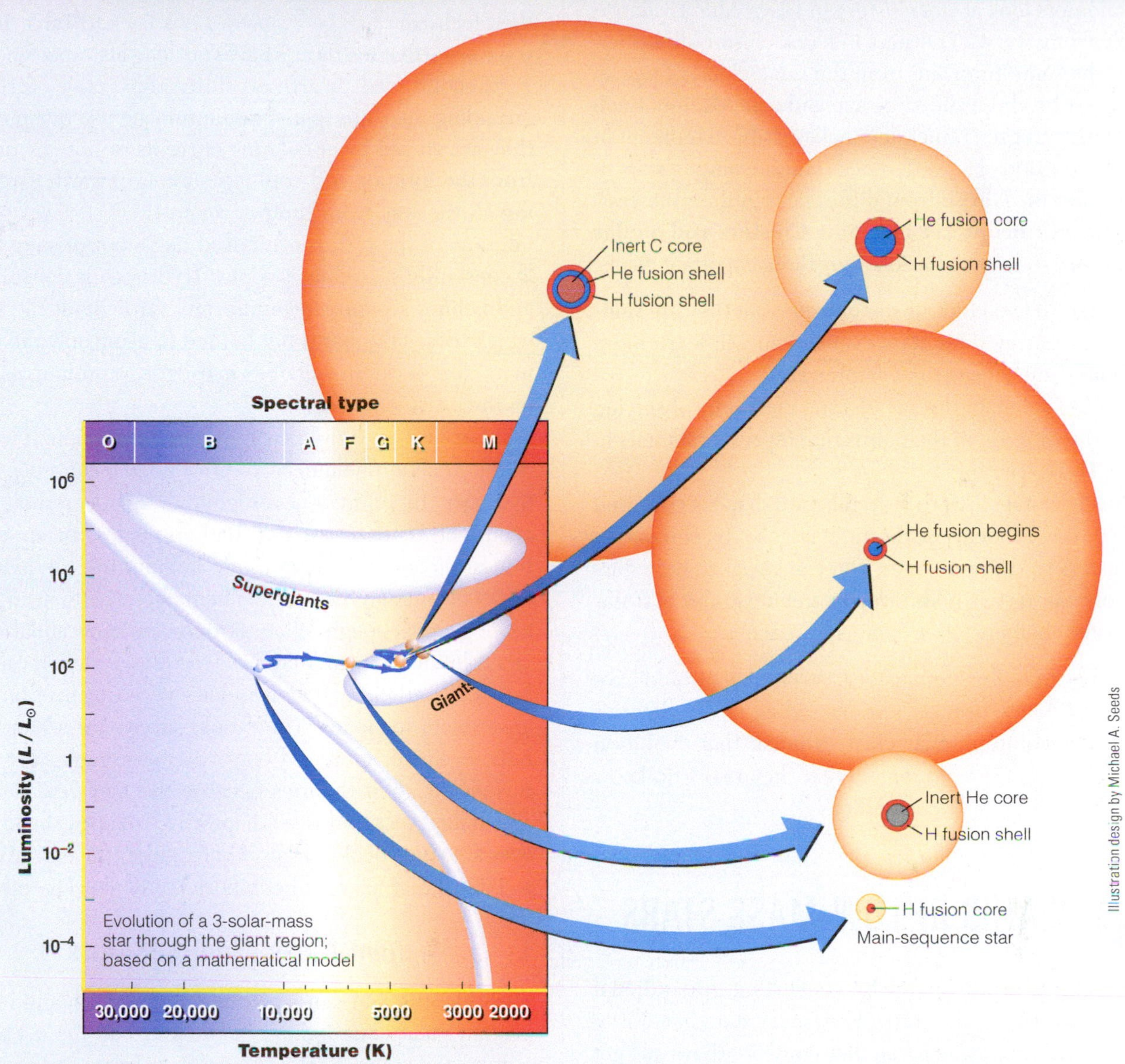

When a main-sequence star exhausts the hydrogen in its core, it evolves rapidly to the right in the H–R diagram as it expands to become a cool giant. It then follows a looping path as it fuses helium in its core and then fuses helium in a shell. (Evolutionary track adapted from the work of Icko Iben.)

surface cools once again. The point that represents the star in the H–R diagram moves back to the right, completing a loop to become a red giant again. The approximate rule is that if the core of a post-main-sequence star is "dead" (has no nuclear reactions), the star is a red giant, and if the core is "alive" (has fusion reactions), the star is a yellow giant.

Now you can understand why giant stars are relatively rare (look back to Chapter 9, **Concept Art 9A**). A star spends about 90 percent of its lifetime on the main sequence and only 10 percent as a giant star. At any moment you look, only a fraction of the visible stars will be passing through the red and yellow giant stages.

What happens to a star after helium fusion depends on its mass, but in any event it cannot survive long: It must eventually collapse and end its career as a star. This chapter will trace the details of this process of stellar death, but before you begin that story, you must ask the most important question in science: What is the evidence? What evidence shows that stars actually evolve as theory predicts? You can find the answers in observations of star clusters.

11-1c Star Clusters: Evidence of Evolution

Astronomers look at star clusters and say, "Aha! Evidence to solve a mystery." The stars in a star cluster all formed about the same time and from the same cloud of gas, so they must be about the same age and composition. Each star cluster freeze-frames and makes visible a moment in stellar evolution. The differences you see among stars in one cluster must arise from differences in their masses.

Study **Concept Art 11A, Star Clusters and Stellar Evolution**, and notice the following three important points:

1. There are two kinds of star clusters, but they are similar in the way their stars evolve. You will learn more about these clusters in Chapter 12.
2. You can estimate the age of a star cluster by observing the distribution of the points that represent its stars in the H–R diagram.
3. Finally, a star cluster's H–R diagram shape is governed by the evolutionary paths the stars take. By comparing clusters of different ages, you can visualize how stars evolve almost as if you were watching a film of a star cluster changing over billions of years.

Were it not for star clusters, astronomers would have little confidence in their understanding of the theories of stellar evolution. Star clusters make that evolution visible and assure astronomers that they really do know how stars are born, live, and die.

11-2 DEATHS OF LOW-MASS STARS

Contracting stars heat up by converting gravitational energy into thermal energy. Low-mass stars have little gravitational energy, so when they contract, they can't get very hot. This limits the nuclear fuels they can ignite. In Chapter 10 Section 10-3b, you saw that objects less massive than 0.08 solar mass cannot sustain hydrogen fusion. Consequently, this section will concentrate on stars more massive than 0.08 solar mass but no more than about four times the mass of the Sun.

Structural differences divide the lower-main-sequence stars into two subgroups, very-low-mass red dwarfs and medium-mass stars such as the Sun. The critical difference between the two groups is the extent of interior convection.

11-2a Red Dwarfs

Stars less massive than about 0.4 solar mass—the red dwarfs—have two important differences from more massive stars. First, because they have very small masses, they have very little weight to support. Their pressure–temperature thermostats are set low, and they consume their hydrogen fuel very slowly. Look back to Table 10-2; you can see that red dwarfs have very long life expectancies.

Second, red dwarfs are fully convective, further extending their life spans. Being fully convective means they are stirred by circulating currents of hot gas rising from the interior and cool gas sinking inward, extending all the way from their cores to their surfaces. The interiors of these stars are mixed like a pot of soup that is constantly stirred as it cooks. Hydrogen is consumed and helium accumulates uniformly throughout the star, which means the star is not limited to using only the fuel in its core. It can use all of its hydrogen to prolong its life on the main sequence.

Because a red dwarf is mixed by convection, it won't develop an inert helium core surrounded by unprocessed hydrogen; therefore it never ignites a hydrogen shell and cannot become a giant star. Rather, nuclear fusion continues to convert hydrogen into helium, but the helium does not fuse into heavier elements because the star can never get hot enough. What astronomers know about stellar evolution indicates that these red dwarfs should use up nearly all of their hydrogen and live very long lives on the lower main sequence. They could survive for a hundred billion years or more. Of course, astronomers can't test this part of their theories because the Universe is only 13.8 billion years old (see Chapter 14), so not a single red dwarf has died of old age anywhere in the Universe. Every red dwarf that has ever been born is still glowing today.

11-2b Medium-Mass (Sunlike) Stars

Stars like the Sun can ignite hydrogen and helium and become giants; but, if they contain less than about 4 solar masses, they cannot get hot enough to ignite carbon, the next fuel after helium. Note that this mass limit is uncertain, as are many of the masses quoted here. The evolution of stars is highly complex, and such thresholds are difficult to determine exactly.

There are two keys to the evolution of these sunlike stars, the lack of complete mixing and mass loss. The interiors of medium-mass stars are not completely mixed because, unlike the red dwarf stars, they are not completely convective. The helium ash accumulates in an inert helium core surrounded by unprocessed hydrogen. As described earlier, when this core contracts, the unprocessed hydrogen ignites in a shell and swells the star into a red giant. During red giant stages, stars tend to lose mass because of strong stellar winds. When the helium core finally ignites, the star becomes a yellow giant. When the core fills up with carbon–oxygen ash, that is the end of fusion for stars in the medium-mass range because they have masses too low to make their cores hot enough to ignite carbon or oxygen fusion. The carbon–oxygen core is the dead end for these stars.

All of this discussion is based on theoretical models of stars and a general understanding of how stars evolve. Does it really happen? Astronomers need observational evidence to confirm this theoretical discussion, and the gas that is expelled from these giant stars gives visible evidence that sunlike stars have gone through these stages of element building and do indeed die in this way.

11-2c Planetary Nebulae

When a medium-mass star like the Sun becomes a distended giant, its atmosphere becomes cool and consequently more opaque. Light has to push against it to escape. At the same time, the fusion shells become thin and unstable, and they begin to flare, which pushes the atmosphere outward. An aging giant can expel its outer atmosphere in repeated surges to form one of the most beautiful objects in astronomy, a **planetary nebula**, so called because the first ones discovered looked like the greenish-blue disks of planets such as Uranus or Neptune. In fact, a planetary nebula has nothing to do with a planet. It is composed of ionized gases expelled by a dying star, like the nebulae in Figure 11-3.

FIGURE 11-3

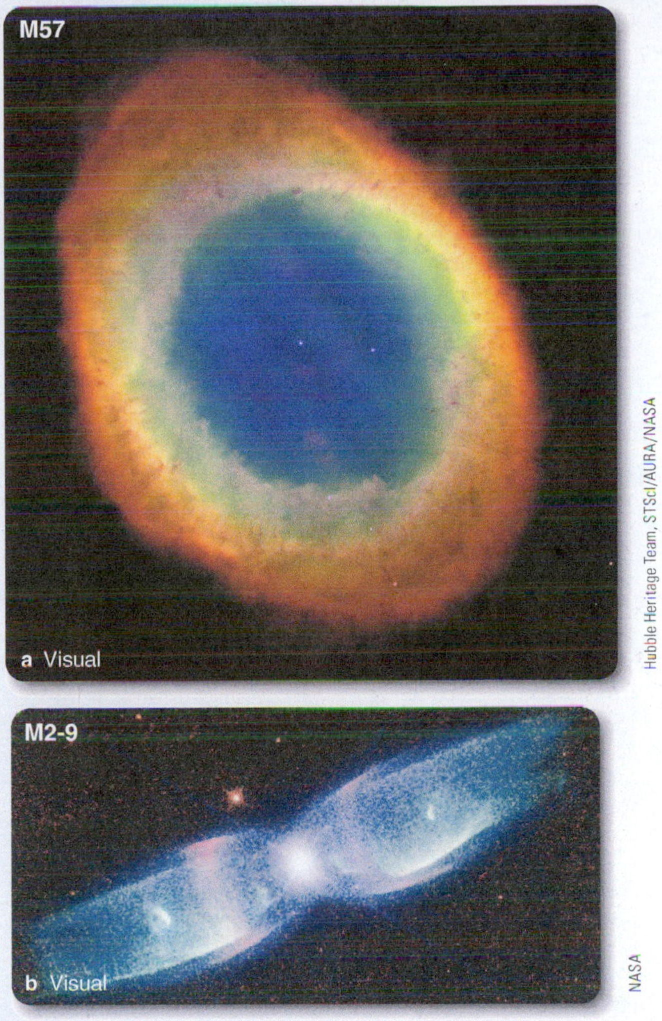

(a) The Ring Nebula (M57) in the constellation Lyra is visible even in small telescopes. Note the hot blue star at its center and the radial texture in the gas that suggests outward motion. (b) Some planetary nebulae such as M2-9 are highly elongated, and it has been suggested that the Ring Nebula in (a) actually has a tubular shape like M2-9 that happens to be pointed approximately toward Earth.

You can understand what planetary nebulae are like by using simple observations and theoretical methods explained in Chapter 5 such as Kirchhoff's laws and the Doppler effect. Although real nebulae are quite complex, the simple model shown in Figure 11-4 of a slow stellar wind followed by a fast wind explains their structures fairly well and provides a way to organize the observed phenomena. The complexities and asymmetries seen in planetary nebulae might be because of repeated expulsions of expanding shells and oppositely directed jets, much like bipolar flows observed coming from protostars. Observations and stellar evolution models indicate that the central star of a planetary nebula star finally must contract and become a white dwarf.

11-2d White Dwarfs

As you have just learned, medium-mass stars die by ejecting gas into space and contracting into white dwarfs. In Chapter 9 Section 9-6b, you surveyed the stars and learned that white dwarfs are the second most common kind of star (look back to **Concept Art 9A**). Only red dwarfs are more abundant. The billions of white dwarfs in our Galaxy must be the remains of medium-mass stars.

The first white dwarf discovered was the faint companion to the well-known star Sirius, the brightest star in the sky. Sirius is a visual binary star, the most luminous member of which is Sirius A. The white dwarf, Sirius B, is 8000 times fainter than Sirius A.

planetary nebula An expanding shell of gas ejected from a medium-mass star during the latter stages of its evolution.

Concept Art 11A

1 STAR CLUSTERS AND STELLAR EVOLUTION

An **open cluster** is a collection of 10 to 1000 stars in a region about 25 pc in diameter. Some open clusters are quite small, and some are large, but they all have an open, transparent appearance because the stars are not crowded together.

a In a star cluster, each star follows its orbit around the center of mass of the cluster.

N.A.Sharp/NOAO/AURA/NSF

1a A **globular cluster** can contain 10^5 to 10^6 stars in a region only 10 to 30 pc in diameter. The term *globular cluster* comes from the word *globe*, although globular cluster is pronounced like glob as in *glob* of butter. These clusters are nearly spherical, and the stars are much closer together than the stars in an open cluster.

Anglo-Australian Observatory / David Malin Images

b Astronomers can construct an H–R diagram for a star cluster by plotting a point to represent the luminosity and temperature of each star.

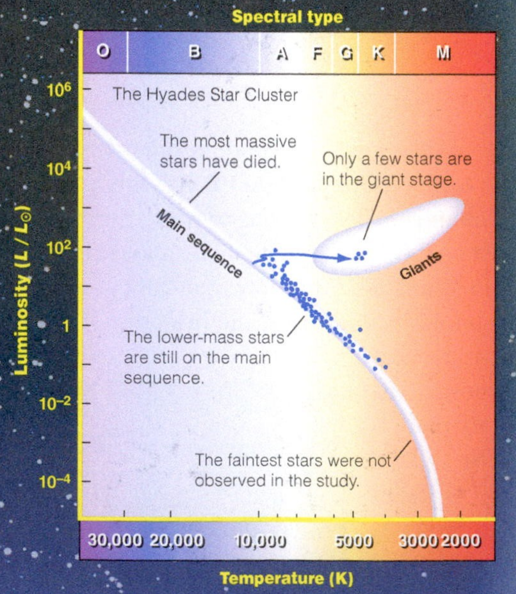

2

The H–R diagram of a star cluster can make the evolution of stars visible. The key is to remember that all of the stars in the star cluster have the same age and composition, differing only in mass. The H–R diagram of a star cluster provides a "snapshot" of the evolutionary state of the stars at the time you happen to observe them. This diagram shows the 650-million-year-old star cluster called the Hyades, easily visible to the unaided eye in the constellation Taurus. The upper main sequence is missing because the more massive stars have died, and this snapshot catches a few medium-mass stars leaving the main sequence to become giants.

As a star cluster ages, its main sequence grows shorter, like a candle burning down. You can judge the age of a star cluster by looking at the turnoff point, the point on the main sequence where stars are currently evolving to the right to become giants. Stars at the **turnoff point** have lived out their lives and are about to die. Consequently, the life expectancy of the stars at the turnoff point equals the age of the cluster. (Evolutionary track adapted from the work of Icko Iben.)

3 From theoretical models of stars, you could construct a film to show how the H–R diagram of a star cluster changes as it ages. You can then compare theory (*left*) with observation (*right*) to understand how stars evolve. Note that the time step for each frame in this film increases by a factor of 10.

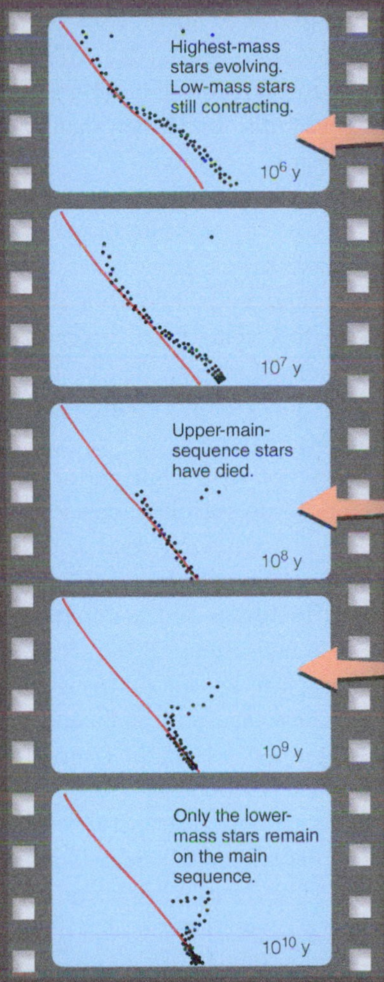

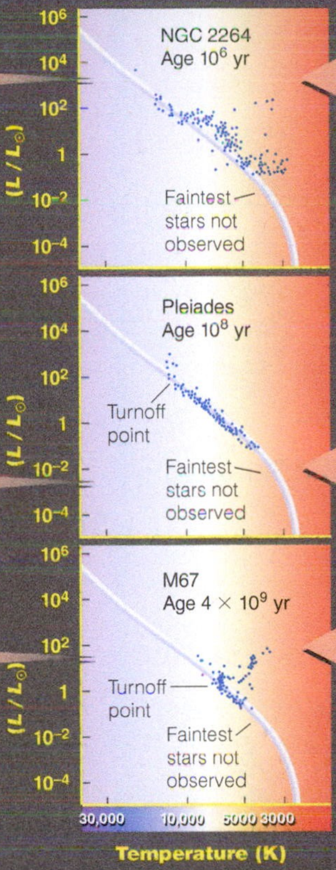

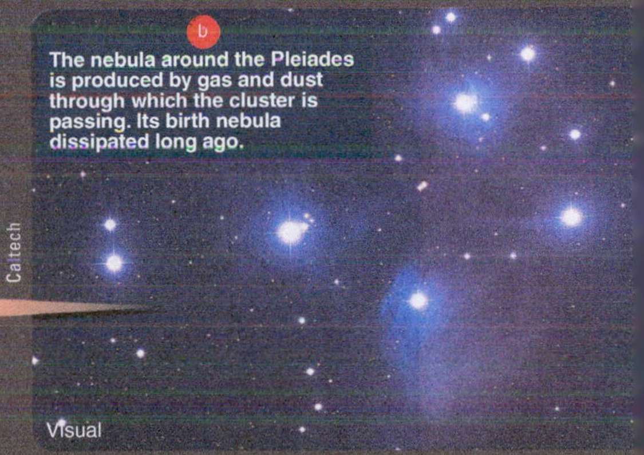

Illustration design by Michael A. Seeds

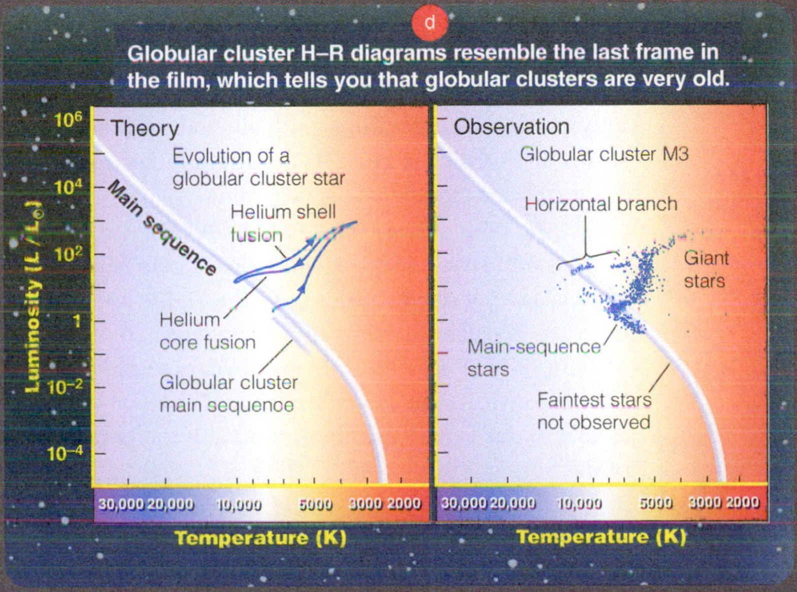

3a The H–R diagrams of globular clusters have very faint (low-luminosity, low-mass) turnoff points showing that they are very old clusters. The most complete analysis suggests these clusters are about 11 billion years old.

The **horizontal branch** stars are giants fusing helium in their cores or in shells. The shape of the horizontal branch outlines the evolution of these stars.

The main-sequence stars in globular clusters are fainter and bluer than the zero-age main sequence. Spectra reveal that globular cluster stars are poor in elements heavier than helium, which means their gases are less opaque. That means energy can flow outward more easily, making these low-metallicity stars slightly smaller and hotter than their high-metallicity counterparts with the same masses.

FIGURE 11-4

The process that produces planetary nebulae involves two stellar winds. First, as an aging giant, the star gradually blows away its outer layers in a slow breeze of low-excitation gas that is not easily visible. Once the hot interior of the star is exposed, it ejects a high-speed wind that overtakes and compresses the gas of the slow wind, like a snowplow, while ultraviolet radiation from the hot remains of the central star excites the gases to glow like a giant neon sign.

The orbital motions of the stars (see Figure 9-12) reveal that the white dwarf's mass is 0.98 solar mass, and its blue-white color tells you that its surface is hot, about 25,000 K. Although it is very hot, it has a very low luminosity, so it must have a small surface area. In fact, it is about the size of Earth. Dividing its mass by its volume reveals that it is very dense—about 2×10^6 g/cm^3. On Earth, a teaspoonful of Sirius B material would weigh more than 10 tons. These basic observations and simple physics lead to the conclusion that white dwarfs are astonishingly dense.

A normal star is supported by energy flowing outward from its core, but a white dwarf cannot generate energy by nuclear fusion. It has exhausted its hydrogen and helium fuel and produced carbon and oxygen. As the star contracts into a white dwarf, it converts gravitational energy into thermal energy, and its interior becomes very hot, but it cannot get hot enough to fuse carbon into heavier elements.

The contraction of a white dwarf compresses the gases in its interior to such high densities that quantum mechanical laws become important, and the electrons in the gas cannot get closer together. Such a gas is termed **degenerate matter**, and it takes on two properties that are important in understanding the structure and evolution of dying stars. A degenerate gas is millions of times harder to compress than solid steel, and the pressure in the gas no longer depends on the temperature. Unlike a normal star, which is supported by ordinary gas pressure, a white dwarf is supported against its own gravity by the resistance to compression of a degenerate gas.

Clearly, a white dwarf is not a true star. It generates no nuclear energy, is almost completely degenerate matter, and, except for a thin layer at its surface, contains no gas. Instead of calling a white dwarf a "star," you could call it a **compact object**. Later sections of this chapter discuss two other types of compact objects, neutron stars and black holes.

A white dwarf's future is bleak. As it radiates energy into space, its temperature gradually falls, but it cannot shrink any smaller because its degenerate electrons cannot get closer together. This degenerate matter is a very good thermal conductor, so heat flows to the surface and escapes into space, and the white dwarf gets fainter and cooler, moving downward and to the right in the H–R diagram. Because the white dwarf contains a tremendous amount of heat, it needs billions of years to radiate that heat through its small surface area. The coolest white dwarfs in our Galaxy are about the same temperature as of the Sun.

Perhaps the most interesting thing astronomers have learned about white dwarfs has come from mathematical

degenerate matter Extremely high-density matter in which, because of quantum mechanical effects, pressure does not depend on temperature.

compact object One of the three final states of stellar evolution, which generates no nuclear energy and is much smaller and denser than a normal star.

models. The equations predict that degenerate electron pressure cannot support an object with more than about 1.4 solar masses. A white dwarf with that mass would have such strong gravity that its radius would shrink to zero. This is called the **Chandrasekhar limit** after Subrahmanyan Chandrasekhar, the astronomer who calculated it. (*Subrahmanyan* was his family name, *Chandrasekhar* his given name.) This seems to imply that a star more massive than 1.4 solar masses could not become a white dwarf unless it got rid of the extra mass in some way.

Can stars lose substantial amounts of mass? Observations provide clear evidence that young stars have strong stellar winds, and aging giants and supergiants also lose mass. This suggests that stars more massive than the Chandrasekhar limit can eventually end up as white dwarfs if they reduce their mass under the limit. Theoretical models show that stars that begin life with as much as 8 solar masses could lose mass fast enough to reduce their mass so low they could collapse to form white dwarfs with masses below 1.4 solar masses. With mass loss, a wide range of medium-mass stars could eventually die as white dwarfs.

11-2e The Fate of the Sun and the End of Earth

Astronomy is about you. Although this chapter has been discussing the deaths of medium-mass stars, it has also been discussing the future of our Earth. The Sun is a medium-mass star and must eventually die by becoming a giant, possibly producing a planetary nebula, and collapsing into a white dwarf. That will spell the end of Earth.

Evolutionary models of the Sun suggest that it might survive as an energy-generating star for another 7 or 8 billion years. In about 6 billion years, it will exhaust the hydrogen in its core, leave the main sequence, begin burning hydrogen in a shell, and swell into a red giant star about 30 times its current radius. Later, helium fusion will ignite in the core and the Sun will become a horizontal branch star. After the helium fuel is exhausted in its core, helium fusion will begin in a shell, and the Sun will expand again. That second red giant version of the Sun will be about as large as the orbit of Earth. Before that, the Sun's increasing luminosity will certainly evaporate Earth's oceans, drive away the atmosphere, and even vaporize much of Earth's crust. Astronomers are still uncertain about some of the details, but computer models that include tidal effects predict that the expanding Sun eventually will engulf and destroy Mercury, Venus, and Earth.

While it is a giant star, the Sun will have a strong wind and lose a substantial fraction of its mass into space. The atoms that were once in Earth will be part of the expanding nebula around the Sun. Your atoms will be part of that nebula. If the white dwarf remnant Sun becomes hot enough, it will ionize the expelled gas and light it (and you) up as a planetary nebula.

Models of the Sun's evolution are not precise enough to predict whether its white dwarf remnant will become hot enough soon enough to light up its expelled gas and create a planetary nebula before that gas disperses. Whether the expelled gas lights up or not, it would include atoms that were once part of Earth. Some research also suggests that a star needs a close binary companion to speed up its spin in order to create a planetary nebula. The Sun, of course, has no close stellar companion. This is an area of active research, and there are as yet no firm conclusions.

There is no danger that the Sun will explode as a nova; it has no binary companion (see the next section of this chapter). And, as you will see, the Sun is definitely not massive enough to die the violent supernova death of the most massive stars.

11-3 THE EVOLUTION OF BINARY SYSTEMS

Stars in binary systems can evolve independently of each other if they orbit at a large distance from each other. In this situation, one of the stars can swell into a giant and collapse without disturbing its companion. Other binary stars are as close to each other as 0.1 AU, and when one of those stars begins to swell into a giant, its companion can be affected in strong and surprising ways.

11-3a Mass Transfer and Accretion Disks

Binary stars can sometimes interact by transferring mass from one star to the other. Of course, the gravitational field of each star holds its mass together, but the gravitational fields of the two stars, combined with the rotation of the binary system, define a dumbbell-shaped volume called the **Roche lobes** around the pair of stars. Matter inside a star's Roche lobe is gravitationally bound to the star, but matter outside the lobe can be transferred to the other star or lost completely from the system. If the stars are close together, the Roche

Chandrasekhar limit The maximum mass of a white dwarf, about 1.4 solar masses. A white dwarf of greater mass cannot support itself and will collapse.

Roche lobes The two volumes of space gravitationally controlled by the stars in a binary system.

lobes are relatively small and can interfere with the evolution of the stars. When an evolving star in a close binary system expands so far that it fills its Roche lobe, matter can flow from that star's lobe into the other lobe and onto the other star, as you can see in the series in Figure 11-5.

Matter flowing from one star to another cannot fall directly into the star. Rather, because of conservation of angular momentum, it must flow into a whirling disk around the star. **Angular momentum** refers to the tendency of a rotating or revolving body to continue that motion. All such objects possess some angular momentum, and in the absence of external forces, an object maintains (conserves) its total angular momentum. A figure skater takes advantage of conservation of angular momentum by starting a spin slowly with arms extended and then drawing them in. As her mass becomes concentrated closer to her axis of rotation, she spins faster. The same effect causes the slowly circulating water in a bathtub to spin faster in a whirlpool as it approaches the drain.

Mass transferred from one star to another in a binary system must conserve its angular momentum. Thus it must flow into a rapidly rotating whirlpool called an **accretion disk** around the second star, as shown in Figure 11-6. The gas in the disk grows hot because of friction and tidal forces (look back to Chapter 3 Section 3-5c) and eventually falls onto the second star. If that second star, the one receiving the matter lost from its companion, is a compact object like a white dwarf, the gas in the accretion disk can become very compressed. The gas temperature can exceed a million Kelvin, producing X-rays. In addition, the matter accumulating on the white dwarf can eventually cause a violent explosion called a nova.

11-3b Novae

At the beginning of this chapter you saw that the word *nova* refers to an apparently new star that appears in the sky for a while and then fades away. Modern astronomers know that a nova is not a new star but an old star flaring up. After a nova fades, astronomers can photograph the spectrum of the remaining faint point of light. Invariably, they find a normal star and a white dwarf in a close binary system. A nova is evidently an explosion involving a white dwarf.

Observational evidence can tell you how nova explosions occur. As the explosion begins, spectra show blueshifted absorption lines

angular momentum A measure of the tendency of a rotating or revolving body to continue that motion. Mathematically, the product of mass, velocity, and radius.

accretion disk The rotating disk that forms in some situations as matter is drawn gravitationally toward a central body.

FIGURE 11-5

The Evolution of a Binary System

A pair of stars orbiting close to each other can exchange mass and modify their evolution.

FIGURE 11-6

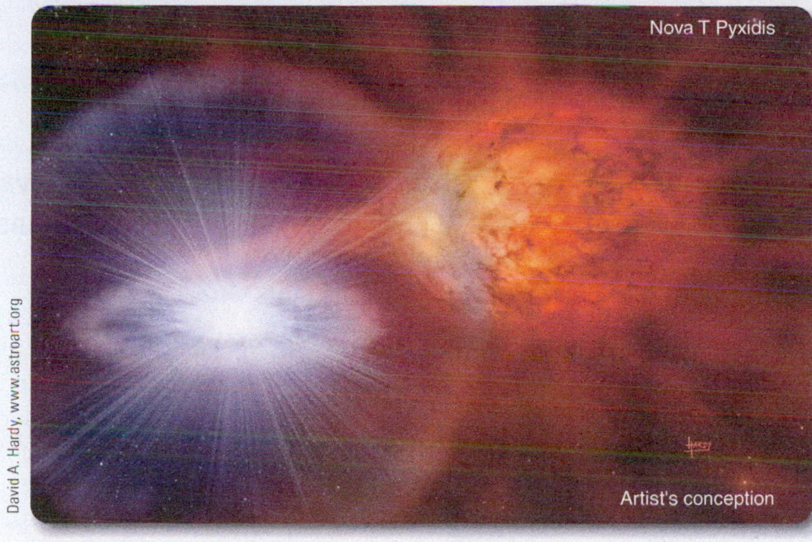

Matter from an evolving red giant falls into a white dwarf and forms a whirling accretion disk. Friction and tidal forces can make the disk very hot. Such systems can lead to nova explosions on the surface of the white dwarf, as shown in this artist's impression of the recurrent nova T Pyxidis.

that tell you the gas is dense and coming toward you at a few thousand kilometers per second. After a few days, the spectral lines change to emission lines, which tells you the gas has thinned, but the blueshifts remain, showing that a cloud of debris has been ejected into space.

Nova explosions occur when mass transfers from a normal star into an accretion disk around a white dwarf. As the matter loses its angular momentum in the accretion disk, it settles inward onto the surface of the white dwarf and forms a layer of unused nuclear fuel—mostly hydrogen. As the layer deepens, it becomes denser and hotter until the hydrogen fuses in a sudden explosion that blows the surface off the white dwarf. Although the expanding cloud of debris contains less than 0.0001 solar mass, it is hot, and its expanding surface area makes it very luminous. Nova explosions can become 100,000 times more luminous than the Sun. As the debris cloud expands, cools, and thins over a period of weeks and months, the nova fades from view.

The explosion of its surface hardly disturbs the white dwarf and its companion star. Mass transfer quickly resumes, and a new layer of fuel begins to accumulate. How fast the fuel builds up depends on the rate of mass transfer. Accordingly, you can expect novae to repeat each time an explosive layer accumulates. Many novae take thousands of years to build an explosive layer, but some take only decades.

11-4 THE DEATHS OF MASSIVE STARS

You have seen that low- and medium-mass stars die relatively quietly as they exhaust their hydrogen and helium, and then some push away their surface layers to form planetary nebulae. In contrast, massive stars live spectacular lives, aspects of which can be seen in Figure 11-7, and destroy themselves in violent explosions.

11-4a Nuclear Fusion in Massive Stars

Stars on the upper main sequence have too much mass to die as white dwarfs, but their evolution begins much like that of their lower-mass cousins. They consume the hydrogen in their cores, ignite hydrogen shells, and become giants or, for the most massive stars, supergiants. Their cores contract and fuse helium first in the core and then in a shell, producing a carbon–oxygen core.

Unlike medium-mass stars, the massive stars finally can get hot enough to ignite carbon fusion at a temperature of about 1 billion Kelvin. Carbon fusion produces more oxygen plus neon. As soon as the carbon is exhausted in the core, the core contracts, and carbon ignites in a shell. This pattern of core ignition and shell ignition continues with a series of heavier nuclei as fusion fuel, and the star develops a layered structure as shown in Figure 11-7, with a hydrogen-fusion shell surrounding a helium-fusion shell surrounding a carbon-fusion shell surrounding … and so on. At higher temperatures than carbon fusion, nuclei of oxygen, neon, and magnesium fuse to make silicon and sulfur, and at even higher temperatures silicon can fuse to make iron.

The fusion of the nuclear fuels in this series goes faster and faster as the massive star evolves rapidly. The amount of energy released per fusion reaction decreases as the mass of the types of atoms involved increases. To support its weight and remain stable, a star must fuse oxygen much faster than it fused hydrogen. Also, there are fewer nuclei in the core of the star by the time heavy nuclei begin to fuse. Four hydrogen nuclei make one helium nucleus, and three helium nuclei make one

FIGURE 11-7

Massive stars live fast and die young. (a) This star is among the most massive known, estimated to contain more than 100 solar masses. It is rapidly ejecting matter into space. (b) The centers of massive stars develop Earth-size cores (magnified 100,000 times in this figure) composed of concentric layers of gases undergoing nuclear fusion. The iron core at the center eventually causes a star-destroying explosion.

TABLE 11-1 HEAVY-ELEMENT FUSION IN A 25-$M_\odot$ STAR

Fuel	Time	Percentage of Lifetime
H	3,000,000 years	88
He	400,000 years	12
C	600 years	0.018
O	0.5 years	0.000015
Si	1 day	0.00000008

carbon, so there are 12 times fewer nuclei of carbon available for fusion than there were hydrogen nuclei. This means the heavy elements are used up, and fusion goes very quickly in massive stars (Table 11-1). Hydrogen fusion can last 7 million years in a 25-solar-mass star, but that same star will fuse its oxygen in 6 months and its silicon in just 1 day.

11-4b Supernova Explosions

Theoretical models of evolving stars combined with nuclear physics allow astronomers to describe what happens inside a massive star when the last nuclear fuels are exhausted. It begins with iron nuclei and ends in cosmic violence.

Silicon fusion produces iron, the most tightly bound of all atomic nuclei (see Figure 10-5). Nuclear fusion releases energy only when less tightly bound nuclei combine into a more tightly bound nucleus. Once the gas in the core of the star has been converted to iron, there are no further nuclear reactions that can release energy. The iron core is a dead end in the evolution of a massive star.

As a star develops an iron core, energy production begins to decline, and the core contracts. Nuclear reactions involving iron begin, but they remove energy from the core, causing it to contract even further. Once this process starts, the core of the star collapses inward in less than a tenth of a second.

The collapse of a giant star's core after iron fusion starts is calculated to happen so rapidly that the most powerful computers are unable to predict the details. Thus, models of supernova explosions contain many approximations. Nevertheless, the models predict exotic nuclear reactions in the collapsing core that should produce a flood of neutrinos (look back to Chapter 10 Section 10-2c). In fact, for a short time the core produces more energy per second than all of the stars in all of the visible galaxies in the Universe, and 99 percent of that energy is in the form of neutrinos. This flood of neutrinos carries large amounts of energy out of the core, allowing the core to collapse further. The models also predict that the collapsing core of the star must quickly become a neutron star or a black hole, the subjects of the next sections of this chapter, while the envelope of the star is blasted outward.

To understand how the inward collapse of the core can produce an outward explosion, you can think about a traffic jam. The collapse of the innermost part of the degenerate core allows the rest of the core to fall inward, and this creates a tremendous traffic jam as all of the nuclei fall toward the center. The position of the traffic jam, called a shock wave, begins to move outward as more in-falling material encounters the jam. The torrent of neutrinos, as well as energy flowing out of the core in sudden violent convective turbulence, help drive the shock wave outward. Within a few hours, the shock wave bursts outward through the surface of the star and blasts it apart.

The supernova seen from Earth is the brightening of the star as its envelope is blasted outward by the shock wave. As months pass, the cloud of gas expands, thins, and fades, but the manner in which it fades tells astronomers more about the death throes of the star. The rate at which the supernova's brightness decreases matches the rate at which radioactive nickel and cobalt decay, so the explosion must produce great abundances of those atoms. The radioactive cobalt decays into iron, so destruction of iron in the core of the star is followed by the production of iron through nuclear reactions in the expanding outer layers.

11-4c Types of Supernovae

In studying supernovae in other galaxies, astronomers have noticed that there are a number of different types (Figure 11-10). **Type I supernovae** have no hydrogen lines in their spectra whereas **Type II supernovae** spectra have hydrogen lines. Type II supernovae appear to be produced by the collapse and explosion of a massive star, the process discussed in the previous section.

A Type Ia supernova evidently occurs when a white dwarf in a binary system receives enough mass to exceed the Chandrasekhar limit and collapse. The collapse of a white dwarf is different from the collapse of a massive star because the core of the white dwarf contains usable fuel. As the collapse begins, the temperature and density shoot up, and the carbon–oxygen core begins to fuse in violent nuclear reactions. In a few seconds, the carbon–oxygen core interior is entirely consumed, and the outermost layers are blasted away in a violent explosion that, at its brightest, is about six times more luminous than a Type II supernova. The white dwarf is destroyed; no neutron star or black hole is left behind. This explains why no hydrogen lines are seen in the spectrum of a Type Ia supernova explosion—white dwarfs contain very little hydrogen.

The less common Type Ib supernova is understood to occur when a massive star in a binary system loses its hydrogen-rich outer layers to its companion star. The remains of the massive star could develop an iron core and collapse, as described in the previous section, producing a supernova explosion that lacked hydrogen lines in its spectrum.

Astronomers working with the largest and fastest computers are using modern theory to try to understand supernova explosions. The companion to theory is observation, so you should ask what observational evidence supports this story of supernova explosions.

11-4d Observations of Supernovae

In the year 1054, Chinese astronomers saw a "guest star" appear in the constellation known in the Western tradition as Taurus the Bull. The star quickly became so bright it was visible in the daytime, and then, after a month, it slowly faded, taking almost 2 years to vanish from sight. When modern astronomers turned their telescopes to the location of the guest star, they found a peculiar nebula now known as the Crab Nebula for its many-legged shape, which you can see in Figure 11-8. In fact, the legs of the Crab Nebula are filaments of gas that are moving away from the site of the explosion at about 1400 kilometers/second. Comparing the radius of the nebula, 1.4 pc, with its velocity of expansion reveals that the nebula began expanding nine or ten centuries ago, just when the guest star made its appearance. The Crab Nebula is clearly the remains of the supernova seen in 1054. In the next section, you will meet the neutron star found at the center of the Crab Nebula.

Type I supernova A supernova explosion caused by the collapse of a white dwarf.

Type II supernova A supernova explosion caused by the collapse of the core of a massive star.

FIGURE 11-8

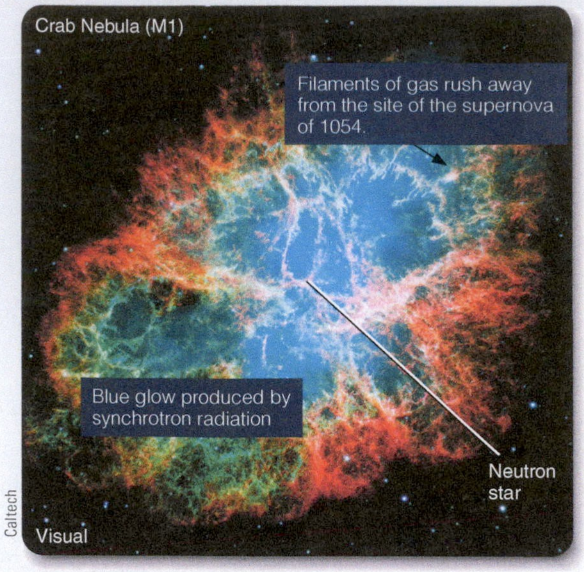

The Crab Nebula is located in the constellation Taurus the Bull, just where Chinese astronomers saw a brilliant guest star in the year 1054. High-speed electrons produced by the central neutron star spiral through magnetic fields and produce the foggy glow of synchrotron radiation that fills the nebula. The star just to the right of the neutron star lies much closer to Earth and is not in the Crab Nebula.

The blue glow of the Crab Nebula is produced by **synchrotron radiation**. This form of electromagnetic radiation, unlike blackbody radiation, is produced by rapidly moving electrons spiraling through magnetic fields and is common in the nebulae produced by supernovae. In the case of the Crab Nebula, the electrons travel so fast that they emit visual wavelengths. In most such nebulae, the electrons move slower, and the synchrotron radiation is at radio wavelengths.

Supernovae are rare. Only a few have been seen with the naked eye in recorded history. Arab astronomers saw one in 1006, and the Chinese saw one in 1054. European astronomers observed two—one in 1572 (Tycho's supernova) and one in 1604 (Kepler's supernova). In addition, the guest stars in the years 185, 386, 393, and 1181 might have been supernovae.

synchrotron radiation
Radiation emitted when high-speed electrons move through a magnetic field.

In the centuries following the invention of the astronomical telescope in 1609, no supernova was bright enough to be visible to the naked eye. Then, in the early morning hours of February 24, 1987, astronomers around the world were startled by the discovery of an unaided-eye supernova still growing brighter in the southern sky, which you can see in Figure 11-9. The supernova, known officially as SN 1987A, is 50,000 pc away in the Large Magellanic Cloud, a small satellite galaxy to our own Milky Way Galaxy. This first unaided-eye supernova in 383 years gave astronomers a ringside seat for the most spectacular event in stellar evolution.

One observation of SN 1987A is critical in that it confirms the hypothesis of core collapse. About 3 hours before the supernova was first noticeable in photographs, a blast of neutrinos swept through Earth. Instruments buried in a salt mine under Lake Erie and in a lead mine in Japan, though designed for another purpose, recorded 19 neutrinos in less than 15 seconds. Neutrinos are so difficult to detect that the 19 neutrinos actually detected mean that some 10^{17} neutrinos must have passed through the detectors in those 15 seconds. Furthermore, the neutrinos were arriving from the direction of the supernova. Thus, astronomers conclude that the burst of neutrinos was released when the iron core collapsed, and the supernova was first seen at visual wavelengths hours later when the shock wave blasted the star's surface into space.

Most supernovae are seen in distant galaxies, and careful observations allow astronomers to compare types, which can be done with charts such as the one in Figure 11-10. Type Ia supernovae, caused by the collapse of white dwarfs, are more luminous at maximum brightness and decline rapidly at first and then more slowly. Type II supernovae, produced by the collapse of massive stars, are not as bright at maximum, and they decline in a more irregular way.

SN 1987A was a Type II supernova, although its light curve shown in Figure 11-10 is not typical. Models indicate that most Type II supernovae are caused by the collapse of red supergiants, but SN 1987A was produced by the explosion of a hot, blue supergiant. Astronomers hypothesize that this star was once a red supergiant but later contracted and heated up slightly, becoming bluer before it exploded.

Although the supernova explosion fades to obscurity in a year or two, an expanding shell of gas marks the site of the explosion. The gas, originally expelled at 10,000 to 20,000 km/s, might carry away a fifth of the mass of the star. The collision of that expanding gas with the surrounding interstellar medium can sweep up even more

FIGURE 11-9

The star that exploded as Supernova 1987A had about 20 times the mass of the Sun. Interactions of the burst of light and gas expanding from the explosion with material lost by the star during previous stages as a blue giant and red supergiant have produced rings around the central glow, as shown in the high-resolution image and the artist's conception. A shock wave from the explosion is now expanding into a 1 light-year-diameter ring of gas ejected roughly 20,000 years before the explosion. As that ring is excited, it is lighting up the region around it, revealing how the star shed mass before it collapsed.

FIGURE 11-10

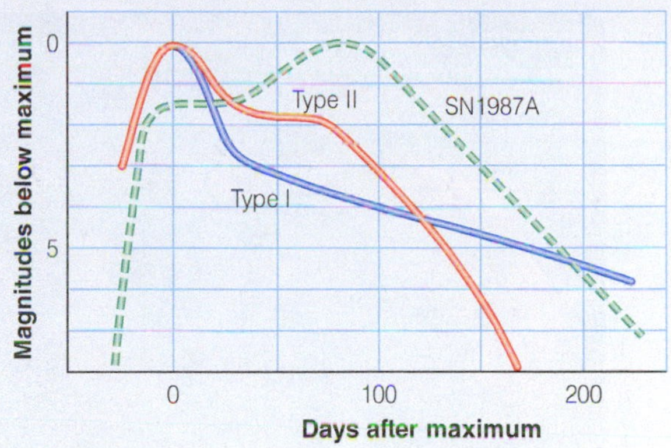

Type I supernovae decline in brightness rapidly at first and then more slowly, but Type II supernovae pause for about 100 days before beginning a steep decline. Supernova 1987A was odd in that it did not rise directly to maximum brightness. These light curves have been adjusted to the same maximum brightness.

gas and excite it to produce a **supernova remnant**, the nebulous remains of a supernova explosion. Figure 11-11 shows images of four supernova remnants; images made at other than visible wavelengths are displayed in false color (look back to Chapter 4 Section 4-4a).

Supernova remnants look quite delicate and do not survive very long—a few tens of thousands of years—before they gradually mix with the interstellar medium and vanish. The Crab Nebula is a young remnant, only about 960 years old and about 9 light-years in diameter. Older remnants can be larger. Some supernova remnants are visible only at radio and X-ray wavelengths. They have become too tenuous to emit detectable light, but the collision of the expanding hot gas with the interstellar medium can generate radio and X-ray radiation. You learned in Chapter 10 Section 10-4b that the compression of the interstellar medium by expanding supernova remnants can also trigger star formation.

Gravity always wins. However a star lives, theory predicts it must eventually die and leave behind one of three types of final remains—a white dwarf, neutron star, or black hole. These compact objects are small monuments to the power of gravity. Almost all of the gravitational potential energy available has been squeezed out of compact objects, and you find them in their final, high-density states.

supernova remnant The expanding shell of gas and dust marking the site of a supernova explosion.

FIGURE 11-11

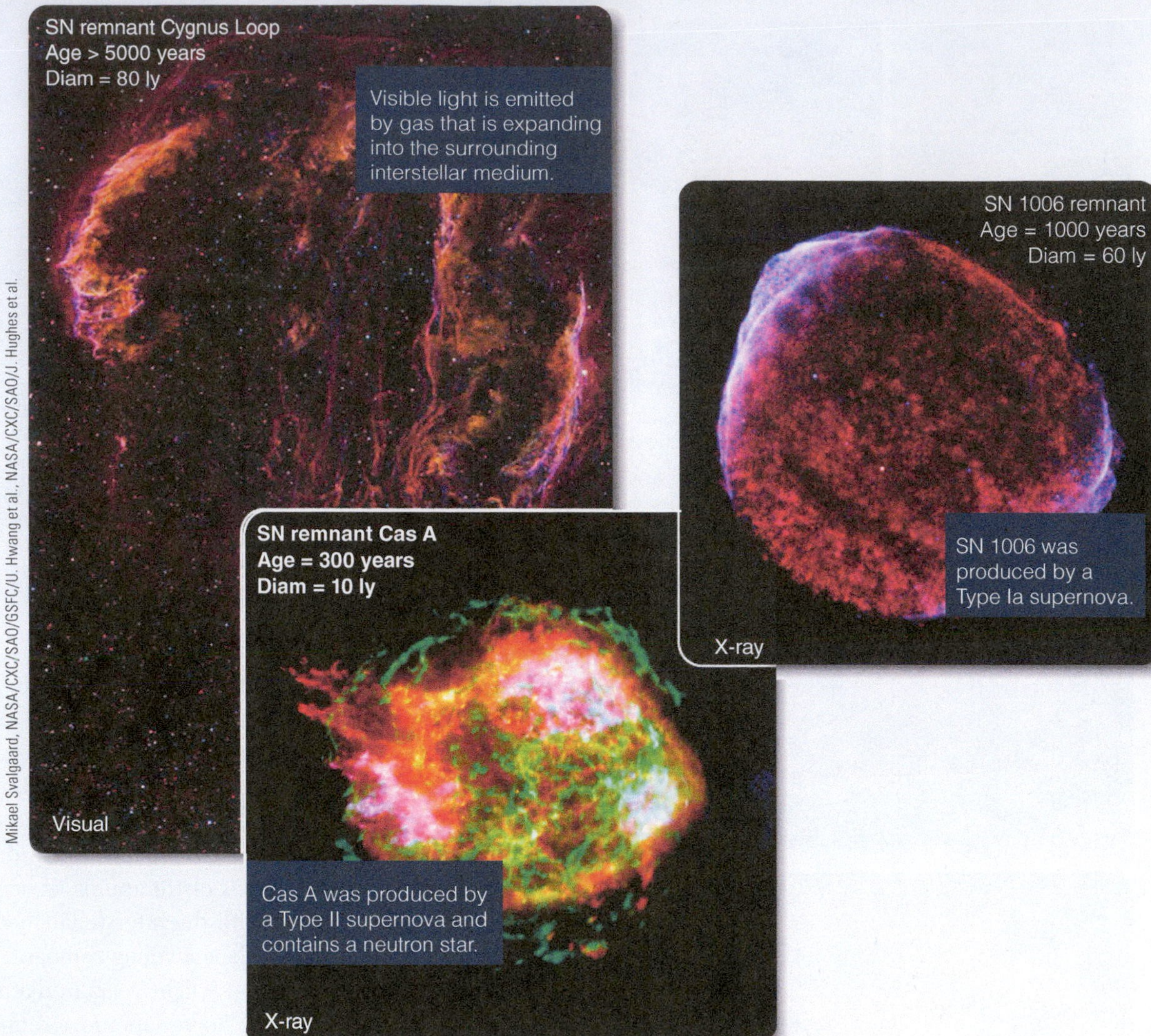

A supernova remnant is an expanding bubble of hot gas created by a supernova explosion. As the remnant expands and pushes into neighboring gas, it can emit radiation at many wavelengths. Jets can be seen extending toward the left and right of Cas A (Cassiopeia A).

11-5 NEUTRON STARS

A **neutron star** contains a little more than 1 solar mass compressed to a radius of about 10 km. Its density is so high that the matter is stable only as a fluid of pure neutrons. Two questions should occur to you immediately. First, how could any theory predict such a strange object? And second, how do we know they exist?

neutron star A small, highly dense star, with radius about 10 km, composed almost entirely of tightly packed neutrons.

11-5a Theoretical Prediction of Neutron Stars

The subatomic particles called neutrons were discovered in a laboratory in 1932, and physicists quickly realized that because neutrons spin much like electrons, a gas of neutrons could become degenerate and therefore nearly incompressible. Just 2 years later, in 1934, two astronomers, Walter Baade and Fritz Zwicky, suggested that some of the most luminous novae in the historical record were not regular novae but were caused by the collapse plus explosion of a massive star in a cataclysm they named a "super-nova." If the collapsing core is more massive than the Chandrasekhar limit of 1.4 solar masses, then the weight is too great to be supported by degenerate electron pressure, and the core cannot become a stable white dwarf. The collapse would force protons to combine with electrons and become neutrons. The envelope of the star would be blasted away in a supernova explosion, and the core of the star would be left behind as a small, tremendously dense sphere of neutrons that Zwicky called a "neutron star."

FIGURE 11-12

Michael A. Seeds

A tennis ball and a road map illustrate the scale of a neutron star. Such an object, containing slightly more mass than the Sun, would fit with room to spare inside the beltway around Washington, DC.

Mathematical models predict that a neutron star will be only 10 or so kilometers in radius (Figure 11-12) and will have a density of almost 10^{15} g/cm^3. That is roughly the density of atomic nuclei, and you can think of a neutron star as matter with all of the empty space squeezed out of it. On Earth, a sugar-cube-sized lump of this material would weigh more than a billion tons, the mass of a small mountain.

Simple physics, the physics you have used in previous chapters to understand normal stars, predicts that neutron stars should (1) spin rapidly, perhaps 100 to 1000 rotations per second; (2) be hot, with surface temperatures of millions of degrees Kelvin; and (3) have strong magnetic fields, up to a trillion times stronger than the Sun's or Earth's magnetic fields. For example, the collapse of a massive star's core would greatly increase its spin rate by conservation of angular momentum. Other processes during core collapse should create high temperature and magnetic field strength. Despite their high temperature, neutron stars should be difficult to detect because of their tiny size.

What is the maximum mass for a stable neutron star? In other words, is there an upper limit to the mass of neutron stars like the Chandrasekhar limit that defines the maximum mass of a white dwarf star? That is difficult to answer, because physicists don't know enough about the properties of pure neutron material. It can't be made in a laboratory, and theoretical calculations in this case are very difficult. The most widely accepted results suggest that a neutron star must be less massive than 2 to 3 solar masses. An object more massive than that can't be supported by degenerate neutron pressure, so it would collapse, presumably becoming a black hole.

What size stars will end their lives with supernova explosions that leave behind neutron star corpses? Theoretical calculations suggest that stars that begin life on the main sequence with 8 to about 15 solar masses will end up as neutron stars. Stars more massive than about 15 solar masses are expected to form black holes when they die.

11-5b The Discovery of Pulsars

In November 1967, Jocelyn Bell, a graduate student at Cambridge University in England, found a peculiar pattern in the data from a radio telescope. Unlike other radio signals from celestial bodies, this was a series of regular pulses, graphed in Figure 11-13. At first she and the leader of the project, Anthony Hewish, thought the signal was interference, but they found it day after day at the same celestial latitude and longitude. Clearly, it was cosmic in origin.

Another possibility, that it came from a distant civilization, led them to consider naming it LGM for Little Green Men. Within a few weeks, the team found three more objects in other parts of the sky pulsing with different periods. The objects were clearly natural, and the team dropped the name LGM in favor of **pulsar**—a contraction of "pulsing

pulsar A source of short, precisely timed radio bursts, understood to be spinning neutron stars.

FIGURE 11-13

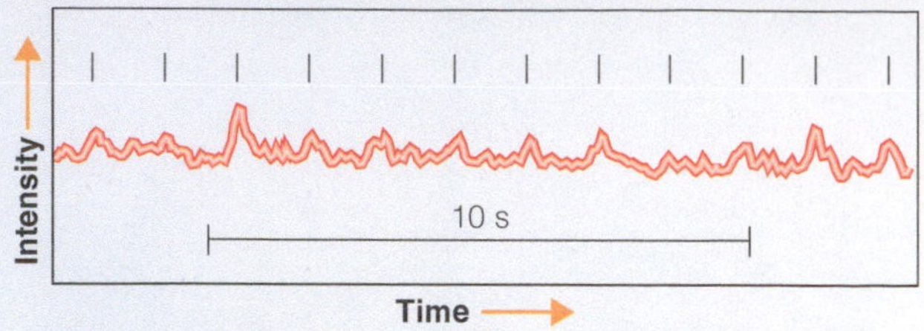

The 1967 detection of regularly spaced pulses in the output of a radio telescope led to the discovery of pulsars. This record of the radio signal from the first pulsar, CP1919, contains regularly spaced pulses (marked by ticks). The period is 1.33730208831 seconds.

star." The pulsing radio source Bell had observed with her radio telescope was the first known pulsar. Hewish received the Nobel Prize in physics for this work.

As more pulsars were found, astronomers argued over their nature. The pulses, which typically last only about 0.001 second, gave astronomers an important clue. The pulse length places an upper limit on the size of the object producing the pulse. This is a very important principle in astronomy—an object cannot change its brightness significantly in an interval shorter than the time light takes to cross its diameter. If pulses from pulsars are no longer than 0.001 second, then the objects cannot be larger than 0.001 light-second, or 300 km (200 miles) in diameter, smaller than white dwarfs, which makes neutron stars the only reasonable explanation.

The missing link between pulsars and neutron stars was found in late 1968, when astronomers discovered a pulsar at the heart of the Crab Nebula (see Figure 11-8). The Crab Nebula is a supernova remnant, which agrees nicely with Zwicky and Baade's prediction that some supernovae should produce a neutron star. The short pulses and the discovery of the pulsar in the Crab Nebula are strong evidence that pulsars are neutron stars.

The modern model of a pulsar has been called the **lighthouse model**, and is illustrated in Figure 11-14 and in **Concept Art 11B, The Lighthouse Model of Pulsars**. The name *pulsar* is, in a sense, inaccurate: A pulsar does not pulse (vibrate) but rather emits beams of radiation that sweep around the sky as the neutron star rotates, like a rotating lighthouse light. The mechanism that produces the beams involves extremely high energies and strong electric and magnetic fields and is not fully understood.

lighthouse model The explanation of a pulsar as a spinning neutron star sweeping beams of electromagnetic radiation around the sky.

More than 1000 pulsars are now known. There might be many more which are undetected because their beams never point toward Earth.

11-5c The Evolution of Pulsars

Neutron stars are not simple objects, and modern astronomers need knowledge about frontier physics to understand them. Nevertheless, the life story of pulsars can be worked out to some extent. When a pulsar first forms, it might be spinning as many as 100 times a second. The energy it radiates into space ultimately comes from its energy of rotation, so as it blasts beams of radiation outward, its rotation slows. Judging from their pulse periods and rates at which they slow down, the average pulsar is apparently only a few million years old and the oldest has an age of about 10 million years. Presumably, neutron stars older than that rotate too slowly to generate detectable radio beams. Supernova remnants last only about 50,000 years before they mix into the interstellar medium and disappear, so most pulsars have long outlived the remnants in which they were originally embedded.

You can expect that a young neutron star should emit especially strong beams of radiation powered by its rapid rotation. The Crab Nebula provides an example of such a system. Only about 960 years old and spinning

FIGURE 11-14

Neutron Star Rotation with Beams

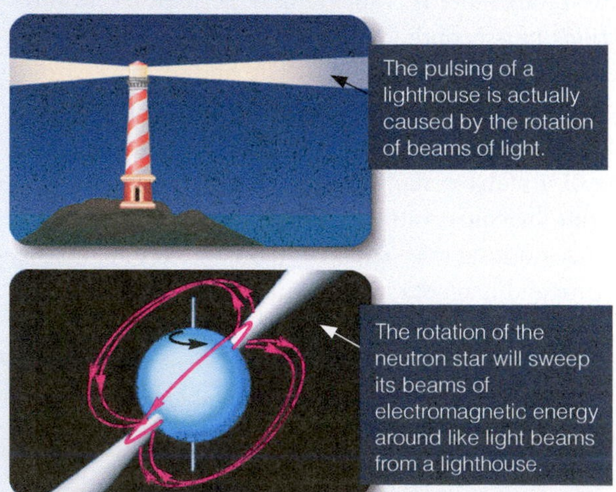

A neutron star contains a powerful magnetic field and spins very rapidly. The spinning magnetic field generates a tremendously powerful electric field, and the field causes the creation of electron–positron pairs that accelerate through the magnetic field, emit photons in the directions of their motion, and thereby produce powerful beams of electromagnetic radiation emerging from the neutron star's magnetic poles.

30 times per second, the Crab pulsar is so powerful that astronomers can detect photons of radio, infrared, visible, X-ray, and gamma-ray wavelengths from it.

The explosion of Supernova 1987A in February 1987 probably formed a neutron star. You can draw this conclusion because a burst of neutrinos was detected passing through Earth, and theory predicts that the collapse of a massive star's core into a ball of neutrons would produce such a burst of neutrinos. The neutron star initially should be hidden at the center of the expanding shells of gas ejected into space by the supernova explosion, but as the gas continues to expand and become thinner, you can expect that astronomers might eventually be able to detect it. As of this writing (October 2016), no neutron star has been detected in the SN 1987A remnant, but astronomers continue to watch the site hoping to find the youngest pulsar known.

11-5d Binary Pulsars

One reason pulsars are so fascinating is the extreme conditions found in spinning neutron stars. To see even more extreme natural processes, you have only to look at the pulsars that are members of binary systems. These pulsars are of special interest because astronomers can learn more about the neutron star and about the behavior of matter in unusual circumstances by studying them.

Binary pulsars can be sites of tremendous violence because of the strength of gravity at the surface of a neutron star. Matter falling onto a neutron star can release titanic amounts of energy. If you dropped an ordinary apple onto the surface of a neutron star from a distance of 1 AU, it would hit with an impact equivalent to a 1-megaton nuclear warhead. Even a small amount of matter flowing from a companion star to a neutron star can generate high temperatures and release X-rays and gamma-rays.

As an example of such an active system, examine Hercules X-1 in Figure 11-15. It emits pulses of X-rays with a period of about 1.2 seconds, but every 1.7 days the pulses vanish for a few hours. Hercules X-1 seems to contain a 2-solar-mass star and a neutron star that orbit each other with a period of 1.7 days. Matter flowing from the normal star into an accretion disk around the neutron star reaches temperatures of millions of degrees and emits a powerful X-ray glow, some of which is in beams that sweep around with the rotating neutron star. Earth receives a pulse of X-rays every time a beam points this way. The X-rays shut off completely every 1.7 days when the neutron star is eclipsed behind the normal star. Hercules X-1 is an intricate system with many different high-energy processes going on simultaneously, and this quick analysis only serves to illustrate how complex and powerful such binary systems are during mass transfer.

FIGURE 11-15

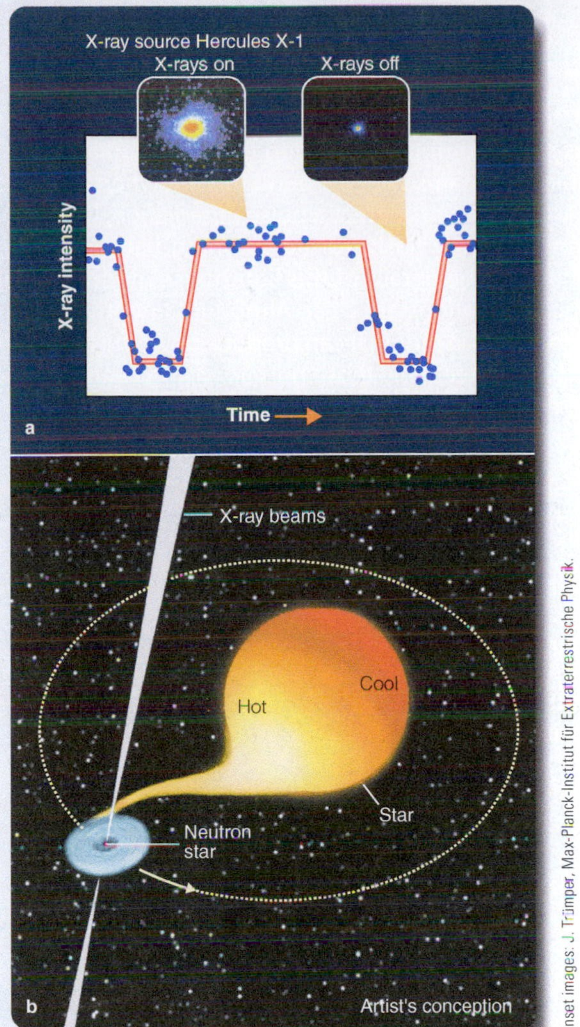

(a) Sometimes the X-ray pulses from Hercules X-1 are on, and sometimes they are off. A graph of X-ray intensity versus time looks like the light curve of an eclipsing binary. (b) In Hercules X-1, matter flows from a star into an accretion disk around a neutron star producing X-rays, which heat the near side of the star to 20,000 K compared with only 7000 K on the far side. X-rays turn off from Earth's point of view when the neutron star is eclipsed behind the star.

A binary system in which both objects are neutron stars was discovered in 1974 when radio astronomers Joseph Taylor and Russell Hulse noticed that the pulse period of the pulsar PSR B1913+16 grew longer and then shorter in a cycle that takes 7.75 hours. Taylor

Concept Art 11B

THE LIGHTHOUSE MODEL OF PULSARS

1

Astronomers understand pulsars not as pulsing objects but rather as spinning objects emitting beams of radiation. As the pulsars spin, the beams sweep around the sky; when a beam sweeps over Earth, observers detect a pulse of radiation, like a mariner seeing a momentary flash from a spinning lighthouse beam. Understanding the details of this lighthouse model is a challenge, but the implications are clear. Although a neutron star is only a few kilometers in radius, it can produce powerful beams. Also, observers can detect only those pulsars whose beams happen to sweep over Earth.

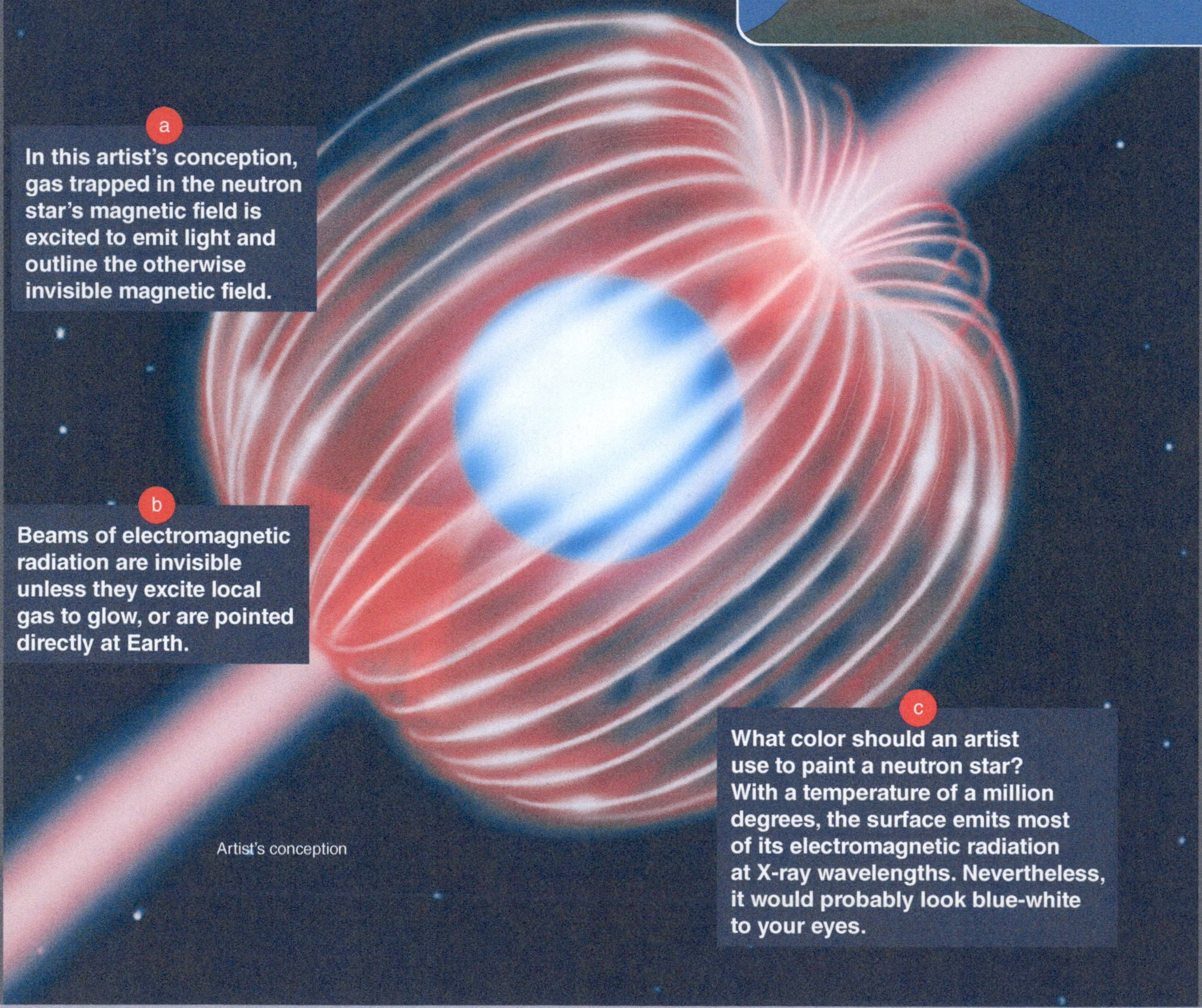

Artist's conception

2

Why a neutron star emits beams of electromagnetic radiation is one of the challenging problems of modern astronomy, but astronomers have a general idea. A neutron star contains a powerful magnetic field and spins very rapidly. The spinning magnetic field generates a tremendously powerful electric field, and the field causes the production of electron–positron (matter–antimatter) pairs. As these charged particles are accelerated through the magnetic field, they emit photons in the direction of their motion, which produce powerful beams of radiation emerging from the magnetic poles of the neutron star.

Neutron Star Rotation with Beams

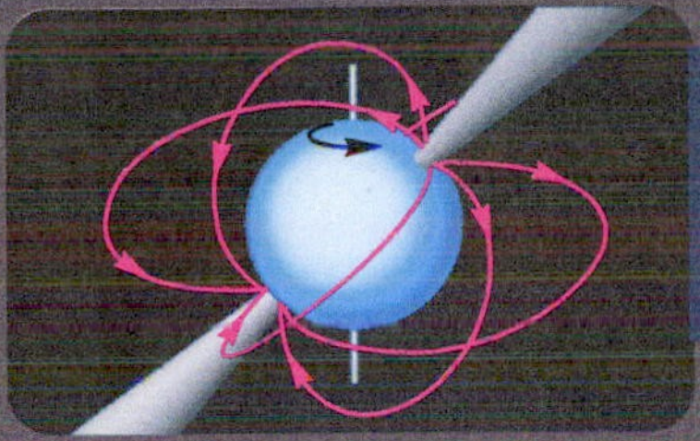

a As is true for Earth, the magnetic axis of a neutron star can be inclined to its rotational axis.

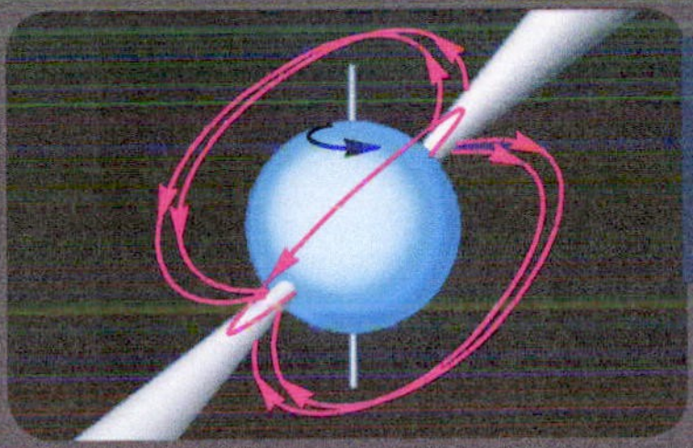

b The rotation of the neutron star sweeps its beams around like lighthouse beams.

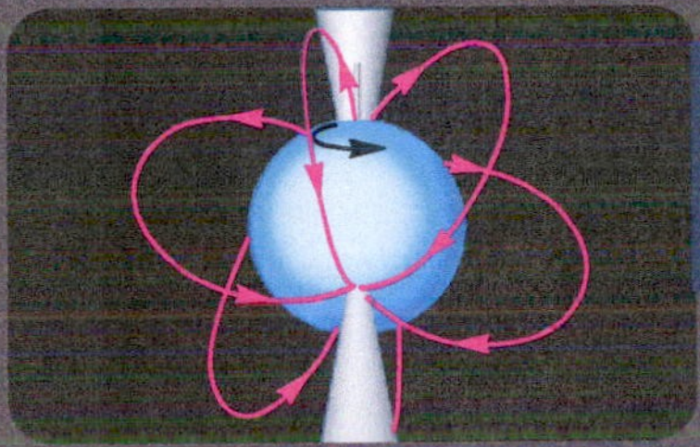

c While a beam points toward Earth, observers detect a pulse.

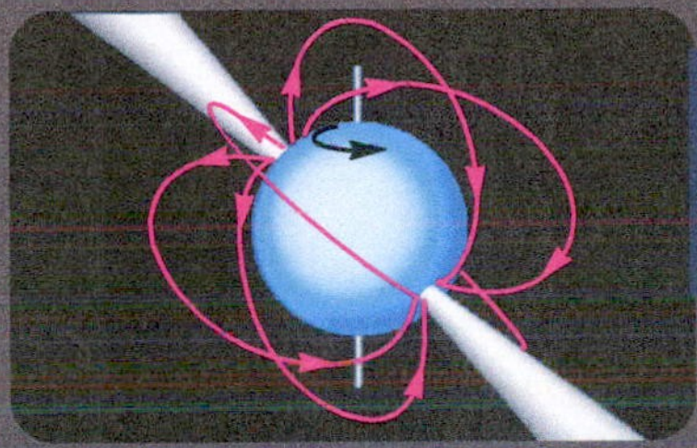

d While neither beam is pointed toward Earth, observers detect no energy.

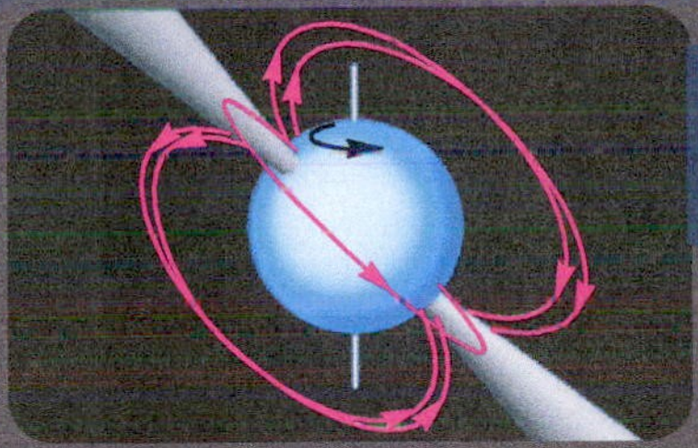

e Beams might not be as exactly symmetric as in this model.

3 X-ray observations of young pulsars show that they are surrounded by disks of excited matter and emit powerful jets of excited gas. The disks and jets are shaped by electromagnetic fields and the jets may curve if they encounter external magnetic fields.

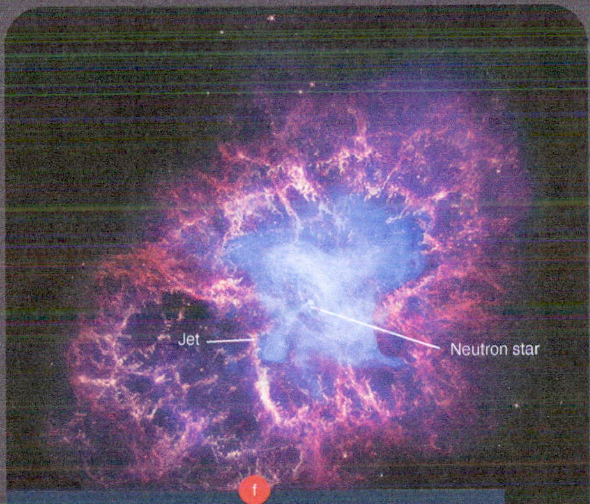

f This visual plus infrared plus X-ray image of the Crab Nebula reveals the neutron star plus a disk of hot gas over 1 ly across and a curving jet.

NASA/CXC/J.Hester (Arizona State Univ.); Visual: NASA/ESA/J. Hester and A. Loll (Arizona State Univ.); Infrared: NASA/JPL-Caltech/R.Gehrz (Univ. of Minnesota)

g Pulsar 3C 58 at right was produced by the supernova seen on Earth in the year 1181. It pulses (rotates) 15 times per second, is surrounded by a disk, and is ejecting jets in both directions.

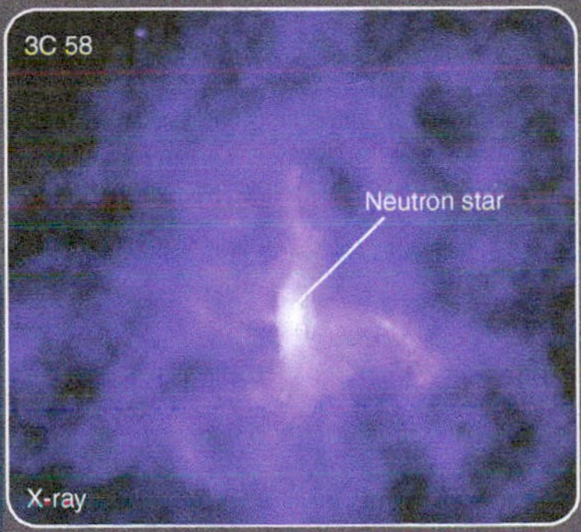

NASA/CXC/SAO/P. Slane et al.

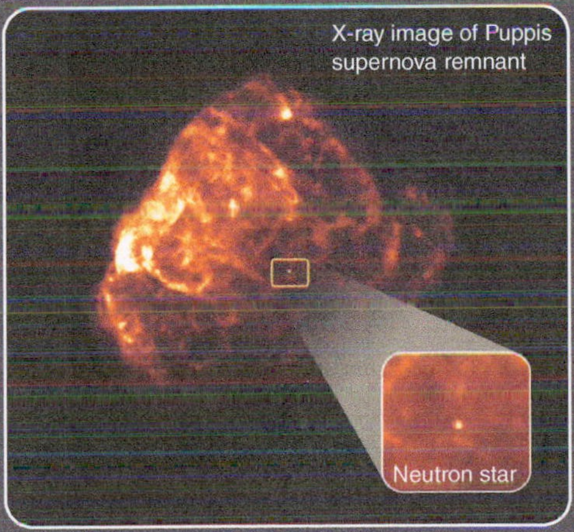

NASA/ROSAT Project/S. Snowden, R. Petre (GSFC), C. Becker (MIT)

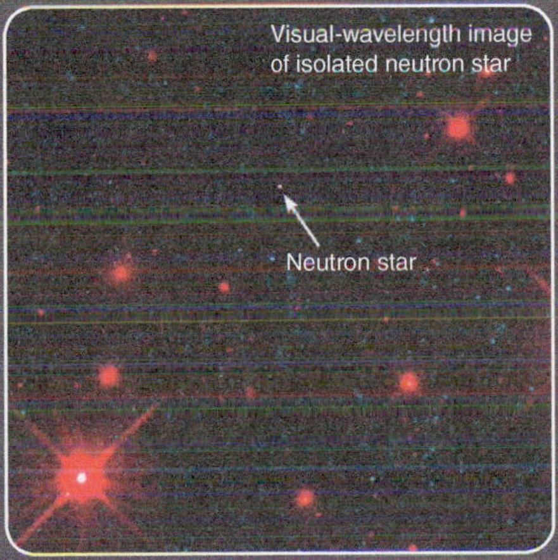

NASA/ESA/STScI/AURA/NSF/F. Walter (SUNY Stony Brook)

3a If a pulsar's beams do not sweep over Earth, observers detect no pulses, and the neutron star is difficult to find. A few such objects are known, however. The Puppis A supernova remnant in the left-hand image is about 4000 years old and contains a point source of X-rays thought to be a neutron star. The isolated neutron star RX J185635-3754 seen in the right-hand image from the *Hubble Space Telescope* has a surface temperature of 700,000 K.

How Do We Know? 11-1

Hypotheses, Theories, and Proofs

Why do astronomers say that a theory is confirmed but never say that it is proven? What scientists mean by the word *theory* is a hypothesis that has "graduated" to being confidently considered a well-tested truth. You can think of a hypothesis as equivalent to having a suspect in a criminal case, and a theory as equivalent to the trial being finished and someone being convicted of the crime.

Of course, no matter how many tests and experiments you conduct, you can never prove that any scientific theory is absolutely true. It is always possible that the next observation you make will disprove the theory. And it is unfortunately sometimes true that innocent people go to jail and guilty people are free, although sometimes, with further evidence, those legal mistakes can be fixed.

There have always been hypotheses about why the Sun is hot. Only a century ago, most astronomers accepted the hypothesis that the Sun was hot because gravity was making it contract. In the late 19th century, geologists showed that Earth was much older than the Sun could be if it was powered by gravity, so the gravity hypothesis had to be wrong. It wasn't until 1920 that Sir Arthur Eddington suggested the Sun is powered somehow by the energy in atomic nuclei. In 1938, the German-American astrophysicist Hans Bethe showed how nuclear fusion could power the Sun. He won the Nobel Prize in 1967.

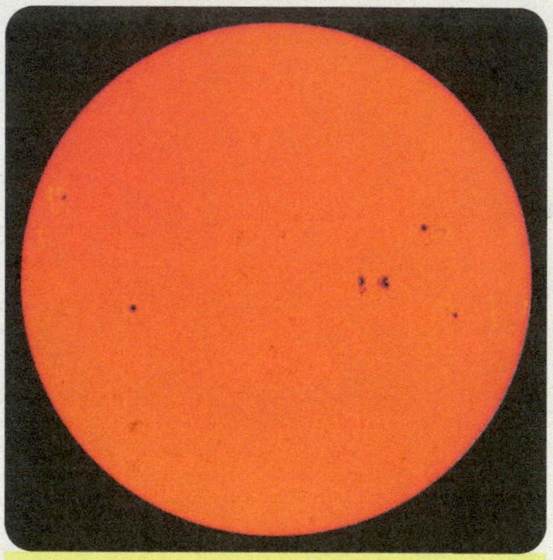

SOHO/MDI

Astronomers have such confidence that the Sun gets its power from nuclear fusion that this idea has "graduated" from being a hypothesis to being considered a theory.

The fusion hypothesis is now so completely confirmed that it is fair to call it a theory. No one will ever go to the center of the Sun, so you can't *prove* the fusion theory is right. Many observations and model calculations support this theory, and in Chapter 10 Section 10-2c, you learned about further evidence provided by the neutrinos that have been detected coming from the Sun's core. Nevertheless, there remains some tiny possibility that all the observations and models are misunderstood and that the theory will be overturned by some future discovery.

There is a great difference between a theory in the colloquial sense of a far-fetched guess and a scientific theory that has undergone decades of testing and confirmation with observations, experiments, and models. But no theory can ever be proven absolutely true. It is up to you as a consumer of knowledge and a responsible citizen to distinguish between a flimsy guess and a well-tested theory that deserves to be treated like truth—at least pending further information.

general theory of relativity Einstein's theory that describes gravity as a result of the curvature of space-time.

gravitational radiation Expanding waves in a gravitational field that transport energy through space at the speed of light, as predicted by general relativity.

and Hulse realized that must be the binary orbital period of the pulsar. They analyzed the system with the same techniques used to study spectroscopic binary stars (look back to Chapter 9 Section 9-5b) to find that PSR B1913+16 consists of two neutron stars separated by a distance roughly equal to the radius of our Sun. The masses of the two neutron stars are each about 1.4 solar masses, in good agreement with models of neutron stars and how they are created.

A nice surprise was hidden in the motion of PSR B1913+16. In 1916, Albert Einstein published his **general theory of relativity** that described gravity as a curvature of space-time. Einstein realized that any rapid change in a gravitational field should spread outward at the speed of light as **gravitational radiation**.

Taylor and Hulse were able to show that the orbital period of the binary pulsar is slowly growing shorter because the stars are radiating orbital energy away at exactly the rate expected for gravitational radiation and so are gradually spiraling toward each other. (Normal binary stars are too far apart and orbit too slowly to emit significant gravitational radiation.) Taylor and Hulse won the Nobel Prize in 1993 for their work confirming general relativity using binary pulsars. In 2016, it was announced that the Laser Interferometer Gravitational Wave Observatory, LIGO, had made the first direct detection of gravitational radiation.

11-5e The Fastest Pulsars

Your knowledge of pulsars suggests that newborn pulsars should blink rapidly, and old pulsars should blink slowly. In fact, the handful that blink the fastest may be quite old. The currently fastest known pulsar is cataloged as PSR J1748-2446ad. It pulses 716 times per second and is slowing down only slightly. The energy stored in the rotation of a neutron star at this rate is equal to the total energy of a supernova explosion, so it seemed difficult at first to explain this pulsar. It now appears that the speedy pulsar is an old neutron star that has gained mass and rotational energy from a companion in a binary system. Like water hitting a mill wheel, the matter falling on the neutron star has spun it up to that fantastic rate. With its weak magnetic field, it slows down only gradually and will continue to spin rapidly for a very long time.

"Show me," say scientists; and, in the case of neutron stars, the evidence seems very strong. Of course, you can never prove a scientific hypothesis or theory is absolutely true, but the evidence for neutron stars is so strong that astronomers have great confidence that they really do exist. (See **How Do We Know? 11-1**.) Other theories that describe how they emit beams of radiation and how they form and evolve are less certain, but continuing observations at many wavelengths are expanding the understanding of these last embers of massive stars. In fact, precise observations have turned up objects no one expected.

11-5f Pulsar Planets

Because a pulsar's period is so precise, astronomers can detect tiny variations by comparison with atomic clocks. When astronomers checked pulsar PSR B1257+12, they found variations in the period of pulsation (Figure 11-16a) analogous to the variations caused by the orbital motion of the binary pulsar but much smaller. When these variations were interpreted as Doppler shifts, it became evident that the pulsar was being orbited by at least two objects with planetlike masses of about 3 to 4 Earth masses. The gravitational tugs of the planets make the pulsar wobble about the center of mass of the system by about 800 km, and that produces the tiny changes that are observed (Figure 11-16b).

Astronomers greeted this discovery with both enthusiasm and skepticism. As usual, they looked for ways to

FIGURE 11-16

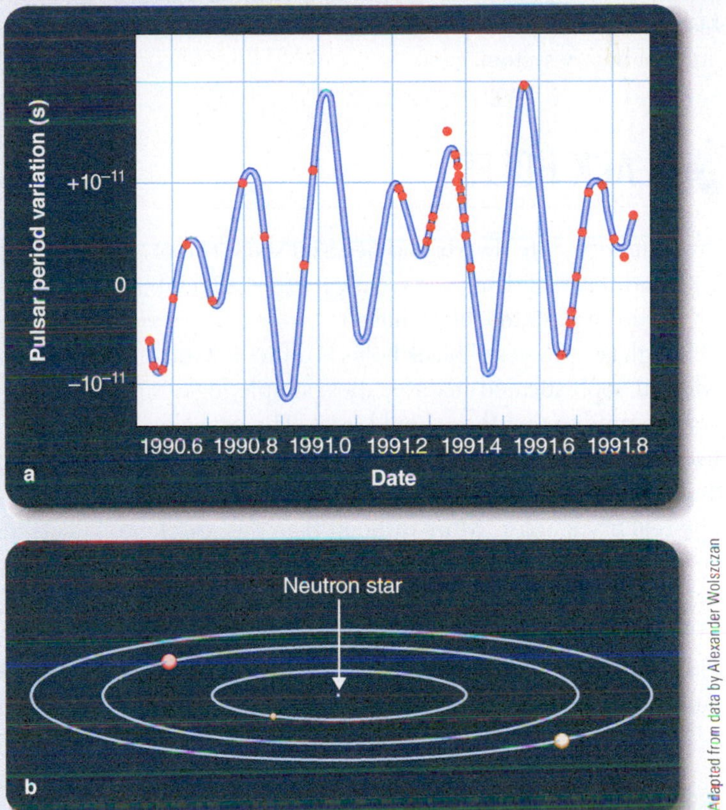

(a) The data points in this graph represent observations showing that the period of pulsar PSR B1257+12 varies from its average value by just ten-trillionths of a second. The blue line indicates the variation that would be produced by the best-fit model for the two largest of the three planets orbiting the pulsar. (b) As the planets orbit the pulsar, they cause it to wobble by less than 800 km (500 mi), a distance that is invisibly small in this diagram. The outermost planet orbits about 0.46 AU from the pulsar, farther than Mercury but closer than Venus is from the Sun.

test the hypothesis. Simple gravitational theory predicts that the planets should interact and slightly modify each other's orbit. When the data were analyzed, that interaction was found, further confirming the hypothesis that the variations in the period of the pulsar are caused by planets. In fact, more observations and analyses revealed the presence of a third planet of about twice the mass of Earth's Moon, and possibly a fourth planet of only 3 percent the mass of Earth's Moon. This illustrates the astonishing precision of studies based on pulsar timing.

Astronomers wonder how a neutron star can have planets. The inner three planets that orbit PSR B1257+12 are closer to the pulsar than Venus is to the Sun. Any planets that orbit a star would be lost or vaporized when the star exploded. Furthermore, a star about to explode as a supernova would be a large giant or a supergiant, and planets only a few AU distant would be inside such a large star and could not survive. It seems more likely that these planets are the remains of a stellar companion that was devoured by the neutron star. In fact, PSR B1257+12 is very fast (161 pulses per second), suggesting that it was spun up in a binary system.

11-6 BLACK HOLES

You have studied white dwarfs and neutron stars, two of the three end states of dying stars. Now it's time to think about the third end state—black holes.

Although the physics of black holes is difficult to discuss without sophisticated mathematics, simple logic is sufficient to predict that they should exist. The problem is to use their predicted properties and attempt to confirm that they exist. What objects observed in the heavens could be real black holes? More difficult than the search for neutron stars, the quest for black holes has nevertheless met with success.

To begin, you can consider a simple question. How fast must an object travel to escape from the surface of a celestial body? The answer will lead to black holes.

singularity An object of zero radius and infinite density.

black hole A mass that has collapsed to such a small volume that its gravity prevents the escape of all radiation. Also, the volume of space from which radiation cannot escape.

11-6a Escape Velocity

In Chapter 3 Section 3-5b you learned that the escape velocity is the initial velocity an object needs to escape from a celestial body. Whether you are discussing a baseball leaving Earth or a photon leaving a collapsing star, the escape velocity depends on two things: the mass of the celestial body and the distance from the center of mass to the escapee's starting point. If the celestial body has a large mass, its gravity is strong, and you need a high velocity to escape; but if you begin your journey farther from the center of mass, the velocity needed is less. For example, to escape from Earth, a spaceship has to leave Earth's surface at 11.2 km/s (25,100 mph), but if you could launch spaceships from the top of a tower 1000 mi high, the escape velocity would be only 10.0 km/s (22,400 mph).

An object massive enough and/or small enough could have an escape velocity greater than the speed of light. Relativity says that nothing can travel faster than the speed of light, so even photons would be unable to escape. Such a small, massive object could never be seen because light could not leave it. This was first noted by British astronomer Reverend John Mitchell in 1783, long before Einstein and relativity.

11-6b Schwarzschild Black Holes

If the core of a star collapses and contains more than about 3 solar masses, no force can stop the collapse. When the object reaches the size of a white dwarf, the collapse continues because degenerate electrons cannot support that much weight. It also cannot stop when it reaches the even smaller size of a neutron star because degenerate neutrons also cannot support that weight. No force remains to stop the object from collapsing to zero radius.

As an object collapses, its density and the strength of its surface gravity increase; and if an object collapses to zero radius, its density and gravity become infinite. Mathematicians call such a point a **singularity**, but in physical terms it is difficult to imagine an object of zero radius. Some theorists think that a singularity is impossible and that the laws of quantum physics must somehow halt the collapse at a subatomic radius roughly 10^{20} times smaller than a proton. Astronomically, it seems to make little difference.

If the contracting core of a star becomes small enough, the escape velocity in a region around it is so large that no light can escape. You can receive no information about the object or about the region of space near it. Such a region is called a **black hole**. Note that the term *black hole* refers to a volume of space, not just the singularity at the region's center. If the core of an exploding star collapsed to create a black hole, the expanding outer layers of the star could produce a supernova remnant, but the core would vanish without a trace.

Albert Einstein's general theory of relativity treats space and time as a single entity—space-time. His equations showed that gravity could be described as a curvature of space-time, and almost immediately astronomer Karl Schwarzschild found a way to solve the equations to describe the gravitational field around a single, nonrotating, electrically neutral lump of matter. That solution contained the first general relativistic description of a black hole, and nonrotating, electrically neutral black holes are now known as Schwarzschild black holes. Theorists such as Roy Kerr and Stephen Hawking have found ways to apply the sophisticated mathematical equations of the general theory of relativity and quantum mechanics to black holes that are rotating and have electrical charges. For this discussion the differences are minor, and you can proceed as if all black holes were Schwarzschild black holes.

FIGURE 11-17

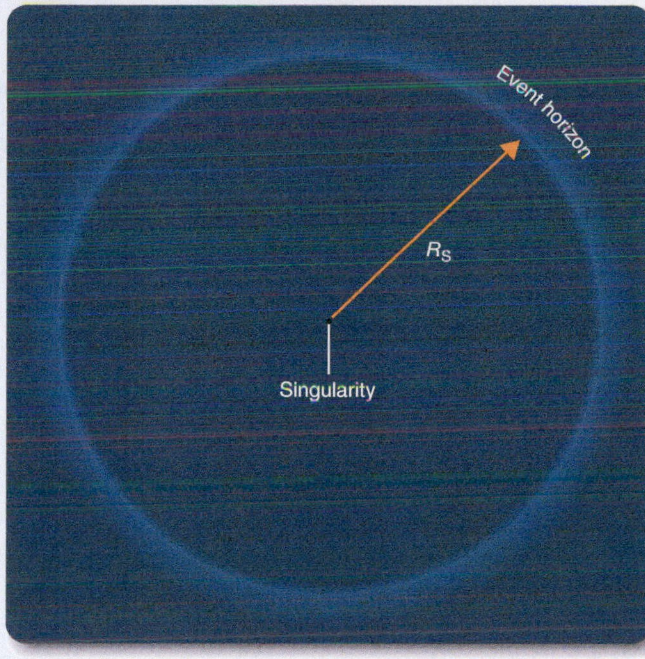

A black hole forms when an object collapses to a small size (perhaps to a singularity) and the escape velocity becomes so great light cannot escape. The boundary of the black hole is called the event horizon because any event that occurs inside is invisible to outside observers. The radius of the black hole R_S is the Schwarzschild radius.

Schwarzschild's solution shows that if matter is packed into a small enough volume, then space-time curves back on themselves. Objects can still follow paths that lead into the black hole, but no path leads out, so nothing can escape, not even light. Consequently, the inside of the black hole is totally beyond the view of an outside observer. The **event horizon** is the boundary between the isolated volume of space-time and the rest of the Universe, and the radius of the event horizon is called the **Schwarzschild radius, R_S**—the radius within which an object must shrink to become a black hole, as in Figure 11-17, and the point of no return for any object falling in later.

Although Schwarzschild's work was highly mathematical, his conclusions were quite simple. The size of a black hole, its Schwarzschild radius, is simply proportional to its mass. A 3-solar-mass black hole will have a Schwarzschild radius of about 9 km, a 10-solar-mass black hole will have a Schwarzschild radius of 30 km, and so on. Note that even a very massive black hole would not be very large—just a few miles across.

It is a common misconception to think of black holes as giant vacuum cleaners that will suck up everything in the Universe. A black hole is just a gravitational field, and at a large distance its gravity is no greater than that of a normal object of similar mass. If the Sun were replaced by a 1-solar-mass black hole, the orbits of the planets would not change at all. The gravity of a black hole becomes extreme only when you approach close to it. Figure 11-18 illustrates this by representing gravitational fields as curvature of the fabric of space-time. Physicists like to graph the strength of gravity around a black hole as curvature in a flat sheet. The graphs look like funnels in which the depth of the funnel indicates the strength of the gravitational field, but black holes themselves are not shaped like funnels; they are spheres or spheroids. In Figure 11-18, you should note that the strength of the gravitational field around the black hole becomes extreme only if you venture too close.

This chapter discusses black holes that might originate from the deaths of massive stars. In later

event horizon The boundary of the region of a black hole from which no radiation can escape. No event that occurs within the event horizon is visible to a distant observer.

Schwarzschild radius, R_S The radius of the event horizon around a black hole.

FIGURE 11-18

If you fell into the gravitational field of a star, you would hit the star's surface before you fell very far. Because a black hole is so small, you could fall much deeper into its gravitational field and eventually cross the event horizon. At a distance, the two gravitational fields are the same.

chapters, you will encounter black holes whose masses might exceed a million solar masses, located in the centers of galaxies.

11-6c Leaping into a Black Hole

Before you can search for real black holes, you must understand what theory predicts about the behavior of a black hole. To explore that idea, you can imagine that you leap, feet-first, into a Schwarzschild black hole.

If you were to leap into a black hole of a few solar masses from a distance of an astronomical unit, the gravitational pull would not be very large, and you would fall slowly at first. Of course, the longer you fell and the closer you came to the center, the faster you would travel. Your wristwatch would tell you that you fell for about 2 months before you reached the event horizon.

Your friends who stayed behind would see something different. They would see you falling more slowly as you came closer to the event horizon because, as described by general relativity, time slows down in curved space-time. This is known as **time dilation**. In fact, your friends would never actually see you cross the event horizon. To them you would fall more and more slowly until you seemed hardly to move. Generations later, your descendants could focus their telescopes on you and see you still inching closer to the event horizon. You, however, would have sensed no slowdown and would conclude that you had crossed the event horizon after about 2 months.

time dilation The slowing of moving clocks or clocks in strong gravitational fields.

gravitational redshift The lengthening of the wavelength of a photon as it escapes from a gravitational field.

Another relativistic effect would make it difficult to see you with normal telescopes. As light travels out of a gravitational field, it loses energy, and its wavelength grows longer. This is known as the **gravitational redshift**. Although you would notice no effect as you fell toward the black hole, your friends would need to observe at longer and longer wavelengths in order to detect you.

While these relativistic effects seem merely peculiar, other effects would be quite unpleasant. Imagine again that you are falling feet-first toward the event horizon of a black hole. You would feel your feet, which would be closer to the black hole, being pulled in more strongly than your head. This is a tidal force, and at first it would be minor. As you fell closer, however, the tidal force would become very large. Another tidal force would compress you as your left side and your right side both fell toward the center of the black hole. For any black hole with a mass like that of a star, the tidal forces would crush you laterally and stretch you longitudinally long before you reached the event horizon (Figure 11-19). Needless to say, this would render you inoperative as a thoughtful observer.

Science fiction books, movies, and TV shows sometimes suggest that you can travel through the Universe by jumping into a black hole in one place and popping out of another somewhere far across space. That might make for good science fiction, but tidal forces would make it an unpopular form of transportation even if it worked. You would certainly lose your luggage.

Your imaginary leap into a black hole is not frivolous. You now know how to find a black hole: Look for a strong source of X-rays that might be from matter compressed and stressed just before disappearing as it approaches the event horizon.

FIGURE 11-19

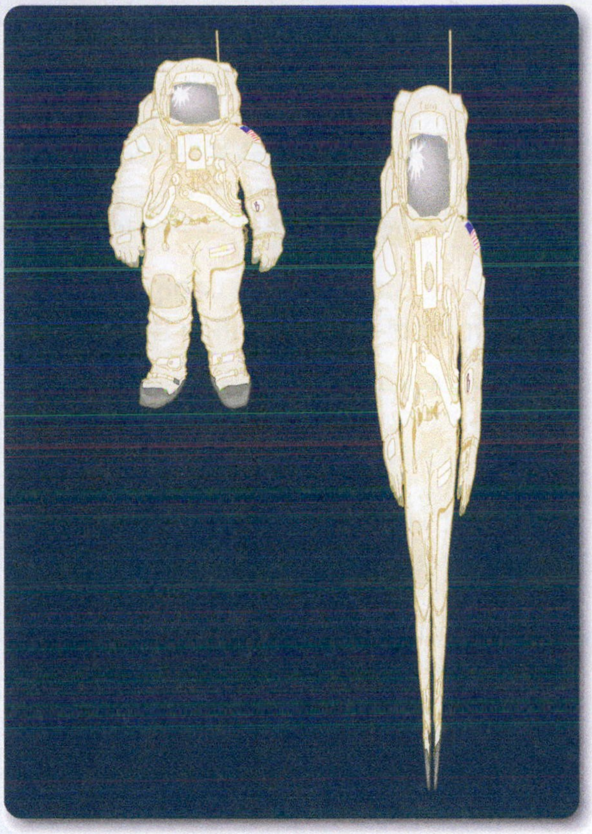

Leaping feet-first into a black hole. A person of normal proportions (left) would be distorted by tidal forces (right) long before reaching the event horizon around a typical black hole of stellar mass. Tidal forces would stretch the body lengthwise while compressing it laterally.

11-6d The Search for Black Holes

An isolated black hole is totally invisible because nothing can escape from the event horizon. But a black hole into which matter is flowing would be a source of X-rays. Of course, X-rays can't escape from *inside* the event horizon, but X-rays emitted by the heated matter flowing into the black hole can escape if the X-rays are emitted *before* the matter crosses the event horizon. An isolated black hole in space will not have much matter flowing into it, but a black hole in a binary system might receive a steady flow of matter transferred from its companion star. This suggests you can search for black holes by searching among X-ray binaries.

You have learned about X-ray binaries such as Hercules X-1 that contain a neutron star, and they emit X-rays much as a binary containing a black hole should. You can tell the difference in two ways. If the compact object emits pulses, you know it is a neutron star. Otherwise, you might check the mass of the object. If the mass of the compact object is greater than about 3 solar masses, the object cannot be a neutron star, and you can conclude that it must be a black hole.

The first X-ray binary suspected of harboring a black hole was Cygnus X-1. It contains a supergiant star of spectral type B0 and a compact object orbiting each other with a period of 5.6 days. Matter flows from the B0 star as a strong stellar wind, and some of that matter enters a hot accretion disk around the compact object (Figure 11-20). The accretion disk is about five times larger in diameter than the orbit of Earth's Moon, and the inner few hundred kilometers of the disk have a temperature of about 2 million Kelvin—hot enough to radiate X-rays. The compact object is invisible, but Doppler shifts in the spectrum reveal the motion of the B0 star around the center of mass of the binary. From the geometry of the orbit, astronomers were able to calculate the mass of the compact object—at least 3.8 solar masses, well above the maximum for a neutron star.

As X-ray telescopes have found more X-ray objects, the list of black hole candidates has grown to a few dozen. Some of these objects are shown in Table 11-2. Each candidate is a compact object surrounded by a hot accretion disk in a close X-ray binary system without regular pulsations. A few of the binary systems are easier to analyze than others, but, in the end, it has become clear that some of these objects are too massive to be neutron stars. Scientists cannot have absolute proof, but the evidence is now overwhelming: Black holes really do exist. (The first LIGO gravity wave detection was of a signal with characteristics signifying the merger of two black holes with 36 and 29 solar masses located 1.3 billion light-years from Earth.)

Another way to confirm that black holes are real is to search for evidence of their distinguishing characteristic—event horizons—and that search also has been successful. In one study, astronomers selected 12 X-ray binary systems, six of which seemed to contain neutron stars and six of which were expected to contain black holes. Using X-ray telescopes, the astronomers could see flares of energy as blobs of matter fell into the accretion disks and spiraled inward. In the six systems thought to contain neutron stars, the astronomers could also detect bursts of energy when the blobs of matter finally fell onto the surfaces of the neutron stars. In the six systems thought to contain black holes, however, the blobs of matter spiraled inward through the accretion disks and vanished without final

TABLE 11-2 SIX BLACK HOLE BINARIES			
Object	Star	Orbital Period	Mass of Black Hole
Cygnus X-1	B0 I	5.6 days	10 $M_{\odot}$
LMC X-3	B3 V	1.7 days	> 8 $M_{\odot}$
A0620-00	K6 V	7.75 hours	11 ± 1.9 $M_{\odot}$
V404 Cygni	K3 III	6.47 days	12 ± 3 $M_{\odot}$
GRO J1655-40	F5 IV	2.61 days	6.9 ± 1 $M_{\odot}$
QZ Vulpecula	K5 V	8.3 hours	10 ± 4 $M_{\odot}$

bursts of energy. Evidently, those blobs of matter had approached the event horizons and become undetectable because of time dilation and gravitational redshift. This is dramatic evidence that event horizons are real.

11-6e Energy from Compact Objects—Jets

As you can see, it is a common misconception to think that it is impossible to get any energy out of a black hole. In a later chapter, you will meet galaxies radiating vast amounts of energy from massive black holes, so you should pause here to see how a compact object and an accretion disk can eject powerful jets and beams of energy.

Whether a compact object is a black hole or a neutron star, it has a strong gravitational field. Any matter flowing into that field is accelerated inward; and, because it must conserve angular momentum, it flows into an accretion disk made so hot by friction that the inner regions can emit X-rays and gamma-rays. Somehow the spinning accretion disk can emit powerful beams of gas and radiation along its axis of rotation. The process isn't well understood, but it seems to involve magnetic fields that get caught in the accretion disk and are twisted into tightly wound tubes that squirt gas and radiation out of the disk and confine it in narrow beams.

This process is similar to the bipolar outflows ejected by protostars (look back to Chapter 10 Section 10-4d), but it is much more powerful. One example of this process is an X-ray binary called SS 433. Its optical spectrum shows sets of spectral lines that are Doppler-shifted by about one-fourth the speed of light, with one set shifted to the red and one set shifted to the blue. Apparently, SS 433 is a binary system in which a compact object (probably a black hole) pulls matter from its companion star and forms an extremely hot accretion disk. Jets of high-temperature gas blast away in beams aimed in opposite directions. SS 433 is a prototype that illustrates how the gravitational field around a compact object can produce powerful beams of radiation and matter.

gamma-ray burst A sudden, powerful burst of gamma-rays.

hypernova Produced when a very massive star collapses into a black hole. Thought to be a possible source of gamma-ray bursts.

11-6f Energy from Compact Objects—Gamma-Ray Bursts

During the 1960s, the United States put a series of satellites in orbit to watch for bursts of gamma-rays coming from Earth indicating nuclear weapons tests that would be violations of an international treaty. The experts were startled when the satellites detected about one **gamma-ray burst** coming from space per day. The *Compton Gamma Ray Observatory* launched in 1991 discovered that gamma-ray bursts were occurring all over the sky and not from any particular region. Starting in 1997, new satellites in orbit were able to detect gamma-ray bursts, determine their location in the sky, and immediately alert astronomers on the ground. When telescopes swiveled to image the locations of the bursts, they detected fading glows that resembled supernovae, and that has led to the conclusion that some relatively long gamma-ray bursts are produced by a kind of supernova explosion called a **hypernova** (Figure 11-21).

FIGURE 11-20

The X-ray source Cygnus X-1 consists of a supergiant B0 star and a compact object orbiting each other. Gas from the B0 star's stellar wind flows into the hot accretion disk around the compact object, and the X-rays astronomers detect come from the disk.

Theoretical calculations indicate that a star more massive than some threshold around 15 or 20 solar masses will collapse and become a black hole when its nuclear fuel is exhausted. Models show that the collapsing star would conserve angular momentum and spin very rapidly, and that would slow the collapse of the equatorial parts of the star. The poles of the star would fall in quickly, and that would focus beams of intense radiation and ejected gas blasting out along the axis of rotation, resulting in a hypernova. If either of those beams happens to point in the right direction, Earth would receive a powerful gamma-ray burst. The evidence seems clear that at least some of the gamma-ray bursts are produced by hypernovae. Massive stars explode as hypernovae only rarely in any one galaxy, but the gamma-ray bursts they produce are so powerful that astronomers can detect these explosions among a vast number of galaxies. Other gamma-ray bursts might be produced by the merger of two neutron stars or a neutron star and a black hole, and yet others by sudden shifts in the crusts of highly magnetized neutron stars.

Incidentally, if a gamma-ray burst occurred only 1000 ly from Earth, the gamma-rays would shower Earth with radiation equivalent to a 10,000-megaton nuclear blast. (The largest bombs ever made were less than 100 megatons in size.) The gamma-rays could create enough nitric oxide in the atmosphere to produce intense acid rain and would destroy the ozone layer, exposing life on Earth to deadly levels of solar ultraviolet radiation. Gamma-ray bursts can occur relatively near the Earth as often as every few 100 million years and could be one of the causes of the mass extinctions that show up in the fossil record.

FIGURE 11-21

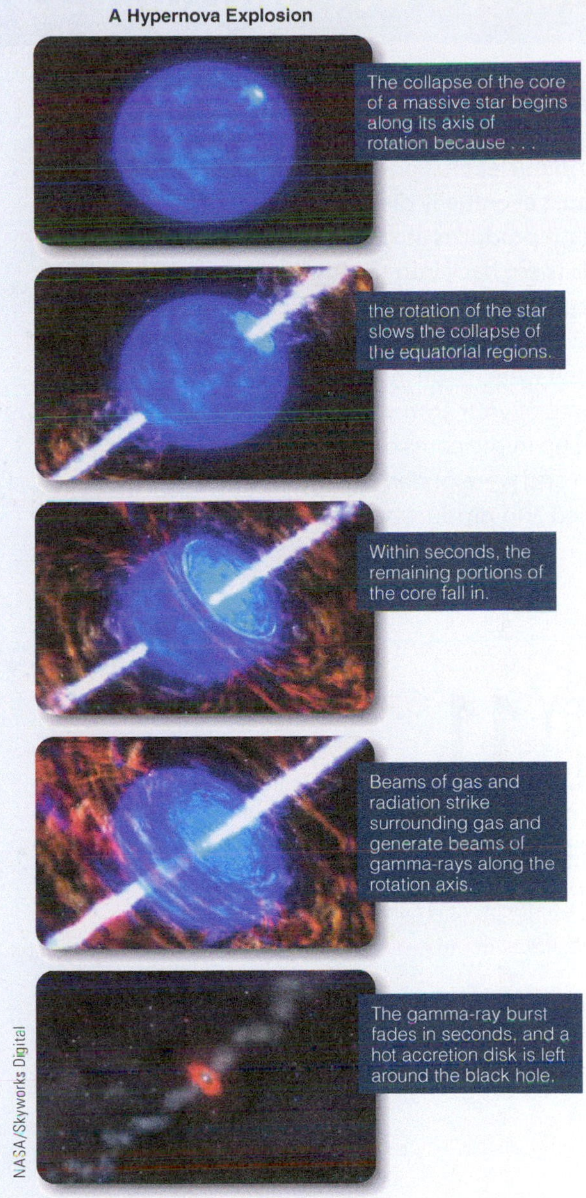

When an extremely massive star collapses in a hypernova explosion, energy from the core is focused into beams of radiation and matter that point along the axis of rotation. This is thought to be the source of GRBs that last longer than 2 seconds.

What Are We? Stardust

You are made of atoms that were cooked up inside stars. Gravity draws matter together to make stars, and although nuclear fusion delays gravity's final victory, stars must eventually die. That process of star life and star death produces atoms heavier than helium and spreads them back into the interstellar medium, where they can become part of the gas clouds that form new stars. All of the atoms in your body except for the hydrogen were made inside stars.

Some of your atoms, such as the carbon, were cooked up in the cores of medium-mass stars like the Sun and were puffed out into space when those stars died and produced planetary nebulae. Some of your atoms, such as the calcium in your bones, were made inside massive stars and were blown out into space during type II supernova explosions. Many of the iron atoms in your blood were made by the sudden fusion of carbon atoms when white dwarfs collapsed in Type Ia supernova explosions. Other heavy atoms, such as iodine in your thyroid gland, selenium in your nerve cells, and gold in your class ring, were also produced in the raging violence of supernova explosions.

You are made of atoms scattered into space long ago by the violent deaths of stars. What are we? We are stardust.

STUDY TOOLS 11

IN THE BOOK

- ☐ Tear Out the Review Card on *The Deaths of Stars*.

ASTRO ONLINE

Visit ASTRO Online at www.cengagebrain.com

- ☐ Interactive Reading
- ☐ Practice Quizzing
- ☐ Videos and Animations
- ☐ Exclusive Features

11 Questions and Problems

REVIEW QUESTIONS

1. Why does helium fusion require a higher temperature than hydrogen fusion?
2. How can the contraction of an inert helium core trigger the ignition of a hydrogen-fusion shell?
3. Why does the expansion of a star's envelope make it cooler and more luminous?
4. Why is degenerate matter so difficult to compress?
5. How can star clusters confirm astronomers' theories of stellar evolution?
6. Why don't red dwarfs become giant stars?
7. What causes an aging giant star to produce a planetary nebula?
8. Why can't a white dwarf have a mass greater than 1.4 solar masses?
9. How can a star of as much as 8 solar masses form a white dwarf when it dies?
10. How can the inward collapse of the core of a massive star produce an outward explosion?
11. What is the difference between Type I and Type II supernovae?
12. What is the difference between a supernova explosion and a nova explosion?
13. How are neutron stars and white dwarfs similar? How do they differ?
14. Why is there an upper limit to the mass of neutron stars?
15. Why do you expect neutron stars to spin rapidly? Why do you expect neutron stars to have a powerful magnetic field?
16. If neutron stars are hot, why aren't they very luminous?
17. What does the short length of pulsar pulses tell you?
18. How does the lighthouse model explain pulsars?
19. What evidence can you cite that pulsars are neutron stars?
20. How can a neutron star in a binary system generate X-rays? How can a black hole emit X-rays?
21. What evidence can you cite that black holes really exist?
22. Discuss the possible causes of gamma-ray bursts.
23. **How Do We Know?** Why do scientists say that a hypothesis or theory is confirmed, but do not say it is proven?

DISCUSSION QUESTIONS

1. You are reading a novel about an advanced alien civilization on a planet that orbits a neutron star. What can you say about the likelihood of such a situation? Is this closer to science-fantasy or science-fact?
2. An eccentric philanthropic billionaire is willing to fund your team in a search for undiscovered black holes. Where do you look, and what kind of instrument do you acquire to perform the search?

PROBLEMS

1. About how long will a 0.4-solar-mass star spend on the main sequence?
2. If the stars at the turnoff point in a star cluster have masses of about 4 solar masses, how old is the cluster?
3. Suppose a planetary nebula is 1 pc in radius. If the Doppler shifts in its spectrum show it is expanding at 30 km/s, how old is it? (*Note:* 1 pc equals 3.1×10^{13} km, and 1 year equals 3.2×10^7 seconds, to 2 significant figures.)
4. If a star the size of the Sun expands to form a giant 20 times larger in radius, by what factor will its average density decrease? (*Note:* The volume of a sphere is $\frac{4}{3}\pi r^3$.)
5. If a star the size of the Sun collapses to form a white dwarf the size of Earth, by what factor will its density increase? (*Note:* The volume of a sphere is $\frac{4}{3}\pi r^3$. See **Celestial Profiles 3** and **9** in Appendix C for the radii of the Sun and Earth.)
6. The Crab Nebula is now 1.35 pc in radius and is expanding at 1400 km/s. About when did the supernova occur? (*Note:* 1 pc equals 3.1×10^{13} km.)
7. If a neutron star has a radius of 10 km and rotates 716 times per second, what is the speed of the surface at the neutron star's equator as a fraction of the speed of light? (*Note:* The speed of light is 3.0×10^5 km/s.)
8. Suppose that a neutron star has a radius of 10 km and a temperature of 1,000,000 K. How luminous is it? (*Hints:* See **Luminosity, Radius, and Temperature** in the Math Reference Cards MR. Make sure you convert the neutron star's radius to meters.)

9. A neutron star and a white dwarf have been found orbiting each other with a period of 11 minutes. If their masses are typical for their respective types of objects, what is their average separation? Compare that separation with the radius of the Sun, 7.0×10^5 km. (*Hints:* See **The Masses of Binary Stars** in the Math Reference Cards MR. Make sure to convert relevant quantities to units of AU, solar masses, and years.) (*Note:* 1 AU is 1.5×10^8 km, and 1 year is 5.3×10^5 minutes.)
10. If the accretion disk around a neutron star with a typical mass has a radius of 2×10^5 km, what is the orbital velocity of a particle at its outer edge? (*Hints:* See **Circular Velocity** in the Math Reference Cards MR. Make sure to convert relevant quantities to units of meters, kilograms, and seconds.)
11. What is the escape velocity at the surface of a typical neutron star? (*Hint:* See **Circular Velocity** in the Math Reference Cards MR.) (*Note:* Escape velocity equals circular orbital velocity times the square root of 2, which is 1.41 to 3 significant figures.)
12. If the inner accretion disk around a black hole has a temperature of 1,000,000 K, at what wavelength will it radiate the most energy? What part of the spectrum is this in? (*Hint:* See **Blackbody Radiation** in the Math Reference Cards MR.)

LEARNING TO LOOK

1. The star cluster in Figure 11-22 contains many hot, blue, luminous stars. Sketch its H–R diagram and discuss its probable age.
2. What processes caused a medium-mass star to produce the nebula in Figure 11-23? The nebula is now about 0.1 ly in diameter and still expanding. What will happen to it?
3. Figure 11-24 combines X-ray (blue), visible (green), and radio (red) images. Observations show that the sphere is expanding at a high speed and is filled with very hot gas. What kind of object produced this nebula? Roughly how old do you think it must be?
4. What is happening in the artist's conception in Figure 11-25? How would you distinguish between a neutron star and a black hole in such a system?

FIGURE 11-22

NASA//ESA/STScI/AURA/NSF/Hubble Heritage Team/The WFC3 Science Oversight Committee/R. O'Connell (Univ. of Virginia), F. Paresce (NIA), E. Young (USRA/SOFIA)

FIGURE 11-23

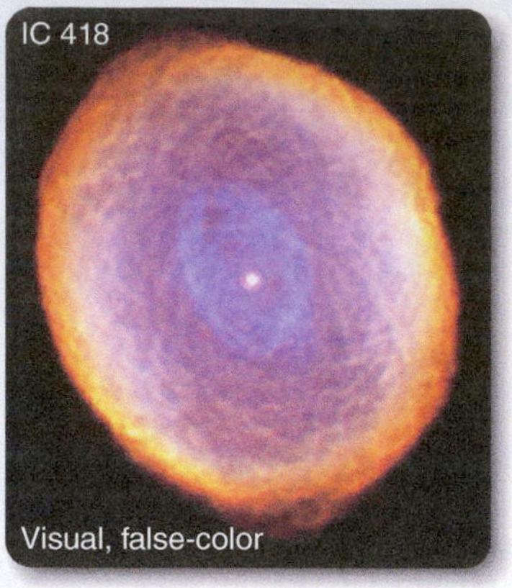

NASA/ESA/STScI/AURA/NSF/The Hubble Heritage Team/ R. Sahai (JPL-Caltech) and A. Hajan (NOAO)

FIGURE 11-24

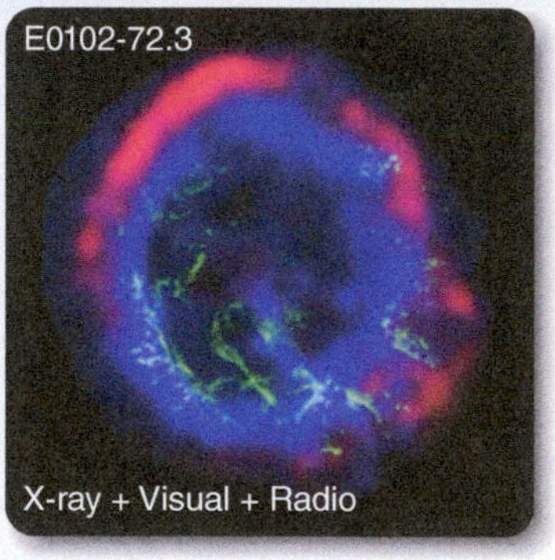

NASA/CXC/SAO/CSIRO/ATNF/ATCA

FIGURE 11-25

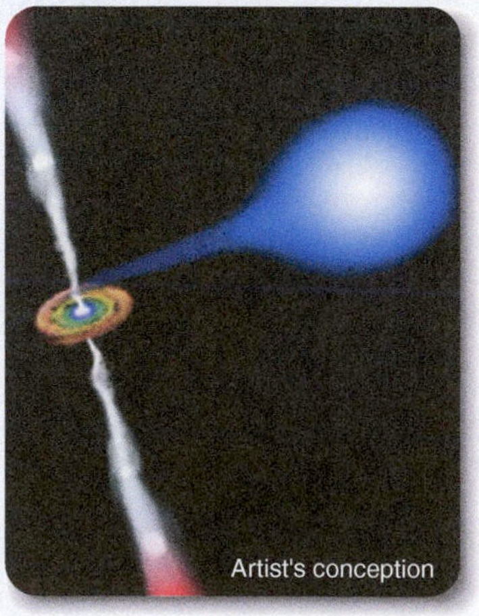

CXC/M. Weiss

ASTRO
ONLINE

STUDY YOUR WAY WITH STUDYBITS!

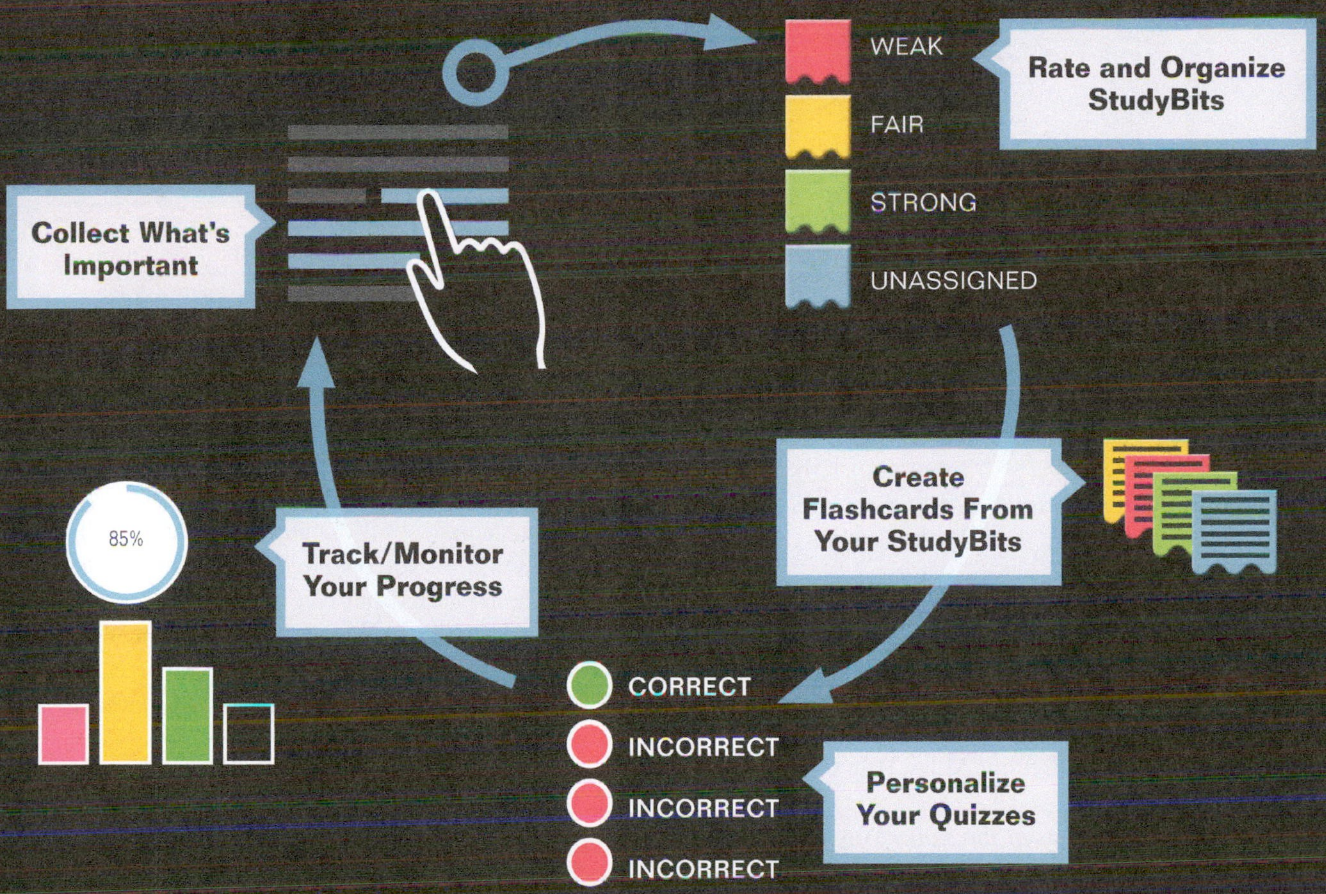

4LTR PRESS

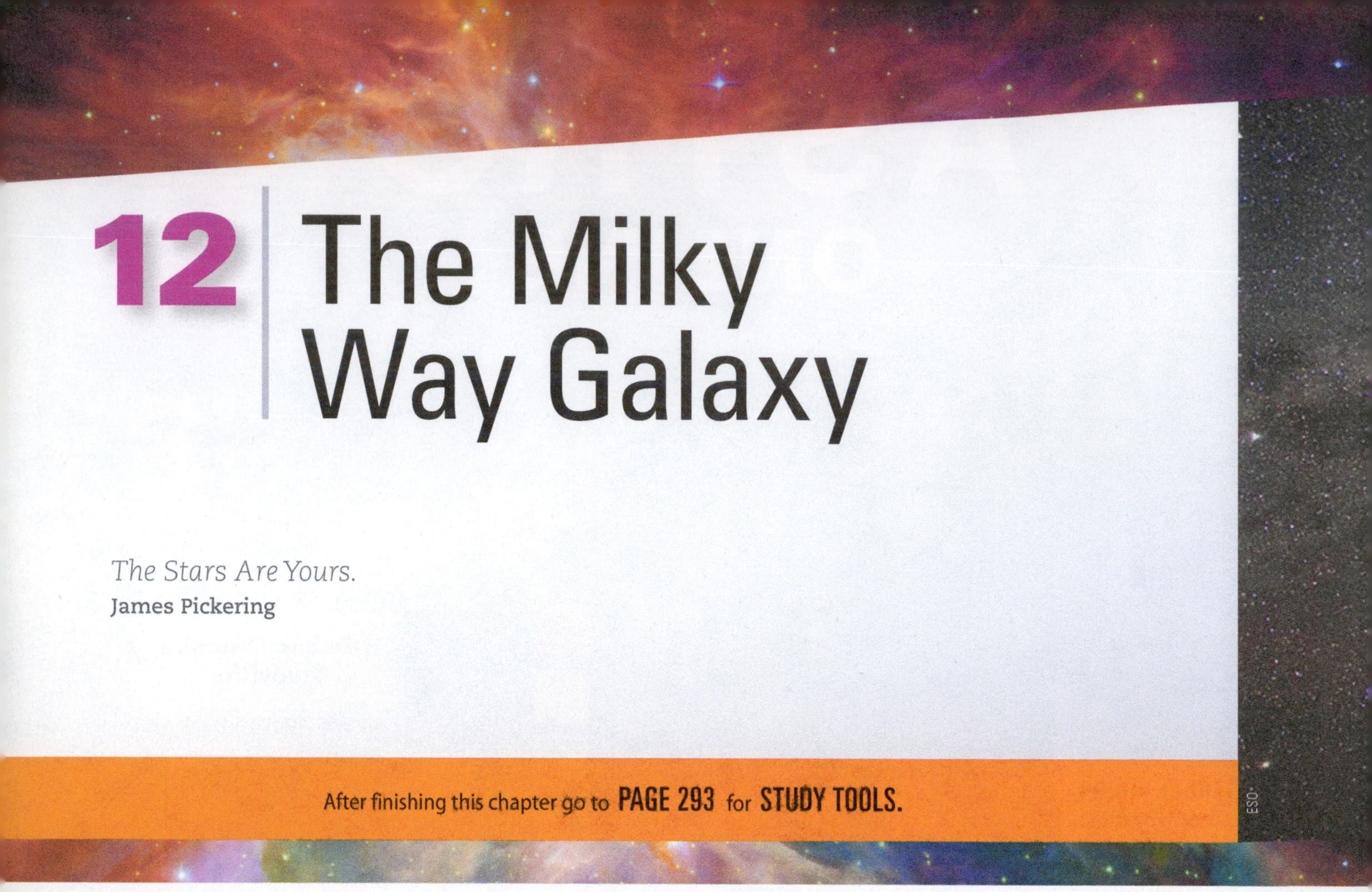

12 The Milky Way Galaxy

The Stars Are Yours.
James Pickering

After finishing this chapter go to **PAGE 293** for **STUDY TOOLS.**

GUIDEPOST

You have traced the life stories of stars from their birth in clouds of gas and dust to their deaths as white dwarfs, neutron stars, or black holes. Now you are ready to step back and view stars in vast communities called galaxies. This chapter focuses on our home galaxy, the Milky Way, and addresses five important questions:

- What is the evidence that we live in a galaxy?
- What is the evidence that our Milky Way Galaxy is a spiral galaxy?
- What are the spiral arms of the Milky Way and other spiral galaxies?
- How did the Milky Way Galaxy form and evolve?
- What is in the nucleus of the Milky Way Galaxy?

Discovering the Milky Way Galaxy by answering these questions will be one more step toward understanding the Universe as a whole. In the chapters that follow, you will leave your home galaxy and voyage out among the billions of other galaxies that fill the depths of the Universe.

The Stars Are Yours was the title of a popular astronomy book written by James S. Pickering. This title expresses the author's view that the stars belong to everyone equally, and you can enjoy the stars as if you owned them. You live inside one of the largest of the star systems that fill the Universe. The Milky Way Galaxy is more than 80,000 light-years in diameter and contains more than 100 billion stars. (See CO 12.) As you read this chapter, you will learn about your home galaxy, but you will also learn how the stars of the galaxy have cooked up the atoms heavier than helium. You know already how the cores of massive stars make atoms heavier than helium and how supernovae blast those atoms back into space and add even heavier atoms made during the supernova explosion. In this chapter you will see how the stars in the Milky Way Galaxy have, generation after generation, made the atoms now in your body.

12-1 DISCOVERY OF THE GALAXY

It seems odd to say astronomers discovered something that is all around you. However, until the early 20th century, no one knew what the Milky Way was.

It isn't obvious that you live in a galaxy. You are inside, and you see nearby stars scattered in all directions over the sky, whereas the more distant clouds of stars in the galaxy make a faint band of light circling the sky, as

CO 12 The Milky Way Galaxy as seen from Cerro Paranal in Chile. A laser beam for adaptive optics imaging points exactly at the galactic center from one of the four 8.2-meter telescopes in the European Southern Observatory's VLT array. The two bright objects below the Milky Way are Antares (Alpha Scorpii; *left*) and Jupiter (*right*).

the artwork in Figure 12-1 illustrates. The ancient Greeks named that band *galaxies kuklos*, the "milky circle." The Romans referred to it as *via lactea*, "milky road" or "milky way." It was not until early in the 20th century that astronomers understood that humans live inside a great wheel of stars and that the Universe is filled with other such star systems. Drawing on the Greek word for milk, astronomers called these star systems *galaxies*.

12-1a The Great Star System

Galileo's telescope revealed that the glowing Milky Way is made up of stars, and later astronomers realized that the great cloud of stars in which the Sun is located, which they called the star system, must be wheel shaped. If the star system were spherical, for example, you would see stars scattered more or less uniformly in all directions over the sky. Only a wheel or disk shape would look, from the inside, like the Milky Way band encircling the sky.

The English astronomers Sir William Herschel and his sister Caroline Herschel attempted to gauge the true shape of the star system by counting stars in 683 different directions in the sky. In directions where they saw more stars, they assumed the star system extended farther into space. As you can see in Figure 12-2, which was published by the Herschels in 1785, they concluded that the star system has a disk shape with some noticeable "holes" lacking stars around its edges. Modern astronomers know that those apparent holes are caused by dense clouds of gas and dust blocking the view of more distant stars. The Herschels counted similar numbers of stars in most directions around the Milky Way and concluded that the Sun and Earth are near the center of the star system.

As the 20th century began, astronomers still believed that the Sun was located near the center of a wheel-shaped star system that they estimated about 10 kiloparsecs in diameter. [A **kiloparsec (kpc)** is 1000 parsecs.] How humanity realized the truth about the size of the galaxy, Earth's location in it, and the galaxy's location in a much larger Universe of other galaxies, is an adventure that begins with a woman studying stars that pulsate and leads to a man studying distant star clusters. The next section will follow that story because it is an important historical moment in human history and because it illustrates how scientists build on the work of their predecessors and step-by-step refine their ideas about the natural world.

kiloparsec (kpc) A unit of distance equal to 1000 pc or about 3300 ly.

FIGURE 12-1

(a) Nearby stars look bright, and people around the world group them into different constellations. Nevertheless, the vast majority of the stars in our Galaxy that are visible from Earth merge into a faintly luminous path that circles the sky: the Milky Way. This artwork shows the location of a portion of the Milky Way near a few bright winter constellations. (See Figure 12-5 and the star charts in Appendix B to locate other sections of the Milky Way in your sky.) (b) The Andromeda Galaxy, also known as Messier 31, a spiral galaxy about 2.5 million ly from Earth, looks somewhat like our Galaxy would appear if you could view it from the same distance. The two oval patches to upper right and lower right of the main galaxy's central bulge are small satellite galaxies.

FIGURE 12-2

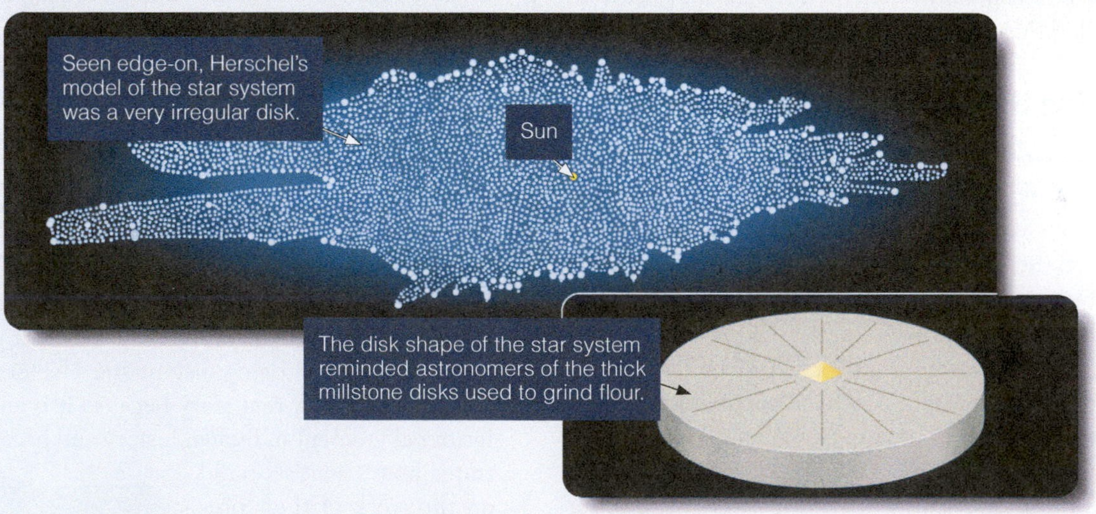

In 1785, William Herschel published this diagram showing the shape of the star system as if it could be viewed edge-on from outside. The Sun is located near the center of this "grindstone" model universe.

FIGURE 12-3

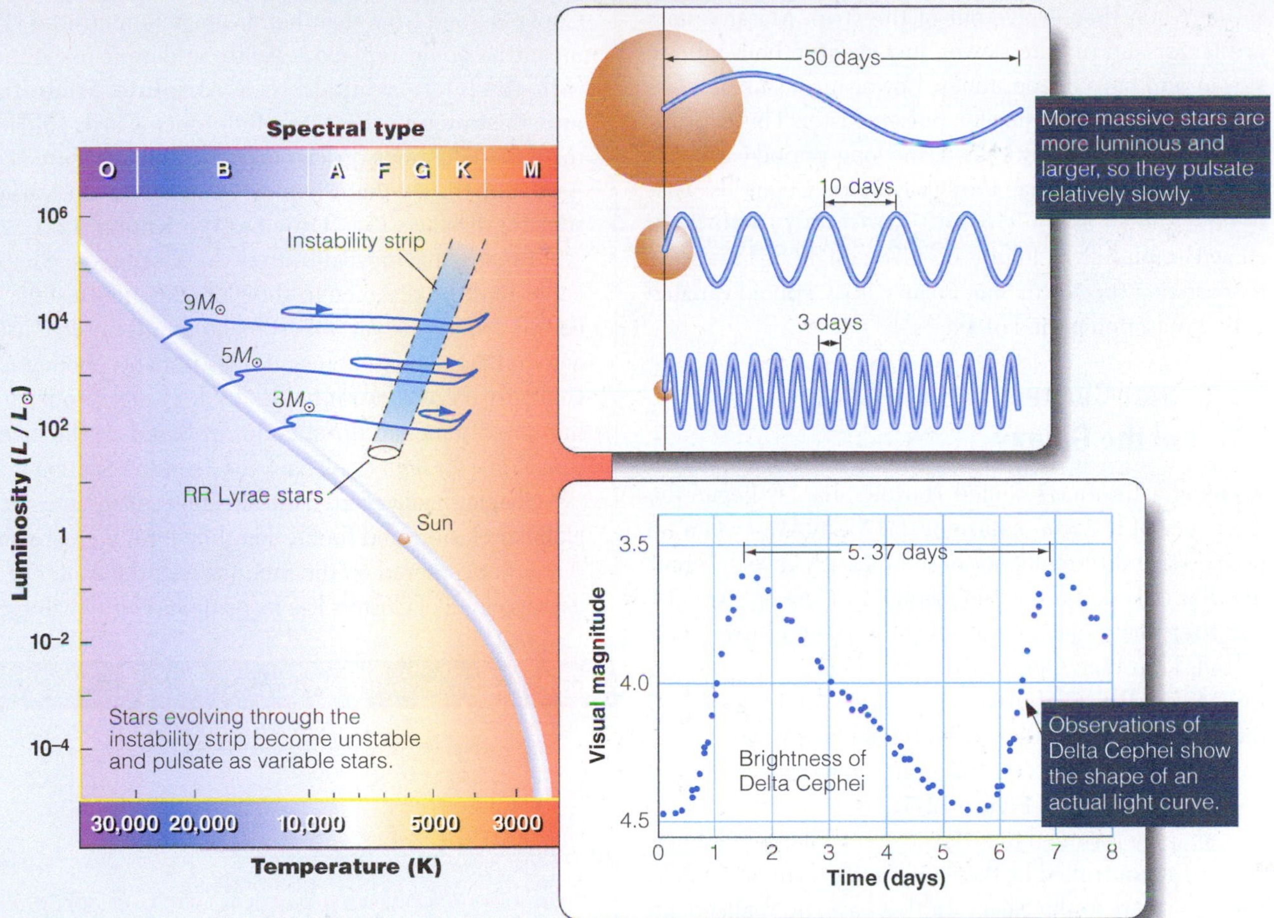

The instability strip on the H–R diagram contains combinations of stellar temperatures and luminosities corresponding to unstable, pulsating internal structures. The more massive a star is, the more luminous it is, and the larger in diameter it becomes when it leaves the main sequence. Those larger stars pulsate with longer periods when they pass through the instability strip. Therefore, because both luminosity and period of pulsation depend on mass, there is a relationship between period and luminosity.

12-1b Size of the Milky Way

It is a common misconception of poets that the stars are eternal and unchanging, but astronomers have known for centuries that some stars change in brightness. Of course, novae and supernovae burst into view, grow brighter, and then fade, but many other variable stars actually pulsate like beating hearts. The period of pulsation is the time it takes a star to complete a cycle from bright to faint to bright again. You will learn next how the properties of some types of variable stars allowed astronomers to measure the size of the galaxy.

In 1912, Henrietta Leavitt (1868–1921) was studying a star cloud in the southern sky known as the Small Magellanic Cloud. On her photographic plates she found many variable stars, and she noticed that the brightest had the longest periods. Because all of the variables were in the same cloud at nearly the same distance, she concluded that there was a relationship between the pulsation periods and intrinsic brightnesses (luminosities, i.e., true total power output) of those variable stars.

The stars Leavitt saw, **Cepheid variable stars**, are named after the first such star discovered, Delta Cephei. They are giant and supergiant stars with pulsation periods of 1 to 60 days with properties that lie in a region of the H–R diagram known as the **instability strip** (Figure 12-3). As stars evolve, and the points that represent their temperatures

Cepheid variable stars Variable stars with pulsation periods of 1 to 60 days for which period of variation is related to luminosity.

instability strip The region of the H–R diagram in which stars are unstable to pulsation. A star evolving through this strip becomes a variable star.

and luminosities move in the H–R diagram, they can cross into the instability strip and start pulsating; they stop pulsating when they evolve out of the strip. Massive stars are larger and pulsate slower, just as large bells vibrate slower and have deeper tones. Lower-mass stars are less luminous and, being smaller, pulsate faster. This explains why, as first noticed by Leavitt, the long-period Cepheids are more luminous than the short-period Cepheids. That is now known as the **period–luminosity relation**, shown graphically in Figure 12-4. You might be interested to learn that the North Star, Polaris, is a Cepheid variable with a pulsation period of 4 days.

12-1c Star Clusters and the Center of the Galaxy

A young astronomer named Harlow Shapley began the discovery of the true nature of the Milky Way when he noticed that different kinds of star clusters have different distributions in the sky. In Chapter 11 Concept Art 11A you met yow types of star clusters: open clusters and globular clusters. Open clusters are concentrated along the Milky Way. Globular clusters are widely scattered, but Shapley noticed that the globular clusters were more common toward the constellations Sagittarius and Scorpius, as illustrated in Figure 12-5a.

Shapley assumed that the concentration of globular clusters is controlled by the combined gravitational field of all the stars in the galaxy. In that case, he realized, he could study the size and extent of the galaxy by studying the globular clusters. To do that, he needed to measure the distances to as many globular clusters as possible.

Globular clusters are much too far away to have measurable parallaxes (refer to **Parallax and Distance** in the Math Reference Cards MR), but they do contain variable stars. Shapley knew about Leavitt's work on these stars, which depended on their relative, rather than true, luminosities. Cepheids are rare, and none is near enough to have measurable parallaxes, so their true luminosities were not then known. Shapley realized that he could calculate the distance to the globular clusters if he found the true luminosities of the Cepheid variable stars in the clusters.

Through a statistical process involving measurements of position shifts because of their motions through space, called **proper motions**, Shapley was able to determine the average distances of a few of the nearest Cepheids, and from that their average luminosities. That meant he could replace Leavitt's apparent magnitudes with absolute magnitudes (see **Absolute Magnitude and Distance** in the Math Reference Cards MR), on the period–luminosity diagram (as shown in Figure 12-4). Astronomers say that Shapley **calibrated** the variable stars for distance. (See **How Do We Know? 12-1**.)

Finally, having calibrated the Cepheids, Shapley could find the distance to the globular star clusters. He could identify the variable stars he found in the clusters and find their apparent magnitudes from his photographs. Comparison of apparent and absolute magnitudes gave him the distance to the star cluster based on the inverse square law for light (look back to Chapter 9 Section 9-2a).

Shapley finally plotted the direction and distance to the globular clusters and found that they form a great swarm that is not centered on the Sun. Instead, the center of the galaxy's cloud of clusters lies many thousands of light-years

period–luminosity relation The relation between period of pulsation and intrinsic brightness among Cepheid variable stars.

proper motion The rate at which a star moves across the sky, measured in arc seconds per year.

calibrate To make observations of reference objects, checks on instrument performance, calculations of units conversions, and so on, needed to completely understand measurements of unknown quantities.

FIGURE 12-4

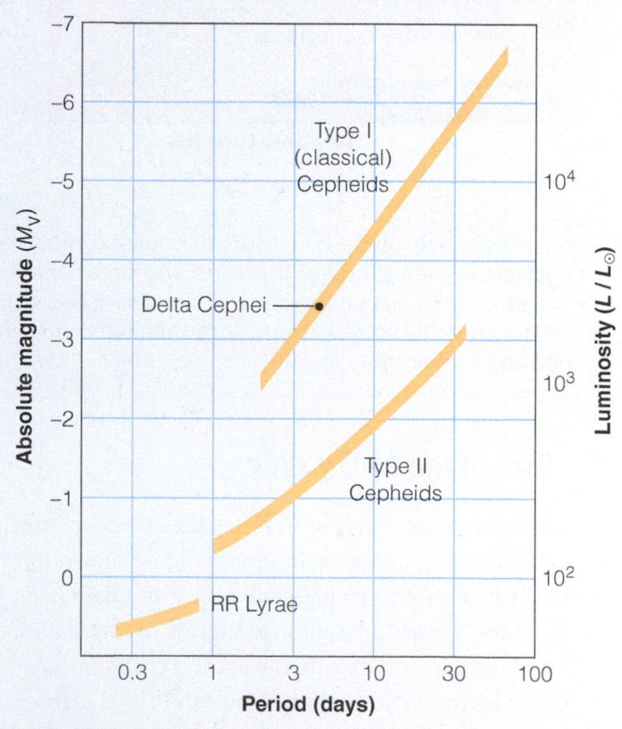

The period–luminosity diagram is a graph of the brightness of variable stars versus their periods of pulsation. You could plot brightness as luminosity, but because the diagram is used in distance calculations, it is more convenient with absolute magnitude on the vertical axis. Modern astronomers know that there are two types of Cepheids, something that astronomers in the early 20th century could not recognize in their limited data.

FIGURE 12-5

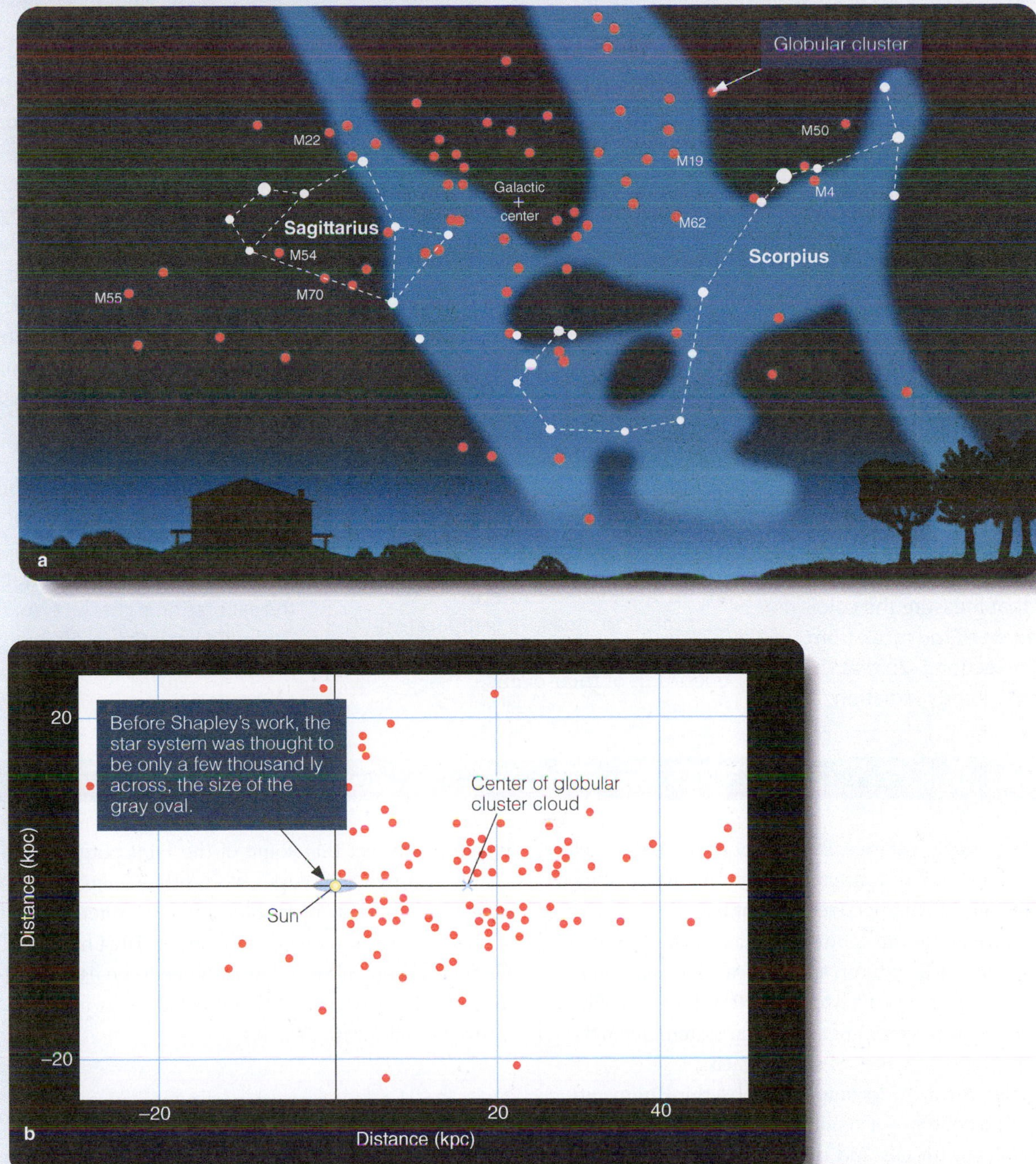

(a) Nearly half of cataloged globular clusters (red dots) are located in or near Sagittarius and Scorpius. A few of the brighter globular clusters, labeled with their catalog designations, are visible in binoculars or small telescopes. Constellations are shown as they appear above the southern horizon on a summer night as seen from latitude 40°N, typical for most of the United States. (b) Shapley's study of globular clusters showed that they were not centered on the Sun, at the origin of this graph, but rather form a great cloud centered far away in the direction of Sagittarius. Distances on this graph are given in thousands of parsecs, corresponding to Shapley's original calibration that produced distances more than two times larger than the modern values.

How Do We Know? 12-1

Calibration

How do you take the temperature of a vat of molten steel? Astronomers often say that Shapley "calibrated" the Cepheids for the determination of distance, meaning that he did all the detailed background work to determine the luminosities of Cepheids. After that, he and other astronomers could use his calibrated period–luminosity diagram to find the distance to other Cepheids without repeating the detailed calibration.

Calibration is actually very common in science because it saves a lot of time and effort. For example, engineers in steel mills must monitor the temperature of molten steel, but they can't dip in a thermometer. Instead, they can use handheld devices that measure the color of molten steel. You recall from Chapter 5 Section 5-2b that the color of blackbody radiation is determined by the temperature of the body emitting it. Molten steel emits visible and infrared radiation that is nearly perfect blackbody radiation, so the manufacturer can calibrate the engineers' devices to convert the measured color to a temperature displayed on digital readouts. The engineers don't have to repeat the calibration every time; they just point their instrument at the molten steel and read off the temperature. Astronomers have made the same kind of color-temperature calibration for stars.

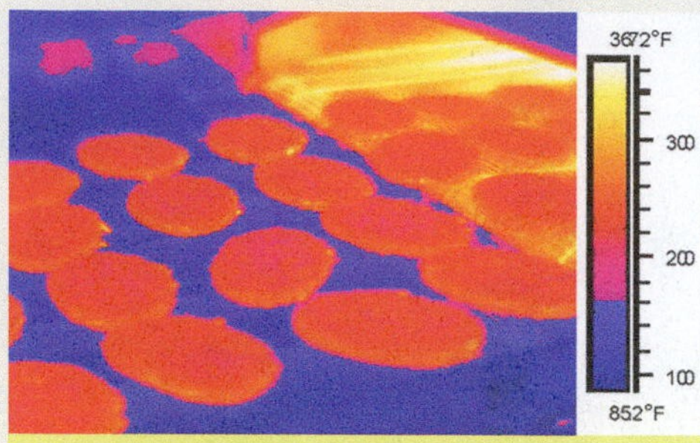

© iStockphoto.com/DuffB

An infrared video camera calibrated to measure temperature allows bakers to monitor the operation of their ovens.

As you read about any science, notice how calibrations are used to simplify common measurements. But notice, too, how important it is to get the calibration right. An error in calibration can throw off every measurement made with that calibration.

in the direction of Sagittarius. By analogy, if you see a bunch of tall buildings, and they are all together in one direction away from you, you might conclude that downtown is over there, and you are in the suburbs. Evidently the center of the star system is in Sagittarius and far away. The star system is much bigger than anyone had suspected (Figure 12-5b). You live not near the center of a small star system but in the suburbs of a very big wheel of stars, a galaxy.

Why did astronomers before Shapley think humanity lives near the center of a small star system? The answer: Space is filled with gas and dust that dims the view of distant stars. When you look toward the band of the Milky Way, you can see only the neighborhood near the Sun. Most of the star system is hidden and, like a traveler in a fog, you seem to be at the center of a small region. Shapley was able to see the globular clusters at greater distances because they lie outside the Milky Way plane and are not dimmed very much by the interstellar dust.

disk component Material confined to the plane of the galaxy.

Building on Shapley's work, other astronomers began to suspect that some of the faint patches of light visible through telescopes were other galaxies like our own. In 1923, Edwin Hubble photographed individual stars in the Andromeda Galaxy, and in 1924 he identified Cepheids there, allowing its distance to be estimated. As a result, it became clear that our Galaxy is just one in a Universe filled with galaxies.

12-1d Components of the Galaxy

Our Galaxy, like many others, contains two primary components—a disk and a sphere. Figure 12-6 shows these components and other features discussed in this section.

The **disk component** consists of all matter confined to the plane of rotation—that is, everything in the disk itself. This includes stars, open star clusters, and nearly all of the galaxy's gas and dust. As you will learn in more detail in the next section, because the disk contains lots of gas, it is the site of most of the star formation in the galaxy. Consequently, the disk is illuminated by recently formed brilliant, blue, massive stars and has an overall relatively blue color.

FIGURE 12-6

An artist's conception of our Milky Way Galaxy, seen face-on and edge-on. Note the position of the Sun and the distribution of globular clusters in the halo. Hot blue stars light up the spiral arms. Only the inner halo is shown here. At this scale, the entire halo would be larger than a dinner plate.

The diameter of the disk and the position of the Sun are difficult to determine accurately. Interstellar dust blocks the view in the plane of the galaxy, so astronomers cannot see to the center or to the edge easily, and the outer edge of the disk is not well defined. Most recent studies suggest the Sun is about 8.3 kpc from the center. The Sun and Earth seem to be about two-thirds of the way from the center to the edge, so the diameter of our Galaxy appears to be about 25 kpc or about 80,000 ly, but that isn't known to be better than 10 percent precision. This is the diameter of the luminous part of our Galaxy, the part you would see from a distance. You will learn later that strong evidence shows that our Galaxy is much larger than that, but the outer parts are not luminous.

Observations made at other than visual wavelengths can help astronomers peer through the dust and gas. Infrared and radio photons have wavelengths long enough to be unaffected by the dust. Thus, a map of the sky at long infrared wavelengths reveals the disk of our Galaxy, which you can see in Figure 12-7.

The most striking features of the disk component are the **spiral arms**—long curves of bright stars, star clusters, gas, and dust. Such spiral arms are easily visible in other galaxies, and you will see later how astronomers found that our own Galaxy has a spiral pattern.

The second component of our Galaxy is the **spherical component**, which includes all matter in our Galaxy scattered in a roughly spherical distribution around the center. This includes a large halo and the central bulge.

The **halo** is a spherical cloud of thinly scattered stars and globular star clusters. It contains only about 2 percent as many stars as the disk of the galaxy and has very little gas and dust. Thus, with no raw material available, no new stars are forming in the halo. In fact, the halo stars are mostly old, cool giants or dim lower-main-sequence stars plus old white dwarfs that are difficult to detect but have been revealed by recent careful studies. Astronomers can map the halo of our Galaxy by studying the more easily detected giant stars.

The **central bulge** is the dense cloud of stars that surrounds the center of our Galaxy. It has a radius of about 2 kpc and is a slightly flattened spheroid. It is hard to observe because thick dust in the disk scatters and absorbs radiation of visible wavelengths, but observations at longer wavelengths can penetrate the dust. The bulge seems to contain little gas and dust, and there is not much star formation there. Most of the stars in the central bulge are old, cool stars like those in the halo.

spiral arms Long spiral pattern of bright stars, star clusters, gas, and dust. Spiral arms extend from the center to the edge of the disk of spiral galaxies.

spherical component The part of the galaxy including all matter in a spherical distribution around the center (the halo and central bulge).

halo The spherical region of a spiral galaxy, containing a thin scattering of stars, star clusters, and small amounts of gas.

central bulge The spherical cloud of stars that lies at the center of spiral galaxies.

FIGURE 12-7

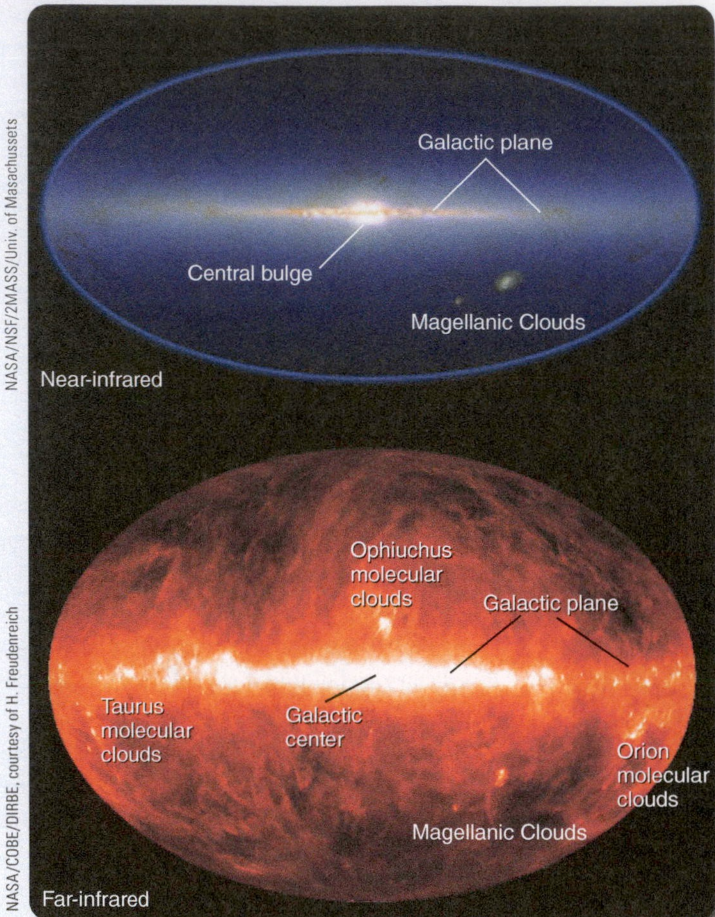

In these infrared images, the entire sky has been projected onto ovals with the center of the galaxy at their respective centers. The central plane of the Milky Way extends from left to right. (a) The central bulge is prominent at near-infrared wavelengths, but dust clouds block the view through the plane. (b) Interstellar dust emits blackbody radiation and glows brightly at longer wavelengths.

12-1e Mass of the Galaxy

The vast numbers of stars in the disk, halo, and central bulge lead to a basic question: How massive is the galaxy?

When you needed to find the masses of stars, you studied the orbital motions of the pairs of stars in binary systems. To find the mass of the galaxy, you must look at the orbital motions of the stars within the galaxy. Every star in the galaxy follows an orbit around the center of mass of the galaxy. In the disk, the stars follow parallel circular orbits, and astronomers say the disk of the galaxy rotates. That rotation can allow an estimate of the mass of the galaxy. Consequently, any discussion of the mass of the galaxy is also a discussion of the rotation of the galaxy and the orbits of the stars within the galaxy.

Astronomers can find the orbits of stars by finding how they move. Of course, the Doppler effect reveals a star's radial velocity (look back to Chapter 5 Section 5-4c). In addition, if astronomers can measure the distance to a star and its proper motion, they can find the velocity of the star perpendicular to the radial direction. Combining all of this information, astronomers can find the shape of the star's orbit.

As you can see in Figure 12-8, the orbital motions of the stars in the halo are strikingly different from those in the disk. In the halo, each star and globular cluster follows its own randomly tipped elliptical orbit. These orbits carry the stars and clusters far out into the spherical halo, where they move slowly, but when they fall back into the inner part of the galaxy, their velocities increase. Motions in the halo do not resemble a general rotation but are more like the random motions of a swarm of bees. In contrast, the stars in the disk of the galaxy move in the same direction in nearly circular orbits that lie in the plane of the galaxy. The Sun is a disk star and follows a circular orbit around the galaxy that always remains within the disk.

You can use the orbital motion of the Sun to find the mass of the galaxy inside the Sun's orbit. By observing the radial velocity of distant galaxies in various directions around the sky, astronomers can tell that the Sun moves about 225 kilometers/second in the direction of Cygnus, carrying Earth and the other planets of our Solar System along with it. Because its orbit is a circle with a radius of 8.3 kpc, you can divide the circumference of the orbit by the velocity and find that the Sun completes a single orbit in about 225 million years.

If you think of the Sun and the center of mass of our Galaxy as two objects orbiting each other, you now have enough information to determine the mass of the galaxy. The Milky Way Galaxy must have a mass of about 100 billion solar masses. This estimate is uncertain for a number of reasons. First, you don't know the radius of the Sun's orbit with great certainty. Astronomers estimate the radius as 8.3 kpc, but they could be wrong by 10 percent or more, and the radius gets cubed in the calculation, which has a large effect. Second, this estimate of the mass includes only the mass inside the Sun's orbit. Mass spread uniformly outside the Sun's orbit will not

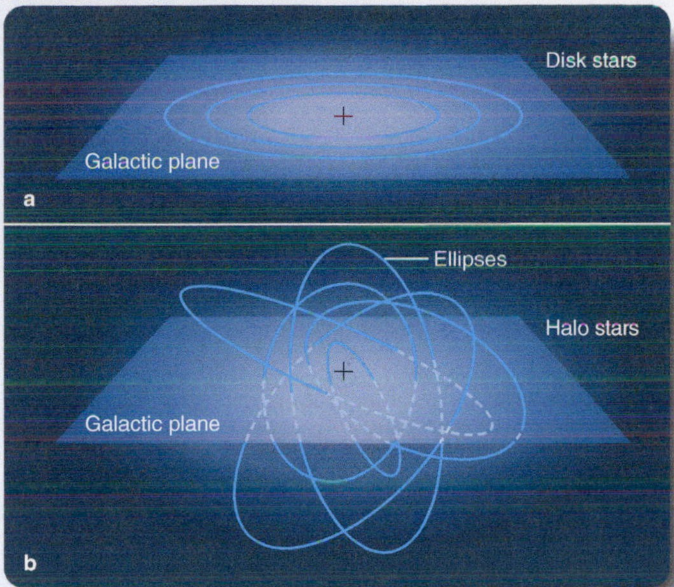

FIGURE 12-8

(a) Stars in the galactic disk have nearly circular orbits that lie in the plane of the galaxy. (b) Stars in the halo have randomly oriented, highly eccentric, elongated orbits.

affect its orbital motion. Thus 100 billion solar masses is a lower limit for the mass of the galaxy, but no one knows exactly how much to increase the estimate to include the rest of the mass in the galaxy that lies outside the Sun's orbit.

The measured motion of the stars shows that the disk does not rotate as a solid body. Each star follows its own orbit, and stars in some regions have shorter or longer orbital periods than the Sun. This is called differential rotation. (Recall that you met the term *differential rotation* in Chapter 5 Section 5-6d where it was applied to the Sun.) The graph in Figure 12-9 is an example of the orbital velocity of stars at various orbital radii in the galaxy, called a **rotation curve**. If all of the mass in the galaxy were concentrated at its center, then the orbital velocity would be high near the center and would decline away from the center. This kind of motion has been called *Keplerian motion* because it follows Kepler's third law, as in the case of our Solar System, where nearly all of the mass is in the Sun.

Of course, you already know the galaxy's mass is not all concentrated at its center, but if most of the mass were inside the orbit of the Sun, then orbital velocities should decline at greater distances. Many observations confirm, however, that velocities do not decline and might actually increase at greater distance; this observation shows that larger and larger orbits enclose more and more mass.

Although it is difficult to determine a precise edge to the visible galaxy, it seems clear that large amounts of matter are located beyond what seems to be the limit of the galaxy's luminous matter.

The evidence is clear that significant extra mass lies in the galaxy's halo, extending up to 10 times farther than the edge of the visible disk and containing up to 2 trillion (2×10^{12}) solar masses. Some small fraction of this mass is made up of low-luminous stars and white dwarfs, but most of the matter is not producing any light. Astronomers call it **dark matter** and conclude that it must be some as yet unknown form of matter. The following two chapters will return to the question of dark matter. It is one of the fundamental problems of modern astronomy.

12-2 SPIRAL ARMS AND STAR FORMATION

The most striking feature of galaxies such as the Milky Way is the system of spiral arms that wind outward through the disk and contain swarms of hot, blue stars; clouds of dust and gas; and young star clusters. These young objects hint that the spiral arms involve star formation. As you try to understand the spiral arms, you face two questions. First, how can you be sure our Galaxy has spiral arms when Earth is embedded inside the galaxy and your view is obscured by gas and dust? Second, why doesn't the differential rotation of the galaxy destroy the arms? The answers to both questions involve star formation.

12-2a Tracing the Spiral Arms

Studies of other galaxies show that spiral arms contain hot, blue stars. Thus, one way to study the spiral arms of our own Galaxy is to locate these stars. Fortunately, this is not difficult, because O and B stars are often found in associations and, being very luminous, are easy to detect across great distances. Unfortunately, at these great distances their parallax is too small to measure, so their distances must be found by other means, usually by spectroscopic parallax (look back to Chapter 9 Section 9-4b).

O and B associations in the sky are not located randomly but reveal parts of three spiral arms near the Sun, which have been named for the prominent

dark matter Nonluminous matter that is detected only by its gravitational influence.

rotation curve A graph of orbital velocity versus radius in the disk of a galaxy.

FIGURE 12-9

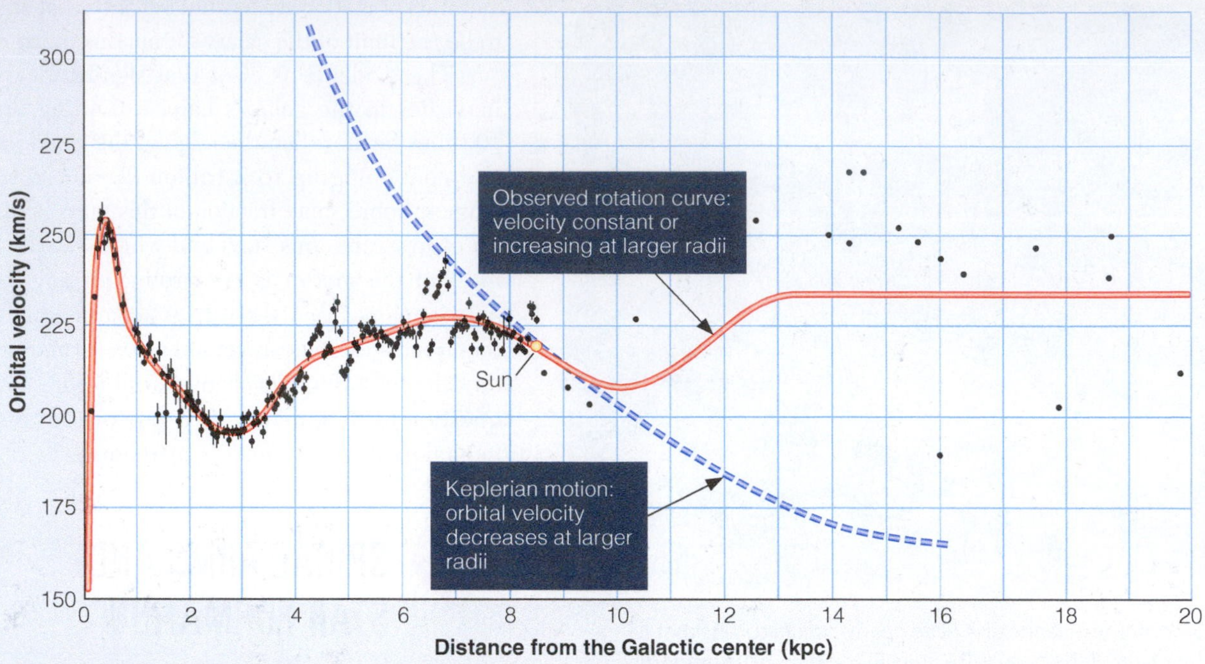

The rotation curve of our Galaxy is plotted here as orbital velocity versus radius. Data points show measurements made by radio telescopes. Observations outside the orbit of the Sun are much more uncertain, and the data points scatter widely. Orbital velocities do not decline outside the orbit of the Sun, as you would expect if most of the mass of the galaxy were concentrated toward the center. Rather, the curve is approximately flat at great distances, suggesting that our Galaxy contains significant mass outside the orbit of the Sun.

constellations through which they pass (Figure 12-10). Objects used to map spiral arms are called **spiral tracers**. O and B associations are good spiral tracers because they are bright and easy to see at great distances. Other tracers include young open clusters, HII, and certain higher-mass variable stars.

Notice that all spiral tracers are young objects. O stars, for example, live for only a few million years. If their orbital velocity is about 200 km/s, they cannot have moved more than about 500 pc since they formed. This is less than the width of a spiral arm. Because they don't live long enough to move away from the spiral arms, they must have formed there. The youth of spiral tracers provides an important clue about spiral arms. Somehow they are associated with star formation.

Radio telescopes can detect emission from clouds of cool neutral hydrogen gas. Radio maps of the galaxy disk, combined with optical and infrared data, allow astronomers to deduce the spiral pattern of our Galaxy. The segments near the Sun are part of a spiral pattern that continues throughout the disk. However, the maps show that the spiral arms are rather irregular and interrupted by branches, spurs, and gaps. The stars in Orion, for example, appear to be a detached segment of a spiral arm, a spur. There are significant sources of error in the mapping methods, but many of the irregularities along the arms seem real, and images of nearby spiral galaxies show similar features (Figure 12-10a). Studies comparing all of the available data on our Galaxy's spiral pattern with patterns seen in other galaxies do not necessarily agree, although the newest models suggest that the central bulge in our Galaxy is elongated into a bar (Figure 12-11). You will see in Chapter 13 that such bars are common in spiral galaxies.

spiral tracers Object used to map the spiral arms—for example, O and B associations, open clusters, clouds of ionized hydrogen, and some types of variable stars.

Spiral tracers show that the arms contain young objects, and that suggests active star formation. The radio maps confirm this suspicion by showing that the material needed to make stars, hydrogen gas, is abundant in spiral arms.

12-2b Star Formation in Spiral Arms

Having mapped the spiral pattern, you can ask, "What are spiral arms?" You can be sure that they are not physically connected structures. Like a kite string caught on a

FIGURE 12-10

(a) Nearby young O and B stars in our Galaxy fall along bands that appear to be segments of spiral arms. (b) Many of the galaxies in the sky are disk shaped like our Galaxy, and most of those galaxies have spiral arms. (c) Images of other galaxies show that spiral arms are marked especially by hot, luminous stars and emission nebulae that must be very young. This strongly suggests that spiral arms are related to star formation.

spinning hubcap, such arms would be wound up and pulled apart by differential rotation within a few tens of millions of years. Yet spiral arms are common in disk-shaped galaxies and must be reasonably permanent features.

FIGURE 12-11

This artist's conception of a two-armed model for the Milky Way is based on observations with the *Spitzer* infrared space telescope. Notice the large central bar.

The most prominent theory about spiral arms is called the **spiral density wave theory**, which proposes that the arms are waves of compression that move around the galaxy triggering star formation. The density wave is a bit like a traffic jam behind a truck moving slowly along a highway. Seen from an airplane overhead, the jam seems a permanent, though slow-moving, feature. But individual cars overtake the jam from behind, slow down, move up through the jam, wait their turn, pass the truck, and resume speed along the highway. Similarly, clouds of gas overtake the spiral density wave, become compressed in the "traffic jam," and eventually move out in front of the arm, leaving the slower-moving density wave behind.

Of course, star formation will occur where the gas clouds are compressed. Stars pass through the spiral arms unaffected, like bullets passing through a wisp of fog, but large clouds of gas slam into the spiral density wave from behind and are suddenly compressed. The spiral density wave series of images demonstrates how this movement occurs (Figure 12-12). You saw in Chapter 10 Section 10-4b that sudden compression could trigger the formation of stars in a gas cloud. Thus, new stars should form along the spiral arms. The spiral arms are not wound up by differential rotation because they are patterns, not physically connected structures.

spiral density wave theory
Theory proposed to account for spiral arms as compressions of the interstellar medium in the disk of the galaxy.

FIGURE 12-12

The Spiral Density Wave

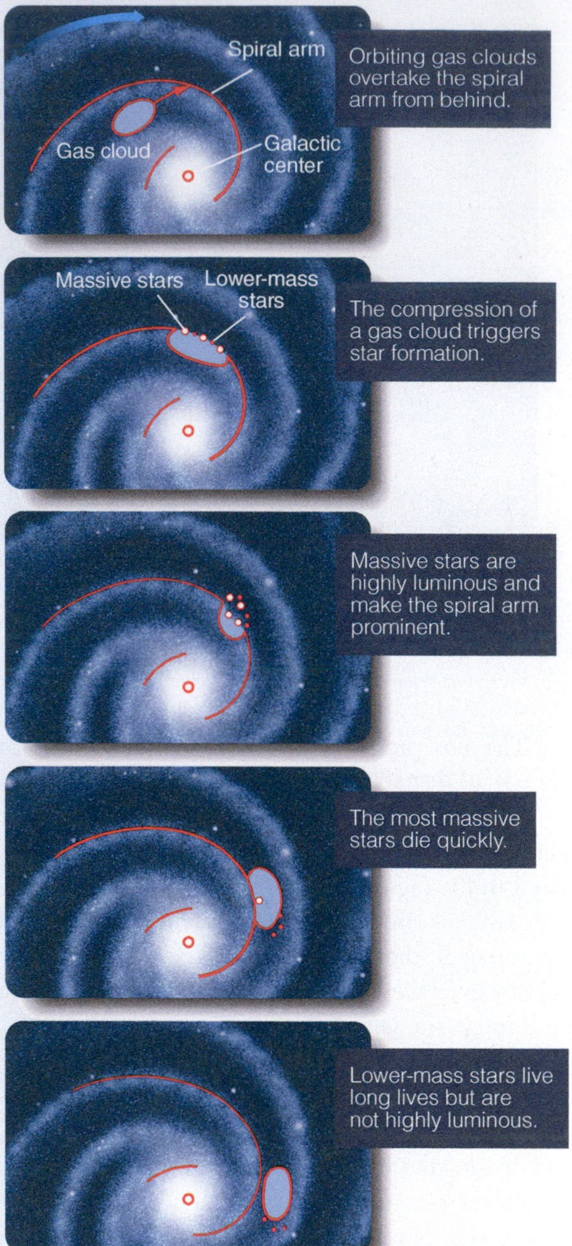

According to the spiral density wave theory, star formation occurs as gas clouds are compressed when they pass through spiral arms.

The brightest stars, the O and B stars, live such short lives that they never travel far from their birthplace and are found only along the arms. Their presence is what makes the spiral arms glow so brightly, because of both their own light and the emission from clouds of gas excited by UV radiation from the stars, as shown in Figure 12-13a. Lower-mass stars, like the Sun, live longer and have time to move out of the arms and continue their journey around the galaxy. The Sun might have formed in a star cluster almost 5 billion years ago when a gas cloud smashed into a spiral arm. Since that time, the Sun has escaped from its birth cluster and made about 20 trips around the galaxy, passing through spiral arms many times.

self-sustaining star formation The process by which the birth of stars compresses the surrounding gas clouds and triggers the formation of more stars.

The spiral density wave theory is very successful in explaining general properties of spiral galaxies, but it has two problems. First, what stimulates the formation of the spiral pattern in the first place? Theorists calculate that minor fluctuations in the galaxy's disk shape or gravitational interactions with passing galaxies might be able to start a density wave. Second, the density wave theory does not account for the branches and spurs in the spiral arms of our own Galaxy and other galaxies. The solution to this second problem might lie in a process that sustains star formation once it begins.

Star formation can control the shape of spiral patterns if the birth of stars in a cloud of gas can cause the birth of more new stars. Massive stars evolve so quickly that their lifetimes are only an instant in the history of a galaxy, and then they explode as supernovae. You learned in Chapter 10 Section 10-4b that bursts of luminosity and jets from newborn stars and the expanding gases of supernova explosions can compress neighboring clouds of gas and trigger more star formation. This process is known as **self-sustaining star formation**. The Orion complex, consisting of the Great Nebula in Orion plus the protostars buried deep in the dark interstellar clouds behind the nebula (look back to Concept Art 10B), seems to be a region of self-sustaining star formation.

Astronomers have calculated models indicating that the differential rotation of the galaxy can drag the inner edge of a star-forming region ahead and let the outer edge lag behind to produce a cloud of star formation shaped like a segment of a spiral arm. Self-sustaining star formation plus differential rotation might produce the branches and spurs so prominent in some galaxies (Figure 12-13b), including our own, but only the spiral density wave can generate the beautiful two-armed grand spiral patterns.

12-3 ORIGIN AND HISTORY OF THE GALAXY

Just as paleontologists reconstruct the history of life on Earth from the fossil record, astronomers try to reconstruct our Galaxy's past from the fossil it left behind as it formed and evolved. That fossil is the spherical

FIGURE 12-13

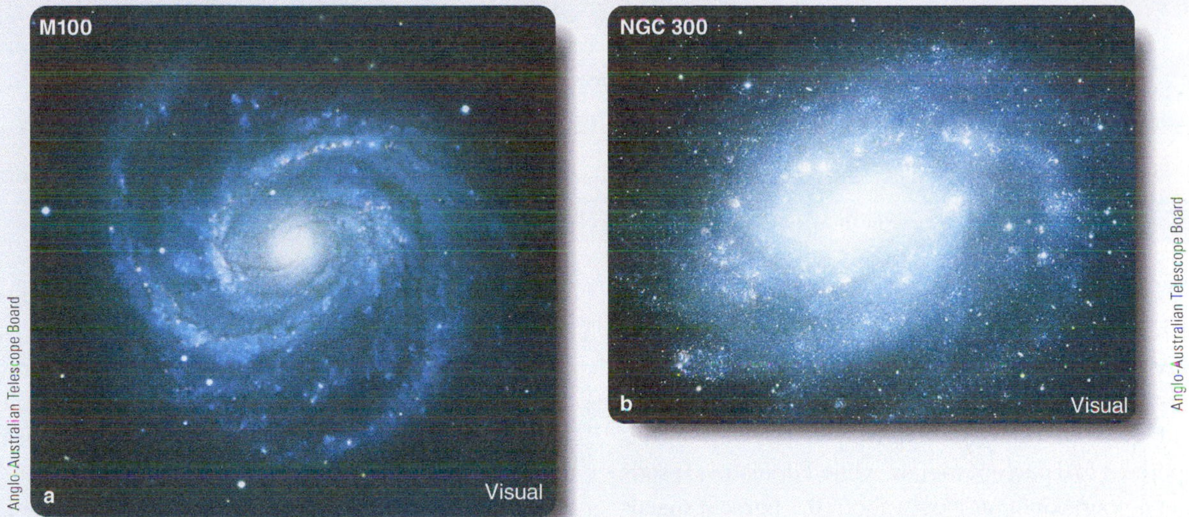

(a) Some galaxies are dominated by two spiral arms; but, even in these galaxies, minor spurs and branches are common. The spiral density wave can generate the two-armed, grand-design pattern, but self-sustained star formation might be responsible for the irregularities. (b) Many spiral galaxies do not appear to have two dominant spiral arms. Abundant spurs and branches nevertheless suggest that star formation is proceeding robustly in such galaxies. Observations indicate that the Milky Way Galaxy's spiral pattern is intermediate between these two examples.

component of the galaxy. The stars in the halo formed when the galaxy was young. The chemical composition and the distribution of those stars can provide clues to how our Galaxy formed.

12-3a Age of the Galaxy

To begin, you should ask yourself how old our Galaxy is. That question is easy to answer because you already know how to find the age of star clusters. But there are uncertainties that make that easy answer hard to interpret.

The oldest open clusters are about 9 to 10 billion years old. These ages are determined by analyzing the turnoff point in each cluster's H-R diagram (look back to Concept Art 11A) and are somewhat uncertain. Nevertheless, from open clusters you can get a rough age for the disk of our Galaxy of at least 9 billion years.

Globular clusters have faint turnoff points in their H–R diagrams and are clearly old, but finding their ages accurately is difficult. Clusters differ slightly in chemical composition, which must be accounted for in calculating the stellar models from which ages are determined. Also, to find the age of a cluster, astronomers must know the distance to the cluster. Precise parallaxes from the *Hipparcos* satellite have allowed astronomers to increase the precision of the Cepheid variable star calibration, and careful studies with the newest large telescopes have refined the cluster H–R diagrams. Globular cluster ages seem to average about 11 billion years, with the oldest being a bit over 13 billion years old. The halo of our Galaxy therefore seems to be at least 13 billion years old, older than the disk.

12-3b Stellar Populations

In the 1940s, astronomers realized that there were two types of stars in the galaxy. The stars they were most accustomed to studying were located in the disk, such as the stars near the Sun. These they called **Population I stars**. The second type, called **Population II stars**, are usually found in the halo, in globular clusters, or in the central bulge. In other words, the two stellar populations are associated, respectively, with the disk and spherical components of the galaxy.

The stars of the two populations fuse nuclear fuels and evolve in nearly

Population I stars Stars with significant amounts of atoms heavier than helium; relatively young stars nearly always found in the galactic disk. The Sun is an intermediate Population I star.

Population II stars Stars poor in atoms heavier than helium; relatively old stars nearly always found in the halo, globular clusters, or the central bulge.

TABLE 12-1 STELLAR POPULATIONS

	Population I		Population II	
	Extreme	Intermediate	Intermediate	Extreme
Typical location	Spiral arms; Thin disk	Thick disk	Central bulge	Halo
Metal content (%)	3	1–2	0.1–1	Less than 0.1
Orbit shape	Circular	Slightly eccentric	Moderately eccentric	Highly eccentric
Range of ages (yr)	0–0.2 billion	0.2–8 billion	8–12 billion	12–13 billion

identical ways. They differ only in their abundance of atoms heavier than helium, atoms that astronomers refer to collectively as **metals**. (Note that this is definitely not the way the word *metal* is commonly used by nonastronomers.) Population I stars are relatively metal rich, containing 2 percent to 3 percent metals, while Population II stars are metal poor, containing only about 0.1 percent metals or less. The metal content of a star defines its population.

Population I stars, sometimes called *disk population stars*, have circular orbits in the plane of the galaxy and are relatively young stars that formed within the past few billion years. The Sun is a Population I star. Population II stars belong to the spherical component of the galaxy and are sometimes called *halo population stars*. These stars have randomly tipped orbits with a wide range of shapes. A few follow circular orbits, but most follow elliptical orbits. The Population II stars are all lower-mass main-sequence stars or giants. They are old stars. The metal-poor globular clusters are part of the halo population. Since the discovery of stellar populations, astronomers have realized that there is a gradation between populations, as illustrated in Table 12-1.

Why do the disk and halo stars have different metal abundances? The two types of stars must have formed at different stages in the life of the galaxy, at times when the chemical composition of the galaxy differed. This is a clue to the history of our Galaxy, but to use the clue, you must understand the cycle of element building.

12-3c The Element-Building Cycle

The atoms of which you are made were created in a process that spanned a number of generations of stars. The process that built the chemical elements over the history of our Galaxy led to the possibility of Earth and life on Earth.

You saw in Chapter 11 Sections 11-2 and 11-4 how elements heavier than helium but lighter than iron are built up by nuclear reactions inside evolving stars. More massive atoms are made only by short-lived nuclear reactions that occur during a supernova explosion. This explains why lower-mass atoms such as carbon, nitrogen, and oxygen are more common and why more massive atoms—such as gold, silver, platinum, and uranium—are so rare and, often, valuable. Figure 12-14a shows the abundance of the chemical elements, but notice that the graph has an exponential scale. To get a feeling for the true abundance of the elements, you should draw this graph using a linear scale as in Figure 12-14b. With that scale you see how rare the elements heavier than helium really are.

metals In astronomical usage, all atoms heavier than helium.

Most of the matter in stars is hydrogen and helium. Other elements, including carbon, nitrogen, oxygen, and the rest of what astronomers call metals, were cooked up inside stars. When the galaxy first formed, there should have been no metals because stars had not yet manufactured any. Judging from the composition of stars in the oldest clusters, the gas from which the galaxy originally condensed must have contained about 90 percent hydrogen atoms and 10 percent helium atoms. (The hydrogen and helium came from the Big Bang that began the Universe, which you will study in Chapter 14.)

The first stars to form from this gas were metal poor, and now, 13 billion years later, their spectra still show few metal lines. Of course, they might have manufactured many atoms heavier than helium in their cores, but because the stars' interiors are not mixed, those heavy atoms stay trapped at the centers of the stars where they were produced and do not affect the spectra (Figure 12-15). The Population II stars in the halo are the survivors of an earlier generation of stars that formed in the galaxy.

Most of the first stars evolved and died, and various types of star death throes, including supernovae, enriched the interstellar gas with metals. Succeeding generations of stars formed from gas clouds that were more enriched, and each generation added to the enrichment with its death. By the time the Sun formed, roughly 5 billion years ago, the element-building process had added about 2 percent metals. Since then, the metal abundance has increased further, and stars forming now incorporate 2 percent to 3 percent metals and become extreme

FIGURE 12-14

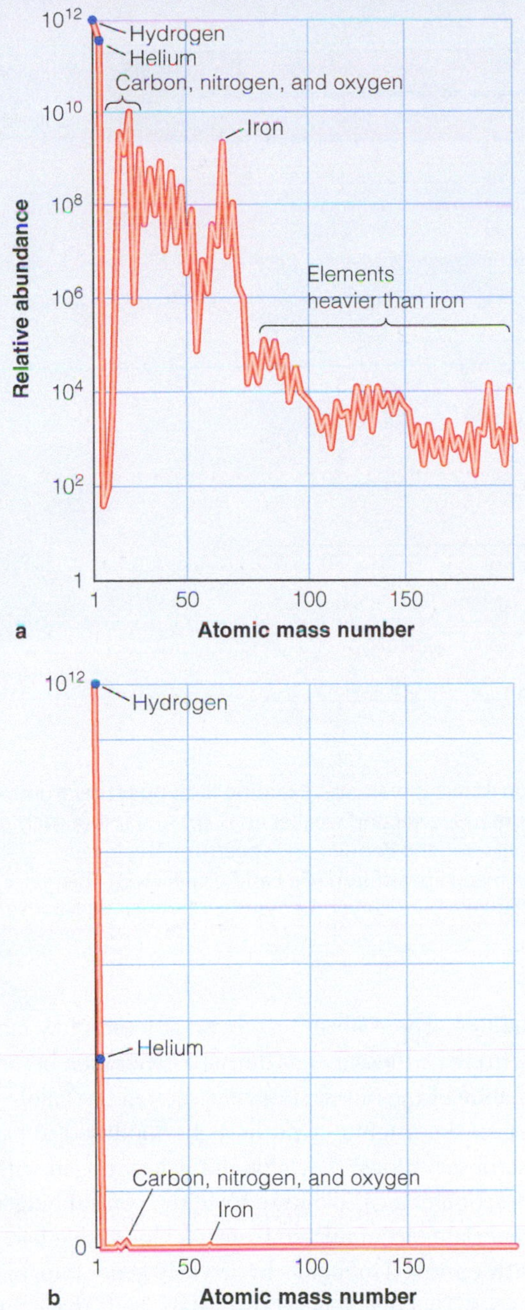

The abundances of chemical elements in the Universe. (a) When the elemental abundances are plotted on an exponential scale, you see that elements heavier than iron are about a million times less common than iron and that all elements heavier than helium (the metals) are quite rare. (b) The same data plotted on a linear scale provide a more realistic impression of how rare the metals are. Carbon, nitrogen, and oxygen make small peaks near atomic mass 15, and iron is just visible in the graph.

Population I stars. Thus metal abundance varies between populations because of the production of heavy atoms in successive generations of stars as the galaxy aged.

The oxygen atoms you are breathing, the carbon atoms in your flesh, and the calcium atoms in your bones were all created in the interiors of red and yellow giant stars hot enough to sustain fusion reactions beyond helium. The gold and silver atoms in your jewelry and dental fillings and the iodine atoms in your thyroid gland were all created in a few moments during supernova explosions, because those are the only places in the Universe hot enough to make these types of atoms. This idea is perhaps the most significant one to take away from an introductory astronomy course and textbook—you, and everything on Earth, are made of stardust.

12-3d History of the Galaxy

The lack of metals in the spherical component of the galaxy tells you it is very old, a fossil left behind by the galaxy when it was young and drastically different from its current disk shape. The study of element building and stellar populations leads to the fundamental question, "How did our Galaxy form?"

In the 1950s, astronomers began to develop a model called the **monolithic collapse**, or **"top-down" hypothesis**, to explain the formation of our Galaxy. Recent observations are forcing a major reevaluation and revision of that traditional hypothesis.

The traditional hypothesis says that the galaxy formed from a single large cloud of turbulent gas more than 13 billion years ago. Stars and star clusters that formed from this material went into randomly shaped and randomly tipped orbits. These first stars were metal poor because no stars had existed earlier to enrich the gas with metals. In this way, the initial contraction of a large, turbulent gas cloud produced the spherical component of the galaxy.

The second stage of this hypothesis accounts for the disk component. The turbulent motions would eventually have canceled out, leaving the cloud with uniform rotation. A rotating, low-density cloud of gas cannot remain spherical. A star is spherical because its high internal pressure balances its gravity, but in a low-density cloud, the pressure cannot support the weight. Like a blob of pizza dough spun in the air, such a cloud must flatten into a disk, as illustrated in Figure 12-16.

This contraction into a disk took billions of years, with the metal abundance gradually increasing as generations of stars were born from the gradually

Monolithic collapse (top-down) hypothesis
An early hypothesis that the galaxy formed from the collapse of a single large cloud of turbulent gas more than 13 billion years ago.

FIGURE 12-15

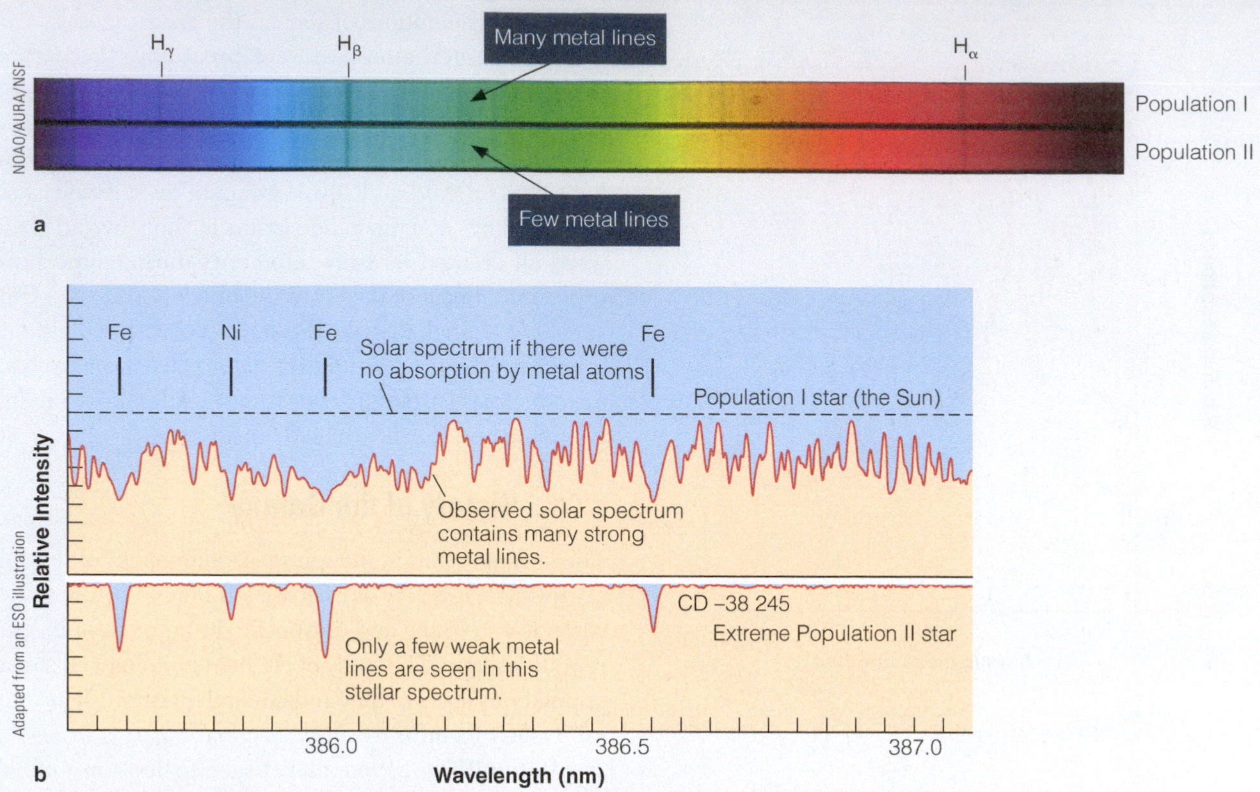

(a) The difference between spectra of Population I stars and Population II stars is dramatic. Examine the upper spectrum here and notice the hundreds of faint spectral lines. The lower spectrum has fewer and weaker lines. (b) A graph of such spectra reveals overlapping absorption lines of metals completely blanketing the Population I spectrum. The lower spectrum is that of an extremely metal-poor star with only a few weak metal lines of iron (Fe) and nickel (Ni) . This Population II star contains about 10,000 times less metal than the Sun.

flattening gas cloud. The stars and globular clusters that formed first in the halo would not have been affected by the motions of the gas and would have been left behind by the cloud as it collapsed and flattened. Later generations of stars formed in flatter distributions. The gas distribution in the galaxy now is so flat that the youngest stars are confined in a thin disk only about 100 pc thick. These stars are metal rich and have nearly circular orbits.

This monolithic collapse hypothesis accounts for many of the Milky Way's properties. Advances in technology, however, have improved astronomical observation; and, beginning in the 1980s, contradictions between theory and observations arose. For example, not all globular clusters have the same age, but, surprisingly, some of the younger clusters seem to be in the outer halo. In contrast, the monolithic collapse hypothesis says that the halo formed first and predicts that the clusters within it should have either a uniform age, or that the most distant clusters should be slightly older. Another problem is that the oldest stars are observed to be metal poor but not completely metal free. There must have been at least a few massive stars to create these metals during a generation before the formation of the oldest stars now seen in the halo.

Can the original hypothesis be modified to explain these observations? Perhaps the galaxy began with the contraction of a gas cloud to form the central bulge, and the halo later accumulated from gas clouds that had been slightly enriched in metals by an early generation of massive stars. That first generation of stars would have formed from almost pure hydrogen and helium gas, which mathematical models indicate would form only stars with very high masses. Those stars would have lived very short lives, made metals, and died in supernova explosions; and none would be left today. This would explain the metals in the oldest stars surviving today.

There is also evidence that entire small galaxies were captured by the growing Milky Way, and that fresh gas was added to the disk in several episodes over the life of the galaxy and not in a single gravitational collapse when the galaxy first formed. An alternative bottom-up hypothesis

FIGURE 12-16

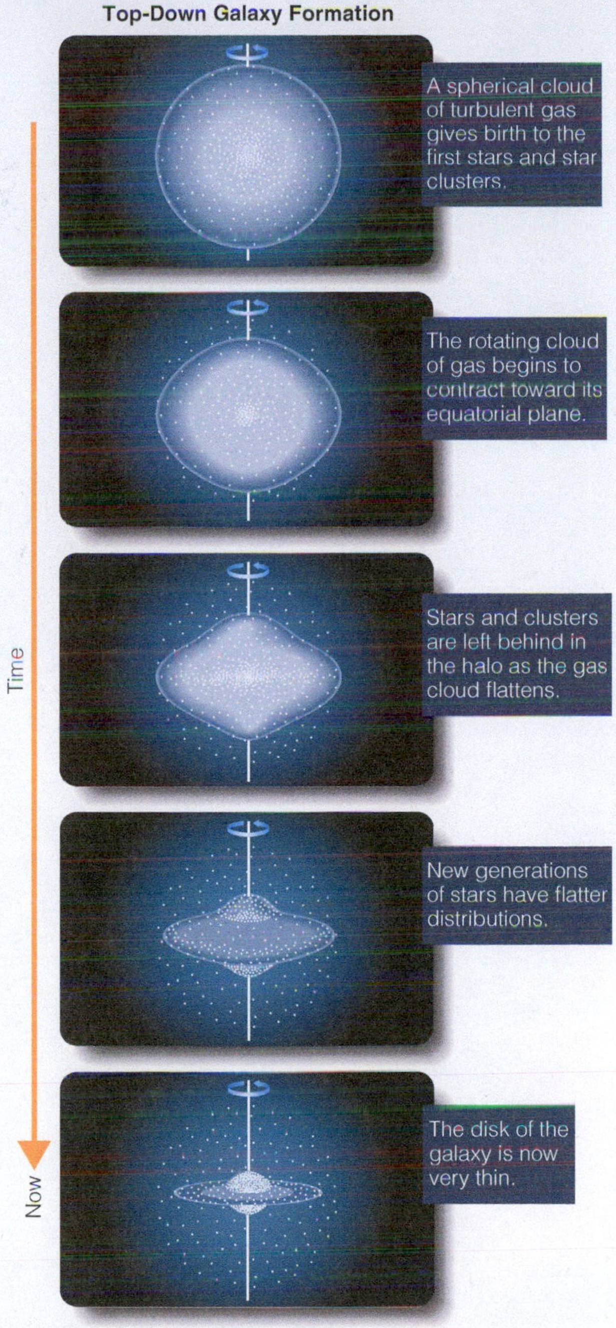

The monolithic collapse, or top-down, model for the origin of our Galaxy begins with a spherical gas cloud that flattens into a disk.

FIGURE 12-17

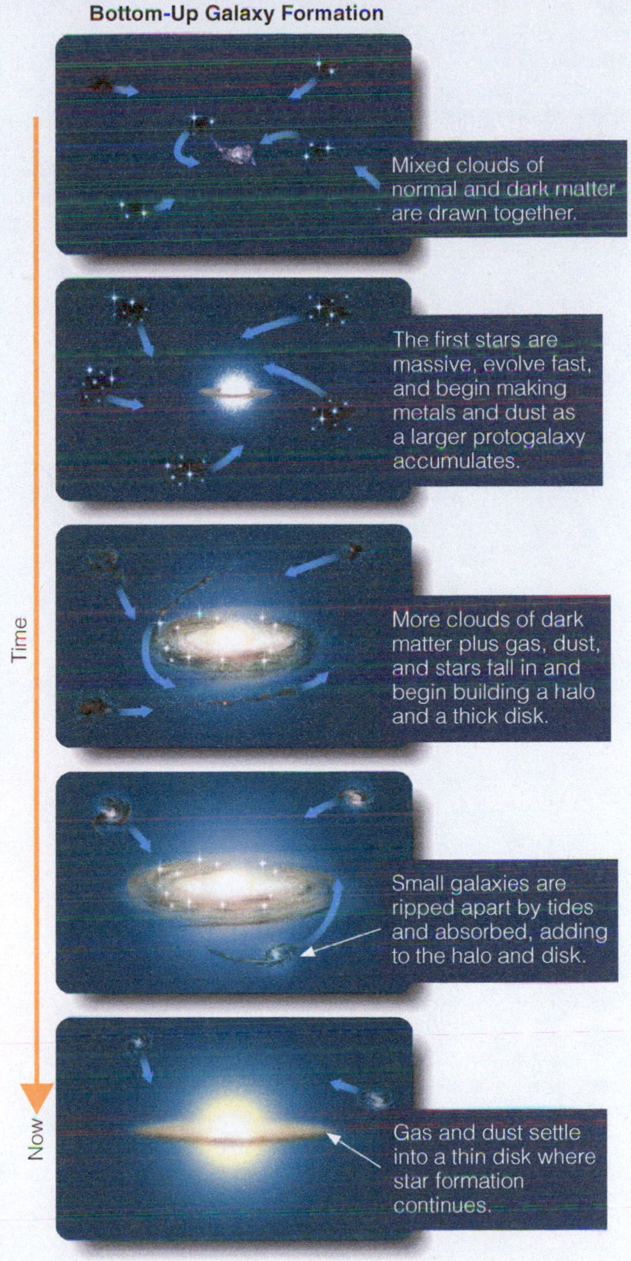

The bottom-up hypothesis for the formation of the Milky Way Galaxy proposes that smaller star systems accumulated to form larger ones. To see how this could have built our Galaxy, start with the top frame, depicting the situation only a few hundred million years after the beginning of the Universe, as small clouds of matter begin accumulating and stars begin forming in them. In the 2nd frame, the central object has grown larger, and in the 3rd frame the galactic halo and disk are forming. By today, represented by the bottom frame, the disk of the galaxy has become very thin.

proposes that our Galaxy was assembled from smaller objects (Figure 12-17). If our Galaxy absorbed a few small but partially evolved galaxies, then some of the globular clusters in the halo might be hitchhikers that originally belonged to the captured galaxies. This could explain the range of globular cluster ages and compositions. (You will see in Chapter 13 that such galaxy mergers do occur.)

Concept Art 12A

SAGITTARIUS A*

1

There is so much interstellar dust in the plane of the Milky Way that you cannot observe the nucleus of our Galaxy at visual wavelengths. The image below is a radio image of the innermost 300 pc (1000 ly). Many of the features are supernova remnants (labeled SNR), and a few are star formation clouds. Peculiar features such as threads, the Arc, and the Snake may be gas trapped in magnetic fields. At the center of the image lies the strong radio source Sagittarius A (Sgr A), the location of the nucleus of our Galaxy.

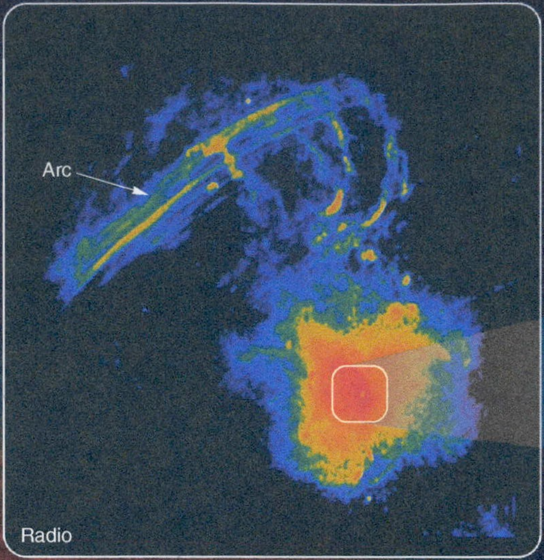

a **The radio map above shows Sgr A and the Arc filaments, 50 pc long. The image was made with the VLA radio telescope. The contents of the white box are shown on the opposite page.**

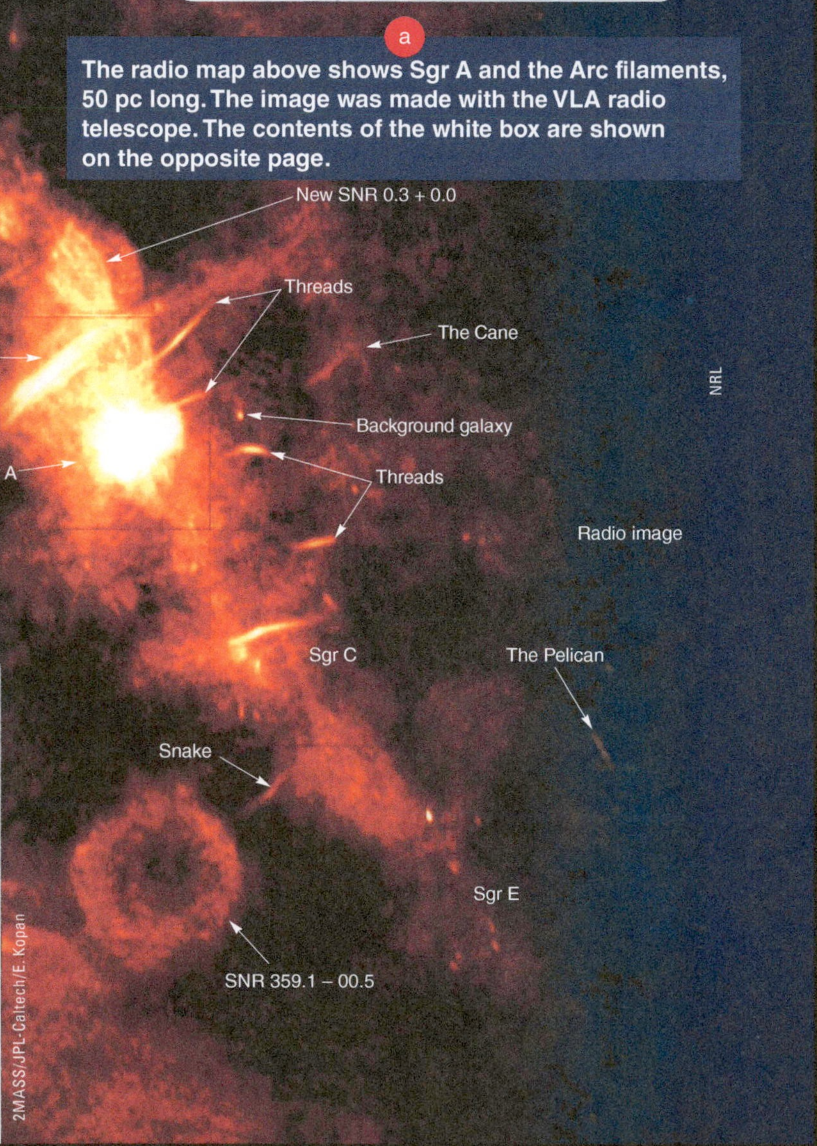

2

Infrared photons with wavelengths longer than 4 microns (4000 nm) are emitted almost entirely by warm interstellar dust. The radiation at those wavelengths coming from Sagittarius is intense, indicating that this region contains lots of dust and is crowded with stars that warm the dust.

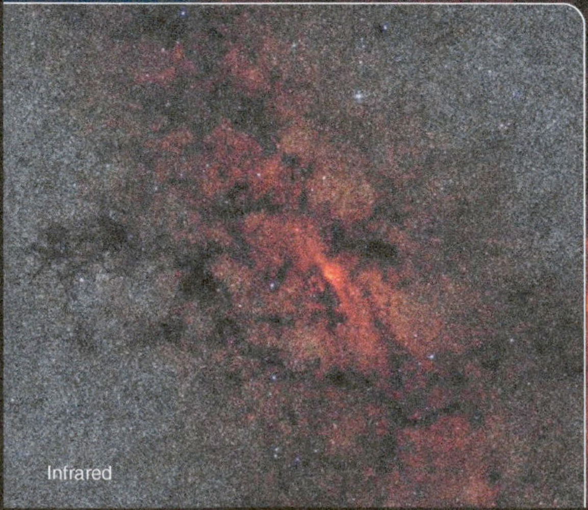

1a This high-resolution radio image of Sgr A (within the white box in the small-scale map on the opposite page) reveals a spiral swirl of gas around an intense point-like radio source known as Sgr A*. About 3 pc (10 ly) across, this spiral lies in a low-density cavity inside a larger disk of neutral gas. The arms of the spiral are thought to be streams of matter flowing into Sgr A* from the inner edge of the larger disk.

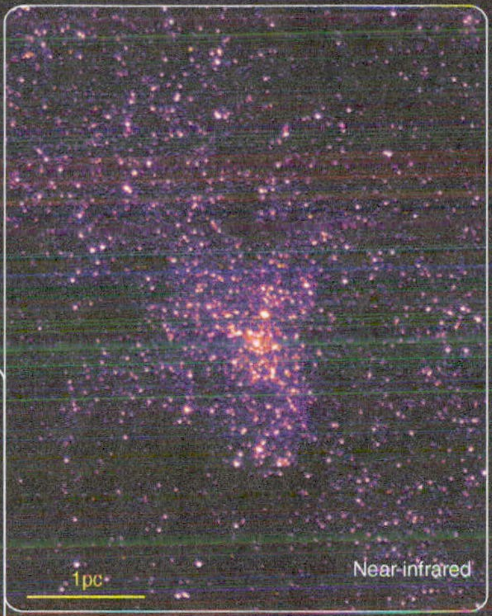

NASA/ESA/STScI/AURA/NSF/NICMOS Team

A near-infrared image from the *Hubble Space Telescope*, showing the same region as the *SOFIA* mid-infrared image on the left. The dense cluster of stars located inside the circumnuclear cloud ring can be seen at this wavelength (1.9 microns), but the ring itself is too cool to emit significant amounts of near-infrared radiation. Nevertheless, especially dense parts of the ring can be discerned in this image as shadowy patches of obscuration. Sgr A* is located in the middle of the central star cluster.

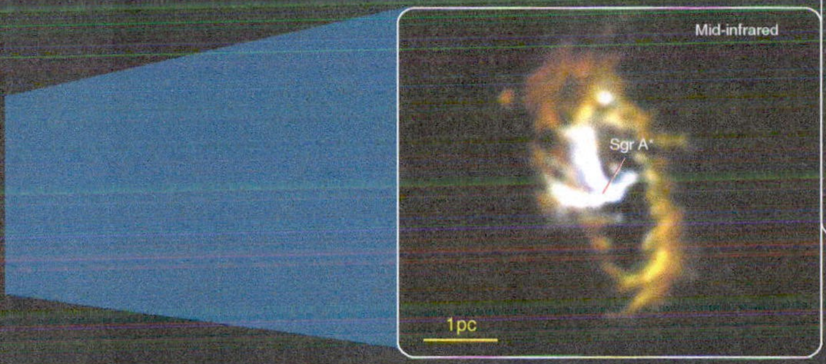

NASA/DLR/SOFIA/FORCAST Team/Lau et al.

An image of the central few parsecs of our Galaxy obtained by astronomers onboard the *Stratospheric Observatory for Infrared Astronomy (SOFIA)*, combining data at mid-infrared wavelengths of 20, 32, and 37 microns. This image contains the region of the white box within the radio image on the opposite page. A ring of gas and dust clouds is observed that surrounds a relatively empty central zone. The bright y shape represents heated material apparently falling from the cloud ring into the center. Sgr A* is located at the intersection of the arms of the y.

3 Evidence of a Black Hole in the Nucleus of Our Galaxy

Astronomers have been able to use large infrared telescopes and adaptive optics to follow the motions of stars orbiting around Sgr A*. A few of those orbits are shown here. The size and period of the orbit allows astronomers to calculate the mass of Sgr A* using Kepler's third law. The orbital period of the star SO-2, for example, is 15.2 years, and the semimajor axis of its orbit is 950 AU. The motions of the observed stars indicate that Sgr A* has a mass of 4 million solar masses.

a **The *Chandra X-ray Observatory* has imaged Sgr A* and detected more than 2000 other X-ray sources in the area.**

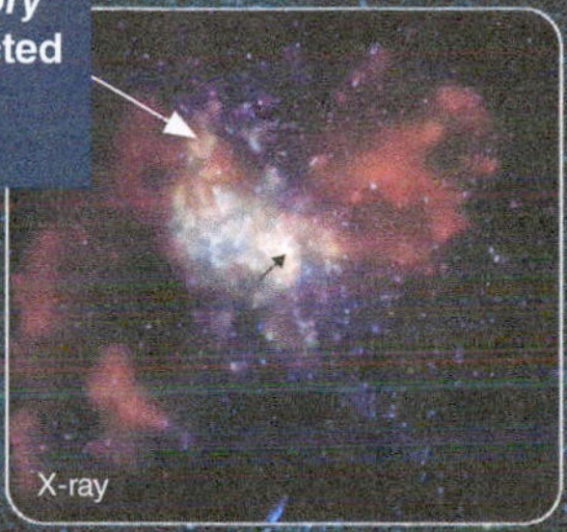

NASA/CXC/SAO/MIT/F. K. Baganoff et al.

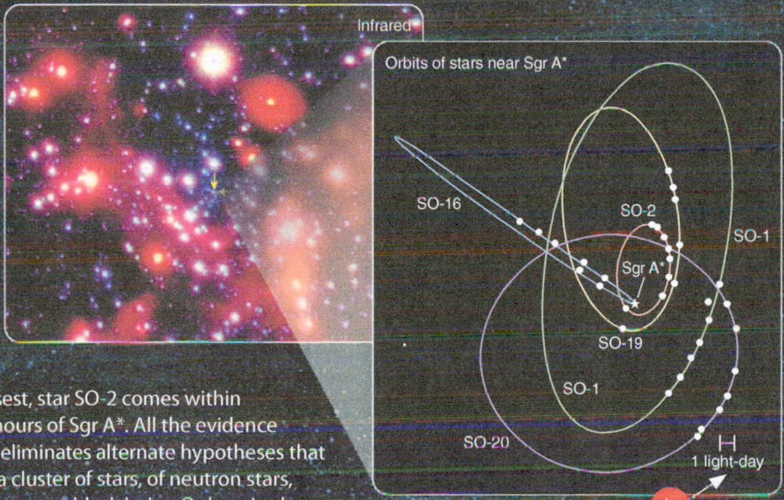

ESO/MPE/S. Gillessen et al.

3a A black hole with a mass of 4 million solar masses would have an event horizon with a size on the scale of this diagram smaller than the period at the end of this sentence. A slow dribble of only 0.0002 $M_\odot$ of gas per year flowing toward the black hole could produce the observed energy. A sudden increase, such as when a star falls in, could produce a violent eruption.

At its closest, star SO-2 comes within 17 light-hours of Sgr A*. All the evidence together eliminates alternate hypotheses that Sgr A* is a cluster of stars, of neutron stars, or of stellar-mass black holes. Only a single supermassive black hole could contain so much mass in such a small region.

b **For comparison, the diameter of the planetary region of our Solar System, defined by Neptune's orbit, is half a light-day.**

The evidence of a massive black hole at the center of our Galaxy seems conclusive. It is much too massive to be the remains of a dead star, however, and astronomers conclude that it probably formed as the galaxy first took shape.

The older monolithic collapse hypothesis has proven inadequate to explain all of the observations. Astronomers are attempting to refine the observations and devise new theories. In the next section you will learn that the nucleus of our Galaxy, like many other galaxies, contains a mysterious and massive object capable of powerful outbursts that could have profoundly influenced galactic evolution.

12-4 THE NUCLEUS OF THE GALAXY

The most mysterious region of our Galaxy is its very center, the nucleus. At visual wavelengths, this region is totally hidden by gas and dust that dim the light by 30 magnitudes. If a trillion (10^{12}) photons of light left the center of the galaxy on a journey to Earth, only one would make it through the gas and dust. The longer-wavelength infrared photons are scattered much less often: One in every 10 makes it to Earth. Consequently, visual-wavelength observations reveal nothing about the nucleus, but it can be observed at longer wavelengths such as in the infrared and radio parts of the spectrum.

12-4a Observations of the Nucleus

Harlow Shapley's study of globular clusters placed the center of our Galaxy in Sagittarius, and the first radio maps of the sky showed a powerful radio source located in Sagittarius. The first infrared map of the central bulge made by Eric Becklin in 1968 showed the location of intense radiation where the stars are most crowded together, identifying the gravitational center of the galaxy. Later high-resolution radio maps of that stellar center revealed a complex collection of radio sources, with one, **Sagittarius A*** (abbreviated **Sgr A*** and pronounced "sadge A-star"), lying at the expected location of the galactic core.

Observations show that Sgr A* is only a few astronomical units in diameter but is a powerful source of radio energy. The tremendous amount of infrared radiation coming from the central area appears to be produced by crowded stars and by dust warmed by those stars. But what could be as relatively tiny as Sgr A* and produce so much radio energy?

Sagittarius A* (pronounced "sadge A-star") The powerful radio source located at the core of the Milky Way Galaxy.

Study **Concept Art 12A, Sagittarius A***, and notice the following three important points:

1. Observations at radio wavelengths reveal complex structures near Sgr A* caused by magnetic fields and by rapid star formation. Supernova remnants show that massive stars have formed there recently and died explosively.
2. The center is very crowded. Tremendous numbers of stars plus radiation from Sgr A* heat the dust, producing strong infrared emission.
3. There is evidence that Sgr A* is a supermassive black hole into which gas is flowing.

Astronomers continue to test the hypothesis that the center of our Galaxy contains a supermassive black hole. Such an object is sufficient to explain the observations, but is it necessary? Is there some other way to explain what is observed? For example, astronomers have suggested that gas flowing toward the center of the galaxy could trigger tremendous bursts of star formation. Such hypotheses have been considered and tested against the evidence, but none appears to be adequate to explain all the observations. So far, the only hypothesis that seems adequate is that our Galaxy's nucleus is home to a supermassive black hole.

Meanwhile, observations are allowing astronomers to improve their models. For instance, Sgr A* is not as bright in X-rays as it should be if it had a hot accretion disk with matter constantly flowing into the black hole. Observations of X-ray and infrared flares lasting only a few hours suggest that mountain-size blobs of matter might occasionally fall into the black hole and be heated and ripped apart by tidal forces (look back to Chapter 11 Section 11-6c). The black hole might be mostly dormant and lack a fully developed hot accretion disk because the rate of matter flow into it is relatively low.

Such a supermassive black hole could not be the remains of a single dead star. It contains much too much mass. It probably formed when the galaxy first formed more than 13 billion years ago. In Chapter 13, you will learn that supermassive black holes are commonly found in the centers of galaxies.

What Are We? Children of the Milky Way

Hang on tight. The Sun, with Earth in its clutch, is ripping along at about 225 km/s (i.e., 500,000 mph) as it orbits the center of the Milky Way Galaxy. We live on a wildly moving ball of rock in a large galaxy that humanity calls home, but the Milky Way is more than just our home. Perhaps "parent galaxy" would be a better name.

Except for hydrogen atoms, which have survived unchanged since the Universe began, you and Earth are made of metals—atoms heavier than helium. There is no helium in your body, but there is plenty of carbon, nitrogen, and oxygen. There is calcium in your bones and iron in your blood. All of those atoms and more were cooked up inside stars or during their supernova death throes.

Stars are born when clouds of gas orbiting the center of our Galaxy collide with the gas in spiral arms and are compressed. That process has given birth to generations of stars, and each generation has produced elements heavier than helium and spread them back into the interstellar medium. The abundance of metals has grown slowly in our Galaxy. About 4.6.billion years ago, a cloud of gas enriched in those heavy atoms slammed into a spiral arm and produced the Sun, Earth, and, eventually, you. You have been created by the Milky Way—your parent galaxy.

STUDY TOOLS 12

IN THE BOOK

- ☐ Tear Out the Review Card on *The Milky Way Galaxy*.

ASTRO ONLINE

Visit ASTRO Online at www.cengagebrain.com

- ☐ Interactive Reading
- ☐ Practice Quizzing
- ☐ Videos and Animations
- ☐ Exclusive Features

Questions and Problems

12

REVIEW QUESTIONS

1. Why is it difficult to specify the dimensions of the disk and halo?
2. Why didn't astronomers before Shapley realize how large the galaxy is?
3. What is the evidence that our Galaxy contains a large amount of dark matter?
4. Contrast the motion of the disk stars and that of the halo stars. Why do their orbits differ?
5. Why are all spiral tracers young?
6. Why couldn't spiral arms be physically connected structures? What would happen to them?
7. What are some hypotheses for mechanisms to trigger or sustain spiral density waves?
8. Why does self-sustaining star formation produce clouds of stars that look like segments of spiral arms?
9. Why are metals less abundant in older stars than in younger stars?
10. Why do metal-poor stars have a wider range of orbital shapes than metal-rich stars like the Sun?
11. What evidence contradicts the monolithic collapse hypothesis for the origin of our Galaxy?
12. Describe the kinds of observations you would make to study the galactic nucleus.
13. Why must astronomers use infrared telescopes to observe the motions of stars around Sgr A*?
14. What evidence can you cite that the nucleus of the galaxy contains a supermassive black hole?
15. **How Do We Know?** Calibration simplifies complex measurements, but how does that make the work of later astronomers dependent on the expertise of the astronomer who did the calibration?

DISCUSSION QUESTIONS

1. How would the history of humanity's discovery of the Milky Way Galaxy have been different if the interstellar medium did not block starlight?
2. The three most important attributes in real estate are said to be: location, location, and location. If you could relocate our Solar System to another region of the Galaxy, where would that be? Why?

PROBLEMS

1. Make a scale sketch of our Galaxy in cross section. Include the disk, Sun, nucleus, halo, and some globular clusters. Try to draw the globular clusters to scale.
2. Because of interstellar dust, astronomers can see at most about 5 kpc into the disk of the galaxy at visual wavelengths. What percentage of the galactic disk does that include? (*Hint:* Consider the area of the entire disk versus the area visible from Earth.)
3. If the fastest passenger aircraft can fly 0.45 km/s (1000 mph), how long would it take to reach the Sun? The galactic center? (*Note:* $1\,\text{pc} = 3.1 \times 10^{13}$ km.)
4. If a typical halo star has an orbital velocity of 250 km/s, how long does it take to pass through the disk of the galaxy? Assume that the disk is 1000 pc thick. (*Note:* $1\,\text{pc} = 3.1 \times 10^{13}$ km and $1\,\text{yr} = 3.2 \times 10^{7}$ s.)
5. If the RR Lyrae stars in a globular cluster have apparent magnitudes of 14, how far away is the cluster? (*Hints:* See Figure 12-4, and the **Magnitude-Distance Formula** in the Math Reference Cards MR.)
6. If interstellar dust makes an RR Lyrae variable star look 1 magnitude fainter than it should, by how much will you overestimate its distance? (*Hint:* Use the **Magnitude-Distance Formula** in the Math Reference Cards MR.)
7. If you assume that a globular cluster 4.0 arc minutes in diameter is actually 25 pc in diameter, how far away is it? (*Hint:* Use the **Small-Angle Formula** in the Math Reference Cards MR.)
8. If the Sun is 4.6 billion years old, how many times has it orbited the galaxy?
9. If astronomers were to find they have made a mistake and our Solar System is actually 7.3 (rather than 8.3) kpc from the center of the galaxy, but the orbital velocity of the Sun is still 225 km/s, what is the minimum mass of the galaxy? (*Hint:* Use Kepler's third law from Chapter 3.)
10. What temperature would interstellar dust have to have to radiate most strongly at 100 μm? (*Note:* $1\,\mu\text{m} = 1000$ nm.) (*Hint:* See **Blackbody Radiation** in the Math Reference Cards MR.)
11. Infrared radiation from the center of our Galaxy with a wavelength of about 2 μm (2×10^{-6} m) comes mainly from cool stars. Use this wavelength as λ_{max} and find the temperature of the stars (*Hint:* See **Blackbody Radiation** in the Math Reference Cards MR.)
12. If an object at the center of the Milky Way Galaxy has a linear diameter of 1.0 AU, what will its angular diameter be as seen from Earth? Assume that the distance to the center of the galaxy is 8.3 kpc. (*Hint:* Use the **Small-Angle Formula** in the Math Reference Cards MR.)

LEARNING TO LOOK

1. Why does the galaxy in Figure 12-18 have so much dust in its disk? How big do you suppose the halo is?

FIGURE 12-18

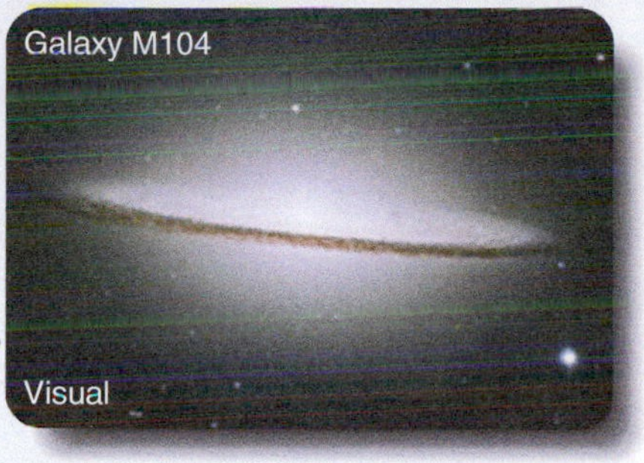

NASA/ESA/STScI/AURA/NSF/The Hubble Heritage Team

2. Why are the spiral arms in the galaxy in Figure 12-19 blue? What color would the halo be if it were bright enough to see in this photo?

FIGURE 12-19

NASA/ESA/STScI/AURA/NSF/Hubble Heritage Team/A. Reiss

13 Galaxies: Normal and Active

X-ray: NASA/CXC/SAO/Univ. of Massachusetts/Q. D. Wang et al.; Visual: NASA/ESA/STScI/AURA/NSF/The Hubble Heritage Team Infrared: NASA/JPL-Caltech/Univ. of Arizona/R. Kennicutt, SINGS Team

Somewhere something incredible is waiting to be known.
Carl Sagan

After finishing this chapter go to **PAGE 319** for **STUDY TOOLS.**

GUIDEPOST

Our Milky Way Galaxy is only one of the many billions of galaxies visible in the sky. (See CO 13.) This chapter will expand your horizon to discuss the different kinds of galaxies, their complex histories, and violent eruptions. Here you can expect answers to the following six important questions:

- What are the different types of galaxies?
- How do astronomers determine distances, sizes, luminosities, and masses of galaxies?
- Do other galaxies contain central supermassive black holes and dark matter, as does our Galaxy?
- How do collisions and interactions affect the evolution of galaxies?
- How do supermassive black holes make some galaxy nuclei active?
- How did supermassive black holes form, evolve, and affect their host galaxies?

As you begin studying galaxies, you will discover they are classified into different types, and that will lead you to insights into how galaxies form, interact, and evolve. In Chapter 14 you will take one last step outward and try to understand the birth and evolution of the entire Universe.

13-1 THE FAMILY OF GALAXIES

Astronomers classify galaxies according to their shape using a system developed in the 1920s by Edwin Hubble, after whom the *Hubble Space Telescope* is named. (See **How Do We Know? 13-1.**) Creating a system of classification is a fundamental technique in science.

Read **Concept Art 13A, Galaxy Classification**, and notice the following three important points:

1. Many galaxies have no disk, no spiral arms, and almost no gas and dust. These *elliptical galaxies* (class E) range from huge giants to small dwarfs.
2. Disk-shaped galaxies (class S) usually contain gas and dust and have spiral arms. Some of these spiral galaxies have a central region shaped like an elongated bar and are called *barred spiral galaxies* (class SB). You learned in Chapter 12 that the Milky Way Galaxy is a barred spiral. A few disk galaxies contain relatively little gas and dust; these are called *lenticular galaxies* (class S0).
3. Finally, notice the *irregular galaxies* (class I), which are generally shapeless and tend to be rich in gas and dust.

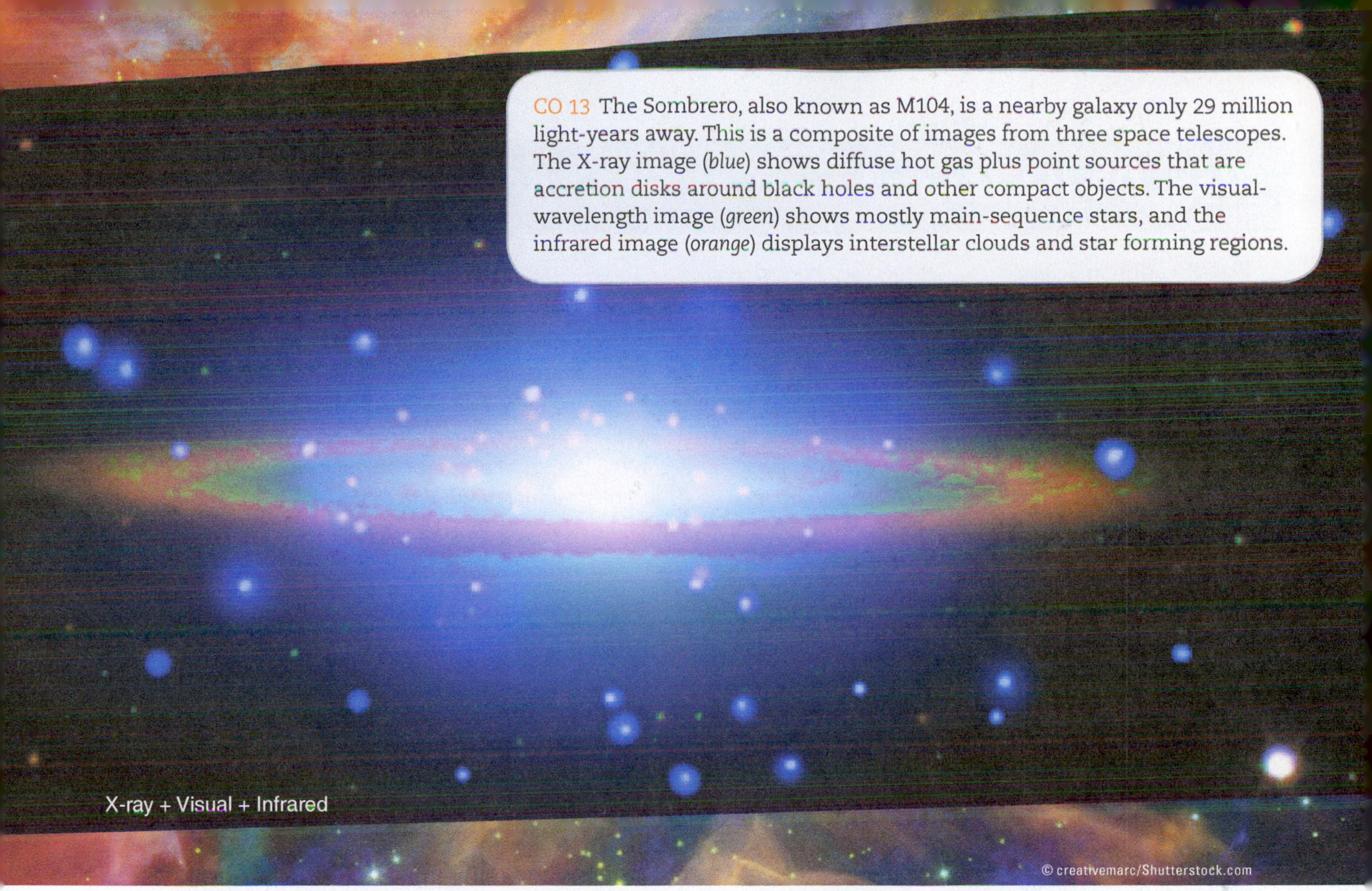

CO 13 The Sombrero, also known as M104, is a nearby galaxy only 29 million light-years away. This is a composite of images from three space telescopes. The X-ray image (*blue*) shows diffuse hot gas plus point sources that are accretion disks around black holes and other compact objects. The visual-wavelength image (*green*) shows mostly main-sequence stars, and the infrared image (*orange*) displays interstellar clouds and star forming regions.

You might also wonder what proportion of galaxies are elliptical, spiral, and irregular, but that is a difficult question to answer. In some catalogs of galaxies, about 70 percent are spiral, but that is misleading because spiral galaxies contain hot, luminous stars and clouds of ionized gas that make them easy to notice (Figure 13-1). From careful studies, astronomers conclude that ellipticals are actually more common than spirals and that irregulars make up about 25 percent of all galaxies. Among spiral galaxies, about two-thirds are barred spirals.

Different kinds of galaxies have different colors, depending mostly on how much star formation is happening in them. Spirals and irregulars usually contain plenty of young stars, including massive, hot, luminous O and B stars. They produce most of the light and give spirals and irregulars a distinct blue tint. In contrast, elliptical galaxies usually have few young stars. The most luminous stars in ellipticals are red giants, which give those galaxies a red tint.

How many galaxies are there? A research effort called GOODS (Great Observatories Origins Deep Survey) has used the *Hubble Space Telescope*, the *Chandra X-ray Observatory*, the *Spitzer Space Telescope*, and the *XMM-Newton X-ray Telescope*, plus the largest ground-based telescopes, to study two selected areas of the northern and southern skies. The GOODS images reveal tremendous numbers of galaxies, which are shown in Figure 13-2. There are good reasons to believe that the two regions of the sky chosen for study are typical, so evidently the entire sky is carpeted with galaxies. The result of the GOODS research is an estimate that at least 100 billion galaxies would be visible if modern telescopes were used for an all-sky census.

13-2 MEASURING THE PROPERTIES OF GALAXIES

Looking beyond the edge of our Milky Way Galaxy, astronomers find many billions of galaxies. What are the properties of these star systems? What are the diameters, luminosities, and masses of galaxies? Just as in your study of stellar characteristics (look back to Chapter 9), the first step in your study of galaxies is to find out how far away they are.

How Do We Know? 13-1

Classification in Science

What does classification tell a scientist? Classification is one of the most basic and most powerful of scientific tools. Establishing a system of classification is often the first step in studying a new aspect of nature, and it can produce unexpected insights.

Charles Darwin sailed around the world from 1831 to 1836 with a scientific expedition aboard the ship *HMS Beagle*. Everywhere he went, he studied the living things he saw and tried to classify them. For example, he classified different types of finches he saw on the Galapagos Islands based on the shapes of their beaks. He found that those that fed on seeds with hard shells had thick, powerful beaks, whereas those that picked insects out of deep crevices had long, thin beaks. His classifications of these and other animals led him to think about how natural selection shapes creatures to survive in their environment, which led him to understand how living things evolve.

Years after Darwin's work, paleontologists classified dinosaurs into two orders, lizard-hipped and bird-hipped dinosaurs. This classification, based on the shapes of dinosaur hip joints, helped the scientists understand patterns of evolution of dinosaurs. It also led to the conclusion that modern birds, including the finches that Darwin saw on the Galapagos, evolved from dinosaurs.

Michael A. Seeds

The careful classification of living things has revealed that birds, including this flamingo, are descended from dinosaurs.

Astronomers use classifications of galaxies, stars, moons, and many other objects to help them see patterns, trace relationships, and generally make sense of the astronomical world. Whenever you encounter a scientific discussion, look for the classifications on which it is based. Classifications are the orderly framework on which much of science is built.

Once you know a galaxy's distance, its size and luminosity are relatively easy to find. Later in this section, you will see that finding the masses of galaxies is more difficult, but the results are intriguing.

13-2a Distance

The distances to galaxies are so large that it is not convenient to express them in light-years, parsecs, or even kiloparsecs. Instead, astronomers use the unit **megaparsec (Mpc)**, or 1 million pc. One Mpc equals 3.3 million ly, or approximately 3×10^{19} kilometers (2×10^{19} miles).

megaparsec (Mpc) A unit of distance equal to 1,000,000 pc.

standard candles Objects of known brightness that astronomers use to find distance—for example, Cepheid variable stars and supernovae.

To find the distance to a galaxy, astronomers must search among its stars, nebulae, and star clusters for familiar objects with known luminosities. Such objects are called **standard candles**. If you can find a standard candle in a galaxy, you can judge its distance.

Cepheid variable stars are reliable standard candles because their period is related to their luminosity (see Figure 12-4). If you know the period of the star's variation, you can use the period–luminosity diagram to learn its absolute magnitude. Then, by comparing its absolute and apparent magnitudes, you can find its distance. Figure 13-3 shows a galaxy in which the *Hubble Space Telescope* detected Cepheids. Even with the *Hubble Space Telescope*, Cepheids are not detectable much beyond 100 million ly (30 Mpc), so astronomers must search for less common but brighter distance indicators and calibrate them using nearby galaxies containing detectable Cepheids.

FIGURE 13-1

A century ago, photos of galaxies looked like spiral clouds of haze. Modern images of these relatively nearby galaxies reveal dramatically beautiful objects filled with newborn stars and clouds of gas and dust.

When a supernova explodes in a distant galaxy, astronomers rush to observe it. Studies show that Type Ia supernovae, those caused by the collapse of a white dwarf (look back to Chapter 11 Section 11-4c), all reach about the same luminosity at maximum (which make them more like standard bombs than standard candles). By searching for Cepheids and other distance indicators in nearby galaxies where Type Ia supernovae have occurred, astronomers have been able to calibrate these supernovae. As a result, when Type Ia supernovae are seen in more distant galaxies, astronomers can measure the apparent brightness at maximum and compare that with the known luminosity of those supernovae to find their distances. Because Type Ia supernovae are much brighter than Cepheids, they can be seen in galaxies at great distances. The drawback is that supernovae are rare, and none might occur during your lifetime in a galaxy you might be studying. You will see in Chapter 14 how these distance indicators have reshaped our understanding about the history of the Universe.

Notice how astronomers use calibration to build a distance scale reaching from the nearest galaxies to the most distant visible galaxies. Often astronomers refer to this as the **distance ladder** because each step depends on the

distance ladder The calibration used to build a distance scale reaching from the size of Earth to the most distant visible galaxies.

Concept Art 13A

GALAXY CLASSIFICATION

1

Elliptical galaxies are round or elliptical, contain no visible gas and dust, and lack hot, bright stars. They are classified with a numerical index ranging from 1 to 7; E0s are round, and E7s are highly elliptical. The index is calculated from the largest and smallest diameter of the galaxy using the following formula, rounding to the nearest integer:

$$\frac{10(a-b)}{a}$$

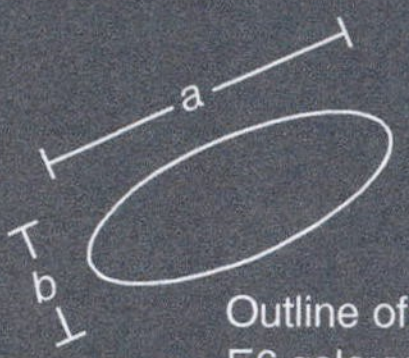

Outline of an E6 galaxy

Australian Astronomical Observatory/ David Malin Images

NOAO/AURA/NSF

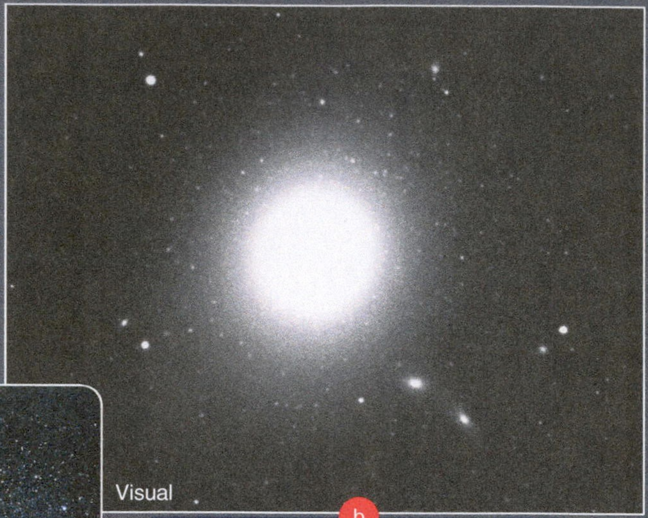

AATB/David Malin Images

a The Leo 1 dwarf elliptical galaxy is not much bigger than a globular cluster.

b M87 is a giant elliptical galaxy classified as E1. It is several times larger in diameter than our own Galaxy and is surrounded by a swarm of more than 500 globular clusters.

2

Spiral galaxies contain a disk and spiral arms. Their halo stars are not visible, but presumably all spiral galaxies have halos. Spirals contain gas and dust and hot, bright O and B stars, as shown at right and below. The presence of short-lived O and B stars alerts us that star formation is occurring in these galaxies. Sa galaxies have larger nuclei, less gas and dust, and fewer hot, bright stars. Sc galaxies have small nuclei, lots of gas and dust, and many hot, bright stars. Sb galaxies are intermediate.

AATB/David Malin Images

AATB/David Malin Images

AATB/David Malin Images

AATB/David Malin Images

2a

Roughly two-thirds of all spiral galaxies are **barred spiral galaxies** classified as SBa, SBb, and SBc. They have an elongated nucleus with spiral arms springing from the ends of the bar, as shown at left. The Milky Way Galaxy is a barred spiral with a spiral form between Sb and Sc, so its complete classification is SBbc.

2b Some disk galaxies are rich in dust, which is concentrated along their spiral arms. NGC 4013, shown below, is a galaxy much like ours, but seen edge-on its dust is readily apparent.

NASA/ESA/STScI/AURA/NSF/Hubble Heritage Team

b Dust in spiral galaxies is most common in the spiral arms. Here the spiral arms of one galaxy are silhouetted in front of a more distant galaxy.

NASA/ESA/STScI/AURA/NSF/Hubble Heritage Team

2c Galaxies with an obvious disk and nuclear bulge but little or no visible gas and dust and no hot, bright stars are called **lenticular galaxies**, classified as S0 (pronounced "Ess Zero"). Compare this galaxy with the edge-on spiral above.

R. E. Schild (Harvard-Smithsonian CfA)

G.J.Jacoby, M.J.Pierce/NOAO/AURA/NSF

a The galaxy IC 4182 is a dwarf irregular galaxy about 4 million pc from our galaxy.

3 **Irregular galaxies** (classified Irr) are a chaotic mix of gas, dust, and stars with no obvious nuclear bulge or spiral arms. The Large and Small Magellanic Clouds are visible to the unaided eye as hazy patches in the Southern Hemisphere sky. Telescopic images show that they are irregular galaxies that are interacting gravitationally with our own much larger Galaxy. Star formation is rapid in the Magellanic Clouds. The bright pink regions are emission nebulae excited by newborn O and B stars. The brightest nebula in the Large Magellanic Cloud is called the Tarantula Nebula.

NOAO/AURA/NSF

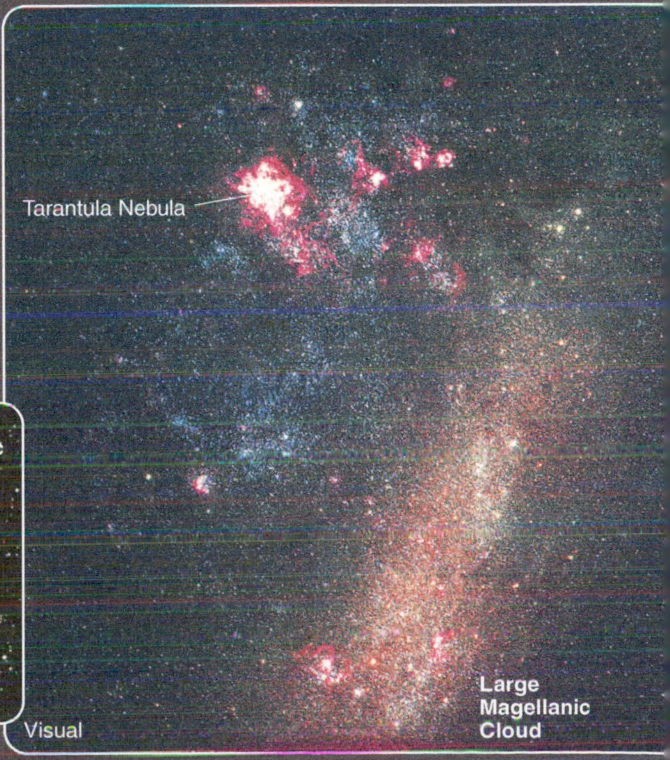

© R. J. Dufour (Rice Univ.)

FIGURE 13-2

NASA/ESA/STScI/AURA/NSF/R. Williams and the Hubble Deep Field Team

An apparently empty spot on the sky only one-thirtieth the diameter of the full moon contains more than 1500 galaxies in this extremely long time exposure known as the Hubble Deep Field North. Only four stars are visible in this image; they are sharp points of light with diffraction spikes produced by the telescope optics. Evidently the entire sky is similarly filled with galaxies.

steps below it. The extragalactic distance scale rests on determining luminosities of stars, which rests on measuring stellar parallax (look back to Chapter 9 Section 9-1).

13-2b Telescopes as Time Machines

The most distant visible galaxies are a little more than 10 billion ly (3000 Mpc) away, and at such distances you see an effect like time travel. When you look at a galaxy millions of light-years away, you do not see it as it is now but as it was millions of years ago when its light began the journey toward Earth. When you look at a distant galaxy, you look into the past by an amount called the **look-back time**, a time in years equal to the distance to the galaxy in light-years.

look-back time The amount by which you look into the past when you look at a distant galaxy, a time equal to the distance to the galaxy in light-years.

Hubble law The linear relation between the distances to galaxies and the apparent velocity of recession.

Hubble constant A measure of the rate of expansion of the Universe, the average value of the apparent velocity of recession divided by distance, about 73 km/s/Mpc.

The look-back time to nearby objects is usually not significant. The look-back time to the Moon is only 1.3 seconds, to the Sun 8 minutes, and to the nearest star about 4 years. The Andromeda Galaxy has a look-back time of about 2 million years, but that is a mere eye blink in the lifetime of a galaxy. When astronomers look at more distant galaxies, the look-back time becomes an appreciable part of the age of the Universe. You will see evidence in Chapter 14 that the Universe began 13.8 billion years ago. When astronomers observe the most distant visible galaxies, they are looking back more than 10 billion years to a time when the Universe was significantly different.

13-2c The Hubble Law

Although astronomers find it difficult to measure the distance to a galaxy precisely, they often estimate such distances using a simple relationship. Early in the 20th century, astronomers noticed that the lines in galaxy spectra are generally shifted toward longer wavelengths—redshifted (look back to Chapter 5 Section 5-4c). These redshifts imply that the galaxies are receding from Earth.

In 1929, the American astronomer Edwin Hubble published a graph that plotted the apparent velocity of recession versus distance for a number of galaxies. The points in the graph fell along a straight line (Figure 13-4). This relation between apparent velocity of recession and distance is known as the **Hubble law**, and the slope of the line is known as the **Hubble constant**, symbolized by H.

The Hubble law is important in astronomy for two reasons. It is evidence that the Universe is expanding, a subject to be discussed in detail in Chapter 14. This chapter discusses the Hubble law because astronomers use it to estimate the distance to galaxies. The distance to a galaxy can be found by dividing its apparent velocity of recession by the Hubble constant. (See **The Hubble Law** in the Math Reference Cards MR.) This is a very useful calculation, because it is usually possible to obtain a spectrum of a galaxy and measure its redshift even if it is too far away to have observable standard candles. Obviously, knowing the precise value of the Hubble constant is important.

Edwin Hubble's original measurement of H was too large because of errors in his measurements of the distances to galaxies. Astronomers worked hard over the following decades to measure this important constant. The most precise measurements of the Hubble constant, made using the *Hubble*, *Spitzer*, and *WMAP* space telescopes yield a value for H of about 70 km/s/Mpc with an uncertainty of just a few percent. This means that a galaxy at a distance of 1 Mpc from the Milky Way is receding

FIGURE 13-3

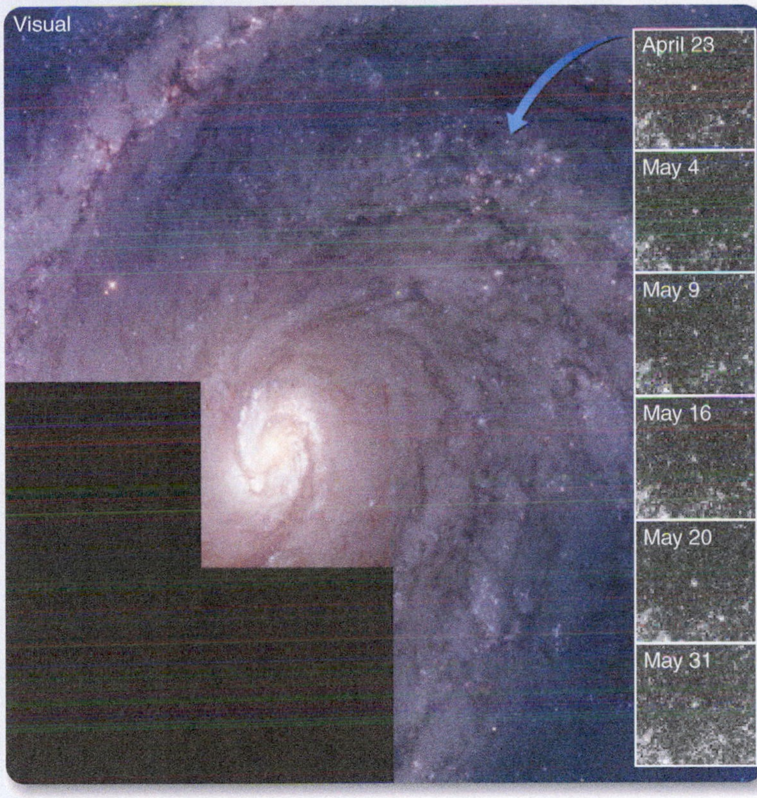

NASA/ESA/STScI/AURA/NSF/J. Trauger (JPL-Caltech) and W. Freedman (CIW)

The vast majority of spiral galaxies are too distant for Earth-based telescopes to detect Cepheid variable stars. The *Hubble Space Telescope*, however, can locate Cepheids in some of these galaxies including in the bright spiral galaxy M100 shown here. From a series of images taken on different dates, astronomers can locate Cepheids (inset), determine the period of pulsation, and measure the average apparent brightness. They can then deduce the distance to the galaxy—16 Mpc (52 million ly) for M100.

from us at a rate of 70 km/s, a galaxy 2 Mpc away is receding at 140 km/s, and so on.

Note that the redshifts of galaxies are not really Doppler shifts, even though astronomers often express the redshifts in kilometers per second as if they were true velocities. In Chapter 14, you will discover a much more sophisticated way of understanding these redshifts. For the moment, you can just keep in mind that modern astronomers interpret redshifts of galaxies as caused by the expansion of the Universe, and the Hubble law allows them to estimate the distance to a galaxy from its redshift.

13-2d Galaxy Diameters and Luminosities

The distance to a galaxy is the key to finding its diameter and its luminosity. With even a modest telescope and a CCD camera, you could photograph a galaxy and measure its angular diameter. If you know the distance to the galaxy, you can then find its linear diameter. Also, if you measure the apparent brightness of the galaxy, you can use the distance to find its luminosity, as you learned regarding stars in Chapter 9 Section 9-2b.

The results of such observations show that galaxies differ dramatically in size and luminosity. Irregular galaxies tend to be small, 5 percent to 25 percent the diameter of our Galaxy, and of low luminosity. Although they are common, they are easy to overlook. Our Milky Way Galaxy is large and luminous compared with most spiral galaxies, though astronomers know about a few spiral galaxies that are even larger and more luminous. Elliptical galaxies cover a wide range of diameters and luminosities. The largest, called giant ellipticals, are 10 times the diameter of our Milky Way Galaxy, but many so-called dwarf elliptical galaxies are only 1 percent the diameter of our Galaxy.

Clearly, the diameter and luminosity of a galaxy do not determine its type. Some small galaxies are irregular, and some are elliptical. Some large galaxies are spiral, and some are elliptical. Other factors must influence the evolution of galaxies.

13-2e Galaxy Masses

Although the mass of a galaxy is difficult to determine, it is an important quantity. It tells you how much matter the galaxy contains, which provides clues to the galaxy's origin and evolution.

The most precise method for measuring the mass of a galaxy is called the **rotation curve method** (Figure 13-5). It requires knowing: (1) the true sizes

rotation curve method A method of determining a galaxy's mass by measuring the orbital velocity and orbital radius of stars in the galaxy.

FIGURE 13-4

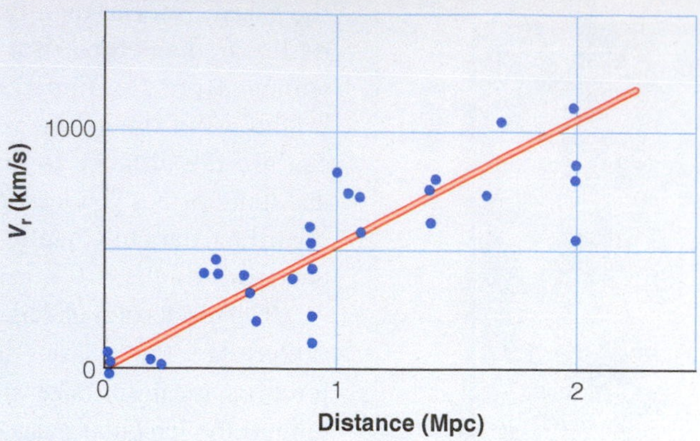

Edwin Hubble and Milton Humason's first diagram of the apparent velocities of recession and distances of galaxies did not probe very deeply into space, and the horizontal axis was later recalibrated. These data did show, however, that the galaxies are receding from one another with speeds proportional to their separation.

of the orbits of stars or gas clouds within a galaxy, which in turn requires knowing the distance of that galaxy; and (2) the orbital speeds of the stars or gas clouds, measured from the Doppler shifts of their spectral lines. That is enough information to use Kepler's third law and find the mass of the part of the galaxy contained within the star orbits with measured sizes and velocities (look back to Chapter 12 Section 12-1e). The rotation curve method works only for galaxies near enough to be well resolved. More distant galaxies appear so small that astronomers cannot measure the radial velocity at different points across the galaxy and must use other, less precise, methods to estimate masses.

The masses of galaxies cover a wide range. The smallest are a million times less massive than the Milky Way, and the largest contain as much as 50 times more mass than the Milky Way.

13-2f Dark Matter in Galaxies

Given the size and luminosity of a galaxy, astronomers can make a rough guess as to the amount of matter it should contain. Astronomers know how much light stars produce, and they know about how much matter there is between the stars, so it should be possible to estimate very roughly the mass of a galaxy from its luminosity. When astronomers measure the masses of galaxies, however, they often find that the measured masses are much larger than expected from the luminosities of the galaxies. You discovered this effect in Chapter 12 when you studied the rotation curve of our own Galaxy and concluded that it must contain large amounts of dark matter, especially in its outer regions (look back to Figure 12-9). Astronomer Vera Rubin found, in observations begun in the 1960s, that this also seems to be true of most nearby galaxies. Measured masses of galaxies amount to 10 to 100 times more mass than you can see.

Dark matter is difficult to detect, and it is even harder to explain. Some astronomers have suggested that dark matter consists of low-luminous white dwarfs and brown dwarfs scattered through the halos of galaxies. Searches for white dwarfs and brown dwarfs in the halo of our Galaxy have found a few but not enough to make up most of the dark matter. The dark matter can't be hidden in vast numbers of black holes and neutron stars, because astronomers don't see the X-rays these objects would emit. The evidence indicates there is 10 to 100 times more dark matter than visible matter in galaxies, and if there were that many black holes they would produce X-rays that would be easy to detect. Furthermore, recent images from the *Chandra X-ray Observatory* indicate that a collision between two galaxy clusters caused their gas and dark matter components to separate.

Because observations imply that the dark matter can't be composed of familiar objects or material, astronomers are forced to conclude that the dark matter is made up of unexpected forms of matter. Until recently, neutrinos were thought to be massless, but studies now suggest they have a very small mass. Thus they can be part of the dark matter, but their masses are too low to make up all of the dark matter. There must be some other undiscovered form of matter in the Universe that is detectable only by its gravitational field.

Dark matter remains one of the fundamental unresolved problems of modern astronomy. Observations of galaxies and clusters of galaxies reveal that 90 percent to 95 percent of the matter in the Universe is dark matter. The Universe you see—the kind of matter that you and the stars are made of—has been compared to the foam on an invisible ocean. You will return to this problem in Chapter 14 when you try to understand how dark matter affects the nature of the Universe, its past, and its future.

13-2g Supermassive Black Holes in Galaxies

Doppler shift measurements show that the stars near the centers of many galaxies are orbiting very rapidly. To hold stars in such small, short-period orbits, the centers

FIGURE 13-5

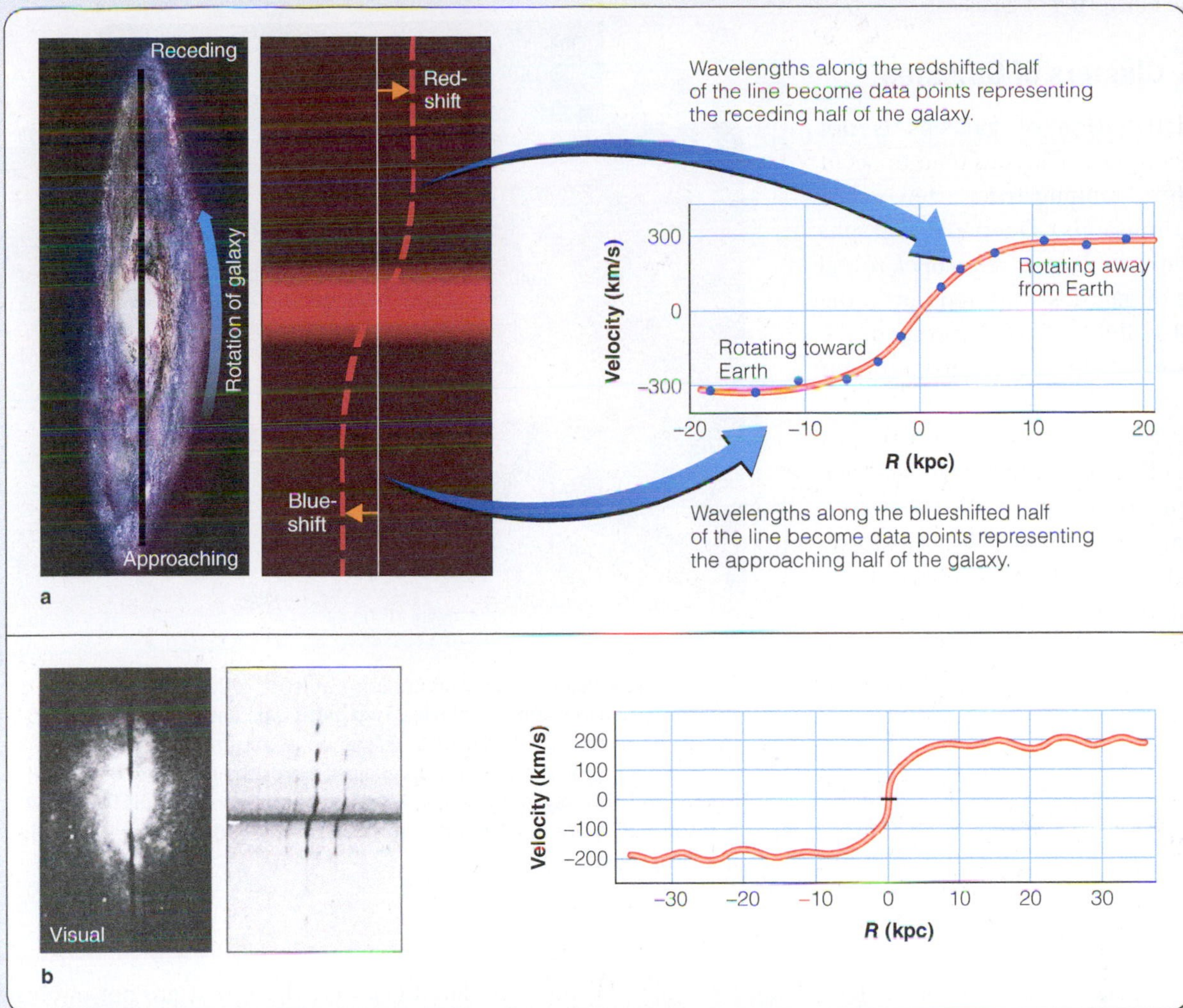

(a) In the upper panel's artwork representation, an astronomer has placed the image of the galaxy over a narrow slit so that light from the galaxy can enter the spectrograph and produce a spectrum. A very short segment of the spectrum shows an emission line redshifted on the receding side of the rotating galaxy and blueshifted on the approaching side. Converting those Doppler shifts into velocities, the astronomer can plot the galaxy's rotation curve (*right*). (b) Real data are shown in the lower panel. Galaxy NGC 2998 is shown over the spectrograph slit, and the segment of the spectrum includes three emission lines.

of those galaxies must contain masses of a million to a few billion solar masses, yet no object is visible. The evidence seems to require that the nuclei of many galaxies contain supermassive black holes. You saw in Chapter 12 Section 12-4 that the Milky Way Galaxy contains a supermassive black hole at its center. Evidently that is typical of galaxies. It is a common misconception that the orbits of stars throughout a galaxy are controlled by the central black hole. The masses of those black holes, large as they might seem, are negligible compared with a galaxy's mass. The 4.3-million-solar-mass black hole at the center of the Milky Way Galaxy contains only a thousandth of 1 percent of the total mass of the galaxy.

13-3 THE EVOLUTION OF GALAXIES

Your goal in this chapter will be to build a theory to explain the evolution of galaxies. In Chapter 12 Section 12-3d, you learned about two models of how our Milky Way Galaxy, a large spiral, might have formed. Presumably, other large spiral galaxies formed similarly. But why did some galaxies become spiral, some elliptical, and some irregular? Clues to that mystery lie in the clustering of galaxies.

13-3a Clusters of Galaxies

The distribution of galaxies is not entirely random. Galaxies tend to occur in clusters, ranging from a few galaxies to thousands. Deep photographs made with the largest telescopes reveal clusters of galaxies scattered out to the limits of visibility. This clustering of the galaxies can help you understand their evolution.

For this discussion, you can sort clusters of galaxies into two groups, rich and poor. **Rich galaxy clusters** contain more than 1000 galaxies, mostly elliptical, crowded into a spherical volume about 3 Mpc (10^7 ly) in diameter. The Coma cluster (located 100 Mpc from Earth in the direction of the constellation Coma Berenices) is an example of a rich cluster, as seen in Figure 13-6. It contains at least 1000 galaxies, mostly E and S0 types. Close to its center are a giant elliptical galaxy and a large S0 galaxy. Rich clusters often contain one or more giant elliptical galaxies at their centers.

FIGURE 13-6

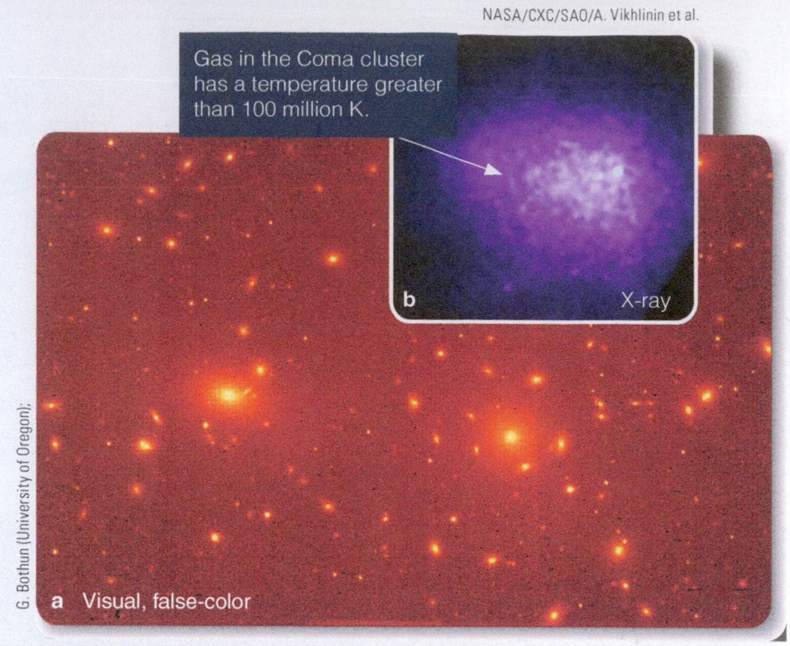

(a) The Coma cluster of galaxies contains at least 1000 galaxies and is especially rich in E and S0 galaxies. Two giant galaxies lie near its center. Only the central area of the cluster is shown in this image. If the cluster were visible in the sky, it would span eight times the diameter of the full moon. (b) This X-ray image of the Coma cluster shows that it is filled and surrounded by hot gas. Note that the two brightest galaxies are visible in the X-ray image.

Poor galaxy clusters contain fewer than 1000 galaxies and are irregularly shaped and less crowded toward the center. Our own **Local Group**, which contains the Milky Way Galaxy, is a good example of a poor cluster (Figure 13-7). It contains a few dozen members scattered irregularly through a volume slightly more than 1 Mpc in diameter. Of the brighter galaxies, 14 are elliptical, 3 are spiral, and 4 are irregular.

Classifying galaxy clusters into rich and poor clusters reveals a fascinating and suggestive clue to the evolution of galaxies. In general, rich clusters tend to contain 80 percent to 90 percent E and S0 galaxies and few spirals. Poor clusters contain a larger percentage of spirals; and among isolated galaxies, those that are not in clusters, 80 percent to 90 percent are spirals. This suggests that a galaxy's environment is important in determining its structure and has led astronomers to suspect that one of the secrets to galaxy evolution lies in the collisions between galaxies.

rich galaxy cluster A cluster containing 1000 or more galaxies, usually mostly ellipticals, scattered over a volume only a few megaparsecs in diameter.

poor galaxy cluster An irregularly shaped cluster that contains fewer than 1000 galaxies, many of which are spiral, and no giant ellipticals.

local group The small cluster of a few dozen galaxies that contains the Milky Way Galaxy.

13-3b Colliding Galaxies

Astronomers are finding more and more evidence to show that galaxies collide, interact, and merge. In fact, collisions among galaxies might dominate their evolution.

You should not be surprised that galaxies collide with each other. The average separation between galaxies is only about 20 times their diameter, so galaxies should bump into each other fairly often, astronomically speaking. In comparison, stars almost never collide because the typical separation between stars is about 10^7 times their diameter. Consequently, a collision between two stars is about as likely as a collision between two gnats flitting about in a baseball stadium.

FIGURE 13-7

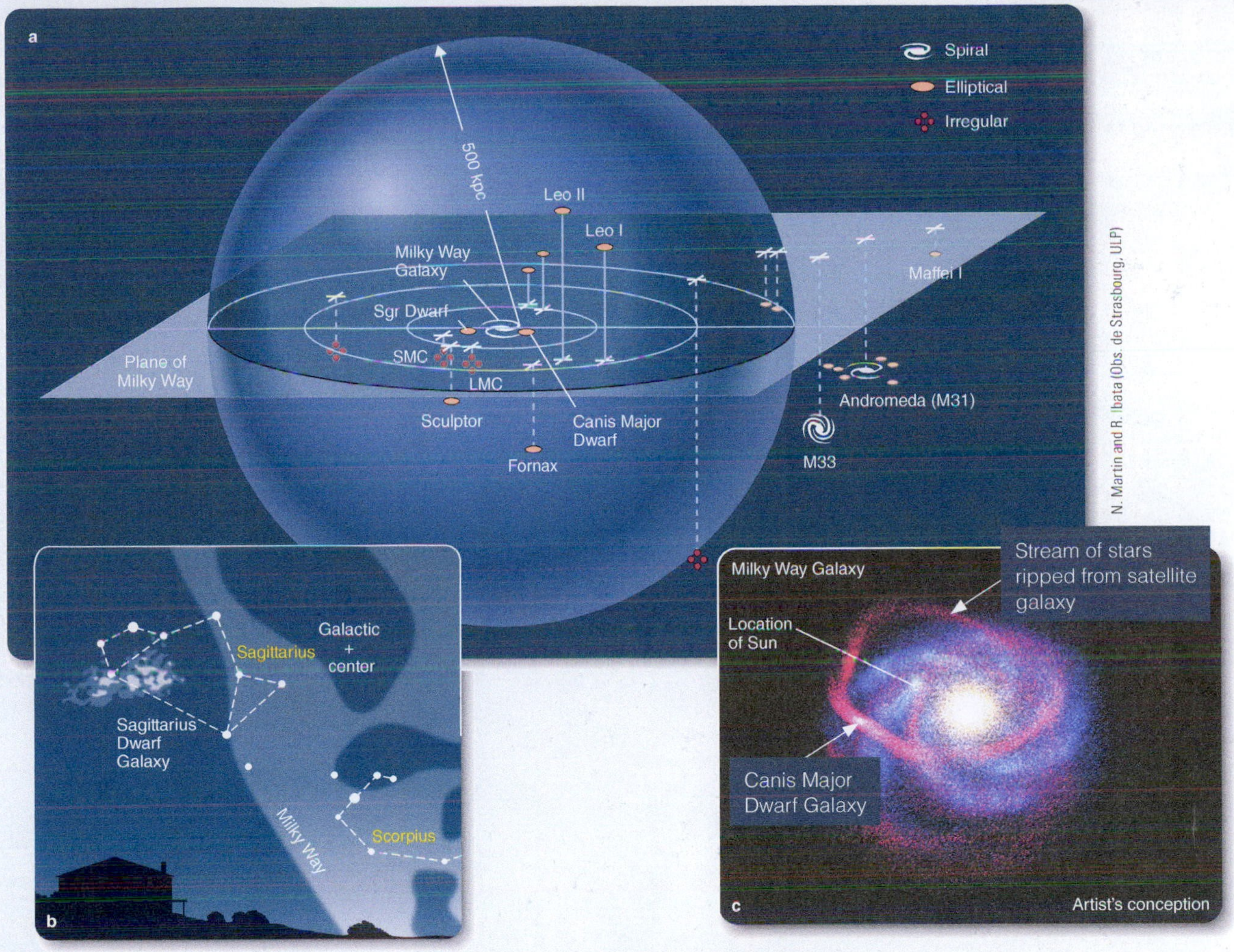

(a) The Local Group. Our Galaxy is located at the center of this diagram. The vertical lines giving distances from the plane of the Milky Way are solid above the plane and dashed below. (b) The Sagittarius Dwarf Galaxy (Sgr Dwarf) lies on the other side of our Galaxy. If you could see it in the sky, its minor axis would appear 17 times wider than the full Moon. (c) The Canis Major Dwarf Galaxy, shown in this simulation, is even closer to the Milky Way Galaxy. It is the remains of a small galaxy that is being pulled apart by our Galaxy and is hidden behind the stars of the constellation Canis Major.

Read **Concept Art 13B, Interacting Galaxies**, and notice the following four important points:

1. Interacting galaxies can distort each other with tides, producing tidal tails and shells of stars. Interactions might help stimulate the formation of spiral arms. Large galaxies can even absorb smaller galaxies.
2. Galaxy interactions can trigger rapid star formation.
3. Evidence within galaxies in the forms of multiple nuclei and conflicting directions of star and gas motions reveals that many have suffered past interactions and mergers.
4. There are beautiful ring galaxies—understood to be bull's-eyes left behind by high-speed perpendicular collisions.

Evidence of galaxy mergers is all around you. Our Milky Way Galaxy is a cannibal galaxy, snacking on the two Magellanic Clouds as they sail past it. Its tides are also pulling apart two other small satellite galaxies, the Sagittarius and Canis Major dwarf galaxies, producing great streamers of stars wrapped around the Milky Way. Almost certainly, our Galaxy has dined on other small galaxies in the past.

Concept Art 13B

INTERACTING GALAXIES

1

When two galaxies collide, they can pass through each other without stars colliding because the stars are so far apart relative to their sizes. Gas clouds and magnetic fields do collide, but the biggest effects may be tidal. Even when two galaxies just pass near each other, tides can cause dramatic effects, such as long streamers called **tidal tails.** In some cases, two galaxies can merge and form a single galaxy.

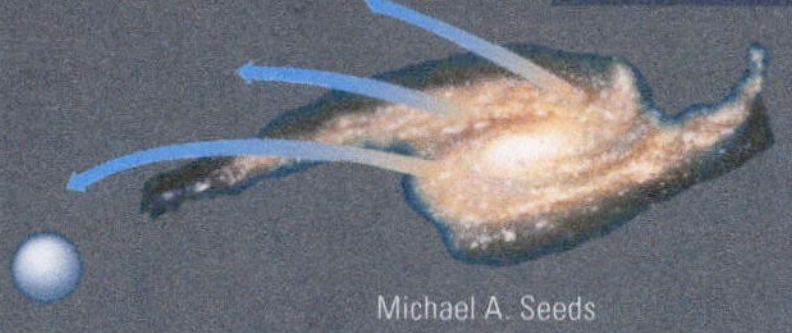

Michael A. Seeds

f — **Small galaxy passing near a massive galaxy.**

g — **Gravity of a second galaxy represented as a single massive object**

1a

When a galaxy swings past a massive object such as another galaxy, tides are severe. Stars near the massive object try to move in smaller, faster orbits whereas stars farther from the massive object follow larger, slower orbits. Such tides can distort a galaxy or even rip it apart.

Galaxy interactions can stimulate the formation of spiral arms.

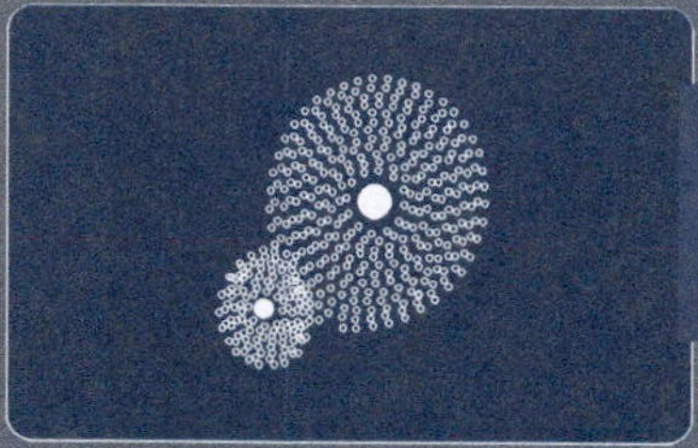

Allen Beechel, CreativeGround.org

a — **In this computer model, two uniform disk galaxies pass near each other.**

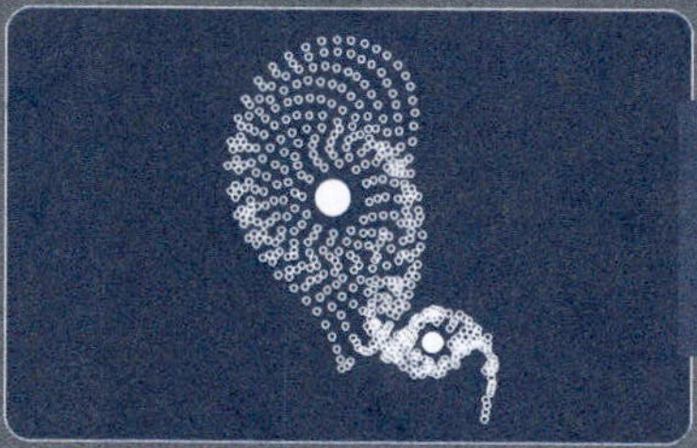

Allen Beechel, CreativeGround.org

b — **The small galaxy passes behind the larger galaxy so they do not actually collide.**

Allen Beechel, CreativeGround.org

c — **Tidal forces deform the galaxies and trigger the formation of spiral arms.**

Allen Beechel, CreativeGround.org

d — **The upper arm of the large galaxy passes in front of the small galaxy.**

NOAO/AURA/NSF

f — **A photo of the well-known Whirlpool Galaxy resembles the computer model.**

NOAO/AURA/NSF

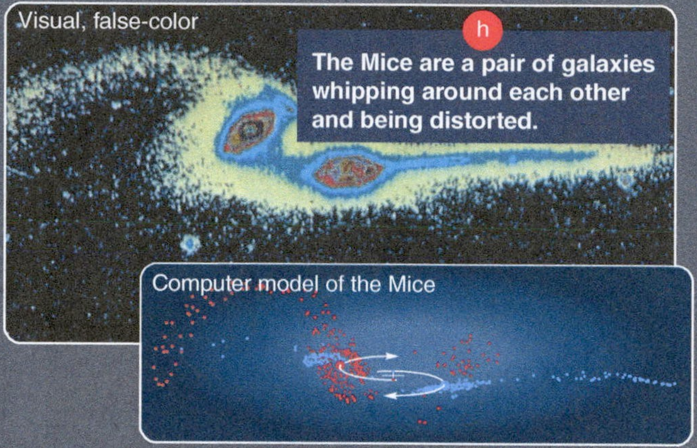

h — **The Mice are a pair of galaxies whipping around each other and being distorted.**

Allen Beechel, CreativeGround.org

1b

The merger of galaxies is called **galactic cannibalism**. Models show that merging galaxies spiral around their common center of mass while tides rip stars away and form shells.

Such shells have been found around elliptical galaxies such as NGC 5128. It is peculiar in many ways and even has a belt of dusty gas. The shells revealed in this enhanced image are evidence that the giant galaxy has cannibalized at least one smaller galaxy. The giant galaxy itself may be the result of the merger of two large galaxies.

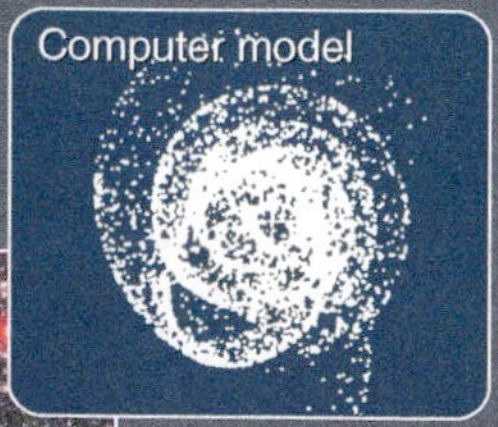

F. Schweizer and A. Toomre

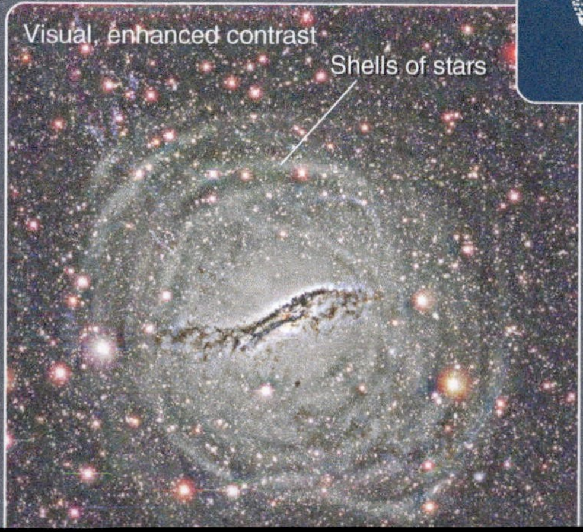

JHU/NOAO/AURA/NSF

2 The collision of two galaxies can trigger firestorms of star formation as gas clouds are compressed. Galaxies NGC 4038 and 4039 have been known for years as the Antennae because the long tails visible in Earth-based photos resemble the antennae of an insect. *Hubble Space Telescope* images reveal that the two galaxies are blazing with star formation. Roughly a thousand massive star clusters have been born.

Spectra show that the Antennae galaxies are 10 to 20 times richer in elements such as magnesium and silicon than the Milky Way. Such metals are produced by nucleosynthesis in massive stars and spread back into space by subsequent supernova explosions.

F. Schweizer (CIW) and B. Whitmore (STScI)

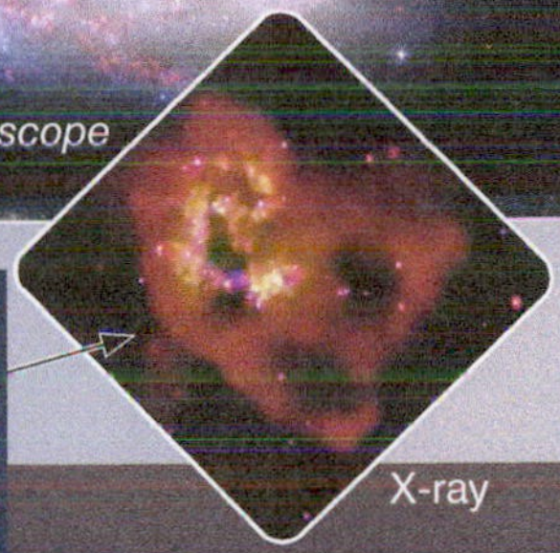

NASA/CXC/SAO/ G. Fabbiano et al.

a **An X-ray image of the Antennae shows clouds of very hot gas heated by supernovae exploding 30 times more often than in our own Galaxy.**

3 Evidence of past galaxy mergers shows up in the motions inside some galaxies. NGC 7251 is a highly distorted galaxy with tidal tails in this ground-based image.

F. Schweizer (CIW) and B. Whitmore (STScI)

a **This *Hubble Space Telescope* image of the core of the galaxy reveals a small spiral spinning backward in the heart of the larger galaxy.**

b **This counter-rotation suggests that NGC 7251 is the remains of two oppositely rotating galaxies that merged about a billion years ago.**

3a Radio evidence of past mergers: Doppler shifts reveal the rotation of the spiral galaxy M64. The upper part of the galaxy has a redshift and is moving away from Earth, and the bottom part of the galaxy has a blueshift and is approaching. A radio map of the core of the galaxy reveals that it is rotating backward. This suggests a merger long ago between two galaxies that rotate in opposite directions.

Rotation of galaxy M64

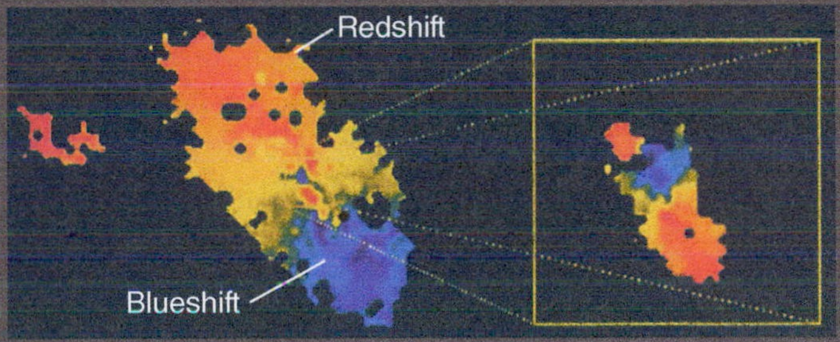

NRAO/AUI/NSF/R. Braun

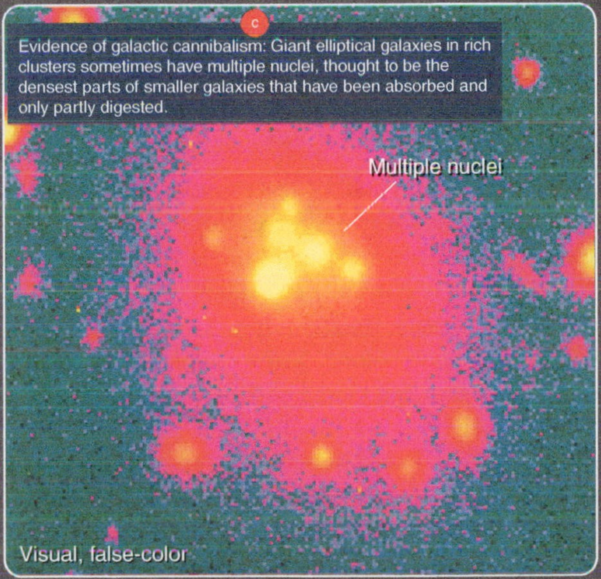

c Evidence of galactic cannibalism: Giant elliptical galaxies in rich clusters sometimes have multiple nuclei, thought to be the densest parts of smaller galaxies that have been absorbed and only partly digested.

ESO/M. J. West (St. Mary's Univ.)

4 The Cartwheel Galaxy pictured below was once a normal spiral or lenticular galaxy but is now a **ring galaxy**. One of its smaller companions has plunged through at high speed almost perpendicular at the Cartwheel's disk. That has triggered a wave of star formation, and the more massive stars have exploded leaving behind black holes and neutron stars. Some of those are in X-ray binaries, and that makes the outer ring bright in X-rays.

NASA/JPL-Caltech/P. N. Appleton (SSC)

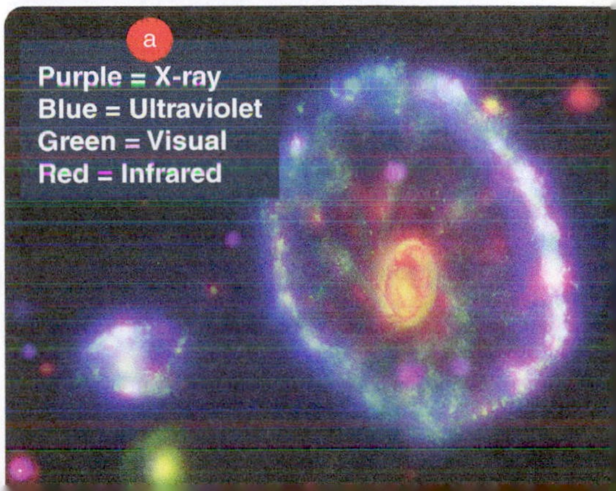

13-3c Origin and Evolution of Galaxies

The test of any scientific understanding is whether you can put all the evidence and theory together to tell the history of the objects studied. Can you describe the origin and evolution of the galaxies? Just a few decades ago, it would have been impossible, but the evidence from space telescopes and new-generation telescopes on Earth combined with advances in computer modeling and theory allow astronomers to outline the story of the galaxies.

Elliptical galaxies appear to be the product of galaxy mergers, which triggered star formation that used up the gas and dust. In fact, as shown in Figure 13-8, astronomers see star formation being stimulated to high levels in many galaxies. These **starburst galaxies** are very luminous in the infrared because a collision has triggered a burst of star formation that is heating the dust. The warm dust reradiates the energy in the infrared. The Antennae Galaxies (see Concept Art 13B) contain more than 15 billion solar masses of hydrogen gas and will become a starburst galaxy as their ongoing merger continues to trigger rapid star formation.

A few collisions and mergers could leave a galaxy with no gas and dust from which to make new stars. Astronomers now suspect that most large ellipticals are formed by the merger of two or more galaxies. The dwarf ellipticals (too small to be formed by mergers) and irregulars might be fragments left over by the merger of larger galaxies.

In contrast, spiral galaxies evidently seem never to have suffered major collisions. Their thin disks are delicate and would be destroyed by tidal forces in a collision with a massive galaxy. Also, they retain plenty of gas and dust and continue making stars. Our Milky Way Galaxy has evidently never merged with another large galaxy, although the Milky Way, as well as its large spiral neighbor the Andromeda Galaxy, both seem to have cannibalized smaller galaxies.

FIGURE 13-8

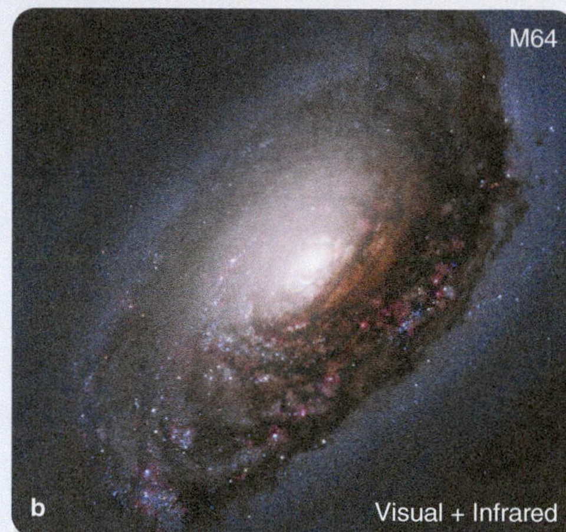

Rapid star formation: (a) NGC 1569 is a starburst galaxy filled with clouds of young stars and supernovae. (b) At least some starbursts are triggered by interactions between galaxies. The inner parts of M64, known as the Black Eye Galaxy, are filled with dust produced by rapid star formation. Radio Doppler shift observations show that the inner part of the galaxy rotates backward compared to the outer part of the galaxy, a product of a merger. Where the counter-rotating parts of the galaxy collide, star formation is stimulated.

Barred spiral galaxies also might be the product of tidal interactions. Mathematical models show that bars are not stable and should eventually dissipate. It might take tidal interactions with other galaxies to regenerate the bars. Because roughly two-thirds of all spiral galaxies have bars, you can suspect that these tidal interactions are common. Our Milky Way Galaxy is probably a barred spiral (look back to Figure 12-11), and interaction with its two Magellanic Cloud companions, or the more distant but very massive Andromeda Galaxy, could be the cause.

starburst galaxy A galaxy undergoing a rapid burst of star formation.

Other processes can alter galaxies. The S0 galaxies, which have disks and bulges like a spiral galaxy but no spiral arms, might have lost much of their gas and dust moving through the gas trapped in the dense clusters to which they belong. For example, X-ray observations show that the Coma cluster contains thin, hot gas between the galaxies (Figure 13-6). A galaxy moving through that gas would encounter a tremendous wind that could strip away its gas and dust.

How Do We Know? 13-2

Statistical Evidence

How can statistics be useful if they can't be specific? Some scientific evidence is statistical. Observations suggest, for example, that Seyfert galaxies are more likely to be interacting with a nearby companion than a normal galaxy is. This is statistical evidence, so you can't be certain that any specific Seyfert galaxy will have a companion. How can scientists use statistical evidence to learn about nature when statistics contain built-in uncertainty?

Meteorologists use statistics to determine how frequently hurricanes of a certain size are likely to occur. Small storms happen every year, but big storms might happen on average only every 10 years. Hundred-year hurricanes are much more powerful but occur much less frequently—on average only once in a hundred years.

Those meteorological statistics can help you make informed decisions—as long as you understand the powers and limitations of statistics. Would you buy a house protected from a river by a levee that was not designed to withstand a hundred-year storm? In any one year, the chance of your house being destroyed would be only 1 in 100. You know the storm will hit eventually, but you don't know when. If you buy the house, a storm might destroy the levee the next year, but you might own the house for your whole life and never see a hundred-year storm. The statistics can't tell you anything about a specific year.

Before you buy that house, there is an important question you should ask the meteorologists: "How much data do you have about storms?" If they have only 10 years of data, then they don't really know much about hundred-year storms. If they have three centuries of data, then their statistical data are significant.

Sometimes people dismiss important warnings by saying, "Oh, that's only statistics." Scientists can use statistical evidence if it passes two tests. It cannot be used to draw conclusions about specific cases, and it must be based on large enough data samples so the statistics are significant. With these restrictions, statistical evidence can be a powerful scientific tool.

Marko Georgiev/Getty Images

Statistics can tell you that a bad storm will eventually hit, but it can't tell you when.

Observations with the largest and most sophisticated telescopes take astronomers back to the age of galaxy formation. At great distances the look-back time is so large that they see the Universe as it was soon after the galaxies began to form. There were more spirals then and fewer ellipticals. Observations also show that galaxies were closer together then; about 33 percent of all distant galaxies are in close pairs, but only 7 percent of nearby (in other words, present-day) galaxies are in pairs. The observational evidence clearly supports the hypothesis that galaxies have evolved by merger.

The evolution of galaxies is not a simple process. A good theory helps you understand how nature works, and astronomers are just beginning to understand the exciting and complex story of the galaxies. Nevertheless, it is already clear that galaxy evolution has some resemblance to a pie-throwing contest and is just about as neat.

13-4 ACTIVE GALAXIES AND QUASARS

Many galaxies have powerful energy sources in their nuclei that in some cases produce powerful jets and other outbursts. These are called **active galaxies**. By looking far away and back in time, astronomers have

active galaxy A galaxy whose center emits large amounts of excess energy, often in the form of radio emission. Active galaxies have massive black holes in their centers into which matter is flowing.

discovered that the origin of active galaxy energy sources and outbursts is closely related to the formation and history of galaxies.

The first type of active galaxy was discovered in the 1950s and named **radio galaxies** because these galaxies are sources of unusually strong radio waves. By the 1970s, astronomers had put space telescopes in orbit and discovered that radio galaxies are generally bright at many other wavelengths. The flood of energy pouring out of active galaxies originates almost entirely in their nuclei, which are referred to as **active galactic nuclei (AGN)**.

13-4a Seyfert Galaxies

In 1943, astronomer Carl Seyfert conducted a study of spiral galaxies in which he noted that about 2 percent of spirals have small, highly luminous nuclei in their bulges. Today, these **Seyfert galaxies** are recognized by the peculiar spectra of these luminous nuclei that contain broad emission lines of highly ionized atoms. Emission lines come from low-density gas; the presence of ionized atoms is evidence that the gas is very hot; broad spectral lines indicate large Doppler shifts produced by high gas velocities. The velocities of gas clouds at the centers of Seyfert galaxies are roughly 10,000 km/s, about 30 times greater than velocities at the center of normal galaxies. Something violent is happening in the cores of Seyfert galaxies.

Astronomers later discovered that the brilliant nuclei of Seyfert galaxies change brightness rapidly, in only a few hours or minutes, especially at X-ray wavelengths. As you learned in Chapter 11 Section 11-5b, an astronomical body cannot change its brightness significantly in a time shorter than the time it takes light to cross its diameter. If a Seyfert nucleus can change in a few minutes, then it cannot be larger in diameter than a few light-minutes. For comparison, the distance from Earth to the Sun is 8 light-minutes. Yet, despite their small size, the brightest Seyfert nuclei emit a hundred times more energy than the entire Milky Way Galaxy. Something in the centers of these galaxies not much bigger than Earth's orbit produces a galaxy's worth of energy.

Seyfert nuclei are three times more common in interacting pairs of galaxies than in isolated galaxies (**see How Do We Know? 13-2**). Also, about 25 percent of Seyfert galaxies have peculiar shapes suggesting tidal interactions with other galaxies, as shown in Figure 13-9. This statistical evidence hints that Seyfert galaxies might have been triggered into activity by collisions or interactions with companions. Some Seyferts are observed to be expelling matter in oppositely directed jets, a geometry you have seen on smaller scales when matter flows into neutron stars and black holes and forms an accretion disk plus jets.

radio galaxy A galaxy that is a strong source of radio signals.

active galactic nuclei (AGN) The centers of active galaxies that are emitting large amounts of excess energy.

Seyfert galaxy An otherwise normal spiral galaxy with an unusually bright, small core that fluctuates in brightness.

FIGURE 13-9

Top: NASA/ESA/STScI/AURA/NSF/The Hubble Heritage Team; center: W. Keel (Univ. of Alabama); bottom: NASA/ESA/STScI/AURA/NSF/The Hubble Heritage Team/A. Evans (Univ. of Virginia/NRAO/Stony Brook)

Seyfert galaxies are spiral galaxies with small, highly luminous nuclei. Some are interacting with nearby companions and appear distorted, with tidal tails and bridges.

The accumulated evidence leads modern astronomers to conclude that the core of a Seyfert galaxy contains a supermassive black hole, with a mass as high as a billion solar masses, plus a correspondingly large accretion disk. Gas in the centers of Seyfert galaxies is traveling so fast it would escape from a normal galaxy, and only large central masses could exert enough gravity to hold the gas inside the nuclei. Encounters with other galaxies could throw matter toward the black hole; and, as you learned in Chapter 11 Section 11-6d, lots of energy can be liberated by matter flowing through an accretion disk into a black hole. In Chapter 12 Section 12-4 you saw that the Milky Way Galaxy contains a massive central black hole, but one that seems to be on a starvation diet and is therefore relatively inactive.

FIGURE 13-10

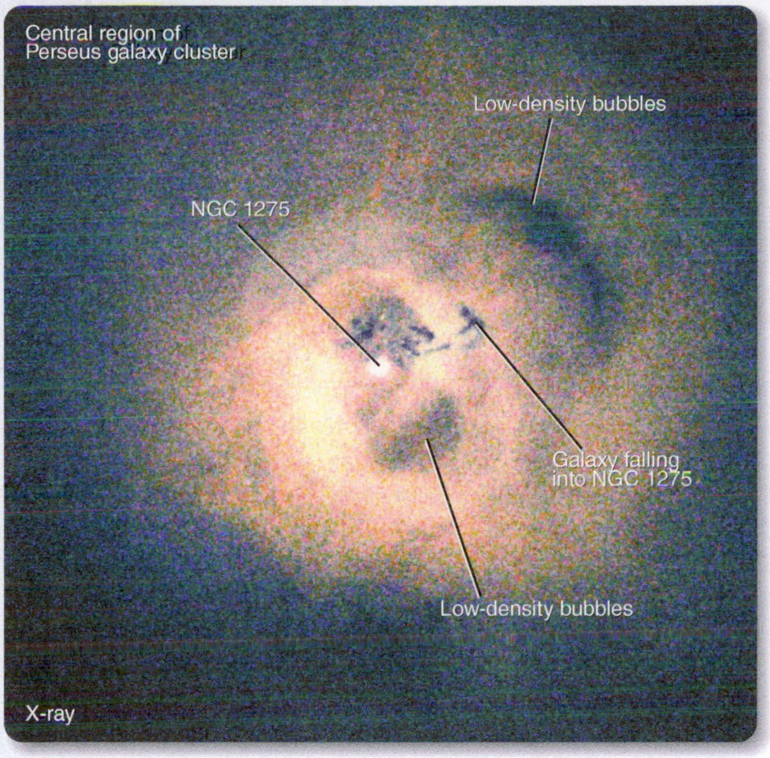

NGC 1275 in the Perseus galaxy cluster is spewing out jets and streamers of high-energy particles that are inflating low-density cavities in the hot gas within the cluster. The entire galaxy cluster is roughly 50 times larger in diameter than this image.

13-4b Double-Lobed Radio Sources

Beginning in the 1950s, radio astronomers found that some sources of radio energy in the sky consist of pairs of radio-bright regions. When optical telescopes studied the locations of these **double-lobed radio sources**, they revealed galaxies located between the two lobes. The geometry suggests that radio lobes are inflated by jets of excited gas emerging in two directions from the central galaxy. Statistical evidence indicates that jets and radio lobes, like Seyfert nuclei, are associated with interacting galaxies. The AGN jets seem to be related to matter falling into a central black hole via an accretion disk, although the details of this process are not understood.

The violence of these active galaxies is so great that it can influence the entire clusters of galaxies. The Perseus galaxy cluster contains thousands of galaxies and is one of the largest objects in the Universe. One of its galaxies, NGC 1275, is among the largest galaxies known. It is pumping out jets of high-energy particles, heating the gas in the galaxy cluster, and inflating low-density bubbles that distort the huge gas cloud (Figure 13-10). The hot gas observed in galaxy clusters is heated to multimillion-degree temperatures as galaxy after galaxy goes through eruptive stages that can last for hundreds of millions of years. NGC 1275 is erupting now, and it is so powerful and has heated the surrounding gas to such high temperatures that the gas can no longer fall in and the galaxy has probably limited its own growth.

13-4c Quasars

The largest telescopes detect multitudes of faint points of light with peculiar emission spectra, objects called **quasars** (also known as *quasi-stellar objects*, or *QSOs*). Although astronomers now recognize quasars as extreme examples of AGN as well as some of the most distant visible objects in the Universe, they were a mystery when they were first identified.

In the early 1960s, photographs of the location of some radio sources that resembled radio galaxies revealed only single starlike points of light. The first of

double-lobed radio sources A galaxy that emits radio energy from two regions (lobes) located on opposite sides of the galaxy.

quasar Small, powerful source of energy in the active core of a very distant galaxy.

these objects identified was 3C 48, and later the source 3C 273 was found. They were obviously not normal radio galaxies. Even the most distant photographable galaxies look fuzzy, but these objects looked like stars. Their spectra, however, were totally unlike stellar spectra, so the objects were called quasi-stellar objects.

For a few years, the spectra of quasars were a mystery. A few unidentifiable emission lines were superimposed on a continuous spectrum. In 1963, astronomer Maarten Schmidt calculated that if hydrogen Balmer lines were redshifted by $z = 0.158$ (meaning each line's change in wavelength divided by its lab wavelength equals 0.158), they would fit the observed lines in 3C 273's spectrum (Figure 13-11). Other quasar spectra quickly yielded to this type of analysis, revealing even larger redshifts.

To understand the significance of these large redshifts and the large velocities of recession they imply, recall the Hubble law from earlier in this chapter stating that galaxies have apparent velocities of recession proportional to their distances. The large redshifts of the quasars imply that they must be at great distances, some farther away than any known galaxy. Many quasars are evidently so far away that galaxies at those distances are very difficult to detect, yet the quasars are easily photographed. This leads to the conclusion that quasars must be ultraluminous, 10 to 1000 times the luminosity of a large galaxy.

Soon after quasars were discovered, astronomers detected fluctuations in their brightness over time scales of hours or minutes. The rapid fluctuations in quasars showed that they are small objects like AGN, only a few light-minutes or light-hours in diameter. Evidence, such as that seen in Figure 13-12, has accumulated that quasars are the most luminous AGN, located in very distant galaxies. For example, some quasars, like AGN, are at the centers of double radio lobes plus jets.

Perhaps you have a skeptical question about quasar distances at this point: "How can you be sure quasars really are that far away?" Astronomers faced with explaining how a small object could produce so much energy asked themselves the same question. In the early 1980s, astronomers were able to photograph faint nebulosity surrounding some quasars, which was called quasar fuzz. The spectra of quasar fuzz looked like the spectra of normal but very distant galaxies with the same redshift as the central quasar. In other cases, quasar light shines through the outskirts of a distant galaxy, and the quasar spectrum has extra absorption lines from gas at the redshift of the galaxy, smaller than the quasar's redshift. This means the quasar is farther away than the galaxy containing the gas that absorbed some of the quasar's light on its way to Earth. Both of these observations confirm that galaxy and quasar redshifts indicate distances in a mutually consistent way. Evidently, quasars are embedded in distant galaxies.

FIGURE 13-11

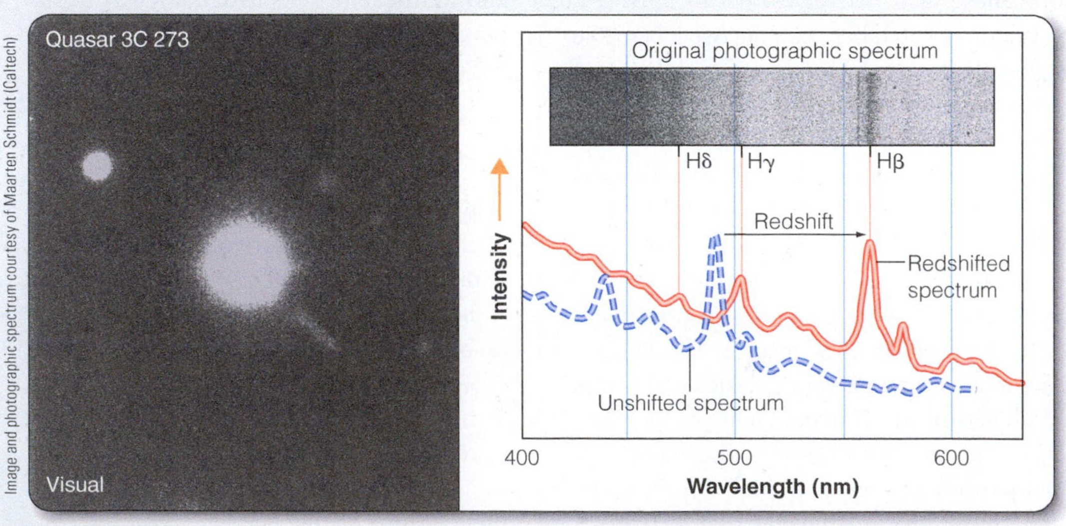

This image of 3C 273 shows the bright quasar at the center surrounded by faint fuzz. Note the jet protruding to the lower right. The original photographic plate containing a spectrum of 3C 273 with three hydrogen Balmer lines, Hδ, Hγ, and Hβ. The spectrum is redshifted by 15.8 percent. The dashed line shows the position of the spectrum without a redshift.

FIGURE 13-12

This radio image of quasar 3C 175 reveals that it is ejecting a jet and is flanked by radio lobes. Presumably astronomers sometimes see only one jet because it is directed approximately toward Earth, and the other jet is invisible because it is directed away from Earth. The presence of jets and radio lobes indicates that quasars are the active cores of distant galaxies.

13-4d A Unified Model of Active Galaxies

Astronomers studying galaxies are now developing a **unified model** of AGN and quasars. A monster black hole is the centerpiece (Figure 13-13).

Even a supermassive black hole is quite small compared with a galaxy. A 10-million-solar-mass black hole would be only one-fifth the diameter of Earth's orbit. This means that matter in an accretion disk can get very close to the black hole, orbit very fast, and grow very hot. Theoretical calculations indicate that the disk immediately around the black hole is "puffed up," thick enough to hide the central black hole from some viewing angles. The jets of gas and radiation seen racing out of the cores of many active galaxies appear to be produced as the spinning accretion disk draws in a magnetic field and winds it up, but the exact process is not fully understood. Mathematical models indicate that the outer part of the disk is a fat, relatively cold torus (doughnut shape) of dusty gas.

According to the unified model depicted in Figure 13-13, what you see when you view the core of an AGN or QSO depends on how its accretion disk is tipped with respect to your line of sight.

1. If you view the accretion disk from the edge, you cannot see the central zone at all because the thick dusty torus blocks your view. Instead you see radiation emitted by gas lying above and below the central disk that is therefore relatively cool and moving relatively slowly, with small Doppler shifts. Thus, you see narrower spectral lines coming from what is called the narrow-line region.
2. If the accretion disk is tipped slightly, you might be able to see some of the intensely hot gas in the central cavity. This is called the broad-line region because the gas is hot and also orbiting at high velocities. The resulting high Doppler shifts spread out the spectral lines.
3. If you look directly into the central cavity around the black hole—down the dragon's throat, so to speak—you see the jet emerging perpendicular to the accretion disk and coming straight at you. Model calculations indicate this would result in a very luminous and highly variable source with few or no emission lines. This is the appearance of AGN known as **blazars**.

Astronomers are now using this unified model to sort out the different kinds of active galaxies and quasars so they can understand how they are related. For example, about 1 percent of quasars are strong radio sources, and the radio radiation might come from synchrotron radiation (look back to Chapter 11 Section 11-4d) produced in the high-energy gas and magnetic fields in the jets. Another example: Using infrared cameras to see through dust, astronomers observed the core of the double-lobed radio galaxy Cygnus A and found an object much like a quasar. Astronomers have begun to refer to such hidden objects as "buried quasars."

unified model An attempt to explain the different types of active galactic nuclei using a single model viewed from different directions.

blazar A type of active galaxy nucleus that is especially variable and has few or no spectral emission lines.

FIGURE 13-13

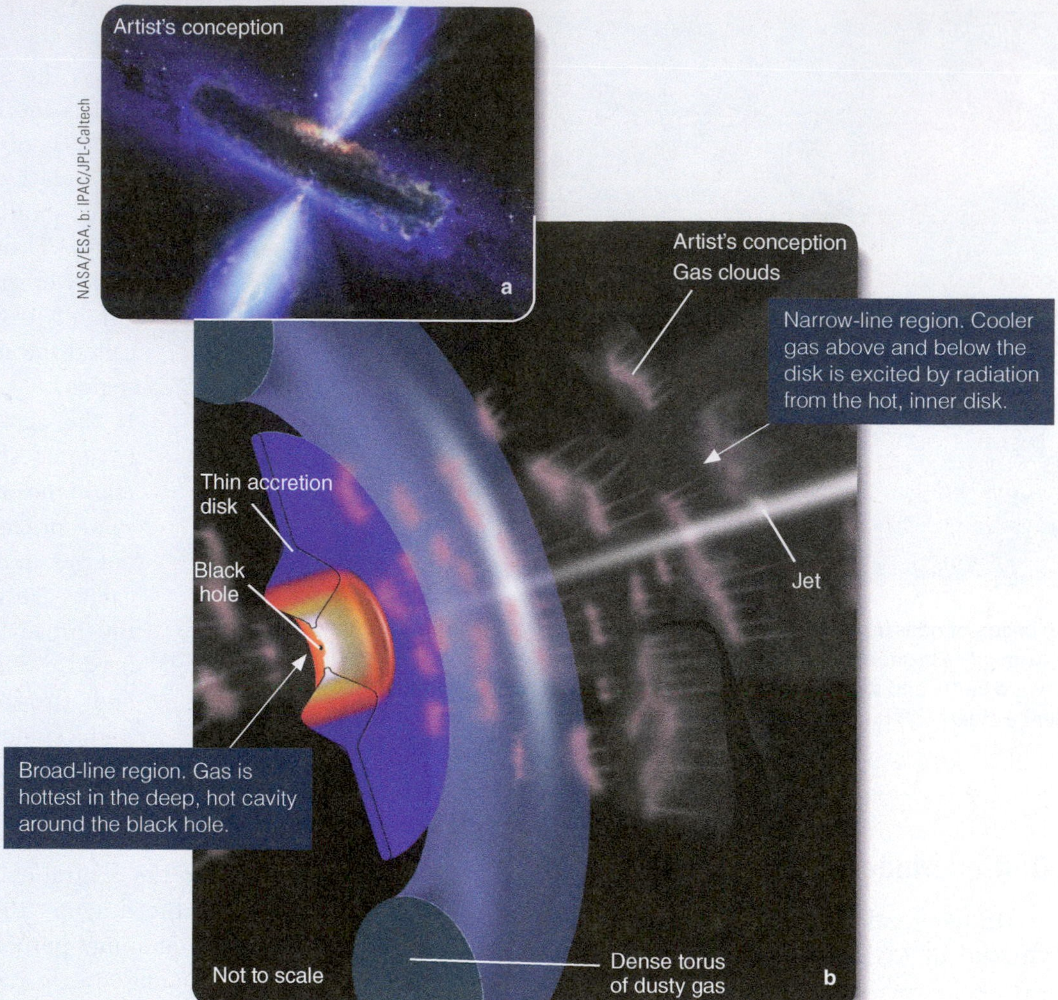

(a) Artist's conception of an AGN viewed edge-on, with the view of the hot accretion disk and central cavity around the supermassive black hole blocked by the outer cool dusty gas torus. (b) Features visible in the spectrum of an AGN depend on the angle at which it is viewed. The unified model, shown in cross section, suggests that matter flowing inward passes first through a large, opaque torus; then into a thinner, hotter disk; and finally into a small, hot cavity around the black hole. Telescopes viewing such a disk edge-on would see only narrow spectral lines from cooler gas, but a telescope looking into the central cavity would see broad spectral lines formed by the hot gas. This diagram is not to scale. The central cavity may be only 0.01 pc in radius, whereas the outer torus may be 1000 pc in radius.

The unified model is far from complete. The detailed structure of accretion disks is poorly understood, as is the process by which the disks produce jets. Furthermore, the spiral Seyfert galaxies are clearly different from the giant elliptical galaxies that have double radio lobes. Unification does not explain all of the differences among active galaxies. Rather, it is a model that provides some clues to what is happening in AGN and quasars.

13-4e The Origin of Supermassive Black Holes

Naturally you are wondering where these supermassive black holes in the nuclei of galaxies came from, and that question is linked to a second question. What makes a supermassive black hole erupt? Answering those questions will help you understand how galaxies form.

Evidence is accumulating that most galaxies developed a supermassive black hole at their center as they formed (Figure 13-14). Even our own Milky Way Galaxy and the nearby Andromeda Galaxy contain central black holes. Only a few percent of galaxies, however, have obviously active nuclei. That must mean that most of the supermassive black holes are dormant. Presumably they are not being fed large amounts of matter. A slow trickle of matter flowing into the supermassive black hole at the center of our Galaxy could explain the relatively mild activity seen there. It would take a larger meal to trigger an eruption such as those seen in active galactic nuclei.

What could trigger a supermassive black hole to erupt? The answer is something that you studied back in Chapter 3 Section 3-5c—tides. You have seen in a previous section of this chapter how tides twist interacting galaxies and rip matter away into tidal tails. Active galaxies are often distorted: They have evidently been twisted by tidal forces as they interacted or merged with another galaxy. Mathematical models show that those interactions can also throw stars plus clouds of interstellar gas and dust inward toward the galaxies' centers. A sudden flood of matter flowing into a supermassive black hole would trigger it into eruption. Figure 13-15 shows how a passing star would be shredded and partially consumed by a supermassive hole. A steady diet of inflowing gas, dust, and an occasional star would keep the supermassive black hole in an AGN active.

A few dozen supermassive black holes have measured masses, and their masses are correlated with the masses of the host galaxies' central bulges. In each case, the mass of the black hole is about 0.5 percent the mass of the surrounding central bulge. Apparently, as a galaxy forms its central bulge, a certain fraction of the mass sinks to the center where it forms a supermassive black hole. All of that matter flowing together to form the black hole would release a tremendous amount of energy and trigger a violent eruption. When the Universe was young—more than 10 billion years ago—when galaxies were actively forming, the birth of the central bulges must have triggered many AGN. Later episodes of AGN activity could be triggered by interactions or mergers with other galaxies.

FIGURE 13-14

Top: NASA/ESA/STScI/AURA/NSF/W. Jaffe (Leiden Obs.) and H. Ford (JHU/STScI); bottom: NASA/L. Ferrarese (JHU)

The elliptical galaxy NGC 4261 is ejecting jets and inflating radio lobes. High-resolution images show that the galaxy contains a small, bright nucleus orbited by a spinning disk. The orbital velocity and size of the disk confirm that the central object is a supermassive black hole.

13-4f Quasars Through Time

The redshifts of some quasars indicate that their light has traveled for more than 10 billion years to reach us. Because of those large look-back times, these quasars appear as they were when the Universe was only about 20 percent of its current age of 13.8 billion years (see Chapter 14). The first clouds of gas that formed galaxies would have also made supermassive black holes at the centers of those galaxies' central bulges. The abundance of matter flooding into those early black holes could have triggered outbursts that are seen as quasars.

You should also note that galaxies were closer together when the Universe was young and had not expanded very much. Because they were closer together, the forming galaxies collided more often, and you have seen how collisions between galaxies could throw matter into central supermassive black holes and trigger eruptions. Quasars are often located in host galaxies that are distorted as if they were interacting with other galaxies (Figure 13-16).

FIGURE 13-15

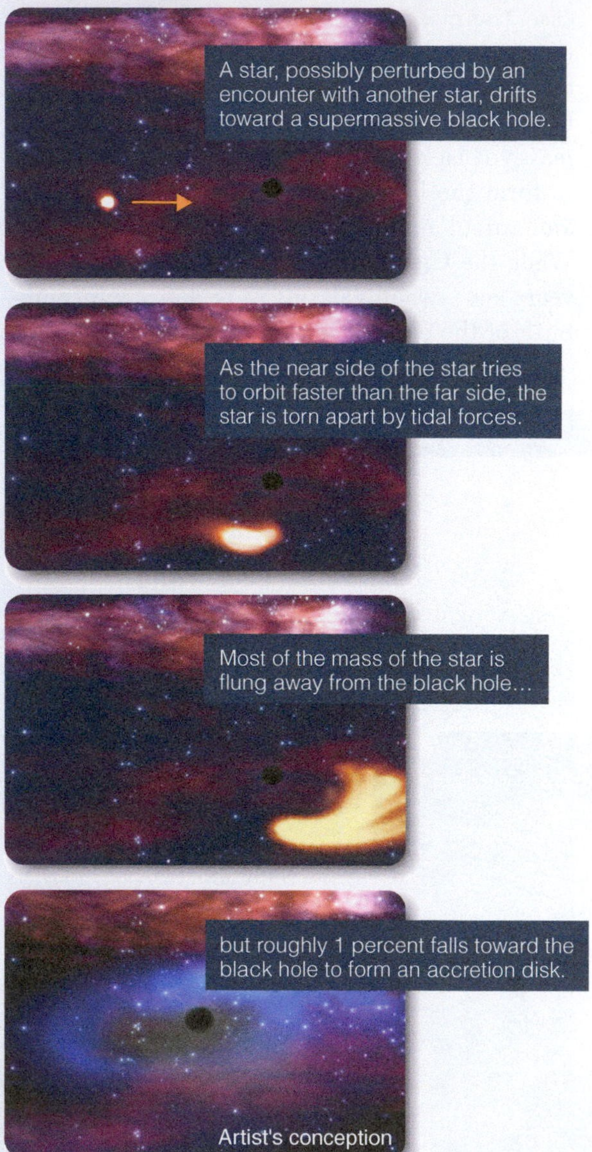

European Space Agency (ESA).

Orbiting X-ray telescopes observing active galaxies sometimes detect X-ray flares equaling the energy of a supernova explosion. Such flares are evidently caused when a star wanders too close to the supermassive black hole at the center of the galaxy and tidal forces rip the star apart.

FIGURE 13-16

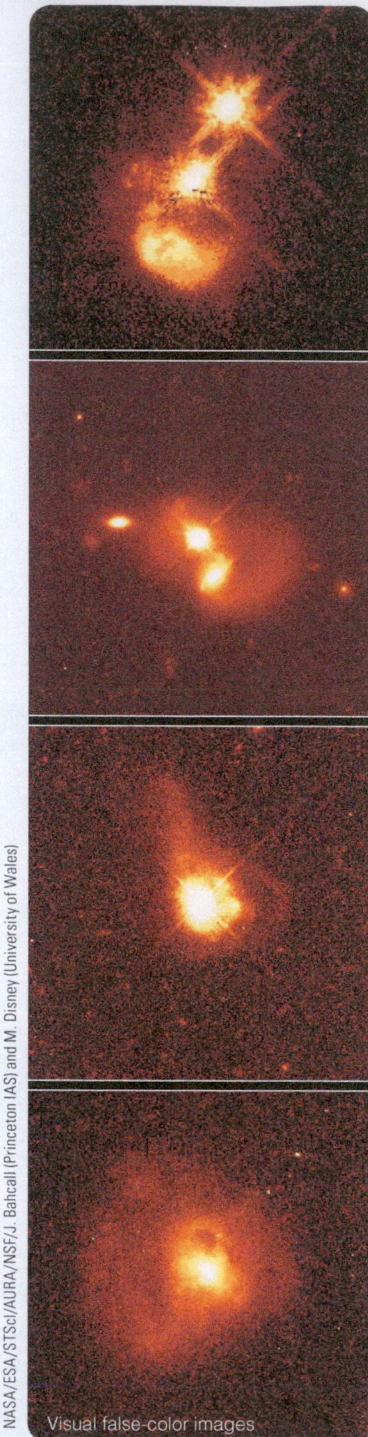

NASA/ESA/STScI/AURA/NSF/J. Bahcall (Princeton IAS) and M. Disney (University of Wales)

The bright object at the center of each of these images is a quasar. Fainter objects near and around the quasars are galaxies distorted by collisions. Compare the ring-shaped galaxy in the top panel and the galaxy with the tidal tail in the third panel respectively with the ring galaxy and the Antennae Galaxies on the right-hand page of Concept Art 13B.

Quasars are most common with redshifts of a little more than 2 and less common with redshifts higher than 2.7. The largest quasar redshifts are more than 7, but such high-redshift quasars are quite rare. Evidently, if you looked at quasars with redshifts a bit more than 2, you would be looking back to an age when galaxies were actively forming, colliding, and merging. In that era, quasars were about 1000 times more common than they are now, although, even so, only a fraction of galaxies had quasars erupting in their cores at any one time. If you looked back to higher redshifts, you would see fewer quasars because you would be looking back to an age when the Universe was so young it had not yet begun to form many galaxies and quasars.

Then where are all the dead quasars? There is no way to get rid of supermassive black holes. Astronomers have discovered that nearly all galaxies contain supermassive black holes, and those black holes might have suffered quasar eruptions when the Universe was younger, galaxies were closer together, and infalling gas and dust were more plentiful. Quasar eruptions became less common as galaxies became more stable and as the abundance of gas and dust in the centers of galaxies was exhausted. Our own Milky Way Galaxy is a good example. It could have been home to a quasar long ago, but today its supermassive black hole is resting. Dormant black holes at the centers of galaxies today can be reawakened to become AGN by galaxy collisions and interactions.

What Are We? *Small and Proud*

Do you feel insignificant yet? You are riding a small planet orbiting a humdrum star that is just one of at least 200 billion stars in the Milky Way Galaxy, and you have just learned that there are more than 100 billion galaxies visible with existing telescopes.

When you look at galaxies, you are looking across voids deeper than human imagination. You can express such distances with numbers and say a certain galaxy is 5 billion light-years from Earth, but the distance is truly beyond human comprehension. Furthermore, looking at distant galaxies also leads you back in time. You see distant galaxies as they were billions of years ago. The realm of the galaxies is deep space and deep time.

Some people say astronomy makes them feel humble, but before you agree, consider that you can feel small without feeling humble. Size is not necessarily the same thing as significance. We humans live out our little lives on our little planet, but our tiny brains are able to figure out some of the most profound mysteries of the Universe. We are exploring deep space and deep time and coming to understand what galaxies are and how they evolve. Most of all, we humans are beginning to understand what we are. That's something to be proud of.

STUDY TOOLS 13

IN THE BOOK

☐ Tear Out the Review Card on *Galaxies: Normal and Active*.

ASTRO ONLINE

Visit ASTRO Online at www.cengagebrain.com

☐ Interactive Reading

☐ Practice Quizzing

☐ Videos and Animations

☐ Exclusive Features

Questions and Problems

13

REVIEW QUESTIONS

1. How do long-exposure photos of two selected areas of the sky reveal that galaxies are very common?
2. What is the difference between an Sa and an Sb galaxy? Between an S0 and an Sa galaxy? Between an Sb and an SBb galaxy? Between an E7 and an S0 galaxy?
3. Explain how the rotation curve method of finding a galaxy's mass is similar to the method used to find the masses of binary stars.
4. Explain how the Hubble law allows you to estimate the distances to galaxies.
5. How can collisions affect the shapes of galaxies?
6. What evidence can you cite that galactic cannibalism really happens?
7. What evidence suggests that Seyfert galaxies have suffered recent interactions with other galaxies?
8. How do the peculiar rotations of M64 and NGC 7251 help explain the histories of these starburst galaxies?
9. What evidence shows that quasars are ultraluminous but must be small?
10. What evidence is there that quasars occur in distant galaxies?
11. Why are there few quasars at low redshifts and at very high redshifts but many at redshifts between about 2 and 3?
12. **How Do We Know?** Classification helped Darwin understand how creatures evolve. Has classification helped you understand how galaxies evolve?
13. **How Do We Know?** How would you respond to someone who said, "Oh, that's only statistics"?

DISCUSSION QUESTIONS

1. From what you know about star formation and the evolution of galaxies, do you think irregular galaxies should be bright or faint in the infrared relative to visible wavelengths? Why or why not? What about starburst galaxies? What about elliptical galaxies?
2. Do you think that our Galaxy is currently an active galaxy? Do you think it ever was in the past? Could it have hosted a quasar when it was young?

PROBLEMS

1. Hubble's first determination of the Hubble constant was too high because the calibration of Cepheid variable stars he used was not very good. Use his original diagram shown in Figure 13-4 to estimate his first determination of the Hubble constant. (*Hint:* See **The Hubble Law** in the Math Reference Cards MR.)
2. If a galaxy contains a type I (classical) Cepheid with a period of 30 days and an apparent magnitude of +20, what is the distance to the galaxy? (*Hints:* See Figure 12-4 and the **Magnitude-Distance Formula** in the Math Reference Cards MR.)
3. If a galaxy contains a supernova that at its brightest has an apparent magnitude of +17, how far away is the galaxy? Assume that the absolute magnitude of the supernova is −19. (*Hint:* See the **Magnitude-Distance Formula** in the Math Reference Cards MR.)
4. If you find a galaxy that is the same mass as our Milky Way Galaxy, what orbital velocity would a small satellite galaxy have if it orbits 50 kpc from the center of the larger galaxy? (*Hints:* See **Circular Velocity** in the Math Reference Cards MR and find the mass of the Galaxy in Chapter 12 Section 12-1e; make sure you convert relevant quantities to units of meters, kilograms, and seconds.) (*Note:* 1 kpc = 3.1×10^{19} m.)
5. If a galaxy has an apparent radial velocity of 2000 km/s and the Hubble constant is 70 km/s/Mpc, how far away is the galaxy? (*Hint:* See **The Hubble Law** in the Math Reference Cards MR.)
6. Suppose you found a galaxy in which the outer stars have orbital velocities of 150 km/s. If the radius of the galaxy is 4.0 kpc, what is the orbital period of the outer stars? (*Note:* 1 kpc = 3.1×10^{19} m, and 1 yr = 3.2×10^{7} seconds.)
7. Among the globular clusters orbiting a distant galaxy, one is moving at 420 km/s and is located 11 kpc from the center of the galaxy. Assuming the globular cluster is located outside most of the mass of the galaxy, what is the mass of the galaxy? (*Hints:* See **Circular Velocity** in the Math Reference Cards MR; make sure you convert relevant quantities to units of meters, kilograms, and seconds. *Note:* 1 kpc = 3.1×10^{19} m.)
8. If the jet in NGC 5128 is traveling at 5000 km/s and is 40 kpc long, how long will it take for gas to travel from the core of the galaxy to the end of the jet? (*Note:* 1 kpc equals 3.1×10^{16} km, and 1 yr = 3.2×10^{7} seconds.)
9. If the active core of a galaxy contains a black hole of 10^6 solar masses, what will the orbital period

be for matter orbiting the black hole at a distance of 0.33 AU? (*Hint:* See **Circular Velocity** in the Math Reference Cards MR; make sure you convert relevant quantities to units of meters, kilograms, and seconds.)

10. If a quasar is 1000 times more luminous than an entire galaxy, what is the absolute magnitude of such a quasar? Note that the absolute magnitude of the Milky Way Galaxy is about −21. (*Hint:* See **Magnitudes** in the Math Reference Cards MR.)
11. The hydrogen Balmer line H_β has a wavelength of 486.1 nm. It is shifted to 563.9 nm in the spectrum of quasar 3C 273. What is the redshift of this quasar? (*Hints:* Evaluate $\Delta\lambda$; see **The Doppler Formula** in the Math Reference Cards MR.)

LEARNING TO LOOK

1. In Figure 13-17 you see two interacting galaxies; one is nearly face-on and the other is nearly edge-on. Discuss the shapes of these galaxies and describe what is happening.
2. Figure 13-18 combines visual (blue) with radio (red) to show the galaxy radio astronomers call Fornax A. Explain the features of this image. Is it significant that the object is a distorted elliptical galaxy in a cluster?

FIGURE 13-17

FIGURE 13-18

14 Modern Cosmology

The Universe, as has been observed before, is an unsettlingly big place, a fact which for the sake of a quiet life most people tend to ignore.

Douglas Adams

The Restaurant at the End of the Universe

ESA/D. Ducros, The Planck Collaboration

After finishing this chapter go to **PAGE 344** for **STUDY TOOLS.**

GUIDEPOST

Since Chapter 1, you have been on an outward journey through the Universe. Now you have reached the limit of your travels in space and time and can contemplate the Universe as a whole. The ideas in this chapter are among the biggest and most difficult in all of science. Can you imagine a limitless Universe, or the first instant of time?

As you explore cosmology, you will find answers to the following three important questions:

- Does the Universe have an edge and a center?
- What is the evidence that the Universe began with a "Big Bang," an expansion from a hot, dense state?
- How has the Universe evolved, and what will be its fate?

When you have finished this chapter, you will have a modern insight into the nature of the Universe, as well as where you are and what you are. After that, with an understanding of the big picture, you can focus in chapter 15 on one final important and intriguing question: Is there life elsewhere in the Universe?

cosmology The study of the nature, origin, and evolution of the Universe.

Look at your thumb. The matter in your thumb was present in the fiery beginning of the Universe. **Cosmology**, the study of the Universe as a whole, can tell you where your matter came from, and it can tell you where your matter is going. Cosmology is a mind-bendingly weird subject, and you can enjoy it for its strange ideas. It is fun to think about space stretching like a rubber sheet, invisible energy pushing the Universe to expand faster and faster, and the origin of vast walls of galaxy clusters. Notice that this is better than speculation—it is all supported by evidence. Cosmology, however strange it might seem, is a serious and logical attempt to understand how the Universe works, and it leads to wonderful insights into how you came to be a part of it. (See CO 14.)

14-1 INTRODUCTION TO THE UNIVERSE

Many people have an impression of the Universe as a vast sphere filled with stars and galaxies. Your vision might be similar to the image in Figure 14-1, but, as you begin exploring the Universe, you need to become aware of your assumptions so they do not mislead you. The first step is to deal with an expectation so obvious that most people, for the sake of a quiet life, don't even think about it.

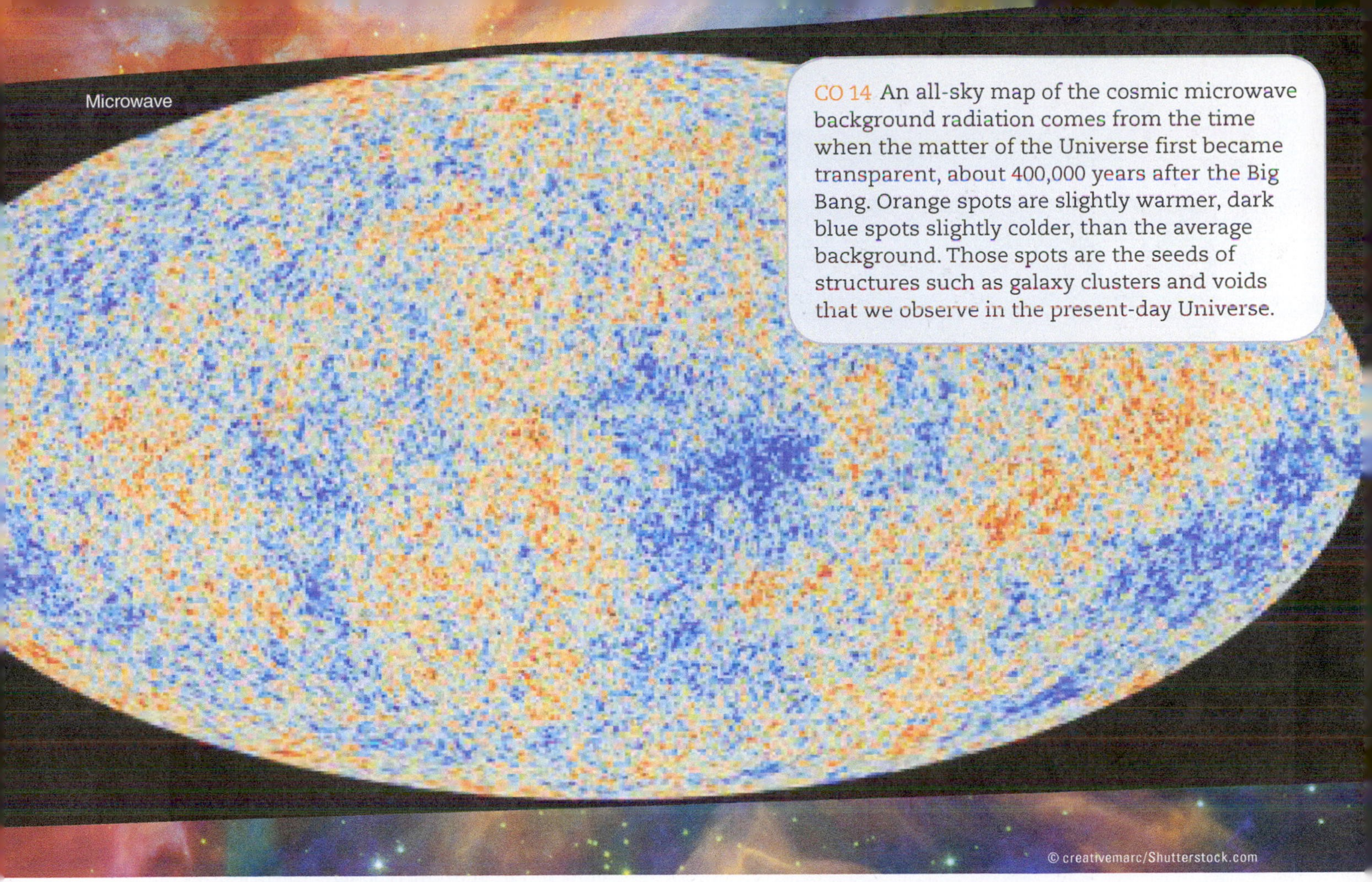

CO 14 An all-sky map of the cosmic microwave background radiation comes from the time when the matter of the Universe first became transparent, about 400,000 years after the Big Bang. Orange spots are slightly warmer, dark blue spots slightly colder, than the average background. Those spots are the seeds of structures such as galaxy clusters and voids that we observe in the present-day Universe.

14-1a The Edge–Center Problem

In your daily life, you are accustomed to boundaries. Rooms have walls, athletic fields have boundary lines, countries have borders, and oceans have shores. It is natural to think of the Universe as having an edge, but that can't be right.

If the Universe had an edge, imagine going to the edge. What would you find there: A wall? A great empty space? Nothing? Even a child can ask: If there is an edge to space, what's beyond it? A true edge would have to be more than just an end of the distribution of matter, it would have to be an end of space itself; but then, what would happen if you tried to reach past or move past that edge?

An edge to the Universe violates common sense, and modern observations (which you will study later in this chapter) indicate that the Universe could be infinite and have no edge. Note that you find the centers of things—galaxies, globular clusters, oceans, and pizzas—by referring to their edges. If the Universe has no edge, then it cannot have a center.

It is a common misconception to imagine that the Universe has a center, but, as you have just learned, that is impossible. As you study cosmology, you need to take care to avoid thinking that there is a center of the Universe.

14-1b The Idea of a Beginning

Of course you have noticed that the night sky is dark. That is an important observation because reasonable assumptions about the Universe can lead to the conclusion that the night sky actually should glow blindingly bright! This conflict between observation and theory is called **Olbers's paradox** after Heinrich Olbers, an Austrian physician and astronomer who publicized the problem in 1826 (*problem* or *question* would be a more accurate word than *paradox*).

Olbers was not the first to pose the question, but it is named after him because modern cosmologists were not aware of the earlier discussions. You will be able to answer Olbers's question and understand why the night sky is dark by revising your assumptions about the Universe.

The point Olbers made seems simple. Suppose you assume that the Universe is infinite and filled with stars. (The clumping of stars into galaxies can be shown mathematically to make no difference.) If you look in any direction, your line of sight must eventually reach

Olbers's paradox The conflict between theory and evidence regarding the darkness of the night sky.

FIGURE 14-1

NASA/ESA/STScI/AURA/R. Williams, HDF-South Team

The entire sky is filled with galaxies. Some lie in clusters of thousands, and others are isolated in nearly empty voids between the clusters. In this image of a typical spot on the sky, most of the bright objects are foreground stars; their "spikes" are caused by diffraction in the telescope. All other objects are galaxies ranging from the relatively close face-on spiral at upper right to the most distant galaxies visible only in the infrared, shown as red in this composite image.

the surface of a star, as shown in Figure 14-2. Think about trying to see out of a forest. When you are deep in a forest, every line of sight ends on a tree trunk, and you cannot see out of the forest. By analogy, every line of sight from Earth out into space should eventually end on the surface of a star, so the entire sky should be as bright as the surface of an average star—like suns crowded "shoulder to shoulder," covering the sky from horizon to horizon. It should not get dark at night.

Today, cosmologists believe they understand why the sky is dark. Olbers's paradox makes an incorrect prediction because it is based on a hidden assumption. The Universe might be infinite in size but it is not eternal—that is, not infinitely old. That answer to Olbers's question was suggested by Edgar Allan Poe in 1848. He proposed that the night sky is dark because the Universe is not infinitely old but instead began at some time in the past. The more distant stars are so far away that light from them has not yet reached Earth. That is, if you look far enough, the look-back time is greater than the age of the Universe. The night sky is dark because the Universe had a beginning.

observable Universe The part of the Universe that you can see from your location in space and time.

This is a powerful idea because it clearly illustrates the difference between the Universe and the observable Universe. The Universe is everything that exists, but the **observable Universe** is the part that you can see. You will learn later that the Universe is 13.8 billion years old. Because of that, the observable Universe is limited by a light travel time of 13.8 billion years. Do not confuse the observable Universe, which is finite, with the Universe as a whole, which could be infinite.

14-1c Cosmic Expansion

In 1929, Edwin P. Hubble published his discovery that the sizes of galaxy redshifts are proportional to galaxy distances (see Figure 13-4). Nearby galaxies have small redshifts, but more distant galaxies have larger redshifts. These redshifts imply that the galaxies are receding from each other. Figure 14-3 shows spectra of galaxies in clusters at various distances. The Virgo cluster is relatively nearby, and its redshift is small. The Hydra cluster is very distant, and its redshift is so large that the two dark lines formed by ionized calcium are shifted from near-ultraviolet wavelengths well into the visible part of the spectrum.

The expansion of the Universe does not imply that Earth is at the center. To see why, think about raisin bread. As the dough rises, it pushes the raisins away from each other uniformly at speeds that are proportional to their distances from each other (Figure 14-4). Two raisins that were originally close are pushed apart slowly, but two raisins that were far apart, having more dough between them, are pushed apart faster. If bacterial astronomers lived on a raisin in your raisin bread, they could observe the redshifts of the other raisins and derive a bacterial Hubble law. They would conclude that their universe was expanding uniformly. It does not matter which raisin the bacterial astronomers lived on, they would get the same Hubble law—no raisin has a special viewpoint. Similarly, astronomers in any galaxy will see the same law of expansion—no galaxy has a special viewpoint.

When you think about that loaf of bread, you see the edge of the loaf, and you can identify a center to the loaf of bread. The raisin bread analogy for the expansion of the Universe no longer works when you consider the crust—the edge—of the bread. Remember that the Universe cannot have an edge or a center, so there can be no center to the expansion.

FIGURE 14-2

Photo courtesy of Janet Seeds

(a) Every direction you look in a forest eventually reaches a tree trunk, and you cannot see out of the forest. (b) If the Universe is infinite and filled with stars, then any line from Earth should eventually reach the surface of a star. This assumption leads to a prediction that the night sky should glow as brightly as the surface of the average star, a puzzle commonly referred to as Olbers's paradox.

14-2 THE BIG BANG

Now you are ready to take a historic step up the cosmology pyramid. The expansion of the Universe must have begun with an event of unimaginable intensity. Popular television shows notwithstanding, you will learn that the Big Bang can be called a "theory" rather than a hypothesis because the supporting evidence supporting it is so solid and comprehensive.

14-2a Imagining the Big Bang

Imagine that you have a video of the expanding Universe, and you run it backward. You would see the galaxies moving toward each other. There is no center to the expansion of the Universe, so you would not see galaxies approaching a single spot. Rather, you would see the space between galaxies disappearing, distances between all galaxies decreasing without the galaxies themselves moving, and eventually galaxies beginning to merge. If you ran your video far enough back, you would see the matter and energy of the Universe compressed into a high-density, high-temperature state. You can conclude that the expanding Universe began with expansion from that condition of extremely high density and temperature, which modern astronomers call the **Big Bang**.

How long ago did the Universe begin? You can estimate the age of the Universe with a simple calculation. If you need to drive to a city 100 miles away and you can travel 50 miles per hour, you divide distance by rate of travel and learn the travel time—in this example, 2 hours. To find the age of the Universe, you can divide the distance between galaxies by the rate at which they are separating and find out how much time was required for them to

Big Bang The high-density, high-temperature state from which the expanding Universe of galaxies began.

have reached their current separation. You will fine-tune your estimate later in this chapter, but for the moment you can conclude that basic observations of the recession of the galaxies require that the Universe began approximately 14 billion years ago. That estimated span is called the **Hubble time**.

Your instinct is to think of the Big Bang as an historical event, like the Gettysburg Address—something that happened long ago and can no longer be observed. But the look-back time makes it possible to observe the Big Bang directly. The look-back time to nearby galaxies is only a few million years, but the look-back time to more distant galaxies is a large fraction of the age of the Universe (look back to Chapter 13 Section 13-2b). Suppose you were to look between and beyond the distant galaxies, back to the time of the Big Bang. You should be able to detect the hot gas that filled the Universe long ago.

Again, do not think of an edge or a center when you think of the Big Bang. It is a very common misconception that the Big Bang was an explosion and that the galaxies are flying away from a center. Although your imagination tries to visualize the Big Bang as a localized event, you must keep firmly in mind that the Big Bang did not occur at a single place but filled the entire volume of the Universe. You cannot point to any particular place and say, "The Big Bang occurred over there."

At the time of the Big Bang, all of the galaxies, stars, and atoms in the observable Universe were confined to a very small volume. That hot, dense state, the Big Bang, occurred everywhere, including right where you are now. The matter of which you are made was part of the Big Bang, so you are inside the remains of that event, and the Universe continues to expand around you. In whatever direction astronomers look, at great distances they can see back to the age when the Universe was filled with dense, hot gas (Figure 14-5).

FIGURE 14-3

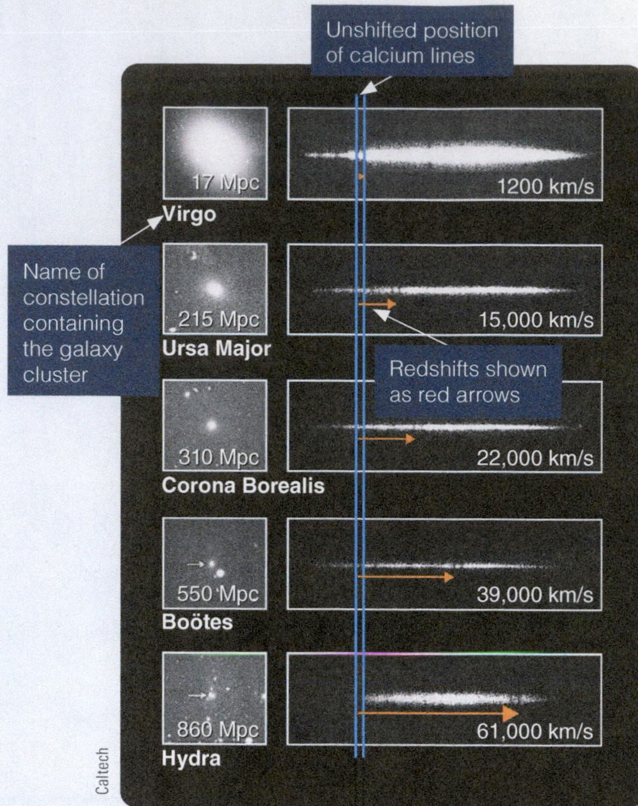

These galaxy spectra extend from the near-ultraviolet at left to the blue part of the visible spectrum at right. The two dark absorption lines of ionized calcium are prominent in the near-ultraviolet. The redshifts in galaxy spectra are expressed here as apparent velocities of recession. Note that the apparent velocity of recession is proportional to distance, which is known as the Hubble law.

FIGURE 14-4

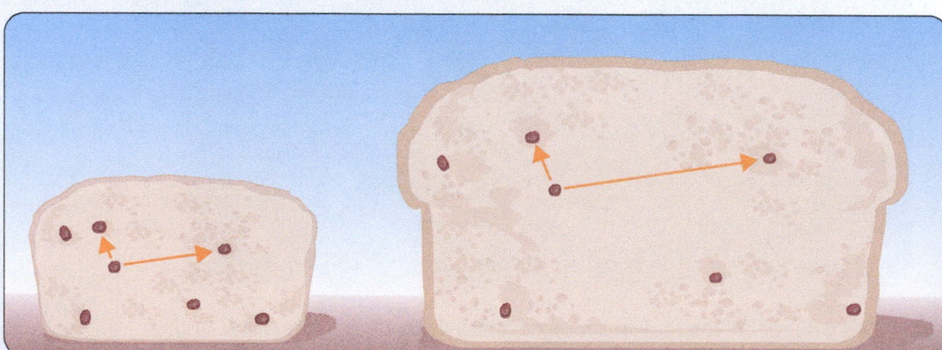

An illustration of the raisin bread analogy for the expansion of the Universe. As the dough rises, raisins are pushed apart with velocities proportional to the distance. A colony of bacteria living on any raisin will find that the redshifts of the other raisins are proportional to their distances.

Hubble time The age of the Universe, equal to the reciprocal of the Hubble constant. The Hubble time is the age of the Universe if it expanded since the Big Bang at a constant rate.

FIGURE 14-5

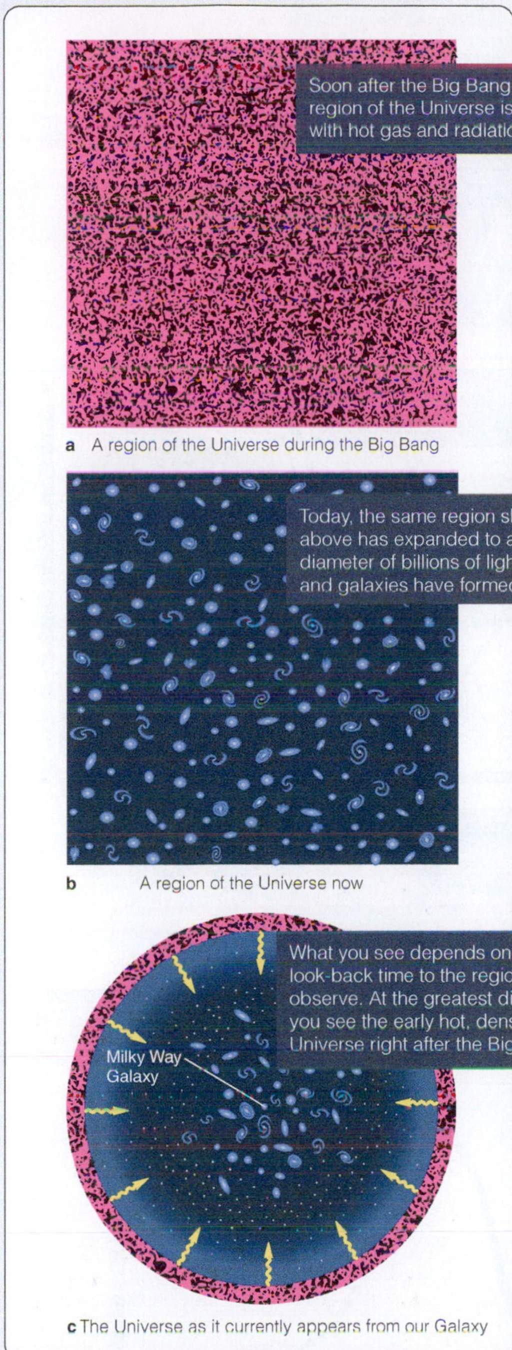

a A region of the Universe during the Big Bang

b A region of the Universe now

c The Universe as it currently appears from our Galaxy

This diagram shows schematically the expansion of a small part of the Universe. Although the Universe is now filled with galaxies, the look-back time distorts what you see. Nearby you see galaxies, but at greater distances the look-back time reveals the Universe at earlier stages, before galaxies formed. At very great distances, the hot gas that filled the Universe soon after the Big Bang is detectable as a source of infrared, microwave, and radio energy arriving from all directions.

14-2b Evidence for the Big Bang—The Cosmic Microwave Background

The radiation that comes from great distance has a tremendous redshift. The most distant visible objects known to date are faint galaxies with redshifts of about 11. In contrast, the radiation from the hot gas of the Big Bang is calculated to have a redshift of about 1100. That means the light emitted by the Big Bang gases arrives at Earth as far-infrared radiation and short-wavelength radio waves. You can't see it with your eyes, but it should be detectable with infrared and radio telescopes. Amazingly, the Big Bang can still be detected by the radiation it emitted.

In the mid-1960s, Bell Laboratories physicists Arno Penzias and Robert Wilson were measuring the radio brightness of the sky (Figure 14-6) when they discovered peculiar radio noise coming from all directions. Physicists George Gamow and Ralph Alpher had predicted in the 1940s that the Big Bang would have emitted blackbody radiation (look back to Chapter 5 Section 5-2b) that should now be in the far-infrared and radio parts of the spectrum. Physicist Robert Dicke and his team at Princeton began building a receiver to detect that radiation in the early 1960s. When Penzias and Wilson learned about the preceding work, they realized the radio noise they had detected, now called the **cosmic microwave background (CMB)**, is actually radiation from the Big Bang. They received the 1978 Nobel Prize in physics for their discovery.

The detection of the background radiation was tremendously exciting, but astronomers wanted confirmation. Critical observations in the far-infrared checking whether the CMB really had a blackbody spectrum could not be made from the ground. It was not until January 1990 that satellite measurements confirmed that the CMB is blackbody radiation with an apparent temperature of 2.725 ± 0.002 K—in good agreement with theoretical predictions.

It might seem strange that the hot gas of the Big Bang appears to have a temperature 2.7 degrees above absolute zero, but recall the tremendous redshift. Observers on Earth see light that has a redshift of about 1100—that is, the wavelengths of the photons are about 1100 times longer than when they were emitted. The gas clouds that emitted the photons had a temperature of about

cosmic microwave background (CMB) Radiation from the hot matter of the Universe soon after the Big Bang. The large redshift makes it appear to come from a blackbody with a temperature of 2.7 K.

3000 K, and they emitted blackbody radiation with a wavelength of maximum intensity of about 1000 nm (see **Wien's Law** in the Math Reference Cards MR). The expansion of the Universe has redshifted the wavelengths about 1100 times longer, so wavelength of maximum intensity is now about 1 million nm (1 millimeter). That is why the hot gas of the Big Bang seems to be 1100 times cooler now, about 2.7 K.

14-2c Particles and Nucleosynthesis—The First Seconds and Minutes

Simple observations of the darkness of the night sky and the redshifts of the galaxies tell you that the Universe must have had a beginning. Furthermore, you have seen that the CMB is clear evidence that conditions at the beginning were hot and dense. Theorists can combine these observations with knowledge from physics about how atoms and subatomic particles behave to work out the story about how the Big Bang occurred.

Cosmologists cannot begin their history of the Big Bang at time zero, because no one understands the behavior of matter and energy under such extreme conditions, but they can come amazingly close. If you could visit the Universe when it was only one-millionth of a second old, you would find it filled with high-energy photons having a blackbody temperature of 20 trillion (2×10^{12}) K and a density (using Einstein's equation $E = mc^2$ to calculate the mass equivalent to a certain amount of energy) of 5×10^{20} kg/m^3, more than a thousand times the density of an atomic nucleus.

If photons have enough energy, two photons can combine and convert their energy into a pair of particles—a particle of normal matter and a particle of **antimatter**. And then, when an antimatter particle meets its matching particle of normal matter, the two particles annihilate each other and convert their mass back into energy in the form of two gamma-rays. In the early Universe, photons had enough energy to produce proton–antiproton pairs or neutron–antineutron pairs. When these particles collided with their antiparticles, they converted their mass back into photons. Thus, the early Universe was filled with a dynamic soup of energy flickering from photons into particles and back again.

antimatter Matter composed of antiparticles, which upon colliding with a matching particle of normal matter, annihilate and convert the mass of both particles into energy.

FIGURE 14-6

In 1965, Arno Penzias (right) and Robert Wilson first detected the cosmic microwave background radiation with the horn antenna behind them.

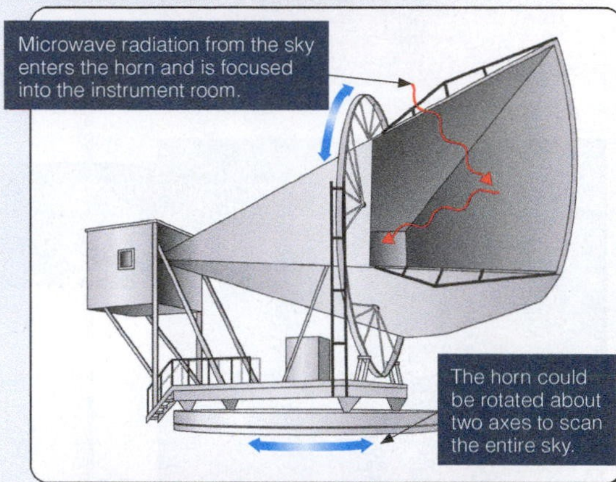

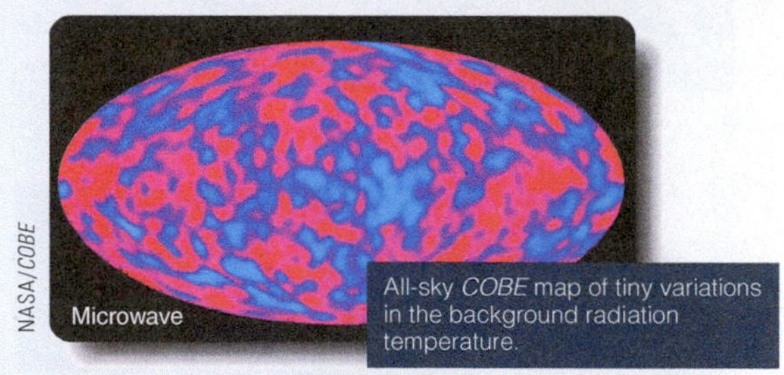

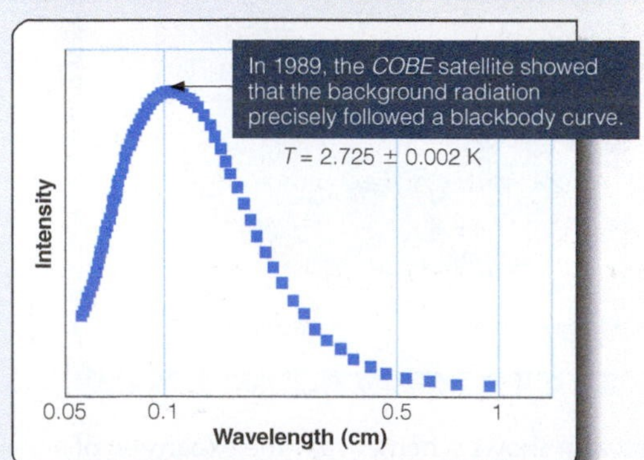

When the cosmic microwave background radiation was first detected in 1965, technology did not allow measurements at many wavelengths. Not until infrared and microwave detectors could be put in space was it conclusively shown that the background radiation has a blackbody spectrum, as predicted by theory.

While all this went on, the expansion of the Universe cooled the radiation. By the time the Universe was ten-millionths of a second old, its blackbody temperature had had fallen to about 2 trillion (2×10^{12}) K. At that point the average energy of the gamma-ray photons fell below the energy equivalent to the mass of a proton or a neutron, so the gamma-rays could no longer produce such heavy particles, and the creation of protons and neutrons stopped. Those particles combined with their antiparticles and quickly converted most of the mass into photons.

You might guess from this that all of the protons and neutrons would have been annihilated with their antiparticles, and the Universe should now consist of nothing but photons. However, for reasons that are poorly understood, a small excess of normal particles apparently existed. For every billion protons annihilated by antiprotons, one survived with no antiparticle to destroy it. Consequently, you live in a world of normal matter, and antimatter is very rare.

Although the gamma-rays did not have enough energy to produce protons and neutrons after the temperature of the Universe fell below about 10^{12} K, electron–positron pairs, having lower mass, could still be produced. (The positron is the antiparticle of the electron, first mentioned in Chapter 10 Section 10-2b in the context of hydrogen fusion.) That continued until the Universe was about 1 minute old, at which time the expansion had cooled the gamma-rays to the point where they could no longer create even electron–positron pairs. Most of the electrons and positrons combined to form photons, and only one in a billion elections survived. The protons, neutrons, and electrons of which our Universe is made were produced during the first 4 seconds of its history.

This soup of hot gas and radiation continued to cool. By the time the Universe was about 2 minutes old, protons and neutrons could link to form deuterium, the nucleus of a heavy hydrogen atom, and not be broken apart again. By the end of the next minute, further reactions began converting deuterium into helium, but almost no heavier atoms could be built because there are no stable nuclei with atomic weights of 5 or 8 (in units of the hydrogen atom)—they fall apart as soon as they are created. Cosmic element building during the Big Bang had to proceed rapidly, step by step, like someone hopping up a flight of stairs. The lack of stable nuclei at atomic weights of 5 and 8 meant there were missing steps in the stairway, and the step-by-step reactions had great difficulty jumping over these gaps. Astronomers can calculate that the Big Bang produced a tiny amount of lithium (atomic weight 7) but no elements heavier than that.

By the time the Universe was 3 minutes old, it had become so cool that most nuclear reactions had stopped, and by the time it was 30 minutes old the nuclear reactions had ended completely. At that time, about 25 percent of the mass was in the form of helium nuclei, and the rest was in the form of hydrogen nuclei (protons). That is the abundance of hydrogen and helium observed today in the oldest stars. The cosmic abundance of hydrogen and helium was essentially fixed during the first minutes of the Universe. The hydrogen nuclei in water molecules in your body have survived unchanged since they formed during the first moments of the Big Bang. Heavier elements were built by nucleosynthesis inside later generations of stars (look back to Chapters 11 and 12).

14-2d Recombination and Reionization—The First Thousands and Millions of Years

At first, the Universe was so hot that the gas was totally ionized, and the electrons were not attached to nuclei. Free electrons interact with photons so easily that a photon could not travel very far before it encountered an electron and was deflected (Figure 14-7a). The radiation and matter interacted continuously with each other and cooled together as the Universe expanded.

As the young Universe expanded, it went through three important changes. First, when the Universe reached an age of roughly 50,000 years, the density of the energy present in the form of photons became less than the density of the gas. Before that time, matter could not clump together because the intense sea of photons smoothed out the gas. When the density of the radiation fell below that of matter, the matter could begin to draw together under the influence of gravity and form the clouds that eventually became galaxy clusters and galaxies.

The expansion of the Universe spread the particles of the ionized gas farther and farther apart. As the Universe reached the age of about 400,000 years, the second important change began. As the density decreased and the falling temperature of the Universe reached 3000 K, protons were able to capture and hold free electrons to form neutral hydrogen, a process called **recombination**, although "combination" (for the first time) would be more accurate. As the free electrons

recombination The stage, about 400,000 years after the Big Bang, when neutral atoms formed and the gas of the Universe became transparent to radiation.

FIGURE 14-7

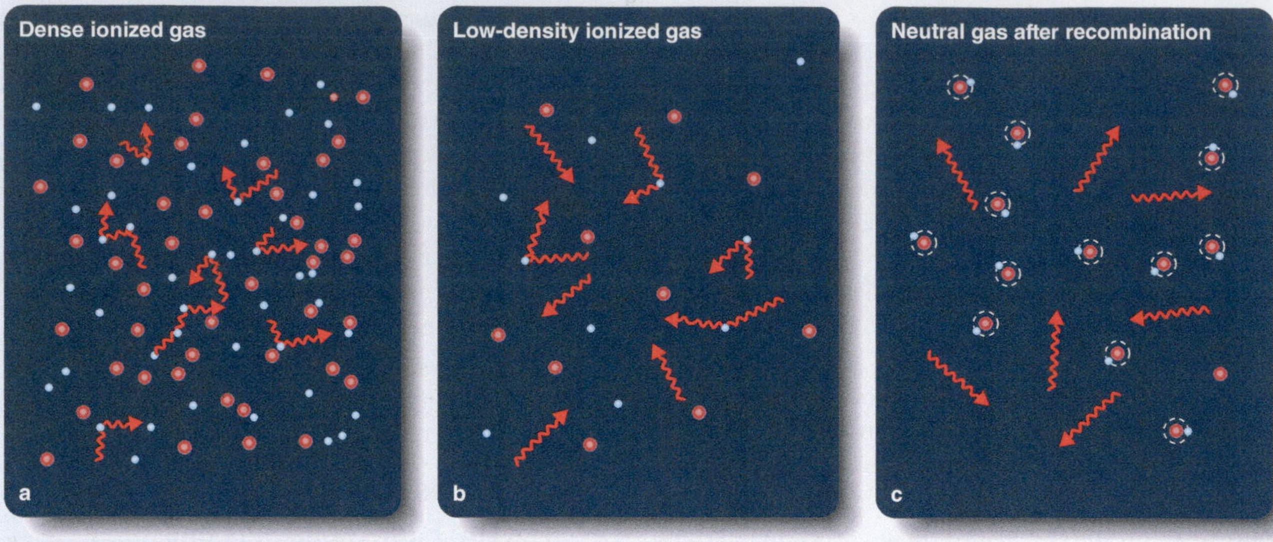

Photons scatter from electrons (blue) easily but hardly at all from the much more massive protons (red). (a) When the Universe was very dense and ionized, photons could not travel very far before they scattered off an electron. This made the gas opaque. (b) As the Universe expanded, the electrons were spread further apart, and the photons could travel farther; this made the gas more transparent. (c) After recombination, most electrons were locked to protons to form neutral atoms, and the gas became completely transparent.

were gobbled up into atoms, they could no longer deflect photons. The photons could travel easily through the gas, so the gas became transparent (Figure 14-7c), and the photons retained the blackbody temperature of 3000 K that the gas and photons together had at the time of recombination. Those photons are what are observed today as the CMB, with a large redshift that makes their temperature now about 2.7 K.

Recombination left the gas of the Big Bang neutral, hot, dense, and transparent. At first the Universe was filled with the glow of the hot gas, which would have been partly detectable at visible wavelengths. As the Universe expanded and cooled, the glow faded and the Universe entered what cosmologists call the **dark age**, a period lasting hundreds of millions of years until the formation of the first stars. During the dark age, there were no sources of visible light in the Universe.

dark age The period of time after the glow of the Big Bang faded into the infrared and before the birth of the first stars, during which the Universe expanded in darkness.

reionization The stage in the early history of the Universe when ultraviolet photons from the first stars ionized the gas of the Universe.

The dark age ended as the first stars began to form. The gas from which those first stars formed contained almost no metals and was consequently quite transparent. Mathematical models show that the first stars formed from this metal-poor gas would have been very massive, very luminous, and very short lived. That first burst of star formation produced enough ultraviolet light to begin ionizing the gas, and today's astronomers, looking back to the most distant visible quasars and galaxies, can see traces of that **reionization** of the Universe (Figure 14-8). Reionization marks the end of the dark ages and the beginning of the age of stars and galaxies that continues today.

Look carefully at Figure 14-9; it summarizes the story of the Big Bang, from the formation of helium in the first 3 minutes through energy–matter equality, recombination, and finally reionization of the gas. It might seem amazing that mere humans trapped on Earth can draw such a diagram, but remember that it is based on evidence and on the best understanding about how matter and energy interact. (See **How Do We Know? 14-1**.)

FIGURE 14-8

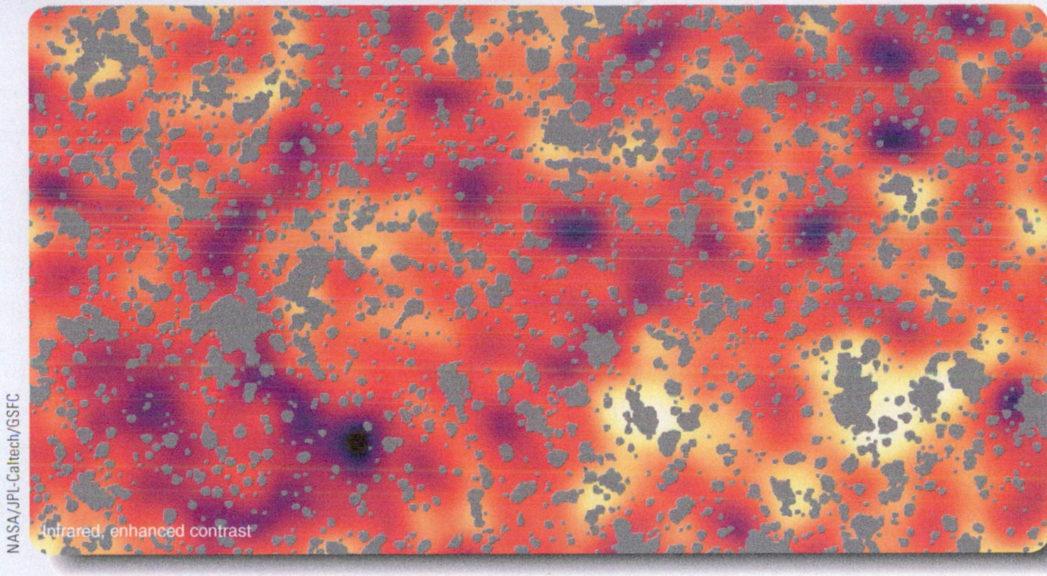

A map of the far-infrared background of a region in the constellation Ursa Major made using the *Spitzer Space Telescope*. Foreground stars, galaxies, and other known sources have been removed by computer processing. The remaining background glow is understood to be from objects that formed in the first billion years of the Universe's history, such as massive stars and voracious black holes at the centers of forming galaxies. Yellow and white indicate the brightest parts of the background.

FIGURE 14-9

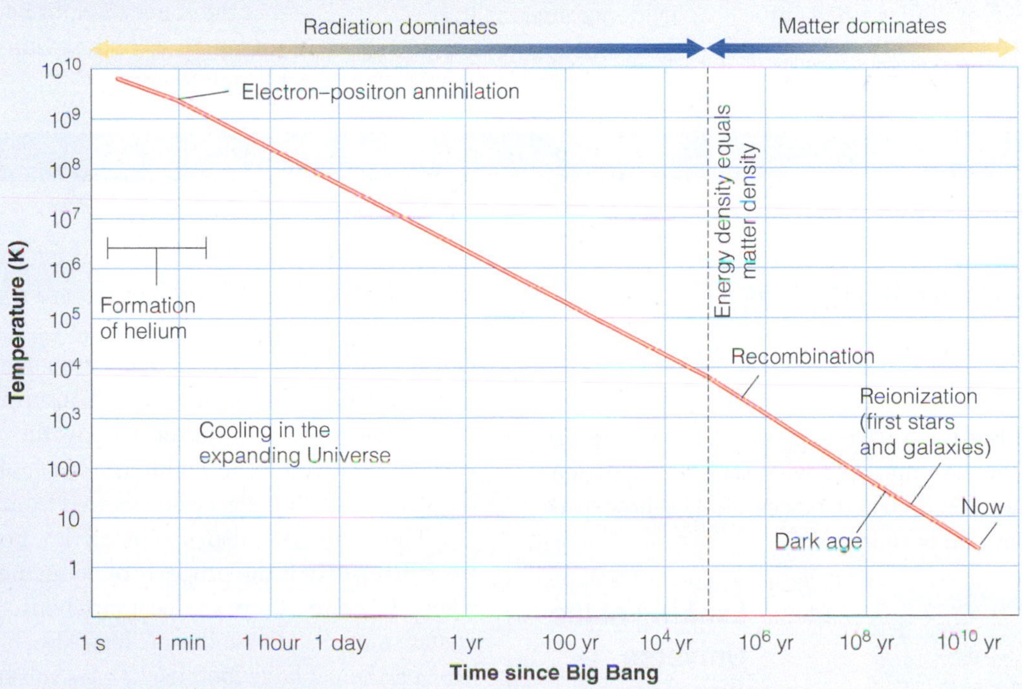

During the first few minutes of the Big Bang, some hydrogen fused to produce helium, but the Universe quickly became too cool for such fusion reactions to continue. The rate of cooling increased as matter began to dominate over radiation. Recombination freed the radiation from the influence of the gas, and reionization was caused by the birth of the first stars. Note how the exponential scale in time stretches early history and compresses recent history.

How Do We Know? 14-1

Science: A System of Knowledge

What is the difference between believing in the Big Bang and understanding it? If you ask a scientist, "Do you believe in the Big Bang?" she or he might hesitate before responding. The question implies something incorrect about the way science works.

Science is a system of knowledge and process but not a system of belief. Science is an attempt to understand logically how nature works and is based on observations and experiments used to test and confirm hypotheses and theories. A scientist does not really believe in even a well-confirmed theory in the way people normally use the word *believe*. Rather, the scientist understands the theory and recognizes how different pieces of evidence support or contradict the theory.

There are other ways to know things, and there are many systems of belief. Religions, for example, are systems of belief that are not entirely based on observation. In some cases, a political system is also a system of belief; many people believe that democracy is the best form of government and do not ask for, or expect, evidence supporting that belief. A system of belief can be powerful and lead to deep insights, but it is different from science.

Scientists try to be careful with words, so thoughtful scientists would not say they *believe* in the Big Bang. They would say that the evidence is overwhelming that the Big Bang really did occur and that they are compelled by a logical analysis of both the observations and the theory to conclude that the theory is very likely correct. In this way, scientists try to be objective and reason without distortion by personal feelings and prejudices.

NASA/*WMAP* Science Team

Scientific knowledge is based objectively on evidence such as that gathered by spacecraft.

A scientist once referred to "the terrible rule of evidence." Sometimes the evidence forces a scientist to a conclusion she or he does not like, but science is not a system of belief, so the personal preferences of each scientist must take second place to the rule of evidence.

Do you believe in the Big Bang? Or, instead, do you have confidence that the theory is right because of your analysis of the evidence? There is a big difference.

14-3 SPACE AND TIME, MATTER AND ENERGY

How can the Big Bang have happened everywhere? To solve the puzzle, you must put what seem to be reasonable expectations on hold and look carefully at how space and time behave on cosmic scales.

isotropy The observation that, in its general properties, the Universe looks the same in every direction.

homogeneity The observation that, on the large scale, matter is uniformly spread through the Universe.

14-3a Looking at the Universe

The Universe looks about the same whichever way you look. That is called **isotropy**. Of course, there are local differences. If you look toward a galaxy cluster you see more galaxies, but that is only a local variation. On the average, you see similar numbers of galaxies in every direction. Furthermore, the background radiation is also almost perfectly uniform across the sky. The Universe is observed to be highly isotropic, almost exactly the same in all directions when viewed from our position.

The Universe also seems to be homogeneous; **homogeneity** is the property of being the same everywhere. Of course there are local variations. Some regions contain more galaxies and some less. Also, if the Universe evolves, then at large look-back times you see galaxies at an earlier stage. If you account for these well-understood variations, then the Universe seems to be, on average, the same everywhere. This is harder to check because you can't actually go to the locations of distant galaxies and check in detail that things are about the same there as here, but all astronomical observations indicate this is so.

Isotropy and homogeneity together lead to the **cosmological principle**, that any observer in any galaxy sees that the Universe has the same general properties, after accounting for relatively minor local and evolutionary variations. The cosmological principle implies that there are no special places in the Universe. What you see from the Milky Way Galaxy is typical of what all intelligent creatures see from their respective home galaxies. Furthermore, the cosmological principle is another way of saying that the Universe has no center or edge. Such locations would be special places, and the cosmological principle means there are no special places.

14-3b Cosmic Redshifts

Distance is the separation between two points in space; time is the separation between two moments. Einstein's theories of special relativity and general relativity (published respectively in 1905 and 1916) describe how space and time are related and can be considered together as the fabric of the Universe, called space-time. As you read **Concept Art 14A, The Nature of Space-Time**, you can think of space-time as the canvas on which the Universe is painted. Einstein's theories predict that the canvas of space-time can potentially expand (or contract), and, amazingly, that has been confirmed by observations.

The stretching of space-time explains one of the most important observations in cosmology—cosmological redshifts. Modern astronomers understand that, except for small local motions within clusters of galaxies, the galaxies are basically at rest and have kept approximately the same "address" in space since the Big Bang. The distances between them increase as space-time expands. Furthermore, as space-time expands, it stretches photons traveling through space to longer wavelengths, as you can see in the vertical sequence of panels on the left-hand page of Concept Art 14A. Photons from distant galaxies spend more time traveling through space and are stretched more than photons from nearby galaxies. That is why redshift depends on distance. Note that objects—such as the Milky Way Galaxy, Earth, and you—that are held together by gravity or electromagnetic forces do not expand as the Universe expands.

Astronomers often express redshifts as if they were radial velocities, but the redshifts of the galaxies are not Doppler shifts. That is why this book is careful to refer to a galaxy's *apparent* velocity of recession. All a cosmological redshift tells you directly is how much the Universe has expanded since the light began its journey to Earth. The formula to calculate the distance a photon has traveled, given its redshift, is complicated, and the parameters are difficult to measure precisely. Nevertheless, the Hubble law does apply, and redshifts can be used to estimate the distances to galaxies.

14-3c Model Universes

Almost immediately after Einstein published his theories, theorists were able to solve the highly sophisticated mathematics to compute simplified descriptions of the behavior of space-time and matter. Those "model Universes" dominated cosmology throughout the 20th century.

The equations allowed three general possibilities. Space-time might be curved in ways that are called an **open universe** or a **closed universe**, or space-time might have no overall curvature at all, a situation that is called a **flat universe**. Most people find the curved space-time models difficult to imagine. Fortunately, modern observations have shown that the flat universe model is almost certainly correct, so you don't have to wrap your brain around the curved models. Note that "flat" does not mean two dimensional. Rather, it means that the familiar rules of geometry you learned in elementary and middle school, for instance that the circumference of a circle is 2π times its radius and that the interior angles of a triangle add up to 180 degrees, are true on the largest scales. Different rules of geometry would apply on large scales if the Universe were curved.

A main criterion separating the three models is the average density of the Universe, which according to general relativity, determines the overall curvature of space-time. If the average density of matter and energy in the Universe equals what is called the **critical density**, calculated to be about 9×10^{-27} kg/m^3 (depending on the exact value of Hubble constant), space-time will be flat. If the average density is more than the critical density, the Universe must be closed; and if it is less, the Universe must be open.

The expansions versus time of the three different models are compared in Figure 14-10. The parameter R on the vertical axis is a measure of the extent to which the Universe has expanded. You could think of it, essentially, as the average distance between

cosmological principle The assumption that any observer in any galaxy sees the same general features of the Universe.

open universe A model of the Universe in which space-time is curved in such a way that the Universe is infinite.

closed universe A model of the Universe in which space-time is curved to meet itself and the Universe is finite.

flat universe A model of the Universe in which space-time is not curved.

critical density The average density of the Universe needed to make its curvature flat.

Concept Art 14A

1 THE NATURE OF SPACE-TIME

In 1929, Edwin Hubble discovered that the redshifts of the galaxies are proportional to distance—a relationship now known as the Hubble law. It was immediately understood to be compelling evidence that the Universe is expanding.

Distance is the separation between two points in space, and interval is the separation between two moments in time. Einstein's theories of special and general relativity show how distances in space and intervals in time are interrelated, so that space and time should be considered together as space-time. You can think of space-time as a canvas on which the Universe is painted, a canvas that can stretch.

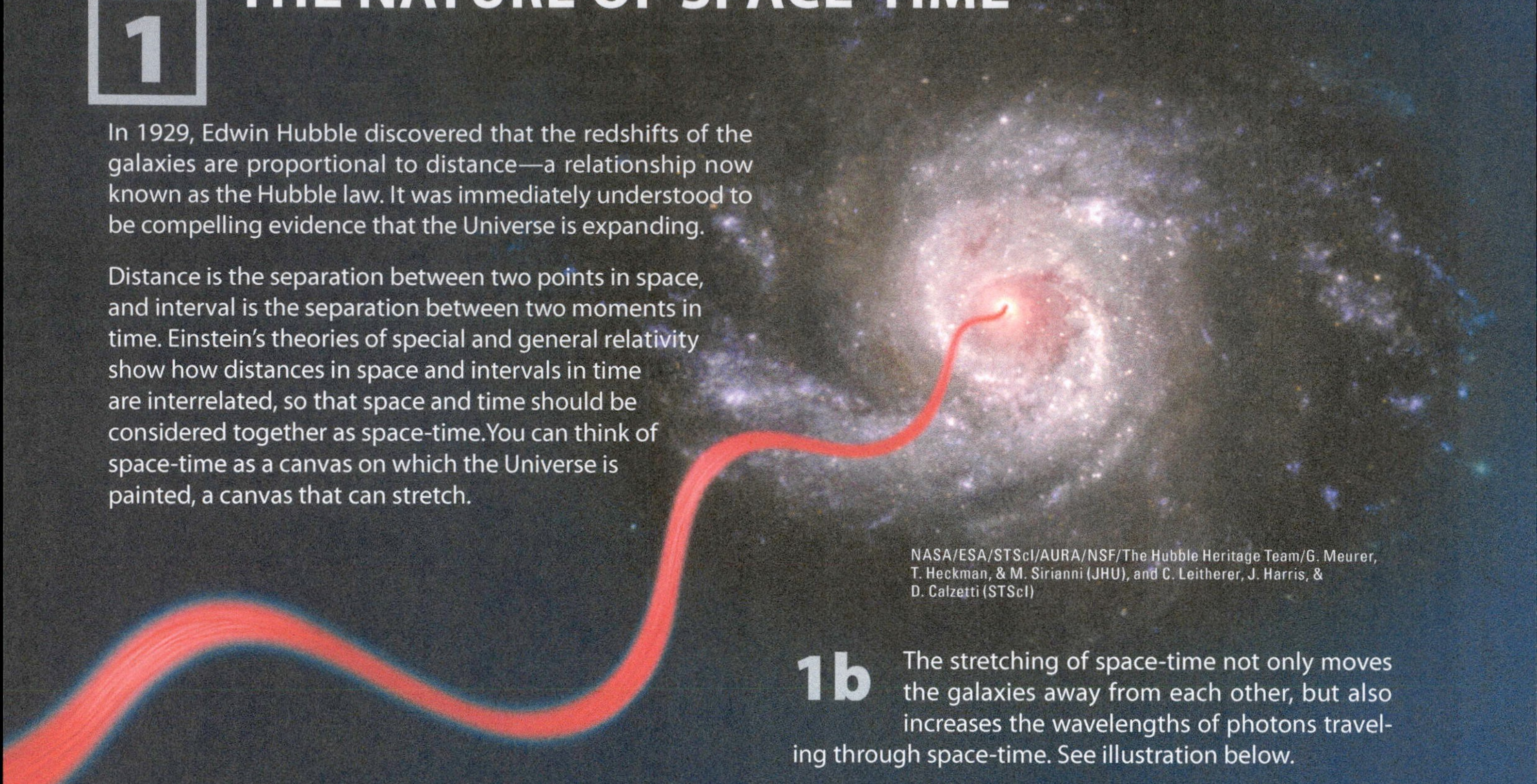

NASA/ESA/STScI/AURA/NSF/The Hubble Heritage Team/G. Meurer, T. Heckman, & M. Sirianni (JHU), and C. Leitherer, J. Harris, & D. Calzetti (STScI)

1a Astronomers often express galaxy redshifts as apparent velocities of recession, but these redshifts are not Doppler shifts. They are caused by the expansion of space-time.

For decades textbooks have described the cosmological redshifts using Einstein's relativisitic Doppler formula, but that formula applies to motion through space and not to the behavior of space itself.

Galaxies are not moving through space as the Universe expands any more than the raisins in the raisin bread analogy are swimming through the dough as the bread rises. Except for small relative orbital motions among neighbors, galaxies are motionless in space and are being carried away from each other by the stretching of space-time.

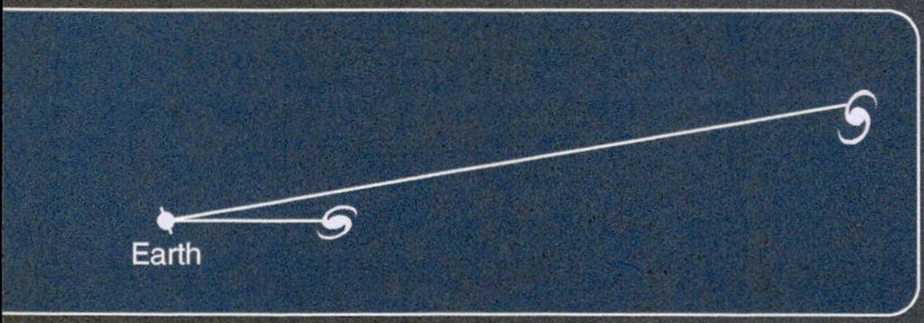

1c A photon traveling to Earth from a more distant galaxy spends more time moving through space, and is therefore stretched more, than a photon from a nearby galaxy. That is why cosmological redshifts are proportional to distance.

1b The stretching of space-time not only moves the galaxies away from each other, but also increases the wavelengths of photons traveling through space-time. See illustration below.

What are the cosmological redshifts?

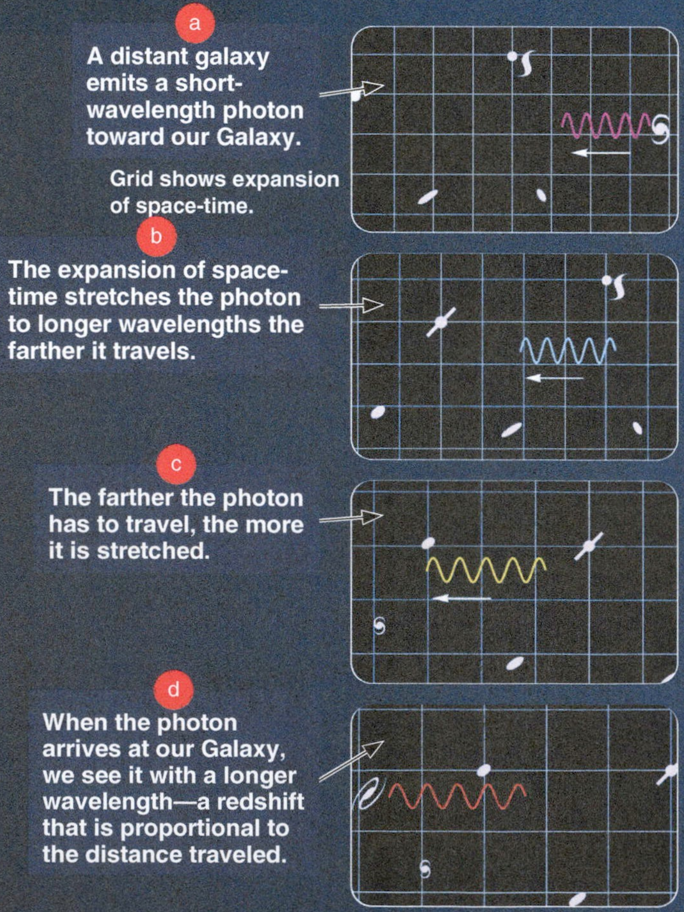

2 To think about space-time curvature, you can use an analogy of a two-dimensional ant on an orange. If it is truly two dimensional, it can travel only forward and back, right and left; it will not perceive up and down. As it walks around this mini-universe, it doesn't realize it is actually the surface of a three-dimensional sphere and will eventually realize it has been everywhere because that universe is covered by its footprints. The ant will conclude that it lives in a finite universe that has no edge and no center.

Similarly, a three-dimensional universe could be curved back on itself so that it could be finite, but an explorer would never find an edge. With no edge, such a universe would also have no center.

SPUTNIK / Alamy Stock Photo

Bygone Collection / Alamy Stock Photo

Nikolai Lobachevsky 1792–1856

Bernhard Riemann 1826–1866

b During the 19th century, two mathematicians, the Russian Nikolai Lobachevsky and the German Bernhard Riemann, developed the mathematics of curved surfaces—non-Euclidian geometry. That, with Einstein's theory of general relativity, showed that space-time might be curved by an amount that depends on the average density of the Universe. Curved space-time dominated 20th-century cosmology.

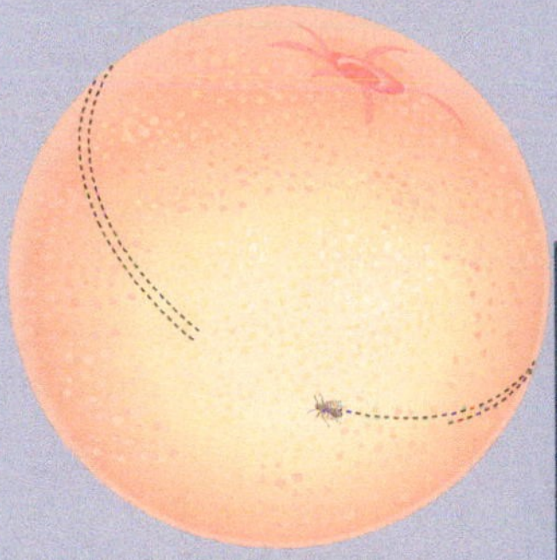

a Notice that the center of the three-dimensional orange is not the center of the ant's two-dimensional universe—the center of the orange does not lie on the surface and is not part of the ant's universe.

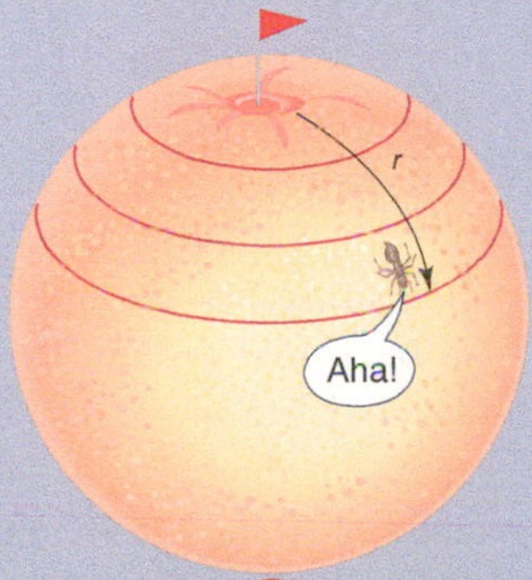

There are three possible types of curvature for a model universe: positive, zero, or negative. A positively curved universe can be represented by a sphere, a zero curvature universe by a flat surface, and a negatively curved universe by a saddle shape.

a Don't forget: These analogies are only two-dimensional. When you think about our Universe, you must think about a three-dimensional universe. It is as difficult for you to imagine curvature in a three-dimensional universe as it is for ants to imagine curvature in a two-dimensional universes.

c A positively curved model universe is termed a **closed universe** because it curves back on itself. It is finite, but it has no edge. In a two-dimensional analogy, ants on an orange could discover that they are in a curved universe because the circumferences of large circles are less than $2\pi r$.

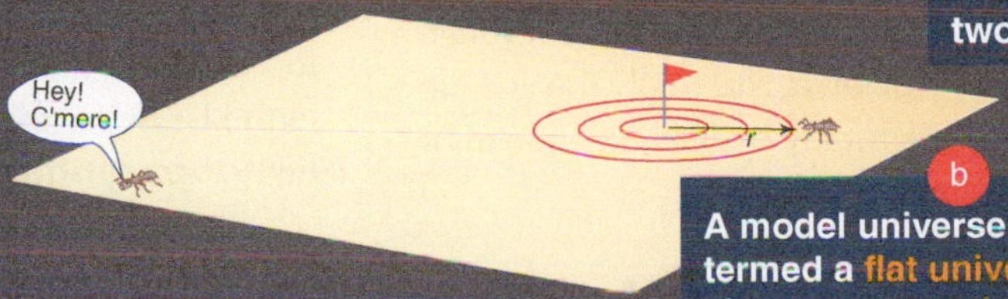

b A model universe with zero curvature is termed a **flat universe**. A flat universe is also an **open universe** because it does not curve back on itself. In a two-dimensional analogy, ants on a flat sheet of paper would find that all circles have circumferences equal to $2\pi r$. If the Universe truly has flat (Euclidean) geometry, then it must be infinite and has no center.

3 Throughout the 20th century, cosmologists tried to determine whether our Universe is actually positively curved, negatively curved, or uncurved (flat) by making difficult and inconclusive measurements analogous to the ants measuring the circumferences of large circles. As you will see later in this chapter, new ways to measure the curvature show that the Universe is close to exactly flat.

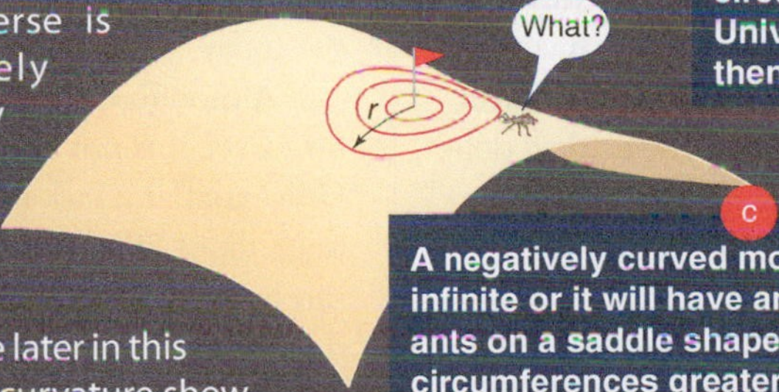

c A negatively curved model is also an open universe. It must be infinite or it will have an edge. In a two-dimensional analogy, ants on a saddle shape will discover that larger circles have circumferences greater than $2\pi r$.

FIGURE 14-10

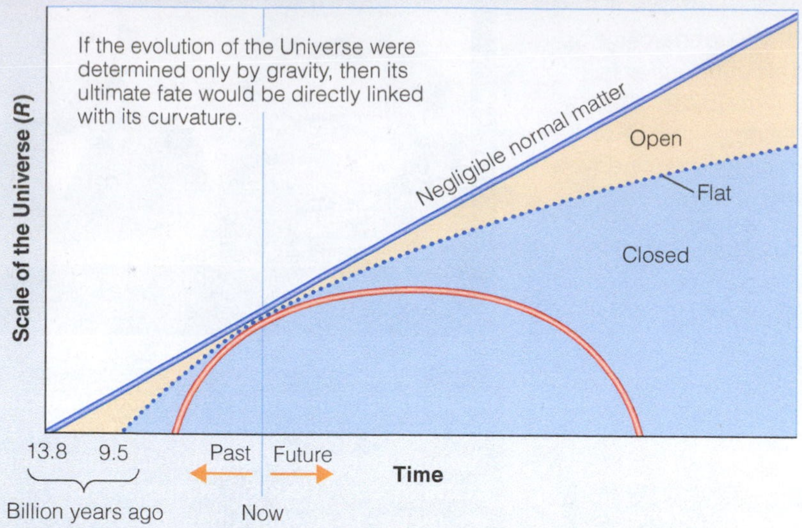

Illustrations of some simple models of the Universe that depend only on the effect of gravity. Open universe models expand without end, and the corresponding curves fall in the region shaded orange. Closed models expand and then contract again (*red curve*). A flat universe (*dotted line*) marks the boundary between open and closed universe models. In these models, the relationship between estimated age and actual age of the universe depends on the curvature of space-time. The age of the Universe for each model is shown on the graph as the horizontal distance from "Now" back to the time when the Universe scale factor R equaled 0.

galaxies. In the figure you can see that closed universes can expand and then contract, and open universes can expand forever. Notice also that you can't find the actual age of the Universe—the distance on the horizontal axis between the present and the beginning, when R was zero—from the expansion rate alone. You also need to know whether the Universe is open, closed, or flat. The Hubble time, discussed earlier in this chapter as a rough estimate of the age of the Universe, is actually the age the Universe would be if it were totally open, which means it would have to contain almost no matter at all.

14-3d Ordinary Matter and Dark Matter

Later in this chapter you will see how observational evidence indicates that the Universe is probably flat. For now, you must solve a different problem. If the Universe is flat, then its average density must equal the critical density, yet when astronomers added up the matter they could detect, they found only a few percent of the critical density. They wondered if the dark matter made up the rest.

nonbaryonic matter Proposed dark matter made up of particles other than protons and neutrons (baryons).

cold dark matter Dark matter which is made of slow-moving particles.

In Chapters 12 and 13, you learned that our Galaxy, other galaxies, and galaxy clusters have much stronger gravitational fields than expected based on the amount of visible matter. Even when you add in the nonluminous gas and dust that you expect to find, their gravitational fields are stronger than expected. Galaxies must contain dark matter—in fact, much more dark matter than normal matter (look back to Figures 12-9 and 13-5).

The protons and neutrons that make up normal matter, including Earth and you, belong to a family of subatomic particles called *baryons*. Modern evidence based on what can be determined about the products of nuclear reactions in the first few minutes after the Big Bang shows that the dark matter is not baryonic. If there were lots of baryons present during those early moments, they would have (1) collided with and destroyed deuterium nuclei and (2) collided with some of the helium to make lithium.

Figure 14-11 shows that the observed amount of deuterium sets a lower limit on the density of the Universe, and the observed abundance of lithium-7 sets an upper limit. Those limits indicate that the baryons you and Earth and the stars are made of cannot add up to more than about 5 percent of the critical density. Yet, observations show that galaxies and galaxy clusters contain as much as 30 percent of the critical density in the form of dark matter. Only a small amount of the matter in the Universe can be baryonic, so the dark matter must be **nonbaryonic matter**.

The true nature of dark matter remains one of the mysteries of astronomy. The most successful models of galaxy formation require that the dark matter be made up of **cold dark matter**, meaning the particles move much slower than light and therefore can clump into structures with sizes that explain the galaxies and galaxy clusters you see today.

Although the evidence is very strong that dark matter exists, it is not abundant enough by itself to make the Universe flat. Dark matter appears to constitute no more than about 30 percent of the critical density. As you will see later in this chapter, there is more to the Universe than meets the eye, more even than dark matter.

FIGURE 14-11

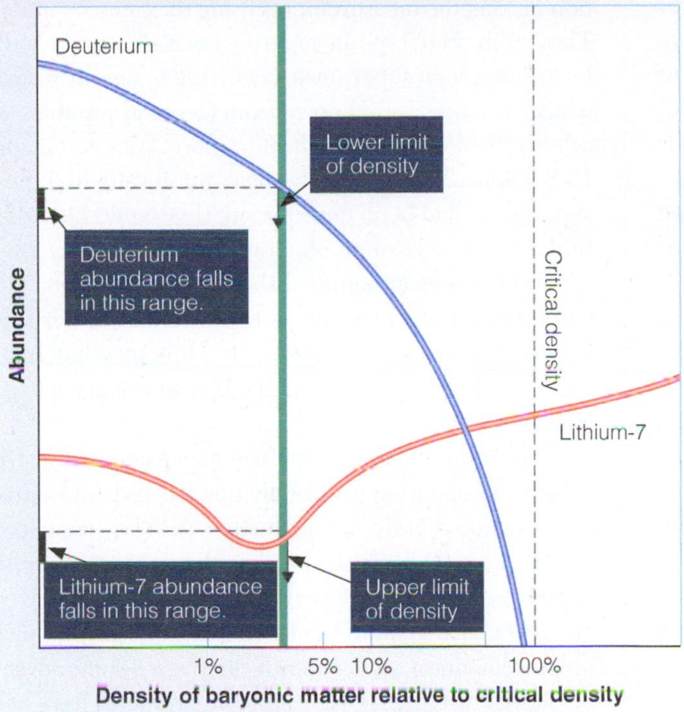

This diagram compares observations with theory. Theory predicts how much deuterium and lithium-7 you would observe for different densities of baryonic matter (red and blue curves). The observed density of deuterium falls in a narrow range shown at upper left and sets a lower limit on the possible density of baryonic matter. The observed density of lithium-7 sets an upper limit. This means the true density of normal matter must fall in a narrow range represented by the green column. Certainly, the density of baryonic matter is much less than the critical density.

14-4 TWENTY-FIRST-CENTURY COSMOLOGY

If you are a little dizzy from the weirdness of expanding space-time and dark matter, make sure you are sitting down before you read further. As the 21st century began, astronomers made a discovery that startled all cosmologists: The expansion of the Universe is actually accelerating. To get a running start on these new discoveries, you'll have to go back a couple of decades.

14-4a Inflation

By 1980, the Big Bang model was widely accepted, but it faced two problems that led to the development of an improved theory—a Big Bang model with an important addition.

One of the problems is called the **flatness problem**. The curvature of space-time seems to be near the transition between an open and a closed Universe. That is, the Universe seems approximately flat. It seems peculiar that the actual density of the Universe is anywhere near the critical density that would make it flat. To be so near critical density now, the density of the Universe during its first moments must have been very close, within 1 part in 10^{49}, of the critical density. So, the flatness problem is: *Why was the Universe so close to exactly flat, with no space-time curvature, at the time of the Big Bang?*

The second problem with the original Big Bang theory is called the **horizon problem**. When astronomers correct for the motion of Earth, they find that the CMB is very isotropic, the same in all directions to a precision of better than 1 part in 1000. Yet, background radiation coming from two points in the sky separated by more than an angle of 1 degree is from two parts of the Big Bang far enough apart that they should not have been connected at any previous time. That is, when the CMB photons were released, the Universe was not old enough for energy to have traveled at the speed of light from one of those regions to the other—the regions should always have been beyond each other's "horizon" and could not have exchanged heat to make their temperatures equal. So, the horizon problem is: *How did every part of the observable Universe get to be so nearly the same temperature by the time of recombination?*

The key to these two problems and to other problems with the simple Big Bang model might lie with a modified model called the **inflationary Big Bang** that predicts there was a sudden extra expansion when the Universe was very young, even more extreme than that predicted by the original Big Bang model. According to the inflationary Universe model, the Universe expanded and cooled until about 10^{-36} seconds after the Big Bang. Then, the

flatness problem The peculiar circumstance that the early Universe must have contained almost exactly the right amount of matter to make space-time flat.

horizon problem The circumstance that the primordial background radiation seems much more isotropic than can be explained by the standard Big Bang theory.

inflationary Big Bang A version of the Big Bang theory, derived from grand unified theories of particle physics, that includes a rapid expansion when the Universe was very young.

Universe became cool enough that the forces of nature (look back to Chapter 10 Section 10-2a), which at earlier extremely high temperatures would have behaved identically, began to differ from each other. Physicists calculate that this would have released tremendous amounts of energy and suddenly inflated the Universe by a factor of 10^{50} or larger. As a result, the part of the Universe that is now visible from Earth, the entire observable Universe, expanded rapidly from 10^{35} times smaller than a proton to roughly a meter across and then continued a slower expansion to its current state.

That sudden inflation can solve both the flatness problem and the horizon problem. The inflation of the Universe would have forced whatever curvature it had toward zero, just as inflating a balloon makes a small spot on its surface flatter. The Universe now has almost perfectly flat space-time geometry because of that sudden inflation long ago. In addition, because the observable part of the Universe was no larger in volume than an atom before inflation, it was small enough to have equalized its temperature by then. The Universe now has almost exactly the same CMB temperature in all directions.

The inflationary theory predicts that the Universe is almost perfectly flat. Observations, however, give evidence that the masses scientists know about (baryonic matter plus dark matter) add up only to about 30 percent of the amount needed to make space-time flat. Can there be more to the Universe than baryonic matter and dark matter? What could be weirder than dark matter? Read on.

14-4b Acceleration of the Expansion

Both common sense and mathematical models suggest that, as the galaxies recede from each other, the expansion should be slowed by gravity trying to pull the galaxies toward each other. How much the expansion is slowed should depend on the amount of matter in the Universe. If the density of matter is less than the critical density, the expansion should be slowed only slightly, and the Universe should expand forever. If the density of matter in the Universe is greater than the critical density, the expansion should be slowing down dramatically, and the Universe should eventually begin contracting.

For decades, astronomers struggled to measure the distance to very distant galaxies directly, compare distances with redshifts, and thereby detect the slowing of the expansion. The rate of slowing would in turn reveal the true curvature of the Universe. This was one of the key projects for the *Hubble Space Telescope*, and two teams of astronomers spent years in competition making the measurements using the same technique. They calibrated Type Ia supernovae as standard candles by locating such supernovae occurring in nearby galaxies whose distances were known from Cepheid variables and other reliable distance indicators (look back to Chapter 13 Section 13-2a). Once the peak luminosity of Type Ia supernovae had been determined, they could be used to find the distances of much more distant galaxies.

cosmological constant A constant in Einstein's equations of space and time that represents a force of repulsion.

Both teams announced their results in 1998. They agreed that the expansion of the Universe is not slowing down. Contrary to expectations, it is speeding up! In other words, the expansion of the Universe is accelerating (Figure 14-12).

The announcement that the expansion of the Universe is accelerating was totally unexpected, and astronomers immediately began testing it. The most likely problem was thought to be that the calibration of the supernovae by the original teams might be incorrect. However, that type of problem has been ruled out by measurements of more recently discovered supernovae at very great distances. Other possible problems have been checked and appear to have been eliminated. The Universe really does seem to be expanding faster and faster.

14-4c Dark Energy and Acceleration

If the expansion of the Universe is accelerating, then there must be a force of repulsion in the Universe, and astronomers are struggling to understand what it could be. One possibility leads back to Albert Einstein.

When Einstein published his theory of general relativity in 1916, he noticed that his equations describing space-time implied that the Universe should contract because of the gravitational attraction of galaxies for each other. In 1916, astronomers did not yet know that the Universe was expanding, so Einstein thought he needed to balance the attractive force of gravity by adding a constant to his equations called the **cosmological constant**, representing a force of repulsion that would balance gravity and make the Universe hold still. Thirteen years later, in 1929, Edwin Hubble announced that the Universe was expanding, and Einstein said that introducing the cosmological constant was his biggest blunder. Modern astronomers aren't so sure.

One explanation for the acceleration of the Universe is that there is a cosmological constant after all, representing a real force that drives a continuing acceleration in the expansion of the Universe. The cosmological constant, as its name implies, would be constant in strength over time. Another possibility is a type of energy that is

FIGURE 14-12

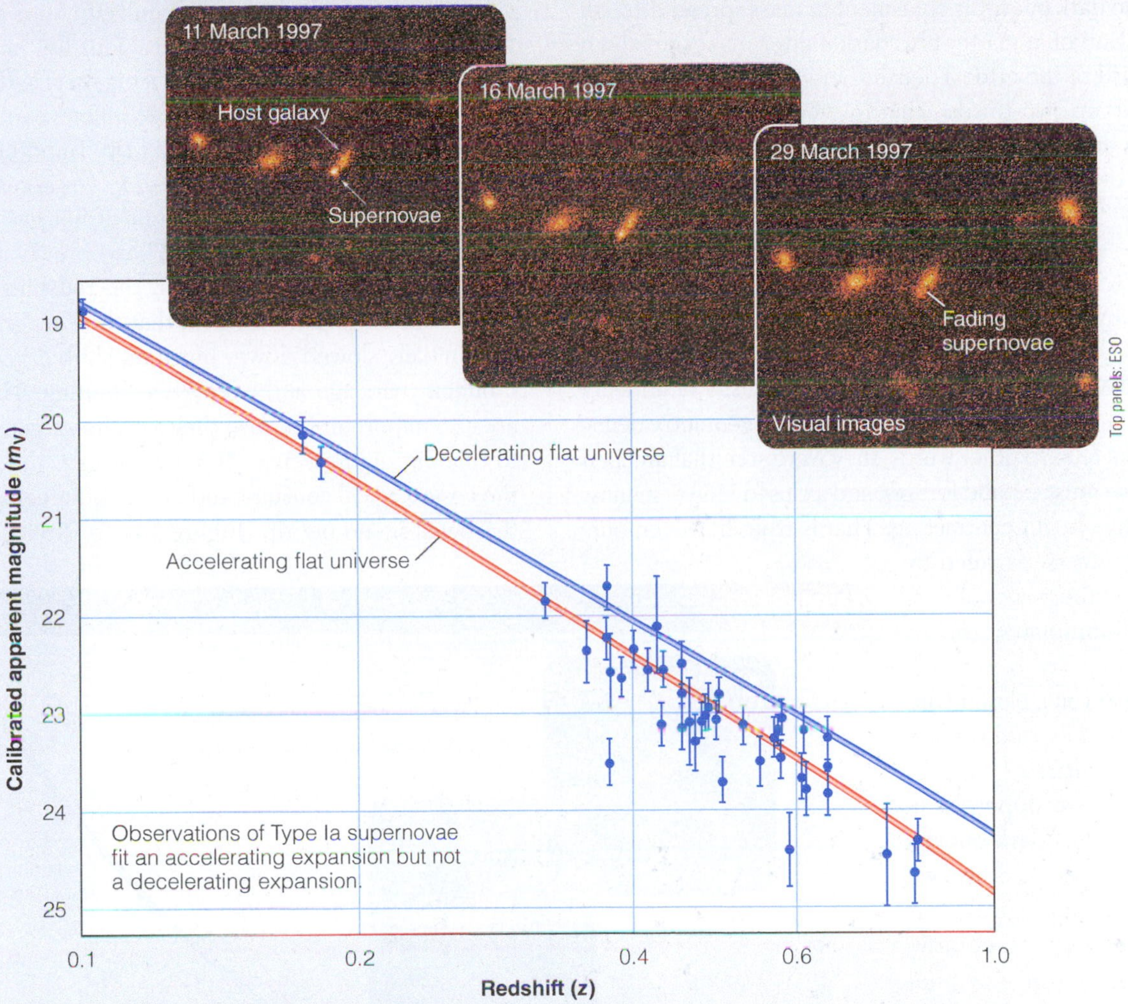

From the way supernovae fade over time, astronomers can identify those that are Type Ia. Once calibrated, the peak brightness of each of those supernovae could be compared with their respective redshifts, revealing that Type Ia supernovae with the range of redshifts plotted were about 25 percent fainter than expected. That must mean they are farther away than expected, given their redshifts. This is strong evidence that the expansion of the Universe is accelerating.

not constant in strength over time. Astronomers refer to this type of energy as **quintessence**. In either case, the observed acceleration is evidence that some form of energy, either a cosmological constant or quintessence, is spread throughout space. Astronomers refer to this as **dark energy**, energy that drives the acceleration of the Universe but does not contribute to starlight or the CMB.

You will recall that acceleration and dark energy were first discovered when astronomers found that supernovae a few billion light-years away were slightly fainter than expected. Since then even more distant supernovae have been determined to be a bit brighter than expected, meaning they are not as far away as the redshifts of their galaxies would seemingly indicate. This means that sometime about 6 billion years ago the Universe shifted gears from deceleration to acceleration. The careful calibration of Type Ia supernovae (look back to Chapter 11 Section 11-4d) allows astronomers to observe this change from deceleration to acceleration. This discovery has the important consequence of increasing previous estimates of the age of the Universe by several billion years.

Dark energy can also help you understand the lack of curvature of space-time. The theory of inflation makes the specific prediction that the Universe is flat.

quintessence A possible form of dark energy that can change in strength as the Universe ages.

dark energy The energy believed to fill empty spaces and drive the acceleration of the expanding Universe.

Dark energy seems to fit with that prediction. As mentioned earlier in this chapter, energy and matter are equivalent, so dark energy is equivalent to mass spread through space. Baryonic matter plus dark matter makes up about one-third of the critical density, and dark energy appears to make up two-thirds. That is, when you include dark energy, the total mass-plus-energy density of the Universe equals the critical density, making the Universe flat.

14-4d The Universe's Future

For many years, cosmologists enjoyed saying: "Geometry is destiny." Thinking about models of open, closed, and flat universes, they concluded that the density of a model Universe determines its geometry, and its geometry determines its fate. In other words, they were sure that an open Universe must expand forever, and a closed Universe must eventually begin contracting. That is true, however, only if the Universe is ruled by gravity. If the power of dark energy dominates gravity, then geometry is not destiny, and even a closed Universe might expand forever.

The ultimate fate of the Universe depends on the nature of dark energy. If dark energy is described by the cosmological constant, then the force driving acceleration does not change with time, and our flat Universe will expand forever. The galaxies will get farther and farther apart, use up their gas and dust making stars, and the stars will ultimately all die, until each galaxy is isolated, burnt out, dark, and alone. If, however, dark energy is described by quintessence, then its strength could increase with time, and the Universe expansion might accelerate faster and faster as space pulls the galaxies away from each other and eventually pulls the galaxies apart, then pulls the stars apart, and finally rips individual atoms apart. This has been called the **big rip**. Don't worry. Even if a big rip is in the future, nothing will happen for at least 30 billion years.

There probably will be no big rip. Important observations made by the *Chandra X-Ray Observatory* have been used to measure the amount of hot gas and dark matter in 25 galaxy clusters. These observations are important for two reasons. First, the redshifts and distance of these galaxies confirm that the Universe expansion initially slowed down, but then shifted gears about 6 billion years ago and is now accelerating. The results are also important because they are almost good enough to rule out quintessence. If dark energy is described by the cosmological constant and not by quintessence, then there will be no big rip (Figure 14-13). Notice that this

FIGURE 14-13

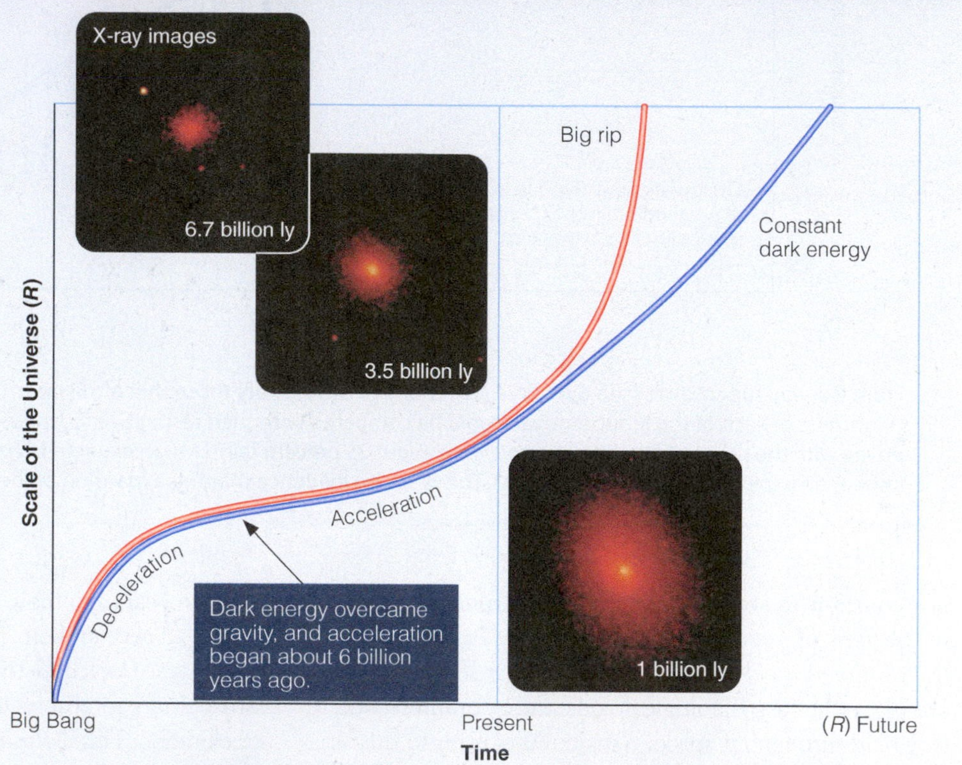

X-ray observations of hot gas in galaxy clusters confirm that in its early history the Universe was decelerating because gravity was stronger than the dark energy. As expansion weakened the influence of gravity, dark energy began to cause acceleration. The evidence is not conclusive, but it most directly supports the cosmological constant form of dark energy and weighs against quintessence, which means the Universe may not undergo a "big rip." This diagram is only schematic, and the two curves are drawn separated for clarity; currently, the two curves have not diverged from each other. Compare this plot with Figure 14-10 that includes only the effects of gravity.

big rip The fate of the Universe if dark energy increases with time and galaxies, stars, and even atoms are eventually ripped apart by the accelerating expansion of the Universe.

method does not depend on Type Ia supernovae at all, so it is independent confirmation that acceleration is real. When a theory is confirmed by observations of many different types, scientists have much more confidence that it is a true description of nature.

14-4e Origin of Structure

On the largest scales, the Universe is isotropic, meaning it looks the same in all directions. On smaller scales, there are irregularities. Galaxies are grouped in clusters ranging from a few galaxies to thousands, and those clusters appear to be grouped into **superclusters**. The Local Supercluster, in which we live, is a roughly disk-shaped swarm of galaxy clusters 50 to 75 Mpc in diameter. By measuring the redshifts and positions of thousands of galaxies in great slices across the sky, astronomers have been able to create maps revealing that the superclusters are not scattered at random. They are distributed in long, narrow filaments and thin walls that outline great voids nearly empty of galaxies, as demonstrated in Figure 14-14. These aggregations and gaps are referred to as **large-scale structure**.

This large-scale structure is a problem because the cosmic microwave background radiation is very uniform, and that means the gas of the Big Bang must have been extremely uniform at the time of recombination. Yet the look-back time to the farthest known galaxy clusters, galaxies, and quasars is about 95 percent of the way back to the Big Bang. How did the uniform gas at the time of recombination coagulate relatively quickly to form galaxy clusters, galaxies, and supermassive black holes in the centers of galaxies so early in the history of the Universe? The answer appears to lie in the characteristics of dark matter.

Baryonic matter is so rare in the Universe that it did not have enough gravity to pull itself together quickly after the Big Bang. As you have read earlier, astronomers have evidence that dark matter is nonbaryonic and therefore immune to the smoothing effect of the intense radiation that prevented normal matter from contracting. Dark matter was able to collapse into clouds and then pull in the normal matter to begin the formation of galaxies, clusters, and superclusters. Mathematical models, such as Figure 14-15, have attempted to describe this process, and cold dark matter does seem capable of jump-starting the formation of structure.

But what started the clumping of the dark matter? Theorists say that space is filled with tiny, random quantum mechanical fluctuations smaller than the smallest atomic particles. At the moment of inflation, those tiny fluctuations would have been stretched to become very large but very subtle variations in gravitational fields that could have later led to the formation of clusters, filaments, and walls. The structure you see in Figure 14-15 might be the ghostly traces of quantum fluctuations in the infant Universe.

FIGURE 14-14

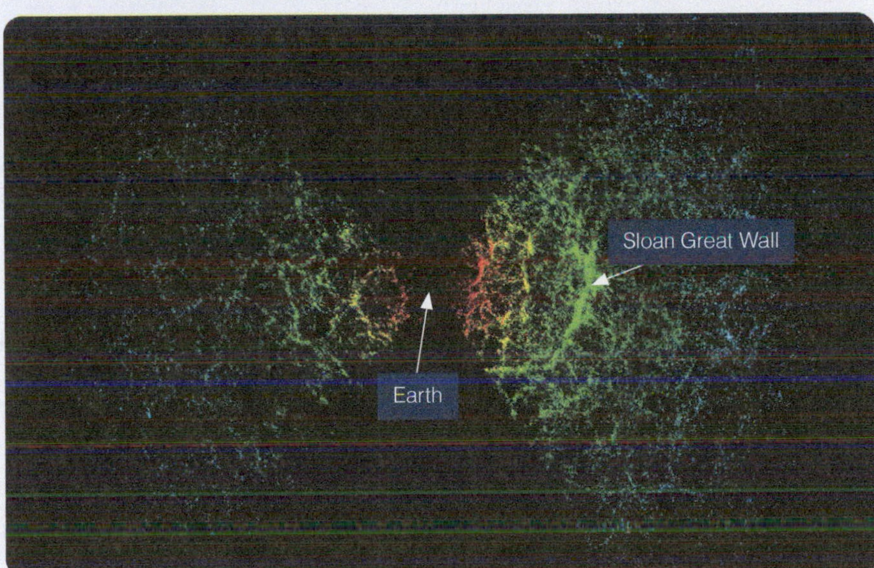

Sloan Digital Sky Survey

Nearly 70,000 galaxies are plotted in this double slice of the Universe extending outward in the plane of Earth's equator. The nearest galaxies are shown in red and the more distant in green and blue. The galaxies form filaments and walls enclosing nearly empty voids. The Sloan Great Wall, discovered by the Sloan Digital Sky Survey (look back to Chapter 4 Section 4-3), is almost 1.4 billion light-years long and is the largest known structure in the universe. The most distant galaxies in this diagram are roughly 3 billion light-years from Earth.

superclusters Cluster of galaxy clusters.

large-scale structure The distribution of clusters and superclusters of galaxies in filaments and walls enclosing voids.

FIGURE 14-15

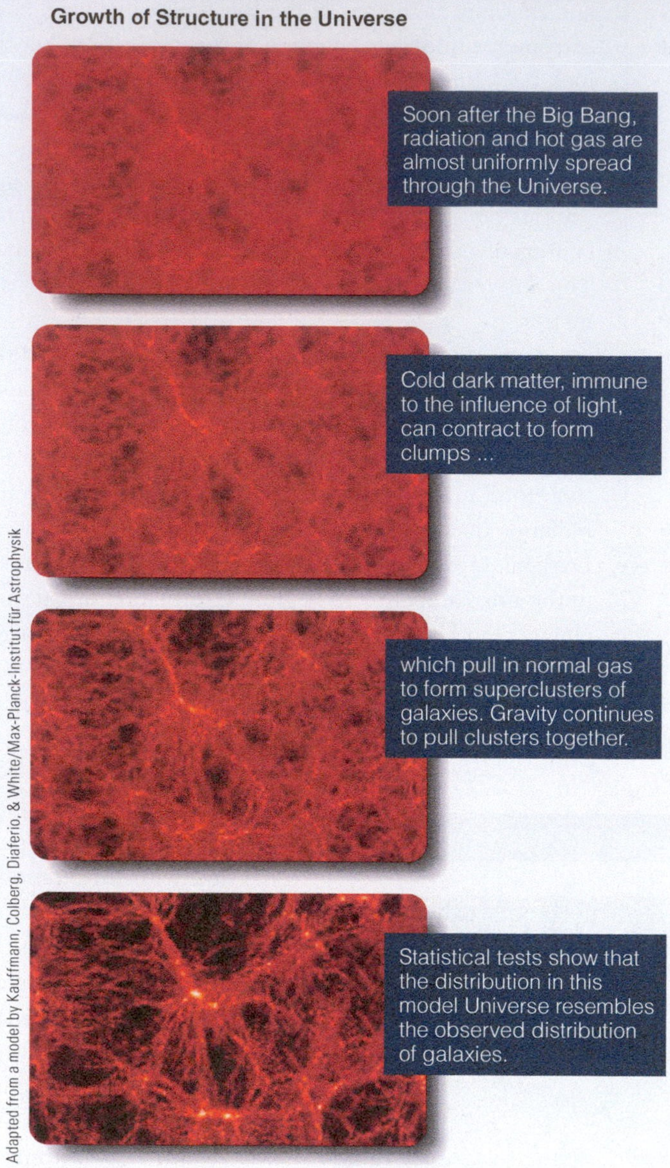

This computer model traces the formation of structure in the Universe from soon after the Big Bang to the present.

14-4f Curvature of Space-Time

Observations of tiny irregularities in the background radiation can also reveal details about inflation and acceleration. In fact, the inflationary theory of the Universe makes very specific predictions about the sizes of the irregularities an observer on Earth should see in the CMB. The *Wilkinson Microwave Anisotropy Probe (WMAP)*, a space infrared telescope, has made extensive observations of the type required to test those predictions.

The background radiation is very isotropic—it looks almost exactly the same in all directions. However, when the average intensity is subtracted from each spot on the sky, small irregularities are evident. That is, some spots on the sky look a tiny bit hotter and brighter than other spots (Figure 14-16; also look back to Figure 14-6).

Cosmologists can analyze those irregularities in the intensity of the CMB using sophisticated mathematics to measure how often spots of different sizes occur. The analysis confirms that spots about 1 degree in diameter are the most common, but spots of other sizes occur as well, and it is possible to plot a graph such as Figure 14-17 to show how common irregularities of various sizes are.

Theory predicts that most of the irregularities in the hot gas of the Big Bang should be about 1 degree in diameter if the Universe is flat. If the Universe were open, the most common irregularities would be smaller. The size of the irregularities in the cosmic background radiation show that the observations fit the flat universe model well, as you can see in Figures 14-16 and 14-17. You learned earlier in this chapter that a flat universe is most easily explained by inflation. Not only is the theory of inflation supported, an exciting result itself, but these data show that the Universe is flat, which indirectly confirms the existence of dark energy and the acceleration of the Universe.

The results from the *WMAP* observations make a complicated curve in Figure 14-17, and details of the wiggles tell cosmologists a great deal about the Universe. The curve shows that the Universe is flat, accelerating, and will expand forever. The age of the Universe derived from the data is 13.8 billion years. Furthermore, the smaller peaks in the curve reveal that the Universe contains 4.5 percent baryonic (normal) matter, 22.7 percent dark matter, and 72.8 percent dark energy. A Hubble constant of 70 km/sec/Mpc is confirmed by combining *WMAP* with *Hubble* and *Spitzer* space telescope data. The inflationary theory is confirmed, and the data support the cosmological constant version of dark energy, although quintessence is not quite ruled out. The dark

FIGURE 14-16

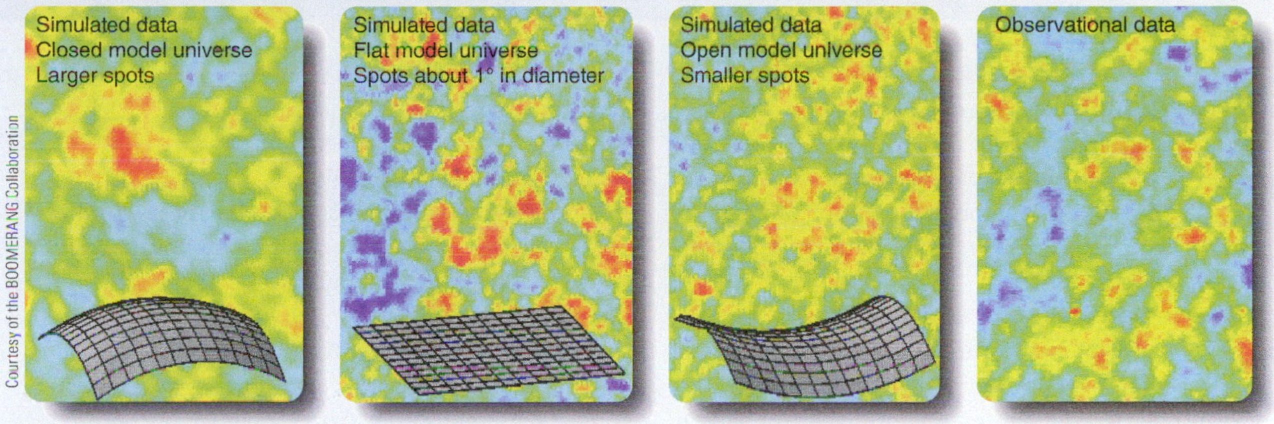

You can see the difference yourself. Compare the observations of the irregularities in the cosmic background radiation in the far right panel with the three simulations starting from the left. The observed size of the irregularities fits best with cosmological models having flat geometry. Detailed mathematical analysis confirms your visual impression: The Universe is flat.

FIGURE 14-17

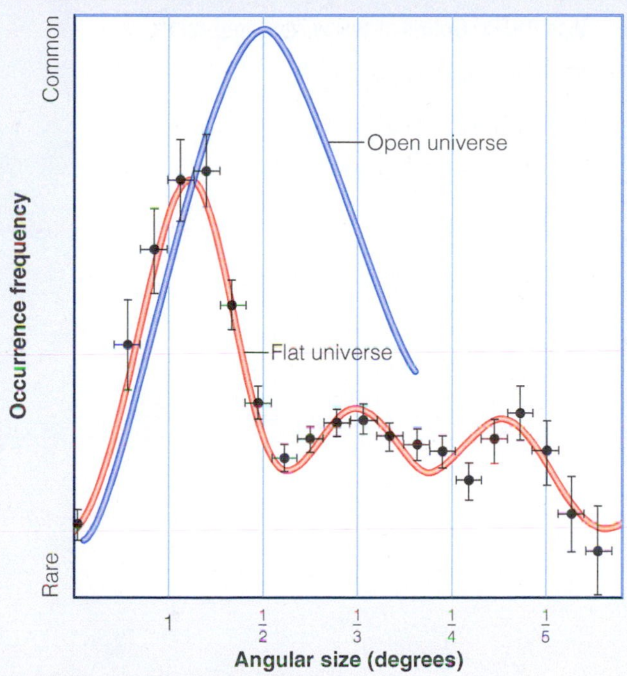

This graph shows how commonly irregularities of different sizes occur in the cosmic microwave background radiation. Irregularities of about 1 degree in diameter are most common. Models of the Universe that are open or closed are ruled out. The data fit a flat model of the Universe very well. Crosses on data points show the uncertainty in the measurements.

matter needs to be "cold" (meaning, in this context, slowly moving) in order to clump together rapidly enough after the Big Bang to make the first galaxies, quasars, and galaxy clusters.

Please reread the preceding paragraph. It is mindblastingly amazing. *WMAP* and other studies of the cosmic microwave background radiation and the distribution of galaxies have revolutionized cosmology. At last, astronomers have accurate observations against which to test theories. The basic constants are known to a precision of a few percent.

On reviewing these results, one cosmologist announced that "Cosmology is solved!"—but that might be premature. Scientists don't understand dark matter or the dark energy that drives the acceleration, so more than 95 percent of the Universe is not understood. Hearing this, another astronomer suggested that a better phrase is "Cosmology in crisis!" Certainly there are further mysteries to be explored, but cosmologists are growing more confident that they can describe the origin and evolution of the Universe.

What Are We? Products

As you climbed the cosmology pyramid, you negotiated many steps. Most were easy, and all were logical. They have carried you up to some sweeping views, and you have traced the origin of the Universe, the formation of the chemical elements, the birth of galaxies, and the births and deaths of stars. You now have a perspective that few humans share.

We are among the products of two cosmic processes, gravitational contraction and nuclear fusion. Gravity created instabilities in the hot matter of the Big Bang and triggered the formation of clusters of galaxies. Further contraction of that material formed individual galaxies and then stars within the galaxies. As stars began to shine through the Universe, nuclear fusion in their cores began to cook hydrogen and helium into the heavier atoms from which humans are made.

As you review the history and structure of the Universe, it is wise to recognize the mysteries that remain; but note that they are mysteries that might be solved rather than mysteries that are unknowable. Only a century ago, humanity didn't know there were other galaxies, or that the Universe was expanding, or that stars generate energy by nuclear fusion. Human curiosity has solved many of the mysteries of cosmology and will solve more during your lifetime.

STUDY TOOLS 14

IN THE BOOK

- ☐ Tear Out the Review Card on *Modern Cosmology.*

ASTRO ONLINE

Visit ASTRO Online at www.cengagebrain.com

- ☐ Interactive Reading
- ☐ Practice Quizzing
- ☐ Videos and Animations
- ☐ Exclusive Features

14 Questions and Problems

REVIEW QUESTIONS

1. How does the darkness of the night sky tell you something important about the Universe?
2. How can Earth be located at the center of the observable Universe if you accept the cosmological principle?
3. Why can't the Universe have a center?
4. What evidence shows that the Universe is expanding? What evidence shows that it began with a Big Bang?
5. Why couldn't atomic nuclei exist when the age of the Universe was less than 2 minutes?
6. Why are measurements of the current density of the Universe important?
7. How does the inflationary Universe theory resolve the flatness problem? How does it resolve the horizon problem?
8. If the Hubble constant were really 100 instead of 70 km/s/Mpc, much of what astronomers understand about the evolution of stars and star clusters would have to be wrong. Explain why. (*Hint:* What would the age of the Universe be?)
9. What is the evidence that the Universe was very uniform during its first 400,000 years?
10. To what does "cold" refer in the term "cold dark matter"? Why do models of large-scale structure agree best with dark matter being "cold"?
11. What evidence can you cite that the expansion of the Universe is accelerating?
12. What evidence can you cite that the Universe is flat?

DISCUSSION QUESTIONS

1. Of the possible geometries for a simple Universe model (open, closed, and flat) which one seems easiest for you to imagine? With that in mind, how do you feel about the growing consensus that the Universe actually has a flat geometry?
2. Since the start of the 20th century humanity has gained a seemingly impressive amount of cosmological knowledge. Do you think we are close to knowing the final truth? Why or why not?

PROBLEMS

1. Use the data in Figure 14-3 to plot a velocity–distance diagram, calculate the Hubble constant H, and estimate the Hubble time. (*Hint:* See **The Hubble Law** and **The Age of the Universe** in the Math Reference Cards MR.)
2. If a galaxy is 9.0 Mpc away from Earth and recedes at 510 km/s, what is H? What is the Hubble time? How old would the Universe be assuming space-time is flat and there has been no acceleration of the Universe's expansion? How would acceleration change your answer? (*Hint:* See **The Hubble Law** and **The Age of the Universe** in the Math Reference Cards MR.)
3. What was the wavelength of maximum intensity for radiation from the gas of the Big Bang at the time of recombination? By what factor is that different from the wavelength of maximum intensity of the cosmic microwave background radiation observed now? (*Hint:* See **Blackbody Radiation** in the Math Reference Cards MR.)
4. Pretend that galaxies are spaced evenly, 2 Mpc apart, and the average mass of a galaxy is 10^{11} solar masses. What is the average density of matter in the Universe? (*Note:* The volume of a sphere is $\frac{4}{3}\pi r^3$, and the mass of the Sun is 2×10^{30} kg.)
5. Figure 14-10 is based on an assumed Hubble constant of 70 km/s/Mpc. How would you change the diagram to fit a Hubble constant of 50 km/s/Mpc? (*Hint:* See **The Hubble Law** and **The Age of the Universe** in the Math Reference Cards MR.)
6. Hubble's first estimate of the Hubble constant was 530 km/s/ Mpc. His distances were too small by a factor of about 7 because of a calibration error. If he had not had that calibration problem, what value

for *H* would he have obtained? (*Hint:* See **The Hubble Law** in the Math Reference Cards MR.)

7. What was the maximum age of the Universe predicted by Hubble's first estimate of the Hubble constant (see Problem 6)? (*Hint:* See **The Hubble Law** and **The Age of the Universe** in the Math Reference Cards MR.)

LEARNING TO LOOK

1. Figure 14-18 shows irregularities in the background radiation. Why isn't the background radiation perfectly uniform? What does the size of these irregularities tell you?

FIGURE 14-18

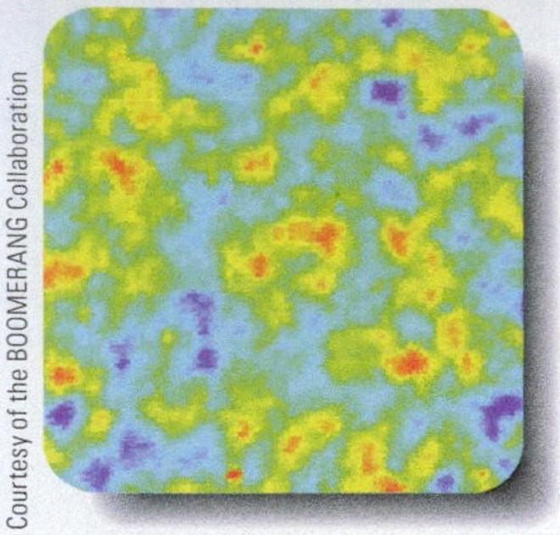

Courtesy of the BOOMERANG Collaboration

15 Life on Other Worlds

"Did I solicit thee From darkness to promote me...?"
ADAM TO GOD
JOHN MILTON, PARADISE LOST

After finishing this chapter go to **PAGE 365** for **STUDY TOOLS.**

Jim Peaco/National Park Service

GUIDEPOST

This chapter is either unnecessary or vital. If you believe that astronomy is the study of the physical Universe above the clouds, then you are done; the previous 14 chapters completed your study of astronomy. But, if you believe that astronomy is the study not only of the physical Universe but also of your role as a living being in the evolution of the Universe, then everything you have learned so far from this book has been preparation for this final chapter.

As you read this chapter, you will encounter four important questions:

- What is life?
- How did life originate on Earth?
- Does life exist on other worlds?
- Can humans communicate with intelligent beings on other worlds?

You won't get more than the beginnings of answers to those questions here, but often in science asking a question is more important than getting an immediate answer.

astrobiology The field of study focused on searches for life and possible habitats for such life on other worlds.

You have explored the Universe from the phases of the Moon to the Big Bang, from the origin of Earth to the death of the Sun. Astronomy is meaningful, not just because it is about the Universe but because it is also about you. Now that you know astronomy, you see yourself and your world in a different way. Astronomy has changed you.

As a living thing, you have been promoted from darkness. (See CO 15.) The atoms of carbon, oxygen, and other heavy elements that are necessary components of your body did not exist at the beginning of the Universe but were cooked up by successive generations of stars. The elements from which you are made are common everywhere in the observable Universe, so it is possible that life began on other worlds and evolved to intelligence there as well. If so, perhaps those other civilizations will be detected from Earth. Future astronomers might discover distant alien species completely different from any life on Earth. (See **How Do We Know? 15-1.**) Your goal in this chapter is to try to understand truly intriguing puzzles—the origin and evolution of life on Earth and what that reveals about the possibility of life on other worlds, a field of study named **astrobiology**.

CO 15 Artist's conception of a scene on the young Earth, 3 billion years ago, with bacterial mats growing near the shores of an ocean. If you landed on an extrasolar planet resembling this, you might not consider it truly Earth-like; for example, the atmosphere had no oxygen at that time.

15-1 THE NATURE OF LIFE

What is life? Philosophers have struggled with that question for thousands of years, and it is not possible to answer it in one chapter or even in one book. An attempt at a general definition of what living things do, distinguishing them from nonliving things, might be: Life is a *process* by which an organism extracts energy from the surroundings, maintains itself, and modifies the surroundings to foster its own survival and reproduction.

One very important observation is that all living things on Earth, no matter how apparently different, share certain characteristics in how they perform the process of life.

15-1a The Physical Basis of Life

The physical basis of life on Earth is the element carbon. Because of the way carbon atoms bond to each other and to other atoms, they can form long, complex, stable chains that are capable of storing and transmitting information. A large amount of information of some sort is necessary to maintain the forms and control the functions of living things.

Carbon might not be crucial to life. Science fiction authors have speculated that silicon could be substituted for carbon because the two elements share many chemical properties, but that seems unlikely because silicon chains are harder to assemble and disassemble than their carbon counterparts and can't be as lengthy. Even stranger life forms have been proposed, based on electromagnetic fields and ionized gas, and none of these possibilities can be ruled out. These hypothetical life forms make for fascinating speculation, but they can't be studied as systematically as life on Earth can be. This chapter is concerned with the origin and evolution of life as it is on Earth, based on carbon, not because of lack of imagination, but because it is the only form of life about which we know anything.

Even carbon-based life has its mysteries. What makes a lump of carbon-based molecules a living thing? An important part of the answer lies in the transmission of information from one molecule to another.

15-1b Information Storage and Duplication

Almost every action performed by a living cell is carried out by chemicals it manufactures. Cells must store recipes for all these chemicals, use them when they need them, and pass them on to their offspring.

How Do We Know? 15-1

The Nature of Scientific Explanation

Must science and religion be in conflict? Science is a way of understanding the world around you, and at the heart of that understanding are explanations that science gives for natural phenomena. Whether you call these explanations stories, histories, hypotheses, or theories, they are attempts to describe how nature works based on evidence and intellectual honesty. While you might take these explanations as factual truth, you can understand that they are not the only explanations that describe the Universe.

A separate class of explanations involves religion. For example, the Old Testament description of creation does not fit well with scientific observations, but it is a way of adding "who" and "why" to the scientific "how" and "when." Religious explanations are based partly on faith rather than on strict rules of logic and evidence, and it is wrong to demand that they follow the same rules as scientific explanations. In the same way, it is wrong to demand that scientific explanations take into account religious beliefs. The so-called conflict between science and religion arises when people fail to recognize that science and religion are different ways of knowing about the Universe.

World History Archive/Alamy Stock Photo

Galileo's telescope gave him a new way to know about the Universe.

Scientific explanations are quite compelling because science has been so successful at producing technological innovations that have changed the world you live in. From new vaccines, to digital music players, to telescopes that can observe the most distant galaxies, the products of the scientific process are all around you. Scientific explanations have provided tremendous insights into the workings of nature.

Many people are attracted to the suggestion, made by evolutionary biologist Stephen Jay Gould and others, that religious explanations and scientific explanations should be considered as "separate magisteria." In other words, religion and science are devoted to different realms of the mystery of existence.

Science and religion offer differing ways of explaining the Universe, but the two ways follow separate rules and cannot be judged by each other's standards. The trial of Galileo can be understood as a conflict between these two ways of knowing.

Study **Concept Art 15A, DNA: The Code of Life**, and notice the following three important points:

1. The chemical recipes of life are stored in each cell as information on DNA molecules that resemble a ladder with rungs that are composed of chemical bases. The recipe information is expressed by the sequence of ladder rungs, providing instructions to guide chemical reactions within the cell.
2. DNA instructions normally are expressed by being copied into a messenger molecule called RNA that causes molecular units called amino acids to be connected into large molecules called proteins. Proteins serve as the cell's basic structural molecules or as enzymes that control chemical reactions.
3. The instructions stored in DNA are genetic information passed along to offspring. The DNA molecule reproduces itself when a cell divides so that each new cell contains a copy of the original information.

To produce viable offspring, a cell must be able to make copies of its DNA. Surprisingly, it is important for the continued existence of all life that not all the copies be exact duplicates.

15-1c Modifying the Information

Earth's environment changes continuously. To survive, species must change as their food supply, climate, or home terrain changes. If the information stored in DNA

FIGURE 15-1

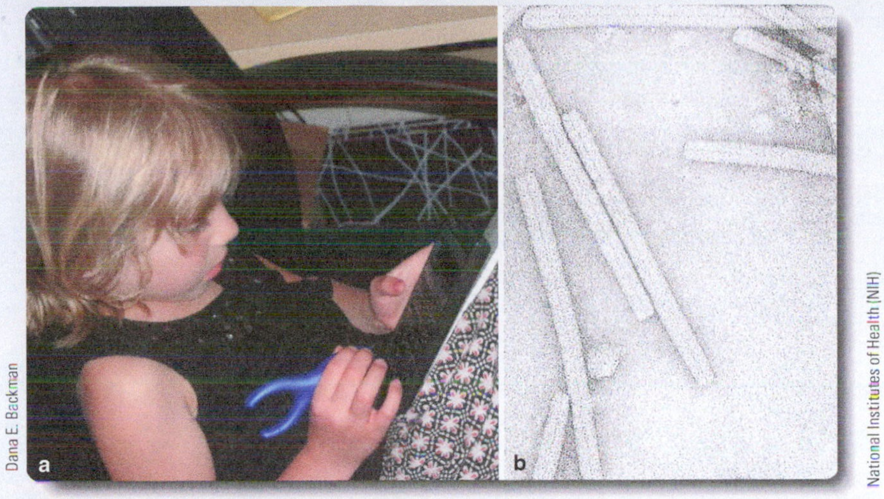

All living things on Earth are based on carbon chemistry in water. Even the long molecules that carry genetic information, DNA and RNA, have a framework defined by chains of carbon atoms. (a) Anastasia, a complex mammal, contains about 150 AU of DNA. (b) Each rodlike tobacco mosaic virus contains a single spiral strand of RNA about 0.01 mm long as its genetic material.

could not change, then life would quickly go extinct. The process by which life adjusts itself to its changing environment is called **biological evolution**.

When an organism reproduces, its offspring receives a copy of its DNA. Sometimes external effects such as radiation alter the DNA during the parent organism's lifetime, and sometimes mistakes occur in the copying process such that occasionally the copy is slightly different from the original. Offspring born with random alterations to their DNA are called **mutants**. Most mutations make no difference, but some mutations are fatal, killing the afflicted organisms before they can reproduce. In rare but vitally important cases, a mutation can actually help an organism survive.

These changes produce variations among the members of a species. All of the squirrels in the park might look the same, but they carry a range of genetic variation. Some might have slightly longer tails or faster-growing claws. These variations make almost no difference until the environment changes. For example, if the environment becomes colder, a squirrel with a heavier coat of fur will, on average, survive longer and produce more offspring than its normal contemporaries. Likewise, the offspring that inherit this beneficial variation will also live longer and have more offspring of their own. These differing rates of survival and reproduction are examples of **natural selection**. Over time, the beneficial variation becomes more common and a species can evolve until the entire population shares the trait. In this way, natural selection adapts species to their changing environments by selecting, from the huge array of random variations, those that would most benefit the survival of the species.

It is commonly believed that evolution is random, but that is not true. The underlying variation within species is random, but natural selection is not random because progressive changes in a species are directed by changes in the environment.

15-2 LIFE IN THE UNIVERSE

It is obvious that the 4.5 billion chemical bases that make up human DNA did not just come together in the right order by chance. The key to understanding the origin of life lies in the processes of evolution. The complex interplay of environmental factors with the DNA of generation after generation of organisms drove some life forms to become more sophisticated over time, until they became the unique and specialized creatures on Earth today.

This means that life on Earth could have begun very simply, even as simple a form as carbon chain molecules able to copy themselves. Of course, this is a hypothesis for which you can seek evidence. What evidence exists regarding the origin of life on Earth?

15-2a Origin of Life on Earth

The oldest fossils are all the remains of sea creatures, and this indicates that life began in the sea. Identifying

biological evolution The process of mutation, variation, and natural selection by which life adjusts itself to its changing environment.

mutants Offspring born with DNA that is altered relative to parental DNA.

natural selection The process by which the best genetic traits are preserved and accumulated, allowing the fittest organisms and species to survive and proliferate.

Concept Art 15A

DNA: THE CODE OF LIFE

1

The key to understanding life is information—the information that guides all of the processes in an organism. In most living things on Earth, that information is stored in a large molecule called **DNA (deoxyribonucleic acid)**.

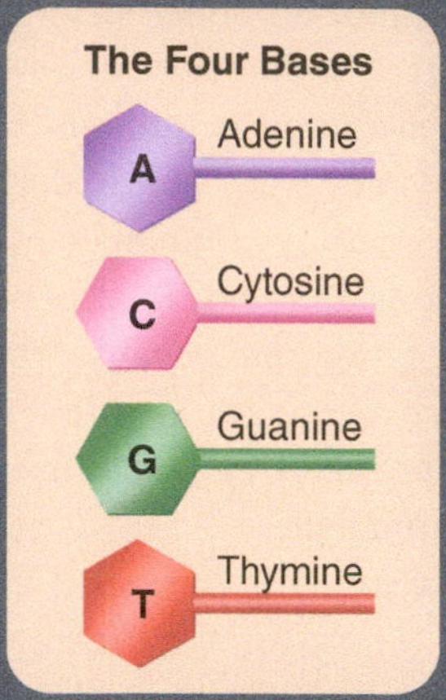

1a The DNA molecule looks like a spiral ladder with rails made of phosphates and sugars. The rungs of the ladder are made of four chemical bases arranged in pairs. The bases always pair the same way. That is, base A always pairs with base T, and base G always pairs with base C.

1b Information is coded on the DNA molecule by the order in which the base pairs occur. To read that code, molecular biologists have to "sequence" the DNA. That is, they must determine the order in which the base pairs occur along the DNA ladder.

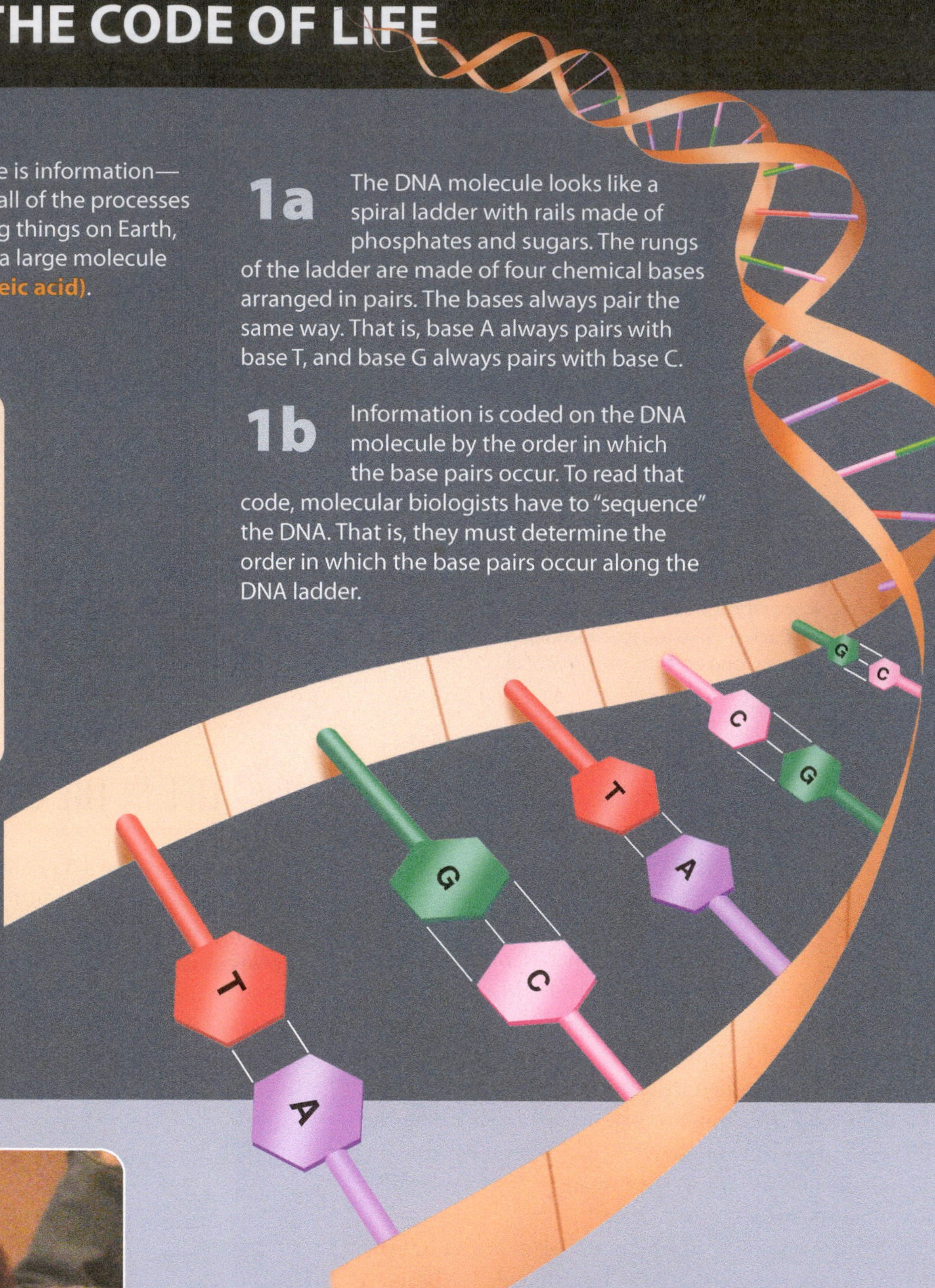

Dana E. Backman

a The traits you inherit from your parents, the chemical processes that animate you, and the structure of your body are all encoded in your DNA. When people say, "You have your mother's eyes," they are talking about DNA codes.

2

The information in DNA is directions for combinations of raw materials to form important chemical compounds. The building blocks of these compounds are relatively simple **amino acids**. Segments of DNA act as templates that guide the amino acids to join together in the correct order to build specific **proteins**, chemical compounds important to the structure and function of organisms. Some proteins called **enzymes** regulate chemical processes. In this way, DNA recipes regulate the production of the compounds of life.

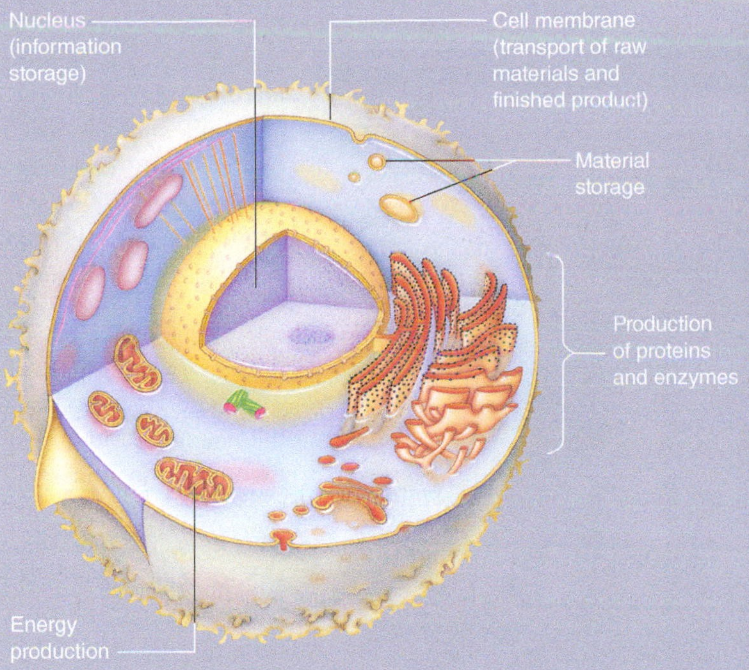

2a A cell is a tiny factory that uses the information encoded in DNA to produce molecules. Most of the DNA remains safe in the nucleus of a cell, and the code is copied to create a strand of **RNA (ribonucleic acid)**. Like a messenger carrying blueprints, the RNA carries the code out of the nucleus to the work site where the proteins and enzymes are made.

2b A single cell from a human being contains about 1.5 meters of DNA with about 4.5 billion base pairs—enough to record the entire works of Shakespeare 200 times. A typical adult human contains a total of about 600 AU of DNA. Yet the DNA in each cell, only 1.5 m in length, contains all of the information to create a new human. A clone is a new creature created from the DNA code found in a single cell.

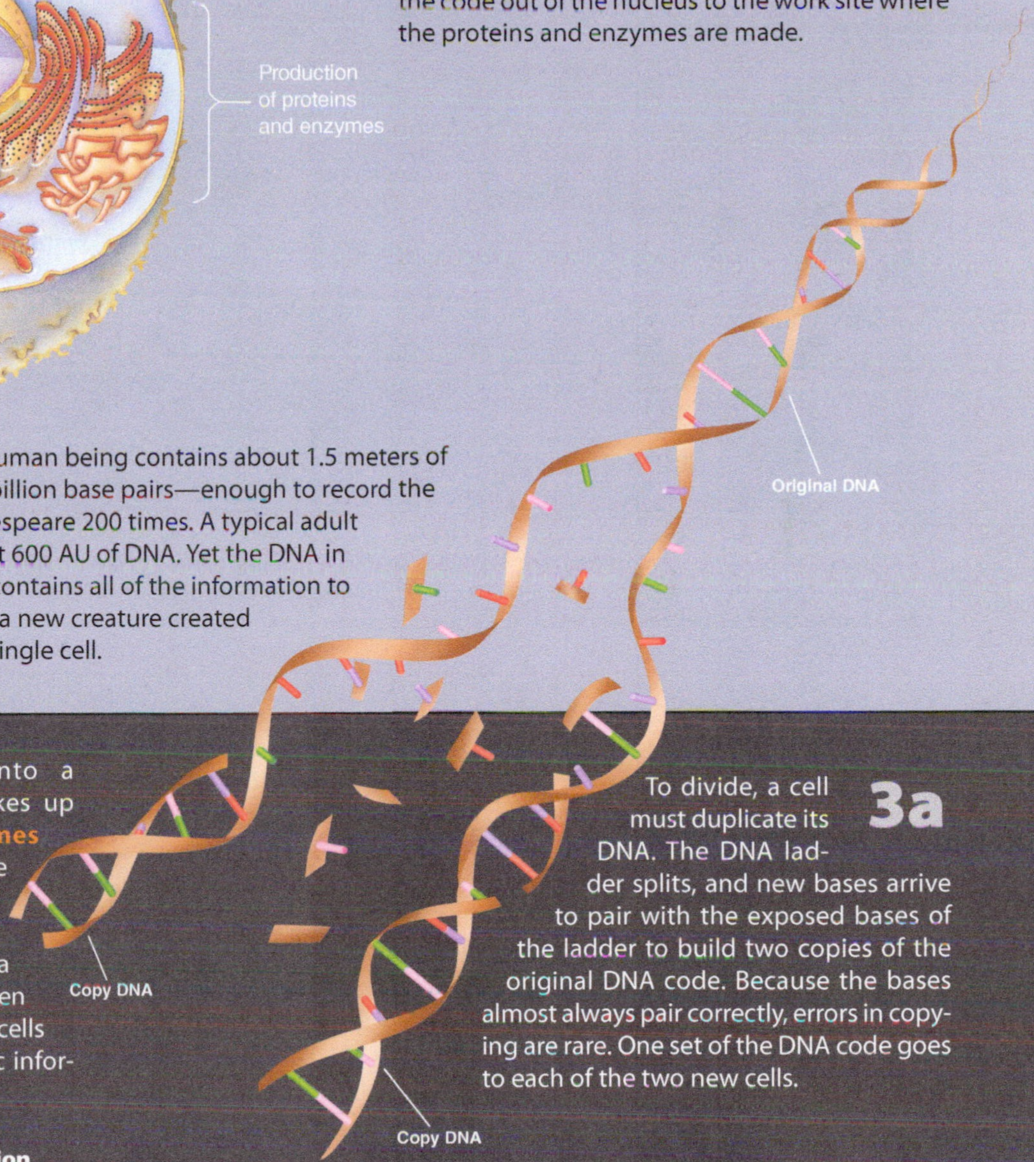

3 DNA, coiled into a tight spiral, makes up the **chromosomes** that contain the genetic material in a cell. A **gene** is a segment of a chromosome that controls a certain product or function. When a cell divides, each of the new cells receives a copy of the genetic information in the chromosomes.

3a To divide, a cell must duplicate its DNA. The DNA ladder splits, and new bases arrive to pair with the exposed bases of the ladder to build two copies of the original DNA code. Because the bases almost always pair correctly, errors in copying are rare. One set of the DNA code goes to each of the two new cells.

Cell Reproduction by Division

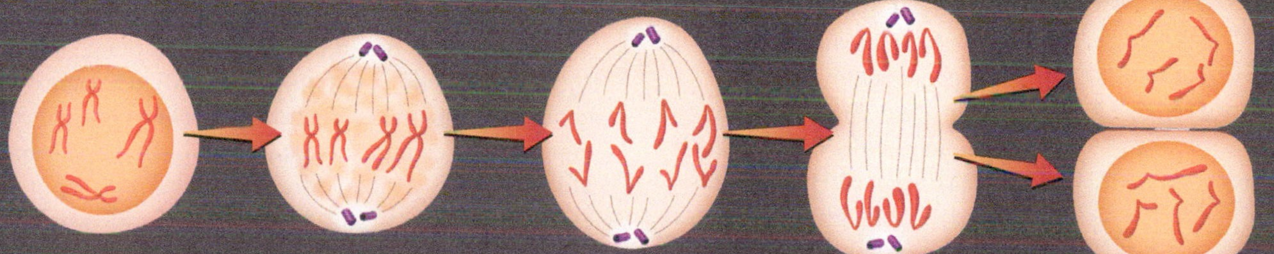

a — As a cell begins to divide, its DNA duplicates itself.

b — The duplicated chromosomes move to the middle.

c — The two sets of chromosomes separate, and . . .

d — the cell divides to produce . . .

e — two cells, each containing a full set of the DNA code.

FIGURE 15-2

Dirk Wiersma/Science Source

Cross section of layers in a fossilized stromatolite. Layers are made of mineralized mats of bacteria covered repeatedly by sediment. Fossilized stromatolites have been found within rocks more than 3.5 billion years old. This sample measures 30 centimeters (1 foot) across.

the oldest fossils is not easy, however. Fossils billions of years old are difficult to recognize because the earliest living things contained no easily preserved hard parts such as bones or shells, and because the individual organisms were microscopic. Some rocks from western Australia that are about 3.5 billion years old contain features that experts identify as fossil **stromatolites**, accumulated layers deposited by colonies of single-celled organisms that lived in shallow water long ago (Figure 15-2). The evidence, though scarce, indicates that simple organisms lived in Earth's oceans 3.5 billion or more years ago, less than 1.1 billion years after Earth formed. Where did these simple organisms come from?

stromatolites A layered formation caused by mats of algae or bacteria combined with sediments.

Miller–Urey experiment An experiment that attempted to reproduce early Earth conditions and showed how easily amino acids and other organic compounds could have formed then.

primordial soup The rich solution of organic molecules in Earth's first oceans.

An important experiment performed by Stanley Miller and Harold Urey in 1952 sought to recreate the conditions in which life on Earth began. The **Miller–Urey experiment** consisted of a sterile, sealed glass container holding water, hydrogen, ammonia, and methane. An electric arc inside the apparatus created sparks to simulate the effects of lightning in Earth's early atmosphere (Figure 15-3).

Miller and Urey let the experiment run for a week and then analyzed the material inside. They found that the interaction between the electric arc and the simulated atmosphere had produced many organic molecules from the raw material of the experiment, including such important building blocks of life as amino acids. When the experiment was run again using different energy sources such as hot silica to represent molten lava spilling into the ocean, similar molecules were produced. Even the amount of UV radiation present in sunlight is sufficient to produce complex organic molecules.

According to updated models of the formation of the Solar System and Earth (look back to Chapters 6 and 8), Earth's early atmosphere probably consisted mostly of carbon dioxide, nitrogen, and water vapor instead of the mix of hydrogen, ammonia, and methane assumed by Miller and Urey. When gases corresponding to the newer understanding of the early Earth atmosphere are processed in a Miller–Urey apparatus, lesser but still significant numbers of organic molecules are created.

The Miller–Urey experiment is important because it shows that complex organic molecules form naturally in a wide variety of circumstances. Lightning, sunlight, and hot lava pouring into the oceans are just some of the energy sources that can naturally rearrange simple common molecules into the complex molecules that make life possible. If you could travel back in time, you would expect to find Earth's first oceans filled with a rich mixture of organic compounds called the **primordial soup**.

Many of these organic compounds would have been able to link up to form larger molecules. Amino acids, for example, can link together to form proteins by joining

FIGURE 15-3

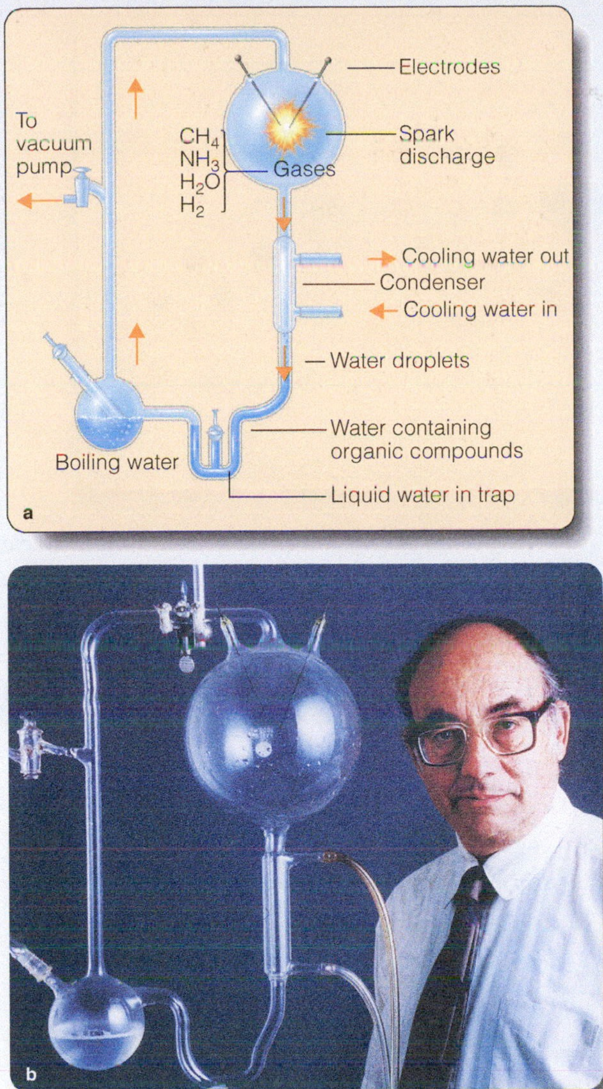

Stanley Miller

(a) The Miller–Urey experiment circulated gases through water in the presence of an electric arc. This simulation of primitive conditions on Earth produced many complex organic molecules, including amino acids, the building blocks of proteins. (b) Stanley Miller with a Miller–Urey apparatus.

ends and releasing a water molecule (Figure 15-4). It was initially thought that this must have occurred in sun-warmed tidal pools where organic molecules were concentrated by evaporation. However, violent episodes of volcanism and catastrophic meteorite impacts would probably have destroyed complex molecules developing at the surface, so it is now thought that successful linkage of complex molecules might have taken place on the ocean floor, perhaps near the hot springs at midocean rises (look back to Chapter 6 Concept Art 6A).

These complex organic molecules were still not living things. Even though some proteins might have contained hundreds of amino acids, they did not reproduce but rather linked and broke apart at random. Because some molecules are more stable than others, and some bond together more easily than others, scientists hypothesize that a process of **chemical evolution** eventually concentrated the various smaller molecules into the most stable larger forms. Eventually, according to the hypothesis, somewhere in the oceans, after sufficient time, a molecule formed that could copy itself. At that point, the chemical evolution of molecules became the biological evolution of living things.

An alternate theory for the origin of life holds that reproducing molecules might have arrived here from space. Radio astronomers have found a wide variety of organic molecules in the interstellar medium, and similar compounds have been found inside meteorites, such as the Murchison meteorite pictured here (Figure 15-5).

The Miller experiment showed how easy it is to create organic molecules in a hydrogen-right environment, so it is not surprising to find them in space. Although speculation is fun, the hypothesis that life arrived on Earth from space is presently more difficult to test than the hypothesis that Earth life originated on Earth.

Whether the first reproducing molecules formed here on Earth or in space, the important thing is that they could have formed by natural processes. Scientists know enough about these processes to feel confident about them, even though some of the steps remain unknown.

The details of the evolution of the first cells are unknown, but the first reproducing molecule to surround itself with a protective membrane must have gained an important survival advantage. Experiments have shown that microscopic spheres the size of cells containing organic molecules form relatively easily in water, as shown in Figure 15-6, so the evolution of cell-like structures is not surprising.

The first cells must have been simple single-celled organisms much like modern bacteria. As you now know, these kinds of cells are preserved in stromatolites (Figure 15-2),

chemical evolution
Chemical processes involving growth and natural selection of complex molecules prior to the origin of life.

FIGURE 15-4

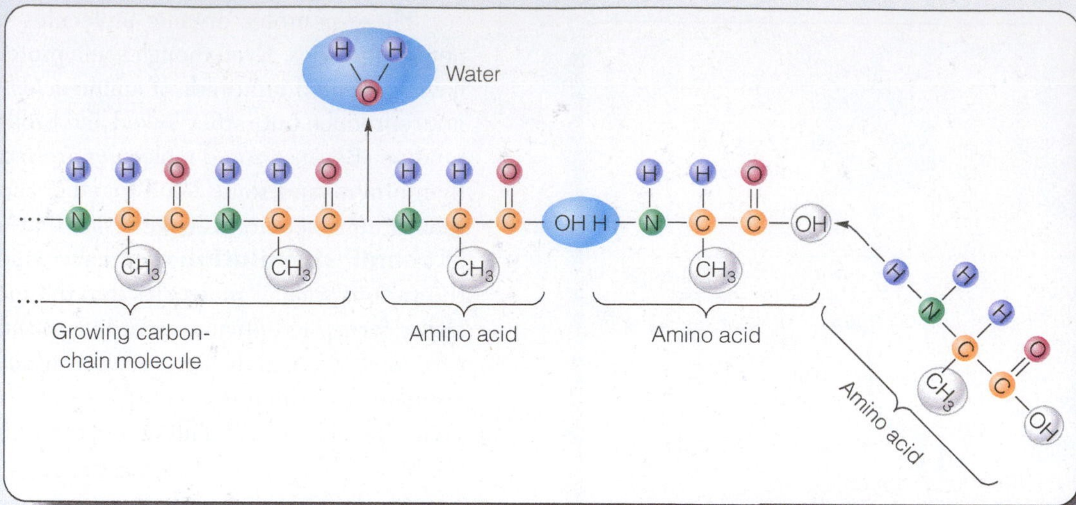

Amino acids can link together by releasing a water molecule to form long carbon-chain protein molecules. The amino acid in this hypothetical example is alanine, one of the simplest.

FIGURE 15-5

Chip Clark/National Museum of Natural History

A piece of the Murchison meteorite, a carbonaceous chondrite that fell near Murchison, Australia, in 1969. Analysis of the interior of the meteorite revealed the presence of amino acids. Whether the first chemical building blocks of life on Earth originated in space is a matter of debate, but the amino acids found in meteorites illustrate how commonly amino acids and other complex organic molecules occur in the Universe, even in the absence of living things.

FIGURE 15-6

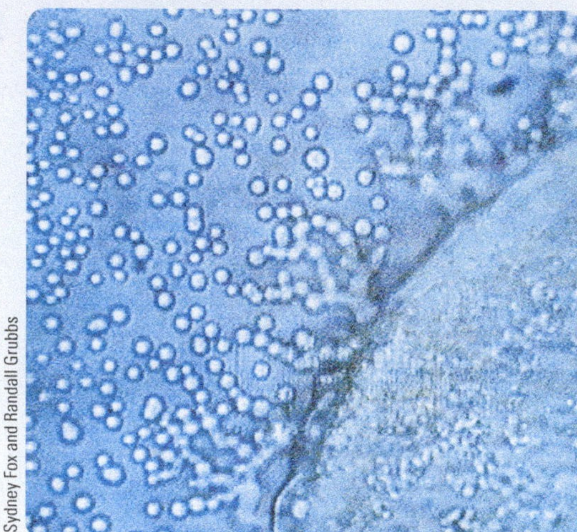

Sydney Fox and Randall Grubbs

Single amino acids can be assembled into long proteinlike molecules. When such material cools in water, it can form microspheres, microscopic spheres with double-layered boundaries similar to cell membranes. Microspheres might have been an intermediate stage in the evolution of life from complex molecules to living cells holding molecules reproducing genetic information.

mineral formations formed by layers of photosynthetic bacteria and shallow ocean sediments. Stromatolites formed in rocks with radioactive ages of 3.5 billion years, and they still form in some places today. Stromatolites and other photosynthetic organisms would have begun

FIGURE 15-7

Panel a: Michael A. Seeds (Franklin & Marshall College); Panel b: Painting by D. W. Miller, appearing in American Scientist, March-April 1997

(a) Trilobites made their first appearance in the Cambrian oceans. The smallest were almost microscopic, and the largest were bigger than dinner plates. This example, about the size of your hand, lived 400 million years ago in an ocean floor that is now a limestone deposit in Pennsylvania. (b) In this artist's conception of a Cambrian sea bottom, Anomalocaris (center-rear and looming at upper right) had specialized organs including eyes, coordinated fins, gripping mandibles, and a powerful, toothed maw. Notice Opabinia at center right with its long snout.

adding oxygen, a product of photosynthesis, to Earth's early atmosphere. An oxygen abundance of only 0.1 percent would have created an ozone screen, protecting organisms from the Sun's UV radiation and later allowing life to colonize the land.

Over the course of eons, the natural processes of evolution gave rise to stunningly complex **multicellular** life forms with their own widely differing ways of life. It is a common misconception to imagine that life is too complex to have evolved from such simple beginnings. It is possible because small variations can accumulate, but that accumulation requires huge amounts of time.

15-2b Deep Time and the Evolution of Life

Life has existed on Earth for at least 3.5 billion years, but there is no evidence of anything more than simple organisms until about 540 million years ago, when life suddenly branched into a wide variety of complex forms. This sudden increase in complexity is known as the **Cambrian explosion**, and it marks the beginning of the Cambrian period (Figure 15-7).

If you represented the entire history of Earth on a scale diagram, the Cambrian explosion would be near the top of the column, as shown at the left of Figure 15-8. The emergence of most animals familiar to you today, including fishes, amphibians, reptiles, birds, and mammals, would be crammed into the topmost part of the chart, above the Cambrian explosion.

If you magnify this portion of the diagram, as shown on the right side of Figure 15-8, you can get a better idea of when these events occurred in the history of life. Humanoid creatures have walked the Earth for about 4 million years. This is a long time by the standard of a human lifetime, but it makes only a narrow red line at the top of the diagram. All of recorded history would be a microscopically thin line at the very top of the column.

To understand just how thin that line is, imagine that the entire 4.6-billion-year history of the Earth has been compressed onto a yearlong video and that you began watching this video on January 1. You would not see any signs of life until March or early April, and the slow evolution of the first simple forms would take up the next 6 or 7 months. Suddenly, in mid-November, you would see complex organisms of the Cambrian explosion.

You would see no life of any kind on land until November 28, but once it appeared it would diversify

multicellular Composed of more than one cell.

Cambrian explosion A geologically brief period about 540 million years ago during which fossil evidence indicates Earth life became complex and diverse. Cambrian rocks contain the oldest easily identifiable fossils.

FIGURE 15-8

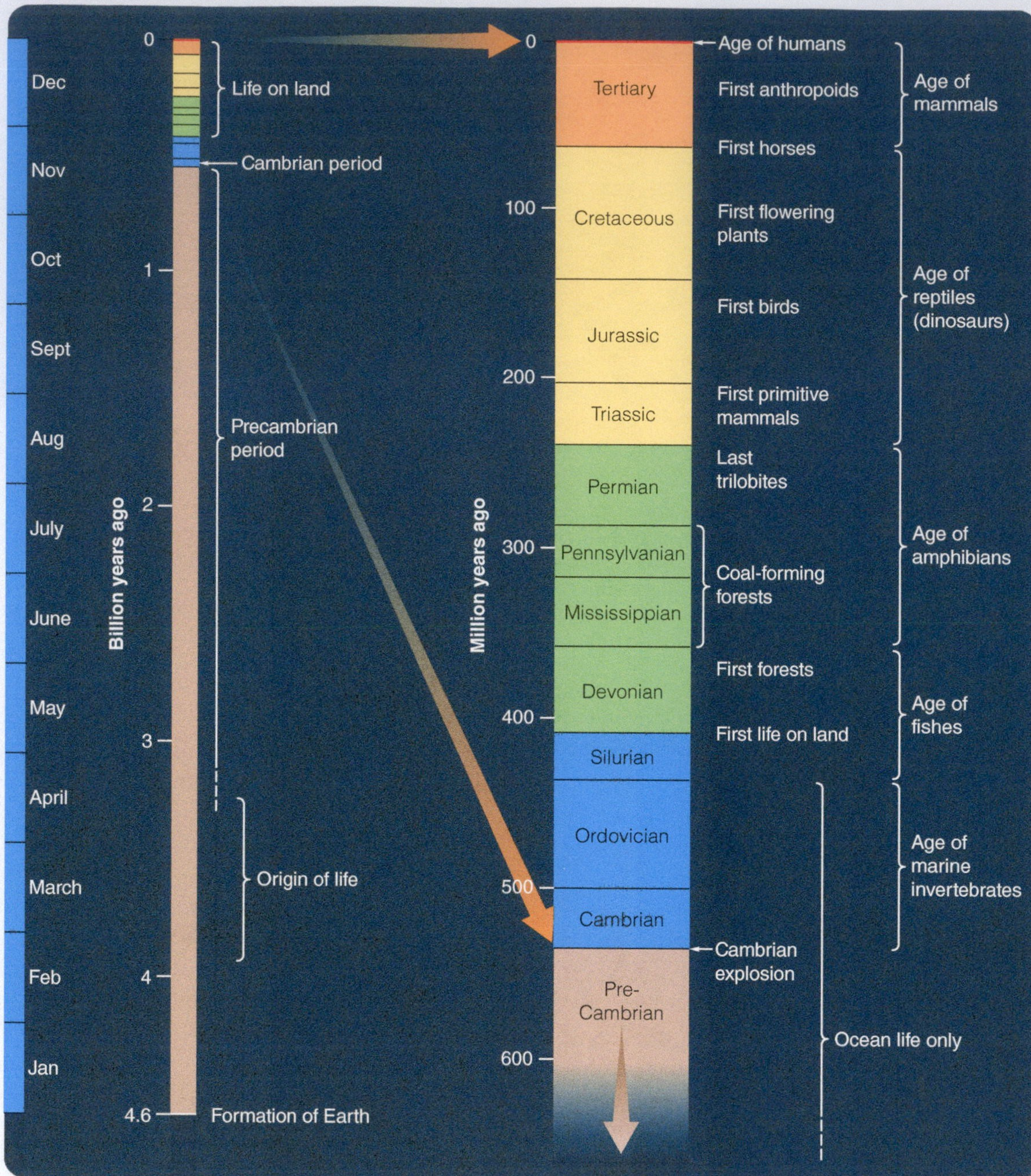

Complex life has developed on Earth only recently. If the entire history of Earth were represented in a time line (left), you have to examine the end of the line closely to see details such as life leaving the oceans and dinosaurs appearing. The age of humans would still be only a thin line at the top of your diagram. If the history of Earth were a yearlong videotape, humans would not appear until the last minutes of December 31.

quickly, and by December 12 you would see dinosaurs walking the continents. By December 26 they would be gone, and mammals and birds would be on the rise.

If you watched closely, you might see the first humanoids around suppertime on New Year's Eve, and by late evening you could see humans making the first stone tools. The Stone Age would last until 11:59 pm, after which the first towns, then cities, would appear. Suddenly things would begin to happen at lightning speed. Babylon would flourish, the pyramids would rise, and Troy would fall.

The Christian era would begin 14 seconds before the New Year. Rome would fall, the Middle Ages and the Renaissance would flicker past. The American and French revolutions would occur 1 second before the end of the video.

By imagining the history of Earth as a yearlong video, you have gained some perspective on the rise of life. Tremendous amounts of time were needed for the first simple living things to evolve in the oceans; but, as life became more complex, new forms arose more and more quickly as the hardest problems—how to reproduce, how to take energy efficiently from the environment, how to move around—were solved. The easier problems, such as what to eat, where to live, and how to raise offspring, were solved in different ways by different organisms, leading to the diversity that is seen today.

Even intelligence—which appears to set humans apart from other animals—might be a unique solution to an evolutionary problem posed to humanity's ancient ancestors. A smart animal is better able to escape predators, outwit its prey, and feed and shelter itself and its offspring, so under certain conditions evolution is likely to select for intelligence. Could intelligent life arise on other worlds? To try to answer this question, you can estimate the chances of any type of life arising on other worlds, then assess the likelihood of that life developing intelligence.

15-2c Life in Our Solar System

Could there be carbon-based life elsewhere in our Solar System? Liquid water seems to be a requirement of carbon-based life, necessary both for vital chemical reactions and as a medium to transport nutrients and wastes. It is not surprising that life developed in Earth's oceans and stayed there for billions of years before it was able to colonize the land.

Scientists are in agreement that any world harboring living things must have significant quantities of some type of liquid. Water is a cosmically abundant substance with properties such as high heat capacity that set it apart from other common molecules that are liquid at the temperatures of planetary surfaces. It is difficult to pin down a range of environments and be sure that life based on carbon and water cannot exist outside those conditions if there is even occasionally some liquid water present. Life has been found on Earth in places previously judged inhospitable such as the bottoms of ice-covered lakes in Antarctica, or far underground inside solid rock, or in pools of nearly boiling water (Figure 15-9).

Many worlds in the Solar System can be eliminated immediately as hosts for life because liquid water is not possible there. The Moon and Mercury are airless, and water would boil away into space immediately. Venus has traces of water vapor in its atmosphere, but it is too hot for liquid water to survive on the surface. The Jovian planets have deep atmospheres, and at a certain level water condenses into liquid droplets. However, it seems unlikely that life could have originated there; the Jovian planets do not have solid surfaces (look back to Chapter 7 Section 7-1b). Isolated water droplets could never mingle to mimic the rich primordial oceans of Earth, where organic

FIGURE 15-9

Jim Peaco/National Park Service

Colonies of thermophilic (heat-loving) single-celled organisms thrive around the edge of Yellowstone National Park's Grand Prismatic Spring at temperatures up to 72°C (162°F), producing the green pigments. The blue water in the center of the pool is too hot even for thermophiles.

molecules grew and interacted. Additionally, powerful currents in the gas giants' atmospheres would quickly carry any reproducing molecules that did form there down into inhospitably hot regions of the atmosphere.

As you learned in Chapter 7 Section 7-2d, at least one of the Jovian satellites could potentially support life. Jupiter's moon Europa appears to have a liquid-water ocean below its icy crust, and minerals dissolved in the water could provide a source of raw material for chemical evolution. Europa's ocean is kept warm and liquid now by tidal heating. Evidently there are also liquid-water layers under the surfaces of Ganymede and Callisto. That can change as the orbits of the moons interact; Europa, Ganymede, and Callisto maight have been frozen solid at other times in their histories, destroying organisms that had developed there.

Saturn's moon Titan is rich in organic molecules. You saw in Chapter 3 Section 3-3c how sunlight converts the methane in Titan's atmosphere into organic smog particles that settle to the surface. The chemistry of any life that could have evolved from those molecules and survived in Titan's lakes of methane is unknown. It is fascinating to consider possibilities, but Titan's extremely low temperature of −180°C (−290°F) could make chemical reactions slow to the point where life processes seem unlikely.

Observations of water venting from the south-polar region of Saturn's moon Enceladus show that it has liquid water below its crust. It is possible that life could exist in that water, but the moon is very small and has been warmed by tidal heating that might operate only occasionally. Enceladus might not have had plentiful liquid water for the extended time thought necessary for the rise of life.

Mars is the most likely place for life to exist in the Solar System because, as you learned in Chapter 6 Section 6-6c, there is a great deal of evidence that liquid water once flowed on its surface. Even so, results from searches for signs of life on Mars are not encouraging. The robotic spacecraft *Viking 1* and *Viking 2* landed on Mars in 1976 and tested soil samples for living organisms. Some of the tests had puzzling semipositive results that scientists hypothesize were caused by nonbiological chemical reactions in the soil. There is no clear evidence of life or even of organic molecules in the Martian soil.

In 1996, there were prominent news stories regarding chemical and physical traces of life on Mars discovered inside a Martian meteorite found in Antarctica (Figure 15-10). Scientists were excited by the announcement, but they immediately began testing the evidence. Their results suggest that the unusual chemical signatures in the rock might have formed by processes that did not involve life. Tiny features in the rock that were originally

FIGURE 15-10

About the size of a compact car, a *Viking* lander model sits in a simulated Martian environment on Earth. A claw for grabbing rock and soil samples is in the foreground at the end of the black arm. In 1976, the two *Viking* landers made measurements possibly indicating some sort of microbial activity in the Martian dirt. (b) Meteorite ALH 84001 is one of a dozen meteorites known to have originated on Mars. Its name means this meteorite was the first one found in 1984 near Antarctica's Allan Hills. (c) A group of researchers claimed that ALH 84001 contained chemical and physical traces of ancient life on Mars, including what appear to be fossils of microscopic organisms. That evidence has not been confirmed, and the claim continues to be tested and debated.

thought to be fossils of ancient Martian microorganisms could be nonbiological mineral formations. This is the only direct evidence yet found regarding potential life on Mars, but it remains highly controversial. The rover *Curiosity* landed on Mars in the fall of 2012 to search for evidence of environments, both ancient and modern, capable of supporting life, but conclusive evidence of life on Mars might have to wait until a geologist from Earth can scramble down dry Martian streambeds and crack open rocks looking for fossils.

There is no strong evidence for the existence of life elsewhere in the Solar System. Now your search will take you to distant planetary systems.

15-2d Life in Other Planetary Systems

Could life exist in other planetary systems? You already know that there are many different kinds of stars and that many of these stars have planetary systems. As a first step toward answering this question, you can try to identify the kinds of stars that seem most likely to have stable planetary systems where life could evolve.

If a planet is to be a suitable home for living things, it seems reasonable that it must be in a stable orbit around its sun. That is simple in a planetary system like our own, but most planet orbits in binary star systems are unstable unless the component stars are very close together or very far apart. Astronomers can calculate that in binary systems with stars separated by distances of a few AU, the planets should eventually be swallowed up by one of the stars or ejected from the system. Half the stars in the galaxy are members of binary systems, and many of them are unlikely to support life on planets.

Moreover, just because a star is single does not necessarily make it a good candidate for sustaining life. Earth required between 0.5 and 1 billion years to produce the first cells and 4.6 billion years for intelligence to emerge. Massive stars that live only a few million years do not meet this criterion. If the history of life on Earth is at all representative, then stars more massive and luminous than about spectral type F5 are too short lived for complex life to develop. Main-sequence stars of types G and K, and possibly some of the M stars, are the best candidates.

The temperature of a planet is also important, and that depends on the type of star it orbits and its distance from the star. Astronomers have defined a **habitable zone** around a star as a region within which planets have temperatures that permit the existence of liquid water. The Sun's habitable zone extends from around the orbit of Venus to the orbit of Mars, with Earth right in the middle. A low-luminous star has a small habitable zone, and a high-luminous star has a large one.

Scientists on Earth are finding life in places previously judged inhospitable, such as the bottoms of ice-covered lakes in Antarctica and far underground inside solid rock. Life has also been found in boiling hot springs with highly acidic water. As a result, it is difficult for scientists to pin down a range of environments and state with certainty that life cannot exist outside those conditions. You should also note that two environments mentioned as locations of liquid water and therefore possible havens for life—Europa and Enceladus—are in the outer Solar System and lie far outside the Sun's conventional habitable zone. Stable planets inside the habitable zones of long-lived stars are the places where life seems most likely, but, given the tenacity and resilience of Earth's life forms, there might be other, seemingly inhospitable, places in the Universe where life exists.

15-3 INTELLIGENT LIFE IN THE UNIVERSE

Visiting extrasolar planets is, for now, impossible. Nevertheless, if other civilizations exist, it is possible humans can communicate with them. Nature puts restrictions on such conversations, but the main question lies in the unknown life expectancy of civilizations.

15-3a Travel Between the Stars

The distances between stars are almost beyond comprehension. The fastest human device ever launched, the *New Horizons* probe, will take about 90,000 years to travel the distance to the nearest star. The obvious way to overcome these huge distances is with tremendously fast spaceships, but even the closest stars are many light-years away.

Nothing can exceed the speed of light, and accelerating a spaceship close to the speed of light takes huge amounts of energy. Even if you travel more slowly, your rocket would require massive amounts of fuel. If you were piloting a spaceship with the mass of 100 tons (roughly the size of a commercial jet airliner) to the nearest star, 4 light-years away, and you wanted to travel at half the speed of light so as to arrive in 8 years, the trip would require 400 times as much energy as the entire United States consumes in a year.

These limitations not only make it difficult for humans to leave the Solar System, but they would also make it difficult for aliens to visit Earth. Reputable

habitable zone A region around a star within which planets have temperatures that permit the existence of liquid water on their surfaces.

scientists have studied unidentified flying objects (UFOs) and have never found any evidence that Earth is being visited or has ever been visited by aliens. (See **How Do We Know? 15-2.**) Humans are unlikely ever to meet aliens face to face. However, communication by radio across interstellar distances takes relatively little energy.

15-3b Radio Communication

Nature puts restrictions on travel through space, and it also restricts astronomers' ability to communicate with distant civilizations by radio. One restriction is based on simple physics. Radio signals are electromagnetic waves and travel at the speed of light. Because of the distances between the stars, the speed of radio waves would severely limit astronomers' ability to carry on normal conversations with distant civilizations. Decades could elapse between asking a question and getting an answer.

So, rather than try to begin a conversation, one group of astronomers decided in 1974 to broadcast a simple message of greeting toward the globular cluster M13, 26,000 ly away, using the Arecibo radio telescope. When the signal arrives 26,000 years in the future, alien astronomers might be able to decode it.

The Arecibo beacon is an anticoded message, meaning that it is intended to be decoded by beings about whom we know nothing except that they build radio telescopes. The message is a string of 1679 pulses and gaps. Pulses represent 1s, and gaps represent 0s. The string can be arranged in only two possible ways: as 23 rows of 73 or as 73 rows of 23. The second arrangement forms a picture containing information about life on Earth (Figure 15-11).

Although the 1974 Arecibo beacon was the only powerful signal sent purposely from Earth to other planetary systems, Earth is sending out many other signals. Shortwave radio signals, such as TV and FM, have been

How Do We Know? 15-2

UFOs and Space Aliens

Has Earth been visited by aliens? If you conclude that there is likely to be life on other worlds, then you might be tempted to use UFO sightings as evidence to test your hypothesis. Scientists don't do this for two reasons.

First, the reputation of UFO sightings and alien encounters does not inspire confidence that these data are reliable. Most people hear of such events in grocery store tabloids, daytime talk shows, or sensational "specials" on viewer-hungry cable networks. You should take note of the low reputation of the media that report UFOs and space aliens. Most of these reports, like the reports that Elvis is alive and well, are simply made up for the sake of sensation, and you cannot use them as reliable evidence.

Second, the few UFO sightings that are not made up do not survive careful examination. Most are mistakes and unintentional misinterpretations, committed by honest people, of natural events or human-made objects. It is important to realize that experts have studied these incidents over many decades and found none that is convincing to the professional scientific community. In short, despite false claims to the contrary on TV shows, there is no dependable evidence that Earth has ever been visited by aliens.

Flying saucers from space are fun to think about, but there is no evidence that they are real.

In a way, that's too bad. A confirmed visit by intelligent creatures from beyond our Solar System would answer many questions. It would be exciting, enlightening, and, like any real adventure, a bit scary. Most scientists would love to be part of such a discovery. But, scientists must professionally pay attention to what is supported by evidence rather than what might be thrilling. There is not yet any direct evidence of life on other worlds.

FIGURE 15-11

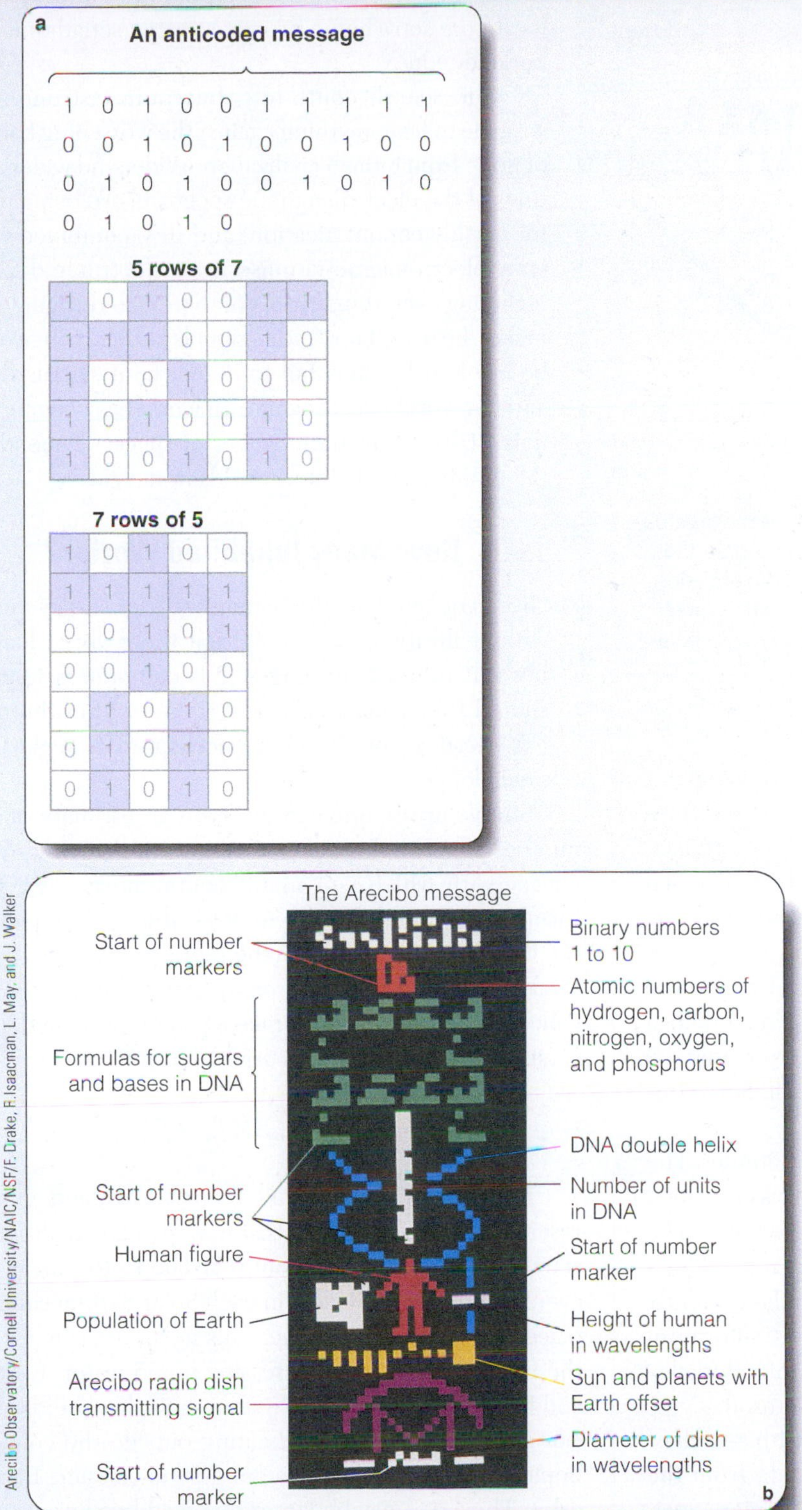

Arecibo Observatory/Cornell University/NAIC/NSF/F. Drake, R. Isaacman, L. May, and J. Walker

(a) An anticoded message is designed for easy decoding. Here a string of 35 radio pulses, represented as 1s and 0s, can be arranged in only two ways, as 5 rows of 7 or 7 rows of 5. The second way produces a friendly message. (b) The Arecibo message describes life on Earth (color added for clarity). Binary numbers give the height of the human figure (1110) and the diameter of the telescope dish (100101111110) in units of the wavelength of the signal, 12.3 cm.

leaking into space for the past 50 years or so. Any civilization within 50 ly could already have detected Earth's civilization. That works both ways: Alien signals, whether intentional messages of friendship or the blather of their equivalent to daytime TV, could be arriving at Earth now. Astronomers all over the world are pointing radio telescopes at the most likely stars and listening for alien civilizations.

Which channels should astronomers monitor? Wavelengths longer than 100 cm would get lost in the background noise of the Milky Way Galaxy, while wavelengths shorter than about 1 cm are absorbed in Earth's atmosphere. Between those wavelengths is a radio window that is open for communication. Even this restricted window contains millions of possible radio-frequency bands and is too wide to monitor easily, but astronomers might have found a way to narrow the search. Within this window lie the 21-cm spectral line of neutral hydrogen and the 18-cm line of OH (Figure 15-12). The interval between those lines has low background interference and is named the **water hole** because H plus OH yields water. Any civilizations sophisticated enough to do radio astronomy research must know about these lines and might appreciate their significance in the same way as do Earthlings.

A number of searches for extraterrestrial radio signals have been made, and some are now under way. This field of study is known as **SETI**, Search for Extra-Terrestrial Intelligence, and it has generated heated

water hole The interval of the radio spectrum between the 21-cm hydrogen emission line and the 18-cm OH emission line; these are likely wavelengths to use in the search for extraterrestrial life.

SETI The Search for Extra-Terrestrial Intelligence.

FIGURE 15-12

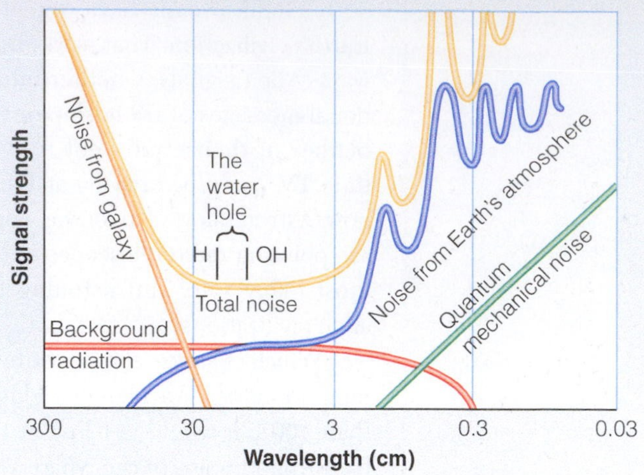

Radio noise from various astronomical sources makes it difficult to detect distant signals at wavelengths longer than 100 cm or shorter than 1 cm. In this range, radio emission lines from H atoms and from OH molecules mark a small wavelength range named the water hole that might be a likely channel for communication.

debate among astronomers, philosophers, theologians, and politicians. Congress funded a NASA search for a short time but ended support in the early 1990s. In fact, the annual cost of a major search is only about as much as a single Air Force attack helicopter, but much of the reluctance to fund searches probably stems from issues other than cost. One point to keep in mind is that the discovery of alien intelligence would cause a huge change in humans' worldview, akin to Galileo's discovery that the moons of Jupiter do not go around the Earth. Some turmoil would inevitably result.

In spite of the controversy, the search continues. The NASA SETI project canceled by Congress was renamed Project Phoenix and completed using private funds. The SETI Institute, founded in 1984, managed Project Phoenix plus several other important searches and is currently operating a radio telescope array in northern California in collaboration with the University of California, Berkeley that was partly funded by Paul Allen of Microsoft.

There is even a way for you to help with searches. The Berkeley SETI team (which is separate from the SETI Institute), with the support of the Planetary Society, has recruited about 4 million owners of personal computers that are connected to the Internet. Participants download a screen saver that searches data files from the Arecibo radio telescope for signals whenever the owner is not using the computer. For information, locate the seti@home project at http://setiathome.ssl.berkeley.edu/.

Drake equation The equation that estimates the total number of communicative civilizations in our Galaxy.

The search continues, but radio astronomers struggle to hear anything against the worsening babble of noise from human civilization. Wider and wider sections of the electromagnetic spectrum are being used for earthly communication, and this, combined with stray electromagnetic noise from electronic devices including everything from computers to refrigerators, makes hearing faint radio signals difficult. It would be ironic if humans fail to detect faint signals from another world because our own world has become too noisy. Ultimately, the chances of success depend on the number of inhabited worlds in the galaxy.

15-3c How Many Inhabited Worlds?

Given enough time, the searches will find other worlds with civilizations, assuming that there are at least a few out there. If intelligence is common, scientists should find signals relatively soon—within the next few decades—but if intelligence is rare, it might take much longer.

Simple arithmetic can give you an estimate of the number of technological civilizations in the Milky Way Galaxy with which you might communicate, N_c. The formula proposed for discussions about N_c is named the **Drake equation** after the radio astronomer Frank Drake, a pioneer in the search for extraterrestrial intelligence. The version of the Drake equation presented here is modified slightly from its original form:

$$N_C = N_* \times f_P \times n_{HZ} \times f_L \times f_I \times f_S$$

N_* is the number of stars in our Galaxy, and f_P represents the fraction of stars that have planets. If all single stars have planets, f_P is about 0.5. The factor n_{HZ} is the average number of planets in each Solar System suitably placed in the habitable zone. This factor actually means the number of planets possessing liquid water. Europa and Enceladus in our Solar System show that liquid water can exist because of tidal heating outside the conventional habitable zone that in our system contains Earth's orbit. Thus, n_{HZ} may be larger than had been previously thought. The factor f_L is the fraction of suitable planets on which life begins, and f_I is the fraction of those planets where life evolved to intelligence.

Notice that the Drake equation is, in a sense, based on an extension of the idea put forth by Copernicus that Earth is not unique. You can put in numbers for the various factors that represent low probabilities or high probabilities, but, either way, the implicit assumption of the equation is that Earth, life on Earth, and the human species that desires to make contact with other intelligences in the Universe are single examples of larger sets of such things.

The first five factors in the Drake equation can be roughly estimated, but the final factor, f_S, the fraction of a star's life during which an intelligent species is communicative, is extremely uncertain. If a society survives at a technological level for only 100 years, the chances of communicating with it are small. But a society that stabilizes and remains technological for a long time is much more likely to be detected. For a star with a life span of 10 billion years, f_S can range from 10^{-8} for extremely short-lived societies to 10^{-3} for societies that survive for a million years. Table 15-1 summarizes what many scientists consider a reasonable range of values for f_s and the other factors.

If the optimistic estimates are true, there could be a communicative civilization within a few tens of light-years of Earth. On the other hand, if the pessimistic estimates are true, Earth might be the only planet that is capable of communication within thousands of the nearest galaxies.

TABLE 15-1 THE NUMBER OF TECHNOLOGICAL CIVILIZATIONS PER GALAXY

Estimates	Variables	Pessimistic	Optimistic
N_*	Number of stars in a typical large galaxy	2×10^{11}	2×10^{11}
f_P	Fraction of stars with planets	0.1	0.5
n_{HZ}	Number of planets per star that orbit in the habitable zone for longer than 4 billion years	0.01	1
f_L	Fraction of habitable zone planets on which life begins	0.01	1
f_I	Fraction of planets with life on which some species evolves to intelligence	0.01	1
f_S	Fraction of star's existence during which a technological civilization survives	10^{-8}	10^{-3}
N_c	Current number of communicative civilizations per galaxy	2×10^{-4}	1×10^{8}

What Are We? Matter and Spirit

The matter you are made of appeared in the Big Bang and was cooked into a wide range of elements inside stars. Eventually, your atoms became part of a nebula that contracted to form our Sun and the planets of the Solar System. They have been recycled many times through dinosaurs, bacteria, grass, and other living things. When the Sun swells into a red giant star and dies in a few billion years, your atoms will become part of future generations of stars and planets.

The message of astronomy is that humans are not just observers: We are participants. If by chance the human race is the only thinking species in the Universe, we bear the sole responsibility to understand and admire it. However, if we are not alone, we might never realize our full potential as humans until we make contact with nonhuman intelligent life.

STUDY TOOLS 15

IN THE BOOK

- ☐ Tear Out the Review Card on *Life on Other Worlds*.

ASTRO ONLINE

Visit ASTRO Online at www.cengagebrain.com

- ☐ Interactive Reading
- ☐ Practice Quizzing
- ☐ Videos and Animations
- ☐ Exclusive Features

Questions and Problems

15

REVIEW QUESTIONS

1. If life is based on information, what is that information?
2. How does the DNA molecule produce a copy of itself?
3. What would happen to a life-form if the genetic information handed down to offspring was copied extremely inaccurately? How would that endanger the future of the life-form?
4. What would happen to a life-form if the information handed down to offspring was always the same? How would that endanger the future of the life-form?
5. Give an example of natural selection acting on new DNA patterns to select the most advantageous characteristics.
6. What evidence do scientists have that life on Earth began in the sea?
7. Why do scientists generally think that liquid water is necessary for the origin of life?
8. What is the difference between chemical evolution and biological evolution?
9. What is the significance of the Miller-Urey experiment?
10. How does intelligence make a creature more likely to survive?
11. Why are upper-main-sequence (high-luminous) stars unlikely sites for intelligent civilizations?
12. Why is it reasonable to suspect that travel between stars is nearly impossible?
13. How does the stability of technological civilizations affect the probability that Earth can communicate with them?
14. What is the water hole, and why is it a good "place" to search for extraterrestrial civilizations?
15. How was the Arecibo message designed to be understood by extraterrestrial recipients?
16. **How Do We Know?** How do science and religion have complementary explanations of the world?
17. **How Do We Know?** Why are scientists confident that Earth has never been visited by aliens?

DISCUSSION QUESTIONS

1. Do you expect that hypothetical alien recipients of the Arecibo message will be able to decode it? Why or why not?
2. How do you think the detection of extraterrestrial intelligence would be received by the public? Would it be likelier to upset, or confirm, humans' beliefs about themselves and the world?
3. What do you think it would mean if decades of careful searches for radio signals for extraterrestrial intelligence turn up nothing?

PROBLEMS

1. A single human cell encloses about 1.5 m of DNA, containing 4.5 billion base pairs. What is the spacing between these base pairs in nanometers? That is, how far apart are the rungs on the DNA ladder?
2. If you represent Earth's history by a line 1 m long, how long a segment would represent the 400 million years since life moved onto the land? How long a segment would represent the 4-million-year history of human life?
3. If a human generation, the average time from birth to childbearing, has been 20 years long, how many generations have passed in the past 1 million years?
4. If a star must remain on the main sequence for at least 4 billion years for life to evolve to intelligence, what is the most massive a star can be and still possibly harbor intelligent life on one of its planets? (*Hints:* See Appendix Table A-3, Properties of Main-Sequence Stars, and also the **Stellar Mass-Lifetime Relationship** in the Math Reference Cards MR.)
5. If there are about 1.4×10^{-4} stars like the Sun per cubic light-year, how many lie within 100 light-years of Earth? (*Note:* The volume of a sphere is $\frac{4}{3}\pi r^3$.)
6. Mathematician Karl Gauss suggested planting forests and fields in gigantic geometric figures as signals to possible Martians that intelligent life exists on Earth. If Martians had telescopes that could resolve details no smaller than 1 arc second, how large would the smallest element of Gauss's signal have to be for it to be visible at Mars's closest approach to Earth? (*Hints:* See Appendix Table A-6, Properties of the Planets, and the **Small-Angle Formula** in the Math Reference Cards MR.)
7. If you detected radio signals with an average wavelength of 20 cm and suspected that they came from a civilization on a distant Earth-like planet orbiting a Sun-like star, roughly how much of a change in wavelength should you expect to see because of the orbital motion of the distant planet?
8. Calculate the number of communicative civilizations per galaxy using your own estimates of the factors in Table 15-1.

LEARNING TO LOOK

1. The star cluster shown in Figure 15-13 contains cool red giants and main-sequence stars from hot blue stars all the way down to red dwarfs. Discuss the likelihood that planets orbiting any of these stars might be home to life. (*Hint:* Estimate the age of the cluster.)

FIGURE 15-13

2. If you could search for life in the galaxy shown in Figure 15-14, would you look among disk stars or halo stars? Discuss the factors that influence your decision.

FIGURE 15-14

Appendix A

Units and Astronomical Data

The Metric System and SI Units

A system of measurement is based on the three fundamental units for length, mass, and time. By international agreement, there is a preferred subset of metric units known as the *Système International d'Unités* (SI units) that is based on the meter, kilogram, and second.

Density should be expressed in SI units as kilograms per cubic meter, but no human hand can enclose a cubic meter, so that unit does not help you easily grasp the significance of a given density. Instead, this book refers to density in grams per cubic centimeter. A gram is roughly the mass of a paperclip, a cubic centimeter is the size of a small sugar cube, and a density of 1 g/cm^3 equals the density of water, all tangible units.

Americans generally use the British system of units. In that system (now official only in Liberia, Myanmar, and the United States, but, ironically, no longer in Great Britain), the fundamental unit of length is the foot, composed of 12 inches. For conceptual purposes, this book expresses some quantities in both metric and English units. Instead of saying that the average adult person would weigh 133 N on the Moon, it might help some readers to see that weight described as 30 pounds. Consequently, this text often gives quantities in metric form followed by the British form in parentheses. For example the equatorial radius of the Moon is 1738 kilometers (1080 miles).

Log in to ASTRO Online to view **Supplemental Tables** including a list of the fundamental SI units and their abbreviations in Table S-1. The metric prefixes and their abbreviations, used to describe multiples or fractions of the fundamental units, are given in Table S-2. Conversion factors between English and metric units are given in Table S-3.

Temperature Scales

In astronomy, as in most other sciences, temperatures are expressed on the Kelvin scale, although the centigrade scale is also used. The centigrade scale is also called the Celsius scale after its inventor, the Swedish astronomer Anders Celsius (1701–1744). Temperatures on the centigrade scale are measured from the freezing point of water (0°C). Temperatures on the Kelvin scale are measured from absolute zero, the temperature of an object that contains no extractable heat (0 K). In practice, no object can be as cold as absolute zero, although laboratory apparatus have reached temperatures less than 10^{-9} K. The Kelvin scale is named after the Scottish mathematical physicist William Thomson, Lord Kelvin (1824–1907). Centigrade and Kelvin degree steps are the same size.

The Fahrenheit scale fixes the freezing point of water at 32°F and the boiling point at 212°F. Named after the German physicist Daniel Gabriel Fahrenheit (1686–1736), who made the first successful mercury thermometer in 1720, the Fahrenheit scale is used routinely only in the United States.

Log in to ASTRO Online to view **Supplemental Tables** including a list of numerical values and formulas for conversion of temperatures from one scale to another in Table S-4.

Astronomy Units and Constants

Astronomy, and science in general, is a way of learning about nature and understanding the Universe. To test hypotheses about how nature works, scientists use observations of nature. The tables that follow contain some of the basic observations that support science's best understanding of the astronomical Universe. Of course, these data are expressed in the form of numbers, not because science reduces all understanding to mere numbers, but because the struggle to understand nature is so demanding that science must use every good means available. Quantitative thinking—reasoning mathematically—is one of the most powerful techniques ever invented by the human brain. Thus these tables are not nature reduced to mere numbers but numbers that support humanity's growing understanding of the natural world around us.

Table A-1 defines units of measure commonly used by astronomers. Table A-2 gives physical constants such as the speed of light, plus basic astronomical data such as the mass and luminosity of the Sun. Table A-3 presents

TABLE A-1 UNITS USED IN ASTRONOMY

1 Angstrom (Å)	= 10^{-8} cm = 10^{-10} m = 10 nm
1 astronomical unit (AU)	= 1.50×10^{11} m = 93.0×10^{6} mi
1 light-year (ly)	= 6.32×10^{4} AU = 9.46×10^{15} m = 5.88×10^{12} mi
1 parsec (pc)	= 2.06×10^{5} AU = 3.09×10^{16} m = 3.26 ly
1 kiloparsec (kpc)	= 1000 pc
1 megaparsec (Mpc)	= 1,000,000 pc

TABLE A-2 ASTRONOMICAL CONSTANTS

Velocity of light (*c*)	= 3.00×10^{8} m/s
Gravitational constant (*G*)	= 6.67×10^{-11} m^3/s^2 kg
Mass of H atom	= 1.67×10^{-27} kg
Mass of Earth ($M_{\oplus}$)	= 5.97×10^{24} kg
Earth equatorial radius ($R_{\oplus}$)	= 6.38×10^{3} km
Mass of the Sun ($M_{\odot}$)	= 1.99×10^{30} kg
Radius of the Sun ($R_{\odot}$)	= 6.96×10^{8} m
Solar luminosity ($L_{\odot}$)	= 3.83×10^{26} J/s
Mass of the Moon	= 7.35×10^{22} kg
Radius of the Moon	= 1.74×10^{3} km

TABLE A-3 PROPERTIES OF MAIN-SEQUENCE STARS

Spectral Type	Absolute Visual Magnitude (M_v)	Luminosity[1,2]	Temp. (K)	λ^{max} (nm)	Mass[1]	Radius[1]	Average Density (g/cm^3)
O6	−5.7	600,000	38,000	76	40	18	0.01
B0	−4.0	40,000	30,000	97	18	7.4	0.06
B5	−1.2	800	16,000	181	5.0	3.8	0.1
A0	0.7	30	10,000	290	2.1	1.8	0.5
A5	2.0	13	8600	337	1.7	1.6	0.6
F0	2.7	4.1	7200	402	1.5	1.3	1.0
F5	3.5	2.3	6500	446	1.3	1.2	1.1
G0	4.4	1.3	5900	491	1.1	1.1	1.2
G2	4.8	1.0	5800	500	1.0	1.0	1.4
G5	5.1	0.7	5600	518	0.9	0.9	1.7
K0	5.9	0.4	5200	557	0.8	0.8	2.2
K5	7.4	0.2	4400	659	0.7	0.7	2.9
M0	8.8	0.1	3900	743	0.5	0.6	3.2
M5	16.3	0.01	3100	935	0.2	0.3	10

[1]Luminosity, mass, and radius are expressed in terms of the Sun's luminosity, mass, and radius. The Sun's luminosity, mass, and radius are given in Table A-2.

[2]Luminosity is computed from radius and temperature.

the characteristics of different spectral types of main-sequence stars. Tables A-4 and A-5, respectively, list the 15 nearest and the 15 brightest stars with their individual properties. The fact that these two star lists have almost no overlap is an important truth about stars that you can read more about in Chapter 9. Table A-6 gives the physical and orbital properties of the major planets in our Solar System. Table A-7 lists the larger moons of each

TABLE A-4 THE 15 NEAREST STARS

Name	Distance (ly)	Distance (pc)	Apparent Visual Magnitude (m_V)	Absolute Visual Magnitude (M_V)	Spectral Type
Sun			−26.7	4.8	G2
Proxima Cen	4.2	1.3	11.0	15.5	M6
α Cen A	4.4	1.3	0.0	4.4	G2
α Cen B	4.4	1.3	1.3	5.7	K1
Barnard's Star	5.9	1.8	9.6	13.2	M4
Wolf 359	7.8	2.4	13.5	16.6	M6
Lalande 21185	8.3	2.5	7.5	10.4	M2
α CMa A (Sirius A)	8.6	2.6	−1.5	1.4	A1
α CMa B (Sirius B)	8.6	2.6	8.4	11.3	White dwarf
Luyten 726-8A	8.7	2.7	12.6	15.5	M5
Luyten 726-8B	8.7	2.7	13.1	15.9	M6
Ross 154	9.7	3.0	10.4	13.1	M4
Ross 248	10.3	3.2	12.3	14.8	M6
ϵ Eri	10.5	3.2	3.7	6.2	K2
Lacaille 9352	10.7	3.3	7.3	9.8	M1

TABLE A-5 THE 15 BRIGHTEST STARS

Star	Name	Apparent Visual Magnitude (m_V)	Distance (pc)	Absolute Visual Magnitude (M_V)	Spectral Type
	Sun	−26.74		4.8	G2 V
α CMa	Sirius	−1.47	2.6	1.4	A1 V
α Car	Canopus	−0.72	96	−5.6	F0 II
α Cen	Rigil Kentaurus	−0.29	1.3	4.1	G2 V
α Boo	Arcturus	−0.04	11	−0.3	K2 III
α Lyr	Vega	0.03	7.8	0.5	A0 V
α Aur	Capella	0.08	13	−0.5	G8 III
β Ori	Rigel	0.12	240	−6.8	B8 Iab
α CMi	Procyon	0.34	3.5	2.6	F5 IV-V
α Eri	Achernar	0.43	44	−2.8	B6 V
α Ori	Betelguese	0.58	200	−5.9	M2 Iab
β Cen	Hadar	0.60	120	−4.8	B1 III
α Aql	Altair	0.77	5.1	2.2	A7 V
α Cru	Acrux	0.81	98	−4.2	B0 IV
α Tau	Aldebaran	0.85	20	−0.7	K5 III

planet and their characteristics. The outer planets have many smaller moons that are not listed here, with more being discovered each year.

Log in to ASTRO Online to view **Supplemental Tables** including Table S-5, which contains the Greek alphabet used in naming the brighter stars. You will also find Table S-6, the periodic chart and list of the chemical elements. One of the primary themes of this book is the history of the atoms in your body: the story of how hydrogen and helium (plus a very small amount of lithium) were produced in the big bang and how the rest of the stable elements up to uranium (atomic number 92) are produced inside stars, either slowly during giant star phases or rapidly during supernova explosions.

TABLE A-6 PROPERTIES OF THE PLANETS

ORBITAL PROPERTIES

Planet	Semimajor Axis (a) (AU)	Semimajor Axis (a) (10^6 km)	Orbital Period (P) (y)	Orbital Period (P) (days)	Average Orbital Velocity (km/s)	Orbital Eccentricity	Inclination to Ecliptic
Mercury	0.387	57.9	0.241	88.0	47.9	0.206	7.0°
Venus	0.723	108	0.615	224.7	35.0	0.007	3.4°
Earth	1.00*	150	1.00*	365.3	29.8	0.017	0°*
Mars	1.52	228	1.88	687.0	24.1	0.093	1.8°
Jupiter	5.20	779	11.9	4333	13.1	0.048	1.3°
Saturn	9.55	1429	29.5	10,759	9.6	0.056	2.5°
Uranus	19.22	2875	84.0	30,688	6.8	0.047	0.8°
Neptune	30.10	4504	164.8	60,182	5.4	0.009	1.8°

*By definition.

PHYSICAL PROPERTIES (Earth = ⊕)

Planet	Equatorial Radius (km)	Equatorial Radius (⊕ = 1)	Mass (⊕ = 1)*	Average Density (g/cm³)	Surface Gravity (⊕ = 1)	Escape Velocity (km/s)	Sidereal Period of Rotation	Inclination of Equator to Orbit
Mercury	2440	0.383	0.055	5.43	0.38	4.3	58.6 d	0.0°
Venus	6052	0.949	0.815	5.24	0.90	10.4	243.0 d	177.3°
Earth	6378	1.00	1.000	5.51	1.00	11.2	23.93 h	23.4°
Mars	3396	0.533	0.107	3.93	0.38	5.0	24.62 h	25.2°
Jupiter	71,492	11.2	318	1.33	2.53	59.5	9.92 h	3.1°
Saturn	60,268	9.45	95.2	0.69	1.06	35.5	10.57 h	26.7°
Uranus	25,559	4.01	14.5	1.27	0.89	21.3	17.23 h	97.8°
Neptune	24,764	3.88	17.1	1.64	1.14	23.5	16.11 h	28.3°

*Earth's mass = 5.97×10^{24} kg

TABLE A-7 PRINCIPAL SATELLITES OF THE SOLAR SYSTEM

Planet	Satellite	Radius (km)	Distance from Planet (10^3 km)	Orbital Period (days)	Orbital Eccentricity	Orbital Inclination**
Earth	Moon	1738	384.4	27.32	0.055	18.3°
Mars	Phobos	14 × 12 × 10	9.4	0.32	0.018	1.0°
	Deimos	8 × 6 × 5	23.5	1.26	0.002	2.8°
Jupiter	Amalthea	135 × 100 × 78	182	0.50	0.003	0.4°
	Io	1820	422	1.77	0.000	0.3°
	Europa	1560	671	3.55	0.000	0.5°
	Ganymede	2630	1071	7.16	0.002	0.2°
	Callisto	2410	1884	16.69	0.008	0.2°
	Himalia	~85*	11,470	250.6	0.158	27.6°
Saturn	Janus	110 × 80 × 100	151.5	0.70	0.007	0.1°
	Mimas	196	185.5	0.94	0.020	1.5°
	Enceladus	260	238.0	1.37	0.004	0.0°
	Tethys	530	294.7	1.89	0.000	1.1°
	Dione	560	377	2.74	0.002	0.0°
	Rhea	765	527	4.52	0.001	0.4°
	Titan	2575	1222	15.94	0.029	0.3°
	Hyperion	205 × 130 × 110	1484	21.28	0.104	~0.5°
	Iapetus	720	3562	79.33	0.028	14.7°
	Phoebe	105	12,930	550.4	0.163	150°
Uranus	Miranda	235	129.9	1.41	0.017	3.4°
	Ariel	580	190.9	2.52	0.003	0°
	Umbriel	585	266.0	4.14	0.003	0°
	Titania	805	436.3	8.71	0.002	0°
	Oberon	790	583.4	13.46	0.001	0°
Neptune	Proteus	205	117.6	1.12	~0	~0°
	Triton	1355	354.59	5.88	0.00	160°
	Nereid	170	5588.6	360.12	0.76	27.7°

*The ~ symbol means "approximately."

**Relative to the planet's equator.

Appendix B

Observing the Sky

Observing the sky with the unaided eye is as important a prelude to studying astronomy as picking up pretty pebbles is to understanding modern geology. The sky is a natural wonder unimaginably bigger than the Grand Canyon, the Rocky Mountains, or any other site that tourists visit every year. To neglect the beauty of the sky is equivalent to geologists neglecting the beauty of the minerals they study. This supplement is meant to act as a tourist's guide to the sky. You analyzed the Universe in the textbook's chapters; here you can admire it.

The brighter stars in the sky are visible even from the centers of cities with their air and light pollution. In the countryside, only a few miles beyond the cities, the night sky is a velvety blackness strewn with thousands of glittering stars. From a wilderness location, far from the city's glare, and especially from high mountains, the night sky is spectacular.

Using Star Charts

The constellations are a fascinating cultural heritage of our planet, but they are sometimes a bit difficult to learn because of Earth's motion. The constellations above the horizon change with the time of night and the seasons.

Because Earth rotates eastward, the sky appears to rotate westward around Earth. A constellation visible overhead soon after sunset will appear to move westward, and in a few hours it will disappear below the horizon. Other constellations will rise in the east, so the sky changes gradually through the night.

In addition, Earth's orbital motion makes the Sun appear to move eastward among the stars. Each day the Sun moves about twice its own diameter, about 1 degree, eastward along the ecliptic. Consequently, each night at sunset, the constellations are about 1 degree farther toward the west.

Orion, for instance, is visible in the evening sky in January, but, as the days pass, the Sun moves closer to Orion. By March, Orion is difficult to see in the western sky soon after sunset. By June, the Sun is so close to Orion that the constellation sets with the Sun and is invisible. Not until late July is the Sun far enough past Orion for the constellation to become visible rising in the eastern sky just before dawn.

Because of the rotation and orbital motion of Earth, you need more than one star chart to map the sky. Which chart you select depends on the month and the time of night. The charts on the following pages show the evening sky for each season, as viewed from the Northern Hemisphere at a latitude typical of the United States or central Europe.

To use the charts, select the appropriate chart and hold it overhead as shown in Figure B-1. If you face south, turn the chart until the words Southern Horizon are at the bottom of the chart. If you face other directions, turn the chart appropriately. Note that hours are in Standard Time; for Daylight Savings Time (spring, summer, and the first half of fall in the United States) add 1 hour. For a more modern experience, there are several star chart apps available on Android and iOS, such as the excellent SkySafari 5 by Simulation Curriculum.

FIGURE B-1

To use the star charts in this book, select the appropriate chart for the season and time. Hold it overhead and turn it until the direction at the bottom of the chart is the same as the direction you are facing. The center of the chart represents the point directly overhead.

Star Charts

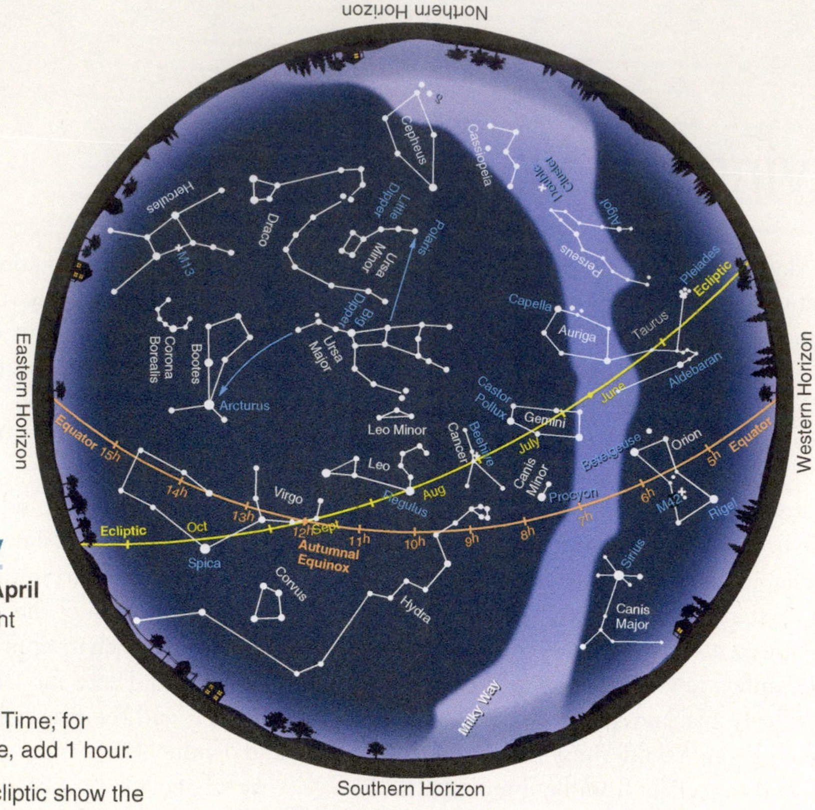

Northern Hemisphere Sky

February–March–April

February	midnight
March	10 PM
April	8 PM

Times are Standard Time; for Daylight Saving Time, add 1 hour.

Months along the ecliptic show the location of the Sun during the year.

Numbers along the celestial equator show right ascension.

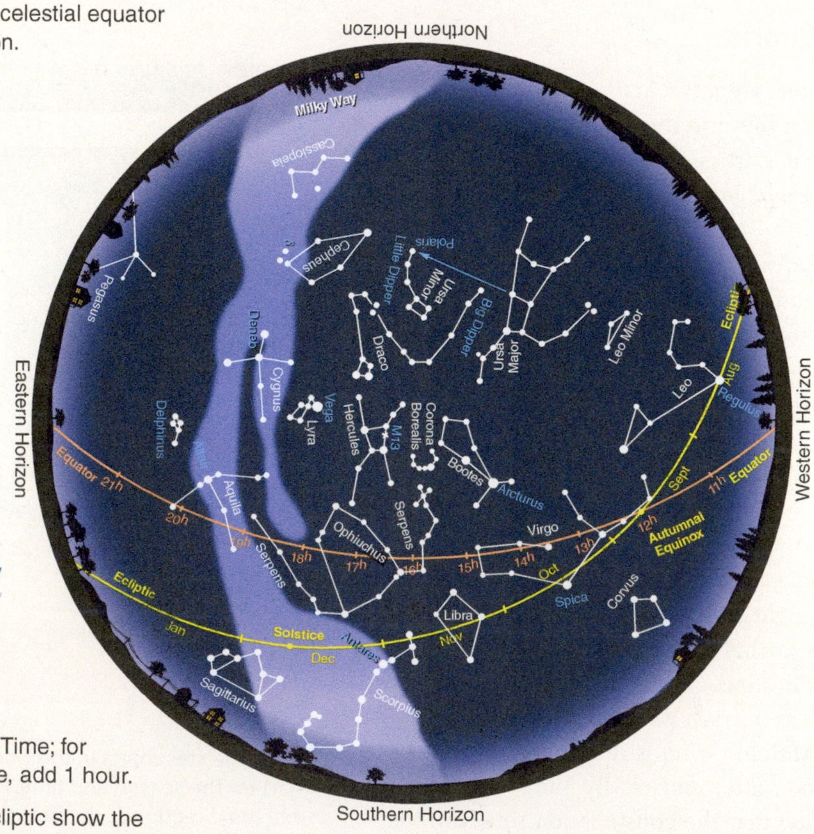

Northern Hemisphere Sky

May–June–July

May	midnight
June	10 PM
July	8 PM

Times are Standard Time; for Daylight Saving Time, add 1 hour.

Months along the ecliptic show the location of the Sun during the year.

Numbers along the celestial equator show right ascension.

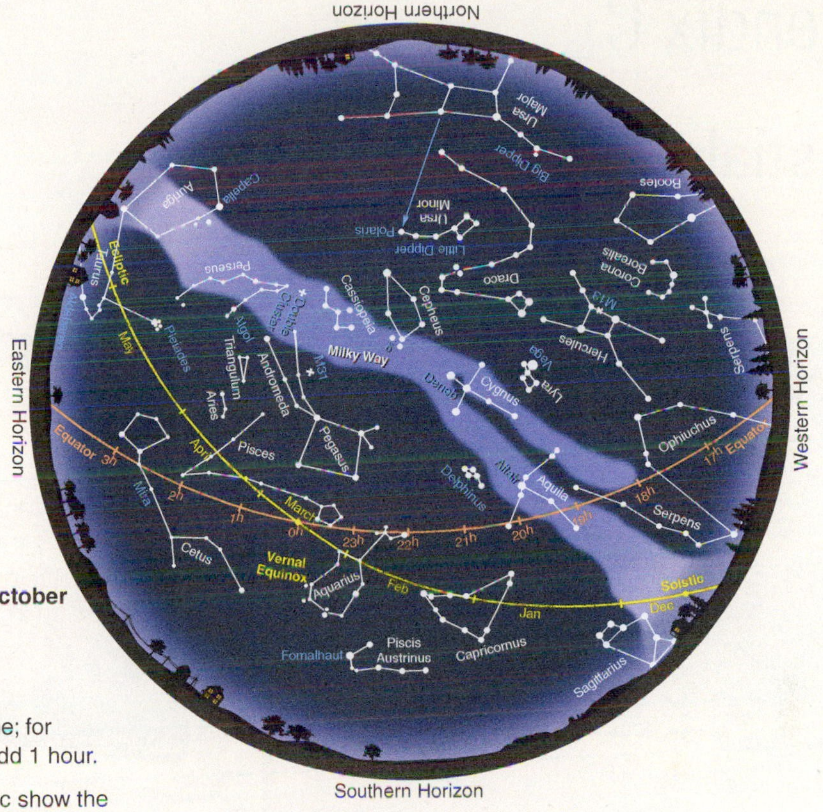

Northern Hemisphere Sky

August–September–October

August	midnight
September	10 PM
October	8 PM

Times are Standard Time; for Daylight Saving Time, add 1 hour.

Months along the ecliptic show the location of the Sun during the year.

Numbers along the celestial equator show right ascension.

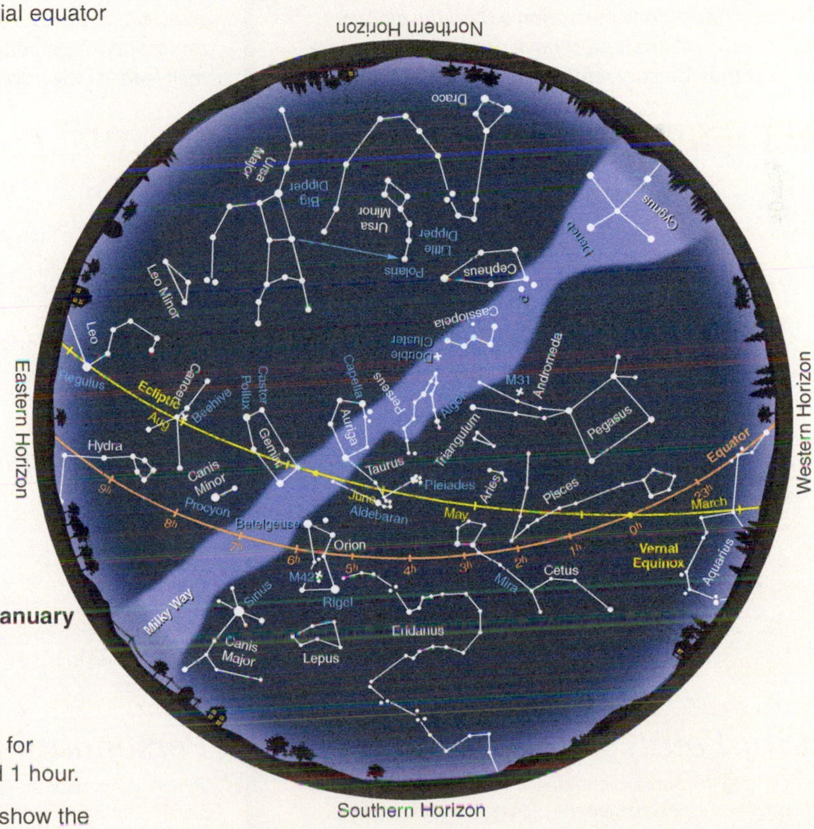

Northern Hemisphere Sky

November–December–January

November	midnight
December	10 PM
January	8 PM

Times are Standard Time; for Daylight Saving Time, add 1 hour.

Months along the ecliptic show the location of the Sun during the year.

Numbers along the celestial equator show right ascension.

Appendix C

Celestial Profiles

NASA

Figure C-1 *Mercury has slightly more than a third the diameter of Earth. Its high density means it must have a very large iron core. The amount of heat that Mercury retains is unknown.*

Celestial Profile 1: Mercury

Motion:

Average distance from the Sun	0.387 AU (5.79×10^7 km)
Eccentricity of orbit	0.206
Inclination of orbit to ecliptic	7.0°
Orbital period	0.241 y (88.0 d)
Period of rotation (sidereal)	58.6 d
Period of rotation (solar)	175.9 d
Inclination of equator to orbit	0.0°

Characteristics:

Equatorial diameter	4.88×10^3 km (0.383 $D_{\oplus}$)
Mass	3.30×10^{23} kg (0.0553 $M_{\oplus}$)
Average density	5.43 g/cm^3 (5.0 g/cm^3 uncompressed)
Surface gravity	0.38 Earth gravity
Escape velocity	4.3 km/s (0.38 $V_{\oplus}$)
Surface temperature	−170° to 430°C (−275° to 805°F)
Average albedo	0.12
Oblateness	0

Personality Point:

Mercury lies very close to the Sun and completes an orbit in only 88 Earth days. For this reason, the ancients named the planet after Mercury, the fleet-footed messenger of the gods. The name is also applied to the element mercury, which is known as quicksilver because it is a heavy, quick-flowing silvery liquid at room temperatures.

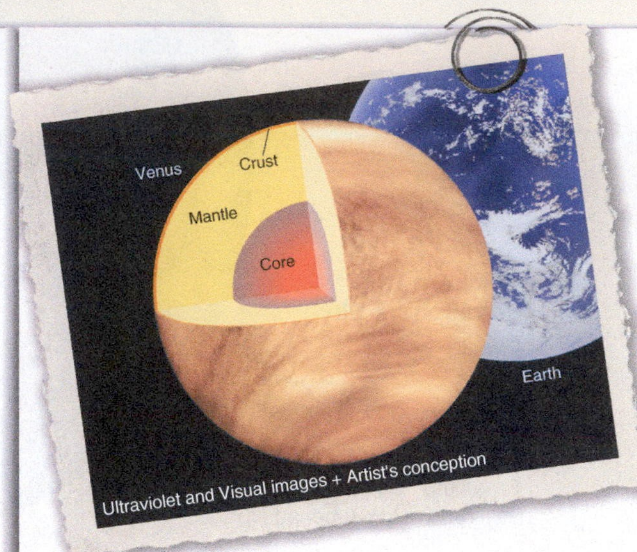

NASA

Figure C-2 *Venus is only 5 percent smaller than Earth, but its atmosphere is perpetually cloudy, and its surface is hot enough to melt lead. It may have a metal core about the size of Earth's.*

Celestial Profile 2: Venus

Motion:

Average distance from the Sun	0.723 AU (1.08×10^8 km)
Eccentricity of orbit	0.007
Inclination of orbit to ecliptic	3.4°
Orbital period	0.615 y (224.7 d)
Period of rotation (sidereal)	243.0 d
Period of rotation (solar)	116.8 d
Inclination of equator to orbit	177.3° (retrograde rotation)

Characteristics:

Equatorial diameter	1.21×10^4 km (0.949 $D_{\oplus}$)
Mass	4.87×10^{24} kg (0.815 $M_{\oplus}$)
Average density	5.24 g/cm^3 (3.9 g/cm^3 uncompressed)
Surface gravity	0.90 Earth gravity
Escape velocity	10.4 km/s (0.93 $V_{\oplus}$)
Surface temperature	470°C (880°F)
Albedo (cloud tops)	0.90
Oblateness	0

Personality Point:

Venus is named for the Roman goddess of love, perhaps because the planet often shines so beautifully in the evening or dawn sky. In contrast, the ancient Maya identified Venus as their war god Kukulkan and sacrificed human victims to the planet when it rose in the dawn sky.

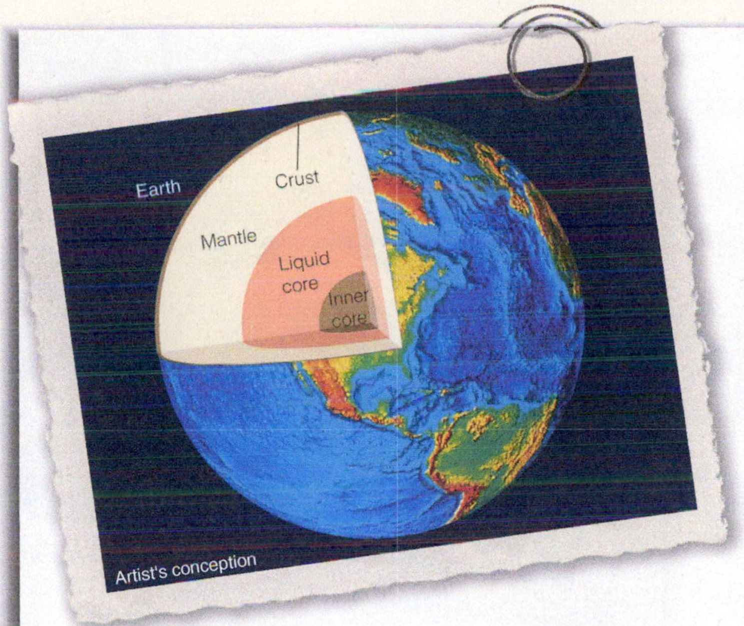

iStockphoto.com/janrysavy

Figure C-3 *Earth's surface is marked by high continents and low seafloors. The crust is only 10 to 60 km (6 to 36 mi) thick. Interior to that are a thick mantle, liquid outer core, and solid inner core.*

Celestial Profile 3: Earth

Motion:

Average distance from the Sun	1.00 AU (1.50×10^8 km)
Eccentricity of orbit	0.017
Inclination of orbit to ecliptic	0° (by definition)
Average orbital velocity	29.8 km/s
Orbital period	1.0000 y (365.26 d)
Period of rotation (solar)	24.00 h
Period of rotation (sidereal)	23.93 h
Inclination of equator to orbit	23.4°

Characteristics:

Equatorial diameter	1.28×10^4 km
Mass	5.97×10^{24} kg
Average density	5.51 g/cm^3 (3.96 g/cm^3 uncompressed)
Surface gravity	1.00 Earth gravity
Escape velocity	11.2 km/s
Surface temperature	−90° to 60°C (−130° to 140°F)
Average albedo	0.31
Oblateness	0.0034

Personality Point:

The modern English word *Earth* comes from Old English eorthe and ultimately from the Indo-European root er-, meaning "ground" or "dirt." Terra comes from the Roman goddess of fertility and growth; thus, *Terra Mater,* Mother Earth.

NASA Images

Figure C-4 *Mars is only half the diameter of Earth and probably retains some internal heat, but the size and composition of its core are not well known.*

Celestial Profile 4: Mars

Motion:

Average distance from the Sun	1.52 AU (2.28×10^8 km)
Eccentricity of orbit	0.093
Inclination of orbit to ecliptic	1.8°
Orbital period	1.881 y (687.0 d)
Period of rotation (sidereal)	24.62 h
Inclination of equator to orbit	25.2°

Characteristics:

Equatorial diameter	6.79×10^3 km (0.533 $D_\oplus$)
Mass	6.42×10^{23} kg (0.107 $M_\oplus$)
Average density	3.93 g/cm^3 (3.3 g/cm^3 uncompressed)
Surface gravity	0.38 Earth gravity
Escape velocity	5.0 km/s (0.45 $V_\oplus$)
Surface temperature	−140° to 15°C (−220° to 60°F)
Average albedo	0.25
Oblateness	0.009

Personality Point:

Mars is named for the god of war. Minerva was the goddess of defensive war, but Bullfinch's Mythology refers to Mars's "savage love of violence and bloodshed." You can see the planet glowing reddish orange from Earth, reminiscent of blood to cultures throughout history, because of iron oxides in its soil.

NASA/JPL/University of Arizona

Figure C-5 *Jupiter's interior is mostly liquid. It may have a small core of heavy elements that is not much larger than Earth.*

Celestial Profile 5: Jupiter

Motion:

Average distance from the Sun	5.20 AU (7.79×10^8 km)
Eccentricity of orbit	0.048
Inclination of orbit to ecliptic	1.3°
Orbital period	11.9 y
Period of rotation	9.92 h
Inclination of equator to orbit	3.1°

Characteristics:

Equatorial diameter	1.43×10^5 km (11.2 $D_{\oplus}$)
Mass	1.90×10^{27} kg (318 $M_{\oplus}$)
Average density	1.33 g/cm^3
Gravity at cloud tops	2.5 Earth gravity
Escape velocity	59.5 km/s (5.3 $V_{\oplus}$)
Temperature at cloud tops	145 K (−200°F)
Albedo	0.34
Oblateness	0.065

Personality Point:

Jupiter, named for the Roman king of the gods (the Greek Zeus), is the largest planet in our Solar System. It can be very bright in the night sky, and its cloud belts and four largest moons can be seen through even a small telescope. Its moons are even visible with a good pair of binoculars mounted on a tripod.

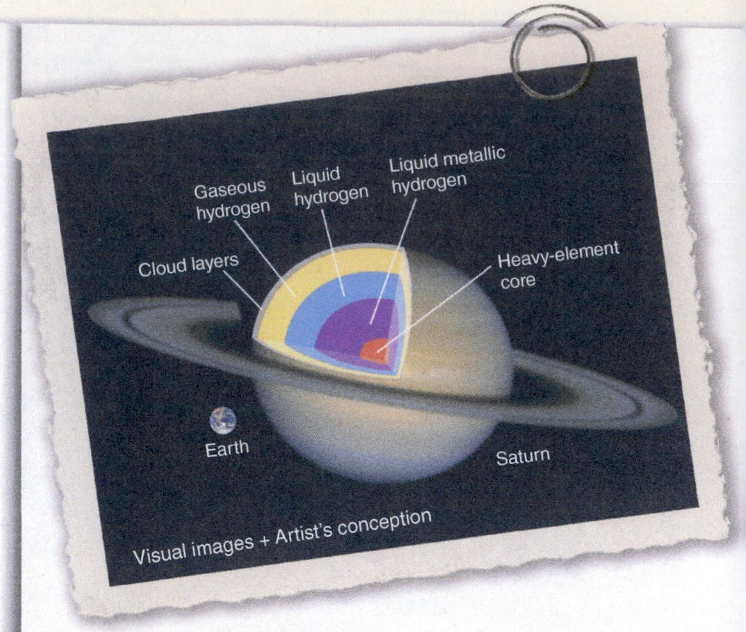

NASA, STScI

Figure C-6 *Saturn's composition, diameter, and internal structure are similar to Jupiter's, although it has less than one-third of Jupiter's mass.*

Celestial Profile 6: Saturn

Motion:

Average distance from the Sun	9.58 AU (1.43×10^9 km)
Eccentricity of orbit	0.056
Inclination of orbit to ecliptic	2.5°
Orbital period	29.5 y
Period of rotation	10.57 h
Inclination of equator to orbit	26.7°

Characteristics:

Equatorial diameter	1.21×10^5 km (9.45 $D_{\oplus}$)
Mass	5.68×10^{26} kg (95.2 $M_{\oplus}$)
Average density	0.69 g/cm^3
Gravity at cloud tops	1.1 Earth gravity
Escape velocity	35.5 km/s (3.2 $V_{\oplus}$)
Temperature at cloud tops	95 K (−290°F)
Albedo	0.34
Oblateness	0.098

Personality Point:

The Greek god Cronus was forced to flee when his son Zeus took power. Cronus went to Italy where the Romans called him Saturn, protector of the sowing of seed. He was celebrated in a weeklong Saturnalia festival at the time of the winter solstice in late December. Early Christians took over the holiday to celebrate Christmas.

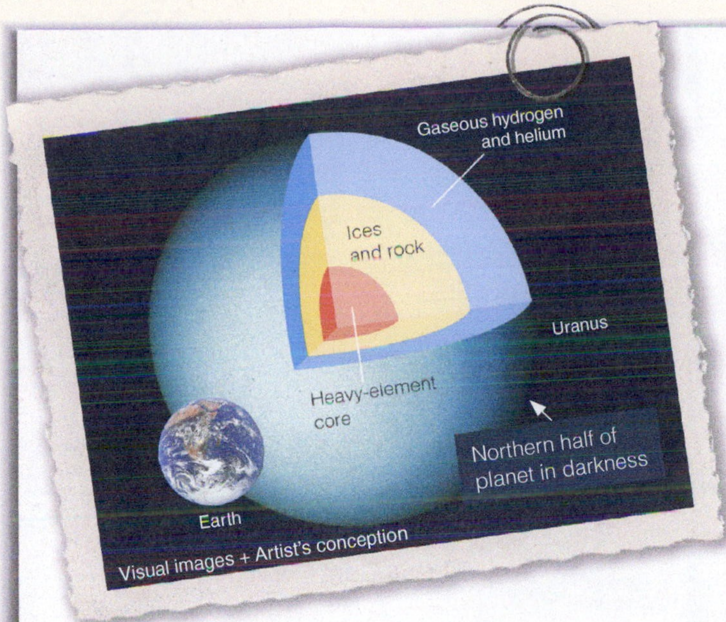

NASA/JPL-Caltech

Figure C-7 *Uranus rotates on its side, and, when Voyager 2 flew past in 1986, the planet's south pole was pointed almost directly at the Sun.*

Celestial Profile 7: Uranus

Motion:

Average distance from the Sun	19.2 AU (2.88×10^9 km)
Eccentricity of orbit	0.047
Inclination of orbit to ecliptic	0.8°
Orbital period	84.0 y
Period of rotation	17.23 h
Inclination of equator to orbit	97.8° (retrograde rotation)

Characteristics:

Equatorial diameter	5.11×10^4 km (4.01 $D_\oplus$)
Mass	8.68×10^{25} kg (14.5 $M_\oplus$)
Average density	1.27 g/cm^3
Gravity	0.9 Earth gravity
Escape velocity	21.3 km/s (1.9 $V_\oplus$)
Temperature at cloud tops	55 K (−360°F)
Albedo	0.30
Oblateness	0.023

Personality Point:

Uranus was discovered in 1781 by William Herschel, a German-born scientist who lived and worked most of his life in England. He named the new planet *Georgium Sidus*, meaning "George's Star" in Latin, after the English King George III. European astronomers, especially the French, refused to accept a planet named after an English king. Instead, they called the planet Herschel. Years later, German astronomer J. E. Bode suggested it be named Uranus after the oldest of the Greek gods.

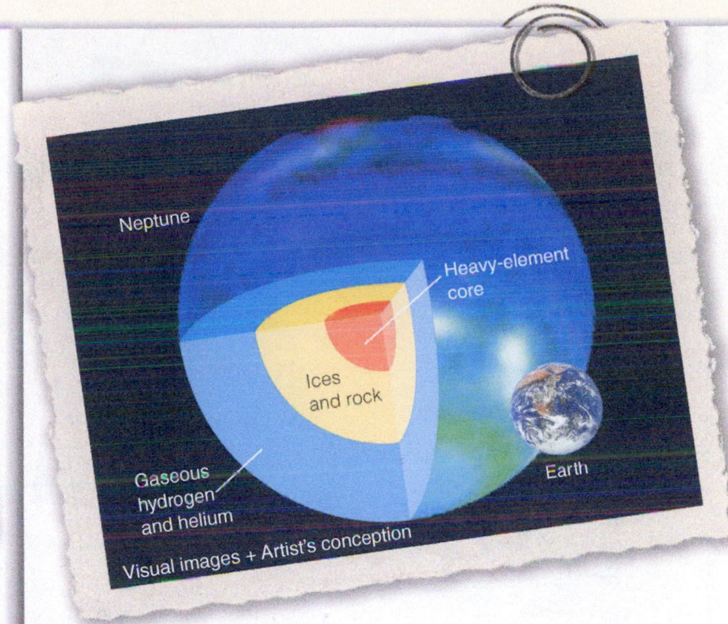

NASA/JPL-Caltech

Figure C-8 *Neptune was tipped slightly away from the Sun when the* Hubble Space Telescope *recorded this image. The interior is much like that of Uranus, but Neptune has more heat flowing outward.*

Celestial Profile 8: Neptune

Motion:

Average distance from the Sun	30.1 AU (4.50×10^9 km)
Eccentricity of orbit	0.009
Inclination of orbit to ecliptic	1.8°
Orbital period	164.8 y
Period of rotation	16.11 h
Inclination of equator to orbit	28.3°

Characteristics:

Equatorial diameter	4.95×10^4 km (3.88 $D_\oplus$)
Mass	1.02×10^{26} kg (17.1 $M_\oplus$)
Average density	1.64 g/cm^3
Gravity	1.1 Earth gravity
Escape velocity	23.5 km/s (2.1 $V_\oplus$)
Temperature at cloud tops	55 K (−360°F)
Albedo	0.29
Oblateness	0.017

Personality Point:

A British and a French astronomer independently calculated the existence and location of Neptune from its gravitational influence on the motion of Uranus. British observers were too slow to act on this information; Neptune was discovered in 1846, and the French astronomer got the credit. Because of its blue color, astronomers named Neptune after the god of the sea.

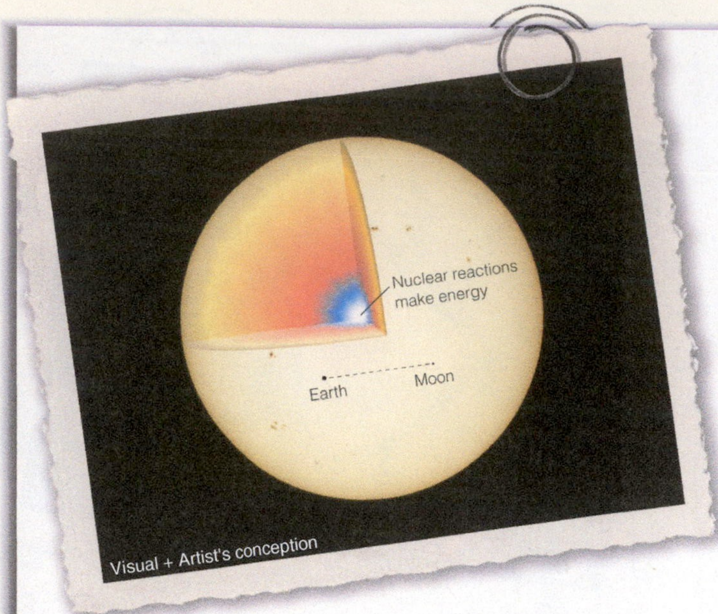

Daniel Good

Figure C-9 *This visible image of the Sun shows a few sunspots and is cut away to show the location of energy generation at the Sun's center. The Earth–Moon system is shown for scale.*

Celestial Profile 9: The Sun

From Earth:

Average distance	1.000 AU (1.496 × 10^8 km)
Maximum distance	1.017 AU (1.521 × 10^8 km)
Minimum distance	0.983 AU (1.471 × 10^8 km)
Angular diameter	0.533° (1920 arc seconds)
Period of rotation (sidereal)	24.5 d at equator
Apparent visual magnitude	−26.74

Characteristics:

Radius	6.96 × 10^5 km
Mass	1.99 × 10^{30} kg
Average density	1.41 g/cm^3
Escape velocity at surface	618 km/s
Luminosity	3.84 × 10^{26} J/s
Surface temperature	5780 K
Central temperature	15.7 × 10^6 K
Spectral type	G2 V
Absolute visual magnitude	+4.83

Personality Profile:

In Greek mythology, the Sun was carried across the sky in a golden chariot pulled by powerful horses and guided by the Sun god Helios. When Phaeton, the son of Helios, drove the chariot one day, he lost control of the horses, and Earth was nearly set ablaze before Zeus smote Phaeton from the sky. Even in classical times, people understood that life on Earth depends critically on the Sun.

NASA

Figure C-10 *Earth's Moon has about a quarter the diameter of Earth. Its low density indicates that it does not contain much iron. The size of its core, if any, and the amount of remaining heat are unknown.*

Celestial Profile 10: The Moon

Motion:

Average distance from Earth	3.84 × 10^5 km
Eccentricity of orbit	0.055
Inclination of orbit to ecliptic	5.1°
Average orbital velocity	1.02 km/s
Orbital period (sidereal)	27.3 d (with respect to the stars)
Synodic period (phase cycle)	29.5 d
Inclination of equator to orbit	6.7°

Characteristics:

Equatorial diameter	3.48 × 10^3 km (0.273 $D_\oplus$)
Mass	7.35 × 10^{22} kg (0.0123 $M_\oplus$)
Average density	3.35 g/cm^3 (3.27 g/cm^3 uncompressed)
Surface gravity	0.17 Earth gravity
Escape velocity	2.4 km/s (0.21 $V_\oplus$)
Surface temperature	−170° to 130°C (−275° to 265°F)
Average albedo	0.12
Oblateness	0

Personality Point:

Lunar superstitions are common. The words *lunatic* and *lunacy* come from *luna*, the Moon. Someone who is *moonstruck* is supposed to be a bit nutty. Because the Moon affects the ocean tides, many superstitions link the Moon to water, to weather, and to women's cycle of fertility. According to legend, moonlight is supposed to be harmful to unborn children, but on the plus side, moonlight rituals are said to remove warts.

H–R Diagrams

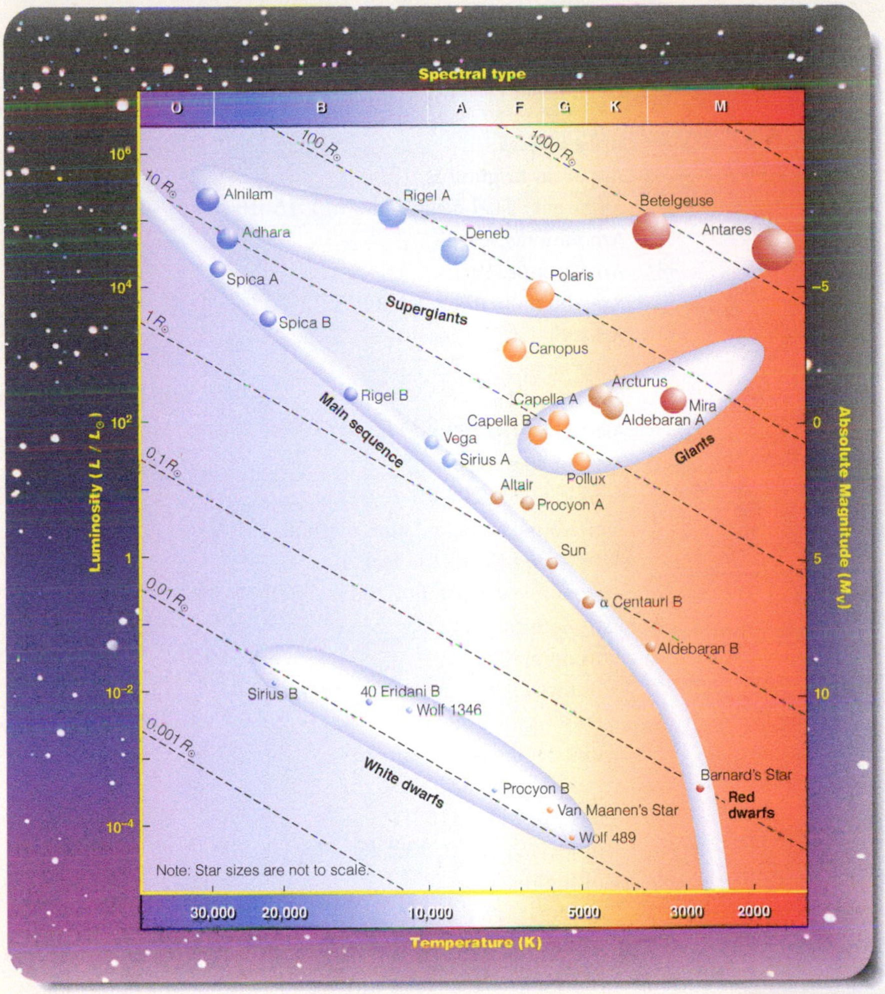

To get a better feel for how our Sun relates to other stars, examine these two H–R diagrams.

This H–R diagram shows the luminosity and temperature of many well-known stars, including our Sun. Now you can compare the Sun with other stars that you can see in the sky.

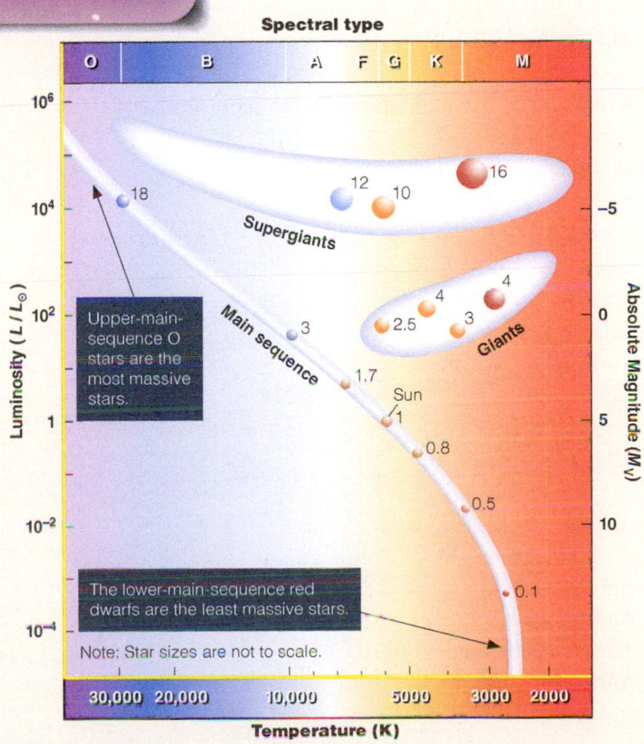

This H–R diagram shows the masses of stars along the main sequence, as well as the masses of some giants and supergiants. Note that the Sun's mass is defined as 1.

INDEX

C

D

E

H

I

J

K

L

M

P

Q

R

S

T

U

V

W

X

Y

Z

ASTRO ONLINE—IT'S ALL IN THERE!

Working with ASTRO Online

To help you take your reading outside the covers of ASTRO, each new text comes with access to ASTRO Online! You can read ASTRO wherever and whenever you're online – ASTRO Online will work on nearly any device, including smartphones!

But you can do more than just read:

- Create StudyBits from text, photos, and figures, and rate your understanding for later test prep.
- Review key term flashcards and even create your own from StudyBits!
- Interact with a variety of digital study tools, including:

Videos covering important chapter concepts

Interactive activities to test your understanding

In-line animated gifs

All-new Pop Astronomy, Featured Astronomer, and Timeline features

Key Terms and Definitions

Assessment

Log in to ASTRO Online with the access code found in the front of your book at www.cengagebrain.com.

CHAPTER 1 TOPIC SUMMARIES/KEY TERMS

1-1 Where Is Here?

- You surveyed the Universe by taking a cosmic zoom in which each **field of view** was 100 times wider than the previous field of view.
- Astronomers use the metric system because it simplifies calculations and use **scientific notation** for very large or very small numbers.
- The **Solar System** includes the Sun at the center, all of the planets that orbit around it—Mercury, Venus, Earth, Mars, Jupiter, Saturn, Uranus, and Neptune—plus the moons of the planets and all other objects bound to the Sun by its gravity.
- You live on a **planet**, Earth, which orbits our **star**, the Sun, once a year. As Earth rotates once a day, you see the Sun rise and set.
- The Moon is only one-fourth the diameter of Earth, but the Sun is about 100 times larger in diameter than Earth—a typical size for a star.
- The **astronomical unit (AU)** is the average distance from Earth to the Sun. Mars, for example, orbits 1.5 AU from the Sun. The **light-year (ly)** is the distance light can travel in 1 year. The nearest star is 4.2 ly from the Sun.
- Many stars seem to have planets, but such small, distant worlds are difficult to detect. Nevertheless, thousands have been found after almost 30 years of searching by astronomers. So far, only a few of those planets have been determined to be Earth-like in terms of size and temperature.
- **Galaxies** contain many billions of stars. Our Galaxy is about 80,000 ly in diameter and contains more than 100 billion stars.
- The **Milky Way**, the hazy band of light that encircles the sky, is the **Milky Way Galaxy** seen from inside. The Sun is just one of the billions of stars that fill the Milky Way Galaxy.
- Some galaxies, including our own, have graceful **spiral arms** bright with stars, but some galaxies are plain clouds of stars.
- The Solar System consists of the Sun plus eight planets, including Earth. Our Galaxy contains our Solar System plus billions of other stars and whatever planets orbit around them. The Universe includes everything that there is: billions of galaxies, each containing billions of stars and, presumably, billions of planetary systems.
- Our Galaxy is just one of billions of galaxies that fill the Universe in great clusters, **superclusters**, filaments, and walls—the largest structures in the Universe.

Astronomical unit Average distance from Earth to the Sun; 1.5×10^8 kilometers (93 million miles).

Field of view The area visible in an image. Usually given as the diameter of the region.

Galaxy A large system of stars, star clusters, gas, dust, and nebulae orbiting a common center of mass.

Light-year (ly) Unit of distance equal to the distance light travels in 1 year.

Milky Way The hazy band of light that circles our sky, produced by the glow of our Galaxy.

Milky Way Galaxy The spiral Galaxy containing our Sun, visible in the night sky as the Milky Way.

Planet A nonluminous body in orbit around a star, massive enough to be spherical and to have cleared its orbital zone of other objects.

Scientific notation A recording method for expressing very large or very small numbers by using powers of 10.

Solar System The Sun and all the objects gravitationally bound to it, including Earth, the other planets plus their moons, asteroids, comets, and so on.

Spiral arms Long spiral pattern of bright stars, star clusters, gas, and dust. Spiral arms extend across the disks of spiral galaxies.

Star A globe of gas, held together by its own gravity and supported by the internal pressure of its hot gases, which generates energy by nuclear fusion.

Supercluster A cluster of galaxy clusters.

1 Here and Now

1-2 When Is Now

- The Sun and planets of our Solar System formed about 4.6 billion years ago.
- Human-like creatures developed on Earth only about 4 million years ago, and human civilizations developed only about 10,000 years ago.
- Although astronomy seems to be about stars and planets, it describes the Universe in which you live, so it is really about you. Astronomy helps you answer the question "What are we?"
- As you study astronomy, you should ask "How do we know?" and that will help you understand how science gives us a way to understand nature.

2-1 The Stars

- Astronomers divide the sky into 88 **constellations** with modern boundaries defined by the International Astronomical Union (IAU). Although the constellations originated in Middle Eastern and Greek mythology, the names are Latin. Named groups of stars that are not constellations are called **asterisms**.
- The names of stars usually come from ancient Arabic, though modern astronomers often refer to a star by its constellation and a Greek letter assigned according to its brightness within the constellation.
- Astronomers refer to the brightness of stars using the **magnitude scale**. First-magnitude stars are brighter than second-magnitude stars, which are brighter than third-magnitude stars, and so on. The magnitude you see when you look at a star in the sky is its **apparent visual magnitude**, m_v, which includes only types of light visible to the human eye and does not take into account the star's distance from Earth.
- **Flux** is the amount of light energy that hits 1 square meter per second, a rigorous definition of brightness. The magnitude of a star is related mathematically to the flux of light received on Earth from the star, in other words, the star's brightness.

Apparent visual magnitude (m_v) A measure of the brightness of a star as seen by human eyes from Earth.

Asterism A named grouping of stars that is not one of the recognized constellations.

Constellation One of the stellar patterns identified by name, usually of mythological gods, people, animals, or objects. Also, the region of the sky containing that star pattern.

Flux A measure of the flow of energy through a surface. Usually applied to light.

Magnitude scale The astronomical brightness scale. The larger the number, the fainter the star.

2-2 The Sky and Its Motions

- The **celestial sphere** is a **scientific model** of the sky, to which the stars appear to be attached. Because Earth rotates eastward, the celestial sphere appears to rotate westward on its axis.
- **Precession** is caused by the gravitational forces of the Moon and Sun acting on the equatorial bulge of the spinning Earth and causing its axis to sweep around in a conical motion like the motion of a top's axis. Earth's axis of rotation precesses with a period of 26,000 years, and consequently the celestial poles and celestial equator move slowly against the background of the stars.
- The **north** and **south celestial poles** are the pivots on which the sky appears to rotate, and they define the four directions around the **horizon:** the **north, south, east,** and **west points**. The point directly overhead is the **zenith**, and the point on the sky directly underfoot is the **nadir**.
- The **celestial equator**, an imaginary line around the sky above Earth's equator, divides the sky into northern and southern halves.
- Astronomers often refer to distances "on" the sky as if the stars, Sun, Moon, and planets were equivalent to spots painted on a plaster ceiling. These **angular distances**, measured in degrees, **arc minutes**, and **arc seconds**, are unrelated to the true distance between the objects in light-years. The angular distance across an object is its **angular diameter**.
- What you see of the celestial sphere depends on your latitude. Much of the Southern Hemisphere of the sky is not visible from northern latitudes. To see that part of the sky, you would have to travel southward over Earth's surface. **Circumpolar constellations** are those close enough to a celestial pole that they do not rise or set.
- The angular distance from the horizon to the north celestial pole always equals your latitude. This is the basis for celestial navigation.

Angular diameter The angle formed by lines extending from the observer to opposite edges of an object.

Angular distance The angle formed by lines extending from the observer to two locations in the sky.

Arc minute 1/60th of a degree.

Arc second 1/60th of an arc minute.

Celestial equator The imaginary line around the sky directly above Earth's equator.

Celestial sphere An imaginary sphere of very large radius surrounding Earth to which the planets, stars, Sun, and Moon seem to be attached.

Circumpolar constellation Any of the constellations so close to the celestial poles that they never set (or never rise) from a given location.

Horizon The circular boundary between the sky and Earth.

Nadir The point on the celestial sphere directly below the observer; the opposite of the zenith.

North and south celestial poles The points on the celestial sphere directly above Earth's north and south poles.

North, south, east, and west points The four cardinal directions; the points on the horizon in those exact directions.

Precession The slow change in orientation of the Earth's axis of rotation. One cycle takes nearly 26,000 years.

Scientific model A scientific metaphor; a concept that organizes thought about an aspect of nature without necessarily being literally true.

Zenith The point in the sky directly above the observer.

2-3 Cycles of the Sun

- The **rotation** of Earth on its axis produces the cycle of day and night, and the **revolution** of Earth around the Sun produces the cycle of the year.
- Because Earth orbits the Sun, the Sun appears to move eastward along the **ecliptic** through the constellations, completing a circuit of the sky in a year. Because the ecliptic is tipped 23.4° to the celestial equator, the Sun spends half the year in the Northern Celestial Hemisphere and half in the Southern Celestial Hemisphere.
- The locations of the Sun and planets along the **zodiac** are diagrammed in a horoscope, which is the basis for the ancient **pseudoscience** known as astrology.
- In the summer, the Sun is above the horizon longer and shines more directly down on the ground, producing warmer weather. In the winter, the Sun is above the horizon for a

(continues)

(continued)

shorter time and shines on the ground at a less direct angle. That produces colder weather. Summer occurs in the United States and Europe in July, August, and September, while the Sun is in the northern sky, but the seasons are reversed in Earth's southern half relative to the Northern Hemisphere. Australia, for example, experiences summer in January, February, and March when the Sun is in the southern sky.

- The beginnings of the seasons are marked by the **vernal equinox**, the **summer solstice**, the **autumnal equinox**, and the **winter solstice**. These names refer to the seasons in the Earth's Northern Hemisphere.
- Earth is slightly closer to the Sun at **perihelion** in January and slightly farther away from the Sun at **aphelion** in July. This has almost no effect on the seasons.
- The planets move generally eastward along the ecliptic, and all but Uranus and Neptune are visible to the unaided eye, looking like stars. Mercury and Venus never wander far from the Sun and are sometimes visible in the evening sky after sunset or in the dawn sky before sunrise.
- Planets visible in the sky at sunset are traditionally called **evening stars**, and planets visible in the dawn sky are called **morning stars**, even though they are not really stars.
- The Moon orbits eastward around Earth once a month and rotates on its axis, keeping the same side facing Earth throughout the month.

Aphelion The orbital point of greatest distance from the Sun.

Autumnal equinox The point on the celestial sphere where the Sun crosses the celestial equator going southward. Also, the beginning of autumn.

Ecliptic The apparent path of the Sun around the sky.

Evening star Any planet visible in the sky just after sunset.

Morning star Any planet visible in the sky just before sunrise.

Perihelion The orbital point of closest approach to the Sun.

Pseudoscience A set of beliefs that appear to be based on scientific ideas but that fail to obey the most basic rules of science.

Revolution Orbital motion about a point located outside the orbiting body.

Rotation Motion around an axis passing through the rotating body.

Summer solstice The point on the celestial sphere where the Sun is at its most northerly point. Also, the beginning of summer.

Vernal equinox The place on the celestial sphere where the Sun crosses the celestial equator moving northward. Also, the beginning of spring.

Winter solstice The point on the celestial sphere where the Sun is farthest south. Also, the beginning of winter.

Zodiac A band centered on the ecliptic and encircling the sky.

2-4 Cycles of the Moon

- Because you see the Moon by reflected sunlight, its shape appears to change as it orbits Earth and sunlight illuminates different amounts of the side you can see.
- The lunar phases wax from new moon to first quarter to full moon and wane from full moon to third quarter to new moon.
- A complete cycle of lunar phases takes 29.53 days, which is known as the Moon's **synodic period**. The **sidereal period** of the Moon—its orbital period with respect to the stars—is 27.32 days, more than 2 days shorter.
- If a full moon passes through Earth's shadow, sunlight is cut off, and the Moon darkens in a **lunar eclipse**. If the Moon fully enters the dark umbra of Earth's shadow, the eclipse is total, but if it only grazes the **umbra**, the region of a shadow that is totally shaded, the eclipse is partial. If the Moon enters the partial shadow of the **penumbra** but not the umbra, the eclipse is penumbral.
- During totality, the eclipsed Moon looks copper red because of sunlight refracted through Earth's atmosphere.
- A **solar eclipse** occurs if a new moon passes between the Sun and Earth and the Moon's shadow sweeps over Earth's surface along the path of totality. Observers inside the path of totality see a total eclipse, and those just outside the path of totality see a partial eclipse as the penumbra sweeps over their location.
- When the Moon is near perigee, the closest point in its orbit, its angular diameter is large enough to cover the Sun's photosphere and produce a total eclipse. But if the Moon is near apogee, the farthest point in its orbit, it looks too small and can't entirely cover the photosphere. A solar eclipse occurring then would be an **annular eclipse**.
- Looking at the Sun is dangerous and can burn the retinas of your eyes. The safest way to observe the partial phases of a solar eclipse is by pinhole projection. Only during totality, when the photosphere is completely hidden, it is safe to look at the Sun directly.
- Because the Moon's orbit is tipped a few degrees from the plane of Earth's orbit, most full moons pass north or south of Earth's shadow, and no lunar eclipse occurs. Also, most new moons cross north or south of the Sun, the Moon's shadow does not sweep over Earth, and there is no solar eclipse.
- Eclipses can occur only when a full moon or a new moon occurs near one of the two nodes of its orbit, where it crosses the ecliptic. These two eclipse seasons occur about 6 months apart, but move slightly earlier each year. By keeping track of the location of the nodes of the Moon's orbit, you could predict which full and new moons were most likely to be eclipsed.
- Eclipses follow a pattern lasting 18 years, 11⅓ days called the **Saros cycle**. If ancient astronomers could have understood that pattern, they could have predicted eclipses.

Annular eclipse A solar eclipse in which the solar photosphere appears around the edge of the Moon in a bright ring, or annulus. Features of the solar atmosphere cannot be seen during an annular eclipse.

Lunar eclipse The darkening of the Moon when it moves through Earth's shadow.

Penumbra The portion of a shadow that is only partially shaded.

Saros cycle The 18-year, 11⅓-day period after which the pattern of lunar and solar eclipses repeats.

Sidereal period The period of rotation or revolution of an astronomical body relative to the stars.

Solar eclipse The event that occurs when the Moon passes directly between Earth and the Sun, blocking your view of the Sun.

Synodic period The period of rotation or revolution of a celestial body with respect to the Sun.

Umbra The region of a shadow that is totally shaded.

CHAPTER REVIEW

CHAPTER 3 TOPIC SUMMARIES/KEY TERMS

3-1 Astronomy Before Copernicus

- Ancient philosophers accepted as a **first principle** that the heavens were perfect, so philosophers such as Plato argued that, because the sphere was the only perfect geometrical form and carried a point on its surface around in a circle, the heavens must move in **uniform circular motion**.
- They also accepted that Earth was the unmoving center of all motion in the middle of a **geocentric universe**. These principles became part of the teachings of the great philosopher Aristotle, who argued that the Sun, Moon, and stars were carried around Earth on rotating crystalline spheres.

First principle Something that seems obviously true and needs no further examination.

Geocentric universe A model universe with Earth at the center, such as the Ptolemaic universe.

Uniform circular motion The classical belief that the perfect heavens could move only by the combination of uniform motion along circular orbits.

3-2 Nicolaus Copernicus

- The lack of any **parallax** in the positions of the stars gave astronomers confidence that Earth could not be moving.
- About the year 140, Ptolemy gave mathematical form to Aristotle's model in the *Almagest*. Ptolemy preserved the principles of geocentrism and uniform circular motion, but he added **epicycles**, **deferents**, and **equants** to better predict the motions of the planets. To account for **retrograde motion**, his epicycles had to be quite large. Even so, his model was not very accurate in predicting the positions of the planets.
- The problem of the place of Earth was solved when Copernicus devised a model that was a **heliocentric universe**. He preserved the principle of uniform circular motion, but he put the Sun at the center and argued that Earth rotates on its axis and circles the Sun once a year. His theory was controversial in part because it contradicted Church teaching.
- Copernicus published his theory in his book *De Revolutionibus* in 1543, the same year he died.
- In Copernicus's model, retrograde motion was explained without epicycles, but because he kept uniform circular motion, he had to include small epicycles, and his model did not predict the motions of the planets well.
- One reason the Copernican model won gradual acceptance was that it was more elegant. Venus and Mercury were treated the same as all the other planets, and the velocity of each planet was related to its distance from the Sun. The shift from the geocentric **paradigm** to the heliocentric paradigm is an example of a scientific revolution.

Deferent In the Ptolemaic model of the Solar System, the large circle around Earth along which the center of an epicycle moved.

Epicycle The small circle followed by a planet in the Ptolemaic theory. The center of the epicycle follows a larger circle (the deferent) around Earth.

Equant In the Ptolemaic model of the Solar System, the point off-center in a deferent from which the center of an epicycle appears to move uniformly.

Heliocentric universe A model universe with the Sun at the center, such as the Copernican universe.

Paradigm A commonly accepted set of scientific ideas and assumptions.

Parallax The apparent change in position of an object due to a change in the location of the observer. Astronomical parallax is measured in seconds of arc.

Retrograde motion The apparent backward (westward) motion of planets as seen against the background of stars.

3-3 Tycho Brahe, Johannes Kepler, and Planetary Motion

- The problem of planetary motion was finally solved through the work of two astronomers, Tycho Brahe and Johannes Kepler.
- Brahe developed his own model in which the Sun and Moon circled Earth and the planets circled the Sun. His great contribution was to compile detailed observations over a period of 20 years, observations that were later used by Johannes Kepler.
- Kepler inherited Brahe's books of observations in 1601 and used them to devise three **empirical** laws of planetary motion. The first law says that the planets follow **ellipses** with the Sun at one focus. According to the second law, planets move faster when nearer the Sun and slower when farther away. The third law says that a planet's orbital period squared is proportional to the **semi-major axis, *a***, of its orbit cubed.
- The **eccentricity, *e***, of an ellipse equals 0 for a circle and grows closer and closer to 1 as the ellipse becomes more and more elongated.

Eccentricity, *e* A number between 1 and 0 that describes the shape of an ellipse; the distance from one focus to the center of the ellipse divided by the semi-major axis.

Ellipse A closed curve around two points called the foci, such that the total distance from one focus to the curve and back to the other focus remains constant.

Empirical Description of a phenomenon without explaining why it occurs.

Semi-major axis, *a* Half of the longest diameter of an ellipse.

3 The Origin of Modern Astronomy

3-4 Galileo Galilei

- A **hypothesis** is a statement about nature that needs further testing, but a **theory** is usually a description of nature that has been tested. Some theories are very well understood and widely accepted. A **natural law** is a fundamental principle in which scientists have great confidence.
- Kepler's final book, *The Rudolphine Tables* (1627), combined heliocentrism with elliptical orbits and predicted the positions of the planets well.
- Galileo used the newly invented telescope to observe the heavens, and he recognized the significance of what he saw there. His discoveries of the phases of Venus, the Galilean moons of Jupiter, the mountains of the Moon, and other phenomena helped undermine the Ptolemaic universe.
- Galileo based his analysis on observational evidence rather than on first principles or on scripture. In 1633, he was condemned before the Inquisition for refusing to halt his defense of Copernicanism.

Hypothesis A conjecture, subject to further tests, that accounts for a set of facts.

Natural law A theory that has been so well confirmed that it is almost universally accepted as correct.

Theory A system of assumptions and principles applicable to a wide range of phenomena that has been repeatedly verified.

3-5 Isaac Newton, Gravity, and Orbits

- Isaac Newton used the work of Kepler and Galileo to discover three laws of motion and the law of gravity. These laws made it possible to understand such phenomena as orbital motion and the tides.
- Newton showed that gravity was mutual and universal. It depends on the **mass** of the bodies and the distance between them according to the **inverse square relation**. **Weight** is the force that Earth's gravity exerts on an object.
- Newton used the image of a cannon on a mountaintop to explain that an object in orbit is falling toward Earth's center and simultaneously moving fast enough to continually miss hitting Earth's surface. To maintain a circular orbit, the object must have **circular velocity**. All elliptical orbits, including circles, are **closed orbits**, but if the object's velocity equals or exceeds **escape velocity**, it will follow an **open orbit** and never return.
- A **geosynchronous satellite** orbits far enough from Earth that its orbital period is 24 hours, and it remains above a single spot on Earth as Earth turns.
- Two objects that orbit each other actually orbit their common **center of mass**.
- Newton's laws gave scientists a unified way to think about nature—cause and effect. Every effect has a cause, and science is the search for those causes.
- Newton's laws also explain that tides are caused by small differences in the gravity acting on different parts of a body. Ocean tides occur because the Moon's gravity pulls more strongly on the near side of Earth than on the center. A tidal bulge occurs on the far side of Earth because the Moon's gravity is slightly weaker there than on the center of Earth.
- Tides produced by the Moon combine with tides produced by the Sun to cause extreme tides, called **spring tides**, at new and full moons. The Moon and Sun work against each other to produce less extreme tides, called **neap tides**, at quarter moons.
- Friction from tides can slow the rotation of a rotating world, and the gravitational pull of tidal bulges can make orbits change slowly.
- The 99 years from the death of Copernicus to the birth of Newton marked the beginning of modern science. From that time on, science depended on evidence to test theories and relied on the analytic methods first demonstrated by Kepler and Newton.

Center of mass The balance point of a body or system of masses. The point about which a body or system of masses rotates in the absence of external forces.

Circular velocity The velocity an object needs to stay in orbit around another object.

Closed orbit An orbit that repeatedly returns to the same starting point.

Escape velocity The initial velocity an object needs to escape from the surface of a celestial body.

Geosynchronous satellite A satellite that orbits eastward around Earth with a period of 24 hours and remains above the same spot on Earth's surface.

Inverse square relation A rule that the strength of an effect (such as gravity) decreases in proportion as the distance squared increases.

Mass A measure of the amount of matter making up an object.

Neap tide Ocean tide of small range occurring at first- and third-quarter moons.

Open orbit An orbit that carries an object away, never to return to its starting point.

Spring tide Ocean tide of large range that occurs at full and new moons. Note that the term "spring" refers to leaping, not to the season of spring.

Weight The force that gravity exerts on an object.

4-1 Radiation: Information from Space

- Light is the visible form of **electromagnetic radiation**, an electric and magnetic disturbance that transports energy at the speed of light. The **wavelength** of visible light is usually measured in **nanometers** (10^{-9} m) or **ångstroms** (10^{-10} m), and ranges from 400 nm to 700 nm (4000 Å to 7000 Å).
- The complete electromagnetic spectrum includes radio waves, microwaves, **infrared (IR)** radiation, visible light, **ultraviolet (UV)** radiation, **X-rays**, and **gamma-rays**.
- A **photon** is a packet of light waves that can act as a particle or as a wave. The energy carried by a photon is inversely proportional to its wavelength. Gamma-rays, X-rays, and ultraviolet radiation have shorter wavelengths and carry more energy per photon than visible light. Infrared rays, microwaves, and radio waves have longer wavelengths and carry less energy per photon than visible light.
- Earth's atmosphere is fully transparent in only two **atmospheric windows**: visible light and radio.

Ångstrom (Å) A unit of distance commonly used to measure the wavelength of light. $1Å = 10^{-10}$ m.

Atmospheric window Wavelength region in which our atmosphere is transparent—at visual, radio, and some infrared wavelengths.

Electromagnetic radiation Changing electric and magnetic fields that travel through space and transfer energy from one place to another; examples are light and radio waves.

Gamma-rays The shortest-wavelength electromagnetic waves.

Infrared (IR) The portion of the electromagnetic spectrum with wavelengths longer than red light, ranging from 700 nm to about 1 mm, between visible light and radio waves.

Nanometer (nm) A unit of distance equaling one-billionth of a meter (10^{-9} m), commonly used to measure the wavelength of light.

Photon A quantum of electromagnetic energy that carries an amount of energy that depends inversely on its wavelength.

Ultraviolet (UV) The portion of the electromagnetic spectrum with wavelengths shorter than violet light, between visible light and X-rays.

Wavelength The distance between successive peaks or troughs of a wave, usually represented by a lowercase Greek lambda, λ.

X-rays Electromagnetic waves with wavelengths shorter than ultraviolet light.

4-2 Telescopes

- **Refracting telescopes** use a **primary lens** to bend the light and focus it into an image. **Reflecting telescopes** use a **primary mirror** to focus the light. The image produced by the telescope's primary lens or mirror can be magnified by an **eyepiece.** Lenses and mirrors with short **focal lengths** must be strongly curved and are more expensive to grind to an accurate shape.
- Because of **chromatic aberration**, refracting telescopes cannot bring all colors to the same focus, resulting in color fringes around the images. Lenses that can partially correct for this effect are difficult to make and expensive.
- Reflecting telescopes are easier to build and less expensive than refracting telescopes of the same diameter. Also, reflecting telescopes do not suffer from chromatic aberration. Most large **optical telescopes** and all **radio telescopes** are reflecting telescopes.
- **Light-gathering power** refers to the ability of a telescope to produce bright images. **Resolving power** refers to the ability of a telescope to resolve fine detail. **Diffraction fringes** in an image, caused by the interaction of light waves with the telescope's apertures, limit the detail visible.
- An **interferometer** is two or more separate telescopes combined to act as a single large telescope that has a resolution equivalent to that of a telescope as large in diameter as the separation between the individual telescopes. The first working interferometers were composed of multiple radio telescopes.
- Turbulence in Earth's atmosphere blurs the image of an astronomical object, a phenomenon that astronomers refer to as **seeing.** Atop a mountain, the air is relatively steady, and the seeing is better. **Adaptive optics** techniques involve measuring seeing distortions caused by turbulence in Earth's atmosphere and partially canceling out those distortions by rapidly altering some of the telescope's optical components.
- **Magnifying power**, the ability to make an object look bigger, is a less important telescope power because it is not a property of the telescope itself but can be changed simply by changing the eyepiece.

Adaptive optics A computer-controlled optical system in an astronomical telescope used to partially correct for seeing.

Chromatic aberration A distortion found in refracting telescopes because lenses focus different colors at slightly different distances. Images are consequently surrounded by color fringes.

Diffraction fringe Blurred fringe surrounding any image, caused by the wave properties of light. Because of this, no image detail smaller than the fringe can be seen.

Eyepiece A short-focal-length lens used to enlarge the image in a telescope. The lens nearest the eye.

(*continues*)

4 Light and Telescopes

(continued)

Focal length The focal length of a lens or mirror is the distance from that lens or mirror to the point where it focuses parallel rays of light.

Interferometer Separated telescopes combined to produce a virtual telescope with the resolution of a much larger-diameter telescope.

Light-gathering power The ability of a telescope to collect light; proportional to the area of the telescope's primary lens or mirror.

Magnifying power The ability of a telescope to make an image larger.

Optical telescope Telescope that gathers visible light.

Primary lens In a refracting telescope, the largest lens.

Primary mirror In a reflecting telescope, the largest mirror.

Radio telescope Telescope that gathers radio radiation.

Reflecting telescope A telescope that forms images by reflecting light with a mirror.

Refracting telescope A telescope that forms images by bending (refracting) light with a lens.

Resolving power The ability of a telescope to reveal fine detail. Depends on the diameter of the telescope's primary lens or mirror.

Seeing Atmospheric conditions on a given night. When the atmosphere is unsteady, producing blurred images, the seeing is said to be poor.

4-3 Observatories on Earth—Optical and Radio

- Observatories are located far from cities to avoid **light pollution.** Radio telescopes are also located far from cities to avoid human-produced radio noise.
- Because Earth rotates, telescopes must use **sidereal tracking** to remain pointed at celestial objects.
- In a reflecting telescope, light first comes to a focus at the **prime focus**, but **secondary mirrors** can direct light to other locations such as the **Cassegrain focus.** The **Newtonian focus** and **Schmidt-Cassegrain focus** locations are used in some telescopes.

Cassegrain focus The optical design in which the secondary mirror reflects light back down the tube through a hole in the center of the objective mirror.

Light pollution The illumination of the night sky by waste light from cities and outdoor lighting, which prevents the observation of faint objects.

Newtonian focus The optical design in which a diagonal mirror reflects light out the side of the telescope tube for easier access.

Prime focus The point at which the objective mirror forms an image in a reflecting telescope.

Schmidt-Cassegrain focus The optical design that uses a thin corrector plate at the entrance to the telescope tube. A popular design for small telescopes.

Secondary mirror In a reflecting telescope, a mirror that directs the light from the primary mirror to a focal position.

Sidereal tracking (pronounced "sih-DARE-ee-al") The continuous movement of a telescope to keep it pointed at a star as Earth rotates.

4-4 Astronomical Instruments and Techniques

- Astronomers in the past used **photographic plates** to record images at the telescope, and **photometers** to precisely measure the brightness of celestial objects. Modern electronic systems such as **charge-coupled devices (CCD)** and other types of **array detectors** have replaced both photographic plates and photometers in most applications.
- Electronic detectors have the advantage that data from them is automatically **digitized** in numerical format and can be easily recorded and manipulated. Astronomical images in digital form can be computer enhanced to produce **false-color images** (also known as "representational-color" images) that bring out subtle details.
- A **spectrograph** is an instrument that uses a prism or a **grating** to spread light into a **spectrum.**

Array detector Device for collecting and recording electromagnetic radiation using multiple individual detectors arrayed on the surface of a chip; for example, a CCD electronic camera.

Charge-coupled device (CCD) An electronic device consisting of a large array of light-sensitive elements used to record very faint images.

Digitized Converted to numerical data that can be read directly into a computer memory for later analysis.

False-color image A representation of graphical data with added or enhanced color to reveal detail.

Grating A piece of material in which numerous microscopic parallel lines are scribed. Light encountering a grating is dispersed to form a spectrum.

Photographic plate The first image-recording device used with telescopes; it records the brightness of objects, but with only moderate precision.

Photometer Sensitive astronomical instrument that measures the brightness of individual objects very precisely.

Spectrograph A device that separates light by wavelengths to produce a spectrum.

Spectrum A range of electromagnetic radiation spread into its component wavelengths (colors); for example, a rainbow; also, representation of a spectrum as a graph showing intensity of radiation as a function of wavelength or frequency.

4-5 Airborne and Space Observatories

- Earth's atmosphere distorts and blurs images. Telescopes in orbit are above this seeing distortion and are limited only by diffraction in their optics.

Earth's atmosphere absorbs gamma-rays, X-rays, ultraviolet, far-infrared, and microwave light. To observe at these wavelengths, telescopes must be carried in aircraft, balloons, rockets, or spacecraft.

Guidepost

- Beginning in the 19th century, the study of features called **spectral lines** in the Sun's spectrum and in spectra of gas samples in laboratories has allowed determination of the composition of the Sun, measurement of motions in the Sun's material, and detection of magnetic fields in its surface and atmosphere.
- **Spectral line:** A feature in a spectrum at a specific wavelength produced by the absorption or emission of light by certain atoms.

5-1 The Sun: Basic Characteristics

- Observations of parallax during rare **transits of Venus** allow measurement of the Sun's distance and thereby determination of its true diameter and volume. The Sun's mass can be determined from the orbits of the planets via Newton's laws. Dividing the Sun's mass by its volume yields its **density**, an important clue to its composition and internal structure.

Density Mass per volume.

Transits of Venus Rare occasions when Venus can be seen as a tiny dot directly between Earth and the Sun.

5-2 The Origin of Sunlight

- An **atom** consists of a **nucleus** made up of positively charged **protons** and uncharged **neutrons**. The number of protons in an atom determines which element it is. Atoms of the same element (i.e., having the same number of protons) with different numbers of neutrons are called **isotopes**. In ordinary Earthly conditions, an atomic nucleus is surrounded by a cloud of **electrons**. Two or more atoms joined together form a **molecule**.
- Energy in the form of agitation of atoms, molecules, and other particles is **heat**. **Temperature** refers to the intensity of the agitation and can be expressed on the **Kelvin temperature scale**, which gives temperature above **absolute zero**.
- Collisions among the particles in a body accelerate electrons and cause the emission of electromagnetic radiation. The characteristic emission from an opaque object is termed **blackbody radiation**. The hotter an object is, the shorter is its **wavelength of maximum intensity**, λ_{max} (known as **Wien's law**, and the more total energy it radiates [known as the **Stefan–Boltzmann law**]).

Absolute zero The theoretical lowest possible temperature at which a material contains no extractable heat energy; zero on the Kelvin temperature scale.

Atom The smallest unit of a chemical element, consisting of a nucleus containing protons and neutrons plus a surrounding cloud of electrons.

Blackbody radiation Spectrum of radiation emitted by an ideal perfect radiator. The spectrum is continuous and its characteristics depend only on the object's temperature.

Electron Low-mass atomic particle carrying a negative charge.

Heat Energy contained in matter in the form of agitation of its particles.

Isotope A form of an atom with the same number of protons but a different number of neutrons.

Kelvin temperature scale A temperature scale using Celsius degrees and based on zero being equal to absolute zero.

Molecule Two or more atoms bonded together.

Neutron An atomic particle with no charge and about the same mass as a proton.

Nucleus The central core of an atom containing protons and neutrons that carries a net positive charge.

Proton A positively charged atomic particle contained in the nucleus of an atom. The nucleus of a hydrogen atom.

Stefan–Boltzmann law Hotter objects emit more energy than cooler objects of the same size, in proportion to the fourth power of temperature.

Temperature A measure of the agitation among the atoms and molecules of a material.

Wavelength of maximum intensity The wavelength at which the radiation emitted by a glowing object is most intense.

Wien's law The hotter a glowing object is, the shorter will be its wavelength of maximum intensity, inversely proportional to its temperature.

5-3 The Sun's Surface

- The Sun's visible surface, the **photosphere**, is the layer in the Sun from which visible photons most easily escape. Dark **sunspots** are observed that seem dark because they are slightly cooler than the rest of the photosphere. The **granulation** of the photosphere is produced by **convection** currents of hot gas rising from below.

Convection Circulation in a fluid driven by heat. Hot material rises and cool material sinks.

Granulation The fine structure of bright grains with dark edges covering the Sun's surface.

Photosphere The bright visible surface of the Sun.

Sunspot Relatively dark spot on the Sun that contains intense magnetic fields.

5-4 Light, Matter, and Motion

- The electrons in an atom are attracted to the nucleus by the **Coulomb force**. A neutral atom is surrounded by a number of negatively charged electrons equal to the number of protons in the nucleus. An atom that has lost or gained an electron is called an **ion** and the process is called **ionization**.
- As described by **quantum mechanics**, the **binding energy** that holds electrons in an atom is limited to certain energies, and that means the electrons may occupy only certain **permitted orbits**.
- The size of an electron's orbit depends on its energy, so the orbits can be thought of as **energy levels**. The lowest possible energy level is known as its **ground state**. An **excited atom** is one in which the atom has more energy than the ground state because one or more electrons have been raised to higher orbits. An atom can become excited by a collision between atoms or the absorption of a photon of the proper energy.
- An atom can emit or absorb a photon when an electron makes a **quantum jump** between orbits. Because orbits of only certain energy differences are permitted in an atom, photons of only certain wavelengths can be absorbed or emitted. Each kind of atom has its own characteristic set of spectral lines.
- The **Doppler effect** can provide clues to the motions of the stars. When a star is approaching, you observe slightly shorter wavelengths, called a **blueshift**, and when it is receding, you observe slightly longer wavelengths, a **redshift**. Measurement of the Doppler effect reveals a star's **radial velocity, V_r**, that part of its velocity directed toward or away from Earth.

Binding energy The energy needed to pull an electron away from its atom.

Blueshift A Doppler shift toward shorter wavelengths caused by a velocity of approach.

Coulomb force The electrostatic force of repulsion or attraction between charged bodies.

Doppler effect The change in the wavelength of radiation because of relative radial motion of source and observer.

Energy level One of a number of states an electron may occupy in an atom, depending on its binding energy.

Excited atom An atom in which an electron has moved from a lower to a higher energy level.

Ground state The lowest permitted electron energy level in an atom.

Ion An atom that has lost or gained one or more electrons.

Ionization The process in which atoms lose or gain electrons.

Permitted orbit One of the unique orbits that an electron may occupy in an atom.

Quantum jump Jumps of electrons from one orbit or energy state to another.

Quantum mechanics The study of the behavior of atoms and atomic particles.

Radial velocity (V_r) That component of an object's velocity directed away from or toward the observer.

Redshift A Doppler shift toward longer wavelengths caused by a velocity of recession.

CHAPTER REVIEW

CHAPTER 5 TOPIC SUMMARIES/KEY TERMS

5 Sun Light and Sun Atoms

5-5 The Sun's Atmosphere

- The solar atmosphere consists of two layers of hot, low-density gas above the photosphere: the **chromosphere** and **corona**. The chromosphere is just above the photosphere and is most easily visible during total solar eclipses when it flashes into view for a few seconds. The corona is the Sun's outermost atmospheric layer, composed of a very low-density, very hot gas extending many solar radii from the visible Sun. The chromosphere and corona can be imaged using a special instrument called a **coronagraph** that blocks light from the photosphere.
- **Kirchhoff's laws** summarize laboratory and astronomical observations that (1) a hot solid, liquid, or dense gas emits electromagnetic radiation at all wavelengths and produces a **continuous spectrum**; (2) a light source viewed through a low-density gas produces an **absorption (dark-line) spectrum** containing **absorption lines**, and (3) an excited low-density gas produces an **emission (bright-line) spectrum** containing **emission lines**.
- The hydrogen atom has a set of spectral lines at visible wavelengths called the **Balmer series** that are caused by quantum jumps (also known as **transitions**) starting or ending in the second orbital. The rose color of the chromosphere is caused by hydrogen Balmer emission lines. The **Lyman series** and **Paschen series** of hydrogen spectral lines are observed, respectively, at ultraviolet and infrared wavelengths.
- The presence of spectral lines of a certain element shows that this element must be present in the Sun, but the strength of spectral lines depends partly on the temperature of gas as well as on abundance. The mathematical techniques for deriving true abundances of elements from the solar (and stellar) spectra were worked out by Cecilia Payne in the 1920s.
- **Filtergrams** of the chromosphere reveal **spicules**, flamelike structures extending upward into the lower corona.
- The corona's high temperature—up to 2 million K—is understood to be maintained by energy transported via motions of the magnetic field extending up through the photosphere—the **magnetic carpet**—and by magnetic waves coming from below the photosphere. Parts of the corona give rise to the **solar wind**, a breeze of low-density ionized gas streaming away from the Sun.

Absorption line A dark line in a spectrum produced by the absence of photons absorbed by atoms or molecules.

Absorption spectrum (dark-line spectrum) A spectrum that contains absorption lines.

Balmer series A series of spectral lines produced by hydrogen in the near-ultraviolet and visible parts of the spectrum. The three longest-wavelength Balmer lines are visible to the human eye.

Chromosphere Bright gases just above the photosphere of the Sun.

Continuous spectrum A spectrum in which there are no absorption or emission lines.

Corona The faint outer atmosphere of the Sun, composed of low-density, high-temperature gas.

Coronagraph A telescope designed to capture images of faint objects such as the corona of the Sun that are near relatively bright objects.

Emission line A bright line in a spectrum caused by the emission of photons from atoms.

Emission spectrum (bright-line spectrum) A spectrum produced by photons emitted by an excited gas.

Filament A solar eruption, seen from above, silhouetted against the bright photosphere.

Filtergram A photograph (usually of the Sun) taken in the light of a specific region of the spectrum—for example, an Hα (hydrogen Balmer-alpha) filtergram.

Kirchhoff's laws A set of laws that describe the absorption and emission of light by matter.

Lyman series Spectral lines in the ultraviolet spectrum of hydrogen produced by transitions whose lowest energy level is the ground state.

Magnetic carpet The network of small magnetic loops that covers the solar surface.

Paschen series Spectral lines in the infrared spectrum of hydrogen produced by transitions whose lowest energy level is the third.

Solar wind Rapidly moving atoms and ions that escape from the solar corona and blow outward through the Solar System.

Spicule A small, flamelike projection in the chromosphere of the Sun.

Transition The movement of an electron from one atomic energy level to another.

5-6 Solar Activity

- Solar astronomers can study the motion, density, and temperature of gases inside the Sun by analyzing the way the solar surface oscillates. Known as **helioseismology**, this field of study requires large amounts of data and extensive computer analysis.
- Sunspots often have diameters larger than Earth. They appear for a month or so and then fade away, and the number of spots on the Sun varies with an 11-year cycle. Early in a sunspot cycle, spots appear farther from the Sun's equator, and later in the cycle they appear closer to the equator. This is shown in the **Maunder butterfly diagram**.
- Astronomers can use the **Zeeman effect** to measure magnetic fields on the Sun. The average sunspot contains magnetic fields a few thousand times stronger than Earth's field. This is part of the evidence that the sunspot cycle is produced by a solar magnetic cycle that is 22 years long.
- Sunspots are the visible components of **active regions** where the Sun's magnetic field is strong. Arches of magnetic field can produce sunspots where the field passes through the photosphere. The Sun's magnetic field is produced by the **dynamo effect** operating at the base of the **convection zone**.
- Alternate sunspot cycles have reversed **magnetic polarity**, which has been explained by the **Babcock model**, in which the Sun's **differential rotation** and convection currents tangle the magnetic field. The magnetic field's tangles arch through the photosphere and cause active regions visible to the eye as sunspot pairs. When the field becomes strongly tangled, it reorders itself into a simpler but reversed field, and the cycle starts over.
- The sunspot cycle does not repeat exactly each cycle, and the decades from 1645 to 1715, known as the **Maunder minimum**, seem to have been a time when solar activity was very low and Earth's climate was slightly colder.
- Arches of magnetic field are visible as **prominences** in the chromosphere and corona. Seen from above in filtergrams, prominences are visible as dark **filaments** silhouetted against the bright chromosphere.
- Magnetic field **reconnection events** can produce powerful **flares**, sudden eruptions of X-ray, ultraviolet, and visible radiation plus high-energy atomic particles. Flares are important because they can have dramatic effects on Earth, such as communications blackouts.
- The solar wind originates in regions on the solar surface called **coronal holes**, where the Sun's magnetic field leads out into space and does not loop back to the Sun. **Coronal mass ejections (CMEs)** occur when magnetic fields on the surface of the Sun eject bursts of ionized gas that flow outward in the solar wind. Such bursts can produce **auroras** and other phenomena if they strike Earth.

Active region A magnetic region on the solar surface that includes sunspots, prominences, flares, and the like.

Aurora The glowing light display that results when a planet's magnetic field guides charged particles toward the north and south magnetic poles, where they strike the upper atmosphere and excite atoms to emit photons.

Babcock model A model of the Sun's magnetic cycle in which the differential rotation of the Sun winds up and tangles the solar magnetic field. This is thought to be responsible for the sunspot cycle.

Convection zone A region inside a star where energy is carried outward as rising hot gas and sinking cool gas.

Coronal hole An area of the solar surface that is dark at X-ray wavelengths, thought to be associated with divergent magnetic fields and the source of the solar wind.

Coronal mass ejection (CME) Matter ejected from the Sun's corona in powerful surges guided by magnetic fields.

Differential rotation The rotation of a body in which different parts of the body have different periods of rotation. This is true of the Sun, the giant planets like Jupiter, and the disk of the galaxy.

Dynamo effect The process by which a rotating, convecting body of conducting matter, such as in Earth's core or in the Sun's convection zone, can generate a magnetic field.

Flare A violent eruption on the Sun's surface.

Helioseismology The study of the interior of the Sun by the analysis of its modes of vibration.

Magnetic polarity Orientation and strength of a magnetic field's north and south poles.

Maunder butterfly diagram A graph showing the latitude of sunspots versus time, first plotted by W. W. Maunder in 1904.

Maunder minimum A period between 1645 and 1715 of less numerous sunspots and other solar activity.

Prominence Eruption on the solar surface. Visible during total solar eclipses.

Reconnection event On the Sun, the merging of magnetic fields to release energy in the form of flares.

Zeeman effect The splitting of spectral lines into multiple components when the atoms are in a magnetic field.

CHAPTER 6 TOPIC SUMMARIES/KEY TERMS

6-1 A Travel Guide to the Terrestrial Planets

- Earth is the standard of **comparative planetology** in the study of the **Terrestrial planets** because you know it best and because it contains all of the phenomena found on the other Terrestrial planets.
- Our discussion of the Terrestrial planets considers Earth, the Moon, Mercury, Venus, and Mars. Earth's Moon is included because it is a complex world and makes a striking comparison with Earth. The **Jovian planets** are giant, low-density worlds in the outer Solar System.
- The Terrestrial worlds differ mainly in size, but they all have low-density crusts, **mantles** of dense rock, and metallic cores.
- The **heavy bombardment** early in the history of the Solar System cratered all solid surfaces. Once cratering declined, geological activity was driven by heat flowing out of a planet's interior. The nature of its atmosphere depends on the size of the planet and its temperature.
- Earth has passed through four stages: (1) **differentiation**, (2) cratering, (3) flooding by lava and water, and (4) slow surface evolution. The other Terrestrial planets and the Moon also passed through the same stages, which had different effects and durations depending on the specific properties of each body.

Comparative planetology Understanding planets by searching for and analyzing contrasts and similarities among them.

Differentiation Separation of planetary material within a planet into layers according to density.

Heavy bombardment The intense cratering that occurred during the first 0.7 billion years of the Solar System's history.

Jovian planet A Jupiter-like planet with a large diameter and low density.

Mantle The layer of dense rock and metal oxides that lies between the molten core and Earth's surface or a similar layer in another planet.

Terrestrial planet An Earth-like planet—small, dense, rocky, and metallic.

6-2 Earth: The Active Planet

- Earth is unique in the Solar System in that it has large amounts of liquid water on its surface, and that water drives strong erosion that alters the surface geology. Earth is also unique in that it is the only known home for life.
- Seismic waves generated by earthquakes can be detected by seismographs all over the world and can reveal Earth's internal structure.
- **Pressure *(P)* waves** can travel through a liquid, but **shear *(S)* waves** cannot. Observations show that *S* waves cannot pass through Earth's core and that is evidence that the core is liquid. Measurements of heat flowing outward from the interior, combined with mathematical models, reveal that the core is very hot and composed of iron and nickel.
- Although Earth's crust is brittle and breaks under stress, the mantle is plastic and can deform and flow under pressure.
- Earth is dominated by **plate tectonics**, which breaks the crust into moving sections. Plate tectonics is driven by heat flowing upward from the interior.
- Tectonic plates are made of low-density, brittle rock that floats on the hotter plastic upper layers of the mantle. **Rift valleys** can be produced where plates begin pulling away from each other.
- New crust is formed along **mid-ocean rises**, where molten rock solidifies to form **basalt.** Crust is destroyed when it sinks into the mantle along **subduction zones**. Volcanism and earthquakes are common along the edges of the plates.
- The motion of a plate across a hot spot can produce a chain of volcanic islands such as the Hawaiian Islands. Hot-spot volcanism is not related to subduction zones.
- The continents are drifting slowly on the plastic mantle, and their arrangement changes with time. Where they collide, they can form **folded mountain ranges**.
- Most geological features on Earth, such as mountain ranges and the Grand Canyon, have been formed recently. The first billion years of Earth's geology have been almost entirely erased by plate tectonics and erosion.
- Because Earth formed in a molten state, its **primary atmosphere** was probably mostly carbon dioxide, nitrogen, and water vapor.
- Most of the carbon dioxide eventually dissolved in seawater and was added to ocean sediments, and plant life has added oxygen to the atmosphere, producing the current **secondary atmosphere**.
- Ultraviolet photons can break up water molecules in a planet's atmosphere, but as soon as Earth had enough oxygen, an ozone layer formed high in Earth's atmosphere. The ozone absorbs ultraviolet photons and protects water molecules.
- The **greenhouse effect** can warm a planet if gases such as carbon dioxide in the atmosphere are transparent to light but opaque to infrared. The natural greenhouse effect warms Earth and makes it comfortable for life, but greenhouse gases added by industrial civilization are responsible for **global warming**.
- Measurement of carbon isotope ratios and carbon dioxide versus oxygen abundances make it clear that the CO_2 added to the atmosphere since 1800 is predominantly from burning of fossil fuels. Observations and model calculations have eliminated other candidate causes for the current warming such as natural climate cycles or variations in the Sun's output. The precise amount of future warming is difficult to predict.

Basalt Dark igneous rock, characteristic of solidified lava.

Folded mountain range A long range of mountains formed by the compression of a planet's crust.

Global warming The gradual increase in the surface temperature of Earth caused by human modifications to Earth's atmosphere.

Greenhouse effect The process by which an atmosphere traps heat and raises the temperature of a planetary surface.

Mid-ocean rise One of the undersea mountain ranges that push up from the seafloor in the center of the oceans.

***P* wave** A type of seismic wave involving compression and decompression of the material through which it passes.

Plate tectonics The constant destruction and renewal of Earth's surface by the motion of sections of crust.

Primary atmosphere A planet's first atmosphere.

Rift valley A long, straight, deep valley produced by the separation of crustal plates.

***S* wave** A type of seismic wave involving lateral motion of the material through which it passes.

Secondary atmosphere A planet's atmosphere that replaces the primary atmosphere, for example by outgassing, impact of volatile-bearing planetesimals, or biological activity.

Subduction zone A region of a planetary crust where a tectonic plate slides downward.

6 The Terrestrial Planets

6-3 The Moon

- Earth's Moon is a small world and so has lost most of its internal heat and is no longer geologically active. It contains little metal and has an overall low density. Because of its low escape velocity, the Moon has no significant atmosphere.
- The **albedo** of the Moon is very low, meaning it is a dark gray world.
- Much of the lunar surface is heavily cratered. Debris blasted out of craters is called **ejecta**. Large impacts have resulted in prominent **multiringed basins**. Constant bombardment by tiny **micrometeorites** continues to erode the Moon, producing a layer of dust on the surface.
- The Moon's highlands are the oldest portions of its surface, heavily cratered by impacts that occurred during the heavy bombardment of all the planets after the formation of the Solar System. There is evidence from crater counts and lunar rock sample ages that, before the bombardment ended, there was a temporary burst of impacts called the **late heavy bombardment**.
- The Moon's lowlands are filled by lava flows that formed smooth **maria** (singular, **mare**) soon after the end of the heavy bombardments.
- Rocks from the Moon are typical of solidified lava such as **anorthosite**, which shows that the Moon was once covered by a **magma ocean**. Many of the rocks are **breccias**, showing that the crust has been repeatedly pounded by impacts.
- The **large-impact hypothesis** suggests the Moon formed when an impact between the proto-Earth and a very large planetesimal left Earth surrounded with a disk of collision debris. The Moon formed from that disk.

Albedo The ratio of the amount of light reflected from an object to the amount of light received by the object. Albedo equals 0 for perfect absorption and 1 for perfect reflection.

Anorthosite Aluminum- and calcium-rich silicate rock found in the lunar highlands.

Breccia Rock composed of fragments of older rocks bonded together.

Ejecta Pulverized rock scattered by meteorite impacts on a planetary surface.

Large-impact hypothesis Hypothesis that the Moon formed from debris ejected during a collision between Earth and another planet-sized object.

Late heavy bombardment A sudden temporary increase in the Solar System's cratering rate that evidently occurred about 4 billion years ago.

Magma ocean The exterior of the newborn Moon, a shell of molten rock hundreds of kilometers deep.

Mare One of the lunar lowlands filled by successive flows of dark lava, from the Latin for "sea." Plural "maria."

Micrometeorite Meteorite of microscopic size.

Multiringed basin Large impact feature (crater) containing two or more concentric rims formed by fracturing of the planetary crust.

6-4 Mercury

- Mercury is smaller than Earth but larger than Earth's Moon. It has an old, heavily cratered surface and only a sporadic, extremely thin atmosphere.
- Mercury is very dense and has a large iron core. The reason for the large core is not well known but one idea suggests that a giant impact may have fractured and driven off some of the rocky surface. As the interior cooled, it shrank and the crust wrinkled to produce curving ridges.

6-5 Venus

- Although Venus is almost as large as Earth, it has a thick, cloudy atmosphere of carbon dioxide that hides the surface from sight. It can be studied by radar mapping.
- Venus's carbon dioxide atmosphere drives an intense greenhouse effect and makes that planet's surface hot enough to melt lead.
- Venus is slightly closer to the Sun than Earth is, too warm for liquid water oceans to dissolve carbon dioxide from the atmosphere easily, and warm enough to start evaporating its oceans, leading to a **runaway greenhouse effect**.
- Geologic activity on Venus is dominated not by plate tectonics but rather by volcanism and vertical tectonics, including **coronae**, large circular uplifted regions. Crater counts show that the entire surface of Venus has been covered over or replaced in the past half billion years.

Coronae On Venus, large round geological faults in the crust caused by the intrusion of magma below the crust.

Runaway greenhouse effect A greenhouse effect so dramatic that it amplifies itself, becoming stronger with time.

6-6 Mars

- Mars is about half the size of Earth. It has a thin atmosphere and has lost much, but not all, of its internal heat. Although its surface is marked by some large **shield volcanoes**, it is no longer very active geologically.
- The loss of atmospheric gases depends on the size of a planet and its temperature. Mars is cold, but it is small and has a low escape velocity, and thus many of its lighter gases have leaked away.
- Some water may have leaked away from Mars as ultraviolet radiation from the Sun broke it into hydrogen and oxygen, but some water is frozen in the polar caps and as **permafrost** in the soil.
- **Outflow channels** and **valley networks** visible from Mars orbiters seem to have been cut by sudden floods or by longer-term drainage. There are many other signs that there has been water on the Martian surface, some of which are geologically recent. When humans reach Mars, they will not need to dig far to find water in the form of ice.
- The Southern Hemisphere of Mars is old cratered terrain, but some large volcanoes lie in the north. The size of these volcanoes strongly suggests that the crust does not move horizontally.
- Robot rovers have found clear signs that the Martian climate was different in the past and that liquid water flowed over the surface in at least some places. The northern lowlands might even have held an ocean at one time.
- The two moons of Mars are probably captured asteroids. They are small, airless, and cratered. They lost their internal heat long ago.

Outflow channels Geological features on Mars and Earth caused by flows of vast amounts of water released suddenly.

Permafrost Permanently frozen soil.

Shield volcano Wide, low-profile volcanic cone produced by highly liquid lava.

Valley network A system of dry drainage channels on Mars that resembles the beds of rivers and tributary streams on Earth.

CHAPTER 7 TOPIC SUMMARIES/KEY TERMS

7-1 A Travel Guide to the Outer Solar System

- The Jovian planets—Jupiter, Saturn, Uranus, and Neptune—are large, massive low-density worlds located in the outer Solar System.

7-2 Jupiter

- Models of planetary interiors can be calculated based on each planet's density and **oblateness**, the fraction by which its equatorial diameter exceeds its polar diameter.
- Jupiter has a core of heavy elements surrounded by a deep mantle of **liquid metallic hydrogen** in which the planet's large and strong magnetic field is generated.
- Both Jupiter and Saturn are composed mostly of liquid hydrogen with liquid metallic hydrogen in deep regions close to their cores. Consequently, they are sometimes referred to as "liquid giants."
- The **magnetosphere** around Jupiter traps high-energy particles from the Sun to form intense radiation belts.
- Atmospheres of the Jovian planets are marked by **belt–zone circulation** that produces cloud belts parallel to their equators.
- Uranus and Neptune contain abundant water in solid form and are sometimes called "ice giants."
- All the Jovian planets have large systems of satellites and rings that have had complex histories.
- Jupiter is observed to have heat flowing out of it at a high rate, indicating that its interior is very hot.
- Jupiter's atmosphere contains three layers of clouds formed of hydrogen-rich molecules, including water and ammonia. The cloud layers are at altitudes corresponding to the condensation temperatures of their respective main constituents.
- The cloud stripes parallel to Jupiter's equator consist of light-colored zones that are high-pressure regions of rising gas plus darker belts that are lower-pressure areas of sinking gas.
- Spots in Jupiter's atmosphere, including the Great Red Spot, are circulating weather patterns.
- The four large **Galilean moons** show signs of geologic activity. Grooved terrain on Ganymede, smooth ice and cracks on Europa, and active volcanoes on Io show that that **tidal heating** driven by orbital resonances has made these moons active.
- Many of Jupiter's moons are small, rocky bodies that are probably captured objects. They are too small to retain heat and are not geologically active.
- Jupiter's rings are composed of dark particles that are bright when illuminated from behind, which is called **forward scattering**. This shows that the particles are very small and are probably dust from meteoroid impacts on moons.
- Jupiter's rings, like all of the rings in the Solar System, lie inside the planet's **Roche limit**, within which tidal stress can destroy a moon or prevent one from forming.

Belt–zone circulation The atmospheric circulation typical of Jovian planets in which dark belts and bright zones encircle the planet parallel to its equator.

Forward scattering The optical property of finely divided particles to preferentially direct light in the original direction of the light's travel.

Liquid metallic hydrogen A form of liquid hydrogen that is a good electrical conductor, inferred to exist in the interiors of Jupiter and Saturn.

Magnetosphere The volume of space around a planet within which the motion of charged particles is dominated by the planetary magnetic field rather than the solar wind.

Oblateness The flattening of a spherical body, usually caused by rotation.

Roche limit The minimum distance between a planet and a satellite that can hold itself together by its own gravity.

Tidal heating The heating of a planet or satellite because of friction caused by flexing due to the gravitational influence of a nearby body.

7-3 Saturn

- Saturn is less dense than water and contains a small core and less metallic hydrogen than Jupiter.
- Cloud layers on Saturn occur at the same temperature as those on Jupiter, but because Saturn is farther from the Sun and colder, the cloud layers are deeper in the hydrogen atmosphere below a layer of methane haze and so are less prominent.
- Saturn's moons are icy and mostly heavily cratered.
- Saturn's largest moon, Titan, has a cold, cloudy nitrogen atmosphere. It might have had methane falling as rain on its icy surface and forming lakes and rivers.
- Sunlight entering Titan's atmosphere can convert methane into complex carbon-rich molecules to form haze and particles that settle out to coat the surface with dark organic material.
- Enceladus has a light surface with some uncratered regions. Geysers of water and ice containing organic molecules vent from the region around its south pole and provide ice particles to the extensive low-density E ring.
- Saturn's rings are composed of icy particles ranging in size from boulders to dust. The composition and brightness of the ring particles vary from place to place in the rings.

(continues)

(continued)

- Grooves in the rings can be produced by orbital resonances, or waves that propagate through the rings, caused by moons near or within the rings.
- Narrow rings as well as sharp ring edges can be produced by **shepherd satellites**.
- The material observed now in the Jovian planets' rings cannot have lasted since the formation of the Solar System. The rings are understood to be replenished occasionally with material produced by meteoroids, asteroids, and comets colliding with moons.

Shepherd satellite A satellite that, by its gravitational field, confines particles to a planetary ring.

7-4 Uranus

- Uranus is much less massive than Jupiter, and its internal pressure cannot produce liquid hydrogen. Models indicate it has a heavy-element core and a mantle of solid or slushy ice and rocky material below a hydrogen-rich atmosphere. Little heat flows out of Uranus, so it cannot be very hot inside.
- Uranus's atmosphere is almost featureless at visual wavelengths with a pale blue color caused by traces of methane, which absorbs red light. Images at selected wavelengths can be enhanced to show traces of belt–zone circulation.
- Uranus rotates on its side, perhaps because of a major impact or tidal interactions with other planets during its early history.
- The larger moons of Uranus are icy and heavily cratered, with signs on some of past geological activity, including **ovoids** on Miranda.
- The rings of Uranus, discovered by stellar **occultations**, are narrow hoops confined by shepherd satellites. The particles appear to be ice with traces of methane darkened by the radiation belt.

Occultation The passage of a larger body in front of a smaller body.

Ovoids The large oval grooves found on Uranus's moon Miranda that indicate past geologic activity.

7-5 Neptune

- Neptune, like Uranus, is an ice giant with no liquid hydrogen. Unlike Uranus, Neptune does have heat flowing outward from its interior.
- The atmosphere of Neptune, marked by traces of belt–zone circulation, is rich in hydrogen and colored blue by traces of methane.
- Neptune's satellite system is odd in that distant Nereid follows an elliptical orbit and Triton orbits backward. These might be signs of catastrophic encounters with other objects early in the Solar System's history.
- Triton is icy with a thin atmosphere and frosty polar caps. Smooth areas suggest past geological activity, and dark smudges mark the location of active nitrogen geysers.
- Neptune's rings are made of icy particles in narrow hoops and contain arcs produced by the gravitational influence of one or more moons.

7-6 Pluto: Planet No More

- Pluto is a small world with five known moons, one of which, Charon, is quite large in relation to Pluto. The moons' orbital planes and Pluto's equator are highly inclined to Pluto's orbit around the Sun. Pluto's composition is mostly rock with a substantial amount of ice. The *New Horizons* spacecraft flew past Pluto in July 2015, radioing back impressive images.
- Pluto is a member of a family of Kuiper Belt Objects orbiting beyond Neptune. At least one of those objects, Eris, is a bit larger than Pluto. The International Astronomical Union (IAU), which decides definitions and criteria for naming celestial objects, redefined Pluto in 2006 as a **dwarf planet**.
- Some Kuiper Belt Objects called **plutinos** follow orbits like Pluto that have an orbital resonance with Neptune.
- Some model calculations suggest that Uranus and Neptune formed closer to the Sun and migrated outward, pushing millions of icy bodies into orbital resonances farther from the Sun to form the Kuiper Belt. Pluto might be one of those objects.

Dwarf planet A body that independently orbits the Sun or another star, is massive enough to pull itself into a spherical shape but not massive enough to clear out other bodies in and near its orbit; for example, Pluto, Eris, and Ceres.

Plutino One of the icy Kuiper Belt Objects that, like Pluto, is caught in a 3:2 orbital resonance with Neptune.

CHAPTER 8 TOPIC SUMMARIES/KEY TERMS

8-1 Solar System Overview

- The Solar System is disk shaped, with all the planets orbiting nearly in the same plane. The orbital revolution of all the planets, the rotation of most of the planets on their axes, and the revolution of most of their moons are all in the same direction, counterclockwise as seen from the north.

8-2 Space Debris: Asteroids, Comets, and Meteorites

- The planets are divided into two groups. The inner four planets are **Terrestrial planets**—small, rocky, dense, Earth-like worlds. The next four outward are Jupiter-like **Jovian planets** that are large and have low density.
- All four of the Jovian worlds have ring systems and large families of moons. The Terrestrial planets have no rings and few moons.
- Most of the **asteroids**, small, irregular, rocky bodies, are located between the orbits of Mars and Jupiter.
- Asteroids are irregular in shape and heavily cratered from collisions. Their surfaces are covered by gray, pulverized rock, and some asteroids have such low densities they must be fragmented rubble piles.
- **Carbonaceous chondrites** are volatile-rich, primitive meteorites.
- A **meteor shower** occurs when Earth moves through the swarm of small solid particles left along the orbit of a comet.
- Most asteroids lie in a belt between Mars and Jupiter. Asteroids that pass near Earth are called **near-Earth objects (NEOs)**.
- **Comets** are icy bodies that pass through the inner Solar System along long elliptical orbits. As the ices vaporize and release dust, effects of the solar wind plus radiation pressure cause the comet to develop a tail that points away from the Sun.
- A visible comet is produced by lump of ices and rock usually about 1 to 10 km in diameter, referred to as the comet nucleus. In long, elliptical orbits, the icy nucleus stays frozen until it nears the Sun. Then, some of the **volatile** ices vaporize and release dust and gas that is blown away to form a prominent head and tail.
- A comet's gas tail is ionized gas carried away by the solar wind. The dust tail is composed of solid debris released from the nucleus and blown outward by the pressure of sunlight. A comet's tail always points away from the Sun, no matter in which direction the comet is moving.
- The **Kuiper Belt** is composed of small, icy bodies called Kuiper Belt Objects (KBOs) that orbit the Sun beyond the orbit of Neptune.
- **Meteoroids** that fall into Earth's atmosphere are vaporized by friction and are visible as streaks of light called **meteors.** Larger and stronger meteoroids might survive to reach the ground, where they are called **meteorites.**

Asteroid Small, rocky world. Most asteroids orbit between Mars and Jupiter in the asteroid belt.

Carbonaceous chondrite Stony meteorite that contains small glassy spheres called chondrules and volatiles. These chondrites may be the least altered remains of the solar nebula still present in the Solar System.

Comet One of the small, icy bodies that orbits the Sun and produces tails of gas and dust when it approaches the Sun.

Jovian planet Jupiter-like planet with a large diameter and low density.

Kuiper Belt The collection of icy objects orbiting in a region from just beyond Neptune out to 50 AU or more from the Sun.

Meteor A small bit of matter heated by friction to incandescent vapor as it falls into Earth's atmosphere.

Meteor shower A display of meteors that appears to come from one point in the sky, understood to be cometary debris.

Meteorite A meteor that survives its passage through the atmosphere and strikes the ground.

Meteoroid A meteor in space before it enters Earth's atmosphere.

Terrestrial planet An Earth-like planet—small, dense, and rocky.

Volatile Easily vaporized.

Widmanstätten pattern (pronounced *VEED-mun-stay-tun*) Bands in iron meteorite sections from large metal crystals.

8-3 The Origin of the Solar System

- The age of a rocky body can be found by radioactive dating, based on the decay half-life of radioactive atoms. The oldest rocks from Earth, the Moon, and Mars are older than 4 billion years. The oldest objects in our Solar System are some meteorites that are 4.6 billion years old. This is taken to be the age of the Solar System.
- An **evolutionary theory** proposes common, gradual causes. A **catastrophic theory** proposes rare, sudden causes. The solar nebula theory is an evolutionary theory.
- The **solar nebula theory** proposes that the planets formed in a disk of gas and dust around the protostar that became the Sun. Hot disks of gas and dust have been detected around many protostars and are believed to be the kind of disk in which planets could form.
- Modern astronomy reveals that all the matter in the Universe, including our Solar System, was originally formed as hydrogen and helium in the big bang. Atoms heavier than helium were cooked up in nuclear reactions in later generations of stars. The Sun and planets evidently formed from a cloud of gas and dust in the interstellar medium.
- **Condensation** in the solar nebula converted some of the gas into solid bits of matter. According to the **condensation sequence**, the inner part of the solar nebula was so hot that only metals and rocky materials could form solid grains. The dense Terrestrial planets grew from those solid particles and did not include many ices or low-density gases. Comparing the **uncompressed densities** of the Terrestrial planets shows that the inner Terrestrial planets have higher densities than the outer ones because of the condensation sequence.
- The outer solar nebula, beyond the **ice line**, was cold enough for large amounts of ices as well as metals and rocky minerals to form solid particles. The Jovian planets grew rapidly and incorporated large amounts

(continues)

(continued)

of low-density ices and gases. Further evidence that the condensation sequence was important in the solar nebula is found in the high densities of the Terrestrial planets relative to the Jovian planets.

- As solid particles became larger they would no longer grow efficiently by condensation but could continue growing by **accretion** to form billions of **planetesimals**.

Accretion The sticking together of solid particles to produce a larger particle.

Condensation The growth of a particle by addition of material from surrounding gas, atom by atom.

Condensation sequence The sequence in which different materials condense from the solar nebula depending on temperature and distance from the Sun.

Half-life The time required for half of the radioactive atoms in a sample to decay.

Ice line Boundary in the forming Solar System beyond which water vapor could freeze to form ice.

Planetesimal One of the small bodies that formed from the solar nebula and eventually grew into a protoplanet. Asteroids and comets are remnants of the planetesimals.

Solar nebula theory The theory that the planets formed from a spinning disk of material around the forming Sun.

Uncompressed density The density a planet would have if its gravity did not compress it.

8-4

The Story of Planet Building

- Planets begin growing by accretion of solid material into **protoplanets**. Once a protoplanet approaches about 15 Earth masses, it can begin growing by **gravitational collapse** as it pulls in gas from the solar nebula.
- The Terrestrial planets might have formed slowly from the accretion of planetesimals of similar composition. Radioactive decay plus **heat of formation** then melted each planet's interior to cause differentiation into layers of differing density. In that scenario, Earth's early atmosphere was probably supplied by a combination of **outgassing** from Earth's interior and planetesimal impacts.
- Icy planetesimals were ejected by gravitational interactions with the Jovian planets to take up orbits in the **Oort cloud**, which is now the source of long-period comets.
- Disks of gas and dust around protostars might not last long enough to form Jovian planets by accretion and then by gravitational collapse. This is referred to as the **Jovian problem**. Some models suggest the Jovian planets could have formed more rapidly by direct collapse, skipping the condensation and accretion steps.
- In addition to evaporation by the light from hot nearby stars and the gravitational influence of passing stars, the solar nebula was eventually cleared away by radiation pressure, the solar wind, and the sweeping up or ejection of debris by the planets.
- All of the old surfaces in the Solar System were heavily cratered during the heavy bombardment by debris that filled the Solar System when it was young, left over from the formation of the planets.

Catastrophic theory An explanation of a phenomenon involving special, sudden, perhaps violent, events.

Evolutionary theory An explanation of a phenomenon involving slow, steady processes of the sort seen happening in the present day.

Gravitational collapse The process by which a forming body such as a planet gravitationally captures gas rapidly from the surrounding nebula.

Heat of formation In planetology, the heat released by infalling matter during the formation of a planetary body.

NEO (near Earth object) A small Solar System body (asteroid or comet) with an orbit near enough to Earth that it poses some threat of eventual collision.

Oort cloud The hypothetical source of comets, a swarm of icy bodies understood to lie in a spherical shell extending to 100,000 AU from the Sun.

Outgassing The release of gases from a planet's interior.

Protoplanet Massive object, destined to become a planet, resulting from the coalescence of planetesimals in the solar nebula.

8-5

Planets Around Other Stars

- Cold dust disks known as **debris disks** around main-sequence stars appear to be produced by dust released by collisions among comets, asteroids, and Kuiper Belt Objects. Such disks might be signs that planets have already formed in those systems.
- Planets orbiting other stars, called **extrasolar planets**, have been detected by the way they tug their stars about, creating small Doppler shifts in the stars' spectra. Planets have also been detected in transits as they cross in front of their star and partly block the star's light. A few planets have been detected when they orbited behind their star and their infrared radiation was cut off. A few more have been detected by gravitational microlensing.
- Nearly all extrasolar planets found so far are massive, Jovian worlds orbiting close to their parent stars, called **hot Jupiters**. Lower-mass Terrestrial planets are harder to detect but might be common. Astronomers speculate that hot Jupiters formed beyond the ice lines of their respective planetary systems and then spiraled inward to their current locations as they swept up protoplanetary disk material.

Debris disk A disk of dust found by infrared observations around some stars. The dust is debris from collisions among asteroids, comets, and Kuiper Belt Objects.

Extrasolar planet A planet orbiting a star other than the Sun.

Hot Jupiter A massive and presumably Jovian-composition planet that orbits close to its star and consequently has a high temperature.

CHAPTER 9 TOPIC SUMMARIES/KEY TERMS

9-1 Star Distances

- Your goal in this chapter was to characterize the stars by finding their luminosities, temperatures, diameters, and masses. Before you could begin, you needed to find their distances. Only by first knowing the distance to a star can you find its intrinsic properties such as luminosity and diameter.
- Astronomers can measure the distance to nearer stars by observing their **stellar parallaxes (*p*)**. The most distant stars are so far away that their parallaxes are immeasurably small. Space telescopes above Earth's atmosphere have measured the parallaxes of huge numbers of stars.
- Stellar distances are commonly expressed in **parsecs (pc)**. One parsec is 2.06 × 105 AU—the distance to an imaginary star whose parallax is 1 arc second. One parsec equals 3.26 light-years.

Parsec (pc) The distance to a hypothetical star whose parallax is 1 arc second.

Stellar parallax (*p*) The small apparent shift in position of a nearby star relative to distant background objects because of Earth's orbital motion.

9-2 Apparent Brightness, Intrinsic Brightness, and Luminosity

- The amount of light received from a star, the light *flux*, is related to its distance by the inverse square law. Once you know the distance to a star, you can use the **Magnitude-Distance Formula** (in the Math Reference Cards MR) to find its **intrinsic brightness** expressed as its **absolute visual magnitude** (M_v)—the apparent magnitude the star would have if it were 10 pc away.
- To find the energy output of a star, astronomers must correct for the light at wavelengths that are not visible to convert absolute visual magnitude into **luminosity (*L*)**, the total energy radiated by the star in 1 second.

Absolute visual magnitude (M_v) Intrinsic brightness of a star. The apparent visual magnitude the star would have if it were 10 pc away.

Intrinsic brightness A measure of the amount of light a star produces.

Luminosity (*L*) The total amount of energy a star radiates per second at all wavelengths.

9-3 Star Temperatures

- The strength of spectral lines depends on the temperature of the star. For example, in cool stars, the Balmer lines are weak because atoms are not excited out of the ground state. In hot stars, the Balmer lines are weak because atoms are excited to higher orbits or are ionized. Only at medium temperatures, around 10,000 K, are the Balmer lines strong.
- A star's temperature **spectral class** is determined by which absorption lines are visible in its spectrum. The **spectral sequence**, OBAF GKM, is important because it is a temperature sequence. By classifying a star, the astronomer learns the temperature of the star's surface.
- Long after the spectral sequence was created, astronomers found the **L dwarfs**, **T dwarfs**, and **Y dwarfs** with temperatures even cooler than the M stars. These are examples of **brown dwarfs**, objects smaller than stars but larger than planets.

Brown dwarf A very cool, low-luminosity star whose mass is not sufficient to ignite nuclear fusion.

L dwarf, T dwarf Spectral classes of brown dwarf stars with lower surface temperatures and luminosities than M dwarfs.

Spectral class A star's label in the temperature classification system based on the appearance of the star's spectrum.

Spectral sequence The arrangement of spectral classes (O, B, A, F, G, K, and M) ranging from hot to cool.

Y dwarf A substellar object with temperature below 500 K, having inferred properties intermediate between brown dwarfs and Jovian planets.

9-4 Star Sizes

- The **Hertzsprung–Russell (H–R) diagram** is a plot of luminosity versus surface temperature. It is an important diagram in astronomy because it sorts the stars into categories by size.
- Roughly 90 percent of normal stars, including the Sun, fall on the **main sequence**, with the more massive stars being hotter and more luminous. The **giants** and **supergiants**, however, are much larger and lie above the main sequence.
- **Red dwarfs** lie at the bottom end of the main sequence. Some of the **white dwarfs** are very hot stars, but they fall below the main sequence because they are so small.
- Observations of a few stars made with interferometers confirm the sizes implied by the H–R diagram.
- The large sizes of the giants and supergiants mean their atmospheres have low densities and their spectra have sharper spectral lines than the spectra of main-sequence stars. In fact, it is possible to assign

(continues)

9 The Family of Stars

(continued)

stars to **luminosity classes** by the widths of their spectral lines. Class V stars are main-sequence stars. Giant stars, class III, have sharper lines, and supergiants, class I, have extremely sharp spectral lines.

- Astronomers can use the locations of the luminosity classes in the H–R diagram to estimate the distances to stars in a technique called **spectroscopic parallax**.

Giant Large, cool, highly luminous star in the upper right of the H–R diagram, typically 10 to 100 times the diameter of the Sun.

Hertzsprung–Russell (H–R) diagram A plot of the intrinsic brightness versus the surface temperature of stars. It separates the effects of temperature and surface area on stellar luminosity.

Luminosity class A category of stars of similar luminosity, determined by the widths of lines in their spectra.

Main sequence The region of the H–R diagram running from upper left to lower right, which includes roughly 90 percent of all stars generating energy by nuclear fusion.

Red dwarf A faint, cool, low-mass, main-sequence star.

Spectroscopic parallax The method of determining a star's distance by comparing its apparent magnitude with its absolute magnitude as estimated from its spectrum.

Supergiant Exceptionally luminous star whose diameter is 100 to 1000 times that of the Sun.

White dwarf Dying star at the lower left of the H–R diagram that has collapsed to the size of Earth and is slowly cooling off.

9-5 Star Masses—Binary Stars

- The only direct way to find the masses of stars is by studying **binary stars**, in which two stars orbit their common center of mass. Astronomers find the masses of the stars by observing the period and sizes of their orbits. In a **visual binary system**, both stars are visible and the orbits can be mapped, but in a **spectroscopic binary system** the stars are so close together they look like a single point of light, and the orbits can't be observed directly. Because the inclination of the orbit can't be found, spectroscopic binaries yield only a lower limit to the masses of the stars.
- In an **eclipsing binary system**, the orbits are edge-on, and the stars cross in front of each other. The resulting brightness changes in the **light curve** can reveal the diameters of the stars as well as their masses.
- A survey in the neighborhood of the Sun shows that lower-main-sequence stars, less luminous than the Sun, are the most common type. Giants and supergiants are rare, but white dwarfs are fairly common, although they are faint and hard to find.

Binary stars Pairs of stars that orbit around their common center of mass.

Eclipsing binary system A binary star system in which the stars cross in front of each other as seen from Earth.

Light curve A graph of brightness versus time commonly used in analyzing variable stars and eclipsing binaries.

Spectroscopic binary system A star system in which the stars are too close together to be visible separately. We see a single point of light, and only by taking a spectrum can we determine that there are two stars.

Visual binary system A binary star system in which the two stars are separately visible in the telescope.

9-6 Typical Stars

- The **mass–luminosity relation** expresses the fact that the more massive a main-sequence star is, the more luminous it is. The most massive main-sequence stars are type O at the upper left of the H–R diagram, and the least massive are type M at the lower right of the diagram. Giants and supergiants do not follow a simple mass–luminosity relation, and neither do white dwarfs.
- Given the mass and diameter of a star, you can find its average density. Main-sequence stars have about the same density as the Sun, but giants and supergiants are very-low-density stars. Some are much thinner than air. The white dwarfs, lying below the main sequence, are tremendously dense.

Mass–luminosity relation The more massive a main-sequence star is, the more luminous it is.

10-1 Stellar Structure

- Four laws describe the insides of stars. The conservation of mass and the conservation of energy laws say that mass and energy cannot arise from nothing nor vanish into nothing. The law of **hydrostatic equilibrium** says that the weight pressing down on a layer of gas in a star must be balanced by the pressure in the gas in that layer.
- The law of energy transport says that energy must flow from hot regions to cool by *conduction, radiation, or convection*. The **opacity** of a gas is a measure of its resistance to the flow of radiation. Where the opacity is high, convection carries the energy.
- Astronomers can use these four laws to compute stellar models that describe the interior structure of stars. **Stellar models** can also be used to predict the future evolution of stars.

Hydrostatic equilibrium Stable balance of forces in a fluid (liquid or gas).

Opacity The resistance of matter to the passage of radiation.

Stellar model A table of numbers representing the conditions in various layers within a star.

10-2 Nuclear Fusion in the Sun and Stars

- Stars generate energy by extracting it from the **nuclear forces** that bind atomic nuclei together. **Nuclear fission** breaks heavy nuclei into lower mass nuclei, while **nuclear fusion** combines lower mass nuclei to form heavy nuclei.
- Because atomic nuclei carry positive charges, they repel each other with a force called the **Coulomb barrier**. To overcome that barrier and combine, they must collide violently, and that requires very high temperatures. The Sun begins making its energy by combining two protons to form a **deuterium**, which is then converted to helium. This **proton–proton chain** ultimately combines four hydrogen nuclei to make one helium nucleus. The **CNO (carbon-nitrogen-oxygen) cycle** operates in more massive stars and does the same thing but uses carbon as a catalyst.
- **Neutrinos** carry a small part of the energy out of the Sun's core, and the detection of these neutrinos confirms the theoretical description of nuclear fusion in stars.
- The relationship between pressure and temperature, the pressure–temperature thermostat, ensures that the star generates just enough energy to support itself. This explains the mass–luminosity relation.

CNO (carbon-nitrogen-oxygen) cycle A series of nuclear reactions that use carbon as a catalyst to combine four hydrogen nuclei to make one helium nucleus plus energy, effective in stars more massive than the Sun.

Coulomb barrier The electrostatic force of repulsion between bodies of like charge, commonly applied to atomic particles.

Deuterium An isotope of hydrogen in which the nucleus contains a proton and a neutron.

Neutrino A neutral, nearly massless atomic particle that travels at or nearly at the speed of light.

Nuclear fission Reactions that break the nuclei of atoms into fragments.

Nuclear forces The two forces of nature that only affect the particles in the nuclei of atoms.

Nuclear fusion Reactions that join the nuclei of atoms to form more massive nuclei.

Proton–proton chain A series of three nuclear reactions that builds a helium atom by adding together protons. The main energy source in the Sun.

10-3 Main-Sequence Stars

- The main sequence has a lower end because stars less massive than 0.08 solar mass cannot get hot enough to begin hydrogen fusion. Such objects become **brown dwarfs**.
- When a star first begins fusing hydrogen into helium, it is located on the H–R diagram along the **zero-age main sequence (ZAMS)**. As hydrogen nuclei are converted to helium nuclei, the total number of nuclei in the star's core declines. As a result, the core slowly contracts and the outer layers gradually expand, making the star gradually move upward and to the right across the band of the main sequence.
- How long a star can remain on the main sequence depends on its mass. The more massive a star is, the faster it uses up its hydrogen fuel. A 25-solar-mass star will exhaust its hydrogen and die in only about 7 million years, but the Sun is expected to last for 10 billion years.

Brown dwarf A stellar object with such low mass that it cannot raise its central temperature high enough to sustain hydrogen fusion.

Zero-age main sequence (ZAMS) The location in the H–R diagram where stars first reach stability as hydrogen-burning stars.

10-4 The Birth of Stars

- The **interstellar medium (ISM)**, the gas and **interstellar dust** between the stars, is spread in irregular clouds separated by lower density gas and dust.
- A **nebula** is a cloud of gas in space, and an **H II region**, also known as an **emission nebula**, is produced when ultraviolet radiation from hot stars ionizes nearby gas, making it glow like a giant neon sign. The red, blue, and violet Balmer lines blend together to produce the characteristic pink-red color of ionized hydrogen.
- A **reflection nebula** is produced by gas and dust illuminated by a star that is not hot enough to ionize the gas. Rather, the dust scatters the starlight to produce a reflection of the stellar absorption spectrum. Because shorter-wavelength photons scatter more easily than longer-wavelength photons, reflection nebulae look blue. The daytime sky looks blue for the same reason.
- A **dark nebula** is a cloud of gas and dust that is noticeable because it blocks the light of distant stars. The irregular shapes of these dark nebulae reveal the turbulence in the interstellar medium.
- **Interstellar reddening** makes distant stars appear too red because dust particles in the interstellar medium scatter blue light more easily than red light. This shows that the scattering dust particles are very small. Further evidence of an ISM comes from very narrow absorption lines in the spectra of distant stars. The lines are produced by the cold gas in the interstellar medium. The ISM can be detected at infrared, radio, and X-ray wavelengths.
- Large dense clouds of gas and dust are called **molecular clouds** because they are so dense that molecules can form inside them. They are the sites of star formation. Radio, infrared, and X-ray telescopes have detected emission from these molecular clouds. Such clouds can begin contracting to form stars if they are compressed by **shock waves**.
- A contracting molecular cloud can form an entire **star cluster** or a **stellar association** of stars.
- A collapsing fragment of a cloud can become a **protostar**. They form deep inside gas and dust and are best detected in the infrared. Small dark nebula called **Bok globules** seem to be clouds in the very first stages of star formation. The detection of **bipolar flows** and **Herbig–Haro** objects is further evidence of star formation.
- **T Tauri stars** are very young objects and are found throughout the region of the Orion Nebula, a nebula well known to contain many protostars just forming.

Bipolar flow Jets flowing in opposite directions away from a central object.

Bok globule Small, dark cloud only about 1 ly in diameter that contains 10 to 1000 solar masses of gas and dust, thought to be related to star formation.

Dark nebula A cloud of gas and dust seen silhouetted against a brighter nebula.

Emission nebula A cloud of glowing gas excited by ultraviolet radiation from hot stars.

H II region A region of ionized hydrogen around a hot star.

Herbig–Haro object A small nebula that varies irregularly in brightness, evidently associated with star formation.

Interstellar dust Microscopic solid grains in the interstellar medium.

Interstellar medium (ISM) The gas and dust distributed between the stars.

Interstellar reddening The process in which dust scatters blue light out of starlight and makes the stars look redder.

Molecular cloud A dense interstellar gas cloud in which atoms are able to link together to form molecules such as H_2 and CO.

Nebula A relatively dense cloud of interstellar gas and dust.

Protostar A collapsing cloud of gas and dust destined to become a star.

Reflection nebula A nebula produced by starlight reflecting off dust particles in the interstellar medium.

Shock wave A sudden change in pressure that travels as an intense sound wave.

Star cluster A group of stars that formed together and orbit a common center of mass.

Stellar association A group of stars that formed together but are not gravitationally bound to one another.

T Tauri star A young medium-mass star surrounded by gas and dust, understood to be contracting toward the main sequence.

CHAPTER REVIEW

CHAPTER 11 TOPIC SUMMARIES/KEY TERMS

Guidepost

- A **nova** appears to be a new star that becomes visible and then fades after a few weeks. **Supernovae** are more luminous and last longer. Both are associated with the deaths of stars.

Nova From the Latin, meaning "new," a sudden and temporary brightening of a star making it appear as a new star in the sky, evidently caused by an explosion of nuclear fuel on the surface of a white dwarf stellar remnant.

Supernova A "new star" in the sky that is up to 50,000 times more luminous than a normal nova and longer lasting, evidently the result of an explosion destroying a massive star.

11-1 Giant Stars

- Main-sequence stars such as the Sun can expand and become **giant stars** when they use all of the hydrogen fuel in their cores. As that happens, the core contracts and heats up. Hydrogen fusion begins in a spherical layer around the core—a hydrogen-fusion shell.
- Energy from the hydrogen-fusion shell swells the star into a cool giant 10 to 100 times larger in diameter than the Sun. Massive stars swell to become **supergiant stars** up to 1000 times larger in diameter than the Sun.
- The contraction of the star's core eventually ignites helium fusion, first in the core and later in a shell. If the star is massive enough, it can eventually fuse carbon and other elements.
- As the giant star fuses helium in its core and hydrogen in a shell, it moves toward the hot side of the H–R diagram. As soon as it exhausts the helium in its core and begins fusing helium in a shell, it moves back toward the cool side producing a loop in its evolutionary path.
- In old clusters, stars fusing helium follow a loop in the H–R diagram that is visible in the diagrams of star clusters as the **horizontal branch**.

Giant star Large, cool, highly luminous star in the upper right of the H–R diagram, typically 10 to 100 times the diameter of the Sun.

Horizontal branch The location in the H–R diagram of giant stars fusing helium in their cores.

Supergiant star Exceptionally luminous star whose diameter is 100 to 1000 times that of the Sun.

11-2 Deaths of Low-Mass Stars

- Red dwarfs less massive than about 0.4 solar mass are completely mixed and will have very little hydrogen left when they die. They can't ignite a hydrogen-fusion shell, so they can't become giant stars. They will remain on the main sequence for many times the current age of the Universe.
- Stars more massive than about 8 solar masses can't lose mass fast enough to reduce their mass low enough to die by ejecting a **planetary nebula** and collapsing into a white dwarf. Such massive stars must die more violent deaths.
- Medium-mass stars such as the Sun become cool giants and fuse helium but can't fuse carbon. They eventually blow away their outer layers and collapse into hot white dwarfs. Ultraviolet radiation from the white dwarfs ionize the gas to produce planetary nebulae.
- You can see evidence of stellar evolution in the H–R diagrams of star clusters. Stars in both **open clusters** and in **globular clusters** evolve in similar ways. The stars begin their evolution at the same time but evolve at different rates, depending on their masses. The most massive stars leave the main sequence first and are followed later by progressively less massive stars. This makes the evolution of stars visible in the H–R diagram.
- You can estimate the age of a star cluster from the **turnoff point** in its H–R diagram.
- In **degenerate matter**, the density is so high that quantum mechanical effects prevent electrons from changing their energies. Such matter is very difficult to compress, and its pressure does not depend on its temperature.
- After a star ceases generating energy by nuclear reactions, it becomes one of three types of **compact objects**, depending on the mass of the star.
- The **Chandrasekhar limit** shows that no white dwarf more massive than 1.4 solar masses can be stable. Presumably more massive stars can become white dwarfs only if they shed mass.

Chandrasekhar limit The maximum mass of a white dwarf, about 1.4 solar masses. A white dwarf of greater mass cannot support itself and will collapse.

Compact object One of the three final states of stellar evolution, which generates no nuclear energy and is much smaller and denser than a normal star.

Degenerate matter Extremely high-density matter in which, because of quantum mechanical effects, pressure does not depend on temperature.

Globular cluster A star cluster containing 100,000 to 1 million stars in a sphere about 75 ly in diameter, generally old, metal-poor, and found in the spherical component of the galaxy.

Open cluster A cluster of 100 to 1000 stars with an open, transparent appearance, usually relatively young and located in the disk of the galaxy. The stars are not tightly grouped.

Planetary nebula An expanding shell of gas ejected from a medium-mass star during the latter stages of its evolution.

Turnoff point The point in an H–R diagram at which a cluster's stars turn off from the main sequence and move toward the red-giant region, revealing the approximate age of the cluster.

11-3 The Evolution of Binary Systems

- Close binary stars evolve in complex ways because they can transfer mass from one star to the other. The two stars are enclosed by a dumbbell region known as the **Roche lobes.**
- As a star becomes a giant, it can expand and fill its Roche lobe, spilling mass to the other star. Mass transfer explains why some binary systems contain a main-sequence star more massive than its giant companion.
- Mass that is transferred from one star to the other must conserve **angular momentum** and can form a whirling **accretion disk** around the receiving star. Accretion disks can become hot enough to emit light and even X-rays.
- Mass transferred onto the surface of a white dwarf can build up a layer of fuel that erupts in a nova explosion. A white dwarf can erupt repeatedly so long as mass transfer continues to form new layers of fuel.

Accretion disk The rotating disk that forms in some situations as matter is drawn gravitationally toward a central body.

Angular momentum A measure of the tendency of a rotating or revolving body to continue that motion. Mathematically, the product of mass, velocity, and radius.

Roche lobes The two volumes of space gravitationally controlled by the stars in a binary system.

11-4 The Deaths of Massive Stars

- The massive stars on the upper main sequence fuse nuclear fuels one after the other producing a layering of fusion shells, but such stars can't fuse iron because iron is the most tightly bound of all atomic nuclei. When an aging massive star forms an iron core, the core collapses and triggers a supernova explosion known as a **Type II supernova**.
- The spectra of Type II supernovae contain hydrogen lines, but the spectra of **Type I supernovae** do not. At least two causes of Type I supernovae are known.

(*continues*)

11 The Deaths of Stars

(continued)

- A Type Ia supernova can occur when mass transferred onto a white dwarf pushes it over the Chandrasekhar limit and it collapses suddenly, fusing all of its carbon at once. A Type Ib supernova occurs when a massive star in a binary system loses its outer layers of hydrogen before it explodes.
- A supernova expels an expanding shell of gas called a **supernova remnant**. The supernova seen in the year 1054 produced a supernova remnant known as the Crab Nebula, which emits **synchrotron radiation**, evidence of a powerful energy source remaining inside the remnant.

Supernova remnant The expanding shell of gas and dust marking the site of a supernova explosion.

Synchrotron radiation Radiation emitted when high-speed electrons move through a magnetic field.

Type I supernova A supernova explosion caused by the collapse of a white dwarf.

Type II supernova A supernova explosion caused by the collapse of the core of a massive star.

11-5 Neutron Stars

- The supernova 1987A is only a few years old, but its expanding gases will eventually form a supernova remnant. Neutrinos observed coming from the direction of the supernova are evidence that the core collapsed and formed a neutron star.
- When a supernova explodes, the core collapses to very small size. Theory predicts that the collapsing core cannot support itself as a white dwarf if its mass is greater than 1.4 solar masses, the Chandrasekhar limit. If its mass lies between 1.4 solar masses and about 3 solar masses, it can halt its contraction and form **a neutron star**.
- A neutron star is supported by the pressure of the degenerate neutrons. Theory predicts that a neutron star should be about 10 km in radius, spin very fast because it conserves angular momentum as it contracts, and have a powerful magnetic field.
- **Pulsars**, rapidly pulsing radio sources, were discovered in 1967. The **lighthouse model** explains pulsars as spinning neutron stars that emit beams of radiation from their magnetic poles. As they spin, they sweep the beams around the sky like lighthouses; if the beams sweep over Earth, astronomers detect pulses. The short pulses and the discovery of a pulsar in the supernova remnant called the Crab Nebula were key evidence that pulsars are neutron stars.
- Theory predicts that a neutron star cannot have a mass greater than about 3 solar masses. Dozens of pulsars have been found in binary systems, and those objects allow astronomers to estimate the masses of the neutron stars. Such masses are consistent with the predicted masses of neutron stars.
- Observations of the first binary containing two neutron stars revealed that the system is losing orbital energy by emitting **gravitational radiation**, as predicted by Einstein's **general theory of relativity**.
- The fastest pulsars, the **millisecond pulsars**, appear to be old pulsars that have been spun up to high speed by mass flowing from binary companions.
- Planets have been found orbiting at least one neutron star. They might be the remains of a companion star that was mostly devoured by the neutron star, or they might have formed from a ring of gas and dust left orbiting the neutron star after the supernova explosion.

General theory of relativity Einstein's theory that describes gravity as a result of the curvature of space-time.

Gravitational radiation Expanding waves in a gravitational field that transport energy through space at the speed of light, as predicted by general relativity.

Lighthouse model The explanation of a pulsar as a spinning neutron star sweeping beams of electromagnetic radiation around the sky.

Millisecond pulsar A pulsar with a pulse period of only a few milliseconds.

Neutron star A small, highly dense star, with radius about 10 km, composed almost entirely of tightly packed neutrons.

Pulsar A source of short, precisely timed radio bursts, understood to be spinning neutron stars

11-6 Black Holes

- If the collapsing core of a supernova has a mass greater than 3 solar masses, then it must contract to a very small size—perhaps to a **singularity**, an object of zero radius. Near such an object, gravity is so strong that not even light can escape, and the region is called a **black hole**.
- The outer boundary of a black hole is the **event horizon**; no event inside is detectable. The radius of the event horizon is the **Schwarzschild radius,** R_S, amounting to only a few kilometers for a black hole of stellar mass.
- If you were to leap into a black hole, your friends who stayed behind would see two relativistic effects. They would see your clock slow relative to their own clock because of **time dilation**. Also, they would see your light redshifted to longer wavelengths because of the **gravitational redshift**. You would not notice these effects, but you would feel powerful tidal forces that would deform and heat your mass until you grew hot enough to emit X-rays. Any X-rays you emitted before reaching the event horizon could escape.
- To search for black holes, astronomers must look for binary star systems in which mass flows into a compact object and emits X-rays. If the mass of the compact object is clearly greater than about 3 solar masses, then the object is presumably a black hole. A number of such objects have been located.
- Clues that an X-ray source is caused by accretion into a black hole rather than onto a neutron star are lack of sustained regular pulsations and lack of flares when blobs of material in the accretion disk arrive at the central compact object.
- Black holes and neutron stars at the center of accretion disks can eject powerful jets of gas and radiation. Such jets have been detected.
- **Gamma-ray bursts** appear to be related to violent events involving neutron stars or black holes. Bursts longer than 2 seconds appear to arise during **hypernovae**, the collapse of massive stars to form black holes.

Black hole A mass that has collapsed to such a small volume that its gravity prevents the escape of all radiation. Also, the volume of space from which radiation cannot escape.

Event horizon The boundary of the region of a black hole from which no radiation can escape. No event that occurs within the event horizon is visible to a distant observer.

Gamma-ray burst A sudden, powerful burst of gamma-rays.

Gravitational redshift The lengthening of the wavelength of a photon as it escapes from a gravitational field.

Hypernova Produced when a very massive star collapses into a black hole. Thought to be a possible source of gamma-ray bursts.

Schwarzschild radius, R_S The radius of the event horizon around a black hole.

Singularity An object of zero radius and infinite density.

Time dilation The slowing of moving clocks or clocks in strong gravitational fields.

12-1 Discovery of the Galaxy

- The hazy band of the Milky Way is our wheel-shaped galaxy seen from within, but its size and shape are not obvious. William and Caroline Herschel counted stars at many locations over the sky to show that our star system seemed to be shaped like a grindstone with the Sun near the center.
- Later astronomers studied the distributions of stars, but because they didn't know that interstellar dust blocked their view of distant stars, they incorrectly concluded the star system is only about 10 **kiloparsecs (kpc)** in diameter with the Sun at the center.
- In the early 20th century, Harlow Shapley observed **Cepheid variable stars** and **other types of variables** to determine the size of the galaxy. Cepheids fall within an **instability strip** on the H–R diagram. The **period–luminosity relation** for Cepheid variables originally discovered by Henrietta Leavitt was **calibrated** by Shapley using the **proper motions** of a few Cepheids. That allowed him to estimate the distance to globular clusters, demonstrating that our Galaxy is much larger than the portion that can easily be observed from Earth and that the Sun is not at the center.
- Modern observations suggest that our Galaxy contains a **disk component** about 80,000 ly in diameter and that the Sun is two-thirds of the way from the center to the visible edge. The **central bulge** around the center and an extensive **halo** containing old stars and little gas and dust make up the **spherical component**.
- The mass of the galaxy can be found from its **rotation curve**. Kepler's third law reveals that the galaxy contains at least 110 billion solar masses. If stars orbited in Keplerian motion more distant stars would orbit more slowly. They do not, and that shows that the halo might contain much more mass than is visible. Because much of the mass in the halo is not emitting or absorbing electromagnetic radiation, astronomers call it **dark matter**.

Calibrate To make observations of reference objects, checks on instrument performance, calculations of unit conversions, and so on, needed to completely understand measurements of unknown quantities.

Central bulge The spherical cloud of stars that lies at the center of spiral galaxies.

Cepheid variable stars Variable stars with pulsation periods of 1 to 60 days for which period of variation is related to luminosity.

Dark matter Nonluminous matter that is detected only by its gravitational influence.

Disk component Material confined to the plane of the galaxy.

Halo The spherical region of a spiral galaxy, containing a thin scattering of stars, star clusters, and small amounts of gas.

Instability strip The region of the H–R diagram in which stars are unstable to pulsation. A star evolving through this strip becomes a variable star.

Kiloparsec (kpc) A unit of distance equal to 1000 pc or about 3300 ly.

Period–luminosity relation The relation between period of pulsation and intrinsic brightness among Cepheid variable stars.

Proper motion The rate at which a star moves across the sky, measured in arc seconds per year.

Rotation curve A graph of orbital velocity versus radius in the disk of a galaxy.

Spherical component The part of the galaxy including all matter in a spherical distribution around the center (the halo and central bulge).

Spiral arms Long spiral pattern of bright stars, star clusters, gas, and dust. Spiral arms extend from the center to the edge of the disk of spiral galaxies.

12-2 Spiral Arms and Star Formation

- You can trace the spiral arms through the Sun's neighborhood by using **spiral tracers** such as OB associations, but to extend the map over the entire galaxy, astronomers must use radio, infrared, and X-ray telescopes to see through the interstellar medium.
- The most massive stars live such short lives that they don't have time to move from their place of birth. Because they are found scattered along the spiral arms, astronomers conclude that the spiral arms are sites of star formation.
- The **spiral density wave theory** suggests that the spiral arms are regions of compression that move around the disk. When an orbiting gas cloud overtakes the compression wave, the gas cloud is compressed and forms stars. A density wave produces a galaxy with a prominent two-armed spiral pattern.
- Another process, **self-sustaining star formation,** might act to modify the arms with branches and spurs as the birth of massive stars triggers the formation of more stars by compressing neighboring gas clouds. This might account for the "wooly" or fragmentary appearance of the spiral pattern in some galaxies.

Self-sustaining star formation The process by which the birth of stars compresses the surrounding gas clouds and triggers the formation of more stars.

Spiral density wave theory Theory proposed to account for spiral arms as compressions of the interstellar medium in the disk of the galaxy.

Spiral tracer Object used to map the spiral arms—for example, O and B associations, open clusters, clouds of ionized hydrogen, and some types of variable stars.

12 The Milky Way Galaxy

12-3 Origin and History of the Galaxy

- The oldest star clusters reveal that the disk of our Galaxy is younger than the halo, and the oldest globular clusters appear to be about 13 billion years old. So our Galaxy must have formed about 13 billion years ago.
- Stellar populations are an important clue to the formation of our Galaxy. The first stars to form, termed **Population II stars**, were poor in elements heavier than helium—elements that astronomers call **metals.** As generations of stars manufactured metals in a process called nucleosynthesis and spread them back into the interstellar medium, the metal abundance of more recent generations increased. **Population I stars**, including the Sun, are richer in metals.
- Because the halo is made up of Population II stars and the disk is made up of Population I stars, astronomers conclude that the halo formed first and the disk later.
- The **monolithic collapse (top-down) hypothesis**—that the galaxy formed from a single, roughly spherical cloud of gas and gradually flatted into a disk—does not match all of the current evidence. A newer bottom-up hypothesis includes mergers with smaller galaxies plus infalling gas contributing to the disk.

Metal In astronomical usage, all atoms heavier than helium.

Monolithic collapse (top-down) hypothesis An early hypothesis that the galaxy formed from the collapse of a single large cloud of turbulent gas more than 13 billion years ago.

Population I star Stars with significant amounts of atoms heavier than helium; relatively young stars nearly always found in the galactic disk. The Sun is an intermediate Population I star.

Population II star Stars poor in atoms heavier than helium; relatively old stars nearly always found in the halo, globular clusters, or the central bulge.

12-4 The Nucleus of the Galaxy

- The nucleus of the galaxy is invisible at visual wavelengths, but radio, infrared, and X-ray radiation can penetrate the gas and dust. These wavelengths reveal crowded central stars and warmed dust.
- The very center of the Milky Way Galaxy is marked by a radio source, **Sagittarius A* (Sgr A*)**. That object must be less than 1 AU in diameter, but the motions of stars around the center show that it must contain at least 4 million solar masses. A supermassive black hole is the only object that could contain so much mass in such a small space.

Sagittarius A* (pronounced "Sagittarius A-star") The powerful radio source located at the core of the Milky Way Galaxy.

CHAPTER 13 TOPIC SUMMARIES/KEY TERMS

13-1 The Family of Galaxies

- Astronomers divide galaxies into three classes—**elliptical, spiral,** and **irregular**—with subclasses specifying the galaxy's shape.
- Elliptical galaxies contain little gas and dust and cannot make new stars, but spiral galaxies contain more gas and dust and make new stars, especially in the spiral arms.
- About two-thirds of spirals, including our Milky Way Galaxy, are **barred spiral galaxies.** Irregular galaxies have no obvious shape but contain gas and dust and young stars.

13-2 Measuring the Properties of Galaxies

- Galaxies are so distant astronomers measure their distances in **megaparsecs (Mpc)**—millions of parsecs.
- Astronomers find the distance to galaxies using **standard candles**, objects of known luminosity such as Cepheid variable stars, the most accurate distance indicators. By calibrating other distance indicators, astronomers have built a **distance ladder**.
- When astronomers look at a distant galaxy, they see it as it was when it emitted the light now reaching Earth. The **look-back time** to distant galaxies can be a significant fraction of the age of the Universe.
- According to the **Hubble law**, the apparent velocity of recession of a galaxy equals its distance times the **Hubble constant.** Astronomers can estimate the distance to a galaxy by observing its redshift, calculating its apparent velocity of recession, and then dividing by the Hubble constant.
- Once the distance to a galaxy is known, its diameter can be found from the **small-angle formula** and its luminosity from the **magnitude-distance formula**. See the Math Reference Cards MR.
- The rotation curve of a galaxy shows the orbital motion of its stars, and astronomers can use the **rotation curve method** to find the galaxy's mass.
- Some dwarf galaxies are only a few percent the size and luminosity of our Galaxy, but some giant elliptical galaxies are five times larger than the Milky Way Galaxy.
- High orbital velocities near the centers of galaxies suggest the presence of supermassive black holes.
- The rotation rates of galaxies reveal they contain large amounts of dark matter, which must be composed of some as-yet undiscovered form of matter.

Barred spiral galaxy A spiral galaxy with an elongated nucleus resembling a bar from which the arms originate.

Distance ladder The calibration used to build a distance scale reaching from the size of Earth to the most distant visible galaxies.

Elliptical galaxy A galaxy that is round or elliptical in outline and contains little gas and dust, no disk or spiral arms, and few hot, bright stars.

Hubble constant A measure of the rate of expansion of the Universe, the average value of the apparent velocity of recession divided by distance, about 73 km/s/Mpc.

Hubble law The linear relation between the distances to galaxies and the apparent velocity of recession.

Irregular galaxy A galaxy with a chaotic appearance, large clouds of gas and dust, and both population I and II stars, but without spiral arms.

Look-back time The amount by which you look into the past when you look at a distant galaxy, a time equal to the distance to the galaxy in light-years.

Megaparsec (Mpc) A unit of distance equal to 1,000,000 pc.

Rotation curve method A method of determining a galaxy's mass by measuring the orbital velocity and orbital radius of stars in the galaxy.

Spiral galaxy A galaxy with an obvious disk component containing gas; dust; hot, bright stars; and spiral arms.

Standard candles Objects of known brightness that astronomers use to find distance—for example, Cepheid variable stars and supernovae.

13-3 The Evolution of Galaxies

- **Rich clusters** of galaxies contain thousands of galaxies with fewer spirals and more ellipticals. **Poor clusters** of galaxies contain few galaxies with a larger proportion of spirals. This is evidence that galaxies evolve by collisions and mergers. Our Galaxy is located in a poor cluster known as the **Local Group**.
- When galaxies collide, tides twist and distort their shapes and can produce **tidal tails**. Large galaxies can absorb smaller Galaxies in what is called **galactic cannibalism**. Shells of stars, counterrotating parts of galaxies, streams of stars in the halos of galaxies, and multiple nuclei are evidence that galaxies can merge.
- **Ring galaxies** are produced by high-speed collisions in which a small galaxy plunges through a larger galaxy perpendicular to its disk.
- The compression of gas clouds can trigger rapid star formation, producing **starburst galaxies**.
- The merger of two larger galaxies can scramble star orbits and drive bursts of star formation to use up gas and dust. Most large ellipticals have evidently been produced by past mergers. Spiral galaxies have thin, delicate disks and appear not to have suffered mergers with large galaxies.
- A galaxy moving through the gas in a cluster of galaxies can be stripped of its own gas and dust and may become an S0 galaxy.

(continued)

Local Group The small cluster of a few dozen galaxies that contains our Milky Way Galaxy.

Poor galaxy cluster An irregularly shaped cluster that contains fewer than 1000 galaxies, many of which are spiral, and no giant ellipticals.

Rich galaxy cluster A cluster containing a thousand or more galaxies, usually mostly ellipticals, scattered over a volume only a few Mpc in diameter.

Starburst galaxy A galaxy undergoing a rapid burst of star formation.

13-4 Active Galaxies and Quasars

- First called **radio galaxies**, **active galaxies** are now known to emit energy at many wavelengths. Because the energy comes from their cores, they are known also as **active galactic nuclei (AGN)** .
- **Seyfert galaxies** are spiral galaxies with small, highly luminous cores and spectra that show the nuclei contain highly excited gas.
- **Double-lobed radio galaxies** emit radio energy from areas on either side of the galaxies. These lobes of excited gas are inflated by high-energy jets of gas and radiation ejected in opposite directions from the center of the galaxies.
- The **quasars** have a starlike appearance but very large redshifts that imply they are very distant. To be visible at such distances, quasars must be ultraluminous. Yet rapid fluctuations in brightness show that quasars must be small, with diameters of only a few light hours, comparable in size to our Solar System.
- The best images show that quasars are embedded in galaxies that are often distorted or have close companions. From this, astronomers conclude that quasars are the highly luminous cores of very distant active galaxies.
- Matter flowing into a supermassive black hole must conserve angular momentum and form an accretion disk. A few such disks have been imaged in the cores of active galaxies, and measurement of their rotation velocities permit determination of the masses of the black holes.
- The spinning accretion disk pulls in and winds up the magnetic field, and by a process not yet fully understood, this magnetic field ejects and focuses two jets of radiation and high-speed gas in opposite directions along the axis of rotation.
- Supermassive black holes erupt only when large amounts of matter flow inward, so most galaxies have dormant cores with little mass falling inward. During galaxy interactions and collisions, tides can throw matter into the galaxy centers and trigger eruptions.
- Because quasars lie at great distances, astronomers see them as they were long ago—more than 10 billion years ago— when the Universe was young and just forming galaxies.
- Astronomers have developed a **unified model** that explains differences among some AGN as differences in the inclination of the accretion disk. If the jet from the AGN points directly at Earth, you see a **blazar**.
- Because the mass of a supermassive black hole is related to the mass of the galaxy central bulge around it, astronomers conclude that the black holes formed when the first gas clouds fell together to form the central bulges.
- Some of the most distant quasars may be erupting because of the formation of supermassive black holes, but most activity is triggered by interactions and collisions with other galaxies. Collisions were more common in the past before the Universe had expanded very far and separated galaxies.
- The age of quasars, at redshifts from about 2 to 3, was a time when galaxies were actively growing and merging.
- The supermassive black holes that once produced quasar eruptions are now mostly dormant because very little mass is flowing into them. A galaxy can be triggered to become an active galaxy if new mass falls into the black hole.

Active galactic nucleus (AGN) The centers of active galaxies that are emitting large amounts of excess energy.

Active galaxy A galaxy whose center emits large amounts of excess energy, often in the form of radio emission. Active galaxies have massive black holes in their centers into which matter is flowing.

Blazar A type of active galaxy nucleus that is especially variable and has few or no spectral emission lines.

Double-lobed radio source A galaxy that emits radio energy from two regions (lobes) located on opposite sides of the galaxy.

Galactic cannibalism The theory that large galaxies absorb smaller galaxies.

Quasar Small, powerful source of energy in the active core of a very distant galaxy.

Radio galaxy A galaxy that is a strong source of radio signals.

Ring galaxy A galaxy that resembles a ring around a bright nucleus, resulting from a head-on collision between two galaxies.

Seyfert galaxy An otherwise normal spiral galaxy with an unusually bright, small core that fluctuates in brightness.

Tidal tail A long streamer of stars, gas, and dust torn from a galaxy during its close interaction with another passing galaxy.

Unified model An attempt to explain the different types of active galactic nuclei using a single model viewed from different directions.

14-1 Introduction to the Universe

- **Cosmology** is the study of the structure and history of the Universe as a whole. Astronomers and physicists who do research in cosmology are called cosmologists.
- Cosmologists conclude that it is impossible for the Universe to have an edge because an edge would introduce logical inconsistencies. In other words, an edge to the Universe does not make sense. If the Universe has no edge, then it cannot have a center.
- The darkness of the night sky leads to the conclusion that the Universe is not infinitely old. If the Universe were infinite in extent, infinite in age, and unchanging overall, then every spot on the sky would glow as brightly as the surface of a star. This problem, commonly labeled **Olbers's paradox**, implies that the Universe had a beginning.
- The **observable universe**, the part you can see from Earth with a large telescope, is limited in size. It might be only a tiny portion of the entire Universe, which could be infinite.
- Edwin Hubble's 1929 discovery that the redshift of a galaxy is proportional to its distance, now known as the Hubble law, shows that the galaxies appear to be moving away from each other. That phenomenon is called the expanding Universe.

Cosmology The study of the nature, origin, and evolution of the Universe.

Observable Universe The part of the Universe that you can see from your location in space and time.

Olbers's paradox The conflict between theory and evidence regarding the darkness of the night sky.

14-2 The Big Bang

- Tracing the expansion of the Universe backward in time brings you to imagine an initial high-density, high-temperature state commonly called the **Big Bang**.
- A rough estimate of the age of the Universe based on the currently observed expansion rate is called the **Hubble time**.
- Although the expanding Universe began from the Big Bang, it has no center. The galaxies do not move away from a single point. Cosmologists understand that galaxies remain approximately in their respective positions and are carried farther away from each other as the space between them expands.
- During the earliest moments of the Universe, matter and **antimatter** particles continually flashed in and out of existence. A slight excess of ordinary matter remained after most of the matter and antimatter particles annihilated each other.
- The **cosmic microwave background (CMB)** radiation is blackbody radiation with a temperature of about 2.73 K, spread nearly uniformly over the entire sky. This radiation is the light from the Big Bang, freed from the gas at the time of **recombination** and now redshifted by a factor of 1100. The cosmic background radiation is clear evidence that the Universe began with a Big Bang.
- During the first few minutes of the Big Bang, nuclear fusion converted some of the hydrogen into helium but was unable to make many other heavy atoms because no stable nuclei exist with weights of 5 or 8. Now, hydrogen and helium are common in the Universe, but heavier atoms are rare.
- After recombination, for a period of hundreds of millions of years called the **dark age**, the Universe expanded in darkness until the first stars came into existence. Astronomers have observed signs of **reionization** of the Universe caused by that first generation of stars.
- The chemical composition of the oldest stars is about 75 percent hydrogen and 25 percent helium, which is what models of the Big Bang nuclear processes would predict. This is further evidence supporting the Big Bang theory.

Antimatter Matter composed of antiparticles, which upon colliding with a matching particle of normal matter, annihilate and convert the mass of both particles into energy.

Big bang The high-density, high-temperature state from which the expanding Universe of galaxies began.

Cosmic microwave background (CMB) Radiation from the hot matter of the Universe soon after the Big Bang. The large redshift makes it appear to come from a blackbody with a temperature of 2.7 K.

Dark age The period of time after the glow of the Big Bang faded into the infrared and before the birth of the first stars, during which the Universe expanded in darkness.

Hubble time The age of the Universe, equal to the reciprocal of the Hubble constant. The Hubble time is the age of the Universe if it expanded since the Big Bang at a constant rate.

Recombination The stage, about 400,000 years after the Big Bang, when neutral atoms formed and the gas of the Universe became transparent to radiation.

Reionization The stage in the early history of the Universe when ultraviolet photons from the first stars ionized the gas of the Universe.

14-3 Space and Time, Matter and Energy

- The Universe has properties of **isotropy** and **homogeneity**, meaning, respectively, that in its general features it looks approximately the same in all directions from Earth and also appears to have approximately the same properties in all locations.
- Isotropy and homogeneity together imply the **cosmological principle**, that there are no special places in the Universe. Except for minor local differences, every place is the same, and the view from every place is the same.
- Einstein's theory of general relativity explains that cosmic redshifts are caused by wavelengths of photons stretching as they travel through expanding space-time.
- **Open universe** models have curved space-time, but it is not curved back on itself. Such universes are infinite.
- **Closed universe** models are finite in size, but their space–time is curved back on itself so they have no edge or center.
- **Flat universe** models have uncurved space-time and are infinite. Modern observations show that the Universe is probably flat.
- Open model universes contain less than the critical density and closed model universes more. If the Universe is flat, then its density must equal the critical density.
- The amounts of deuterium and lithium-7 in the Universe show that normal baryonic matter can make up only about 5 percent of the critical density. Dark matter must be **nonbaryonic matter** and makes up less than 30 percent of the critical density.
- Dark matter could be composed of as-yet undiscovered types of subatomic particles. The observed ranges of masses and diameters of galaxy clusters agree with the predictions of models that assume the dark matter is slow-moving, readily-clumping **cold dark matter**.

14 Modern Cosmology

Closed universe A model of the Universe in which space-time is curved to meet itself and the Universe is finite.

Cold dark matter Dark matter that is made of slow-moving particles.

Cosmological principle The assumption that any observer in any galaxy sees the same general features of the Universe.

Critical density The average density of the Universe needed to make its curvature flat.

Flat universe A model of the Universe in which space-time is not curved.

Homogeneity The observation that, on the large scale, matter is uniformly spread through the Universe.

Isotropy The observation that, in its general properties, the Universe looks the same in every direction.

Nonbaryonic matter Proposed dark matter made up of particles other than protons and neutrons (baryons).

Open universe A model of the Universe in which space-time is curved in such a way that the Universe is infinite.

14-4 Twenty-First-Century Cosmology

- The **inflationary Big Bang** modification to the Big Bang theory proposes that the Universe briefly expanded dramatically, just a tiny fraction of a second after the Big Bang.
- The energy to drive inflation would have been released when the four forces of nature changed their respective properties to become different as the Universe cooled in its earliest moments. This "separation" of forces is predicted by physics theories that explain the forces of nature as being aspects of a single force that are unified only in particle interactions with very high energies.
- Inflation explains the **flatness problem** because the large expansion forced the Universe to become flat, just as a spot on an inflating balloon becomes flatter as the balloon inflates.
- Inflation explains the **horizon problem** because what is now the observable part of the Universe was so small before inflation that energy could move and equalize the temperature everywhere in that volume.
- Observations of Type 1a supernovae reveal the surprising fact that the expansion of the Universe is speeding up. Cosmologists propose that this acceleration is caused by energy present in empty space labeled **dark energy**.
- The nature of dark energy is unknown. It might be described by Einstein's **cosmological constant**, or its strength might change with time, which cosmologists label **quintessence**. Some models of quintessence predict an ever-accelerating expansion leading to a **big rip** that eventually would tear all the objects, even atoms, apart.
- Corrected for acceleration, the observed value of the Hubble constant implies that the Universe is 13.7 billion years old.
- The sudden inflation of the Universe is hypothesized to have magnified tiny quantum mechanical fluctuations in the density of matter and energy. These very wide but very weak differences in density caused dark matter, followed by baryonic matter, to draw together and produce the current **large-scale structure** consisting of galaxy **superclusters** arranged in great walls and filaments outlining enormous voids.
- Statistical observations of the large-scale structure of the Universe and the size of irregularities in the cosmic microwave background confirm that the Universe is flat and contains 4.6 percent baryonic matter, 23.3 percent dark matter, and 72.1 percent dark energy.
- The mass equivalent of dark energy added to dark matter and baryonic matter makes the observed density of the Universe equal to the critical density, thereby confirming the prediction made by inflation theory that the Universe is flat.

Big rip The fate of the Universe if dark energy increases with time and galaxies, stars, and even atoms are eventually ripped apart by the accelerating expansion of the Universe.

Cosmological constant A constant in Einstein's equations of space and time that represents a force of repulsion.

Dark energy The energy believed to fill empty spaces and drive the acceleration of the expanding Universe.

Flatness problem The peculiar circumstance that the early Universe must have contained almost exactly the right amount of matter to make space-time flat.

Horizon problem The circumstance that the primordial background radiation seems much more isotropic than can be explained by the standard Big Bang theory.

Inflationary Big Bang A version of the Big Bang theory, derived from grand unified theories of particle physics, that includes a rapid expansion when the Universe was very young.

Large-scale structure The distribution of clusters and superclusters of galaxies in filaments and walls enclosing voids.

Quintessence A possible form of dark energy that can change in strength as the Universe ages.

Supercluster Cluster of galaxy clusters.

15-1 The Nature of Life

- **Astrobiology** is the scientific study of the likelihood that life exists elsewhere in the Universe, and the study of life and its origin on Earth as it relates to life on other worlds.
- Life can be defined as a process that extracts energy from the surroundings, maintains an organism, and modifies the surroundings to promote the organism's survival.
- Living things have a physical basis—the arrangement of matter and energy that makes life possible. Life on Earth is based on carbon chemistry occurring in tiny bags of water (cells).
- Living things must have a controlling unit of information that can be passed to successive generations.
- Genetic information for life on Earth is stored in long carbon-chain molecules such as **DNA (deoxyribonucleic acid)**.
- The DNA molecule stores information in the form of chemical bases linked together like the rungs of a ladder. Copied by the **RNA (ribonucleic acid)** molecule, the patterns of bases act as recipes for connecting together **amino acid** subunits to construct **proteins**, including **enzymes**, which are the main structural and control components of the life process.
- The unit of heredity is a **gene**, a piece or several pieces of DNA specifying construction of one or more protein molecules corresponding to one inheritable trait. Genes are connected in structures called **chromosomes**, which are essentially single long DNA molecules.
- When a cell divides, the chromosomes split lengthwise and duplicate themselves so that each of the new cells can receive a copy of the genetic information.
- **Biological evolution** is the process by which life adjusts itself to changing environments.
- Errors in duplication or damage to the DNA molecule can produce **mutants**, organisms that contain new DNA information and therefore have new properties. Variation in genetic codes can become widespread among individuals in a species. **Natural selection** determines which of these variations are best suited to survival, and the species evolves to fit its environment.
- Evolution is not random. Genetic variation is random, but natural selection is controlled by the environment.

Amino acid Carbon-chain molecule that is the building block of protein.

Astrobiology The field of study focused on searches for life and possible habitats for such life on other worlds.

Biological evolution The process of mutation, variation, and natural selection by which life adjusts itself to its changing environment.

Chromosome A body within a living cell that contains genetic information responsible for the determination and transmission of hereditary traits.

DNA (deoxyribonucleic acid) The long carbon-chain molecule that records information to govern the biological activity of the organism. DNA carries the genetic data passed to offspring.

Enzyme Special protein that controls processes in an organism.

Gene A unit of DNA—or sometimes RNA—information responsible for controlling an inherited physiological trait.

Mutant Offspring born with DNA that is altered relative to parental DNA.

Natural selection The process by which the best genetic traits are preserved and accumulated, allowing the fittest organisms and species to survive and proliferate.

Protein Complex molecule composed of amino acid units.

RNA (ribonucleic acid) Long carbon-chain molecules that use the information stored in DNA to manufacture complex molecules necessary to the organism.

15-2 Life in the Universe

- The oldest definitely identified fossils on Earth, structures called **stromatolites** that are composed of stacks of bacterial mats and sediment layers, are at least 3.4 billion years old. Those fossils provide evidence that life began in the oceans.
- The **Miller experiment** shows that the chemical building blocks of life form naturally under a wide range of circumstances.
- Scientists hypothesize that **chemical evolution** occurred before biological evolution. Chemical evolution concentrated simple molecules into a diversity of larger stable organic molecules dissolved in the young Earth's oceans, but those molecules did not reproduce copies of themselves. The hypothetical organic-rich water is sometimes referred to as the **primordial soup**. Biological evolution began when molecules developed the ability to make copies of themselves.
- Fossil evidence indicates that life began on Earth as simple single-celled organisms such as bacteria and much later evolved into more complex, **multicellular** creatures.
- Life-forms did not become large and complex until about 0.5 billion years ago, during what is called the **Cambrian explosion**. Life emerged from the oceans only about 0.4 billion years ago, and human intelligence developed over the past 4 million (0.004 billion) years.
- Life as it is known on Earth requires liquid water and thus a specific range of temperatures. Some organisms have adapted to environments at the hot or cold ends of the liquid water temperature range or are extreme in other ways.

(continues)

(continued)

- No other planet in our Solar System appears to harbor life currently. Most are too hot or too cold, although life might have begun on Mars before the planet became too cold and dry. If so, life conceivably could persist today on Mars in limited hospitable environments.
- Liquid water exists, and therefore Earth-like life is at least possible, under the surfaces of Jupiter's moons Europa and Ganymede and Saturn's moon Enceladus. Saturn's moon Titan has abundant organic compounds but does not have liquid water.
- Because the origin of life and its evolution into intelligent creatures took so long on Earth, scientists do not consider middle- and upper-main-sequence stars, which shine for astronomically short time spans, as likely homes for life.
- Main-sequence G and K stars and possibly some of the M stars are thought to be likely candidates to host planets with life.
- The **habitable zone** around a star, within which planets can have liquid water on their surfaces, might be larger than scientists had expected, given the wide variety of living things now found in extreme environments on Earth. Tidally heated moons orbiting large planets could have liquid water at any distance from a star.

Cambrian explosion A geologically brief period about 540 million years ago during which fossil evidence indicates Earth life became complex and diverse. Cambrian rocks contain the oldest easily identifiable fossils.

Chemical evolution Chemical processes involving growth and natural selection of complex molecules prior to the origin of life.

Habitable zone A region around a star within which planets have temperatures that permit the existence of liquid water on their surfaces.

Miller-Urey experiment An experiment that attempted to reproduce early Earth conditions and showed how easily amino acids and other organic compounds could have formed then.

Multicellular Composed of more than one cell.

Primordial soup The rich solution of organic molecules in Earth's first oceans.

Stromatolite A layered formation caused by mats of algae or bacteria combined with sediments.

15-3 Intelligent Life in the Universe

- Because of distance, speed, and fuel, travel between the stars seems almost impossible for humans or for aliens who might visit Earth.
- Communication between planetary systems using electromagnetic signals such as radio waves or laser beams might be possible, but a real conversation would be difficult because of long travel times for such signals.
- Broadcasting a radio (or light) beacon of pulses would distinguish the signal from naturally occurring emissions and identify the source as a technological civilization. It might be possible to design a signal such that an extraterrestrial civilization could decode it.
- One good part of the radio spectrum for communication is called the **water hole**, the wavelength range from the 21-cm spectral line of hydrogen to the 18-cm line of OH. Even so, millions of radio wavelengths need to be tested to fully survey the water hole for a given target star.
- Sophisticated searches are now under way to detect radio transmissions from civilizations on other worlds, but such **SETI** (Search for Extra-Terrestrial Intelligence) programs are hampered by limited computer power and radio noise pollution from human civilization.
- The number of civilizations in our Galaxy that are at a technological level and able to communicate while humans are listening can be estimated by the **Drake equation**. This number is limited primarily by the lifetimes of their and our civilizations.

Drake equation The equation that estimates the total number of communicative civilizations in our Galaxy.

SETI The Search for Extra-Terrestrial Intelligence.

Water hole The interval of the radio spectrum between the 21-cm hydrogen emission line and the 18-cm OH emission line; these are likely wavelengths to use in the search for extraterrestrial life.